Computer Applications in Production and Engineering

Proceedings of CAPE '95

Edited by

Qiangnan Sun
Beijing Information Technology Institute
Indian Institute of Technology
Beijing
China

Zesheng Tang
Tsinghua University
China

and

Yijun Zhang
Beijing Research Institute of Automation for MBI
Beijing
China

Published by Chapman & Hall on behalf of the
International Federation for Information Processing (IFIP)

CHAPMAN & HALL

London · Glasgow · Weinheim · New York · Tokyo · Melbourne · Madras

Published by Chapman & Hall, 2–6 Boundary Row, London SE1 8HN, UK

Chapman & Hall, 2–6 Boundary Row, London SE1 8HN, UK

Blackie Academic & Professional, Wester Cleddens Road, Bishopbriggs, Glasgow G64 2NZ, UK

Chapman & Hall GmbH, Pappelallee 3, 69469 Weinheim, Germany

Chapman & Hall USA, 115 Fifth Avenue, New York, NY 10003, USA

Chapman & Hall Japan, ITP-Japan, Kyowa Building, 3F, 2-2-1 Hirakawacho, Chiyoda-ku, Tokyo 102, Japan

Chapman & Hall Australia, 102 Dodds Street, South Melbourne, Victoria 3205, Australia

Chapman & Hall India, R. Seshadri, 32 Second Main Road, CIT East, Madras 600 035, India

First edition 1995

© 1995 IFIP

Printed in Great Britain by T.J. Press, Padstow, Cornwall

ISBN 0 412 70770 5

A catalogue record for this book is available from the British Library

CONTENTS

PREFACE

CAPE'95 is the fifth International Conference on Computer Applications in Production and Engineering, which is held once in three years. The conference is sponsored by IFIP (International Federation for Information Processing) Technical Committee 5 (Computer Applications in Technology) and the Computer Engineering and Application Society of CIE (Chinese Institute of Electronics). The first CAPE conference was taken place in Amsterdam in 1983. It was follows by a series of conferences in Copenhagen, Tokyo and Bordeaux. The relay baton is now handed to Beijing.

The purpose of CAPE'95 is in accordance with the IFIP basic rules to promote the exchange of advanced experiences in academic research and industrial practice of application of information technology in various fields in industry and to bridge the gap between research and implementation. For there have been many specialized conferences organized by IFIP TC5's working groups, and in line with the above mentioned purpose, papers presented in CAPE'95 are of a wider range in computer applications in technology, from basic theory, architecture, methodology to implementation and case study. Topics covered in this proceedings include Computational Geometry and its Application, Modeling Technology, Theory and Methodology of CAPM, Production Management Techniques, CIM system, Theory and Methodology of CIM, Modeling and Simulation in CAM, Neural Network Applications, AI Applications, Computer Application Techniques, Robotics, and other new technologies in computer application in production and engineering.

The framework of CAPE'95 is planned to raise issues with keynote papers at plenary session and invited and submitted papers at the parallel sessions, along with panel discussions. The submitted papers were offered from over twenty countries/regions. They were selected after undergoing a intensive review process with three reviewers for each paper according to the normal IFIP practice.

Geographically, the authors of papers were widely dispersed over the world. One of the attractive features of CAPE'95 is the considerable number of papers from Asia-Pacific Rim, an area with great potential. For in recent years the growth of economy in that area bas shown an extraordinarily high rate. So we do wish that the proceedings of CAPE'95 will be a good medium for the sharing of experiences in all aspects of computer applications in production and engineering, and be beneficial to broad audience all over the world, both in developed and in developing countries.

We would like to thank all the authors for their valuable contributions, and the members of the International Program Committee for their efforts in reviewing the papers. We would like specially to send our deepest gratitude to Mr. Marco Tomljanovich, the former TC5 chairman, Dr. Toru Mikami, the TC5 chairman, and Prof. Asbjorn Rolstadas, the IFIP president, for their kind supports in preparing the conference. The supports from IFIP and many others involved in CAPE'95 are also highly appreciated.

Qiangnan Sun
Zesheng Tang
Yijun Zhang

Beijing

PROGRAM COMMITTEES AND SPONSORS
INTERNATIONAL PROGRAM COMMITTEE

S. Bologna ITALY

J. Browne, IRELAND

L. Camarinha-Matos, PORTUGAL

E. Cheung, HONG KONG

J. Crestin, FRANCE

Z. Deng, CHINA

G. Doumeingts, FRANCE

E. Eloranta, FINLAND

J. Encarnacao, GERMANY

P. Falster, DENMARK

R. Gan, CHINA

R. Goebl, AUSTRIA

M. Goyal INDIA

G. Guariso ITALY

G. Halevi, ISRAEL

Ho Nai Choon, SINGAPORE

D. Hu, CHINA

F. Kimura, JAPAN

D. Kochan, GERMANY

W. Loeve, THE NETHERLANDS

T. Mikami, JAPAN

K. Muller, SWITZERLAND

L. Nemes, AUSTRALIA

G. Olling USA

C. Park, KOREA

B. Li, CHINA

J. Reinfelds, USA

J. Riis, DENMARK

A. Rolstadas, NORWAY

Q. Sun, CHINA

Z. Tang, CHINA, Chairm

M. Tomljanovich, ITALY

S. Vanichseni, THAILAND

M. Villabo, NORWAY

T. Williams, USA

M. Wozny, USA

G. Yang, CHINA

J. Zabala, SPAIN

Sponsorship

Sponsored by

Technical Committee 5 of International Federation
    for Information Processing
Computer Engineering and Application Society of
    Chinese Institute of Electronics
Co-sponsored by
Mechanical Industry Automation Institute of Chinese
    Mechanical Engineering Society
National Natural Science Foundation of China
IEEE Computer Society Beijing Chapter
Chinese National Hi-Tech. Expert Group of CIMS

CHAIRPERSONS
Honorary Co-Chairmen
Dr. Jun-ren Sun
Mr. Marco TomlJanovich
General Co-Chairmen
Mr. Youming Liao, China Huaneng Group
Dr. Toru Mikami, NEC Corporation
Dr. Mike Wozny, Rensselaer polytechnic Institute

NATIONAL ORGANIZING COMMITTEE
Prof. Qiangnan Sun, Beijing Information Tech. Inst. (Chairman)
Mr.   YiJun Zhang, Beijing Research Institute.   of Automation
for MBI (Vice-chairman)

PART ONE

Keynote Papers

1

CAD In China: Applications and Industrialization

Yang Tianxing
Director of Computer Department, Electronic Ministry, Professor

Gong Bingzheng
6th Institute, Electronic Ministry, Professor

Abstract

The current status of CAD applications and industrialization in China is concisely described in the paper. Then, the potentiality of an extensive market of CAD applications is analyzed. Finally, models of developing CAD applications and industrialization in China are related.

1 THE CURRENT STATUS OF CAD APPLICATIONS AND INDUSTRIALIZATION IN CHINA

1.1 Chinese government supports the development of CAD applications and industrialization

Computer aided design is a critical high technology to accomplish design automation, expedite the transformation from scientific and technological achievements to production forces, and to accelerate national economy development and four modernizations. The application of CAD technology is not only the main aspect of reformation of traditional industrial technologies but also an important leverage to raise product and engineering design to a higher level, reduce cost, shorten design cycle, and to improve the labor productivity, as well a significant requirement for enterprises to enhance their competitive power and adaptive faculty in market. The level of applying CAD technology is also one of the major marks indicating the level of national industrial technologies.

In order to impel the development of CAD applications and industrialization, Chinese government has been including the development of CAD in the National Science and Technology Key Projects and Torch Program for a long time, encouraging enterprises to adopt CAD technology and carrying out preferential policies for them. The national developing plan of CAD applications has been formulated, in addition, a coordinating and directing group of CAD application engineering, which was sponsored by the National Science Committee and participated by eight state ministries and committees, has been established.

1.2   The current status of CAD applications and industrialization in China

Starting in 1960s and passing through its developing period mainly in 1970s and 1980s, CAD in China is now beginning its preliminary applications in engineering designs and product designs in diverse fields such as machinery, electron, aviation, aerospace, architecture, shipbuilding, light industry, textile industry, and so on, and has obtained remarkable technical and economic benefits. For example, in National Designing Institutes, more than 90 percent of calculation, 50 percent of project design and 30 percent of drawings were finished by using CAD technology. It raised working efficiency as 3 to 10 times as before, and saved as much as 2 percent of capital construction investment. As a result, 2 billion yuan of engineering investment has been cut down on only during the 7th. five-year plan period.

According to the results of our sample survey, the installation situation of computers and CAD systems in 1992 is listed as follows.

	Aviation Industry	Shipping Industry	Weapon Industry	Oil Industry & Capital Construction Bureau	Railway Ministry	Machinery	Electron	Construction
Installed Computers Number	5280	2000	6464	577	2352	21500	25453	16069
CAD System Number	1160	515	1006	499	949	5427	10555	9000
Percentage (%)	21.96	25.75	15.6	86.13	40.37	25.24	41.34	55.9

Among CAD systems in our country, microcomputer based CAD systems account for more than 90 percent, whereas workstation based CAD systems are increasing year by year. Hundreds of CAD software have been developed so far. However, a few of CAD systems have been transformed into merchandise.

1.2.1   General situation of CAD hardware

The general situation of microcomputers made in our country is as follows. They have already reached the engineering scale production with a yearly yield of 250,000 sets including 286, 386 and 486 microcomputers produced by computer groups such as Great Wall, Changjiang,

Langchao, Legend, and so on, and supply users with microcomputer based CAD hardware platforms.

The general situation of workstations made in our country is as follows. Huasheng series workstations, as scientific and technological achievements during the 7th and 8th five-year plan periods, have begun its batch production, among which Huasheng 4065, 4075 SPARC workstations have already been produced in a small scale production. Huapu HP workstations are at the initial stage of small scale assembling, while Huaqi SGI workstations have gone into assembling.

Taiji series computers, small-sized computers made in our country, are produced in a small scale production.

The general situation of peripheral devices is as follows. CRT, printer, plotter, floppy, hard disk, and compact disk drive, etc., are produced in batch production by more than 10 factories in our country, respectively.

1.2.2 General situation of CAD software

During more than 10 years CAD software in our country have obtained fairly great development, got a large number of achievements up to now, and started carrying out industrialization. The majority of these software have been developed by universities, colleges, and scientific research institutions.

Nowadays, we have gotten some fairly successful CAD software, listed as follows.

Machinery	ZDMCD, MGPS	Zejiang University
	MSD, BSURF-GI	Nanjing University of Aeronautics and Astronautics
	GEMS, GWCAD	Tsinghua University, Northern CAD Company
	JBZ-CADM, PC-MECADS	Automation Institute of Mechanical Ministry
	APT-X System, CAMS	No. 625 Institute of Aviation and Aerospace Ministry
	NPU-CAD/CAM	Northwest Industrial University
	CAED, PDA	Huazhong University of Science and Technology
	PICAD	Beijing Software Engineering Developing Center of Academia Sinica
Electron	PANDA System	Beijing Integrated Circuit Design Center
	ECADS	CAD Laboratory of Institute of Computing Technology of Academia Sinica
	Printed Circuit CAD	Shanghai Institute of Computing Technology
Architecture	AEDS	Beijing Engineering Design Software Company
	Architecture CAD	Chinese Architectural Scientific Research Academy
	JT-HBCADS	Tongji University

In the mechanical CAD aspect, ZD-MACD system of Zejiang University with SUN workstation as its hardware can accomplish many tasks such as geometry modeling, structural finite element analysis, numerical controlled manufacturing, and so on, meanwhile its software performance/price ratio meets the requirements of domestic users fairly well. GEMS, a geometry modeling system of Department of Computer Science and Technology of Tsinghua University, has already been implemented in varied hardware platforms including PC, HP/APOLLO and SUN workstations, and used by dozens of domestic units. PICAD, a two-dimensional drawing system developed by Beijing Software Engineering Developing Center of Academia Sinica, has already been sold in batches.

In the electronic CAD aspect, PANDA Version 1.0, an IC CAD software of Beijing Integrated Circuit Design Center, which possesses all of the functions of an overseas software, DIASY, was formally promoted on August 17, 1992. PANDA system is presently being used by more than 20 domestic users, and has been used to accomplish several IC design tasks. ECADS software of CAD Laboratory of Institute of Computing Technology of Academia Sinica runs on UNIX workstation and has got fairly good applications in our country.

In the architectural CAD aspect, the CAD software of Chinese Architectural Scientific Research Academy, including architectural CAD, structural CAD, and high-rises CAD, have been developed and used by as many as thousands of domestic users, and have extensive application areas and strong influences.   Running mainly on PC microcomputers, these software are all fairly practical and can be directly served in architectural constructions.

2   Prediction of Chinese computer CAD system market in 1990s

2.1   Prediction of national computer market

In 1990s, the computer market in our country will be expanded steadily. According to our prediction of national computer market, the aggregate sales of commodities of computer market will increase with a rising rate of 20 to 25 percent, meanwhile it will account for a bigger portion in the social fixed investment.

	1995	1996	1997	1998	1999	2000
Aggregate Sales in Market (100 million yuan)	500-520	620-640	740-800	900-1000	1100-1200	1300-1500
Yearly Rising Rate	25-30%	23-25%	20-25%	20-25%	20-22%	20-25%

2.2   Prediction of national CAD demands

To summarize what are mentioned above, during the 8th five-year plan period, there will be a demand of 60,000 or 70,000 to 100,000 sets of CAD systems, including 50,000 or 60,000 to 85,000 sets of microcomputer based CAD systems and 15,000 to 20,000 sets of workstation based CAD systems. During the 9th five-year plan period, there will be a requirement of 310,000 to 360,000 sets of CAD systems, including 250,000 to 280,000 sets of microcomputer based CAD systems and 60,000 to 80,000 sets of workstation based CAD systems.

2.3 Characteristics of the national CAD market

The first characteristic is that the CAD market in our country is still in its primary stage of developing and growing, and needs supporting, fostering and further developing.
Only 10 percent of enterprises are currently using CAD technology. The workstation based CAD systems constitute a low proportion among CAD systems, for example, 7.6 percent in mechanical and electronic enterprises presently, but in developed foreign countries, it occupies 65 percent in mechanical enterprises and 86 percent in electronic enterprises, respectively.

The second characteristic is that the CAD market in our country is a component part of international market, in which mostly CAD software and hardware products from foreign companies have a dominant position currently, whereas CAD software independently developed by our country are of fairly low level, few of which have been industrialized. Forty to 50 percent of microcomputer based CAD systems are developed in China, but 80 percent of CAD products on workstations, small-sized computers and superior computers are products of foreign companies. During 1990s, the national market will still be an international market dominated by foreign products. However, CAD software in China will gain a fairly rapid development.

The third characteristic is that there is a potential extensive market in China.

China has about a million industrial enterprises and a potential extensive market which provides lots of excellent opportunities for China to develop CAD industry and software industry. CAD products developed by our country will undoubtedly share the market, but we will strive for the dominant position in the next century.

The forth characteristic is that the policies of our government such as the social economy developing strategy, the industry policy, the equipment policy, the investment policy, and the import policy, have great influence powers on the market. The general and specific policies of our country have determinant impact on the market in our country.
The computer and software market are presently forming. The national policies are directive for the development and growth of the CAD market.

The fifth characteristic is that laws and regulations of administering on the market are not perfect enough while the CAD market are currently in its growth period. There exist some chaotic phenomena in the market. The software protection rules still need propagating and carrying out continuously. There also exist CAD products of low quality, and counterfeit and inferior CAD systems. It is necessary to enhance administration on the CAD market so that a good circumstance of developing CAD technology could be formed.

3. Models and ways of developing CAD industry in China

3.1 Developing models

In order to develop CAD industry in our country, we should take domestic and overseas markets as guidance to adjust the product structure and the industry structure continuously.

We should select finite objects and give prominence to what is significant. We should vigorously develop CAD software and applications so as to give an impetus to hardware manufacturing industry. We should develop specialized CAD with diversified sectors in the national economy, encourage cooperation among domestic enterprises and introductions of foreign advanced technologies. We should enhance the combination of human resource, financial resource and information resource and the integration of factories, colleges and institutes. We should establish hardware manufacturing industry, software industry and application industry, and ensure them develop harmoniously. We should establish a strip-block-integrated, rationally distributed CAD industry system with specialized CAD companies as its core.

3.1.1   Carrying out the policy of impelling industrialization by applications, meeting needs of overseas and domestic markets, and of being directed by the market

It is urgent to spread and apply CAD achievements and continuously promote industrialization in this process.

Mechanical and electronic industry should close integrate with users and departments in various national economic fields which are engaged in developing, producing, applying CAD technology, and offering CAD technical service, so that we can develop, then produce, and finally spread a large number of CAD products, continuously adjust the product structure, and advance in a rolling style.

We should select finite objects and give prominence to what is significant, for example, machinery, electron, shipping, automobile, and architecture CAD in the near future. The integrated product of CAD application supporting software with CAD system will be selected as a breach to develop CAD systems and products with Chinese characteristics.

3.1.2   Diversified developing models incorporated with national situation

The first model is that the state supports and organizes developing and spreading, impels developing and spreading industrialized achievements of science and technology, meanwhile these achievements can be continuously adjusted and perfected and become more practical and maintainable.

Beijing Integrated Circuit Design Center and CAD Software Center of Institute of Computing Technology of Academia Sinica have made beneficial attempt on CAD industrialization.

The second model is that industries sponsor and organize developing CAD application systems and products which meet their own urgent needs.

With general CAD supporting software as emphases, industrial departments organize a combined group consisting of research   units, colleges and universities, and application demonstration factories, to jointly develop CAD basic database and engineering database in their own industries, and to develop CAD software with their own industrial characteristics. Upon this basis, they jointly establish an integrated entity to spread products, which includes industrialized CAD software development, production, sell, management, and technical service. Chinese Architectural Scientific Research Academy of Construction Ministry has made remarkable achievements in industrializing architectural CAD software and spreading them, and gains beneficial experience.

The third model is to organize diversified specialized CAD companies.

Diversified specialized CAD system companies and software companies will be established with scientific research units, colleges and universities, and enterprises as their main body, which aim at developing CAD application software and application systems which will be extensively used by lots of people in many practical areas and meet large amount of market demands. Specialized CAD companies may also be established by powerful users and key enterprises as main force together with computing centers, institutes and product design sectors of enterprises as auxiliary force.

As a successful instance, Hangzhou Xidebao Electronic Engineering Company, jointly established by Zejiang University and Hangzhou Printing and Dyeing Mill, developed printing and dyeing CAD software.

The forth model is to establish joint ventures so as to absorb foreign investment and technologies and drive national CAD software infiltrate international market through overseas sales and service channels.

Northeast Arpai Software Company, a joint venture established by Software Center of Northeast University and Japan, has exported software to Japan.

Models	Market Object	Investment Source	Achievements Source	Organization Form
The state organizes developing and spreading	CAD supporting software suitable for national industries	State allotments, State loans	National scientific and technological achievements	Enterprise group stock company ECR, State laboratory
Industries organize developing and spreading	CAD supporting software suitable for specific industries	Loans (mainly)	Scientific and technological achievements in Ministries and Committees	Enterprise group CAD company
Enterprises and public institutions organize developing	middle or small sized CAD system, CAD application system	Loans, Enterprises-raised funds	Achievements accomplished by enterprises independently	Specialized CAD company of collective ownership, Scientific and Technological enterprise run by the local people
Joint venture companies	Export (mainly)	Domestic and overseas joint investment	Further development of overseas achievements	Joint venture

3.2 Developing ways

3.2.1 CAD hardware manufacturing industry

To develop CAD hardware manufacturing industry, we should depend on existing state-enterprises, which are engaged in manufacturing microcomputers, workstations, small-sized computers and peripheral devices, as main force, and unite joint ventures and new high technology enterprises run by the local people in new technology developing zones. Nowadays, we should vigorously enhance development of workstations and peripheral devices, cooperate with interested enterprises in USA and Japan, etc., and heighten performance/price ratio of hardware product and market occupying rate.

3.2.2 CAD software industry

Existing scientific research academies and institutes are encouraged to derive specialized CAD software companies, specialized service companies and enterprises run by the local people, which are specially engaged in developing, producing, managing, and technically serving CAD software product.

Joint ventures will be established in new technology developing zones to go in for developing and maintaining CAD application software.
Powerful users take charge of developing, selling and spreading special purpose CAD application software of their own industries.

3.2.3 CAD application service industry

Application service key enterprises will be established by existing factories, institutes and computer enterprises in various industries that are engaged in CAD application system integration. They are encouraged to close cooperate with users, develop typical CAD application system, and to provide complete set of service.

Through technical cooperation with overseas CAD firms, we will absorb technologies and investment, contract domestic and overseas application system engineering, and provide domestic and overseas markets with application systems and specialized service.

In all large areas, technical service nets will be established, including specialized service net and maintenance net concerning selling, maintaining, training, and so on. They will promptly provide various specialized service to vast numbers of users.

3.3 Accelerating the transformation from CAD scientific and technological achievements to products and industrialization

3.3.1 Enhancing developing and spreading CAD achievements

All industries should formulate their programs of spreading and applying CAD technology, and form a large number of typical demonstration enterprises of applying CAD technology.

The combination of enterprises leaders, specialized technicians and operators, and the cooperation of CAD researching and developing units, producing enterprises and users, can ensure spreading and applying CAD technology efficiently.

3.3.2 Drawing up plans in an integrated and overall way, and enhancing CAD application system integration

Through drawing up plans in an integrated and overall way, industrial departments should carry out technical standards and policies, enhance CAD software evaluation and test, and strengthen macroscopic direction for spreading, applying, and industrializing CAD technology.

By introducing more flexible policies and deepening reform, Chinese government will establish a flexible circumstance for developing CAD software, application and service industries, and support and promote the growth of a large number of CAD software companies, application system companies and technical service companies.

REFERENCES

Yang Tianxing (Sept. 27, 1994) Accelerating national economy informationization, and establishing computer industry with Chinese characteristics. China Computerworld.

Gong Bingzheng, et al. (March 6, 1994) Research of prediction and analysis of Chinese CAD market. China Computerworld.

Gong Bingzheng, et al. (March 6, 1994) Research of CAD industrialization. China Computerworld.

Gong Bingzheng, et al. (July 11, 1994) Making an inquiry into models of developing computer software industry in China. International Electronics.

2

Tasks Facing the Manufacturing Industry and NEC's Production Innovation

Makoto Tachibana, Vice President
NEC Corporation
3-484,Tukagosi Saiwai-ku,Kawasaki,Kanagawa 210,JAPAN

Abstract

Drastic changes in the conventional paradigm of mass production have ushered in the era of individualized masses. We at NEC are innovating our productive activities to achieve agile production as a solution to the needs of the era of individualized masses. NEC Yonezawa has achieved drastic cuts in notebook-type personal computer development leadtimes through its pursuit of concurrent engineering. At NEC Nagano and NEC Saitama, lean production has cut manpower requirements, narrowed leadtimes drastically, and greatly reduced materials in process. We are now committed to the goal of building information systems that takes full advantage of the IT (information technology) concept to achieve an optimal production system encompassing the entire flow of supply, from planning to scrapping.

Keywords

Production Innovation, Lean Production, Concurrent Engineering

1. Introduction

The paradigm of mass production has been changing drastically, holding major changes in immediate store for us. As our lives mature and we are more concerned with the betterment of individuality and self-support, the market has been shifting to an era of split masses keeping in pace with diversified market needs. In the world of mass production, standardized products, homogeneous markets, and long product life cycles and development leadtimes were taken for granted. But now, the market has been shifting to more diversified product characteristics and shorter product life cycles. These changes are slowly having an impact on the electronics industry as well. Firms that are unable to adapt to these market trends inevitably incur declining profits.

Requirements for production systems have been changing accordingly. Now that the bubble economy boom is over, building a streamlined production systems with integrated engineering and production, utmost simplicity, and slim facility management is urgently sought to achieve so-called lean production in smaller lots and at lower costs. The need has been mounting for a production system that is capable of addressing changes, including product diversification. It is anticipated that there will come a time calling for exceptionally diversified products more closely tailored to the characteristics of individuals than in the era of split masses. This is the era of individualized masses, or the era of mass customization. The production system that fits into this era will be an agile one capable of manufacturing greater assortments of products in short times.

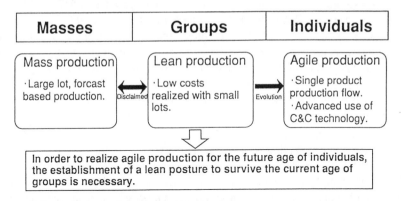

Figure 1 The Shift in the Paradigm of Production

2. Present Status of Product Development and Manufacturing

First among the tasks of agile production from the viewpoints of product development and manufacturing, we have product planning that is precisely geared at diversified and true needs. The second task is a significantly shorter product development leadtime. Realizing these tasks urgently calls for the practice of concurrent engineering (CE), or the organization of parallel execution of increasingly sophisticated and complicated engineering activities through divisional cooperation. We need to standardize engineering activities in response to product diversification and to pursue a global policy of component sharing to achieve low-cost, high-quality designs.

The task in manufacturing, in its narrowest sense, is to deliver diversified customer orders in the required quantities and at the required times. To this end, we need to eradicate inefficiencies from the manufacturing site and also keep ourselves ready to manufacture what has been sold out. In specific terms, efficient production is a scheme of production that minimizes leadtimes, manhours spent, equipment investment, quality defects, and inventories (work in process). As an example, machining time that increase value added accounts for only a small fraction of the total leadtime. Non-processing time in typical companies is said to be 10 thousand times longer than machining time. Japan's leading auto manufacturer has reduced this non-processing time down to three hundred times the machining time. Leadtimes can be cut to a sizable extent by shortening inefficient non-processing time. This concept of eliminating inefficiencies may be extended to the goals of cutting manhours spent, checking equipment investment, reducing quality defects, and running down inventories to build a scheme of efficient production. An information system would also have to be maintained to gain prompt insight into sales results to manufacture products in pace with their movement while holding down work in process to a minimum.

3. Production Innovation Activities

NEC's production innovation activities aime at a complete overhaul of the production structure to meet the change in the pace of production from rapid growth to moderate growth. Rather technology-oriented so far, we at NEC have tended to build costly production systems line after line and maintained a complicated production management system. The production innovations that we are now pursuing are geared toward reviewing the production systems in their narrowest sense, such as engineering, manufacturing, and materials procurement, to rebuild an efficient, streamlined process flow. The activities involved in product development, manufacturing, and materials procurement processes may be summarized as follows:

3.1 Product development process

Concurrent engineering is promoted and developed in the product development process. While the present practice of engineering is characterized by the concepts of task execution and management, and emphasis on technology, it is necessary to renew the designers' ideas, define the engineering processes, maintain an engineering infrastructure, and develop a drastic breakthrough in the attainment of higher engineering efficiency. We will be pursuing better processes through the development of tools for investigating, introducing, promoting, and developing engineering process innovations achieved both in house and outside.

3.2 Manufacturing process

A review and renewal of the flow of production by eliminating inefficiencies is sought in the manufacturing process. Other activities involved include planning and following up lean production promotion activities, streamlining manufacturing processes, downsizing equipment, and minimizing administrative needs.

3.3 Materials procurement process

How information is disseminated is of key importance to the materials procurement process. Various kinds of information flow into the materials purchasing department from suppliers and markets. What is important is to maintain such information in a database to ensure its ready availability to the production line for use. Equally important is the establishment of an optimal purchasing system, including global procurement of materials at low costs to meet the rising yen.

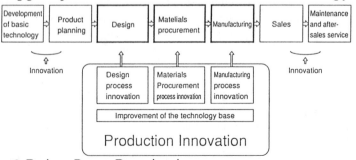

Figure 2 Business Process Re-engineering

To promote the implementation of these processes, three workshops will be organized under the C&C Committee directed by the President. In addition, model shops will be appointed to impart the thorough improvements and process innovations accomplished at the model plants to the rest of the branches of the company. This goal will be approached in step with indexes set forth at the administrative or managerial, as well as field, levels. Innovations will be pursued on a process basis for the time being but will gradually be consolidated to break through the departmental barriers.

4. Successful Practices of Product Development Innovations

Concurrent engineering (CE) is the approach to integrated product development whereby product development capabilities are enhanced throughout the flow of product development work, from the preliminary development stage to materials procurement, manufacturing, sales, and maintenance. CE has come into positive action as a powerful means of improving and innovating engineering processes.

The concept of CE is said to have been developed based on the concept of joint development which is itself derived from the Japanese notion of *wa*, or harmony, and was formulated in the U.S. to bring about the recovery of its manufacturing industry. Successful implementations of the CE concept under powerful leadership have been reported in rapid succession by manufacturers, including Boeing in the aircraft industry, Chrysler and Ford in the automotive industry, and NCR, DEC, and IBM in the electronics industry. Support measures, such as development team organization, information system (CAD) utilization, global product development efforts by way of a network, have been pursued intensively to formulate the bits and pieces of product development techniques learned from Japan into a successful theory of concurrent engineering.

4.1 Considerations for Promoting Concurrent Engineering

4.1.1 Front loading

A firm's total capabilities are directed at the initial stages of engineering to achieve high-quality yet low-cost designs that do not require retouches at a later time, together with representatives of the downstream processes, such as production engineering and purchasing. Action in the upstream processes can be of great benefit to the downstream processes.

4.1.2 Parallel engineering

Engineering activities are conducted in parallel, without waiting for the preceding process to be completed, through inter-divisional sharing of information to shorten the development leadtime.

4.1.3 Standardized engineering

Faster deliveries and high-quality yet low-cost designs are achieved through standardized engineering. Positive standardization allows serial product designs to be developed as early as in the planning stage.

4.1.4 Information system utilization

Higher efficiencies are attained by making effective use of information systems that support the performance of the successive processes, such as CAD/CAM and engineering databases.

4.1.5   Management innovation

Management innovation in the areas of development team organization, engineering progress management, quality appraisal, and product planning decision-making is also an important factor.

It is important to review engineering activities from these viewpoints and to establish standardized engineering processes that are not dependent on individual designers for success by defining the input and output to and from each activity, prescribing work rules, and optimizing the timing of activity execution. Gaining an objective and clear insight into the status of product development contributes to considerable progress in engineering process innovation.

4.2 Promotion of Concurrent Engineering at NEC Yonezawa

NEC Yonezawa has long implemented the CE method in the development of the PC-9800 series of notebook-type personal computers, the 98 Note, with success. The personal computer field is characterized by rapid technological advances and very short product life cycles, dictating us to develop and manufacture new products matched to user needs using good timing.

Under these circumstances, NEC Yonezawa has been pursuing the concept of concurrent engineering in a systematic manner by dividing it into three categories, systems, organization, and support technologies, and 14 elements.

Some of the major measures put into action so far include:

4.2.1   Interdepartmental project team

Key men responsible for design, production engineering, production management, materials procurement and so on are organized under the leadership of project managers to develop processes allowing for the downstream processes from the initial stages of engineering. Project teams are classified into three classes, tiger, heavyweight, and lightweight, according to the priority and urgency of the products of interest and the authority of the project leaders, to meet the characteristics of the projects.

4.2.2   Co-location

The members of a project work in the same place or in places close to each other to promote better interdepartmental communication. Particularly, the members of a tiger team (highest-priority project) get together at the same place to concentrate on their tasks in a short time in one accord, thereby sharing a sense of achievement through enhanced communication.

4.2.3   Utilization of information systems (CAD)

CAD has demonstrated its exceptional usefulness by penetrating all circuit design and structural design fields, including ASIC design. Particularly, three-dimensional CAD using a high-speed engineering workstation has made for greater efficiency and better quality in complicated high-density physical designs. Prototyping using three-dimensional data (stereo lithography) has been used in development applications to enable development team members to share product images and check mechanical units.

Implementation of these and other measures has cut the new product development leadtime from six months to three and a half months.

4.3 Development of a Concurrent Engineering Workbench

In addition to process innovations, such as standardization and parallel engineering activities, information system utilization, which is more advanced in the U.S., is being pursued. As an example, the CE Workbench (CEWB) has been proposed as an instruction system that provides a solution to engineering distributed across Japan as well as the world. Functions provided by the CE Workbench include:

1   Application programs, such as CAD/CAM
2   Groupware that implements virtual co-location by means of communication.
3   Process management functions that make parallel and cooperative development possible by allowing one design process to monitor the performance of other processes.
4   Product data management (PDM) functions that manage product data throughout the life cycles of the products.

The system is beginning to be implemented at various development locations, in addition to NEC Yonezawa and NEC Niigata.

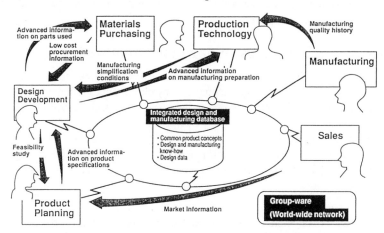

Figure 3   Image of a Concurrent Engineering Workbench

4.4 CE Promotion Guide and CE Diagnostics Method

In parallel with the implementation of the information system mentioned above, a CE Promotion Guide has been prepared that identifies requirements for various product development measures, including their priorities, and defines attainment steps.

The guide classifies CE measures into five categories defining attainment steps: organization, consciousness, process management, information sharing, and tool technology. The tasks or problems that confront the development department are mapped into CE measures to allow the systematic attainment of CE. Product development stories are covered to illustrate ongoing in-house approaches to concurrent engineering as an aid to future product development.

Further, a CE diagnostic method has been developed that measures the degree of concurrent engineering attainment within each development department by a quantitative scale, discusses it from a qualitative viewpoint, and proposes innovation points in the engineering processes. The concurrent engineering diagnostic method has been put to use in the in-house development departments. This method clearly defines four levels. The best practice selected from among those observed in house and outside is rated as level 3. An extension of that level is rated as the highest level, or level 4.

With this CE diagnostic method, all the workers in the department in question are invited to ask certain questions at each level. Their answers are compiled according to the respondents' tasks and functions and assigned levels according to the aforementioned five categories of organization, consciousness, process management, information sharing, and tool technology. The resultant levels are expressed in a radar chart to represent the degree of CE attainment. When the problems involved in the attainment of concurrent engineering by the department in question are thus identified, a detailed hearing is conducted and improvements suggested are, which are put into action with the cooperation of responsible personnel in that department.

This method provides an objective measure of the degree of concurrent engineering attainment by each department, along with mean values, thereby defining the improvement goals to approach.

5. Manufacturing Process Innovations Achieved

Lean production is a plan of continual activity aimed at streamlining successive manufacturing processes into a continuous flow by eliminating inefficient work steps. It thus drastically cuts manpower, space, equipment, time, and the overall cost thus expediting manufacturing.

We set model plants at NEC Nagano and NEC Saitama to proceed with a three-step improvement program in their manufacturing and production management departments involving improved worker morale, removal of inefficiencies, and balanced production.

5.1 Step 1:   Improved Worker Morale

Improved worker morale creates a background that supports the creation of a conceptual breakthrough in the traditional scheme of production. It stimulates the dismantling of established concepts, the location of inefficiencies in the manufacturing site, and their speedy removal.

5.2 Step 2:   Removing Inefficiencies from the Manufacturing Site

Unnecessary shelves, fixture, and apparatus are removed from the manufacturing site first. This action alone should create a space of several thousands of square meters within the plant. Next, inventories and work in process stacked around are reduced to a minimum based on the view that all that is being moved or is stagnant as an inefficiency.

Further, the physical space intervals between workers in a production line are compressed by reviewing their process workload to ensure an equal rate of workflow, taking into account their work capacity. Compression can cut a production line to 1/10 or even 1/40 of its original length, and also enables worker reductions.

Next, parts that had been stored in an external rental warehouse are laid in space created above the production line to allow workers to see at a glance where parts are stored, who manufactured them, and what these parts are. This is called a "store."

The workers in each production line each have a compact part storage yard, called "refrigerator, beside them, in which they keep a set of parts they use. The maximum quantity that can be placed in the refrigerator is fixed for each part. This maximum is indicated on "Kanban". "Kanban" also indicates in which process and for which model each part is required, on which floor and in which corner they are stored, and what is the minimum stock level below which parts must be retrieved from the store. The store, refrigerator, and "Kanban" combine to establish JIT (just in time) in the plant.

5.3 Step 3:   Balanced Production

Suppose that you have three different products, A, B, and C. Their monthly production quantities are 4,800, 2,400, and 1,200 units respectively. Assume that each month has 21 working days. These products are manufactured in descending order of output. Product A is manufactured in the first 12 days of the month, product B in the next six days, and product C in the last three days. This production plan may appear efficient because products A, B, and C require only one setup change each, but dealers who want all three products A, B, and C must wait for one month. The manufacturer, on the other hand, must keep finished units of product A in stock for one month at the longest. To remove these inefficiencies, we need to manufacture all of products A, B, and C in the same day. This is the concept of balanced production.

Since balanced production has the disadvantage of frequent setup changes, it is essential to take measures to cut the setup change time. The advantages offered by balanced production by far offset this demerit, however. They include being able to offer of products A, B, and C to dealers every day, cutting inventories of finished units, and saving storage space and administrative expenses. In step 2, "Kanban" is circulated to represent the flow of parts in the plant. Balanced production would also require JIT production involving the manufacturers of external parts and material.

NEC Nagano and NEC Saitama have proceeded to this step and achieved a certain degree of success each. The total production line length at NEC Nagano has been cut from 1780 m to 440 m. Now a one-worker production line is in existence, in which one worker assembles and inspects more than one hundred parts alone. At NEC Saitama, on the other hand, the total production line length has been reduced from 200 m to 9 m. The line is shaped like a spider to accommodate frequent lot changes and to allow each worker to accomplish a complete job at his or her pace. Floor space has been cut by 12,400 m^2 at NEC Nagano and by 3400 m^2 at NEC Saitama. Added productivity has created 160 excess workers at NEC Nagano and 64 at NEC Saitama. These excess workers have now been assigned to different jobs. Further, sizable cuts in both the leadtimes and work in process have been achieved.

5.4 Single-Unit Production

We are committed to developing the concept of balanced production in Step 3 into small-lot balanced production, in which we are manufacturing products in smaller lots or in smaller quantities, ultimately in lots of one unit. In other words, the ultimate goal of JIT is to produce only the number of units that has been sold. In addition to the cooperation of the external parts and material manufacturers, balanced order taking is essential to achieving this goal. Innovations in the manufacturing processes inevitably demand innovations in the upstream sections (parts and material manufacturers, and even engineering and development departments) and the downstream sections (product logistics, marketing, and sales departments) of the plant as well.

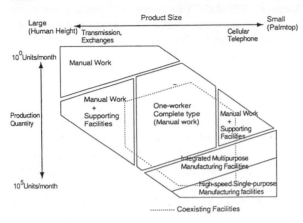

Figure 4 Clasification of Equipment Assenbly
Facilities / Line

These improvements demand the development of manufacturing and testing facilities tailored to the concept of lean production. This goal involves defining the smallest unit of assembly work and developing facilities made up of a combination of basic modules, as well as clearly distinguishing between single-purpose and multi-purpose facilities to cut equipment investment.

6. Conclusion

The history of production began with craftsmanship, in which craftsmen manufactured only small quantities of products craftsmanship gave way to mass production, as examplified by Ford's production lines. Mass production enabled phenomenal leaps in industry. Later, new concepts of production that meets the needs of multi-item low-volume production, such as the Toyota production plan, came into existence.

The scheme of production with which NEC is concerned calls for building information systems that take full advantage of IT (information technologies) to ensure optimal production throughout the flow of supply, from planning to scrapping, as a way to address the upcoming era of mass customiza-

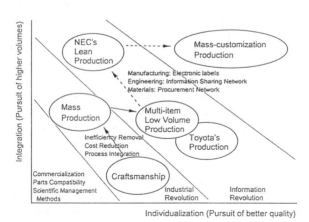

Figure 5 Evolution in the History of Production

tion. Using IT allows the production system to handle and manage vast amounts of custom-specific information in the era of the so-called individualized masses.

A concept advocated and promoted in and around the U.S., CALS (Continuous Acquisition and Life Cycle Support) will greatly aid in future system implementation. CALS assimilates the world's standards thoroughly to realize the exchange of information between firms, as well as within the departments of a single firm. These standards include SGML, which standardizes the practices of documentation management, STEP, which focuses on product data relating to CAD/CAM, and EDI, which computerizes commercial transactions.

We will be realizing on-demand production in an optimal environment to provide a prompt solution to customer needs while accommodating these and other CALS-like approaches.

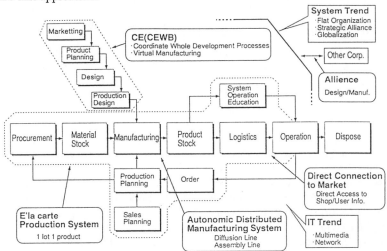

Figure 6 System Image in Late '90

The scope of innovative activities we are proceeding with is shown here. We will be aiming at optimizing the manufacturing processes by expanding this scope of activities further to create a fully integrated system that covers the entire flow of supply.

Biography

Makoto Tachibana received B.E. and M.E. degree from University of Tokyo in 1963 and 1965 respectively. He joined Microwave and Satellite Communications Divisions in NEC Corporation in 1965. From 1970 to 1987 he was engaged in hardware design such as line printers and hard disk drives in Information Processing Group. Since 1987 he has been in Production Engineering Group and in charge of production engineering development. Since 1991 he has been a vice president.

3

Contribution to renovation of a camera factory by intelligent production technology

Sir John H. Noble *Kamera Werke Noble GmbH*
Dresden
Prof. Dr. Kochan *SFM GmbH Dresden*

Abstract

The new technologies for the direct generation of a geometrical object of any complexity, based on 3 D-CAD-Data, are of general importance of the technical progress. Broadly-based and applied in leading industries like airplane and automotive factories, these technologies, named Solid Freeform Manufacturing, Layer Manufacturing, Rapid Prototyping and otherwise, are playing more and more an essential role in small and medium-sized companies.
The following paper presents experiences of product development and manufacturing technologies, also in a small enterprize.

Keywords

Solid Freeform Manufacturing, Rapid Prototyping, Concurrent Engineering

1   INTRODUCTION

The competitiveness of small and medium sized companies depends on the ability for utilization of advanced manufacturing technologies.
The new technologies for the direct generation of geometrical objects, of any complexity based on 3 D-CAD-data (named Solid Freeform Manufacturing, Rapid Prototyping or otherwise) plays an important role in Leading companies of the

airplane and automotive industries. These state of the art technologies are also of increased importance for small and medium sized enterprizes. Because of the high investment cost, new types of innovative service centers and specific kinds of workdivision are requested.

The Noble Camera factory is a midsized corporation of the high tech-optical electronic industry. One of the main product-lines is the NOBLEX-Panorama camera for professional photographers and demanding amateurs.
Based on the international well-known professional panorama camera NOBLEX PRO for roll-film with a rotating lens (picture angle 146°), the goal was formulated in 1993 for the development of a 35 mm panorama camera with a rotating lens (picture angle 136°) and also high functionality.
For the development of the overall housing and mechanical components, a cooperation with the new-founded small company for fast manufacturing. of models incorp. was started. This high-tech company (in german: SFM-Gesellschaft zur schnellen Fertigung von Modellen) deals with all engineering tasks along the process chain for Solid Freeform Manufacturing, especially

- fast digitizing, scanning and reverse engineering
- 3 D-Modelling
- SFM-building processes
- follow up processes, like vacuum casting, investment casting a.o.

The following paper demonstrates the remarkable advantages by using the new technologies for the accellerated product development and manufacturing.

2   CHARACTERISTIC of the new SFM-technologies

The very general principle is shown in Fig. 1, which demonstrates the following points.

1.   The essential prerequisite of 3-D Data models
2.   The application of specific materials such as fluids, powder, wire or laminates. This is currently different from traditional manufacturing processes, a decisive restriction
3.   The use of newly developed highly sophisticated equipment employing different physical principles
4.   The physical objects are currently limited in size depending on the working ranges at the SFM equipment, but without restrictions concerning complexity and geometrical features.

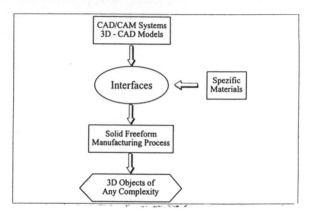

Fig. 1: Simplified Principle of Solid Freeform Manufacturing

A subdivision of the commercially available system, from the application point of view, is possible based on the usable materials and the used physical effects. (see Fig. 2) The stereolithography process is currently the most industrially applied principle and in a leading position on accuracy requirements.

For more detailed explanations, some specific literature is available / 1, 2, 3 /

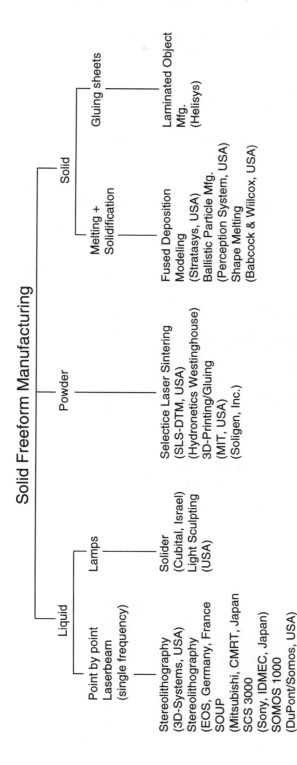

Fig. 2

3   Product development and manufacturing by utilization of advanced manufacturing technology.

The new technologies are not only procedures which allow to replace traditional processes, the SFM-process chain is becoming an increasingly valuable tool in the Integrated Product and Process Development environment.
The advantages and effects can be utilized starting from the product design up to the different follow-up processes depending on the requested number of parts.

3.1   Design and decision-Making of design variants

The market success of a new product is decisively influenced by the quality of the design. Normally some design variants in foam material will be prepared.
Some examples for the new panorama camera are shown in figure 3.

In consideration to the utilization of new SFM-technologies, there was given no restriction concerning freeform shapes and other aestetically-shaped elements.

The result of the decision process by the management of the Noble camera factory is shown in fig. 4.

3.2   Fast digitizing

For some design patterns, the point net was digitized and by the principle of reverse engineering /4/ prepared for further computerized processing
Digitizing is only justified for some selected shape elements, e.g. the frontside of the camera. The other parts must be included into the general task for 3 D modeling.

3.3   3 D-Modeling

As it was emphasized before, the 3D-Modeling is an absolute "must" for the SFM-processes. In the case of the development of the panorama camera the closed cooperation between the constructor and 3D-modeling expert was most useful.
The camera is a very complicated technical device, which includes optical, mechanical and electronical components.

The startpoint for the construction was the functional principle of the camera, the structure, the measurement of the essential function elements and the design drafts for the outside form.

Therefore a closed cooperation between designer, constructor and 3D-Modeling expert was necessary. The 3D-Modeling of the outside shape was realized in consideration to the necessary subdivision of single parts related to functional requirements and manufacturing aspects. For the construction of the irregular single parts of the camera housing with many freeform shapes the utilization of the 3D-surface modeling system STRIM 100 (Cisigraph) was very useful.
Essential advantages are:

- multisided possibilities for the description of complicated shape elements.
- unification of surfaces to surface units by determination of different connection conditions between the single surfaces
- different and independent description of outside and inside contours of objects
- extensive systemsupport for the rounding of surface boundaries and the determination of lift out angles for casting procedures.
- simple and fast possibilities for the local and global changing of surface description.

After the successful construction and 3D-modeling, the CAD-data are available for the following processes.

3.4.   SFM-building procedures

Additional to the 3D-CAD-modeling the availibility of real objects of all technical objects and components is very necessary. The real relationships and proportion in complex and complicated devices or assembly groups can not be simulated by computer support. Therefore the building of design or function pattern for many investigations is a compelling need. The rise of any product-development can be minimized if in an early stage test objects are available.

For this reason the SFM-procedure stereolithography was applied for all seven housing parts. The manufacturing of the models was realized with the stereolithography-system STEREOS 400. For the produced prototypes the following demands was requested

- checking of the design of the outside shape
- checking of the construction of the single parts related to productibility and
  functionality
- suitability as basic models for vacuum casting parts, which can be used for
  assembling of applicable test cameras
- suitability as basic model for manufacturing of forms for injection molding by
  metal spray procedures. This parts must be usable for midsize production
  (up to 5000 pieces).

Because of the necessity for parallelization of product-development tasks and preparation of the manufacturing processes several experimental samples of the camera has to be put ready. For the multiplication of single stereolithographic models into 20 plastic parts the vacuum casting procedure was successfully used.

3.5.   Application of prototypes, testmodels and preparation of the medium sized production.

Without expensive tools it was possible to produce complete camera housing in an early developmental phase. So it was possible to make a lot of necessary corrections at the single parts without additional expenditures for the tool making for the midsize production. Two typical examples will be used for the explanation of the advantages. Selection of the suitable motor for this specific requirements.

Investigations concerning fitting and handlibility of the camera was possible with the parts, which was manufactured by vacuum casting. Additionally, it was possible to provide for the production of the packaging and the casing in an early stage for this specific construction and manufacturing tasks. Furthermore it was useful to prepare prospects by specific painted models. For the necessary marketing activities the prospects was available of the same time when the serial production starts.

The preparation of the serial production starts also simultaneously to the development and construction of the new camera.

The selection of suitable manufacturing procedures must be in agreement with the complexity and number of the part. In this relation has to be considered the relatively small number of parts.

That means it is necessary to apply manufacturing procedures which allow a low price level also for small and medium sized production.

An essential cost factor are the manufacturing cost for the injection molding tools. These manufacturing procedures for complicated freeform shapes requires multi-axis milling and EDM processes for the quality justified production. In this cases the NC-programming was possible direct based on the 3D-CAD-data.

This is also a opportunity for the reduced amount.

Some parts are suitable for the new metal-spray procedure. This procedure reduces the cost up to 20% in comparision with the traditional NC-manufacturing.

The tools, manufactured by the metall-spray procedure, can be used up to 5000 injection moulding parts.

4.   Economical effects and results

The application of the described innovative methods and procedures for the product development and manufacturing of the new panorama camera leads to a lot of effects. Currently the exact qualification is not easy, because there doesn`t exist analogues traditional processes for such complicated parts. Most important is the drastical reduction of the development time for the camerahousing from usually 18 month up to 7 month. In this case has to be considered the increased degree of complexity of the parts.

The consequent realization of the methods of Simultanious Engineering including extented and early tests at real models allows the following advantages and principles.

- investigation and evaluation of constructive and technical variants.
- optimization of application features
- reduction of construction failures up to nearly "Zero"
- multiple usage of optimized CAD-data for follow up and parallel processes
- preparation of serial manufacturing and tooling in parallelization with the product testing.

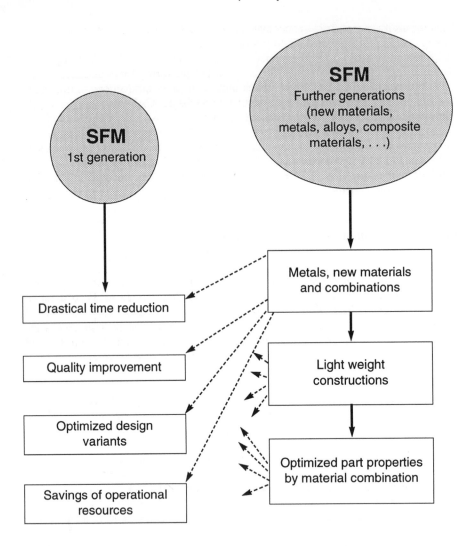

Fig. 5 Essential effects of the new epoch of Production Technology

5.   Summary and concession:

The new technologies of Solid Freeform Manufacturing are still in the first phase of industrial application. But if this possibilities are not only applied as specific procedures for determined requirements, f.e. show and tell models, the entire advantages are most important. If the new technologies are understood and applied as a strategic tool along the entire process chain from product design till manufacturing, the effects can be enormous. The broad utilization of all given advantages guarantees highest quality, shortest time and lowest cost.
The successful development of the new NOBLEX Panorama camera is one evidence of the new development stage in direction of the new epoch of advanced production technology.
In general, by application of the new technologies, time and cost savings in the range of 50% up to 90% can be expected. The explanation for such tremendous advantages is given in Fig 5. This figure demonstrates also some essential trends for further developments.

References

/1/   Kruth, I.P.   Material Incress Manufacturing by Rapid
                                 Prototyping Technologies;
                                 Annals of the CIRP, Vol. 40/2 1991

/2/   Jacobs, P.   Rapid Prototyping & Manufacturing
                                 Fundamentals of Stereolithography;
                                 Dearborn, SME 1992, USA

/3/   Kochan, D.   Solid Freeform Manufacturing -
                                 Advanced Rapid Prototyping
                                 ELSEVIER Amsterdam 1993

/4/   Kochan, D.   Intelligent Production Systems-
                                 Solid Freeform Manufacturing
                                 Proceedings International Conference
                                 Sept. 29.30.1994 GFaJ Sachsen

PART TWO

Modeling Technology

4

Cutting and pasting constrained boundary features

M. Ranta[1], M. Mäntylä[1], M. Inui[2], and F. Kimura[3]

[1]Laboratory of Information Processing Science
Helsinki University of Technology
Tekniikantie 12, Innopoli, 02150 Espoo 15, Finland
tel +358-0-4354 3967, fax +358-0-4354 3965
email mra@cs.hut.fi

[2]Department of System Engineering, Ibaraki University
Naka-narusawa-cho 4-12-1, Hitachi-shi, Ibaraki 316, Japan
tel +81-249-35-6101 ext 409, fax +81-294-32-1546,
e-mail inui@sumire.dse.ibaraki.ac.jp

[3]Department of Precision Machinery Engineering,
University of Tokyo
Hongo 7-3-1, Bunkyo-ku, Tokyo 113, Japan
tel +81-3-3812-2111 ext 6455, fax +81-3-3812-8849,
e-mail kimura@cim.pe.u-tokyo.ac.jp

Abstract

This paper presents a feature modelling approach that allows the designer to reuse former design solutions in the definition of new design features. Instead of the conventional approach of providing the designer a pre-defined set of basic volumetric features to build parts with, the proposed system offers utilities for *cutting* portions of former models as new boundary features that can be *pasted* into new product models. The boundary features are based on an ordinary boundary representation model that has been extended, or relaxed, in order to allow the manipulation of open boundary portions, i.e., incomplete solids that lack some faces and thus have no volume. The size and position dimensions of the boundary features are presented and managed with a constraint system that is attached to control the geometry information of the boundary representation model. A small prototype system has been implemented to demonstrate the suggested approach.

Keywords

Design features, boundary features, reuse of former solutions, presentation of incomplete geometry, dimensioning, constraints

1 INTRODUCTION

The difficulty of feature modelling lies in defining a feature concept that falls between generally applicable geometric modelling and application dependent feature hierarchies. Pre-defined feature libraries restrict the designers too much and there is an obvious need for interactive feature definition tools. This paper goes even further and understands design features to be a method for supporting reuse of former design solutions. In addition to a common library of standard features, each designer may collect his/her own feature catalogue by selecting, or cutting, portions of existing product models to be pasted into new products.

Boundary representation models have been proven to be a suitable basis for feature modelling since they provide easy access to faces and their attributes. However, the potential of boundary manipulation is so far not properly exploited as the emphasis has been on volumetric operations related to manufacturing features. Various non-manifold models, for example (Rossignac 90) and (Gursoz 91), have been presented to extend modelling capabilities from conventional solid models towards tools for managing mixed-dimensional models. The proposed cut and paste approach requires utilities for manipulating a particular type of non-manifolds that correspond to open boundary portions or sets of faces that are the basis for boundary features.

An essential characteristic of feature based systems is parametric modelling. According to a common approach a feature is pre-defined with a certain set of size parameters that are passed as arguments to a procedure that generates the volumetric model. The utilisation of constraints allows more flexible dimension driven geometry such as suggested in (Gossard 88), which proposes relative position operators for dimensions so that changes in dimension values can be automatically translated as changes in geometry and topology. Other examples of constraint based approaches are (Kimura 86) and (Suzuki 88) which propose a declarative method for generating geometry and features according to a constraint representation of a higher level. In this paper constraints are proposed as a methodology for managing the sizing and positioning of boundary features.

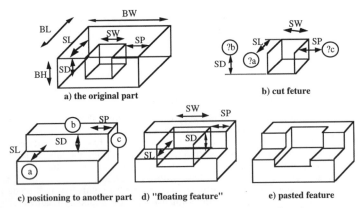

a) the original part   b) cut feture

c) positioning to another part   d) "floating feature"   e) pasted feature

Figure 1 An example of cutting and pasting a constrained boundary feature.

The rest of this paper is organized as follows. First, we give an overview of the design process supported by constrained boundary features. Section 3 gives an outline of the architecture of the implemented prototype system. Next, the constraint model used in this work and its connection with the boundary representation model is described. Section 5 describes the cut and paste operations for constrained boundary features. Finally, we give our conclusions and discuss topics for further research.

2 DESIGN WITH CONSTRAINED BOUNDARY FEATURES

A new boundary feature is defined by selecting a set of faces and cutting the boundary portion as shown in the example of defining a slot-like feature in Figures 1a and b. If the definition is based on a plain boundary model the designer must first explicitly add the dimension constraints to the model. The resulting constrained boundary feature is an incomplete solid with broken constraints as shown in Figure 1b. The open edges and broken constraints maintain default information on the neighbour and reference faces, which allows an unattached feature be modified by changing its size dimensions such as the width of the slot.

Figure 1c and 1d demonstrate the pasting of a feature. First, the designer repairs the positioning constraints by replacing the default reference faces with proper ones according to the new environment and assigns proper values to the positioning dimensions. Figure 1c shows how the slot length is set to refer to the front face of the part and slot depth to the top face and finally the position is related to the right hand side face of the part. Next, the designer accepts the positioning and the system adjusts the boundary and constraint models together. As Figure 1d shows the border edges of the feature are replaced by new ones calculated according to the intersection of the surfaces of the meeting faces.

Figure 2 shows how a feature structure is useful in maintaining a constrained boundary model. When the designer removes the slot of Figure 2a the hole becomes "floating" since it was originally positioned with constraints referring to the faces of the removed slot. The designer may now either delete the hole or repair the positioning by attaching the constraints to the remaining faces of the part as Figure 2c shows.

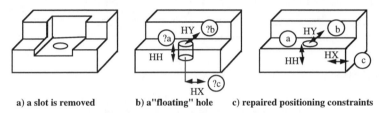

a) a slot is removed   b) a "floating" hole   c) repaired positioning constraints

Figure 2 Repairing constraints of the hole after removing the slot.

The explained cut and paste based design process is a simplified one and only intends to give a clear overview of the design style. In real applications the constrained boundary features are typically much more complicated and larger entities which shows the powerfulness of the reuse approach as new products are created by attaching together just a few large design and function oriented features.

3 CUT AND PASTE BASED FEATURE MODELLING SYSTEM

The architecture of the prototype system consists of a form feature modeller, a dimension feature modeller, and a feature catalogue as shown in Figure 3. The presented applications correspond to the current implementation; however, a realistic feature modelling environment requires tools for also other aspects such as assembly modelling and manufacturing.

The form feature modeller offers access to feature-based and geometric modelling operations. The geometric modelling operations have been built upon the ACIS modeller. This boundary representation modeller has been extended to allow manipulation of boundary portions that do not necessarily have to be solids. The boundary manipulation operations are in turn the basis for cut and paste tools for boundary representation of shapes.

The dimension feature modeller facilitates manipulation of dimension values and definition of constraints for maintaining dimensions. The DeltaBlue algorithm (Freeman-

Benson 88) has been used as a foundation for developing a dimensioning system that controls constraints relating geometric entities, i.e., mainly distances of surface variables. The constraint model controls the dimensions of the boundary representation model and thus the models are in frequent interaction.

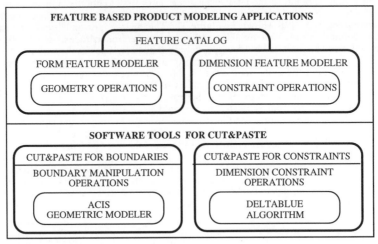

Figure 3 A feature modelling environment.

4 ATTACHING CONSTRAINS TO A BOUNDARY MODEL

The cut and paste approach was first implemented for boundary features as explained in (Ranta 93). The designer may select portions of former boundary representation models and reuse them in new designs. Boundary features are usually not volumetric entities but rather correspond to a sets of faces that form an incomplete or open solid. They can be pasted in a quite flexible manner together by adjusting the open faces and edges according to the intersections of surfaces as shown in Figure 1. However, the scaling possibilities of boundary features are poor and, moreover, their locations are determined by clumsy reference coordinate systems. Both of the shortcomings emphasise the need for proper facilities for managing engineering dimensions.

In order to make boundary features parametric and positioning easier a constraint based dimension feature modeller is suggested. The constraint modeller is founded on the DeltaBlue Algorithm (Freeman-Benson 88, 90) that is a fast incremental constraint solver for satisfying dynamically changing constraint hierarchies. Its successor SkyBlue has been improved to support multi-output methods and allow cycles of constraints to be constructed, although it may not be able to solve the cycles (Sannella 92).

Constraints allow the designer to state declaratively a relation that is to be maintained. Figure 4 shows the operations for expressing the slot length constraint (SL) of Figure 1a. First, two variables are created to store references to the two faces that are to be constrained, i.e., the end face of the slot and the front face of the blank. A third variable is created for representing the value of the constrained length dimension. The `ControllerVariable` and `LongVariable` classes demonstrate how SkyBlue allows the variables to contain arbitrary data, not just numbers. Both of them are subclasses of `Variable` class that records the information needed by the constraint solver, for example a list of all constraints that reference to the variable and the constraint that presently defines the variable.

Next, a constraint is created as an instance of the DistanceConstraint class to bind the distance between the two faces to the length value. Each constraint is labelled with a strength indicating how important it is to satisfy it; in the example the distance constraint is given the strongest S_required priority. The constraint hierarchy guides the constraint satisfaction algorithm to leave weaker constraints unsatisfied if needed for keeping stronger ones satisfied when the model is overconstrained. In the case of an underconstrained hierarchy, the user can add weak constraints to suggest which solution is chosen. Similarly to the Variable class, also the parent class Constraint of DistanceConstraint records other information needed by the constraint satisfaction algorithm, such as a list of variables that are constrained by the constraint. Furthermore, each constraint class provides access to the possible methods for enforcing the constraint. For instance, the methods of DistanceConstraint provide procedures for calculating the LongVariable value according to the distance of the ControllerVariable surfaces, or for moving the ControllerVariable surfaces to a distance determined by the LongVariable value. The appropriate method is chosen incrementally by the constraint satisfaction algorithm.

```
// Create variables for the reference faces and for the dimension values
ControllerVariable *SlotEndFace = new ControllerVariable ( "SlotEndFace", slot_end_face_pointer );
ControllerVariable *BlankFrontFace = new ControllerVariable ( "BlankFrontFace", front_face_pointer );
LongVariable *SlotLenght = new LongVariable ("SlotLength", 100.0);

// Create a constraint that manages the distance of the two faces
DistanceConstraint *SlotLengthConstraint = new DistanceConstraint (SlotEndFace, BlankFrontFace,
                                                     SlotLenght, S_required);
// Assign a new dimension value and solve constraints accordingly
SlotLenght->Assign (150);
// Update the boundary model according to changed controller surfaces
geometry.update();
```

Figure 4 Creating a constraint on slot length and modifying the length value.

In this way the dimensioning constraint operations allow the construction of a model that controls the relative locations of the surfaces of some faces. The system is able to maintain a solution to the constraints incrementally when constraints are added or deleted, or when new values are assigned to the dimensions. As the last row of Figure 4 suggests, it is necessary to integrate the constraint model to the geometric model to make the dimensioning affect it.

Figure 5 shows how the connection between a boundary representation model and a constraint model is built through a surface interface. A ControllerVariable of the constraint model contains a pointer to a surface record that belongs to the data structure of the geometric model which thus changes according to the dimensioning constraints.

(Kimura 86) proposes a natural way to present dimensions of 3-D objects by setting constraints on surfaces. The approach uses solid models for shape description and logical formulae for representing geometric constraints on the solid model elements. We propose a similar approach that is extended to support dimensioning of incomplete solids and to utilize feature-based presentation that makes distinction between sizing and positioning constraints.

The left hand side of Figure 5 illustrates the data structure of a boundary representation model (Mäntylä 88) which represents the topology of a part in terms of shells, faces, loops, coedges, edges, and vertices. Geometric information is represented by attaching surface information to faces, curve information to edges, and point information to vertices. We assume that the topology of a solid remains unchanged and that dimension modifications affect only the geometry of a part. This assumption is acceptable since topological changes would change the character and functionality of a part or a feature in a crucial way. Thus, changes in dimensioning affect only geometric information, and in the current state of our implementation, particularly surface information. When the constraint modeller has propagated some dimension changes to the surfaces, the rest of the geometry is updated accordingly. Curves are updated by recalculating the intersection of the surfaces of the faces that meet at the edges. Accordingly points are updated by recalculating the intersection of the curves of the edges that meet at the vertices. Thus, the constraint model and boundary representation model are integrated into a constrained boundary model.

It is possible to extend the interface to other entities carrying geometric information, such as curves and points. At any rate, the principle is that dimensioning affects the geometric information while the topology is maintained.

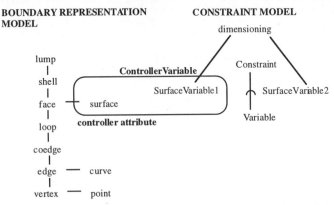

Figure 5 Interface of the boundary representation model and constraint model.

5 CUT AND PASTE OPERATIONS

The cut and paste operations are implemented in an analogous way for different types of models according to the pattern shown in Figure 6. In the case of the constraint model the intersection relation corresponds to a constraint and the entity references are surface variables. The cutting (6a->b) of a constraint occurs when its surface variables are separated into different boundary models as a result of the cutting of a boundary model. During the paste (6b->a) of a constrained boundary feature the constraints are treated first to determine the positions for the pasting of boundary models.

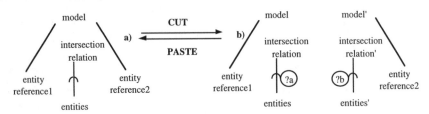

Figure 6 Cutting operation a->b and pasting operation b->a.

In the case of the boundary representation model, the intersection relation of Figure 6 corresponds to an edge which relates two halfedges corresponding to the two faces that meet at it. When an edge is cut (6a->b), the result is two open edges that belong to different models and lack one halfedge as well as a face from one side. The paste operation (6b->a) of two boundary models includes a adjustment that modifies the neighbouring edges to fit each other.

Figure 7 explains the stages of the cutting algorithm for integrated boundary representation and constraint models. In the case of cutting, the boundary model is the basis for user interaction and thus precedes and controls the cutting of the constraint model. Details of the cutting of a boundary feature are explained in (Ranta 93). After the boundary feature has been separated the constraints are classified according to their surface variables. If both controller surfaces of a constraint belong to the boundary feature the constraint classified local to the feature, i.e., it is a size definition. Accordingly when both of the controller surfaces belong to the product the constraint is local to the part and sets a constraint on its size. Such local constraints are simply attached to the appropriate boundary model.

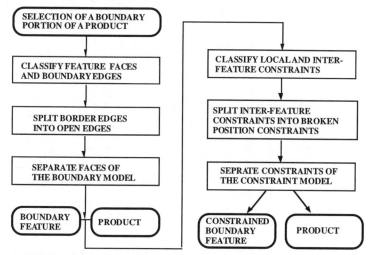

Figure 7 Cutting first the boundary model and then constraint model accordingly.

If one of the surface variables of a constraint belongs to the boundary feature and the other to the product it is an inter-feature constraint, i.e., it is a position definition that defines a relation across feature limits. Such inter-feature constraints are cut according to Figure 6 into two broken constraints that are classified so that one belongs to the feature and the other to the product. The ?a and ?b symbols denote that the cutting left the constraints with default surfaces on the place of one of their controller variables. The a? and b? default variables do no longer have pointers to any actual faces, but just carry a copy of the former surface definition.

```
// The slot lenght constraint was broken by the cutting of the slot.
// Create a default controller variable using a copy of the surface of the front face
ControllerVariable *DefaultFrontSurface = new ControllerVariable ( "DefaultSurface", surface_copy );
// The slot length constraint is now broken and refers to the default front surface
DistanceConstraint *SeparateSlotLengthConstraint = new DistanceConstraint (SlotEndFace,
                                                    DefaultFrontSurface, SlotLenght, S_required);
```

Figure 8 Generating a default surface controller when the constraint model is cut.

Figure 8 shows operations related to the cutting the length constraint that takes place when the slot of Figure 1 is cut from the blank. First, a default controller variable is created to store a copy of the surface of the blank's front face. Then a length constraint of the separated slot is defined to refer to the default surface variable.

Finally the constraints are separated into groups that are attached to the appropriate boundary model. The result is a constrained boundary feature with open edges and broken positioning constraints that are to be repaired while the feature is pasted.

Figure 9 demonstrates the pasting algorithm for an integrated boundary representation and constraint model. In case of pasting the constraint model is the basis for user interaction and thus precedes and controls the cutting of the boundary representation model. First, the designer may change the size of the constrained boundary feature by assigning values to its size constraints and see the effect on the updated boundary model.

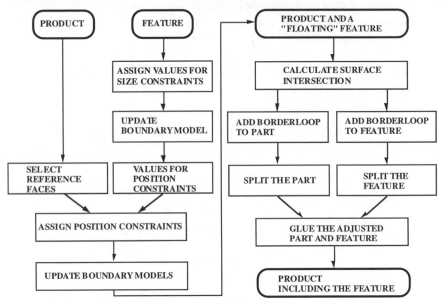

Figure 9 Pasting first position constraints and then boundary models accordingly.

Next, the designer repairs the positioning constraints by selecting new reference faces to the positioning constraints and setting appropriate values to position dimensions. This attaches the constraint models of the feature and the product to into one constraint hierarchy. Figure 10 shows how the default controller variable is assigned with a front face from the new blank and the geometry is updated to be able to display the slot floating at its planned position.

```
// Position the slot according to a face from its new environment by assigning it to the controller variable
DefaultFrontSurface ->Assign(new_front_face_pointer);
// Update the boundary model according to changed controller surfaces
geometry.update();
```

Figure 10 Assigning a new reference face to position the feature in a new environment.

At this stage, the boundary representation models are still separate. The constrained boundary feature is positioned according to its new environment and "floats" in its future location. The designer may now iterate the editing of feature size and position until he/she is happy and ready to finally let the system to paste the boundary models into one product model wholeness. The right hand side of figure 9 demonstrates how the boundary feature and product are adjusted to each other and then glued together. Details of the pasting of boundary features are presented in (Ranta 93).

6 CONCLUSIONS

The proposed cut and paste approach provides the designer an interaction method that is more obvious and produces less surprising results than the set operations on volumetric entities. The constraint based sizing and positioning allows the designer to apply the familiar practices of engineering dimensioning in a straightforward manner instead of constructing less practical systems of reference coordinates. Thus, the cut and paste interaction style was found suitable both boundary representation and constraint feature modelling.

The proposed constraint feature model extends former approaches to allow the management of incomplete solids. Furthermore, it experiments the presentation of broken constraints. Although the current prototype offers a limited set of possible constraints it suggests the advantages of the capability to represent incomplete states during the design process. Product models that are incomplete as solids and features that float according to constraints allow the designer to postpone design decisions to an appropriate time.

If beforehand standardisation is considered to be the crucial characteristic of features it is questionable whether the proposed approach can be regarded as feature modelling. However, particularly in the case of design features, the advantages of user defined features and an evolving feature library seem to be well justified, since designers find it easier to work with large function related features than basic general ones.

Constrained boundary modelling does not necessarily require any feature concept, but may be applied as an interaction method as such without any recording of the feature structure at all. However, the recording of the feature structure is practical for manipulating relationships of features, maintaining meaning of shapes, and so on.

The cut and paste approach has so far been implemented for a boundary representation model and a constraint model. The analogy in cutting and pasting these models was closer than expected which is a promising result when further extensions of the cut and paste approach are considered. In the future the system should manage other types of product information as well, for example tolerance, material and assembly information. Such variety of viewpoints is essential as an ultimate goal of feature based product modelling.

7 REFERENCES

Freeman-Benson, B.N. and Maloney, J. (1988) The DeltaBlue Algorithm: an incremental constraint hierarchy solver. Technical report 88-11-09, Department of Computer Science, University of Washington, Seattle, Washington, 98195, USA.

Freeman-Benson, B.N., Maloney, J. and Borning, A, (1990) An incremental constraint solver. Communications of the ACM, January 1990, 33(1), 54-63.

Gossard, D.C., Zuffante, R.P. and Sakurai, H. (1988) Representing dimensions, tolerances, and features in MCAE systems. IEEE Computer Graphics and Applications, 8(8), 51-59.

Gursoz, E.L., Choi, Y. and Prinz, F.B. (1991) Boolean set operations on non-manifold boundary representation objects. CAD 23(1), 33-39.

Kimura, F., Suzuki, H. and Wingård, L. (1986) A uniform approach to dimensioning and tolerancing in product modelling. In Computer Applications in Production and Engineering, CAPE '86, (ed. K. Bø, L. Estensen, P. Falster and E.A. Warman), 165-178. IFIP, Elsevier Publishers (North-Holland).

Mäntylä, M. (1988) An introduction to solid modelling. Computer Science Press, College Park, Maryland.

Ranta, M., Inui, M., Kimura, F. and Mäntylä, M. (1993) Cut and paste based modelling with boundary features. Proceedings of the Second Symposium on Solid Modelling and Applications (ed. J. Rossignac, J. Turner and G. Allen), ACM, Montreal, Canada.

Rossignac, J. and O'Connor, M. (1990) Selective geometric complex: a dimension-independent model for pointsets with internal structures and incomplete boundaries. Geometric Modelling for Product Engineering (ed. M. Wozny, J.U. Turner and K. Preiss), North-Holland.

Sannella, M. (1992) The SkyBlue constraint solver. Technical Report 92-07-02, Department of Computer Science, University of Washington, Seattle, Washington, 98195, USA.

Suzuki, H., Ando, H. and Kimura, F. (1988) Synthesizing product shapes with geometric design constraints and reasoning. Proceedings of IFIP WG 5.2 2nd Workshop on Intelligent CAD, North-Holland, Amsterdam.

BIOGRAPHY

Mervi Ranta is a research scientist at the Laboratory of Information Processing Science in the Helsinki University of Technology. She is currently working on a design feature approach that promotes the reuse of product knowledge. She received her MS in Computer Science from Helsinki University of Technology in 1987. During 1988-1992 she was a visiting researcher at the Kimura Laboratory in the Department of Precision Machinery Engineering of the University of Tokyo. Her research interests include feature based product modeling, geometric modeling, early design, and concurrent engineering.

Martti Mäntylä received his Dr.Eng. in 1983 at the Helsinki University of Technology. In 1983-1984 he was a visiting scholar with the Computer Systems Laboratory at the Stanford University. In 1989 he was a World Trade Visiting Scientist at the IBM Thomas J. Watson Research Center. Currently he is a Professor of Information Technology with the Laboratory of Information Processing Science at the Helsinki University of Technology, Finland, where he is the head of research teams in product modeling and realization, rapid prototyping, assembly robotics, usability engineering, and analysis and measurement of CIM systems. Mäntylä's research interests include the full range of computer applications in engineering, such as CAD, CAM, computer graphics, user interfaces, and data base management. He is a member of the editorial board of Computer-Aided Design (Butterworth), an associate editor of the ACM Transactions on Graphics and a member of the ACM, the IEEE Computer Society, and the Eurographics Association, where he is also a Member of the Executive Committee. He is also a member of IFIP working groups 5.2, 5.3, and 5.10.

Masatomo Inui is an associate professor in the Department of System Engineering of the Ibaraki University. He was a lecturer in the Department of Precision Machinery Engineering of the University of Tokyo from 1991 to 1993. His interests include solid modelling machining and process-planning automation, design and manufacturing process modelling, product modelling, and geometric reasoning. He received his MSEng. degree in information engineering from the University of Tokyo in 1986, and a Dr.Eng. in precision machinery engineering from the University of Tokyo in 1991.

Fumihiko Kimura is a professor in the Department of Precision Machinery Engineering of the University of Tokyo. He was a research associate at the Electrotechnical Laboratory of the Ministry of International Trade and Industry from 1974 to 1979. He then moved to the University of Tokyo, and was an associate professor from 1979 to 1987. He has been active in the fields of solid modeling, free-form surface modelling and product modelling. His research interests now include the basic theory of CAD/CAM and CIM, concurrent engineering, engineering simulation and virtual manufacturing. He is involved in the product model data exchange standardization activities of ISO/TC184/SC4, and is a member of IFIP WG 5.2 and 5.3, and a corresponding member of CIRP. He graduated from the Department of Aeronautics, the University of Tokyo, in 1968, and received a Dr. Eng. Sci. degree in aeronautics from the University of Tokyo in 1974.

5

Downsizing the geometric model for production engineering

Xiujuan FENG, Jie GAO and Rongxi TANG
Manufacturing Engineering Department
Beijing University of Aeronautics and Astronautics
Beijing 100083, China

Abstract

In our country, 2D drawings are still the predominating form of engineering documentation, so we explored the feasibility of using 2D parametric drafting system to create 3D product model capable of supporting the CAD/CAPP/CAM integration. We have built a feature library to create a parametric feature-based part model, translate a parameterized form feature description file to STEP protocol CAPP module, generate operation card and NC code.

Keywords

Form feature, form feature database, CAD/CAPP/CAM integration

1 MOTIVATION

Contemporary CAD/CAM systems emphasize the role of solid modeler in building up an integrated product information model. Design objects are being constructed in 3D mode, while 2D drawings are generated from the 3D model with bidirectional associativity. But in many countries such as China, 2D drawings are still the predominating form of engineering documentation. Most factories fulfil the customer's order by putting together some old drawings with minor modifications. Professor Zhang has put forward a new idea of making 3D conceptual design by 2D parametric drafting system based on super—2D modeling techniques. [1]In a super—2D modle the geometric elements in different viewports are associated with each other. The system maintains a global coordinate system with serveral viewport coordinate systems derived from it. Super—2D data in different viewports are related each other by the projection rules. Through transformation matrices, a design parameter in one viewport can control not only the geometric elements in that viewport, but also those related geometric elements in other viewports as well. Similar association are set between geometric elements and super — 2D dimensions. In such a way, the new system is able to understand and figure out the non — existing 3D modle from the 2D modle as long as the modle obeys the projection rules of descriptive geometry and engineering drowing standards. For those usres who do not need the 3D modeling capability, such a 2D approach to realize parametric conceptual design is much easier and less expensive than the 3D approach.

In our case, we are undertaking a project of developing a feature—based CAD/ CAPP/CAM integrated system for parts running on the FMS production line at Chang Chun FMS Experimental Center. Two machining centers installed on the production line are of the $2\frac{1}{2}$ D category, this fact predetermines that the overall shape of most parts to be machined on the line is $2\frac{1}{2}$ D. Stimulated by such kind of production requirements, we explored the feasibility of using 2D parametric drafting system to create 3D product modle capable of supporting the CAD/CAPP/CAM integration. The main difference with Professor Zhang's idea is that we implement the system on the platform of a conventional parametric drafting package and the association among different views of a part is built up interactively from the definition of form features. The drafting package does not understand the functional meaning of stored geometric entities, functionality and manipulation rules of form features are implemented in the design—with—feature module.

2   UNDERLYING PLATFORM OF DRAFTING PACKAGE

We use a self — developed parametric drafting system called VCAD to support the exploration. VCAD makes use of construction lines to drive the whole geometry. There are four types of construction lines; horizontal, vertical, inclined lines and circle which are defined by a reference datum and a real number, representing the ordinate, abscissa, inclination angle and radius respectively. Geometry and dimensioning of designed parts are built up on the basis of construction lines. Since construction lines extend over the full screen as shown in Fig. 1, they can be used to correlate different views. All construction lines are positioned relative to the part coordinate system. When some dimenson values are being changed, related construction lines change their position and modify the part geometry. A complete part drawing by its nature contains all the information necessary for its production such as material specification, surface finish, shape and location tolerances etc. , for integrating the CAD/CAPP/CAM activities the key point lies in reorganizing the presentation of part design data in such a way that they can be accessed and processed by an automated CAPP/CAM system. Here the design — with — feature module implemented on the top of conventional parametric drafting package is used to reach this goal.

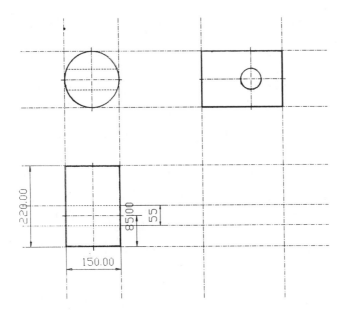

Figure 1   The underlying principle of variational drafing.

3. CASE STUDY OF TYPICAL PART

Fig. 2 shows an oil passage plate—a typical part of moderate complexity being produced on the FMS line. Its constituent form features are quite clear, there is no difficulty to list all of them: contoured plate, holes, slots, threads etc. Contour of the plate should be defined by the users, while the other form features may be retrieved directly from the built—in library. Before entering into the FMS line, the workpiece has been finishing — machined to the required thickness with two diagonally spaced outermost holes having been precisely worked out for locating and fixing on the machining center. Operations planned to be finished on the line are drilling, threading, milling slots and at last milling the peripheral contour of plate. Two setups are needed: upright and bottom up.

In general, solving setup generation and feature sequencing is a crucial step in process planning. [2] We take a simplified approach to generate setups interactively during the drafting stage. In the case of oil passage plate, all the form features located on the toop face of part will be naturally machined in one setup, while the features located on the bottom face will be machined in another setup. In 3D modle, the surface to which a feature is attached is called an attachment surfaces or joint—face,[3]its normal usually represents the tool approach direction to access the feature.

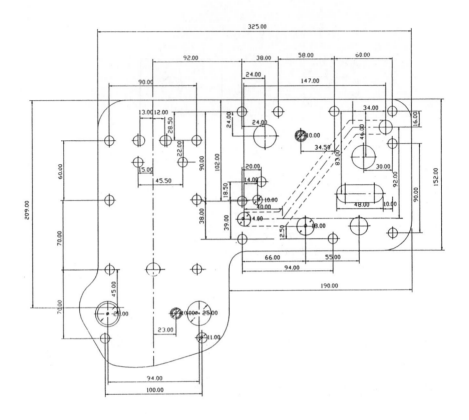

Figure 2   Oil passage plate—a typical part running on FMS line.

So from the point of view of machining, such surfaces may be called operating faces, see Fig. 3. Generating a setup means to choose the appropriate supporting face, operating faces and the nomenclature of features to be machined at that position. In a 2D modle, especially in a conventional parametric drafting system, all the faces are shown as lines, no semantics are involved in the geometric data. The correlation of groups of features to be machined in a setuup with their operating faces can be established only interactively during the part layout period. Return back to the oil passage plate, Fig. 2 shows its top view, all visible features in that view are machined in a setup, so when these features are being constructed one after another using VCAD package, they are linked together and pointed to the common operating face—top face. In the data file top face is a symbolic name, it is in fact the synonym of top view. Once a setup is generated, CAPP module make the detailed process planning for machining these features, i. e. choosing cutting tools, determining number of paths, sequencing the machining order, working out the operation card etc.

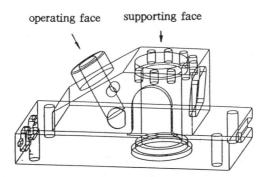

Figure 3   Operating and supporting faces in a setup.
(The figure is taken from [4])

4. SYSTEM ORGANIZATION

Fig. 4 shows the overall structure of our experimental system.

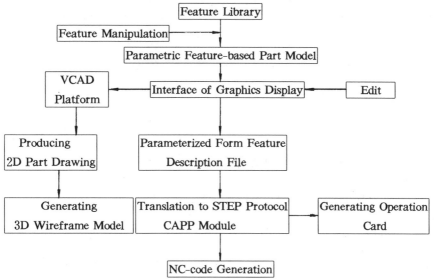

Fig. 4   System organization of experimental software.

Feature library contains comprehensive information about commonly used form features: name, internal code, classification category, geometric parameters, graphic representation, joint face, origin and orientation of local coordinate system, location datum, location dimensions, shape and location tolerances, suface finish etc. The description of features is 3D, so it is possible in principle to generate the 3D wireframe modle of the part as well as ite shaded picture by itself. [3]

Part description file contains administrative information, material specification, heat treatment requirements etc. Features are grouped together under the common heading of operating faces and setup number. Since setups are generated interactively during the part layout stage, we use internal codes to represent the supporting face, locating datum. fixturing method and operating faces for eash satup.

5 CONCLUSION

To our understanding, in the specific conditions in China 2D approach of realizing feature—based CAD/CAPP/CAM integration is the easiest way of creating a STEP compliant product information modle. The technology is rather simple, human intelligence is relied on to solve the most difficult problems such as creating the parametric feature — based part model, generating setups, and let the computer software fulfil trivial and tedious jobs such as formatting the operation card, printing etc. In the case of Chang Chun FMS Experimental Center, an alternative solution is using a 3D solid modeler such as SDRC/I—DEAS as the integration tool. In such a case, people has to create in the first place a 3D model of the part from existing drawinge, recognize and extract invovled form features in certain way, regenerate the 2D drawings through orthographic projections and partial cuts of the 3D model for the purposes of documentation, and finally prepare the STEP physical file of feature—based application protocol for input into the CAPP module. Since I—DEAS stores faceted model of the part, it is very difficult to find the original definition of curved surfaces. We belive that in certain enviroments 2D drawing are still valuable, they are able to offer some effficient shortcuts to solve real problems in factories.

6 ACKNOWLEDGEMENT

This project is sponsored by the National Science Foundation of China, grant No. 69274032.

REFERENCES

1   Zhang Shensheng, Wang Zhonghui and Hou Xiaoling. Supper—2D concept and parametric design. Pacific Graphics' 94/CADDM' 94, Aug. 26—29, Beijing.
2   Chen C L Philip and LeClair S R. Integration of design and manufacturing: solving setup generation and feature sequencing using an unsupervised — learning approach. CAD, 1994,26(1).
3   Tang Rongxi, Wu Hongming and Yang Dong. Framework of a prectical feature

modeling system. CADDM, 1993, 3(1).

4   Miller J R. Incremental boundary evaluation using inference of edge classification. IEEE CC&A, Jan. 1993.

BIOGRAPHY

Xiujuan Feng——Lecturer in the Department of Manufacturing Engineering at the university of BUAA, got Ph. D at BUAA in 1994. Her research interests include parametric design, object — oriented programming, implementation of engineering database.

Jie Gao——Engineer from ChangChun FMS Experimental Center, responsible for CAD/CAPP/CAM integration. Currently engaged in Ph. D. study at BUAA.

Rongxi Tang: professor.

6

Representing Dimensions and Features in a Product Model

Tsuzuki, M. S. G.; Miyagi, P. E.; Moscato, L. A.
Escola Politécnica da Universidade de São Paulo
Departamento de Engenharia Mecânica/Mecatrônica
Av. Prof. Mello Moraes, 2231
CEP 05508-900 - São Paulo - SP - BRAZIL
Tel: +55-11-818-5565
Fax: +55-11-813-1886
email: mtsuzuki@cat.cce.usp.br

Abstract

This work is based on two believes about the next generation of Computer Aided Design: first, it is necessary to *explicitly store features and dimensions* in a Product Model; second, it is necessary to *support user defined features*. The Product Model is represented as a hierarchical structure where it is possible to define two kinds of dimensions: *local dimensions* and *relative dimensions*. Relative dimensions are constraints that associate two different nodes in the hierarchical structure. It is also proposed an algorithm to satisfy a set of relative dimensions.

Keywords

Product Model, Form Feature, Parametric Design, Dimension Representation

1 INTRODUCTION

The state of research of Computer Aided Design can be identified as a transition from design based on geometry to design based on features. In spite of several works related to feature representation, a common sense has not been found. Several works have discussed the usability of the CSG representation and/or the B-Rep representation. The B-Rep representation is preferred because vertices, edges and faces are explicitly represented (Wilson, 1990). However, the CSG modeling approach is most suitable for dimension driven geometry approach (Gossard, 1988).

We are working to create a Design by Feature System in which an user can generate a design by attaching features to the product. This system is based on two believes about the next generation of Computer Aided Design: first, it is necessary to **explicitly store features and dimensions in a Product Model**; second, it is necessary to **support user defined features**.

Ovtcharova (1992) proposed that a Design by Feature Systems has four levels of abstraction: application level, feature definition level, feature representation level and geometric modeller level. This work is part of the implementation of the third level: representation of features.

2 PRODUCT MODEL

The Product Model is represented through a hierarchical tree structure with several levels of complexity (see Figure 1). The Product Model Node represents the root node. Every node has access to all information related to the nodes immediately below. In this way, the Assembly has access to all information related to the Parts immediately below it. In this structure there is a difference between the leftmost node and the others. All information related to the leftmost node is transferred to the upper node directly. All information related to the other nodes is transferred to the upper node based on transformations. Because of this property, we say that the leftmost node is fixed and the other nodes are moveable. Every node has a set of primitive constituent elements that can be a set of **points, lines** or **planes**.

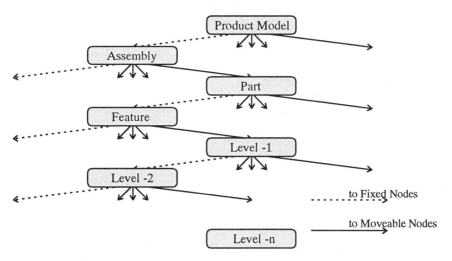

Figure 1 The hierarchical representation of the Product Model.

According to this structure, it is possible to define two types of dimensions: **local dimensions** and **relative dimensions**. Dimensions associated to one unique node are local dimensions. A relative dimension constrains primitive constituent elements from two different nodes, one immediately below and another immediately above. The fixed node has no relative dimension constraining it to the immediately above node as its information is transferred without any transformation.

A node is symbolically represented by:

$$A = F(B_1, t_2 \cdot B_2, t_3 \cdot B_3, ..., t_n \cdot B_n) \tag{1}$$

where A is the upper node, B_i is an inferior node of node A and t_i is a transformation applied to node B_i. B_1 is the leftmost node. Only node B_1 does not have a transformation associated with. Function F represents the transferring of information from the inferior node to the upper node. This work will define an algorithm that will find all the transformations t_i such that the nodes are correctly positioned, i.e., all the relative dimensions are satisfied.

3 REPRESENTING FEATURES

Our objective is to define a mechanism to facilitate the manipulation of features by the user. Such that the user selects a feature from the menu and just provides some primitive constituent elements to attach a feature in the Product Model. This effect can be implement by **associating some relative dimensions to features**. Figure 2 illustrates an example of feature definition.

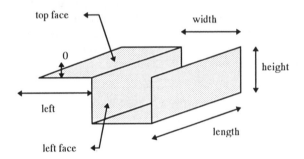

Figure 2 Example of feature definition.

Where width, height and length are feature parameters, and left and 0 are relative dimensions. The relative dimension 0 represents that the top face must have distance 0 (zero) to a Part's face when the feature is attached to the Part. The value of the relative dimension left must be supplied when the feature is attached to a part. The relative dimension left represents that the left face must be at some distance from a Part's face.

4 SPECIAL SETS OF TRANSFORMATIONS

4.1 Relative Dimensions

A set of relative is represented by:

$$\Delta = \{ d_1, d_2, \cdots, d_n \} \tag{2}$$

where, Δ is a set of relative dimensions and d_i is a relative dimension. A relative dimension can be in two different states: satisfied and not satisfied. The inferior node is positioned relatively to the upper node such that all the relative dimensions are satisfied. We will represent the state of the set of relative dimensions by:

$$A <\Delta> B = \delta \tag{3}$$

where, A is the upper node, B is an inferior node of node A and δ is the set of satisfied relative dimensions. In this way, if $\delta = \{ \ \}$ then no relative dimension is satisfied. If $\delta = \Delta$ then all relative dimensions is satisfied. All the fifteen types of relative dimensions are listed in Table 1.

Table 1 All the fifteen types of relative dimensions

relative dimensions type	code
plane-plane distance, with coincident normals	DisCN
plane-plane distance, with opposite normals	DisON
plane-plane angle	AngPP
plane-line distance (and vice versa)	DisPR (DisRP)
plane-line angle (and vice versa)	AngPR (AngRP)
plane-point distance (and vice versa)	DisPp (DispP)
line-line distance, with coincident directions	DisCD
line-line distance, with opposite directions	DisOD
line-line angle	AngRR
line-point distance (and vice versa)	DisRp (DispR)
point-point distance	Dispp

4.2 Set of Transformations

The transformation of a node can be represented by:

$$E' = t \cdot E \tag{4}$$

where, t is the transformation, E represents a node and E' represents the transformed node. Now, it is possible to define two operations over sets of transformations: **intersection** and **distribution**. The intersection operation is defined as:

$$t \in \varphi_1 \ and \ t \in \varphi_2 \Rightarrow t \in \varphi_1 \cap \varphi_2 \tag{5}$$

where, φ_1 and φ_2 represent two sets of transformations, t is a transformation and $\cap$ is the intersection operator. The distribution operation is defined as:

$$t_1 \in \varphi_1 \ and \ t_2 \in \varphi_2 \Rightarrow t_1 \cdot t_2 \in \varphi_1 * \varphi_2 \tag{6}$$

where, φ_1 and φ_2 represent two sets of transformations, t_1 and t_2 are transformations and $*$ is the distribution operator. Note that this operation is not commutative.

4.3 Arbitrary Coordinate

Two nodes A and B are **equivalent**, if every internal point of A is internal of B and if every external point of A is external of B and if every point in the boundary of A is in the boundary of B. Given a transformation t_{ST} such that:

$$E' = t_{ST} \cdot E. \tag{7}$$

Then, if the node E is equivalent to node E' and the transformation t_{ST} is not the identity, then it is said that E has a set of arbitrary coordinates. The set of all transformations t_{ST} is called set of symmetric transformations. This can be better understood through an example: a cylinder can be rotated through its axis of symmetry that the result of this transformation is equivalent to the original cylinder. Particularly, the primitive constituent elements plane, line and point have sets of symmetric transformations that will play an important role in our algorithm:

- **plane**: it is possible to rotate a plane through any axis parallel to its normal vector or to translate it in any direction normal to its normal vector that the result of the transformation is equivalent to the original plane. The plane's set of symmetric transformations is represented by PLN;
- **line**: it is possible to rotate a line through itself or to translate it according to its direction that the result of the transformation is equivalent to the original line. The line's set of symmetric transformations is represented by RET;
- **point**: it is possible to rotate a point through any axis that pass through itself that the result of the transformation is equivalent to the original point. The point's set of symmetric transformations is represented by PNT.

4.4 Degrees of Freedom Associated with Relative Dimensions

A relative dimension has degrees of freedom in the sense that the inferior node can be transformed and the satisfied relative dimensions will remain satisfied. Consider that a transformation $t^{\{dl\}\,1}$ satisfies the relative dimension d_l. Symbolically represented by:

$$A < \Delta > (t^{\{dl\}} \cdot B) = \{ d_l \}. \tag{8}$$

The set of all transformations t_{DF} that maintain the relative dimension d_l satisfied is called set of degrees of freedom associated to the relative dimension d_l. Symbolically represented by:

$$A < \Delta > (t_{DF} \cdot t^{\{dl\}} \cdot B) = \{ d_l \}. \tag{9}$$

The Greek letter τ will be used to represent the sets of degrees of freedom. The sets of degrees of freedom are associated to the set of symmetric transformations of the primitive constituent elements constrained by the relative dimensions. As a relative dimension d_i constrains two primitive constituent elements, one is associated to the upper node and the other is associated to the inferior node. The sets of symmetric transformations are going to be represented by τ_S^{di} and τ_I^{di}, where the first is associated to the upper node and the second is associated to the inferior node. The set of degrees of freedom associated to the relative dimension d_i is represented by:

$$DF(d_i) = \{ \tau_S^{di}, \tau_I^{di} \}. \tag{10}$$

In this way, a relative dimension has two subsets of degrees of freedom. If

$$A < \Delta > B = \{ d_i \} \tag{11}$$

then transforming the inferior node B by any $t_I \in \tau_I^{di}$, the relative dimension d_i will remain satisfied. Symbolically:

---

[1] The index represent that the relative dimension d_l is satisfied by this transformation.

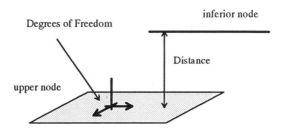

Figure 3 A relative Dimension of type plane-line distance.

Table 2 Sets of Degrees of Freedom associated with the Relative Dimensions

Relative Dimension	Set of Degrees of Freedom
DisNC	{ PLN , PLN }
DisNO	{ PLN , PLN }
AngPP	{ PLN * 3DT , PLN *3DT }
DisRP	{ RET , PLN }
DisPR	{ PLN , RET }
AngRP	{ RET *3DT , PLN *3DT }
AngPR	{ PLN *3DT , RET *3DT }
DisPp	{ PLN , PNT }
DispP	{ PNT , PLN }
DisOC	{ RET , RET }
DisOO	{ RET , RET }
AngRR	{ RET *3DT , RET *3DT }
DispR	{ PNT , RET }
DisRp	{ RET , PNT }
Dispp	{ PNT , PNT }

$$A < \Delta > (t_I \cdot B) = \{ d_i \}. \tag{12}$$

If the upper node is transformed by any $t_S \in \tau_S^{di}$ then the relative dimension d_i will remain satisfied too. However, it is important to note that the relative positioning between the upper node and the inferior node is not altered if the upper node is transformed by t_S or if the inferior node is transformed by t_S^{-1} (inverse transformation). In this way, it is possible to represent the transformation of the upper node by the following expression:

$$A < \Delta > (t_S^{-1} \cdot t_I \cdot B) = \{ d_i \}. \tag{13}$$

This property creates a **hierarchy** between the subsets of degrees of freedom. When, the transformation t_S^{-1} is applied to the inferior node, then the inferior node's subset of symmetric transformations is modified by this transformation. Suppose that a relative dimension of type plane-line distance was satisfied (see Figure 3). In this case, it is possible to transform the line according to its symmetric transformations that the relative dimension will remain satisfied, and it is possible to transform the line according to the plane's symmetric transformations that the relative dimension will remain satisfied. The definition of the set of degrees of freedom associated to a relative dimension depends on its type and the type of the primitive constituent elements that it constrains.

Before listing all sets of degrees of freedom associated with relative dimensions it is necessary to define a special set of transformations that represent all the possible translations in the three dimensional space. This set of transformations is represented by *3DT*. Table 2 defines all the sets of degrees of freedom associated with the relative dimensions.

4.5 Degrees of Symmetry

Both upper and inferior nodes have a set of symmetric transformations. This set will be called set of degrees of symmetry. It is represented by:

$$X = \{ \sigma_s, \sigma_I \} \tag{14}$$

where, σ_s is the set of degrees of symmetry associated to the upper node and σ_I is the set of degrees of symmetry associated with the inferior node. To satisfy the ith relative dimension, it is necessary to determine the set of degrees of freedom $\varphi_{i+1} = \{ \tau_s, \tau_I \}$. The set of degrees of symmetry must be updated every step, such that at the ith step the set of degrees of symmetry is represented by $X_{i+1} = \{ \sigma_s, \sigma_I' \}$. Now, if the expression

$$\tau_s * \tau_I \subset \sigma_s * \sigma_I' \tag{15}$$

is true then the inferior node is fixed to the upper node.

5 SATISFYING A RELATIVE DIMENSION

We are interested in finding a transformation t^Δ such that:

$$A < \Delta > (t^\Delta \cdot B) = \Delta. \tag{16}$$

The inferior node B is relatively positioned by the transformation t^Δ to the upper node A such that all the relative dimensions are satisfied. In this work, we will define an algorithm that will find all these transformations that satisfy a set of relative dimensions. The algorithm will search for the transformations step by step, satisfying one relative dimension each time. In other words, we will find firstly a transformation $t^{\{d1\}2}$; secondly a transformation $t^{\{d2\}}{}_{\{d1\}}{}^3$ and continuing.

The relative dimensions are satisfied step by step. At the ith step a relative dimension d_i will be satisfied. The relative dimensions d_1, d_2, $\cdots$, d_{i-1} are satisfied already. Associated to the ith step there is a set of degrees of freedom φ_i that maintain all the satisfied relative dimensions satisfied. Symbolically represented by:

---

[2]The index represent that the relative dimension d_1 is satisfied by this transformation.
[3]The indexes represent that the relative dimension d_2 is satisfied by this transformation and the relative dimension d_1 remains satisfied.

$$A < \Delta > B' = \{ d_1, d_2, \cdots, d_{i-1} \} \tag{17}$$

and,

$$\forall t \in \varphi_i \Rightarrow A < \Delta > (t \cdot B') = \{ d_1, d_2, \cdots, d_{i-1} \}. \tag{18}$$

Suppose that the relative dimension d_i is satisfied by a transformation:

$$T^{\{di\}}{}_{\{di-1, \dots, d1\}} = \{ t^{\{di\}}{}_{\{di-1, \dots, d1\},S}, t^{\{di\}}{}_{\{di-1, \dots, d1\},I} \} \in \varphi_i \tag{19}$$

where the meaning of the subindexes s and I are, respectively, superior and inferior. In this way,

$$A < \Delta > (t^{\{di\}}{}_{\{di-1, \dots, d1\},S} \cdot t^{\{di\}}{}_{\{di-1, \dots, d1\},I} \cdot B') = \{ d_1, d_2, \cdots, d_{i-1}, d_i \}. \tag{20}$$

The set of degrees of freedom associated with the inferior node must be actualized by the transformation $t^{\{di\}}{}_{\{di-1, \dots, d1\},S}$. Then,

$$\varphi_i' = \{ \tau_S, t^{\{di\}}{}_{\{di-1, \dots, d1\},S} \cdot \tau_I \} \tag{21}$$

The set of degrees of freedom φ_{i+1} is the intersection of φ_i' and $DF(d_i)^4$. Symbolically:

$$\varphi_{i+1} = \varphi_i' \cap DF(d_i). \tag{22}$$

At this moment it is necessary to verify whether the expression 15 is true or not. If it is true then the algorithm is finished. If there is any remaining relative dimension then the nodes were overconstrained. According to equation 22 the intersection between sets of degrees of freedom plays an important role in the algorithm. All the possible cases of intersection are represented in table 4.

Table 4 Intersection between sets of symmetric transformations

	TOT	PLN	RET	PNT	ROT	TRN	R3D	T3D	T2D
TOT	TOT	PLN	RET	PNT	ROT	TRN	R3D	T3D	T2D
PLN		PLN	ROT	ROT	ROT	TRN	PLN	T2D	T2D
		TRN	TRN		EMP	EMP	T2D		TRN
RET			RET	ROT	ROT	TRN	RET	TRN	TRN
			TRN	EMP	EMP	EMP	TRN		EMP
			EMP						
PNT				PNT	ROT	EMP	ROT	EMP	EMP
				ROT	EMP				
ROT					ROT	EMP	ROT	EMP	ROT
					EMP		EMP		
TRN						TRN	TRN	TRN	TRN
						EMP			EMP
R3D							R3D	T3D	T2D
							T3D		
T3D								T3D	T2D
T2D									T2D
									TRN

---

[4] $DF(d_i)$ is the set of degrees of freedom associated with the relative dimension d_i.

where *TOT*, *ROT*, *TRN*, *R3D*, *T2D* and *EMP* are sets of degrees of freedom. Their meaning are explained below:

- *TOT*, totally free;
- *PLN*, plane's set of symmetric transformations;
- *RET*, line's set of symmetric transformations;
- *PNT*, point's set of symmetric transformations;
- *ROT*, it is possible to rotate the node through an axis of rotation that the result of the transformation will be equivalent to the original;
- *TRN*, it is possible to translate the node through a direction that the result of the transformation will be equivalent to the original;
- *R3D*, it is possible to rotate the node through an axis of rotation and to translate it in any direction that the result of the transformation will be equivalent to the original;
- *T2D*, it is possible to translate the node through any direction perpendicular to an specific direction that the result will be equivalent to the original;
- *EMP*, totally fixed.

5.1 Solution Tree

To determine the transformations that satisfy a set of relative dimensions, we will define a hierarchical tree structure. In this tree there will be two kinds of level (see Figure 4). Nodes associated with odd levels represent relative dimensions to be satisfied, and nodes associated with even levels represent transformations that satisfy the relative dimensions associated to the upper node.

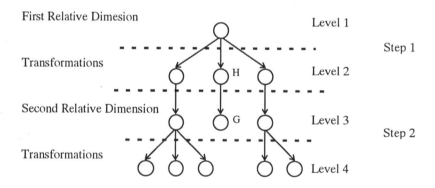

Figure 4 Solution Tree (the relative dimension associated with node *G* could not be satisfied).

The *i*th relative dimension to be satisfied is represented in level $2 \cdot (i-1)+1$. The transformations that satisfy a relative dimension d_i are represented in level $2 \cdot i$. This structure will allow to determine all the transformations that satisfy a given set of relative dimensions.

Figure 4 shows an example with a set of two relative dimensions where the algorithm found five transformations that can satisfy the set. It is not guaranteed that all the transformations are different. In this way, it is necessary to compare them and select only the different transformations. It is possible to observe that the relative dimension associated with node *G* could not be satisfied, so the transformation associated with node *H* is discarded.

6 FUTURE WORKS

We did not show, however it is possible to represent user defined features and to explicitly represent features and dimensions using the concept of relative dimensions (Tsuzuki, 1995). Another implication of the use of relative dimensions is that it is possible to *rationalize* the definition of dimensions in a project, as a consequence of the *encapsulation* of the information in the nodes of the structure shown in Figure 1. It is necessary to define a new representation where points, lines and planes are explicitly represented. We are working in the definition of some heuristics for satisfying relative dimensions. The method is efficient because only the branches in the tree that are affected by a given change need to be reevaluated. Underconstrained nodes can be identified by the method and they can be partially evaluated.

7 REFERENCES

Gossard, D.C.; Zuffante, R.P. and Sakurai, H (1988) Representing Dimensions, Tolerances and Features in MCAE Systems, *IEEE Computer Graphics & Applications*, March, 51-59.

Ovtcharova, J.; Pahl, G. and Rix, J. (1992); A Proposal for Feature Classification in Feature-Based Design, *Computer & Graphics*, 16(2):187-195.

Tsuzuki, M.S.G. (1995), A contribution to parametric feature representation, *Doctor Thesis (in Portuguese)*, Escola Politécnica da USP.

Wilson, P.R. (1990); Feature Modeling Overview, *Proceedings of the 17th International Conference on Computer Graphics and Interactive Techniques SIGGRAPH'90*, Course Notes 12, Dallas, XI.1-XI.56.

TSUZUKI, M. S. G., born in 1963. Graduated in Electrical Eng. from Escola Politécnica da USP (EPUSP) - Brazil in 1985. Received the M.Sc. from the Dep. of Information Processing Eng. from Yokohama National University - Japan in 1988. Assistant of the Dep. of Mechanics/Mechatronics Eng. of EPUSP since 1990.

MIYAGI, P. E. Ph.D., born in 1959, graduated in Electrical Eng. from EPUSP - Brazil in 1981. Received the M.Sc. and Ph.D. degrees from Tokyo Institute of Technology - Japan in 1985 and 1988 respectively. Associate Professor of Dep. of Mechanics/Mechatronics Eng. of EPUSP since 1993.

MOSCATO, L. A. Ph.D., born in 1946, graduated in Electrical Eng. from EPUSP - Brazil in 1969. M.Sc. and Ph.D. from Electrical Eng. Dep. from EPUSP in 1971 and 1978 respectively. Full Professor of Dep. of Mechanics/Mechatronics Eng. of EPUSP since 1988.

7

Product Modeling Technology for Computer Aided Concurrent Design

S.S. Zhang   *X.L. Hou*
Professor   Associate Professor
Department of Computer Science
Shanghai Jiao Tong University
Shanghai 200030 P.R.China
Tel: 86-21-4310310-3536
Fax: 86-21-4315111

Abstract

This paper started from problem analysis of existing CAD systems and proposed a set of new features for Computer Aided Concurrent Design (CACD) --- a new generation CAD system. It focuses on concurrent design support, product modeling, design parameters inheritance and information propogating among different hierarchies of an integrated design process. A prototype of a CACD system for plastic encapsulation mold is implemented at Shanghai Jiao Tong University to demonstrate these new features.

Key Words:

CAD, CIMS, Concurrent Design/Concurrent Engineering, Product Modeling

1. INTRODUCTION

Most machine design starts from requirements specification, goes through concept design, detail design and ends up with a new machine product. The design process is, ingeneral, very complex. Figure 1 shows the different phases of a traditional design process. Here, We want make a clear distinguish among a product, an assembly, a sub-assembly and parts. A product is something can perform certain functions that customer demands; like cars for transpotation, machine tools for fabrication and TV sets for entertainment. Product and assembly are terms used interchangeably in this paper. Sub-assembly is one of the components decomposed from an assembly. It can perform certain function that a product needs such as an engine or wheels for a car. Parts are individual components designed to make these sub-assembly and/or product functional. By looking at the cost distribution of different design phases (Figure 2), it is found that most of the design cost was determined in concept design phase (which includes assembly design and decomposing it into sub-assembly, checking assembly conditions and making necessary modifications). But most of the CAD systems today can support parts design only. For rest of the design process such as product performance specification and concept design, they are just not designed for. They are parts oriented rather than products oriented. Due to lack of assembly hierarchical information, they cannot associate an assembly with different comoponents and with other assemblies/sub-assemblies in diffenrent hierarchies. Therefore, they can not support the entire process of a product design and concurrent design, which are definitely product orientated [1, 2].

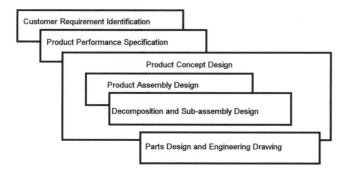

Figure 1 Different Phases of a design process

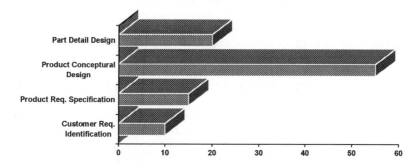

Figure 2 Costs Distribution in Different Design Phases

2. NEW FEATURES OF CACD

Computer Aided Concurrent Design (CACD) proposed in this paper is designed for the integration of entire process of product design. It focuses on concurrent design support, product modeling, design parameters inheritance and information propagating among different hierarchies of an integrated design process.

2.1 Assembly model of in a CACD system

Instead of working on a part model as most CAD systems do today, A CACD system works on a product model -- an assembly model. One of the two major processes of a CACD system is to build a assembly model which could represent the assembly relationship and constrains among different assembly components as well as different assembly hierarchies. Another is to define a set of methods which would enable design parameters and assembly constraints to

propagate among different assembly hierarchies. The key here is to construct an appropriate modeling schema to support product modeling, which should be enable us to:
* represent the assembly relationship between components of an assembly or sub-assembly;
* represent the assembly relationship between upper and lower assembly hierarchies;

2.2 Information flow in a CACD system

Keeping a product model unique and consistant during an entire design process is a key factor for concurrent design support. This is important specially when some changes are made during a design process. The propagation effect of these changes has to be considered and handlled correctly to keep the model correct. The design information (the parameters and constraints) of an assembly model is propagated within the CACD system in the following way:
* A concept design is always started from a product in assembly level. According to the original design constrains imposed by customers, a designer can sketch out the whole system and determine design parameters and assembly relationship among sub-assemblies.
* By such assembly relationship, a product can be decomposed into several sub-assemblies and parts. Using the same design parameters and constraints passed from the upper level hierarchy, these sub-assemblies and parts can be designed simultaneously. New design parameters added in these processes will form the design constrains for the next level design activity.
* All design parameters and constrains are the driving parameters of the design process. By simply adjust these parameters a designer could easily modify his design or even make a new design.

3. AN EXAMPLE OF ASSEMBLY MODEL AND INORMATION FLOW

Let's take a shaft assembly design as an example to illustrate the discussion above. Start from the shaft assembly shown in Figure 3a, which is subtracted from its upper hierarchy and has some initial design constrints (Table 1). The designer could then refine the design and add more design information (parameters and constraints) into the initial design (Figure 3b). It is noted in Table 4 that some of those new added design parameters may be related to those initial design parameters according to design rules and constraints.

Table 1 The initial design parameters

Maximum torque	M	Shaft material	H
Distance between bearings	L	Shaft diameter for gear mounting	D

Table 2 Relationship of current design parameters vs initial design parameters

current design parameter	express by initial and other design parameter	others
d1	f (M, H)	rounding
d2	D	rounding
l1	L + 2 x bearing width + bearing cover +2	rounding
2	> l1 + pulley width + 2	rounding
l3	f (M, H)	rounding
d3	> d12 + 10	rounding

After we finish the design at this level, we can further decomposed it into 5 parts and let more engineers to join the design team. The parameter passing from this level to next level is listed in Table 3.

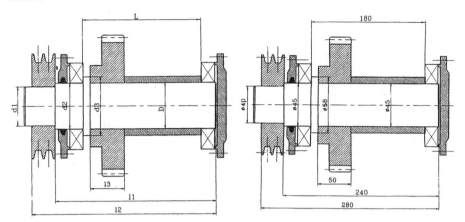

(a) The shaft assembly to be designed   (b) The refined shaft assembly

Figure 3 The shaft assembly design

Table 3 The inheritance relationship between design parameters

Parts	d1	d2	d3	l1	l2	l3	D	L	bearing type
shaft	☐	☐	☐	☐	☐	☐	☐	☐	☐
bearing cover		☐			☐		☐		☐
pulley	☐			☐	☐				
gear						☐	☐		
axle sleeve						☐	☐		

It is clear fron the discussion above that every design step gets some inherited information from its upper hierarchies and adds some new information to the design. These infomation together will be passed to the next hierarchy as initial design constraints. This kind of association ensues the assembly model to be unique and consistent in the entire design and modification process.

In this example, if the design parameter "maximum torque" is changed, not only the calue of D in the same assembly hierarchy will be affected (Table 1), but also all the related parameters d1, d2, d3, l1, l2, l3 at the shaft assembly level will be changed accordingly (Table 2). This propagation mechanism guarantees the consistence of the assembly model.

4. IMPLEMENTATION CONSIDERATION

Plastic encapsulation mold is used to encapsulate IC chips. It consists thousands of components. Its high manufacturing and assembly accuracy requirement increases both the design cost and time. To ensure the accuracy requirement, the mold is designed in a top-down fashion so that all related dimensions of different components in different assembly hierarchies is defined only once and referenced as many times as it is needed. According to this requirement, we designed a particular CACD system with the following features:

• It has an assembly model to represent the layout design of the mold. The layout design is the critical step of the mold design. All the assembly dimensions and constraints are specified in this global model for further reference.
• The layout assembly model is composed of a set of sub-assemblies and parts. Every dimension in the layout assembly is stored in a global model. All other parts constrained by this dimension will reference it.
• The design is started from the layout design. When the layout design is finished, a set of tests is performed on the layout design to check the manufacturability and assembly condition.
• After the layout design is completed, it is decomposed into sub-assemblies for detail design. All the design work thereafter are forced to use the data stored in the global model if they are available. This ensure the design consistence across the entire design process.
• Super-2D parametric design and drafting technology [3, 4] was implemented to handle parameter changes at any design stage. Once a design parameter in the global model is changed for example, all the reference will be changed accordingly, as well as those derived dimensions associated with this dimension.

The system structure of the plastic encapsulation mold CACD system is shown in Figure 4

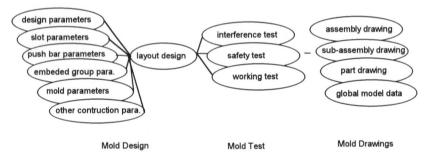

Mold Design   Mold Test   Mold Drawings

Figure 4 Plastic encapsulation mold CACD system

5. CONCLUSIONS

From the discussion above, We conclude that:

• The part oriented CAD systems today can not support product designs; therefore can not meet the demands of CE and CIMS.
• Product oriented CACD technology proposed in this paper, starting with an assembly model, can support the entire process of a product design. Together with other technologies like product model data management, process management & control and an inteoprable integration framework, it can support concurrent design for CIMS.

Concurrent design and product modeling are two hot research subjects today. They are closely related to CIMS and CE implementation. In the future study, we will focus more closely on the assembly model construction and the inheritance mechanism among different assembly hierarchies and different assembly components and associate our study together with the international study on STEP.

REFERENCE

1. Jiang, X.S. and Li B.H. (1994) Integration of Product Development Process --- Concurrent Engineering: Proceedings of the 3rd CIMS Conference of China, Wuhan, pp. 1-26
2. D.D.Bedworth ed. (1991) Computer Integrated Design and Manufacturing. McGraw - Hill Inc.
3. Zhang, S.S. and Wang Z.H. (1994) Descriptive Geometry Based Parameteric Design Technology: Proceedings of 6th ICECGDG, Tokyo, pp. 703-707
4. Zhang, S.S. and Wang Z.H. (1994) Super-2D Concept and Parametric Design Technology: Proceedings of CADDM'94, Beijing, pp. 629-633

8

A methodology for multiple domain mapping of features

S.S. Lim, Ivan B.H. Lee
Gintic Institute of Manufacturing Technology, Nanyang Technological University, Nanyang Avenue, Singapore 2263.

Lennie E.N. Lim, Bryan K.A. Ngoi
School of Mechanical and Production Engineering, Nanyang Technological University, Nanyang Avenue, Singapore 2263.

Abstract

This paper proposes a methodology for mapping features in different application domains. The generic reasoning processes required by various types of mapping processes have been effectively classified and captured. An experimental feature mapping shell has been developed using this approach. The shell is highly flexible and can be easily customized for different products and applications. The shell has been implemented with a blackboard architecture and a case-based reasoning system is used to represent previous mapping cases. These cases are indexed in a library in the feature mapping shell. An explicit semantics representation scheme has been developed to represent deep models of features. A deep model improves the reasoning capability of the mapping process so that the identity of a feature can be established more effectively.

Keywords

Feature mapping, feature modeling, feature-based design, deep model, blackboard, case-based reasoning

1 INTRODUCTION

Feature modeling has been introduced to bridge the information gap between computer-aided design (CAD) and computer-aided manufacturing (CAM). The end result of a CAD system is usually in the form of pure geometrical data. Other attributes such as tolerances can be attached to a design in some of the more advanced CAD systems. These data do not form a complete product definition and therefore, cannot be used effectively by downstream reasoning applications, such as process planning that views a design at feature level. This level of abstraction contains more than geometry alone. The integration between design and manufacturing is necessary for concurrent engineering as well as computer integrated manufacturing (CIM).

To produce a feature-based representation of a design, two main approaches have been proposed. The first, known as feature recognition, attempts to extract feature information from a geometric model by using algorithms, rules, graphs, syntactic patterns or convex hull decomposition techniques (Woodwark, 1988).

Feature-based design constitutes the second main approach. It advocates the use of features during the design phase so that a feature-based model is produced (Dixon *et al.*, 1990). Normally, design features or neutral features are employed to construct such a model. A design feature is related to a product and its functionality. It tends to be product dependent and company specific. On the other hand, a neutral feature is an application independent generic shape but it still provides a feature oriented representation within a model. Design features can often be constructed from neutral features. Examples of feature modeling systems implemented using this approach are GeoNode (Broonsvoort and Jansen, 1993) and ASU Feature Testbed (Shah, 1988).

However, such an approach requires feature mapping as the design features used do not necessary have equivalent representations in other application domains. Mapping is a process of extracting information, or knowledge relevant to a particular application and transforming it to possibly different representations, from a model composed of features. It is also known as feature transformation, feature conversion and feature transmutation (Pratt, 1993). Please refer to Lim *et. al.* (1994) for more information on the various approaches in feature mapping and their limitations.

As the main focus of this paper is on feature mapping, feature recognition techniques will not be reviewed. For further information on the various recognition methodologies, reference is made to Woodwark (1988).

2 FEATURE MAPPING

In this research, feature mapping is defined as the process of identifying emergent features for a target domain from an initial product model represented by design features or neutral features. The emergent features are often application dependent.

The process of identification itself may vary depending on the circumstances of each situation. In certain cases, there is a direct correspondence between the features used in design and those application dependent features. This type of mapping, known as identity mapping, can be handled easily. However, if an initial feature model is represented by features that do not have direct correspondences, conjugate mapping is involved. The process of mapping is made more complicated when feature interaction occurs. Such interactions can change the semantics of the affected design features so drastically that initial assumptions about these features may no longer be valid. It may further affect the efficiency and performance of mapping systems that employ simplifying assumptions in their implementation of the conversion processes. Figure 1 shows an example of features that interact closely with each other, necessitating the use of conjugate mapping techniques.

In the next section, a feature mapping approach that has been developed in this research is described in detail.

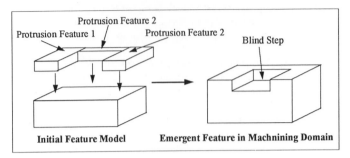

Protrusion Feature 2
Protrusion Feature 1   Protrusion Feature 2   Blind Step

Initial Feature Model   Emergent Feature in Machnining Domain

Figure 1   An example of conjugate mapping involving interaction.

2.1   The Proposed Approach

The proposed approach for feature mapping consists of the following stages of reasoning:
    i) filtration and selection,
    ii) construction of enhanced representation and
    iii) matching.
In the first stage, an initial model of a design has to be systematically searched to determine collections of faces that form a certain pattern of interest to a particular application domain. At first glance, the search space can be very large depending on the initial model. In addition, a large amount of knowledge is required to determine whether any group of connected faces will form a meaningful combination that will yield features in the targeted domain of mapping. However, the search can be constrained by utilizing information contained in the features used to construct the initial design. The relationships among these features can provide some clues about faces that are meaningfully related.

The first stage of reasoning is common for both feature mapping and feature recognition in that both techniques use face information for analysis. However, feature recognition suffers a major drawback caused by the fact that its search space is unconstrained as feature information is not available in the initial model. Failure to identify related faces will render the whole process of recognition useless in later stages. Feature mapping, therefore, has a better opportunity of finding a solution during filtration and selection as compared to feature recognition or feature extraction methodologies.

In the second stage of reasoning, an enhanced representation of the selected faces is constructed. This is necessary as the face description in a solid model, usually represented as Boundary Representation (Brep), does not render itself easily for the determination of emergent features from a collection of faces. The main purpose of this stage is to enhance the representation, making the relationships among related faces explicit. This will increase the efficacy of the subsequent matching process, which is responsible for establishing the identities and for extracting the relevant parameters associated with the emergent features. The enhanced representation developed in this research is described in the next section.

In the final stage, the newly constructed representation of a set of faces is used to match against a pre-defined library of features within the system. It is necessary that the same representation techniques be used for the pre-defined features in the library. The pre-defined library contains not only the descriptions of individual features in the target domain, but also the descriptions of collections of faces that can be decomposed into more than one feature. This extension is particularly important for mapping of complex features.

3 FEATURE SEMANTICS REPRESENTATION

Researchers have attempted to relate form, function and intent through the concept of features but with limited success. Form can be explicitly represented with geometrical elements including faces, edges, surfaces, lines and curves in boundary representation or constructive geometry. Most implementations can associate a feature with function, intent and relevancy implicitly. These semantics dictate the interpretation of features in downstream reasoning processes. Since the implied semantics can be interpreted differently, this can potentially create confusion among different application programs dealing with the same set of features. Semantics is also necessary during mapping as it plays a significant role in verifying the identity of a feature. In order to avoid such limitations, there is a need to represent the semantics of a feature explicitly.

Within the context of feature mapping, two disparate types of semantics can be identified. The first, known as geometric semantics, expresses the form of a feature. Although solid modeling is capable of representing the geometry of a part design unambiguously, the procedural representation is not suitable for matching in the mapping process. An alternate representation that captures the relationships among faces in a solid model is needed. A face connectivity graph (FCG) has been developed in this research to overcome the limitations of conventional solid modeling. Relationships that are represented explicitly in a FCG include convexity, concavity, parallelism and perpendicularity between faces. The type of faces and edges are also captured in the graph. Similar graph techniques have been used by Kyprianou (1980) and Joshi and Chang (1988). The ability of a FCG to represent face relationships compactly is a primary reason for selecting this knowledge representation technique. Figure 2 illustrates how a FCG captures the geometric knowledge of a slot concisely. FCG in essence complements boundary representation (Brep). This is due to the fact that Brep is not suitable to support reasoning in mapping at symbolic level, especially during matching.

The non-geometric knowledge associated with a feature constitutes the second type of semantics. Such knowledge tends to be domain dependent and product specific. In the machining domain, for example, feature classifications are often directly related to knowledge concerned with machining processes and tools that can be used to produce those features. On the other hand, features used in the design of an engine block are related to its functions. Ribs, for instance, provide the function of dissipating heat from an engine block. Constraints on the dimensions of a feature can also form part of this knowledge. The implementation of non-geometric knowledge thus depends on the specific application and product domain. In general, rules and other knowledge representation techniques can be used together with FCGs to specify the non-geometric semantics of a feature. The two types of knowledge described above represent a deep model of a feature.

Figure 3 summarizes the types of knowledge needed for the representation of semantics of a feature.

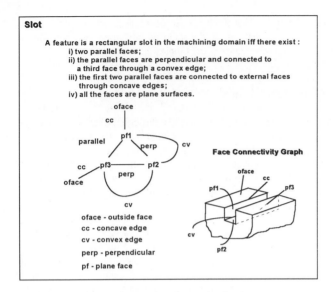

Figure 2 Face Connectivity Graph (FCG) of a slot.

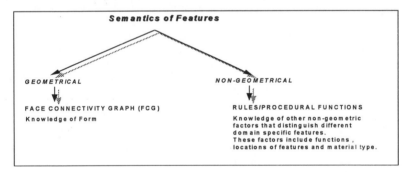

Figure 3 Semantics of a feature.

4 IMPLEMENTATION

The mapping approach described in this paper has been embodied in a prototype mapping shell developed in this research. The shell has been implemented with a blackboard system called Generic Blackboard (GBB) (Gallagher *et al.*, 1988) that integrates a case-based reasoning system (Riesebeck and Schank, 1988), and a feature modeling system. The CBR system has been enhanced to index pre-defined features as cases within a case library. The feature modeling system is developed using Wisdom Systems Concept Modeller (The Concept Modeller, 1992). It is fully integrated with a geometric kernel called CV DORS through the CVkey module provided by Concept Modeller. Common lisp is the main programming language used in these systems.

The overall architecture of the mapping shell is illustrated in Figure 4. It has been implemented with four layers in the blackboard system to model the three phases of reasoning as described in Section 2.1. Each layer contains its own information unit relevant to the associated reasoning processes. The blackboard architecture is employed to take advantage of its ability to integrate and coordinate the activities of different knowledge sources during problem solving. Another advantage of blackboard is that it allows the utilization of different reasoning techniques for problem   solving. Case-based reasoning, graph technique and algorithmic analysis constitutes the three main reasoning techniques in the shell.

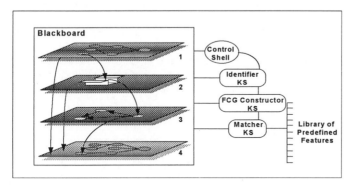

Figure 4 Overall architecture of the feature mapping shell.

Within the shell, three knowledge sources have been developed. The first is the identifier knowledge source that contains knowledge required for filtration and selection processes. The second is the FCG constructor knowledge source that construct a FCG for a collection of faces. The last knowledge source has been developed for matching. This knowledge source is used to find relevant cases in the case library that matches a current situation.

A part design, represented in terms of   design features or neutral features, is initially posted onto the first blackboard layer. The identifier knowledge source uses these features to analyze the part design. Features are sequentially selected and examined to detect any interaction with other features in the design. If there is no interaction, faces in the evaluated boundary model that reside in the same space as the faces in the original features will be grouped together. In the case that involves interactions, certain heuristics are used to guide the analysis process. These heuristics uses the effect of interaction on the faces in the evaluated boundary model to determine faces that should be grouped together. The heuristics are shown in Table 1.

When interactions involve feature that require conjugate mapping, the situation becomes more complex. In the present system, faces of the feature will be grouped together with other faces of features that it is interacting. The effect of the interactions on faces of other features is not considered in such an situation. Those faces that have been included as a group is then posted onto the second layer of the blackboard.

A collection of faces on the second layer will activate the FCG Constructor knowledge source. The knowledge source examine all the faces systematically to extract information such as the type of faces and relationship between these faces. The graph is next sent to the third layer which will invoke a knowledge source containing the necessary algorithms for the examination and creation of data structures for describing the graph. A FCG is implemented as a structure in this research. The various components of the structures are illustrated in Figure 5.

Table 1 Heuristics used to guide the filtration and selection process for feature interactions

i) if inner-loop is created on a face, then consider the feature separately and set up dependency.

ii) if additional edges are created on the outer loop of a face, then consider the feature separately and set up dependency.

iii) if any faces that are sub-divided by another feature, then group the features together.

Finally, the matcher knowledge source in the shell inspects the graph. Based on such information as the number and type of features involved, this knowledge source, implemented as a case-based reasoning system, will attempt to find cases in the case library that partially match the descriptions. The case-based reasoning system is employed to reduce the search space during this phase of reasoning. Schematic diagram of the matcher knowledge source is shown in Figure 6.

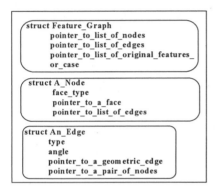

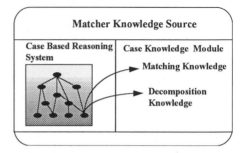

Figure 6   Schematic Diagram of Matcher Knowledge Source.

Figure 5   Components of structures used to represent a FCG graph.

A matching knowledge source and a decomposition knowledge source are always associated with each case in the library. These two type of knowledge are further breakdowns of the non-geometric knowledge in Section 3.0. The cases indexed in the library can represent individual features or past cases that involves interaction or conjugate mapping. The knowledge associated with each case captures the knowledge necessary to perform matching with another case and determine the relevant parameters to describe the emergent features. The case based reasoning system thus provides a systematic method of managing each case and its associated knowledge. It also functions as a repository of application dependent mapping knowledge in the system.

Exhaustive matching then follows to find an exact match by utilizing the matching knowledge of each case. The relevant geometric and non-geometric parameters of emergent features will then be extracted from the corresponding FCGs together with the collection of faces described by the FCGs.

5 CASE STUDY

As the processes of filtration, selection, and FCG construction are based on geometric information alone, the same knowledge sources can be applied to several different applications. The domain dependent aspects of knowledge are employed only during the matching stage when emergent features need to be clearly identified. Therefore, the mapping shell can be easily customized to different domains as only the case base has to be modified.

In this research, the shell has been customized with knowledge to map features into the machining domain. A library of cases that can map neutral features directly into machining features has been developed in the case-based reasoning system. The case base is developed as an abstraction hierarchy known as Memory Organization Packages (MOP). Detail explanation of MOP is beyond the scope of this paper due to limited space.

An example test part, as shown in Figure 7, has been designed with eight typical features to illustrate the mapping approach as implemented in the customized mapping shell. The rectangular depression 1 has a negative volume type and interacts with the rectangular depression 2. Due to the interaction, three machining features can be identified: one through slot and two through steps. Please refer to Table 2 for the results of mapping the example part. This is an example of interaction involving two simple neutral features. In this case, the identifier knowledge source groups together the faces belonging to the two interacting features. It can be seen clearly in Figure 8 that the faces belonging to the rectangular depression 1 have been sub-divided. The identifier knowledge source is able to recognize the effect of interaction on faces and perform appropriate grouping of these faces belonging to different features. In this case, as sub-division of faces occurs, remnant faces of these two features are grouped together. Geometric reasoning is used extensively at this stage.

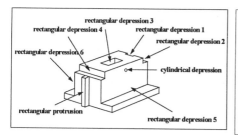

Figure 7 An example test part.

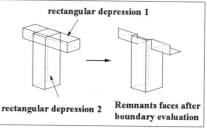

Figure 8 Interaction between rectangular depression 1 and rectangular depression 2, and their evaluated faces.

This group of faces are posted onto the second layer so that the FCG constructor knowledge source can create a FCG that captures the relevant geometric semantics of these faces. Finally, relevant cases of FCGs are then identified from the case base by the matcher knowledge source. The knowledge associated with these cases are applied systematically to determine a specific case that exactly matches the current FCG. The emergent features and their parameters are next established.

Another type of interaction, as illustrated by the rectangular depression 3 and the cylindrical depression, are also taken into consideration in the mapping process. This is clearly shown in Figure 9. The cylindrical depression feature has been decomposed into two through hole features in the evaluated boundary model.

Table 2 Results of mapping the example part shown in Figure 7

	Neutral features used in design	Emergent machining features after mapping
1	rectangular depression 1 & rectangular depression 2	2 through steps & 1 through slot
2	rectangular depression 3	1 rectangular pocket
3	rectangular depression 4	1 through step
4	rectangular depression 5	1 through step
5	rectangular depression 6	1 through step
6	rectangular protrusion	2 through steps
7	rectangular depression	2 through holes

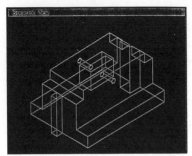

Figure 9 Screen dump of the evaluated boundary model of the test part.

All other features, except for the rectangular protrusion that has a positive volume type, has   direct equivalence in the machining domain. Such mapping cases represent identity mapping as described in Section 2.0. Conjugate mapping, however, is required for the protrusion feature shown in the design. Faces that are related to this feature with concave edges have to be examined by the mapping process to determine emergent features. This results in two through steps being produced owing to the existence of the protrusion feature. This type of mapping can be handled effectively by the feature mapping system.

6   CONCLUSIONS

A mapping methodology has been proposed to meet the requirements imposed by identity mapping, conjugate mapping and mapping involving feature interactions. The methodology has been developed to be applicable for mapping into various application domains. An explicit representation of feature semantics has been adopted to enhance the reasoning capability of the approach. This graph-based representation forms a deep model of a feature.

Based on the proposed approach and the   explicit semantic representation, a feature mapping shell has been implemented. The shell is highly customizable for mapping into different domains by virtue of the separation of domain specific knowledge from the mapping process itself. A specific mapping system can thus be easily developed by adding the domain dependent knowledge into the shell. Such knowledge includes the geometric and non-geometric semantics of features and the construction of a case library.

Although the mapping shell has been proven to be able to handle simple cases of conjugate mapping, further research is still needed to obtain a better understanding of all the issues involved. These issues include the base part enlargement problem due to the existence of protrusion features in the domain of process planning and re-allocation of tolerances originally attached to depression features.

7 REFERENCES

Broonsvoort, W.F. and Jansen, F.W. (1993), Feature Modeling and Conversion - Key Concepts to Concurrent Engineering. *Computers in Industry*, **21**(1) 61-86.

Dixon, J.R., Libardi, E.C., Nielsen, E.H. (1990), Unresolved research issues in development of design-with features systems, in *Geometric Modeling for Product Engineering*, (eds. Wozny M.J., Turner J.U., Preiss K.), IFIP WG 5.2/NSF Working Conference on Geometric Modeling, Rensslaerville, USA, 1988.

Gallagher, K.Q., Corkill, D.D., Johnson, P.M. (1988), GBB Reference Manual GBB Version 1.2, COINS Technical Report 88-66.

Joshi, S. and Chang, T.C. (1988), Graph-based Heuristics for Recognition of Machined Features from a 3-D Solid Model. *Computer Aided Design*, **20**(2), 58-64.

Kyprianou, L.K. (1980), Shape classification in Computer Aided Design, Ph.D. Thesis, University of Cambridge, England.

Lim, S.S., Lee I.B.H., Lim L.E.N. and Ngoi B.K.A. (1994), Multiple Domain Feature Mapping: A Methodology Based on Deep Models of Features, paper accepted by *Journal of Intelligent Manufacturing*, Chapman and Hall, UK.

Pratt, M.J. (1993), Automated Feature Recognition and its Role in Product Modeling, in *Geometric Modelling*, (eds. Farin G., Hagen H., Noltemeir H.), Computing Supplementum **8**, 241-250.

Riesbeck, C.K. and Schank, R.C. (1989), Inside Case-Based Reasoning, Lawrence Erlbaum Associates, New York, USA.

Shah, J.J. (1988), Feature Transformations Between Application-Specific Features Spaces. *Computer-Aided Engineering Journal*, **5**(6), 247-255.

The Concept Modeller (1992), Concept Modeller, Release 2.0, Reference Manual, Wisdom Systems, Pepper Pike, Ohio, USA.

Woodward, J.R. (1988), Some Speculation on Feature Recognition. *Computer-Aided Design*, **20**(4), 189-196.

8　BIOGRAPHY

S.S. Lim holds a BEng(Hons) in Mechanical Engineering from Universiti Teknologi Malaysia. He is currently working as a research engineer at the Gintic Institute of Manufacturing Technology.

Ivan B.H. Lee holds a PhD from the National University of Singapore. He is presently a group manager at the Gintic Institute of Manufacturing Technology.

Lennie E.N. Lim graduated from Surrey University with BSc and PhD degrees in Mechanical Engineering. He is presently a professor and the Vice Dean of the School of Mechanical & Production Engineering, Nanyang Technological University.

Bryan K.A. Ngoi received his BEng from the National University of Singapore in 1985 and his PhD from University of Canterbury, New Zealand in 1990. He is presently a senior lecturer at the Nanyang Technological University.

PART THREE

Computational Geometry and its Application

9

Local and global schemes for interpolating fair curves by $B2/B3$-splines

Mitsuru Kuroda[†],   Fumihiko Kimura[‡],   Susumu Furukawa[§] and   Kang J. Chang[¶]*

[†]*Department of Information and Control Engineering, Toyota Technological Institute, 2-12 Hisakata, Tempaku, Nagoya 468, Japan, Phone: +81-52-802-1111, Fax: +81-52-802-6069, M.Kuroda@toyota-ti.ac.jp*

[‡]*Department of Precision Machinery Engineering, The University of Tokyo, 7-3 Hongo, Bunkyo, Tokyo 113, Japan*

[§]*Department of Mechanical System Engineering, Yamanashi University, 4-3 Takeda, Kofu, Yamanashi 400, Japan*

[¶]*Planning and Organizing Committee for the National Taipei University, Chien Kuo North Road Sec. 2, Taipei, Taiwan, R.O.C.*

Abstract

The local and global schemes are proposed for obtaining fair planar curves useful in the field of smooth curve fitting, computer graphics and computer aided geometric design. These schemes are based on the $B2/B3$-splines, which are embeded in the cubic interpolating B-splines and have 2 and 3 cubic segments by knot-insertion in each span respectively. The local scheme dependent on data points generates span-by-span a C^2 curve with reasonable locality and fairness. The relevant linear equation system is newly derived in symbolic expression for practical use by a symbolic manipulation system. The corresponding equation system is also given for obtaining a quartic C^2 interpolating fair curve (based on the S-splines). A global scheme generates a highly continuous curve of low degree polynomial, because the $B2/B3$-splines are more flexible than the cubic B-splines. The other global scheme generates a C^2 curve like a clothoid splines with piecewise linear curvature, based on the fact that the numerator of the curvature of a cubic segment is not cubic but quadric.

Keywords

Fairness, local behavior, interpolation, cubic B-splines, B2-splines, CAGD, CG

---

*on leave from California State University, Northridge, U.S.A.

1   INTRODUCTION

We develop local and global schemes for obtaining C^2 interpolating fair curves useful in the field of smooth curve fitting, computer graphics and computer aided geometric design. The schemes are based on the $B2/B3$-splines with 2 and 3 cubic segments in each span respectively. The local scheme derives a C^2 curve with reasonable fairness and locality. The global scheme obtains a highly continuous curve of low degree polynomial with curvature consisting of the fewer monotone pieces.

Historically, the $B2$-splines were introduced to bring the local behavior in the cubic C^2 interpolating curve (Woodward, 1987). The $B3$-splines were studied for cubic C^2 interpolating curve with most local behavior (Chu, 1990). Then, the $B\lambda$-splines with more flexibility were developed (Krokos et al., 1992). All these splines were proposed to be used under the interactive environment. However, we analyzed the $B2$-splines and found that there is an instability of shape control in them for interactive operation (Kuroda et al., 1994b). Therefore, we have developed a locality control system for C^2 interpolating curves (Kuroda et al., 1993, Kuroda et al., 1994a, Kuroda et al., 1994b). It makes the curve converge to the conventional cubic interpolating splines, as the influence range of each data point is expanding over the curve shape. The curve (with the global behavior) has *"the minimum norm property"* at the limit, known as the basis of fairness of the cubic interpolatory splines (S.Bu-Qing, 1989). On the contrary, the curves with full locality do not always have satisfactory fairness (Kuroda et al., 1993), because the emphasis is on simple and easy control of the local behavior. Then, we have introduced a *local* minimum property into the quartic C^2 interpolating curve, each span of which is determined by 6 or 8 neighbouring data points (Kuroda et al., 1994c).

In this paper, we derive the linear equation system for a local minimum property in the $B2$-splines by a symbolic manipulation system, in Chapter 2. The equations in symbolic expression are based on minimizing the integral of square of the second derivative over the neighboring 3 spans, and their derivation is quite helpful for practical use. In Chapter 3, we develop a global scheme for highly continuous curves, based on minimizing the summation of square of the third derivative. We also propose the other global scheme for obtaining a C^2 curve like a clothoid splines with piecewise linear curvature, based on the fact that the numerator of the curvature of a cubic segment is not cubic but quadric,

2   LOCAL SCHEMES

To start our development, we first review the interpolating curve with controllable locality (Kuroda et al., 1993, Kuroda et al., 1994a, Kuroda et al., 1994b) briefly, as the basis of our new schemes.

We use the $B2$-splines with an additional control point obtained by knot-insertion in each span of the uniform cubic interpolating B-splines. The notations are as follows.

- Given data points: $p_0, \ldots, p_i, \ldots, p_n$,
- B-spline control points: $d_{-1}, \ldots, d_i, \ldots, d_{2n+1}$,
- Additional control points: $d_{-1}, \ldots, d_{2i-1}, d_{2i+1}, \ldots, d_{2n+1}$,

- B2-spline control points: $d_{-1}, p_0, d_1, p_1, \ldots, d_{2n-1}, p_n, d_{2n+1}$,
- Knots: $u_0, \ldots, u_i, \ldots, u_{2n}$.

The points p_i, d_{2i} and knot u_{2i} correspond each other. d_{2i+1} corresponds to u_{2i+1}. We set the end conditions with zero curvature by adopting the point-symmetric phantom data points around each endpoint. Namely,

$$\begin{cases} d_0 = p_0, & d_{-1} = 2p_0 - d_1, \\ d_{2n} = p_n, & d_{2n+1} = 2p_n - d_{2n-1}. \end{cases}$$

The additional (odd-numbered) control points are expressed as follows, (Hosaka et al., 1976) when the corresponding knots are "pseudo-knots" (Riesenfeld, 1973).

$$d_{2i+1} = \frac{1-\alpha}{1-\alpha^{m+1}} \sum_{j=0}^{m} q_{i,j} \alpha^j, \qquad \alpha = -2 + \sqrt{3} \doteq -0.267949, \tag{1}$$

$$q_{i,j} = \frac{p_{i-j} + p_{i+1+j}}{2}.$$

The other (even-numbered) control points are derived from the following C^2 interpolation conditions.

$$d_{2i-1} + 4d_{2i} + d_{2i+1} = 6p_i, \qquad i = 0, \ldots, n. \tag{2}$$

Therefore, each span is expressed as the uniform cubic B-splines.

$$r_i(t) = \sum_{j=0}^{3} f_j(t) d_{i-1+j}, \qquad 0 \le t \le 1, \qquad i = 0, \ldots, 2n-1. \tag{3}$$

Since $|\alpha| \ll 1$, we can approximate the additional control points with their neighboring data points by cutting off higher order terms and get a curve with the local behavior. This is the idea of the locality control system of the cubic C^2 interpolating curve. Unfortunately, this system does not always give a desired curve with the strong locality, because the curve lost the minimum norm property for getting the local behavior. Then, we developed a new curve (Kuroda et al., 1994c), additional control points of which are determined by their neighboring 4 or 6 data points based on the alternative minimization conditions. We describe the schemes for obtaining the following parameters λ_i, μ_i in Section **2.1** and Section **2.2**.

$$d_{2i+1} = q_{i,0} + \lambda_i a_{i,1} + \mu_i a_{i,2}, \tag{4}$$

$$a_{i,j} = q_{i,j} - q_{i,0}, \quad j = 1, 2, \qquad i = 0, \ldots, n-1.$$

2.1 Scheme independent of data points

The parameters λ, μ independent of data points are determined by minimizing the whole integral of square of the second derivative of the blending function with respect to a certain data point (Kuroda et al., 1994c). The following table lists λ and μ with the corresponding values in the conventional cubic curve with controllable locality.

Table 1 Lists of parameters λ, μ

Schemes	New Scheme		Conventional Scheme	
Parameters	λ	μ	λ	μ
Based on 4 points	-0.295455	0	-0.36603	0
Based on 6 points	-0.344011	0.0821727	-0.3333376	0.0893164

2.2   Scheme dependent on data points

We try to localize the minimum norm property of the conventional cubic C^2 interpolating curve and develope a local scheme for determining additional control points adaptively to the neighboring data points. Namely, we derive the parameters λ_i and μ_i from minimizing the whole integral of square of the second derivative of the neighboring 3 spans (Kuroda et al., 1994c). A symbolic manipulation system leads to the following linear equation system (5), whose coefficient matrix is symmetric and diagonal dominant. We take a pair of $\{\lambda_i,\ \mu_i\}$ and discard $\{\lambda_{i-1},\ \lambda_{i+1}\}$ after solving these simultaneous equations. Now we have got the practical method to obtain a fair curve with the local behavior. The resultant curve is not only as fair as the conventional cubic interpolating curve with the minimum norm property as shown in Figure 1 later, but also locally controllable. In the Appendix for practical use, we give the (quartic C^2) S-spline (Kawai et al., 1988) version of the linear equation system based on the same kind of minimization condition, which generates a shape similar to the $B2$-spline one.

$$
\begin{bmatrix}
46a_{i-1,1} \cdot a_{i-1,1} & 10a_{i-1,1} \cdot a_{i,1} & 10a_{i-1,1} \cdot a_{i,2} & -a_{i-1,1} \cdot a_{i+1,1} \\
\cdots & 54a_{i,1} \cdot a_{i,1} & 54a_{i,1} \cdot a_{i,2} & 10a_{i,1} \cdot a_{i+1,1} \\
\cdots & \cdots & 54a_{i,2} \cdot a_{i,2} & 10a_{i,2} \cdot a_{i+1,1} \\
\cdots & \cdots & \cdots & 46a_{i+1,1} \cdot a_{i+1,1}
\end{bmatrix}
\begin{bmatrix}
\lambda_{i-1} \\
\lambda_i \\
\mu_i \\
\lambda_{i+1}
\end{bmatrix}
$$
$$
= -
\begin{bmatrix}
a_{i-1,1} \cdot (5p_{i-2} + 3p_{i-1} - 22p_i + 15p_{i+1} - p_{i+2})/2 \\
a_{i,1} \cdot (14a_{i,1} - a_{i,2}) \\
a_{i,2} \cdot (14a_{i,1} - a_{i,2}) \\
a_{i+1,1} \cdot (-p_{i-1} + 15p_i - 22p_{i+1} + 3p_{i+2} + 5p_{i+3})/2
\end{bmatrix} . \tag{5}
$$

Figure 1 shows example curves and their curvature plots. An outlined thick line is a cubic C^2 interpolating curve for comparison. It is clear that the new schemes improve the improper part of (a) the conventional curve with the controllable locality. (b) The new scheme based on the neighboring 4 points generates upper-right 3 spans with the smaller curvature, but their curvatures do not consist of more monotone pieces than the cubic interpolatory splines. (c) The new scheme based on the neighboring 6 points generates a shape closer to the cubic interpolating splines, preserving its local behavior. (d) The new scheme dependent on data points generates a shape almost as same as the cubic interpolating splines, preserving its local behavior yet.

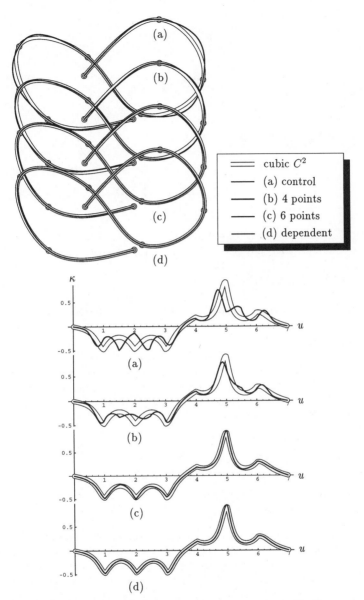

Figure 1   Example curves and their curvatures, comparing with the conventional cubic C^2 curves (outlined line), (a) curve by locality control system, (b) curve by scheme on 4 points, (c) curve by scheme on 6 points, (d) curve by scheme dependent on data points.

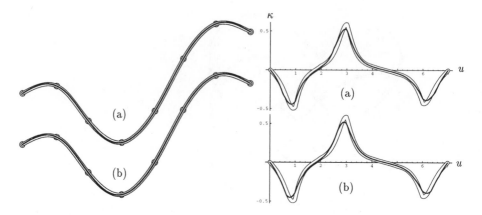

Figure 2   Example curves subject to equation (8) and the curvature plots, comparing with ones (outlined line) subject to equation (6), (a) $B2$-splines, (b) $B3$-splines.

3   GLOBAL SCHEMES

When we determine additional control points by the following condition,

$$\min \int_0^1 \sum_0^{2n-1} \ddot{r}_i^2(t)dt, \tag{6}$$

the corresponding knots degenerate to the "pseudo-knots". The resultant curve becomes the conventional cubic interpolating curve with the minimum norm property, even if we use the $B3$-splines (Chu, 1990, Krokos et al., 1992), which have 3 cubic segments in each span, being subject to the following notations.

- Given data points: $p_0, \ldots, p_i, \ldots, p_n$,
- B-spline control points: $d_{-1}, \ldots, d_i, \ldots, d_{3n+1}$,
- Additional control points: $d_{-1}, \ldots, d_{3i-1}, d_{3i+1}, d_{3i+2}, d_{3i+4}, \ldots, d_{3n+1}$,
- $B3$-spline control points: $d_{-1}, p_0, d_1, d_2, p_1, \ldots, d_{3n-2}, d_{3n-1}, p_n, d_{3n+1}$,
- Knots: $u_0, \ldots, u_i, \ldots, u_{3n}$.

The points p_i, d_{3i} and knot u_{3i} correspond each other. The points d_{3i+1} and d_{3i+2} correspond to the knots u_{3i+1} and u_{3i+2} respectively. We set the end conditions with zero curvature by adopting the point-symmetric phantom data points around each endpoint. As in the $B2$-splines, we eliminate control points d_{3i} by the C^2 interpolation conditions.

$$d_{3i-1} + 4d_{3i} + d_{3i+1} = 6p_i, \qquad i = 0, \ldots, n. \tag{7}$$

The third derivative of the $B2/B3$-splines is constant in each span. When we use the following condition to minimize the difference between the 3rd derivatives of 2 consecutive spans,

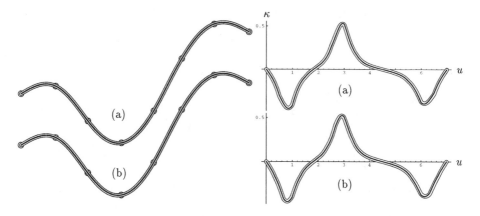

Figure 3   Example curves subject to equation (9) and the curvature plots, comparing with ones (outlined line) in Figure 2, (a) $B2$-splines, (b) $B3$-splines.

$$\min \int_0^1 \sum_0^{\beta n-2} \{\ddot{\boldsymbol{r}}_i(t) - \ddot{\boldsymbol{r}}_{i+1}(t)\}^2 dt, \qquad \beta = 2: \ B2\text{-splines, or } 3: \ B3\text{-splines,} \qquad (8)$$

the condition leads to the linear equation system with respect to the unknown additional control points and we can solve it easily and stably. Figure 2 shows an example curves with the curvatures, comparing with the curves subject to the minimization condition (6).

By the way, the resultant curve is similar to the curve subject to the following condition (9), because $\{\ddot{\boldsymbol{r}}_i(t) - \ddot{\boldsymbol{r}}_{i+1}(t)\}^2 = \ddot{\boldsymbol{r}}_i^2(t) + \ddot{\boldsymbol{r}}_{i+1}^2(t) - 2\ddot{\boldsymbol{r}}_i(t) \cdot \ddot{\boldsymbol{r}}_{i+1}(t)$.

$$\min \int_0^1 \sum_0^{\beta n-1} \ddot{\boldsymbol{r}}_i^2(t) dt, \qquad \beta = 2: \ B2\text{-splines, or } 3: \ B3\text{-splines,} \qquad (9)$$

Figure 3 shows example curves, comparing with the curves subject to the condition (8) in Figure 2.

The curvature κ of the parametric curve is expressed as,

$$\kappa = \frac{det[\dot{\boldsymbol{r}}, \ddot{\boldsymbol{r}}]}{\dot{\boldsymbol{r}}^3}. \qquad (10)$$

Actually, the numerator of κ in a piecewise cubic curve is quadric because the cubic term is zero as follows. Any polynomial is expressed by Bernstein-Bézier basis and,

$$\begin{cases} \boldsymbol{r}(t) = \boldsymbol{b}_0 + 3t\Delta\boldsymbol{b}_0 + 3t^2\Delta^2\boldsymbol{b}_0 + t^3\Delta^3\boldsymbol{b}_0, \\ \dot{\boldsymbol{r}}(t) = 3(\Delta\boldsymbol{b}_0 + 2t\Delta^2\boldsymbol{b}_0 + t^2\Delta^3\boldsymbol{b}_0), \qquad \Delta^i\boldsymbol{b}_0 = \Delta^{i-1}\Delta\boldsymbol{b}_0 = \Delta^{i-1}(\boldsymbol{b}_1 - \boldsymbol{b}_0), \quad (11) \\ \ddot{\boldsymbol{r}}(t) = 6(\Delta^2\boldsymbol{b}_0 + t\Delta^3\boldsymbol{b}_0), \end{cases}$$

$$\Delta^3\boldsymbol{b}_0 \times \Delta^3\boldsymbol{b}_0 = \boldsymbol{o}.$$

We here try to determine additional control points of the $B2/B3$-splines by the following condition.

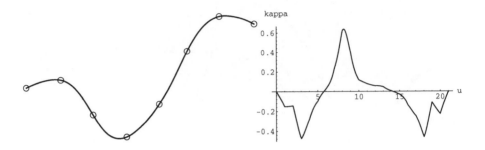

Figure 4 Example curve subject to equation (12) and the curvature plot.

$$\min \int_0^1 \sum_0^{\beta n-1} \{\text{coefficient of } t^2 \text{ of } det[\dot{r}, \ddot{r}]\}^2 dt, \quad \beta = 2: \ B2\text{-splines, or } 3: \ B3\text{-splines}. \quad (12)$$

This condition leads to the simultaneous cubic equation system and it is solved easily by Newton-Raphson method although it is not always stable. Figure 4 shows an example of the $B3$-spline curve with the curvature. The curve is like a piecewise clothoid curve with the curvature continuity, the curvature of which is piecewise linear.

4   CONCLUDING REMARKS

We tried to localize the minimum norm property and have proposed the local scheme for obtaining a C^2 fair curve. Examples showed that the scheme improves improper behavior of the curves with controllable locality and seems to establish a marriage of two opposite-like properties, *fairness* and *local behaviour*. The new curves are defined span-by-span and locally adjustable without undesired undulations and so these are useful in various fields.

It was shown that one of the global schemes derive a curve with curvature of the few monotone pieces and the other gets a curve like a clothoid splines with piecewise linear curvature. Further analysis and development of the latter scheme is for future research.

REFERENCES

[Chu, 1990] Chu, K.-C.(1990) B3-splines for interactive curve and surface fitting. *Computer & Graphics*, **14**, 2, 281 – 288.

[Hosaka et al., 1976] Hosaka, M. and Kuroda, M. (1976) A synthesis theory of curves and surfaces for CAD. *Journal of Information Processing Society of Japan*, **17**, 12, 1120 – 1127 (in Japanese).

[Kawai et al., 1988] Kawai, T., Fujita, T. and Omura, K. (1988) An composition of spline basis with knots of multiplicity 2. *The Transactions of the Institute of Electronics, Information and Communication Engineering of Japan*, **J71-D**, 6, 1149–1150 (in Japanese).

[Krokos et al., 1992] Krokos, M.A. and Slater, M. (1992) Interactive shape control of interpolating B-splines. *Computer Graphics Forum*, in *EUROGRAPHICS '92* (ed. A.Kilgour and L.Kjelldahl), **11**, 3, C-435 – C-447.

[Kuroda et al., 1993] Kuroda, M., Furukawa, S. and Kimura, F. (1993) *S*-splines and *B*2-splines as interpolating curves with controllable locality. *Journal of Information Processing Society of Japan*, **34**, 11, 2294–2301 (in Japanese).

[Kuroda et al., 1994a] Kuroda, M., Furukawa, S. and Kimura, F. (1994) C^2 interpolating *B*2-spline curve with controllable locality. *Journal of the Japan Society for Precision Engineering*, **60**, 1, 65 – 69 (in Japanese).

[Kuroda et al., 1994b] Kuroda, M., Furukawa. S. and Kimura, F. (1994) Controllable locality in C^2 interpolating curves by *B*2-splines / *S*-splines. *Computer Graphics Forum*, **13**, 1, 49 – 55.

[Kuroda et al., 1994c] Kuroda, M., Kimura, F. and Furukawa, S. (1994) Minimization conditions for C^2 interpolating curve with local behavior. *Journal of the Japan Society for Precision Engineering*, **60**, 6, 842 – 846 (in Japanese).

[Riesenfeld, 1973] Riesenfeld, R.F. (1973) Applications of B-spline approximation to geometric problems of computer-aided design. PhD thesis, Syracuse University.

[S.Bu-Qing, 1989] Bu-Qing, S. and Ding-yuan, L. (1989) –Computational Geometry– Curve and Surface Modeling. Academic Press.

[Woodward, 1987] Woodward, C.D. (1987) B2-splines –A local representation for cubic spline interpolation–. in *Proc. CG International '87* (ed. T.L.Kunii), Springer, New York, 197 – 206.

Appendix LINEAR EQUATION SYSTEM FOR THE *S*-SPLINES

The linear equation system for the interpolating *S*-spline curve (Kuroda et al., 1994b) is as follows, for determining additional control points based on the minimization condition dependent on neighboring data points in 3 spans.

$$
\begin{bmatrix}
18a_{i-1,1} \cdot a_{i-1,1} & 2a_{i-1,1} \cdot a_{i,1} & 2a_{i-1,1} \cdot a_{i,2} & -a_{i-1,1} \cdot a_{i+1,1} \\
\cdots & 22a_{i,1} \cdot a_{i,1} & 22a_{i,1} \cdot a_{i,2} & 2a_{i,1} \cdot a_{i+1,1} \\
\cdots & \cdots & 22a_{i,2} \cdot a_{i,2} & 2a_{i,2} \cdot a_{i+1,1} \\
\cdots & \cdots & \cdots & 18a_{i+1,1} \cdot a_{i+1,1}
\end{bmatrix}
\begin{bmatrix}
\lambda_{i-1} \\
\lambda_i \\
\mu_i \\
\lambda_{i+1}
\end{bmatrix}
$$

$$
= - \begin{bmatrix}
a_{i-1,1} \cdot (p_{i-2} + 3p_{i-1} - 10p_i + 7p_{i+1} - p_{i+2})/2 \\
a_{i,1} \cdot (7a_{i,1} - a_{i,2}) \\
a_{i,2} \cdot (7a_{i,1} - a_{i,2}) \\
a_{i+1,1} \cdot (-p_{i-1} + 7p_i - 10p_{i+1} + 3p_{i+2} + p_{i+3})/2
\end{bmatrix}.
\tag{13}
$$

10

Free-form Surface Design by Model Deformation and Image Sculpturing

C.T.Poon, S.T.Tan and K.W.Chan
Department of Mechanical Engineering,
The University of Hong Kong,
Pokfulam Road, Hong Kong.
Email: p91ctp@hkumea.hku.hk,
sttan@hkucc.hku.hk,
kwchan@hkumea.hku.hk

Abstract

Conventional computer aided design systems have been well-developed for the design of general prismatic parts and simple free-form objects. However the available modelling tools and techniques are inadequate for highly "sculptured" objects. This paper presents a new approach for local surface design that provides a rapid and intuitive way to create surface features on a parametric surface. Embossed or depressed patterns can be added to a surface via a 2D grey-level image function. The image function corresponds to a 2D elevation map of the surface. The patterns may included some simple line drawings or 2D contours designs. It allows the user to create peaks and ridges, valleys and pits or other similar shapes by simple sketching over the surface. This module may be used as an additional detailing tool following a global surface design process.

Keywords

Free-form surface, deformation, sculpturing, sketching

1 INTRODUCTION

Many engineering tasks require complex and highly detailed shapes that are not readily describable by traditional modelling systems using either parametric surface representation or Boolean combinations of simple primitives. This is especially true in the design of aesthetic surfaces with intricate shapes and patterns, such as those in the biscuit mold design (Wainwright, 1992), jewellery surface design (Ettagale, 1991) and many other works resulting from highly skillful engraving or carving. This illustrates that the modelling tools and the design environment in existing systems are not sufficient for modelling complex objects.

Many attempts have been made to providing more efficient and effective design methods for 3D modeling systems, A growing trend is to develop some highly intuitive modelling tools and design interfaces that can provide a familiar way of thinking for a designer. The most widely accepted concept in CAD is sculpting in clay, where the model is to be shaped with flexibility similar to a lump of clay in a sculptor's hand. Two design processes should be considered in a more general and sophisticated modelling system:

- Global Solid Design, which allows the user to deform an object smoothly during the initial phase of modelling. This can be limited to simple, global changes, including bending, tapering, twisting, squeezing and so forth. The process may be regarded as deforming a lump of clay with forces applied by the user.
- Local Surface Design, which allows the user to add surface details on the model, producing embossed or depressed forms (e.g. peaks, ridges, valleys, pits, etc.) over it. This module may provide a simple and friendly design interface to get inputs or actions from the user as if he/she is sculpturing on an object's surface with engraving tools.

These two design modules form the basis of a general computer aided system for object modelling. It is different from the conventional B-rep or CSG systems which are good at defining the nominal shape of a solid, but provide limited control of detailed geometry over the object surfaces. These traditional systems are capable of handling many general machining parts and simple free-form shapes in castings but are weak in handling objects with intricate shapes or carvings resulting from skillful engraving work. This kind of work are commonly found in the mold making and handicraft industry which still relies heavily upon experienced and skilled craftsmen.

2 OVERVIEW OF SOME USEFUL TECHNIQUES

There are a number of useful geometric modelling techniques available for global and local design. A common practice in designing spline surfaces is by *Control Point Manipulation* (Piegl, 1989). It is a simple way of interactively modifying the geometry of free-form surfaces. The basic operation includes moving (and refining if necessary) the control vertices of the spline surface. However the modification on the surface is restricted largely by the position and the number of the control vertices. Although this point-based techniques is widely implemented in many systems, it is only sufficient in modelling simple shapes and shows limited success in handling details.

Another more advanced modelling techniques is known as *Solid Deformation* (Barr, 1984). Barr in his paper suggested a set of new hierarchical solid modelling operations which simulate sketching, bending, twisting, tapering and other similar transformation of geometric models. These operations extend the conventional operations of scaling, rotation, translation, Boolean operations, and can be incorporated into the traditional modelling systems, so that users may apply different operations on simple primitives to form a complex model.

The solid deformation technique was further extended by Sederberg and Parry (1986) to form a more general deformation technique, referred to as *Free-Form Deformation (FFD)*. In FFD, an object to be deformed is enclosed by a 3D parallelepipedical lattice. A deformation of the lattice causes a transformation of all interior points of the enclosed object. The deformation effect of the lattice is transferred to the object by some 3D transformation functions such as the trivariate Bernstein polynomial or trivariate B-splines (Griessmair and Purgathofer, 1989). FFD is a highly intuitive modelling tool, allowing the object to be modelled as if it is a real solid being shaped in his hand. An extended free-form deformation (EFFD) is also introduced by Coquillart (1990). This method used non-parallelepipedical 3D lattices for model deformation.

FFD has made a significant change from the control point manipulation method, which relies too much on the underlying surface definition. However, FFD is mainly designed for global shape design, and is less useful and less efficient for local surface design. The aim in local design is to find rapid and intuitive methods that can raise or imprint a surface into a desired pattern. Detailed design on a smooth surface, including arbitrarily shaped pits and valleys, peaks and ridges, like the patterns found in carving works are targets for this work.

A modelling technique known as *3D painting* (Williams, 1990) is an innovative design method to sculpture details on a surface. The paper proposed a means by which the conventional methods used in digital painting and image manipulation be extended into the third dimension. The basic approach is to generate a 3D surface from a 2D image function that is designed interactively by the user with different "painting" tools. The 2D image is essentially a functional (single-valued) surface. The canvas on which the painting program operates is a unit rectangular parametrization of the image. The limitation to surfaces which can be represented by such an image is less restrictive than it seems initially because the image may be a visualization aid to surface features. It may be easier for a user to specify the local surface features in a 2D space and subsequently map the specified surface requirements to the surface model. The basic modelling principle of 3D painting is that the intensity on the image is related to the elevation of surface point sets.

The painting tools for designing sculptured surfaces are commonly used in a general painting system. For example, an arbitrarily-shaped rough surface may be created by a seed filling operation that fills a 2D region with specified intensity distribution. Other tools, like blurring or filtering, may be used to soften and smoothen particular regions of the surface, and are effective in blending and filleting sections together. For a more detailed design, arbitrarily-sized and different style of painting brushes can be used to create embossed shapes over the surface description by conventional interpolation techniques. All these operations and painting tools provide a rapid, effective and interactive procedure for designing details on object surfaces through small refinements.

A new approach for surface sculpturing by an image function is presented in the next section. It is a local surface design method which allows a user to create embossed patterns over a surface from a grey-level image. The pattern may include some simple line drawings or 2D contours. Simple surface features, such as peaks and ridges, valleys and pits or other similar shapes may be embossed or depressed over a surface. The module may be used as an additional detailing tool after using the FFD or other similar shape design techniques. The method is derived from traditional 2D painting system (Smith, 1982), and from displacement mapping (Cook, 1984).

3 SURFACE SCULPTURING

Painting methods have been shown (Williams, 1990) to be a simple way of adding small sculptured details on a smooth surface. The basic approach is to generate a sculptured surface via a 2D image function that is designed interactively by a user with different "painting" tools. The image function corresponds to a 2D elevation map of the surface. it may be used as a canvas of a traditional painting system. The user may paint on it with various shades of grey value, with brighter shades of grey producing peaks and ridges, and darker shades producing valleys and pits. The pattern on the surface is subsequently created by "image sculpturing". It uses grey-level value as a measure of the depth, which is to be added or subtracted from the smooth surface. It is a rapid way for a user to create surface sculpture from a 2D sketch.

There are a number of ways to design a grey-level image function, including all the standard methods used in painting systems and image processing techniques. The aim is to create a grey-level image that is smooth enough to be used as depth values on the surface. A contour-based design method (Carlsson, 1989) which makes use of a diffusion algorithm to smooth out the image is attempted here. The method is used for sculpturing simple patterns. For more complex ones, some other techniques on image manipulation need to be explored.

The role of the *image function* for surface detailing is similar in some ways to the *deformation function* defined by a FFD lattice. In FFD, the lattice is used to define a 3D transformation that modifies the global shape of an object, while in image sculpturing, the image is used to define a 2D function that displaces the surface points on the sculptured surface.

Definition of image function
An image function is a 2D function which defines the depth values to be added or subtracted from the parametrically defined surface. For each point (i, j) in the image space, there will have a corresponding point (u, v) in the parametric space of the smooth surface. It is used to evaluate the displacement of each point in the sculptured surface.

Let $I(i, j)$ be an image function. In the object space, each point $\mathbf{P}$ of the smooth surface would be displaced in the direction of the surface normal $\mathbf{N}$ by a magnitude equal to the value of $I(i, j)$. The new position vector may be written as:

$$\mathbf{P'}(u, v) = \mathbf{P}(u, v) + I(i, j) \times \frac{\mathbf{N}(u, v)}{|\mathbf{N}(u, v)|} \qquad (1)$$

where, (i, j) is index to the image function,

(u, v) is the parametric values of a surface point.

To create a surface with sculptured details, an image function needs to be established and mapped to the smooth surface. Drawing and painting techniques form the basis for the establishment of this image function.

The image $I(i, j)$ could, of course, be defined analytically as a bivariate polynomial function as that found in traditional CAD surfaces. However, in order to generate a function which embraces a sufficient amount of surface complexity, a large number of coefficients of high degree terms may be required. A much simpler way to define complex functions is by a look-up table. The image function is thus represented as a two dimensional array of intensity values.

By increasing the density of the grid, a greater resolution of the image function, and thus the sculptured details, may be obtained.

There are two stages to add surface details on a smooth surface. First, a 2D sketching canvas, like that used in traditional painting system, would be provided to the user. With the canvas, a grey-level image that gives the user a sense of the sculpture geometry may be created. Second, the user is required to attach the image to a smooth surface by mapping the image space to the parametric space of the surface. Adaptive subdivision and surface triangulation, like that used in FFD, are needed for the display of the resulting sculptured surface.

In short, *Image Sculpturing* techniques for surface detailing pose two problems to be considered: i) the ways to design a grey-level image function, and ii) the algorithms to "sculpture" the image on a smooth surface and the method to render it.

Design of image function

In painting systems, a user may freely design an image by different interactive painting tools, such as painting brushes, and various filtering and blurring operators. The painting canvas is usually a region of the frame buffer memory that allows the user to sketch the image and stores the intensity values of the image in it. By utilizing the frame buffer memory as the defining table for an image function $I(i, j)$, a user may directly define the function values by simple painting on the canvas. Dark areas in the canvas would be interpreted as having small values of $I(i, j)$ whilst bright areas as large values of $I(i, j)$. The variation of the intensity values represents an image function for modelling the wrinkled geometry of a sculptured surface.

A simple design interface for generating grey-level images has been implemented. This is a *contour-based image design method*. In this method, a grey-level image is generated from a simple drawing of contours. The contours are created from common two dimensional geometric entities, such as lines and curves. The contour design represents the salient geometric details of a sculptured surface.

The construction of grey-level image from the contour information is an idea derived from an image compression technique, known as the *sketch-based representation of grey-value image* (Carlsson, 1989). It is an image coding technique that encodes a grey-level image by some important details such as the contour information. The image is subsequently reconstructed as a solution to a constrained optimization problem by a diffusion algorithm. The diffusion algorithm determines the maximally smooth image consistent with the given contour information. The emphasis in this research is, however, not on how precise the contour information may reconstruct the original image, but rather, on how the contour-based design method and the diffusion algorithm may be used to create a grey-level image for modelling sculptured surfaces.

To convert a contour design drawing to a grey-level image is basically a two-dimensional interpolation problem. It is analogous to the surface interpolation problems in CAD. For example, a Coons surface patch is interpolated from four initial boundary curves by a mathematical formulation. In the case of interpolating a 2D grey-level image, it is necessary to input the image contour information. Grey values between contours may be interpolated so as to obtain a smooth grey-level image.

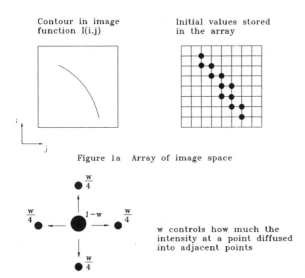

Figure 1a   Array of image space

w controls how much the
intensity at a point diffused
into adjacent points

Figure 1b   Diffusion of contour point in the
iterative algorithm

In the implementation, the initial contour definition contains the significant geometry of a sculptured surface. Each line of the contours may either represent a carving, or an embossing action to be exerted on it. The contours may include combination of two dimensional geometric entities, filled regions and free-hand sketches. In addition to the geometry information, each contour also contains some initial attributes, including:

1. initial intensity value,
2. width of the contour,
3. diffusion rate.

The three attributes for a contour shown above are some of the ways to control the geometry of a sculptured edge. The *initial intensity value* is a measure of depth of cut. For example, a bright intensity may correspond to an embossed edge, and a dark intensity may correspond to a deep cut to the surface. The *width of the contour* is a way of representing the width of the cutting tool used. It controls the width of a trench cut out from the surface or the width of an embossed region added to it. The *diffusion rate* is a parameter used in a smoothing algorithm. It controls the "sharpness" of a cut. A larger diffusion rate would lead to a flatter and smoother cut, while a smaller rate would produce a sharp edge.

The above listed attributes only provide an intuitive way of modifying the geometry of sculptured shapes. They do not provide a precise definition for the sculptured geometry, like the surface definitions used in conventional CAD systems. However, it may be sufficient for some surface detailing applications.

To interpolate 2D contours in the image space, a diffusion algorithm by Carlsson (1989) is used. It may be considered as an algorithm to interpolate given data values over a two dimensional space, just like the case of interpolating contour intensity values over the image space.

The input to the diffusion algorithm is a two dimensional image function defined on the image space. Initial function values are assigned by the user by way of contour design on a drawing canvas. A 2D array representing the image function is constructed from the pixel values on the canvas. The initial intensity values are stored in the array. The output from the algorithm is a diffused image function representing the sculptured geometry.

The interpolation problem is formulated as follows: Suppose there is a two dimensional image space in which an image function I is defined, such that, for any $i, j, N, M \in \mathbf{I} / \{0\}^*$,

$$I \rightarrow I(i,j) \in \mathbf{R}, \text{ for } 0 \le i < M \text{ and } 0 \le j < N \tag{2}$$

The image function (2) is defined as a two dimensional array $M \times N$ entries. The initial function values are set to be the intensity values at contour points (Figure 1). An image function based on the initial contour points is to be determined by two dimensional interpolation.

A successive diffusion process (Carlsson, 1989) may be used to generate a smooth grey level image. The corresponding image function $I(i,j)$ is obtained iteratively as:

$$I_{i,j}^{(n+1)} = I_{i,j}^{(n)} + \frac{\omega}{4} \Delta I_{i,j}^{(n)} \tag{3}$$

where ω defines the diffusion rate and $\Delta I_{i,j}^{(n)}$ is computed as:

$$\Delta I_{i,j}^{(n)} = I_{i-1,j}^{(n+1)} + I_{i,j-1}^{(n+1)} + I_{i+1,j}^{(n)} + I_{i,j+1}^{(n)} - 4I_{i,j}^{(n)} \tag{4}$$

Note that $I_{i,j}$ notes $I(i,j)$ and $i \in [0, M-1], j \in [0, N-1]$.

The algorithm starts with the given intensities at the contour points. The function values at non-contours (i, j) is then obtained in the $(i+1)$**th** iteration by a five point operator centred at (i, j). The values at the contour points are kept during all iterations. The algorithm may be regarded as equivalent to a recursive smoothing operation applied to the image starting at the upper left and proceeding down to the lower right. The repeated smoothing operation may be interpreted as a diffusion process whereby grey values at the contour points are diffused into the regions between contours. Formulae (4) and (5) can be shown to be the heat diffusion equation describing how a quantity evolves by diffusion over time with a constant rate of diffusion.

Thus, by using the diffusion algorithm, a grey level image represented as an image function created from the input contours. The image function may be considered as a look-up table to define the complex pattern to be added on to a 3D smooth surface.

---

* $I/\{0\}$ is set of integers excluding zero.

Image mapping and sculpturing

To create a sculptured surface from a smooth one, the image function is mapped to the continuous parametric space of the surface. The image function $I(i, j)$, which is defined on the image space, represents the complex sculpture details by a look-up table. For convenience, a continuous 2D function of normalized parametric values is used, just like defining a conventional CAD surface in a parametric space.

The continuos image function may be obtained by interpolating values between the image array entries. A simple interpolation technique is, for example, the bilinear interpolation. Such an algorithm would look like:

$$
\begin{aligned}
&\textit{Function I\_val (u , v)}\\
&\quad IV = int (N * v);\\
&\quad IU = int (M * u);\\
&\quad I\_val = I (IU, IV) * (1 - u) * (1 - v)\\
&\qquad + I (IU, IV + 1) * (1 - u) * v\\
&\qquad + I (IU + 1, IV) * u * (1 - v)\\
&\qquad + I (IU + 1, IV + 1) * u * v ;\\
&\quad return (I\_val);
\end{aligned}
$$

This gives a continuos image function representing the geometry of a sculpture pattern. A sculptured surface would then be obtained by applying the image function onto the parametric space of a 3D surface. To evaluate a point $\mathbf{P}'(u, v)$ on the sculptured surface, the original surface $\mathbf{P}(u, v)$ is displaced in the direction of the surface normal $\mathbf{N}(u, v)$ by an amount defined by the image function $I(i, j)$. The new surface point may be calculated from (1).

To display the sculptured surface, an adaptive subdivision technique (Von Herzen, 1987) may be applied and the steps involved are similar to those used to display a deformed surface by the FFD lattice. However, in image sculpturing, a new surface point is determined from a 2D image function instead of using a 3D lattice as in FFD. Both tools are functionally the same for re-mapping the original surface points to new positions.

Examples

Plates 1 and 2 show some of the surfaces created by the image sculpturing technique. In Plate 1a, a simple line pattern drawing is drawn on a 2D canvas. The pattern is mapped to a 3D parametric surface as shown in Plate 1b. Plate 1c displays the resulting surface with sculptured details. Another surface created with a larger diffusion rate is shown in Plate 1d. Plate 2 gives another example of the technique, and the corresponding 2D pattern design is also shown.

4 DESIGN METHODOLOGY FOR IMAGE SCULPTURING: A SUMMARY

Drawing and painting techniques as a basis for surface detailing were explored in the previous section. The basic principle is to sketch on a 2D elevation map of the surface with brighter shades representing peaks and ridges and darker shades representing valleys and pits. The coordinates of this map roughly correspond to the lines of principle curvature on the surface so that details that blend naturally with the overall shape of the object can be easily modelled. The design techniques may be conducted in four steps:

1. Designing a sketch - The user is provided with a set of drawing tools to design a 2D sketch. The sketch is converted into a grey-level image by the diffusion algorithm. All design sketches may be stored as grey-level images into a library.
2. Associating the sketch with a surface - A surface may have a list of sketches associated with it. This step involves extracting the sketches from the library and adding them to a list.
3. Attaching the sketches to the surface - After adding the selected sketches to the list, they can be applied to the surface one by one. This process involves positioning, resizing and fixing a sketch on the surface.
4. Raising/Impressing the sketch - After the sketch is attached onto the surface, each surface point may be raised or impressed along the corresponding normal by a magnitude proportional to the color intensity at that point in the grey-level image of the sketch.

The creation of image data may be supported and supplemented by other display rendering methods. For instance, a 3D rendering system which outputs an image and Z-buffer may be used to create an unlimited number of painting primitives.

For a complete sculpturing system, a roughing stage and a detailing stage of design should be considered. The image sculpturing technique may be used as a detailing tool in conjunction with other conventional modelling techniques, such as the free-form deformation technique.

5 CONCLUSIONS

A design interface for surface detailing is described. The method is based on grey-level image design and mapping. This involves a contour-based design interface, a diffusion algorithm for image creation, and also a subdivision algorithm for rendering the sculptured surface.

Since the contour-based design method is a rapid way of designing grey-level images, it has limitations in covering the design of a wide range of images, as compared to the conventional painting system. It is sufficient for producing embossed edges and contours, or adding simple embossed shapes on a surface. However, only simple geometric pattern designs can be sculptured on a surface. A more versatile system for creating a wide range of grey-level images is necessary. For example, image synthesis algorithms which use Z-buffers or depth buffers to perform hidden surface removal may be considered. The actual Z values left in the depth buffer after executing the algorithm may be used to define the table entries for an image function. A large set of grey-level images may be created in this way, providing an unlimited number of primitives to be sculptured to the surface.

Emphasis also needs to be put on investigating and using a general grey-level image for 'sculpting' a surface and on combining the image sculpturing technique with conventional global design techniques. The subdivision algorithm is important for rendering the deformed solids and sculptured surfaces. Efficient algorithms for surface triangulation require further detailed investigation.

6 ACKNOWLEDGMENT

The authors would like to thank the Department of Mechanical Engineering, the University of Hong Kong for providing the computing facilities and supports.

7 REFERENCES

Barr, A.H. (1984) Global and local deformation of solid primitives. *Computer Graphics,* **18(3)**, 21-30.

Blauer Ettagale (1991) Contemporary American jewellery design. Van Nostrand Reinhold.

Carlsson, S., Reillo, C. & Zetterberg, H. (1989) Sketch based representation of grey value and motion information, in *From Pixels to Features* (ed. Simon, J.C.), Elsevier Science Publishers, North-Holland.

Cook, R.L. (1984) Shade trees. *Computer Graphics,* **18(3)**, 223-31.

Coquillart, S. (1990) Extended free-form deformation: a sculpturing tool for 3D geometric modelling. *Computer Graphics,* **24(4)**, 187-96.

Griessmair, J. & Purgathofer, W. (1989) Deformation of solids with trivariate B-splinee, in *EUROGRAPHICS'89*, Elsevier Science Publishers.

Piegl, L. (1989) Modifying the shape of rational B-splines Part 2: Surfaces. *Computer Aided Design,* **21(9)**, 538-46.

Sederberg, T.W. & Parry, S.R. (1986) Free-form deformation of solid geometric models. *Computer Graphics,* **20(4)**, 151-60.

Smith, A.R. (1982) Paint. *Tutorial: Computer Graphics* (ed. John, C.B. & Booth, K.S.), IEEE Computer Society Press.

Von Herzen, B. (1987) Accurate triangulations of deformed, intersecting surfaces. *Computer Graphics,* **21(4)**, 103-10.

Wainwright, C.E.R., Harrison, D.K. & Leonard, R. (1990) The application of surface modelling techniques to reduce time to market. *Spooner-Vicars,* 3 14-22.

Whitted, T. (1983) Anti-aliased line drawing using brush extrusion. *Computer Graphics,* **17(3)**, 151-6.

Williams, L. (1990) 3D Paint. *Computer Graphics,* **24(2)**, 225-34.

8 BIOGRAPHY

POON Chiu-tak received his BEng in Mechanical Engineering in 1992 from the University of Hong Kong. He is reading for his MPhil in computer aided design and manufacture. His research interests include 3D geometric modelling and CAD/CAM applications.

TAN Sooi-thor is a senior lecturer in the Department of Mechanical Engineering, The University of Hong Kong. He has a BSc and a PhD in mechanical engineering from the University of Leeds, UK. He is a Chartered Engineer, a Member of the Institution of Mechanical Engineers in UK, the American Society of the Mechanical Engineers, and a senior

Member of the Society of Manufacturing Engineers. His research interests include geometric modelling, and the application of computers to mechanical engineering.

CHAN kit-wah is a lecturer in the Department of Mechanical Engineering of The University of Hong Kong. He has a BSc in Mechanical Engineering from Brighton University, a MSc in Engineering Production and Management from the University of Birmingham, and a PhD from Loughborough University of Technology. His research interests include features modelling and CAM/CAD applications.

Plate 1a   A simple line pattern
drawing on a 2D canvas.

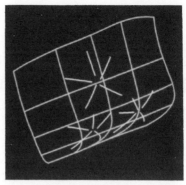

Plate 1b A parametric surface mapped
with the line pattern.

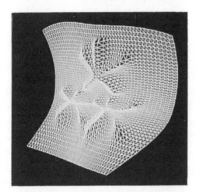

Plate 1c   The resulting surface with
sculptured details.

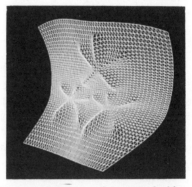

Plate 1d The same surface created with
a larger diffusion rate.

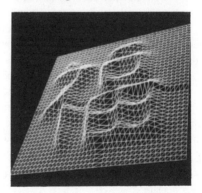

Another example showing how a Chinese character is sculptured on
a smooth surface.

11

Design system of optimizing contour shape to minimize stress concentration in flat plate

Junko Fukuda. and Jiro Suhara.
Faculty of Engineering,Kyushu Kyoritsu University
1-8 Jiyugaoka,Yahatanishi-ku,Kitakyushu 807,Japan
TEL(093)691-3331,FAX(093)603-8186

Abstract

In this paper, the authors present the Shape Optimization Method to minimize the maximum stress around discontinuous parts of plate instead of inserting thick reinforcement plate (Fukuda,1987). The finite element method (FEM) using the constant strain triangles is applied for calculating the displacement and stress of the structure. The improved values of the stresses at nodes on the external boundary of the domain are needed in the process of the shape optimization, but the irregular and dispersed values of FEM results include the errors which are influenced sensitively by the mesh pattern. The accuracy of stresses at a node on the external boundary of domain which are approximated by simple averages of stresses of the adjoined elements having the node in common, is generally known as having lower than those of internal ones.

The authors proposed a smoothing method(Fukuda,1994) furnishing more accurate stresses at a node on the boundary of domain and applied this smoothing method to the optimization system for improving the accuracy of the stress distribution obtained by FEM.

Further, the authors proposed the method of automatic mesh(Suhara and Fukuda,1972, Cavendish,1974) generation which is possible to effectively adapt to transformation of a shape of domain through the process of optimization. Consideration is taken to avoid the formation of elements having acute angled corners and to control the size or distribution density of finite elements in order to avoid the loss of numerical accuracy of result of the analysis. The Mesh Generation Method by means of random numbers is capable of generating mesh of triangular elements in the arbitrarily shaped and multiply connected planar domain.

It is shown in this paper that this system is highly effective in reducing the stress concentration by adapting the optimized shape around the discontinuous part of ship structure and hatch opening.

Keywords

Optimization, finite element method, stress concentration

1 CONCEPTS OF THE SHAPE OPTIMIZATION METHOD

As shown in Figure 1, S_{Opt}^{k} is the boundary of domain to have the shape optimized, P_i^{k} is the nodal point on the boundary and σ_i^{k} is the tangential stress at the node. Where i indicates the nodal number and k indicates iteration steps of the optimization procedure for determining the optimum contour S_{Opt}^{k}. Then ρ_i^{k} is the radius of curvature to S_{Opt}^{k} at P_i^{k} and $\delta\rho_i^{k}$ is the variation of shifting the position(x_i ,y_i) of P_i^{k}. The node is made to shift to the n direction. Where n and t are normal and tangential directions of oriented S_{Opt}^{k} as shown in Figure 1.

The S_{Opt}^{k} is separated into two parts, which is subjected to tensile stresses and compressive stresses as shown in Figure 2. The position of P_i^{k} is shifted as in the following method so that the value of maximum stress in each part, $|\sigma_{max}^{k}|$ or $|\sigma_{min}^{k}|$ is made as small as possible and the distribution of stress is flattened. The stress concentration is made to reduce by enlarging the radii of curvature, of which values can be counted from σ_i^{k} and σ_{max}^{k} (σ_{min}^{k}) around the node occurring the maximum stress as shown in Figure 3. The optimization procedure is repeated and the nodes on the S_{Opt}^{k} are shifted until the value of maximum stress at the node on S_{Opt}^{k} is almost unchanged .

The value of $\delta\rho_i^{k}$,i.e. the new position of the nodes on S_{Opt}^{k} is caluculated from the following equations which has taken the above mentioned opinion into cosideration.

$$\rho_i^{k+1} = \rho_i^{k} + \delta\rho_i^{k}. \quad (i = 1, 2, \cdots, m) \tag{1}$$

where

$$\delta\rho_i^{k} = \alpha^{k}\Delta\sigma_i^{k}.$$

$$\Delta\sigma_i^{k} = \begin{cases} |\sigma_{max}^{k} - \sigma_i^{k}|. & \text{for } \sigma_i^{k} \geq 0 \\ |\sigma_{min}^{k} - \sigma_i^{k}|. & \text{for } \sigma_i^{k} < 0 \end{cases}$$

$$\sigma_{max}^{k} = \underset{i}{\text{Max }} \sigma_i^{k}. *$$

$$\sigma_{min}^{k} = \underset{i}{\text{Min }} \sigma_i^{k}.$$

$$\Delta\sigma_{P,max}^{k} = \underset{i}{\text{Max }} \Delta\sigma_i^{k}. \quad \text{for } \sigma_i^{k} \geq 0$$

$$\Delta\sigma_{M,max}^{k} = \underset{i}{\text{Max }} \Delta\sigma_i^{k}. \quad \text{for } \sigma_i^{k} < 0$$

$$\alpha^{k} = 2\ell_{mean}^{k}\gamma/(\Delta\sigma_{P,max}^{k} + \Delta\sigma_{M,max}^{k}).$$

$$\ell_{mean}^{k} = \ell^{k}/(m-1).$$

$$\ell^{k} = \sum_{i=1}^{m-1} \ell_i^{k}.$$

$$\ell_i^{k} = \sqrt{(x_{i+1} - x_i)^2 + (y_{i+1} - y_i)^2}.$$

$$0 < \gamma \leq 1.$$

* $\underset{i}{\text{Max }} A_i$ stands for finding the maximum value of A_i (i=1,2,3, $\cdots$) and $\underset{i}{\text{Min }} A_i$ for the minmum value.

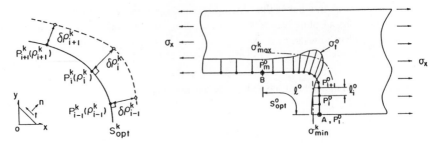

Figure 1 Definition of design variables.

Figure 2 Distribution of tangential stress around the hole.

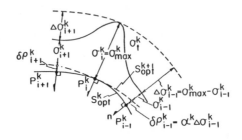

Figure 3 Refining scheme of the corner shape of a perforated plate.

2   ACCURACY OF THE SHAPE OPTIMIZATION METHOD

Examples of a square plate with a hole are used for testing the Shape Optimization Method. In the first example the optimum shape of the hole is to be found in a large square plate which is exposed to equal boundary stresses along two axes(See Figure 4). The ratio of length a and width b of the hole is equal to 1. In case the profile of a hole is circular, it is found, analytically, that a constant tangential stress distribution occurs along the hole. As shown in Figure 5, the configuration S_{opt}^{k} of the hole(k: iteration steps) between two fixed points C and E is optimized by iteration procedure within a limitation S_{LIM} (See Figure 6) of a variable boundary S_{opt}^{k} so that the occuring maximum tangential stress is reduced to a minimum. For every iteration procedure, a mesh is generated over again in a alterable domain Ω_{opt}^{k} of mesh pattern. As may be seen from Figures 7 and 8, in spite of starting the shape optimization of the hole from very different stress distribution and initial configuration, it is confirmed that both constant stress distribution converge around the hole and optimize the circle as the optimum shape of the hole after about 10 iteration steps.

In the second example, boundary stresses σ_x and σ_y have a ratio of 3:2. A and b

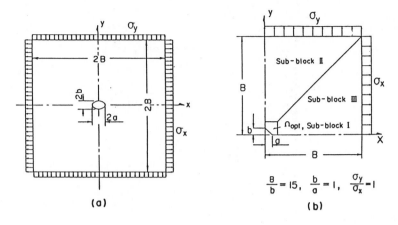

Figure 4 Square plate with a hole at the center subjected to tensile stresses in two perpendicular directions.

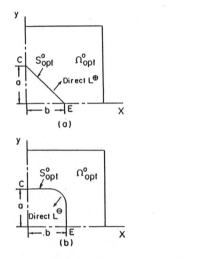

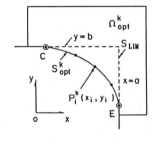

Figure 5 Initial configurations and allowable zones of modifying mesh.

Figure 6 Constraint condition for contour curve.

also have a 3:2 ratio. When an elliptical shape with a semi-axis ratio of 3:2 emerges as the optimum shape, a constant stress distribution along the hole is obtained.

By comparing the numerical with the analytical solutions in these examples, the high degree of accuracy of the Shape Optimization Method has been demonstrated(Fukuda,1984).

$\sigma_x = 0.133$ kg/mm²

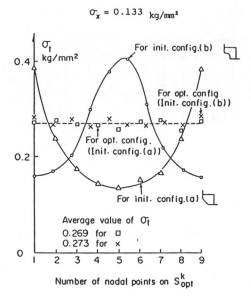

Figure 7 Distributions of tangential stress around the hole.

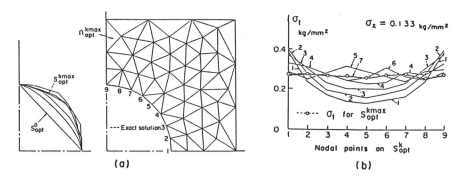

Figure 8 Convergence pattern of shape and distribution of tangential stress in example 1.

3   EXAMPLES OF APPLICATION

The above-mentioned Shape Optimization Method was applied to the optimization of hatch opening and the discontinuous part of an actual ship(Fukuda,1984,1987). In this procedure, geometrical constraints for hatch shape to keep the opening zone and shape clear are taken into account.

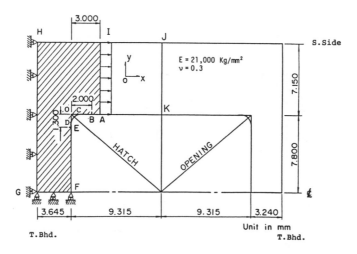

Figure 9 Dimensions of hatch opening and load condition assumed for stress analysis.

In the first example, without the geometrical constrain not undercutting an upper deck, the configurations of the hatch corner between two fixed points B and E are optimized as to the various intial configurations as shown in Figures 9 and 10.   As may be seen from Figure 11, the values of maximum stress around the hatch corner differ with initial configurations, but the values for optimized configurations after about 7 ~ 8 iteration steps converge much the same.   And in spite of the optimized configuration depending on the initial configuration, the constant stress distribution converges around the hatch corner as shown in Figures 12 and 13. It is confirmed by means of Nishida's photoelastic experiment that the configuration which undercuts the inside is effective to reduce the stress concentration.   The value of maximum stress for the optimum configuration has dropped by 33.3 per cent compared with the circular corner(Example for model 350R).

Even in the case of optimizing under the geometircal constraint, that do not make the upper deck to undercut, the value of maximum stress has dropped by 19 per cent compared with the circular corner(Example for model 950R).

Based on the above analystic examples it is confirmed that this shape optimization system is useful in deciding an optimum shape to reduce stress concentration.

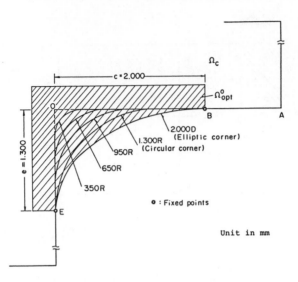

Figure 10   Initial configurations of hatch corner for optimization.

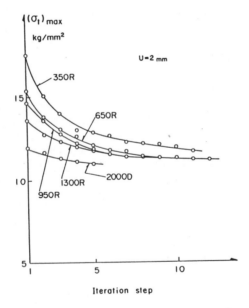

Figure 11   Convergence of maximum stresses around hatch corner in terms of iteration
steps.

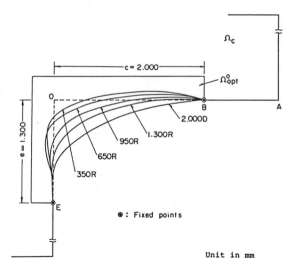

Figure 12 Optimized configurations of hatch corner.

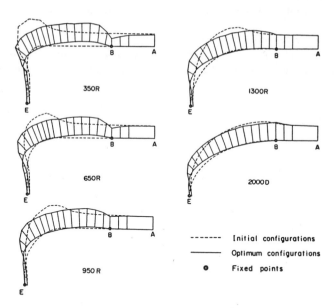

Figure 13 Comparison of stress distributions around hatch corners.

4   REFERENCES

Cavendish,J.C.(1974)Automatic triangulation of arbitrary planar domains for the finite element method,International Journal for Numerical Methods in Engineering,vol.8,679-696.

Suhara,J. and Fukuda,J.(1972)Automatic mesh generation for finite element analysis, Advances in Computational Methods in Structural Mechanics and Design, Univ. Alabama, 607-624.

Fukuda,J.(1984)Shape optimization of perforated plate to reduce stress concentration.,The West-Japan Society of Naval Architects,No.68,191-206.

Fukuda,J.(1987)Effectiveness and optimization of inserted plate around the discontinuous parts of ship structures.,The West-Japan Society of Naval Architects,No.73,162-174.

Fukuda,J.(1994)Accuracy of stress smoothing methods in finite element analysis.,The West-Japan Society of Naval Architects,No.87,239-262.

5   OUR BIOGRAPHY

Fukuda Junko is currently a professor of Faculty of Engineering, Kyushu Kyoritsu University, Kitakyushu, Japan. She took a Ph. D. Engineering from Kyushu University. Her research interests are finite element analysis, information processing and CAD.

Suhara Jiro is a professor emeritus of Kyushu University and Kumamoto University of Tecnology. He took a Ph. D. Engineering from Kyushu University. His field of study is the naval architecture. He has also been the president of the Japan Society of Naval Architects.

12

A Convex Hull Algorithm and Its Application to Shape Comparison of 3-D Objects

Susumu FURUKAWA*, Shinji MUKAI** and Mitsuru KURODA+
* *Department of Mechanical Systems Engineering, Yamanashi University, Takeda 4-3-11, Kofu 400, Japan, Tel. (+81) 552 20 8576, Fax (+81) 552 20 8771, email : furukawa %sun3.esb.yamanashi.ac.jp@yu-gate.yamanashi.ac.jp*
** *Maebashi City College of Technology, Kamisadori-machi 460, Maebashi 371, Japan, Tel. (+81) 272 65 0111, Fax (+81) 272 65 3837*
+ *Toyota Technological Institute, Hisakata 2-12, Tempaku, Nagoya 468, Japan, Tel. (+81) 52 802 1111, Fax (+81) 52 802 6069, email : M.Kuroda@toyota-ti.ac.jp*

Abstract

A new method for constructing the convex hull of a concave polyhedron represented by B-Reps is presented as a basic tool for feature processing of 3-D objects. The algorithm is based on the idea of taking out faces which are used for constructing the convex hull and making new faces to cover gaps between the faces taken out from the polyhedron. It is shown that constructing convex hulls of polyhedra can be applicable to deciding similarities of 3-D objects numerically. Various examples of concave polyhedra for determining similarities have been tested. The results show that the theoretical considerations are reasonable and matches with our feeling.

Keywords

Convex Hull Algorithm, Convex Decomposition, Feature Recognition, Shape Comparison, Solid Model, CAD/CAM

1 INTRODUCTION

At the final stage of manufacturing processes, checking whether products are made within specified accuracy is required to make high quality products. It is then necessary to detect the difference between the design specification and the manufactured products[Gra 92, Furu 94]. Specifically, deciding similar and dissimilar parts between a solid model and its product is required. Thus, developing methods for dealing with features, tolerances and similarities of solid models is necessary. Treating such geometrical properties are also required in process planning, parts assemblies in CAD/CAM processes. Therefore, there are many research works in this field.

Approaches to specifying the features of objects explicitly during the design process have been proposed[Cut 88, Sha 88]. The main advantage of this approach is that the designer can immediately build models which include the necessary information in design, manufacturing and checking. However, the problem with this approach is that users have to define every features prior to each process.

Another way of treating geometrical properties of objects is by feature recognition[Fal 89, Saku 90, Kim 92]. In this case, the designer need not take into account form features at his design work. The main problem with this approach is that only features which are implicitly stored in the data base can be derived.

Combining these two approaches and the system architecture have been proposed[Dix 90, Mar 93]. However, better methods for recognizing features of objects is required in order to reduce the user's load.

Algorithms for determining congruencies of polyhedra[Furu 90] and deciding simlarities of convex polyhedra[Muka 94-1] have been developed and an idea for determining similarities of concave polyhedra[Muka 94-2] has been proposed by the authors. The idea of computing similarities of concave polyhedra proceeds as follows. First, the concave polyhedra to be compared are decomposed into convex components. Then each convex component of one polyhedron is compared with the corresponding component of another polyhedron, and their similarities are computed numerically.

In this paper, a new method for constructing the convex hull of a concave polyhedron represented by B-Reps is presented first as a basic tool for decomposing a polyhedron into convex components. The algorithm utilizes the fundamental properties of a convex polyhedron.

Next, with this convex decomposition, a data structure called two layer data structure of convex components with relationships is discussed briefly. As an example of its applications, similarity tests have been achieved for polyhedra. The results show that the theoretical considerations on this problem are reasonable and matches with our feeling.

2   PREVIOUS RESEARCH WORK

The convex hull algorithm may be applied not only to feature extraction but also in many other fields such as computer graphics, pattern recognition and operations research.

Various methods for computing the convex hull have been developed for 2-dimensional space[Gra 72, Jar 73, Sei 86]. Preparata[Pre 77] and Furukawa[Furu 86] have discussed the methods for given points in 3-D space.

Preparata presented a method which uses the "divided and conquer" technique, recursively merging two nonintersecting convex hulls for points in 3-D space. The convex hulls can be constructed in $O(n \log n)$, where n represents the number of points.

Furukawa developed the method which is based on the idea of determining the outline loop of a convex polyhedron viewing from a point and connecting the point with each vertex of the outline loop. A convex hull polyhedron can be constructed with $O(n)$ operations, if the points are distributed uniformly.

However, methods for constructing convex hulls from solid models already described by a data structure such as B-Reps are not well approached. One method has been developed by Appel[App 77] for a polyhedron for which the data structure is already constructed. This is based on the idea of testing whether a vertex is inside the specified polygon or not, on a projection plane. The estimation of time requirement, however, was not discussed.

3 CONVEX HULL ALGORITHM FOR POLYHEDRA

As the basis for constructing a data structure called "two layer structure" of a polyhedron which can be used for extracting form features, a method for constructing the convex hull of a concave polyhedron represented specified data structure such as B-Reps is required. At the first process for constructing the two layer structure, the convex hull of the concave polyhedron is constructed and faces of the polyhedron are classified into the following three types, i.e., the used and unused faces, and the faces of which parts are used.

3.1 Fundamental properties of convex polyhedra

Before discussing the convex hull algorithm, fundamental properties of convex polyhedra are discussed.

Consider the convex polyhedron shown in Fig.1. Let G_c be a point inside polyhedron P, and R_i be a open region like to a bottomless pyramid surrounded by faces S_j constructed by point G_c and edge line L_j of face F_i for $j=1,2,...,m_i$, where m_i represents the number of edges of face F_i.

[Property 1]
If polyhedron P is convex, regions R_i and R_k have no intersection for all i and k (i, $k=1,2,...,n_f$; $i{\neq}k$), where n_f represents the number of faces of the polyhedron.

[Property 2]
Every interior angles between two faces sharing an edge is less than π.

3.2 Visibility test of faces

Consider a fundamental principle for constructing the convex hull of a polyhedron P in this section.

We discuss visibilities of faces of the polyhedron viewing from a point p on a face F_i of

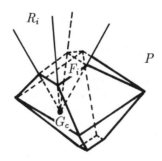

Figure 1: A convex polyhedron and open region.

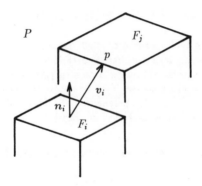

Figure 2: Visibility test of a face.

P. Let n_i and v_i be an outward normal vector of a face F_i and a vector starting with a point on F_i to point p, respectively. The visibility of F_i can be easily determined as follows. If $n_i \cdot v_i > 0$, then F_i is visible from point p. Otherwise, F_i is invisible.

If F_i is visible from a point on a face F_j of P, then face F_i is not used to construct the convex hull of P, because F_i should be inside the resulting convex hull of P[Furu 86](Fig.2).

3.3   Intersection check of regions

We designate the resulting convex hull polyhedron $P^{(c)}$.

Consider the case where two regions R_i and R_j sharing an edge, which are constructed by a face F_i and point G_c, and F_j and G_c respectively, intersect each other. We can classify the relationship between faces F_i and F_j into the following two cases.

Case (1) The interior angle θ of two faces F_i and F_j is less than π.

Case (2) Angle θ is greater than π.

Let g_i be a vector starting with a point on face F_i to point G_c.

In case (1), one of the inequalities

$$n_i \cdot g_i > 0, \quad n_j \cdot g_j > 0 \tag{1}$$

is valid(Fig.3), where n_i represents an outward normal vector of F_i defined in the previous discussions. Assuming $n_i \cdot g_i > 0$, then F_i is not used for constructing the convex hull $P^{(c)}$. Hereafter, such a face is called unused face. In this case, we need not further test of intersections between R_i and the other regions.

In case (2), two faces F_i and F_j are both unused faces.

The edge sharing by F_i and F_j is called a folded edge in this case.

Consider the case where R_i and R_j sharing an edge have no intersection. There are two cases :

Case (3) $n_i \cdot g_i < 0$ and $n_j \cdot g_j < 0$, $\qquad\qquad$ (2)

Case (4) $n_i \cdot g_i \geq 0$ and $n_j \cdot g_j \geq 0$. $\qquad\qquad$ (3)

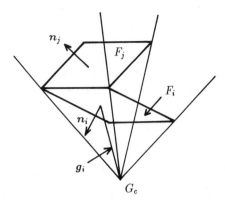

Figure 3: The case where two succesive regions intersect and the interior angle of the faces is less than π.

In case (3), there are possibilities that both faces F_i anf F_j are used to construct the convex hull.

In case (4), two faces F_i and F_j are both unused faces.

Hereafter, this intersection test of regions is called "region check".

3.4 Algorithm for constructing convex hulls of polyhedra

We consider a new algorithm for constructing the convex hull from a polyhedron P.

The faces of the resulting convex polyhedron are classified into the following three types.

(a) Real face : the same face also exists in polyhedron P.

(b) Partial real face : a face including a face of P.

(c) Imaginary face : a face not existing in P.

The procedure for constructing the convex hull $P^{(c)}$ of P is as follows.

(1) Outward normal vectors of all faces are calculated.

(2) A point inside the polyhedron G_c is determined.

(3) Find the most distant vertex v_0 of P from point G_c. This vertex becomes to a vertex of $P^{(c)}$.

(4) Construct a face $F_0^{(c)}$ including vertex v_0 by using so called "gift wrapping" method[Cha 70], where $F_0^{(c)}$ represents a face of $P^{(c)}$. At this stage, whether $F_0^{(c)}$ is a real face or not is detected.

(5) Make the region of "bottomless pyramid" R_0 defined in section 3.1, by using G_c and $F_0^{(c)}$.

(6) Select a face F_i sharing an edge of $F_0^{(c)}$ and make region R_i from G_c and F_i. If the edge is a folded edge, the relation between F_i and $F_0^{(c)}$ corresponds to case (1) in

section 3.3. Then, face F_i is unused face. Otherwise, there is possibility that face F_i is used for constructing $P^{(c)}$.

This procedure is performed for all edges of face $F_0^{(c)}$. Then faces $F_j(j=1,2,...,k)$ which have possibilities to become faces of $P^{(c)}$ and corresponding region R_j are obtained, where k represents the number of such faces.

Those regions are stored in a list, which is called "region table", for intersection tests in the following step.

(7) The following procedures are performed for all faces sharing edges with faces $F_0^{(c)}$, $F_1,...,F_k$. Let F_i be one of the faces F_1, $F_2,...,F_k$ and F_j be a face sharing an edge with F_i. If the relation between F_i and F_j corresponds to case (1), then F_j is an unused face. If it corresponds to case (2), both faces F_i and F_j are unused faces. In this case, region R_i corresponding to face F_i is removed from the region table. At the same time, region checks for region R_j and the regions written in the region list have to be achieved.

The procedure for removing regions from the region table is continued till no intersection is detected.

If the relation corresponds to case (3), then region R_j is added in the region table. At this time, the "angle test" whether the interior angle of faces F_i and F_j is less than π or not is performed. If the angle is greater than π, faces F_i and F_j are unused faces and region R_i is removed from the region table.

(8) Consider whether faces sharing edges with unused faces may be used to construct the convex hull or not.

Let F_i be an unused face and F_j be a face to be tested.

Consider the case where case (1) is valid for F_i and F_j. If F_j is nearer than F_i to point G_c, then F_j is an unused face. Otherwise, region R_j is added into the region table.

In case (2), face F_j is an unused face.

If the relation between F_i and F_j corresponds to case (3) or (4), further checking of intersections between region R_j and regions written in the region table is necessary.

First, we consider the case where region R_j has intersection with region R_l corresponding to face F_l. In both case (3) and (4), F_j is an unused face. In case (4), if F_j is nearer than F_l to point G_c, region R_l is removed from the table. The same discussion is valid for the faces of which regions intersect with R_j.

Next, consider the case where the regions have no intersection. If case (3) is valid for F_i and F_j, region R_j is stored in the region table. Otherwise, face F_j is an unused face.

For faces sharing edges with the faces of which regions are stored in the region table, the same procedure as (7) is applicable.

This procedure (8) is continued until all faces of polyhedron P are tested.

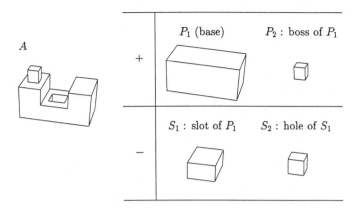

Figure 4: An example of a two layer structure of object A.

(9) At the last step, visibilities are tested for each couple of faces corresponding to neigh-bouring regions stored in the region table. If both faces are invisible each other, then an imaginary face is construct to connect these two faces. On the contrary, in the case where the faces are visible or one of them is visible, the faces become unused faces and an imaginary face is also constructed.

4   SHAPE COMPARISON OF TWO OBJECTS

Before discussing similarities of two polyhedra, a concave polyhedron is decomposed into convex components[Furu 85, Furu 86], using convex hull algorithm described in the previ-ous section, and a data structure called "two layer structure of convex components with relationships" is constructed for each object.

An example of such a data structure is shown in Fig.4, wherein convex polyhedra be-longing to the first layer represent the parts to be added, while those in the second layer represent the parts to be subtracted. Thus object A is given by $P_1 + P_2 - (S_1 + S_2)$.

Although this data structure seems to be same to the hierarchical tree structure presented in [Furu 85], it is more flexible and easier to understand shapes of objects.

The data structure includes the relationships between those convex components too. For example, object P_2 belongs to P_1 as an added part and object S_1 is subtracted from P_1.

By using this type of data structure, we can discuss how to determine similarities of two concave polyhedra. The basic algorithm of shape comparison is as follows.

Suppose that the two layer data structures of given concave objects A and A' are both represented as shown in Fig.4 and only shapes of components to be compared are slightly different from each other.

At first, the bases P_1 and P_1' of objects A, A' are compared and the similarities are calculated according to the procedure described in [Muka 94-1].

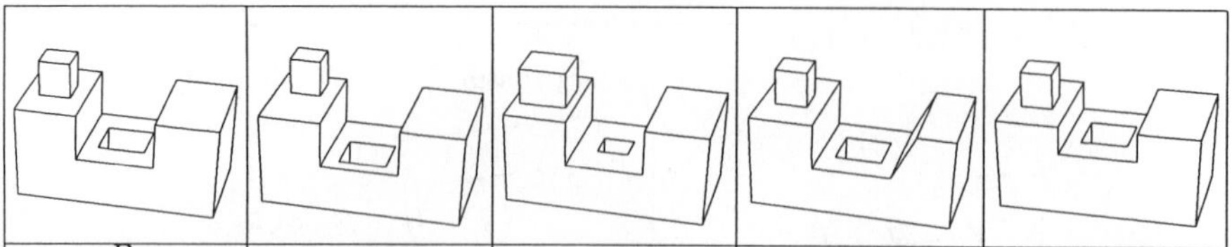

P_1	$\mu = 0.000$ $\sigma = 0.000$	$\mu = 0.000$ $\sigma = 0.000$	$\mu = 0.000$ $\sigma = 0.000$	$\mu = 0.000$ $\sigma = 0.000$
P_2	$\mu = 0.000$ $\sigma = 0.000$ $\delta_G = 0.179$	$\mu = 0.390$ $\sigma = 0.097$ $\delta_G = 0.000$	$\mu = 0.000$ $\sigma = 0.000$ $\delta_G = 0.000$	$\mu = 0.000$ $\sigma = 0.000$ $\delta_G = 0.000$
S_1	$\mu = 0.000$ $\sigma = 0.000$ $\delta_G = 0.000$	$\mu = 0.000$ $\sigma = 0.000$ $\delta_G = 0.000$	$\mu = 0.134$ $\sigma = 0.063$ $\delta_G = 0.045$	$\mu = 0.189$ $\sigma = 0.099$ $\delta_G = 0.089$
S_2	$\mu = 0.000$ $\sigma = 0.000$ $\delta_G = 0.119$	$\mu = 0.283$ $\sigma = 0.087$ $\delta_G = 0.000$	$\mu = 0.000$ $\sigma = 0.000$ $\delta_G = 0.000$	$\mu = 0.294$ $\sigma = 0.074$ $\delta_G = 0.089$

Figure 5: Example of similarity tests.

Next, shapes of P_2, P_2', S_1, S_1' and S_2, S_2' are compared respectively. Then, for example, we may see that two bases P_1 and P_1' are equal, P_2 and P_2' are equal too, but the slots S_1, S_1' and the holes S_2, S_2' are slightly different.

In the case where the data structures of two concave objects are different, we need not compare each pair of convex components, because it is clearly understandable that they are quite different shapes.

An example of this similarity test is shown in Fig.5. The parameters written in the diagram represent numerals of similarities, i.e., if μ and σ equal zero, then corresponding components are congruent, and if $\delta_G=0$, the position of the components are coincident. Therefore, objects tend to exhibit higher similarities, those parameters decrease.

5   CONCLUSIONS

A new algorithm for constracting the convex hull of a polyhedron represented by B-Reps is presented.

Using this algorithm, a data structure called two layer structure which consists of only convex components is constructed and it is applied to deciding similarities of concave polyhedra.

The main features of the paper are as follows.

1. The convex hull algorithm uses fundamental properties of convex polyhedra. Therefore, it is very simple and easy to understand.

2. After constructing the convex hull polyhedron, we need not to check whether faces of the convex hull exist also in the original polyhedron, because necessary informations on those faces are stored while the convex hull is constructed.

3. The results of similarity tests are reasonable and match with our feeling.

REFERENCES

[Cha 70] Chand D.R. and Kapur S.S. : *An Algorithm for Convex Polytopes*, J. ACM, vol.17, no.1, pp.78-86 (1970)

[Cut 88] Cutkosky M.R., Tenenbaum J.M. and Muller D. : *Features in Process-Based Design*, Proc. Comp. Eng., ASME, pp.557-562 (1988)

[Dix 90] Dixon J.R., Libardi E.C. and Nielsen E.H. : *Unresolved Research Issues in Development of Design-with-Features System*, In Geometric Modeling for Product Engineering, North-Holland, pp.183-196 (1990)

[Fal 89] Falcidieno B. and Giannini F. : *Automatic Recognition and Representation of Shape-Based Features in a Geometric Modeling System*, Computer Vision, Graphics, and Image Processing, vol.48, pp.93-123 (1989)

[Furu 86] Furukawa S. : *An Efficient Convex Hull Algorithm for Finite Points Set in 3-D space*, Technical Report of M41 Group in Yamanashi University, no.1, pp.1-7 (1986)

[Furu 85] Furukawa S. : *An Algorithm for Constructing Hierarchical Tree Structures of Polyhedra*, Journal of JSPE., vol.51, no.11, pp.2071-2076 (1985) (in Japanese)

[Furu 90] Furukawa S., Kimura F., Sata T. and Obi M. : *An Algorithm for Deciding Congruence of Polyhedra*, Bulletin of the JSME., vol.56, no.528, pp.2311-2317 (1990) (in Japanese)

[Furu 94] Furukawa S. : *Computer Aided Checking of Products*, Proceedings of Intellectual Facilitation of Creative Activities, SCF., pp.52-57 (1994)

[Gra 72] Graham R.L. : *An Efficient Algorithm for Determining the Convex Hull of a Finite Planar Set*, Information Proc. Letter, pp.132-133 (1972)

[Gra 92] Grabowski H., Ander R., Geiger K. and Schnitt M. : *Vision Based On-Line Inspection of Manufactured Parts*, Human Aspects in Computer Integrated Manufacturing, ed. by G.J.Olling and F.Kimura, Elsevier Science Pub., pp.593-607 (1992)

[Jar 73] Jarvis R.A. : *On the Identification of the Convex Hull of a Finite Set of Points in the Plane*, Information Processing Letter, no.2, pp.18-21 (1973)

[Kim 92] Kim Y.S. : *Recognition of Form Features Using Convex Decomposition*, Computer-Aided Design, vol.24, no.9, pp.461-476 (1992)

[Mar 93] Martino T., Falcidieno B., Giannini F., Haßinger S. and Ovtcharova J. : *Integration of Design by Features and Feature Recognition Approaches Through a Unified Model*, Modeling in Computer Graphics, Springer-Verlag, pp.423-437 (1993)

[Muka 94-1] Mukai S., Furukawa S., Kimura F. and Obi M. : *An Algorithm for Determining Similarities of Convex Polyhedra*, Computers & Graphics, vol.18, no.2, pp.171-176 (1994)

[Muka 94-2] Mukai S., Furukawa S. and Kuroda M. : *A Method for Determining Similarities of Polyhedra using Form Feature Decomposition*, The 49th Annual Conference of Japan Society for Information Processing, pp.2·393-2·394 (1994) (in Japanese)

[Pre 77] Preparata F.P. : *Convex Hulls of Finite Sets of Points in Two and Three Dimensions*, CACM., vol.20, no.2, pp.87-93 (1977)

[Pre 85] Preparata F.P. and Shamos M.I. : *Computational Geometry*, An Introduction, Springer-Verlag, (1985)

[Saku 90] Sakurai H. and Gossard D.C. : *Recognizing Shape Features in Solid Models*, IEEE Computer Graphics & Applications, vol.10, no.5, pp.22-32 (1990)

[Sei 86] Seidel R. : *Constructing Higher-Dimensional Convex Hull at Logarithmic Cost per Face*, ACM, pp.404-413 (1986)

[Sha 88] Shah J.J. and Rogers M.T. : *Functional Requirements and Conceptual Design of the Feature-Based Modeling System*, Computer-Aided Engineering Journal, pp.9-15 (1988)

13

Parallel mesh generation based on a new label-driven subdivision technique

K. T. Miura and *F. Cheng*[†]
*Shape Modeling Laboratory, University of Aizu,
Aizu-Wakamatsu, Fukushima, 965-80, Japan
Phone:+81-242-37-2605, Fax:+81-242-37-2728, E-mail:ktmiura@u-aizu.ac.jp
[†]*Department of Computer Science, University of Kentucky,*
Lexington, KY 40506, U.S.A.
Phone:+1-606-257-6760, Fax:+1-606-323-1971, E-mail:cheng@ms.uky.edu

Abstract

A mesh generation algorithm based on a new label-driven subdivision technique is presented. The algorithm generates 2D mesh of quadrilaterals/triangles on a regular quadrilateral network such as the parameter space of a piecewise surface. Each face (patch) of the quadrilateral network has to be assigned a subdivision level (mesh density) first, which is in tern used to assign labels to the vertices of the quadrilateral network. The mesh is generated on the basis of individual faces by performing label-driven subdivision on each face separately. Parallel processing is achieved by performing label-driven subdivision on all the faces simultaneously. The new label-driven subdivision technique improves a previous label-driven subdivision technique in that it does not require the labels of the vertices to satisfy certain requirement and, consequently, the generation of an admissible extension of the given label assignment is not necessary.

Keywords

mesh generation, vertex label assignment, label-driven subdivision, parallel mesh generation

1 INTRODUCTION

Mesh generation is the process of generating finite element models for simulated structural analysis. Since the accuracy of the finite element solution depends directly on the mesh layout, and the cost of the analysis becomes prohibitively expensive if the number of elements in the mesh is too large, a good mesh generating method should let the user generate a mesh that just fine enough to give an adequate solution accuracy and satisfies the mesh conformity requirement(Cheng et al.,1989).

Several different mesh generation methods are available in the literature. These methods can be categorized as follows:

1. interpolation mesh generation(Gordon and Hall,1973, Harber et al.,1981),

2. automatic triangulation(Rivara,1987, Sadek,1980),

3. quadtree/octree approach(Baehmann et al.,1987),

4. mesh generation based on constructive solid geometry(CSG)(Lee,1984).

Among others, one of the problems in many of the algorithms is that they all require special checking steps to ensure the conformity of the meshes, to allow variable densities and independent local refinement in the surface.

A mesh generation technique which does not require the confirmity checking process has been published recently (Cheng et al., 1989). This mesh generation technique generates 2D mesh of quadrilaterals/triangles on a regular quadrilateral network (such as the parameter space of a piecewise surface) by performing a special subdivision technique, called label-driven subdivision, on the basis of individual faces (patches). Each vertex of the neweork has to be assigned a label first. The label is determined by the subdivision levels (mesh densities) of the adjacent faces. Confirmity requirement is achieved by ensuring that the vertices generated on the common edge of adjacent faces depend on the labels of its labels only. Parallel processing is achieved by performing label-driven subdivision on the faces of the quadrilateral network simultaneously. Unfortunately, the label-driven subdivision process can be performed for faces whose labels satisfy certain requirement only. Otherwise, an admissible extension of the label assignment has to be constructed first.

In this paper, we will present a parallel mesh generation algorithm based on a new label-driven subdivision technique. The new subdivision technique does not require the vertex labels to satisfy any special requirement. Hence, an admissible extension is not required for any given label assignment. Nevertheless, the label-driven subdivision technique has most of the advantages of the previous label-driven subdivision technique, i.e., it is performed on the basis of individual faces and, hence, making parallel processing of the mesh generation process possible; it automatically ensure the conformity of the resultant mesh, no back tracking is required; it allows selective local refinement of the mesh layout. One potential disadvantage of the new subdivision technique is that it would generate more elements in the final mesh than the previous subdivision technique.

2   DEFINITIONS AND NOTATIONS

In this section basic definitions and notations will be summerized, following the same ones used in Cheng et al(1989). We will start with a standard definition of polyhedral networks.

Consider a surface in 3 dimensional space. A *network* on the surface is a finite set of points(called nodes) and curve segments such that

1. each node is an endpoint of a curve segment,

2. each endpoint of a curve segment is a node,

3. two curve segments intersect only at their endpoints.

If the curve segments are segments of straight lines then the corresponding network is called a *polyhedral network*. Polyhedral networks can be represented in a plane by planar graphs. A standard graph terminology (such as vertices, faces, edges) will be used to refer to objects of polyhedral networks. If all of the bounded faces of a polyhedral network are convex quadrilaterals then the network is refered as a *quadrilateral network*. In addition, such a network is called a *regular quadrilateral networks* if the degree of each vertex (in the corresponding planar graph) belonging only to bounded regions is equal to four.

Let F be the set of all faces of a regular quadrilateral network P. A *subdivision level assignment* S of P is a function defined on F, $S : F \rightarrow N \cup \{0\}$, where N is the set of all positive integers. $S(f)$ is called the *subdivision level* of f for $f \in F$.

The problem we consider here is defined as follows which is slightly different from the one dealt with by Cheng at al(1987). Given a regular quadrilateral network P and a subdivision level assignment S on P, our task is to construct a *subdivision mesh* P^* of P such that

(R1) Each specified face f of P is subdivided into at least $9^{S(f)}$ subquadrilaterals.

(R2) The shape of faces generated in P^* is regular, i.e. faces of P^* are not too long or too narrow.

(R3) The resultant subdivision mesh P^* is amenable to local modification, i.e. changing the size or shape of some of the faces without affecting the remainder.

(R4) The number of faces generated in P^* is minimal over all subdivision meshes of P satisfying the goal(R1).

In the next section, we will present a solution to this problem that meets the goals (R1)-(R4).

3   PARALLEL MESH SUBDIVISION ALGORITHM

The algorithm consists of the following two phases:

1. vertex label assignment based on subdivision levels,

2. mesh subdivision based on vertex labels.

3.1   Vertex label assignment

Let P be a regular quadrilateral mesh, and V and F be the sets of vertices and faces, respectively. S is a subdivion level assignment of P. A vertex label assignment L of P with respect to S is a function $L : F \rightarrow N \cup \{0\}$ such that $L(v) = \max\{S(f)|f \in F$ and v is a vertex of $f\}$. Figure 1 shows a subdivision level assignment and the corresponding vertex label assignment.

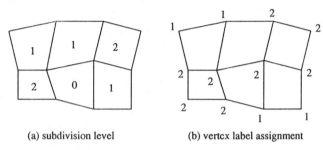

(a) subdivision level   (b) vertex label assignment

Figure 1 Vetex label assignment from subdivision level

3.2   Mesh subdivision based on vertex labels

The algorithm is based on five types of elementary subdivision procedures and they are applied repeatedly according to the relative positions of the vertices labeled 0 of a given quadrilateral until the label of every vertex becomes 0.

The types are classidied into five cases except that all labels are 0:

(1) only one label is non-zero,

(2) two adjacent labels are non-zero,

(3) two diagonal labels are non-zero,

(4) three labels are non-zero,

(5) all labels are non-zero.

For each case, one of the elementatary procedures shown in Figures 2-6 will be applied. For example, in case (1):only one label is non-zero, a given quadrilateral $f = v_1 v_2 v_3 v_4$ shown in Figure 2(a) is subdivide into three smaller subquadrilaterals $f_1 = q_1 q_2 q_3 q_4$, $f_2 = r_1 r_2 r_3 r_4$, and $f_3 = s_1 s_2 s_3 s_4$. New labels are assigned as shown in Figure 2(b). For instance, the label of vertex v_1 is originally i. After the subdivision, it becomes $i - 1$. The coordinates of vertices are calculated as the following expressions:

$$q_1 = v_1,\ r_2 = v_2,\ r_3 = s_3 = v_3,\ s_4 = v_4,$$
$$q_2 = r_1 = \frac{2v_1 + v_2}{3},\ q_4 = s_1 = \frac{2v_1 + v_4}{3},$$
$$q_3 = r_4 = s_2 = \frac{3v_1 + v_2 + v_3 + v_4}{6}.$$

In Figures 3-6, original labels and newly assigned ones are shown in (a) and (b), respectively.

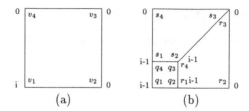

Figure 2 Only one label≥1 (i:vertex label).

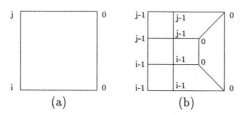

Figure 3 Two labels≥1 (adjacent type).

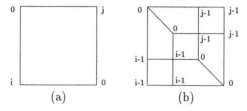

Figure 4 Two labels≥1 (diagonal type).

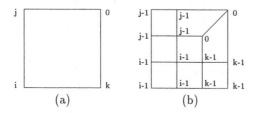

Figure 5 Three labels≥1.

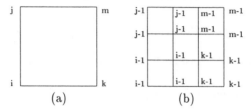

Figure 6 All labels≥1.

3.3   Structure of the algorithm

The overall structure of the algorithm is given in Figure 7. The algorithm has two phases, vertex label assignment and mesh subdivision algorithm explained in the previous sections. We use parallel processing in each phase.

A procedure, subdivide() in Phase 2 is defined in Figure 8. Procedure one_v() is called to subdivide a quadrilateral face one level down if the face has only one non-zero label. It corresponds to the mesh subdivision illustrated in Figure 2.

Similar to one_v(), procedures adjacent_v(), diagonal_v(), ... are all subdivision procedures conducting subdivision in one level deeper, corresponding to different subdivision types illustrated in Figures 3-6.

Algorithm PMS:Parallel Mesh Subdivision
{input:a regular quadrilateral network P and a subdivision level assignment S on P}
{output:a subdivision mesh P^* of P}

Phase 1:[Construct the vertex label assignment L of P with respect to S.]
    PARDO for each vertex v of P do
    $L(v):=\max\{S(f)|f\in F,v$ is a vertex of $f\}$
    DOPAR

Phase 2:[Subdivide the faces of P in parallel.]
    PARDO for each face f of P do
    $subdivide(f)$;
    DOPAR

Figure 7 Parallel mesh subdivision.

$subdivide(f$:quadrilateral);
    begin
      if(only one label of $f > 0$)**then**
        begin
          $one\_v(f, g_1, g_2, g_3)$;
          $subdivide(g_1)$; $subdivide(g_2)$; $subdivide(g_3)$;
        end
      else if(two adjacent labels of $f > 0$)**then**
        begin
          $adjacent\_v(f, g_1, g_2, g_3, g_4, g_5, g_6, g_7)$; ...
        end
      else if(two diagonal labels of $f > 0$)**then**
        ...
    end;{$subdivide$}

Figure 8 Procedure $subdivide()$.

Figures 9-12 illustrate the mesh generation for 2-D grids with different number of faces and subdivision levels. Each resultant quadrilateral is divided into two triangles by one of the diagonals.

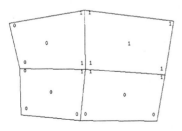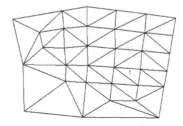

Figure 9 Example no.1.

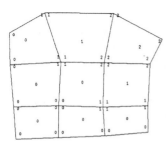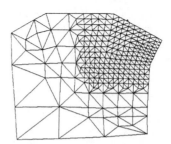

Figure 10 Example no.2.

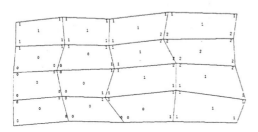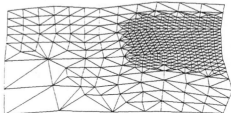

Figure 11 Example no.3.

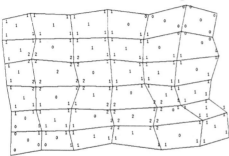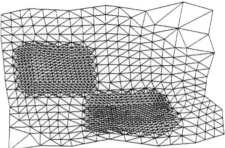

Figure 12 Example no.4.

4 PERFORMANCE ANALYSIS

The algorithm has been implemented in sequential and parallel on Sequent Balance 8000/21000 computer with 26 processors. The performance data for parallel and sequential mesh subdivision recorded from the test are shown in Table 1. The comparison graph bewteen the parallel and sequential version is shown in Figure 13.

In our algorithm, each face can be processed independently. Since the computer has 26 processors, we can process 26 different faces in parallel. As we see from Table 1, when the number of face reaches 25, the ratio of parallel vs sequential is the lowest. The data here is consistent with our expectation.

In Figure 13, the execution time collected for the parallel and the sequential versions is shown. For a regular quadrilateral network of m faces with randomly assigned subdivision levels(maximam subdivision level = 3), the sequential version is implemented using one processor only, the parallel version is implemented using m processors (when the number of processors is greater than the number of patches); one processor per face. According to the data we have collected, parallel version appears well suited to the mesh generation process.

Table 1 Performance data for parallel and sequential mesh subdivision

No. of faces	Parallel(10^{-4}sec)	Sequential(10^{-4}sec)	ration(P/S)
4	3.35	8.93	1:2.67
9	3.40	9.81	1:2.89
16	3.43	22.01	1:6.42
25	4.20	33.94	1:8.08
36	5.11	39.37	1:7.70
49	4.61	24.07	1:5.22

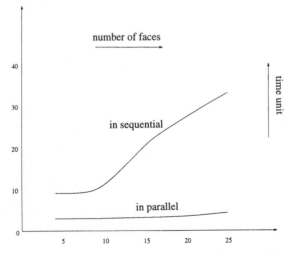

Figure 13 Comparison between the parallel and sequential versions.

5 CONCLUSIONS

A parallel mesh generation algorithm based on a new label-driven subdivision technique is presented. The new label-driven subdivision technique does not impose any restrictions on the labels of the vertices. Hence, it does not require the construction of an admissible extension of the vertex label assignment as does the previous approach. The new algorithm has been implemented in C on a Sequent Balance 21000 computer with 26 processors. The results are quite satisfactory and reliable. Some of the test cases are shown in Figures 9-12.

6 ACKNOWLEDGEMENTS

The authors would like to thank Feng Luo for the implementation of the algorithm presented in this paper. The work of the second author is partially supported by IBM and NSF grant DMI-94000823.

7 REFERENCES

Baehmann,P.L., Wittchen,S.L., Shephard,M.A., Grice,K.R. and Yerry,M.A.(1987) Robust, geometrically based, automatictwo-dimensional mesh generation. *International Journal for Numerical Methods in Engineering*, **24**, 1043-1078.

Cheng,F., Jaromczyk,J.W., Lin,J.R., Chang,S.S. and Lu,J.Y.(1989) A parallel mesh generation algorithm based on the vertex label assignment scheme. *International Journal for Numerical Methods in Engineering*, **28**, 1429-1448.

Gordon,W.F. and Hall,C.A.(1973) Construction of curvilinear coordinate systems and applications to mesh generation. *International Journal for Numerical Methods in Engineering*, **7**, 461-477.

Haber,R.H., Shephard,M.S., Abel,J.F., Gallagher,R.H. and Greenberg,D.P.(1981) A general two-dimensional finite element preprocessor utilizing discrete transfinite mappings. *International Journal for Numerical Methods in Engineering*, **17**, 1015-1044.

Lee,Y.T., De Pennington,A and Shaw,N.K.(1984) Automatic finite-element mesh generation from geometriuc models-A point based approach. *ACM Transactions on Graphics*, **3**, 287-311.

Rivara,M.-C.(1987) A grid generator based on 4-triangles confroming mesh-refinement algorithms. *International Journal for Numerical Methods in Engineering*, **24**, 1343-1354.

Sadek,E.A.(1980) A scheme for the automatic generation of triangular finite elements. *International Journal for Numerical Methods in Engineering*, **15**, 1813-1822.

8   BIBLIOGRAPHY

Kenjiro T. Miura is an associate professor of the Department of Computer Software at the University of Aizu, Japan. He worked for Canon Inc., Japan as a research engineer from 1984 to 1992. His research interests include computer-aided geometric design, computer graphics, and mesh generation. He received his BEng and MEng in precision machinery engineering from the University of Tokyo in 1982 and 1984, respectively and his PhD in mechanical engineering from Cornell University in 1991. He is a member of ACM, ASME, and SIAM.

Fuhua (Frank) Cheng is an associate professor of computer science at the University of Kentucky. He received a BS and an MS in mathematics from the National Tsing Hua University in Taiwan, and an MS in mathematics, an MS in computer science, and a Ph.D. in applied math from the Ohio State University, USA. His research interests include finite-element mesh generation, geometric /solid modeling, computer graphics, and parallel computing in geometric modeling and computer graphics. He is a member of ACM, SIGGRAPH, SIAM, SIAM Activity Group on Geometric Design, and IEEE Computer Society.

PART FOUR

Reverse Engineering

14

An effective method to analyse chronological information aspects in actual engineering processes

J.S.M. Vergeest, E.J.J. van Breemen, W.G. Knoop and T. Wiegers
Delft University of Technology, Faculty of Industrial Design Engineering
Jaffalaan 9, NL 2628 BX Delft, Netherlands
Telephone +31 15 783765, Fax +31 15 787316
Email j.s.m.vergeest@io.tudelft.nl

Abstract

Offering effective computer support during the early stages of the engineering process is a longstanding problem. Research into actual engineering processes is an essential contribution to establishing objective criteria for effective tools. We propose to derive such criteria from observed engineering processes and to test these requirements experimentally. We describe the first series of observations that resulted in data, serving as a reference for subsequent experiments. The control design project was bidisciplinary (mechanical and electrical) and significantly involved conceptual decision making. The experiment focussed on information needs by the designer, information retrieval and information production. Preliminary analysis shows a steep distribution of the time delay between information request and the answer to the request, suggesting a strong influence of information access time on several aspects of the engineering process.

Keywords

Engineering processes, empirical analysis, design information, computer support

1   INTRODUCTION

Computer-aided conceptual design is still known as an extremely difficult goal to achieve. This contrasts the situation for other engineering stages such as mechanical embodiment design, design optimization, analysis and production planning, where the role of computer systems is dominant. The main reason for the lack of computer support during conceptual design is the seemingly chaotic nature of the activities. The designer (or design team) is in the process of generating product ideas and concepts, resulting in many different alternatives based on different technical principles. These concepts are usually quickly evaluated, altered, communicated to partners, or rejected. Once a concept has been selected for more detailed consideration and for technical evaluation, more elaborated models need to be created. Not until that stage of design, CAD/CAE techniques come in effectively.

It is recognized that the industry needs adequate computer support during the concept phase because:

1. The early design phase significantly determines the total costs and the success of the new product, and therefore needs much augmentation.

2. The barrier between the non-electronic world of concept design and the world of computerised product models must be lifted to shorten the time-to-market

3. Computer support enables systematic recording of the history of the early design phase, including the process of decision making. This in turn facilitates design information reuse. Recent evaluations have shown that present commercial and research systems fail to achieve this (Hennessey 1993). To obtain a better understanding of this problem is generally one of the motivations for empirical research into the engineering process.

This problem is central in the Desys project, described in this paper. The main goal is to develop procedures to derive requirements for conceptual CAD from empirical studies, and to apply these procedures in practical situations. In the next section of this paper we give a brief overview of empirical studies into engineering process and the conclusions about supporting computer tools as drawn by the respective authors. In section 3 we motivate and describe the experiment in which we observed engineers at work. Section 4 contains preliminary results from data analysis. In section 5 we propose to use these data for comparative studies with the purpose to determine the impact of computer tools on the engineering process.

2   PAST STUDIES INTO ACTUAL ENGINEERING PROCESSES

Actual engineering projects have been studied in detail and the outcomes generally confirm that engineering design is a complex process consisting of many types of activities, information use and creation, documentation, reasoning and decision making. In addition it turned out that making those studies is very time consuming and expensive. Hence, many published results are rather fragmentary and of qualitative nature.

Stauffer has made a comparison of six empirical studies into the mechanical design process (Stauffer 1988), including his own research carried out at Oregon State University. Additionally, a recent literature survey can be found in (Knoop 1994). It is important to realize that studies of design projects have been made in very different ways on different project scales and on different levels of abstraction. Although these studies can be considered as being complementary, globally aiming at the same goal, they serve distinct research purposes.

Two types of empirical research methods can be distinguished. The first captured design activities in real-time, either by direct observation or by participation. Data was

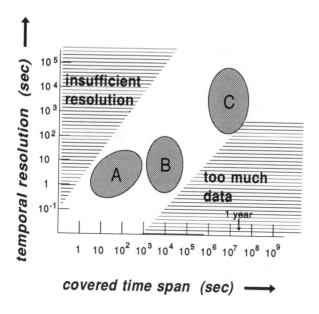

Figure 1: Simplified spectrum of published empirical research in design. The horizontal axis represents the duration of the design process covered by the studies. The vertical axis is the time interval to which a typical empirical datum corresponds, i.e. the temporal resolution of the analysis. The dashed areas are hardly accessible by research, either because the data would too global or because the collected amount of data would become too vast. The research types A, B and C are explained in the text.

usually protocollized and transcribed for further analysis. Due to the large amount of generated data, this type of method was only suited for small size design projects. The other type of method was analysis of retrospective data (interviews, notebooks, design documentation) about past or ongoing design activities. Since retrospective data can be analyzed at very different levels of detail, both small and large design projects can be investigated with this type of method. Figure 1 indicates the clustering of past research with regard to project duration and temporal resolution of the data. Studies of large projects, lasting up to over a year, tend to be based on data integrated over $10^3 \ldots 10^5$ seconds intervals (cluster C), while research based on detailed data $(1 \ldots 10^2$ sec) seems restricted to small design projects (cluster B). Actual measurements of computer tool impacts are mostly in the region of short projects (activities) which are observed in great detail, down to 10^{-1} sec (cluster A). It should be noted that this picture is not strict; for example, the work described in (Blessing 1994) is based on global as well as on some detailed empirical data.

In a few projects the dependence of the data on the subject that carries out the design task was determined. The main reason for this was to detect differences between novice and experienced designers. In addition, an indication could be obtained about the spread of data obtained with supposedly similar subjects. Obviously, subject dependence cannot be measured unless in experimental design projects.

Research projects that report conclusions and recommendations about computer tools for design could be put in three, not completely disjoint categories:

1. The first covers research that takes actual design *projects* as empirical starting points. From research in this category, reported conclusions are relevant for the design project (or task or activity) itself, i.e. without anticipating the introduction of any computer tool. It is this category of research that we consider in this paper. However it is useful to realize that there are two other categories:

2. Research that takes a design *process model* as starting point. Design is described at a certain level of abstraction. In principle, the validity of statements made about the effectiveness of computer tools is restricted to that level and to the situations in which the process model is valid. These statements are sometimes supported by implementation and evaluation of the tools in practice. Obviously, the planning and the conduction of such evaluation is almost inevitably biased by the particular type of tool under consideration.

3. Research that takes existing computer tools as a starting point. Improvements on one or multiple tools are developed and evaluated. Also new types of tools are proposed, developed and demonstrated. If the positive (or negative) impact of the new tools on design is shown, it is usually done by comparison of their performance with that of other tools.

We expect research in the first category to be the most appropriate for deriving requirements for conceptual CAD. We have collected publications of about 16 research projects of this type. Only 6 papers contained comments about CAD or KBS tools. For example Lewis (1981) lists some factors of the design process that should be taken into account for future CAD development (such as allowing user intervention and informal decision making). Kuffner (1991) argues that CAD must be able to make supplemental, non-traditional design information available to designers. Stauffer (1988) indicates the need of making CAD intelligent and Waldron (1988) states that "a flexible tool which can interact, communicate and reason like another designer would be ideal". Stauffer has found evidence for the fact that mechanical design relies little on problem-solving techniques, but much on richness of domain knowledge. He expects that this has implications for future research on design automation.

We did not find any references to empirical studies that resulted in data from which requirements for supporting tools can be derived. This motivated us to set up an experiment with that very purpose.

3   DESCRIPTION OF METHOD AND EXPERIMENT

In the previous section we found that published empirical studies of engineering processes offer global conclusions about requirements for support methods. Our research aims at methods to extract from empirical data such requirements in detail and to test experimentally the validity of these requirements e.g. by showing evidence that a tool made or simulated according to the requirements has the predicted impact on the design process. This sets specific conditions on the way engineering processes must be observed and represented. To gain insight in these conditions and in preparation of the experiment we have made a retrospective study of 6 months of engineering activity by a team of 12 people working on the same project (van Breemen 1994). We then decided to focus our experiment on external information flow, i.e. on information represented by directly observable entities (drawings, written or spoken results). It was also necessary to record, as a function of time, the current engineering activity. No attempt was made to trace the subject's way of thinking, reasoning or mental state.

To capture these data it was not necessary to let the subject think aloud, as is done in some protocol studies. However, the prevailing source of information for the experiment was the subject; he or she can express what the current activity is and which information is needed. We therefore encouraged the subjects to speak only if they had a new information request and/or started a new activity. In addition, at regular time intervals during the session a clock signal reminded the subject to briefly mention his or her current activity and any need for information. For this experiment the time interval was set to 5 minutes. Information requests were handled by an experimenter located near the subject's desk. The experimenter had a set of general technical documentation and some specimen available, and was well prepared to accept questions about the assignment. The role of the experimenter was deliberately restricted to providing simple answers; no advises, hints or judgements about the design itself were given. All questions, answers, activities and results (either pronounced or directly observable) were listed during the session by a second experimenter, watching the session on a video monitor, in a separate room. After the session the drawings and texts produced by the subject were merged with the list. Then, following a well-defined procedure, it was determined which activities were continuations of previous similar activities. This led to the identification of activities and subactivities, discussed further below. Also a search was made for information links, defined as explicit uses of earlier results.

An essential decision for the experiment that no *a priory* categorizations for activities, requests or answers were made. All observables were recorded purely without any attempt to classify them. We did this to avoid the risk of unnecessary data clustering into (perhaps irrelevant) categories. Furthermore this leaves the possibility to discover actual categories in addition to the ones frequently mentioned in the literature.

The engineering assignment in this experiment involved the conception and technical specification of an alternative electrical lighting system for a new type of bicycle. Six technical requirements were listed in the assignment, and the subject was asked to explicitly show that the proposed design could meet those requirements. Since the maximal duration of the session was fixed to 160 minutes, several restrictions were built in to avoid that much time was spent on making too detailed descriptions or illustrations. The technical and functional aspects were emphasized.

10 different subjects, all near-graduate engineering design students, worked on this assignment independently. In advance each subject was asked to work on a smaller assignment, to let him or her get used to the working environment. Before and after the session each subject was interviewed following a list of questions. All subjects declared that they did not feel hampered in any way during their engineering task.

An extensive report about the experimental method is in preparation.

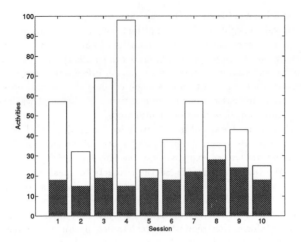

Figure 2: The number of activities (solid bars) shows much less variation over the 10 sessions than does the number of subactivities.

4   INITIAL FINDINGS FROM THE DATA

The design assignment contained very strict directions about the engineering activities; two technical aspects of the new lighting system (power production and mechanical structure) needed to be addressed explicitly by the subjects. This helped to avoid that the 10 engineering processes would become totally different in character and hence incomparable. Although the 10 produced designs are very different (ranging from simple redesign of the existing generator to rather creative inventions), the activities that happened in the 10 processes are largely consistent.

The number of different activities within one session varies between 15 and 28 (Figure 2) corresponding to average durations between approximately 11 and 6 minutes. Very often an activity did not occur in a session as a contiguous activity but it was interrupted by other activities. Hence, a session can be viewed as a sequence of subactivities, where each subactivity is part of exactly one activity. If the number of subactivities in a session is large compared to the number of activities then there have been many interruptions and corresponding switches between activities. In Figure 2 this comparison is shown. Whereas the number of activities remains in the interval 19 ± 5 for all but one of the sessions, the number of switches between activities shows a lot of variation. We find that the 5 longest activities in each session take between 42% and 68% of the total duration of the session. However, only 3 activities occurred in 5 sessions or more. These activities are documenting the design (in 8 sessions), calculating the electric power of the generator (5 sessions) and general study of collected information (5 sessions). This implies that there are significant differences between the sets of dominating activities.

The number of information requests in each session varies between 27 and 89; for eight of the sessions this number stayed within the range 55 ± 14. A significant spread is also found for the number of produced outputs. This is partly due to non-strict interpretation rules during the initial encoding of the data. For some sessions an information output was identified only if the subject explicitly expressed the result. However, on basis of the produced drawings and notes and on the video tape, many more information outputs can be identified, as has been done for some of the sessions. A similar inconsistency has occurred for the information links, representing observed usage by the subject of data that was the result of an earlier activity.

An extensive presentation of the data is in preparation.

In this paper we focus on one aspect of the observations, the information requests of the subjects and the information received upon those requests. As mentioned, a member of the research team acted as an expert and attempted to answer each question as quickly as possible. This effected a steep distribution of the time delay between question and answer, see Figure 3. This articulates the strong influence that access speed to engineering information must have on the engineering process. In a situation that the designer had to find the answers by him or herself, the delay distribution is expected to be much flatter, corresponding to a much longer time to obtain the same design result. Of course it should be taken into account that the presence of the experimenter, in our study, stimulated the subject to ask many questions. Summed over all 10 sessions, of the 557 answers, 299 (54%) were given within 10 seconds and 487 (87%) within 100 seconds. 31 information requests (6% of 480) were never satisfied. It was verified that all these requests did not require an answer or a confirmation by the experimenter. The difference between the number of requests (480) and the number of answers (557) is due to multiple answers (at different times) to the same question. Careful study of the information requests will reveal which of the questions are specific for the session (i.e. dependent on the particular design solution) and which ones are not. A first categorization of the 480 requests showed that only 21 questions (4%) are solution specific, while the remaining questions could have been expected for any design solution. This is a low fraction compared to the 36% found for the redesign study in Kuffner (1991). This might indicate that the designers at work in Kuffner's experiment did, on average, more detailed engineering than the designers in our experiment, where a lot of time was spent on gathering general information.

5   ON MEASURING THE IMPACT OF COMPUTER TOOLS

In view of the findings above, one of the basic questions is whether a contemporary knowledge-based system would be able to play the role of the assisting experimenter, and to which extent. But this is not the only imaginable support for the engineering task of our experiment. Many questions were about the geometrical lay-out of the rear side of the bicycle and about the mechanical interfacing between the (housing of the) new product and the bicycle. Here, a 3D computer simulation system could be more effective than the static drawings, tables and texts available during the experiment. Although tools to support 3D modeling or decision making do exist, two fundamental points of concern remain.

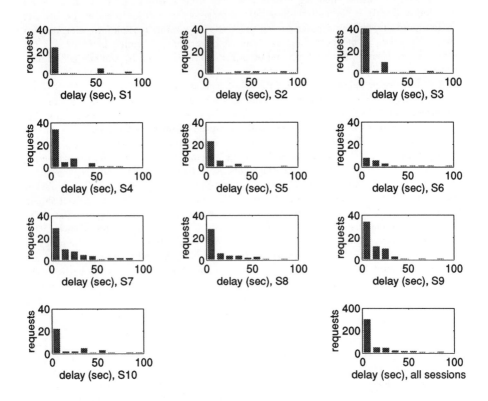

Figure 3: Time delay between information requests and answers. 52% of all answers were provided within 10 seconds, 87% within 100 seconds.

The first is that most existing tools have been developed to perform their specific task and they are often optimized and evaluated as such. This corresponds to research type A in Figure 1. It is the type of research located in the lower-left corner of figure 1, and as we stated before, this research is not sufficient to detect the overall impact of a tool on a design process. What is needed instead is an optimization of tools in a (perhaps simulated) practical situation, rather than retrospective evaluations (often over a many-years' period) by companies that had the tool in use. Unfortunately this is often the only source of information about the effectiveness of support systems in industry. The second concern is that, even if it has been determined that a specific tool is effective in a practical context, the inclusion of another tool could cancel this unless sufficient attention is paid to unify the tools. This topic is being studied as an information technological problem, see e.g. (Smith 1994), but again, any proposal for an integrated system should be tested on its overall impact.

We are still far away from a situation that multiple tools (and for some domains even a single tool) can be effectively used during conceptual design. A contribution to achieve that goal can be made by measuring the impact of (new or existing) tools on the design process in experiments as described above. In its simplest form a number of engineering sessions with and without the tool(s) can be run, of course with different subjects for each session.

6   CONCLUSIONS AND FURTHER RESEARCH

We have presented an experiment procedure to extract timed data about information requests and information flow from actual engineering processes. The method has been applied for a 2 1/2 hour design task, carried out by 10 different subjects. Compared to more conventional protocol studies the data acquisition procedure is fast and non-disturbing. Preliminary analysis shows that among the 10 sessions the spread in the number of activities is much smaller than the variation of other quantities, such as number of activity interruptions, information requests and information outputs. A very small fraction (4%) of the information requests is directly related to a particular design solution.

The main objective of the experiment is to verify that the type of obtained data has the potential to serve as a reference for comparative experiments, for the purpose of tool impact measurement. This verification is one of our current research issues. Initial interpretations suggest that the current series of observations offer a firm grip on the dependence of information access time and engineering progress.

By focussing the observation procedures to even more specific quantities, we expect that larger engineering processes of industrial teams can be traced as well. This will then systematically bridge the gap in Figure 1 between the type B and the type C research methods.

7   REFERENCES

Blessing, L.T.M. (1994) A process-based approach to computer-supported engineering design, Cambridge, ISBN 9523504-0-8.

Breemen, E.J.J. (1994) Characterisation of information in product development processes. Report TUD-IO-K311, Faculty of Industrial Design Engineering, Delft University of Technology, Delft.

Hennessey, J.M. (1993) Computerized conceptualizing in industrial design, in *Proceedings Interface '93, humanizing technology*, The North Carolina State University Student Chapter of the Human Factors & Ergonomics Society, 112-117.

Knoop, W.G. (1994) Empirical research in design. Report TUD-IO-K310, Faculty of Industrial Design Engineering, Delft University of Technology, Delft.

Kuffner, T.A. and Ullman, D.G. (1991) The information requests of mechanical design engineers. *design Studies*, **12**, Nr. 1, 42-50.

Lewis, W.P. (1981) The role of intelligence in the design of mechanical components. *Man-machine communication in CAD/CAM*, IFIP, North-Holland.

Smith, G.L. and Muller, J.C. (1994) PreAmp - a pre-competitive project in intelligent manufacturing technology: an architecture to demonstrate concurrent engineering and information sharing, *Concurrent Engineering: Research and Applications*, **2**, Nr. 2, 107-115.

Stauffer, L.A. and Ullman, D.G. (1988) A comparison of the results of empirical studies into the mechanical design process. *Design Studies*, **9**, Nr. 2, 107-114.

Waldron, M.B. and Waldron, K.J. (1988) Conceptual CAD tools for mechanical engineers. In *Proc. of the 1988 ASME Int. Computers in Engineering Conference* (eds. V.A. Tipnis and E.M. Patton), Vol. 2, ASME, New York, 203-209.

15

Extraction and reuse of partial forms in form-feature modeling

A. Tajima,   M. Numao
tajima@trl.ibm.co.jp   numao@trl.ibm.co.jp
Tokyo Research Laboratory, IBM Research
1623-14, Shimotsuruma, Yamato-shi, Kanagawa, 242, Japan
Telephone: +81-462-73-4886   Fax: +81-462-73-7413

Abstract

This paper presents a method for reusing flexible and complicated form data. Based on a form-feature modeler that has constructive solid geometry and geometric constraints as its form description, it handles form data as flexible and reusable components. To build a data management mechanism that registers, preserves, and searches for form data, a concept-classification method is adopted from knowledge engineering. A more complex form is defined by combining simpler forms, and can be used to define further forms. As designers proceed, they obtain composite form data that can be preserved and reused later.

Keywords

reuse, classification, CSG, feature-based, geometric constraints

1 INTRODUCTION

Three-dimensional solid modeling is an indispensable technology for Computer-Aided Design (CAD), Computer-Aided Manufacturing (CAM), and Computer-Aided Engineering (CAE), being a method of form representation whose applications range from product design to evaluation and process design. Recent advances in computational ability and memory capacity have enabled computers to represent and handle very complicated products as solid models. Consequently, the most pressing problem is the modeling method; that is, how to construct efficiently on computers the forms intended by designers.

Reuse of form data is an important factor in this problem. In practice, forms are not always designed from scratch. Data are often diverted from old designs, standard parts are used, and so on. This not only makes design processes efficient and reduces production costs, but also reduces the frequency of design faults, because the data have already passed through the evaluation process. In other words, actual design processes involve reuse, and it is natural to introduce reuse into computer-aided design.

Some existing form-feature modeling systems can handle limited reuse of form data. But reusable forms that have flexibility — for example, a "through hole" that retains its "through" feature regardless of the size and shape of the base model — are limited to simple ones built into the system library. Users who want to define their own flexible and reusable forms are required to perform some special tasks apart from modeling, such as programming in some specific language [Shah,1988]. When users reuse the forms that they have constructed on modelers, they can handle only the whole model as a unit, or they must handle the forms as non-flexible data whose sizes and positions are fixed. Further, when many form data are stored, users have to name all the data and specify the form they need by specifying its name. This means that the greater the quantity of data and the number of users who have stored data become, the less reusable the data are.

We present a method, based on a form-feature modeler that has constructive solid geometry (CSG) and geometric constraints as its form description, for reusing flexible and complicated form data. It packages form data into components, reuses them, and makes the input process on a form-feature modeler more efficient.

The unit of packaging ranges from a simple form such as a "hole" or "step" to a complicated shape that may be a complete part. Thus, the design process is regarded as "a process in which forms are combined continuously." A more complex form is defined by combining simpler forms, and can be used to define further forms. As designers proceed, they can obtain composite form data that can be preserved and reused later.

2   COMPOSITE FORM

2.1   Semantics

A designer specifies and extracts a part of the current model, and apply it to other parts of the model. It can be preserved by the system if it is useful. Thus, as the designer proceeds, he/she can obtain composite form data that can be preserved and reused later. Hereafter, we call the composite form data that are generated during the design process, "composite forms." These composite forms have the following variations:

- Form-features such as "hole" and "step."

- Partial forms such as "head of a piston."

- Transformational operations such as "make a corner round."

- Macros for applying the same operation at several locations.

2.2   Base Technologies

For handling of composite forms, two characteristics are desirable. First, a composite form should be definable simply by "specifying a part of the model." Second, the data should have enough flexibility to be reusable. To realize these characteristics, we used Constructive Solid Geometry (CSG) and geometric constraints methods for representing solid models.

CSG

In [Wilson,1990], Wilson concluded that Boundary Representation is preferable to CSG for features that include both shape and attributes. However, we adopted CSG, for the following reasons:

- We focus solely on generating forms.

- In our approach, a large number of composite forms are stored in the library, and form and inner representation need not always coincide.

- By combining CSG with the following method, which uses geometric constraints, we can handle data at a abstract level and reuse them more easily.

Geometric constraints

Adopting geometric constraint information simplifies the specification of size, orientation and position of each solid primitive, which is one of the main difficulties in modeling with CSG. It also facilitates later modification. The user; for instance,specifies the mutual constraints on geometric elements of constituent solid primitives during modeling "a surface of a primitive coincides with a surface of another primitive." On the basis of symbolic information received from the user, the system executes geometric inference and obtains the size, orientation, and position of each solid primitive. The main merits of this method are as follows:

- Data are preserved at quite an abstract level as geometric constraint information.

- Manipulation for modeling is simple.

- The model is easy to modify, because the partial modification propagates through geometric inference.

2.3   Extracting and applying composite forms

We present a method for preserving composite form data in a reusable format by extracting only abstracted information from a solid model.

In the previously mentioned form-feature modeler based on CSG and geometric constraints, a solid model has the following two kinds of data as its description:

1. CSG-tree: A tree structure that describes the order and types of set operations among solid primitives.

2. Geometric constraints: Constraints that describe structural relations or sizes of geometric elements.

Here, if geometric constraints are regarded as links, and geometric elements of solid primitives as nodes, data of a solid model can be represented in the form of a graph such as Figure 1-a.

Extracting composite forms

First, the user specifies a part in the current model by selecting solid CSG primitives. Then, the system transforms the CSG tree and isolates the part as a sub-tree. Here, the root of the subtree represents the form to be extracted (Figure 1-b). The geometric constraints within the sub-tree and those extending beyond the sub-tree are also extracted as a sub-graph. When extracted, the attributes (such as coincidence and size) of the latter constraints are preserved as they are, and the external geometric constraints – namely, the counterparts of the links – are replaced with variables, which become the interface for determining the composite form (Figure 1-c). If the interface variables have real values when the composite form is applied to a new model, the geometric inference engine instantiates the composite form.

Thus, we adopted geometric elements as the interface between a solid model and composite forms. A composite form is defined as the combination of a sub-tree and a sub-graph. In other

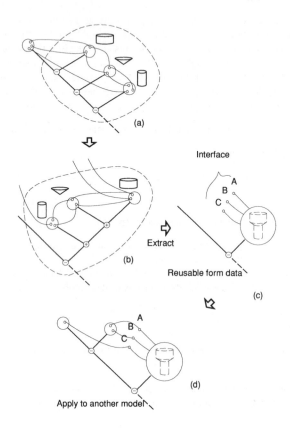

Figure 1: Extraction and application of form data.

words, composite form data do not hold any geometric data except constraints — data at a more abstract level — and the flexibility is always preserved as the form becomes more complicated.

Applying composite forms

First, the user specifies where the composite form is applied to by selecting in the current model the geometric elements that correspond to the interface of the composite form. The system then executes geometric inference and obtains the position, orientation, size of all the solid primitives that constitute the composite form. Next, the solid modeler re-evaluates the CSG-tree and obtains a solid model.

As internal geometric constraints and solid primitives are encapsulated, the user has to be conscious of only the interface of the composite form when applying it to the current model, and the manipulations are simple. The encapsulated information is enough to determine the form, and additional geometric constraints are not necessary. As the whole model is still described by geometric constraints after the part has been applied, it can be modified by changing the attributes or value of geometric constraints.

3 DATA MANAGEMENT SYSTEM

For the reuse of composite forms to be practical, it is indispensable to provide a mechanism that registers composite forms in a database, and queries and presents candidates in response to various requests. In this section, we describe a data management system that meets these conditions.

3.1 Concept-classification

To fulfill its engineering purpose, a data management mechanism should accept various abstract requests such as "I want a hole for a screw," "I want a form like this," and "I want to process this corner." To cope with such requests, we adopted the concept-classification method from knowledge engineering.

The method was first presented in KL-ONE [Schmolze,1983]. It manages concepts at various levels of abstraction. By defining a rule to judge the subsumption relation between two classes, it constructs a tree structure — a lattice, to be precise — of classes (Figure 2). In this tree, classes closer to the root are more abstract, and if there is a link between two classes, the upper class subsumes the lower.

To preserve composite form data as instances of form-concept classes allows the management system to handle data linked with concepts. A lot of calculation is needed to register new data, but only a little calculation is needed for queries that use concepts.

Abstraction-classification strategy

On registration of a new instance, first the instance is abstracted and a virtual class description that can be compared with real concept classes is generated. The virtual class finds its place in the tree, by checking the subsumption relation from the root. The parent of the virtual class is then the class that the instance belongs to.

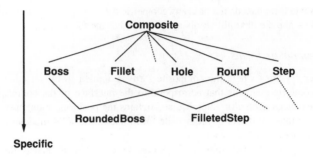

Figure 2: Hierarchy of concept classes.

A newly defined concept class inserts itself into the tree after finding its place in the same way.

As the purpose is data management, we can apply this strategy to query. The required data reside around the class that is arrived at by generating a virtual class description from the query conditions and locating it on the tree.

3.2   Handling composite form data

To use the mechanism described in Section 3.1 for composite forms, it is necessary to have a definition of the class description and a rule to specifying the subsumption relation between classes. In most research on knowledge engineering, concept classes are described by using predicate logic [Villain,1985], but in this case, an instance consists of data extracted from a solid model, and a description must be generated automatically from this instance. Therefore, the expressiveness is limited.

Here, balancing the expressiveness and the ease of extraction from data, we represent the composite form concept with (1) constituent solid primitives and (2) geometric constraints necessary to fix the composite form. For example, the concept of a "through hole" can be represented as (1) "cylinder with a minus set operation" and (2) two parallel planes and an axis.

Simple classes such as step, hole, and fillet are built in as defaults. For forms that are frequently used and/or significant to designers, new classes are defined and inserted. For example, a stepped hole (consisting of two holes with the same axis and different diameters) is classified as a child of the hole class.

Composite form data are preserved as instances of such classes. When a composite form is registered, information on its constituent solid primitives and geometric constraints is extracted from the data, and the virtual class made from it is classified in the tree. The instance belongs to the parent of the virtual class. Thus, data belongs to the most specific class that subsumes it. When a lot of data are registered and classes are defined, the tree grows large and the system becomes more useful.

Query

The following kinds of query are possible with this mechanism:

- Query by concept
  Only data with the characteristics of "through holes" are included in the class "through hole" and its descendant classes. In other words, data can be queried by concept.

- Query by condition
  By generating a virtual class from query conditions and classifying it, one can find the data surrounding it.

- Query by form
  By generating a virtual class from an item of form data, one can find other similar forms.

4 IMPLEMENTATION

We implemented the above system on a UNIX workstation, basing it on a solid modeling system that handles geometric constraints [Shimizu,1991]. Our system visualizes the extracted composite form and the constraints that are necessary to restore it to the model (Figure 4). The data management system is built as a server that uses a commercial object-oriented database, so that it can be simultaneously accessed by multiple users.

5 DISCUSSION

We presented a framework that packages form data into components, reuses them, and makes the design process more efficient by regarding the design process as "a process in which composite forms are continuously combined."

The usefulness of the data management system depends on how it grows, and we will have to consider the methodology for using it. The composite-form class description should be more expressive. The classification mechanism and the subsumption rule are separated from each other and the system is easily expandable.

One demerit of using CSG is that the same form can have different representations. This will affect the usefulness of the method if insufficient data are available. The solution to this problem is to compile form data when registering it.

6 REFERENCES

[Schmolze,1983] Schmolze, J. and Lipkis, T. (1983) Classification in the KL-ONE Knowledge Representation System. *Proceedings of the Eighth International Joint Conference on Artificial Intelligence*, IJCAI.

[Shah,1988] Shah, J. J. and Rogers, M. T. (1988) Expert Form Feature Modelling Shell. *computer-aided design*, volume 20, number 9.

[Shimizu,1991] Shimizu, S., Inoue, K., and Numao, M. (1991) An ATMS-Based Geometric Constraint Solver for 3D CAD. *Proceedings of the Third International Conference on Tools for Artificial Intelligence*, San Jose, California.

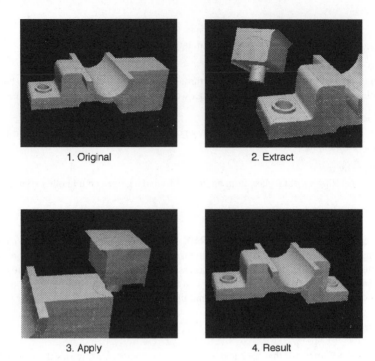

1. Original   2. Extract

3. Apply   4. Result

Figure 3: Example – the height is adjusted automatically.

[Villain,1985] Villain, M. (1985) The Restricted Language Architecture of a Hybrid Representation System. *Proceedings of the Ninth International Joint Conference on Artificial Intelligence*, Los Angeles.

[Wilson,1990] Wilson, P. R. (1990) Feature Modeling Overview. *Tutorial*, SIGGRAPH.

7   BIOGRAPHY

Akira Tajima received BS and MS degrees in aeronautics from the University of Tokyo, Japan. He is an associate researcher at Tokyo Research Laboratory, IBM Research. His research interests include applications of knowledge and databases in engineering.

Masayuki Numao received the BS and MS degrees in electrical engineering from the University of Tokyo in 1981 and 1983, respectively. His research interests include programming environments, program visualization, constraint-based programming, and AI applications to actual scheduling and design problems. He is currently manager of the Advanced Modeling and Analysis group at IBM Research, Tokyo Research Laboratory.

16

New Techniques in TH-DAIMS 2.0

Xinyou Li, Zesheng Tang, Hanwen Huang
CAD center, Tsinghua University,
Beijing 100084, P.R.China
Tel 01-2571980, Fax 01-2556935
Email dcstzs@tsinghua.edu.cn

Abstract

TH-DAIMS is a drawing renewal and management system which is based on binary image process. New techniques in TH-DAIMS 2.0, which are picking primitives from image data, smoothing edges of lines, mapping memory to hard disk for large drawings, adding Chinese characters to user interface, and making database management system suitable for both image and text data, are described.

Key words

Drawing understanding, Drawing renewal system

1 BASIC FUNCTIONS OF TH-DAIMS 1.x

TH-DAIMS is an engineering drawing renewal system which is essentially based on binary image process technique. It first scans drawings into binary image data by use of optical scanner, such as CONTEX FSS5200. In actual application, for poor contrast blueprints, It first scans drawings in grey image, and then converts the grey image into binary image according to a threshold which is a local lowest grey value selected from grey distributions of the grey image. Drawings can be scanned piece by piece in a smaller scanner and then merged into a single image.

TH-DAIMS provides a tool to clean image, which includes removing spots and filling up holes. It can aid engineers to interactively modify contents of drawings in binary image format without any vectorization by providing them many useful interactive design facilities. Available interactive design tools in TH-DAIMS are line, polyline, arc, circle, rectangle, polygon, arrow, ellipse, text, and Chinese character.

The image data of drawings and some attributes of the drawings can hierarchically be stored into FOXBASE compatible databse by the types of projects and can be accessed through image database management system. Vector data from other 2D CAD systems can also be accepted. For example, DXF, the data exchange file of AUTOCAD 11.0, can be read

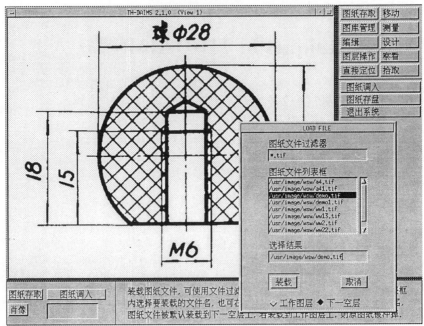

Figure 1 TH-DAIMS user interface with kanji menus and messages.

and interpreted into image data. DXF, the data exchange file of AUTOCAD 11.0, can be read and interpreted into image data. New drawings can be easily and rapidly printed by image data printer, such as HP DesignJet 600.

TH-DAIMS has a well defined graphic user interface which consists of one message area, two level menus, two view regions, four level drawings and temporal dialogues, as shown in Figure 1.

2   TECHNICAL OVERVIEW

The motivation to develop TH-DAIMS is that in Beijing heavy electrical machinery plant, houndreds of drawings have to be drawn when a new product is designed. But most of the drawings are much similar to the one of some old products. Only a little modification is needed. The designers have to spend a lot of time on drawing the drawings again. They need a system which can read the drawings to computer, provide them some interactive design tools for making a little modification, and finally print out new drawings.

Vectorization (Nagasamy, 1990 and Parker, 1988) is one way to solve the drawing renewal problem. Vectorization system converts binary image into vector automatically or manually. Automatical vectorization is to extract vector information from image by use of characteristic extraction, AI recognition, and thinning techniques. Manual vectorization is first to display the image on screen or put up the drawing on digitizer as background, and then interactively draw vector primitives, for example, line, circle and text, according to what displayed in the screen or digitizer. The resulted vectors are then sent to a 2D CAD system in hand to do some modification or further design, and new drawings can be got by printing the vector data.

Although there are not a few vectorization system in sale, the systems can not reach the level of actual usage, because there still remain some complicated problems, such as curve recognition, Chinese character recognition, dimension recognition, to be solved in auto-matical vectorization system, and the manual vectorization process is so time_consuming and terrible bother that one does not like to use it.

TH-DAIMS is another way to solve the drawing renewal problem. TH-DAIMS is a binary image system. The most simple idea in TH-DAIMS is that if we can provide user an interactive design tool in image data, we do not need to do any vectorization for drawing renewal. But it is not a simple thing. It at least needs to solve the following problems:

(1) Picking graphic primitives from image data
Picking primitives such as line, arc, text is one of four basic interactive taskes (Foley, 1990). The picked primitives can be manipulated by user interactively. For example, copy or move a line segment from one place to another, delete a text string, specially, smooth edges of lines. In graphic system, picking primitives is easily implemented if the inputed primitives are well organized in a data structure. The system searches the primitive in the data structure which is the nearest primitive to a given point on the screen. In image data system, there is no primitive information. Given a point on screen, the system should scan the image data from the point, and extract the primitive. Picking primitives from image data is new idea and new work.

(2) Allocating memory for large size drawings
Large memory space is required by image data system for manipulating image data. For an A0 size drawing with 300dpi, about 16 MByte memories are essential. If a user wants to load four A0 size drawings at a time, it will be impossible in a memory limited PC486. The system should allocate memory space available for large size drawings by mapping memory into disk.

(3) Smoothing edges of lines
Smoothing edges of lines is a distinct task in image data system. The edges of lines in original image data are not smooth enogh to pass the qualitative verification of engineering drawings.

TH-DAIMS (was named IMCAD) is one of the 8th national 5 years projects and has been finished in last summer (Li, 1993). From that time, we began to put TH-DAIMS 1.x in market, and it was well received by users in many application fields. Till now, TH-DAIMS 1.x got more than 20 users to use it.

At the same time, we developed a new version of the system, TH-DAIMS 2.0. TH-DAIMS 2.0 will be better than TH-DAIMS 1.x in the following aspects:

(1) Picking line segments from image data;
(2) Smoothing edge of lines;
(3) Allocating memory for large drawings;
(4) Adding Chinese characters to user interface;

(5) Making database management system suitable for both image and text data.

In the rest of this paper, we will describe some techniques used by TH-DAIMS 2.0 in detail.

3   PICKING LINE FROM IMAGE DATA

Although Hough transformation (Illingworth, 1988) can be used to extract line from image data, we would like to use local scanline algorithm to pick lines from image data. Given a point (x_s, y_s) on screen, the algorithm will scan image data line by line from the point until a line segment is found.

3.1 Find the parameters which define a line

A line is defined by
$(k,\ b,\ mean\_line\_width,\ start\_po\text{int},\ end\_po\text{int})$
where $y = kx + b$.

We scan image data from point (x_s, y_s) in X direction. If the inclination of the line is less than $45°$, we will exchange X and Y in image data.

For each scanline, we will have a line width L_w and a middle point (x_m, y_m) as shown in Figure 2. The next scanline will be started from $(x_m, y_m + 1)$, and the scanline width will be three times of L_w. The line equation will be given by

$$\frac{y - y_{m1}}{y_{m2} - y_{m1}} = \frac{x - x_{m1}}{x_{m2} - x_{m1}}$$

where (x_{m1}, y_{m1}) is the middle point of the first scanline, and (x_{m2}, y_{m2}) is the middle point of current scanline. The line parameters will be calculated as followings:

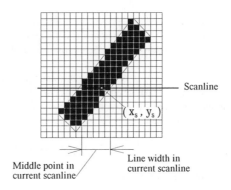

Figure 2 Scanline in an image.

$$k = \frac{y_{m2} - y_{m1}}{x_{m2} - x_{m1}} \ , \ \ b = \frac{x_2 y_1 - x_1 y_2}{x_2 - x_1}$$

The mean *line_width*, k and b of the line will be computed from the mean sum of L_w, k and b for each scanlines except the one whose slope is far difference from the mean slope.

The start point and end point of the line will be detected if there is no line width in some successive scanlines.

3.2 Find and recover all cross points

The purpose of picking lines is to manipulate them. After a move or deletion, the cross points of the picked line with other lines will be destroyed. Figure 3 shows this case. To find all cross points and to recover these cross points after interactive manipulation are necessary.

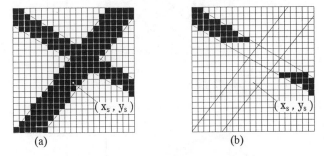

(a) (b)

Figure 3 (a) Two lines intersected (b) Cross point destroyed.

When two segments in one scanline within the scanline width are detected, there may be a cross point. See Figure 4 where two lines are intersected. We can also use scanline method presented in section 3.1 to find the parameters which define the intersection line.

If the cross point is destroyed after interactive manipulation, we will redraw the intersection line to recover the cross point.

An example is given in Figure 6.

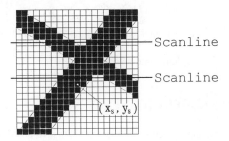

Figure 4 Cross point detection.

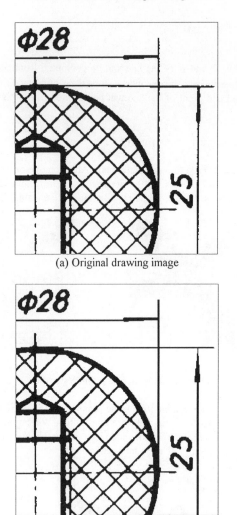

(a) Original drawing image

(b) After a line deleted and some lines smoothed

Figure 5 Example of line deletion and smooth.

4   SMOOTHING EDGES OF LINES

We use two methods to smooth edge of lines. One is first to pick lines and then to redraw the lines. Another is to tracing edge of lines. The first method is described in section 3.1. Here we will address the second one.

In tracing edge algorithm, we scan total image two times started from the first horizontal line and the first vertical line, respectively, to record the runlength codes of the image. From horizontal runlength code, we can trace edges of the lines whose angle from vertical line is less than 45°. From vertical runlength code, we can trace edges of the lines whose angle from horizontal line is less than 45°.

By tracing edge, we can have the parameters which define the edge by the same method as mentioned in section 3.1. Using the parameters, we can smooth the edge.

5   ALLOCATING MEMORY FOR LARGE DRAWINGS

The method we used to allocate memory for large drawings is dynamic memory-disk swapping. Here, What we mean by the large drawing is that the size of drawing is so large that it can not be loaded into memory even through all virtual memory are used.

We divide an image data space into $N \times M$ blocks and store them in a mapping file which has an index table. Each element of the index table has a pointer to a head of a block and a structure to record the location of the block in image space. We also have a memory block queue to record the blocks which are loaded into memory.

Block swapping between mapping file and memory block queue is carried out when an image manipulation is applied to a block which is just not in memory block queue. The mechanism of dynamic swapping is shown as Figure 6 where block swapping first reads image manipulation commands from user interface, then decides whether the block swapping is needed or not, and swaps blocks between mapping file and memory block queue when needed. Finally, block swapping calls X_lib API to carry out image manipulation on memory block queue. The strategy of block swapping is that the blocks which are the farthest to current block are swapped to mapping file when block queue is overflow.

Speed for dynamic swapping is critical problem. Table 2 with Table 1 is a test report of dynamic swapping and non_swapping system.

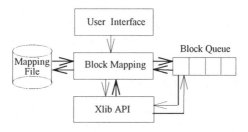

Figure 6 Mechenism of dynamic swapping.

Table 1 Test conditions

Computer	Compaq 386/33m	AST 486/33m
Memory size	8MB	10MB
CUP speed	36	72.1
Disk speed	740KB/s	601KB/s

Table 2 Test results

Manipulation	Seconds for swap/nonswap in PC386	Seconds for swap/nonswap in PC486
Load drawing (A2, 300DPI)	90/60	44/20
Store drawing (A2, 300DPI)	90/70	23/17
Size extend	8/20	2/1
Draw line (70cm, 35°)	25/35	7/2
Draw circle (R=18cm)	40/40	15/2
Move (When swap needed)	2/1	<1/<1

6   ADDING CHINESE CHARACTERS TO USER INTERFACE

Two methods can be used to add Chinese characters to user interface. One is to change UNIX and MOTIF kernel so that an application can have a Chinese interface without or with a little modification. Another is to change all text functions provided in MOTIF toolkit to the functions which can display Chinese characters. We adopted the second method because by the first one the system properties are deeply down while by the second these have not been obviously influenced.

Adding Chinese characters to user interface of an application includes three tasks:

(1) Create Chinese character font(CCF)
CCF can be got from any CC-DOS or CC-Windows but it can not be directly used by X_Window. We first translate CCF into BDF format which is a readable ASCII file and then use *bdftosnf* command to convert it to SNF format.

(2) Display Chinese characters
We use 16-bit text function to display Chinese characters. First we create a multi_font string using *XmStringCreate* function. And then we send the multi_font strings to widgets such as label, title, or other resources.

(3) Receive Chinese character input
Three tasks should be finished. The firest is to grap keyboard to catch user's input, provide user an echo and translate the input into Chinese character standard code. The second is to display Chinese characters in input area. The third is to control cursor movement while user adds or deletes characters or moves cursor.

7   IMAGE DATABASE MANAGEMENT SYSTEM

IDBMS is compatible to FOXBASE in file structure. In order to management both image and

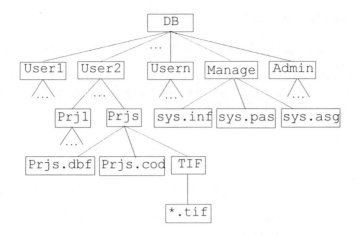

Figure 7 Logical structure of IDB.

text information in a same database, we designed a logical database structure as shown in Figure 7.

A database system has at least a directory called manage in which there are three manage files:

 sys.inf information about database system;
 sys.pas password for each user;
 sys.asg assignment of access right to DBF, DBF record, and TIF file.

A database system may have one or more users. A special user is admin who has a right to access all three manage files. A user may have one or more projects. In each project, there is a DBF file which records text information, a COD file which records image file status, and a TIF directory in which all image files used by the project are stored.

DBF is a relationship database whose structure can be defined by user. DBF is same as the DBF of FOXBASE except that TIF field type and DB field type. TIF field is a pointer to TIF file and DB field is a pointer to another DBF file so that a hierarchical database structure can be constructed.

When user wants to add a drawing to one of his projects, what he needs to do are first to store image data of the drawing into TIF directory and then to add a record about the drawing into DBF file. IDBMS automatically adds a pointer from the record to the TIF file. By the pointer, system can easily find the image data of the drawing from the DBF record when retrieval.

A user can only access his own projects. He can also assign others an access right to one or more his own DBF files, records of a DBF, or TIF files. He can not access DBF file, record of DBF, or TIF file of others if he has no right. A user can add his own crypogram to his TIF files so that others can not understand the means of his drawings recorded in the files even through the others have a way to get the TIF files of his drawings.

A plant or institute can create one or more such image database system in network environment with optical disks. IDBMS can access these database systems through ethernet and can keep concurrence by locking access right to a DBF or TIF file.

8 CONCLUSION

We have presented some techniques of TH-DAIMS2.0. Used these techniques, TH-DAIMS 2.0 becomes more powerful.

There still remain some problems in these methods to be solved. For example, in picking primitives, we only described picking line. Other primitives, such as arc or circle, arrow, symbol, are not addressed. In smoothing edge of lines, how to speed up the process is what we concentrated on. We expect these problems will be solved in the near future.

9 REFERENCES

Foley, J. D. (1990) Computer Graphics Principles and Practice. Addison Vesley Publishing Company.

Illingworth, J. and Kittler, J. (1988) A survey of the Hough transform. *Computer Vision, Graphics & Image processing,* **44**, 87-116.

Li, X.Y. (1993) IMCAD — A Drawing Reading and Interactive Design System Based on Binary Image Process, in *New Advances in CAD&CG* (ed. Z.S.Tang), 809-814, International Academic Publishers.

Nagasamy, V. (1990) Engineering Drawing Processing and Vectorization System. *Computer Vision, Graphics & Image Processing.*

Parker, J. R. (1988) Extraction Vectors From Raster Images. *Computer and Graphics,* **12(1)**, 75-79.

17

Drawing processing -- Toward facilitating the reuse of old engineering paper drawings in new CAD/CAM systems

Liu Wenyin   Tang Long   and   Tang Zesheng
Department of Computer Science and Technology
Tsinghua University, Beijing 100084, PRC
Tel. :(861)2552451-2052
Fax: (861)2562463
E-mail: dcstzs@tsinghua.edu.cn

Abstract

Drawing processing has been researched but far beyond satisfaction for many years. An orientational objective is set in this paper as facilitating the reuse of old paper drawings in new **CAD/CAM** systems, and several ways to achieve it are also presented, in which primitive recognition and 2-D drawing understanding are still main problems.

Keywords

Engineering drawing, engineering drawing understanding, vectorization

1 INTRODUCTION

The reuse of existing drawings is very common in the engineering. They may be either consulted in maintenance or modified for new designs. There are a vast amount of paper drawings having been accumulated by now. The amount of paper drawings in service is still growing so rapidly that the storage and the management of them has become a real problem. The introduction of CAD technology has brought forth a reform in the engineering as well as a new problem--how to make use of the old paper drawings in the new CAD systems.

The two problems have initiated the research on the technology of engineering drawing processing, which includes several aspects such as automated paper drawing input, vectorization(line detection) and engineering drawing understanding. It is a new research subject that needs the powerful backing of great progresses in the high-tech field.

Drawing processing has been researched but far beyond satisfaction for many years. An orientational objective is set in this paper as facilitating the reuse of old paper drawings in new **CAD/CAM** systems, and several ways to achieve it are also presented, in which primitive recognition and 2-D drawing understanding are still main problems.

2 SIGNIFICANCE AND SITUATION

In order to make use of old paper drawings by means of CAD/CAM techniques, the paper drawing must be inputted into computer systems at first. This procedure can be performed in two ways--interactive or automatic. The interactive way is to draw the drawings manually once again but in the 2-D interactive drawing systems, such as **AutoCAD**, therefore is something like to make new designs in the CAD systems and will not be discussed in this paper. The automatic way of paper drawing input is to scan the paper drawings through scanners. The techniques in scanning and binary processing of image have been improved very much in recent years. Now the paper drawings can be easily scanned into and stored in the computer systems. But the efficient redesign of them remains unsolved.

The redesign on paper drawings through CAD/CAM systems can be operated at three levels, as we understand, raster level, vector level and 3-D model level.

The developments of the high-volume secondary storage technology and the image compression technology have made it possible to store the drawings in computer systems in the raster form in which the drawings usually consumed large spaces before. The modification on the drawings can be performed in a raster image editing system in spite of its low efficiency. The **THDAIMS** system(Li,1993) published by our unit is of this kind. In this system, the drawings are stored and managed as compressed raster data after scanned. A raster image editing system is developed to implement the editing operations such as *erasing*, *moving* and *copying*, as well as the drawing operations. Redesigns over the drawings are performed in this form. This system is serving many users, and meets their demands in some extents.

Not only do vector drawings take fewer spaces than raster ones, but also are convenient to be modified. In vector drawings, the binary bits from raster ones have been grouped into vector data, therefore can be picked and edited efficiently as meaningful groups--vectorial primitives. A paper drawing is stored in the raster form after scanned. It needs a so-called-vectorization processing to convert the raster form to the vector form. After vectorization, the drawing can be edited through a 2-D interactive drawing system, such as **AutoCAD**. But the vectorization is often error prone due to the noise in the raster image of the paper drawings, and the employed method of thinning and tracking usually results in distorted vector primitives. The errors and distortions in vector drawings require so much correction that the efficiency of modification at vector level is offset. Although many commercial products, such as **THRV** (Liu, 1993) published by our group, have already been developed, the performance of vectorization needs much more improvements. Many researchers are still working toward this target.

While the research on vectorization was being undertaken, another research work was placed on the agendas of researchers. It is engineering drawing understanding. It has several levels of processing, as Dov Dori mentioned in one of his paper(Dori, 1993), from the lexical level to the semantic level. Its final objective is 3-D reconstruction of objects through 2-D planar drawings. It is a critical procedure to complete the 3-D reuse of old paper drawings. The conversion from 2-D drawings to 3-D models not only facilitates making new designs over old ones but also makes it possible to use old drawings in 3-D CAM systems.

For more than a decade, researchers have not found a better way to achieve the target. In early years, researchers tried to construct the object model only through its orthographical projections(Haralick, 1982) but failed because the solution is not exclusive. This problem is attracting more and more researchers' attentions these years. The understanding process is divided into several procedures, among which the dimension understanding of 2-D drawings has been well researched(Dori, 1988, 1989 and Min, 1993) and therefore provides a strong support for further understanding of drawings. Knowledge-based understanding is also employed to get a better understanding of special components in the drawings(Vaxiviere, 1992). But the main problem of 3-D reconstruction remains unsolved. This problem is so difficult that many researchers shrink back before it.

Therefore, the state of art is: although drawing scanning and drawing processing at raster level are better solved, the efficiency of raster level modification can not be accepted by users in most cases; the performance of vectorization is so far beyond the users' satisfaction that users can not undergo so much vector level modifications; and the 3-D reconstruction is so difficult to implement that the 3-D redesign is just a dream. The researchers are now in such an awkward predicament.

Currently, There are still strong desires from the engineering field for making use of the old paper drawings accumulated over a long period of time in the new CAD/CAM systems. Therefore the reuse techniques of paper drawings, that is, effective and efficient processing of engineering paper drawings must be solved as soon as possible.

3 WAYS OUT

As we mentioned above, improving the performance of vectorization and undertaking understanding of engineering drawings are still many researchers' objectives. But their research subjects are limited to it unfortunately. There is little progress having been made in this field for many years because of its high complicacy. Here we suggest that new subjects in this field should be created, selected and researched to extricate us from the predicament.

(1) Facilitate the 2-D interactive redesign.

The commonly used methods are only but unfortunately either interactive or automatic to perform the modifications over those drawings that exist in computer systems. Why not try to use them simultaneously? The raster level editing is always considered as an interactive or non-automatic operation, while the vectorization processing is usually thought to be an automatic or non-interactive process. To combine the two methods, there may be several ways to improve the efficiency of interactive modification of drawings, as presented hereinafter:

Perform local vectoriaztion interactively in the raster level editing system.

Among the three levels of reuse of paper drawings, as we mentioned above, the raster level is easiest to achieve in spite that the manipulations at this level are always boring. So the way to solve this problem is making it easy to perform the operation at this level. But how can we change this situation? We must go beyond the traditional ways of performing raster operations such as *erasing, moving, copying and drawing with paintbrush.* The new method suggested here includes two aspects: vectorizing interactively and drawing on raster background. The techniques of locally automated vectorization should be used to perform interactive selection of groups of bits. That is, to carry out automatic vectorization on those raster pixels which are stretched to from a selected point or within a local area that is given by the user interactively. In this way, the user can pick those meaningful groups of bits not in regular areas, such as lines and arcs, therefore the editing operations of them can be performed efficiently. And drawing vectorial graphical primitives on a raster background can help user to position quickly, therefore can speed up the 2-D interactive drawing process. The vectorial graphical primitives can be saved onto the raster background. Imposing these two aspects of improvements into the raster editing system can facilitate the modifications on raster drawings.

Perform primitive recognition interactively in the vector editing system.

Now that the situation is that the bars are always too fragmentary to represent meaningful primitives after vectorization, and they are inconvenient to be picked and modified as primitives, effective and efficient ways to group them to meaningful primitives should be found and used to facilitate the modifications. The way suggested here is that adding interactive primitive recognition functions into the 2-D vector editing system. A group of scattered bars should be selected and transformed to meaningful primitives by user interactively, such as longer straight line segments with different line styles. For example, In the following piece of drawing, the parts(1, 2, 3 and 4) of the central line are scattered and one part(the dot between part 1 and 2) is discarded in vectorization, it will be easier to recognize it interactively by means of giving the clue of its two end points or its all parts than it will be to recognize it automatically. Perhaps circular arcs are the most difficult primitives to be extracted in the procedure of vectorization. But the interactive way to find them is very easy because centers and radiuses of these circular arcs or circles given interactively by users are always more precise than those ones that are automatically fitted.

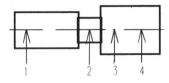

Figure 1 A sample of interactive primitive recognition.

Although it is interactive, this method is different from re-drawing the paper drawing once again. First of all, the interactive recognition is based on the results of automatic recognition which inevitably have some errors in them. Secondly, these interactive operations can be accepted by users because of their small amount.

(2) Endeavor to 2-D understanding of engineering drawings and its applications.

Now that the path toward 3-D understanding is out of our current sights, why not just do our best in 2-D understanding? In fact, 2-D understanding is easier than 3-D understanding and has some promising applications. Some researchers have done some parts such as dimension understanding and 2-D component/block recognition. These fruitful basic works make it possible to do 2-D semantic understanding. Actually, as the basis of 3-D understanding, the research on 2-D understanding will help to approach this final goal.

Similarly, the automatic 2-D understanding can also be helped by means of users' interactions. While the detail of 2-D understanding will be discussed in section 4, here we propose two new applications of 2-D engineering drawing understanding which will enrich the research work in this filed.

Perform 2-D parameterized redesign after 2-D understanding.

A semantic description of the 2-D object will be obtained after 2-D understanding, therefore constrained relations among the dimensions of all its parts can be calculated. The constrained relations make it possible to redesign by modifying the dimensions. This is a higher level reuse of old paper drawings and will improve the utility of old paper drawings in 2-D CAD systems effectively.

Turn to 2-D CAM.

Some engineering drawings are descriptions of flaky objects and for 2-D manufacturing. After 2-D understanding of these drawings which is not so difficult as 3-D understanding, the 2-D objects' shapes are recognized. 2-D manufacturing instructions can be calculated according to the automatically recognized 2-D shape. Then 2-D CAM can be realized.

These two proposed applications will change the current situation in this field. It is an attracting idea to set up an automatic workshop which can complete every operations from drawing input to 2-D dimension-driven redesign and 2-D CAM. But how can we make it true? Automatic 2-D engineering drawing understanding is a critical problem and should be solved first. To solve it, here also we propose a route on which automatic 2-D understanding is not too far to be reached but also needs large amount of research work under the current conditions.

4. TOWARD AUTOMATIC 2-D ENGINEERING DRAWING UNDERSTANDING--A PROPOSAL

4.1 What Is Understanding?

To perform engineering drawing understanding, we should first understand what its meaning is. What kind of results should be obtained to show the computer has understood the drawings.

No standard has been set yet. How can we do it if we do not know what it needs to do? In this section, we will first try to give it a lowest standard and then work toward it.

An engineering drawing is created by one man for other men to understand what object it describes. How does a trained person understand it? After some kind of analyses, the exact object will be constructed in his/her mind finally. But how does a computer understand the drawing? If the 3-D model of the object is constructed and displayed, we surely can say the drawing is understood by the computer. Maybe this is the highest standard and the final stage work of engineering drawing understanding. There are several stages of pre-work before the 3-D model of the object is constructed.

The binary image should be vectorized first. Graphical primitives should be recognized then. These processings are very easy for men, even the untrained men. They can recognize these primitives even with the first look. Therefore these processings can not be called understanding. Dimension recognition is a further processing of these primitives. Special syntactic and semantic analyses should be applied to do it. But it just adds dimension attributes to these primitives, and therefore it can not be called understanding either. What can be called understanding, then? We believe that at least the functions of vectorial graphical primitives and the logical relations among them should be known if we say the drawings have been understood. For example, It should be clear which lines are representing for contours, what relations (connecting to, containing, or of other kinds) they are among these contour lines, and which contour lines are making up parts of the object. After these information has been obtained, dimension attributes can be attached to them to get their shapes, even 3-D shapes from only one projection. This is the lowest level of understanding of engineering drawings, as we understand. It is still a 2-D processing. We just call it 2-D understanding. Although there are steps before 3-D reconstruction, It is enough for 2-D dimension-driven redesign and 2-D CAM. We should fight for this goal first.

4.2 How to Achieve It?

In the current situation, the performance of vectorization and primitive recognition is still perplexing us. The difficulty of improving it is still very high because the vectorial parts of primitives are always inexactly extracted out and usually intervened with others. Although interactive ways can be optional, automatic ways are still preferred. Therefore researchers should endeavor to it. Moreover researchers should expect that the problem of 2-D engineering drawing understanding be solved on the basis of good performance of vectorization and primitive recognition.

In Figure 2, we propose a route of performing 2-D engineering drawing understanding, as well as related works. On this route, the text/graphics separation module is always at the beginning point. The new idea suggested hear is adding a module of primitive recognition after vectorization as well as imposing iteration into primitive recognition. The task of vectorization module is reduced as just extracting the short bars(straight line segments) which will precisely simulate the original drawings after thinning and tracking. The primitive recognition module is expected to extracted the meaningful primitives such as circular arcs, contour lines, hatching lines, central lines, dash lines, arrow heads, etc. Iteration is also suggested to be embedded into character recognition while the text patterns are being recognized. Recognized dimensioned primitives are also helpful for the detection of errors and unrecognized primitives(both graphic primitives and text primitives). 2-D understanding then can be performed on the basis of the

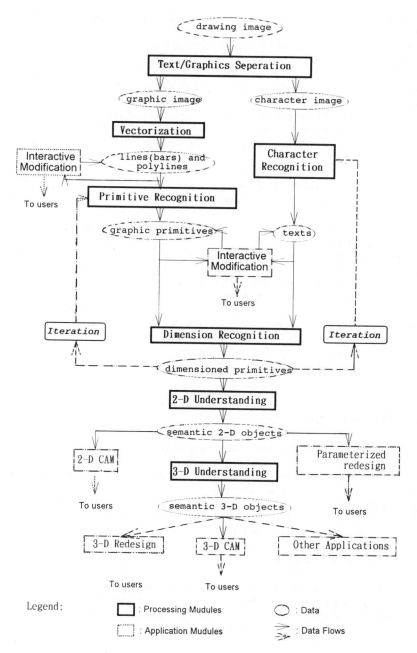

Figure 2 Engineering drawing processing procedures and their applications

well recognized primitives. In the figure, we also propose some practical applications based on different stages of achievements that will satisfy the users at different levels.

As a low level procedure, vectorization should not be expected too much. Fragmentary bars are enough results from it. And this procedure is enough for practical use for the processing of those drawings on which there are only free curves, such as contour maps. Primitive recognition should be performed as its post-processing procedure on the basis of its results if there are high level vectorial primitives in the drawings. In the primitive recognition procedure, domain knowledge should be employed to guide the recognition of primitives. For example, dimension set patterns in the engineering mechanical drawings are always different from those ones in engineering architectural drawings, therefore the recognition processes of them should be different. The drawings of contour lines, hatching lines, central lines, dash-dot lines, and arrow heads are under more strict rules in engineering mechanical drawings than in other drawings. Even the engineering mechanical drawings have some different types. The assembling drawings consist of more complex primitives than other kinds of mechanical drawings.

In the primitive recognition, the iteration of recognition is as important as the knowledge-based method, because the rules must be applied to the data which are generated from the recognition process and its pre-process. And In the iteration, the non-monotonous reasoning method can also be applied to it to get the most possible results when there are so few evidential data that the recognition process can not continue.

As a higher level procedure of drawing processing, 2-D understanding can be carried out on the basis of the recognized primitives. To perform this procedure, It is necessary to make some basic definition of syntactic and semantic rules that guide the description of 2-D objects/blocks. But unfortunately, only can the formalized description of the grammar of dimensions in mechanical drawings be found in the literary. The understanding process of 2-D engineering drawings is the process that detects the rules of 2-D object description and obtains the semantics in that description.

5 MAIN PRESENT PROBLEMS

As mentioned above, The main problems perplexing us currently are still the methods of primitive recognition and 2-D understanding. As long as they are not solved well, the efficient and effective reuse of paper drawings and automatic 3-D understanding can not be carried out and other following procedures can not be realized.

In primitive recognition, the syntactic and semantic descriptions of primitive constructions should be formalized as that of the dimension sets have been done. Perhaps the grammars of these primitives are not so complex than that of the dimension sets. But it is not so easy to implement them on the real data in the practical systems. On the contrast, there is no syntactic or semantic descriptions of 2-D objects/blocks, even less the realizing method to serve the understanding of 2-D objects/blocks.

Because of its high complexity, we suggest that the knowledge-based method rather than the traditional method of pattern recognition should be used to achieve the goal. For completing the automatic primitive recognition and 2-D drawing understanding, here we list some primary problems that should be solved first.

- knowledge representations of primitive descriptions;
- iterative reasoning(non-monotonous reasoning in particular) on them;
- acquisition and representation of syntactic and semantic descriptions of 2-D objects/blocks.

Moreover, it is not too early to list the problems of 2-D CAM and 2-D parameterized design on the researchers' agendas.

6 CONCLUSION

In the current conditions of research works, the reuse of engineering drawing at vector level is still the most preferable choice. Therefore improving the performance of vectorization and primitive recognition is still our main objectives. But it is very utilitarian to make use of current capability in applications. For examples, as shown in Figure 2, after the procedures of vectorization, lines and polygonal lines are enough for some users, while other users are satisfied with (dimensioned) primitives which result from the procedure of primitive recognition or dimension recognition.

But for higher level of applications, 2-D engineering drawing understanding is still taking a critical role. It is at once the basis of 3-D understanding and the sole way to 2-D CAM and 2-D parameterized redesign, therefore deserves more attentions.

References

Dori, D. and Kombre, T. (1993) Paper Drawings to 3-D CAD: A Proposed Agenda, Proc. of 2nd ICDAR, Japan, 866-869.

Dori, D and Pnueli A. (1988) The Grammar of Dimensions in Machine Drawings, CVGIP, **42**, 1-18.

Dori, D (1989) A Syntactic/Geometric Approach to Recognition of Dimensions in Engineering Machine Drawings, CVGIP, **47**, 271-291.

Haralick, M. and Queeney, D. (1982) Understanding Engineering Drawings, CVGIP, **20**, 244-258.

Li Xinyou, *et al* (1993) IMCAD--A Drawing Reading and Interactive Design System Based on Binary Image, Proc. of 3rd International Conference on CAD and Computer Graphics, Beijing, 828-832 .

Liu Wenyin, Tang Long and Tang Zesheng (1993) Intelligent Processing of Dimension Texts in Engineering Drawings, Proc. of "863 High-Tech Project" Conference on Intelligent Interface & Application, Mudanjiang, China, 162-169.

Liu Wenyin (1993) Knowledge-Oriented Recognition of Characters in Engineering Drawings, Proc. of 3rd International Conference on CAD and Computer Graphics, Beijing 828-832.

Min Weidong, Tang Zesheng and Tang Long (1993) Using Web Grammar to Recognize Dimensions in Engineering Drawings, Pattern Recognition, **26-9**, 1407-1416.

Vaxiviere, P. and K. Tombre (1992) Celesstin: CAD Conversion of Mechanical Drawings, **25-7**, 46-54.

Biographies

Liu Wenyin is a lecturer in the Department of Computer Science and Technology, Tsinghua University, Beijing, People's Republic of China. He obtained both his BE degree in 1988 and his ME degree in 1992 from Tsinghua University. His current interests include engineering drawing understanding and artificial intelligence.

Tang Long is an associate professor in the Department of Computer Science and Technology, Tsinghua University, Beijing, People's Republic of China. After he graduated from the Department of Automatic Control at Tsinghua University in 1963, he has been working in education and research on computer science and application for the past 31 years.

Tang Zesheng is a professor of Computer Science and Technology at Tsinghua University, Beijing, People's Republic of China. He graduated from the Department of Electrical Engineering at Tsinghua University in 1953. His current research interests include geometric modeling, volume visualization and computational geometry. Tang Zesheng is the vice chairman of the CAD and Computer Graphics Society of China Computer Federation and the Computer Engineering and Application Society of China Electronic Institute. He was co-chairman of IFIP 1991 Conference on Modeling in Computer Graphics and program chairman of the 3rd International Conference on CAD and Computer Graphics held in Beijing in 1993.

18

Reverse Engineering and Its Application in Rapid Prototyping and Computer Integrated Manufacturing

S.G. Zhang MSc & A. Ajmal BSc MSc PhD CEng MBCS MIIE
School of Engineering, Systems & Design, South Bank University
103 Borough Road, London SE1 0AA, UK Tel: 071-8157632,
Fax: 071-8157699, zhangsa@vax.sbu.ac.uk, ajmala@vax.sbu.ac.uk

S.Z. Yang ACAS
Department of Mechanical Engineering, Huazhong University of
Science and Technology, Wuhan, Hubei, P.R.China

Abstract

Reverse engineering is playing an important role in generating and modifying CAD model, especially for the creation of 3D CAD model of sculptured surface. This paper made a survey of reverse engineering and its applications in rapid prototyping and computer integrated manufacturing. It divided reverse engineering into three main parts: data acquisition, surface reconstruction and data transfer. The recent research results, their merits and restrictions about surface digitizing, reconstruction of surface and data transfer technology are studied. A primary conclusion about reverse engineering and its applications in rapid prototyping and computer integrated manufacturing is drawn up.

Keywords

reverse engineering, rapid prototyping, computer integrated manufacturing, data acquisition, surface digitizing, surface reconstruction, data transfer

1 INTRODUCTION

The basic concept of producing a part based on an original or physical model without the use of engineering drawing is called "Reverse Engineering" (Abella,1994). Reverse engineering has changed from a skilled manual process to an engineering tool using sophisticated computer software and modern measuring instrument. It has rapidly expended its origin functions in maintenance into the areas of design and production. Now it is defined as "the creation of a CAD model from an existing part or a prototype"(Motavalli, 1994). Now the reverse engineering is playing a the role of constructing and modifying CAD model in rapid prototyping (Hosni 1993). In addition it is also an important tool in Computer Integrated Manufacturing (CIM) for rapid part modelling.

In the rapid prototyping environment, 3D CAD model can be utilised to drive sintering, layering, depositing or sculpturing equipment to create a usable prototype within hours much as the ways a computer printer uses a word processor file to fashion a printed page. As the input data of rapid prototyping the CAD files should be constructed with faster speed and higher accuracy than that it does in other applications. The rapid prototyping process is by definition capable of building volume, hence 3D CAD modelling or surface modelling is required. Alternatively, the CAD modelling (for example 2D wire frame modelling) must be transferred into 3D CAD and then this will not benefit from the advantage of time saving of rapid prototyping. Therefore, two demands must be met by CAD system: (1) 3D solid modelling or surface modelling, (2) Create CAD model with faster speed and higher accuracy. Fortunately reverse engineering is exactly a faster surface modelling technique if a manufactured part exists. It greatly simplifies the process of creating a CAD model and meets the demands mentioned above.

In CIM environment, it generally starts from a CAD model. Then CAD files are interpreted as NC machining codes according to which the part is finally manufactured. In addition, the robot movement programs and coordinate measuring machine automatic inspection program can be developed with the CAD model. However, in many cases, the CAD model of a part is not always available to begin the CIM system. Especially when no existing CAD file is usable and a manufactured part has already existed, conversion part data to CAD system is necessary. Reverse engineering involves three main phases: data acquisition, surface reconstruction, and data transfer.

Recent developments in rapid prototyping and CIM have increased part complexity. It is necessary to provide a fast tool for complex part or free-form surface modelling. Therefore reverse engineering has been becoming an important tool and has been utilising extensively in manufacturing context in rapid prototyping and CIM. For example, in rapid prototyping environment, creating model is used for evaluating form, fit, ergonomic study and function test. Generally it is modified such that satisfying some design or functional demands. After modifying it is necessary to digitize the model and to update the CAD database. In automobile industry, the development of new car model takes shorter time from concept to model with reverse engineering than before that development model with CAD surface geometric modelling or tedious manual digitization process. In the die and mould industry, the existing design is often modified on the shop floor due to manufacturing limitations. The product geometric change with such modifications that are usually not reflected in the CAD model. If the dies or moulds are failure for some reason then the experimental results are lost and the test works must be executed again. With the help of reverse engineering the parts, moulds or dies which have been modified can be digitized and the CAD database can be updated. In repair service, when a tooling, die or mould is failed partially because of incorrect operating or collision it can be repaired faster and more efficient with reverse engineering than with other methods.

2   DATA ACQUISITION

Digitizer include two categories: non-contact scanner and touch probe. Non-contact scanners use laser, optics and CCD camera, etc. They can produce huge amounts of point data in a very short time and without physically touching with the object, therefore, no destructive to instrument and part. In addition, they also can digitize an unknown surface automatically. Non-contact 3D surface digitizing technology has been widely researched in rent years for available commercial object. Laser range-finding probe (Che 1992), HYSCAN 3D laser digitizer (Steger 1994), 4D-laser scanner (Ioannides 1994), multi-laser displacement sensors (Smith 1993), laser triangulation probes (Stevenson 1993), vision-based coordinate measurement system (Elhakim 1993), machine vision system (Huang 1994), CCD camera (Bao 1994), etc. are researched.

Non-contact scanner even though there exist obvious advantage over the touch probe in some respects such as high speed and no destructive, there exist some serious drawbacks: for example, due to different work principles the calibration is very difficult, accuracy is low and even lower when wider measurement range is demanded. Until today it is difficult to obtain a commercial usable digitizer within 10 microns. Therefore, it generally is not associated with precision manufacturing applications. In addition non-contact digitizer often has difficulty of reaching concavities and undercuts that extends beyond the probing angle. It is also sensitive to shiny and dark surface. If the surface is shiny reflection it can cause a shift in the location of the point. If the surface is dark the sensor cannot adequately see the surface. An other disadvantage is that most CAD/CAM system is difficult to process enormous amount of point data up to millions. Therefore a data thinning procedure must be carried out before transferring them into CAD/CAM system.

The touch probe can be used in coordinate measuring machines (CMM) to digitize 3D part surface. As contrast with non-contact scanner it has high accuracy but low speed. It usually has high accuracy within microns even sub-micron range. In addition even though part is more complicated a well-developed probe series can be conveniently elected for the different parts and the measurement objects. Manual measurement is tedious and an experienced inspection operator is needed. For complex part inspection accuracy even depends heavily on the experiments of inspection operator. In automatic measurement a detail path measuring planning must be designed in advance such that any collision can not occur during measuring. Therefore, developing a collision-free automatic measurement path planner is necessary. Of late years more studies are concentrated in resolving the questions. Automated dimensional inspection intelligent planning environment (Menq 1992 Yau 1992), automatic inspection planning fragment (Merat 1992), softgauge (Smolky 1992), interactive and integrated inspection system (Pahk 1993), Object-Oriented inspection planner (Chan 1993), integrated CMM within design and manufacturing environment (Medland 1993), optimal CMM position controller (Katebi 1993), etc. are studied. Even though so much more researches are carried out in CMM as mentioned above, it is difficult to meet the all demands of reverse engineering when no existing CAD files can be used to direct automatic inspection planning.

For unknown parts, automatic digitizing unknown free-form curve by CMM (Chen 1992), manually generating initial model then refining the surface model (Hsieh 1993), manually "outline" the part geometry shape of surface, then, high dense measurement (Yau 1993), manual coordinate measuring machine (Abella 1994), etc. methods are researched.

The touch probe has high accuracy that satisfied the digitizing demands of precision manufacturing part but low measurement speed and complicated programming efforts limit its extended applications. Especially under unknown sculptured surface automatic digitizing becomes complexity and difficulty. The questions of speed and automation remain unsolved.

3 REPRESENTATION OF SURFACE

Computer representations of 3D object shape fall into two categories: volume representations and boundary representations. Volume representation can be obtained by Boolean operations such as union, intersection, difference from its solid primitive, like cylinders, cubes, spheres and parallelepipeds, etc. In rapid prototyping, manufacturing process is to be defined by capability of building volume. Therefore, if volume representation is elected and the model is built on primitive solids, the model itself is a true volume. This makes it easier to create the facets and post process is generally not necessary. However, volume representation is not effective in constructing sculptured surface. Boundary representation defines a solid by its bounding surface. Describing curved surfaces requires a large number of polygons, hence parametric surface, such as splines, are often employed. In reverse engineering we will be concerned only with surface representation since measurements are taken at discrete points on the object surface. For reconstruction, quadric can be used to represent the shapes of a wide scope using only few parameters (Bradley 1993). Investigation shown the part surface of 85 per cent of manufactured part can be well approximated by sub-set of quadric or superquadric surface (Hakala 1981). For representation of sculptured surface representation, Bezier surface, B-spline surface and Non Uniform Rational B-Splines (NURBS) often are applied because of their intuitive and local control properties than other parametric surfaces.

The B-spline surface patch has capabilities which in the other systems would require the use of either a composite surface or of a single patch based on polynomials of degree higher than cubic. This advantage is reinforced by the local modification property and the automatic second-order continuity of the B-spline surface. NURBS is a more general case of nonrational B-spline surface. The advantages to use the rational forms are that it is possible to modify a surface by a given set of defining data by adjust the weights and it is also possible to adjust the weights so that the geometry of the surface is unaffected but its parameter is changed. NURBS is a superset of many other types of parametric surfaces such as Bezier, B-spline (Joe 1994), and it also can approximate the COONS surface (Lin 1994). NURBS has become standard tools for the representation and design of geometry in CAD/CAM and geometry community. It has great local control capability and is intuitive to designer modification. It can represent a complex geometry in just one single surface as opposed to many surface patches (Boulanger 1994). This helps reduce the problems of maintaining surface position and tangent continuity. In order to further modify the surfaces for creating a complex model, NURBS has better local control properties than other parametric surfaces.

4 RECONSTRUCTION OF SURFACE

Reconstruction of surface is to fit the selected surface to point data to be measured. An enormous amount of literature about surface fitting can be found (Bolle 1991, Rogers 1989). The choice of surface-fitting technique depends on data source, surface representation and applications. For non-contact scanner, large amounts of point data are collected in order to ensure that the accuracy and resolution of fitting surface are acceptable. A large amount of point data up to millions is very difficult to fit. It must be eliminated to a limited quantity. Keeping in mind, for each of representations a minimum number of point data is desired to uniquely define a surface of nonundulation. When non-contact scanner is used, smoothing of point data must be used in order to eliminate the unwanted point data and to filter the point data fluctuations (Motavalli 1994). For composite surface, the object surface is divided into several regions (Sarkar 1991). Each of these regions is devoid of sharp change in shape and consists of a single feature. Surface-fitting is implemented only on the given region. Therefore, the effect of boundary detection error on accuracy is very large. Especially when the boundary has not discontinuity, it is much difficult to detect. Generally it is recomputed with the intersection of individual surfaces after surface-fitting is carried out. When a single surface is fitted into the point data, this reduces the problems of the surface position and tangent continuity. The amounts of point data are very large and the calculation is very complicated since matrix order is high when fitting with least square method. In addition interpolation method can be used, but surface undulation is unavoidable.

For different applications surface fitting methods could be different (Bolle 1993). In the most of cases least square method is suitable. In addition non-linear least square, maximum-likelihood estimates and variable principle can be used. The surface representations can be classified into explicit and implicit functions. For explicit function, variable principle can be used. The results are invariance to rigid transformation, robust fit and no effect of occlusion on the fit. For implicit function, least square, maximum likelihood estimate can be used. Least square method is invariance to rigid transformation, but estimates are not robust and very sensitive to occlusion. Maximum likelihood estimates are asymptotically unbiased, but estimates are not unique, not robust and not invariance to rigid transformation. The detail fitting formulations of quadric surface, B-spline surface and NURBS surface can found in related references (Chivate 1994, Rogers 1989, Laurent-Gengoux 1993).

5 DATA TRANSFER

Reverse engineering results can be transferred into CAD system by different file formats like IGES, STEP (PDES), VDA-FS, DXF, etc. IGES is the most convenient one of data exchange formats for reverse engineering. It is a neutral file format. IGES only defines each entity to an entity class and a sequence number. Since the entity types vary quite a lot from one CAD system to another, it used to be problems with file transfer. Especially drafting entities like arrow heads and diameter symbols could easily disappear. VDA-FS resolve this problem differently. In VDA-FS file format, all curves (including lines) are converted to B-spline curves and all surfaces are converted to B-spline surfaces. This makes it simple to use. A further development will lead to the STEP (PDES) standard, which is assumed to become the ISO standard for graphical file transfer (Eastman 1994). DXF file format is used by AutoCAD. For special applications, G-code can be directly produced by reverse engineering. It is also possible to produce directly STL file or SLI file. Then the interactive format software between CAD and SL is not necessary and drawbacks of STL format disappear.

6 CONCLUSION

As an efficient tool of generating and modifying CAD model, reverse engineering is playing an important role in rapid prototyping and CIM. Reverse engineering include three mainly phases: data acquisition, reconstruction of surface and data transfer.

For data acquisition, non-contact scanners and touch probes can be used. Non-contact scanner uses laser, optics and CCD camera. It has high speed and no destructive to part and instrument. It is especially suitable to digitize the unknown sculptured surface. However low accuracy limited its applications in precision manufacturing part. It is also sensitive to shiny and dark surface and often has difficulty to reaching concavities and undercuts that extends beyond the probing angle. Touch probe has high accuracy that satisfied the digitizing demands of precision manufacturing part but low measurement speed and complicated programming efforts limit its extended applications. Especially under unknown sculptured surface, automatic digitizing becomes complexity and difficulty.

In representation of surface, volume representation and boundary representation can be used. Volume representation is especially suitable to rapid prototyping and post process is generally not necessary. However, it is not effective to sculptured surface. Boundary representation represents a solid by its bounding surface. Thus it is especially suitable to sculptured surface. For simple part, quadric surface can be used to describe the shapes of a wide scope using only a few parameters. For sculptured surface Bezier patch, B-spline patch and NURBS surface can be used because of their intuitive and local control properties.

In reconstruction of surface, local fitting and global fitting can be used. For local fitting object surface is divided into several regions. Surface-fitting is implemented only on the given region. In this case, surface-fitting is simple but boundary detection is difficulty. For global fitting, object surface is fitted into a simple surface. This reduces the problems of maintaining surface position and tangent continuity but the amounts of point data are very large and the calculation is complicated. With respect to fitting, variable principle, least square method and maximum likelihood estimate can be used. Variable principle is invariance to rigid transformation, robust fit and no effect of occlusion on the fit. Least square method is invariance to rigid transformation, but fitting is not robust and it is very sensitive to occlusion. Maximum likelihood estimates are asymptotically unbiased but estimates are not unique, not robust and not invariance to rigid transformation.

In data transfer, IGES neutral file format is used extensively in data exchange between CAD systems. However it is also problems with file transfer. In addition STEP (PDES), VDA-FS, DFX file format can be used.

7   REFERENCE

Abella, R.J. Daschbach, J.M. and McNichols, R.J. (1994) Reverse engineering industrial applications. Computers & Industry Engineering, **26**, 2, 381-5.

Bao, H.P. Soundar,P. and Yang, T. (1994) Integrated approach to design and manufacture of shoe lasts for orthopaedic use. Computers & Industry Engineering, **26**, 2, 411-21.

Bolle, R.M. and Vemuri, B.C. (1991) On three-dimensional surface reconstruction methods. IEEE transactions on pattern analysis and machine intelligence, **13**, 1, 11-3.

Boulanger, P. Roth, G. and Godin, G. (1994) Applications of 3D active vision to rapid product development. Proceedings of International Conference on Rapid Product Development, IMS Intelligent Manufacturing Systems, 148-156.

Chan, K. And Gu, P. (1993) Object-oriented knowledge-based inspection planner. Proc IEEE Pac Rim Conf Commun Comput Signal Process, 646-9.

Che, C. (1992) Scanning compound surface with no existing CAD model by using laser probe of a coordinate measuring machine. Proceedings of SPIE - The International Society for Optical Engineering, **1779**, 56-67.

Chen, Y.D. Tang, X.J. Ni, J. and Wu, X.M. (1992) Automatic digitization of free-form curve by coordinate measuring machine. American Society of Mechanical Engineers, Production Engineering Division (Publication) PED: Engineering Surfaces, **62**, 113-25.

Chivate, P.N. and Jablokow, A.G. (1994) Solid-model generation from measured point data. Computer-Aided Design, **26**, 2, 587-600.

Eastman, C. (1994) Out of STEP, Computer-Aided Design, **26**, 5, 338-40.

Elhakim, S.F. (1993) Application and performance evaluation of a vision-based automated measurement system. Proceedings of SPIE - The International Society for Optical Engineering, **1820**, 181-195.

Hakala, D.G. Hillyard, R.C. (1981) Malraison,P.F and Nource,B.F Natural quadrics in mechanical design. SIGGRAPH/81, Seminar on Solid modelling, Dallas, Texas.

Hosni, Y. Ferreira, L. and Burjanroppa, R. (1993) Rapid prototyping through reverse engineering. Proceedings of the Industrial Engineering Research Conference, 420-424.

Hsieh, Y.C. Drake, S.H. and Riesenfeld, R.F. (1993) Reconstruction of sculptured surfaces using coordinate measuring machines. American Society of Mechanical Engineers, Design Engineering Division (Publication) DE: Advances in Design Automation, **65_2**, 35-46.

Huang, C. and Motavalli, S. (1994) Reverse engineering of planar parts using machine vision. Computers & Industry Engineering, **26**, 2, 369-79.

Ioannides, M. and Wehr, A. (1994) Reverse engineering, rapid prototyping: generation of our surface information by using a 4D-laser scanner for digitization. Proceedings of International Conference on Rapid Product Development, IMS Intelligent Manufacturing Systems, 388-401.

Joe, B. Wang, W. and Cheng F. (1994) Reduced-knot NURBS representations of rational G1 composite Bezier curves. Computer-Aided Design, **26**, 5, 393-9.

Katebi, M.R. Lee, T. and Grimble, M.J. (1993) Optimal control design for fast coordinate measuring machine. Control Engineering Practice, **1**, 5, 797-806.

Laurent-Gengoux, P. and Mekhilef, M. (1993) Optimization of a NURBS representation. Computer-Aided Design, **25**, 11, 699-710.

Lin, F. and Hewitt, W.T. (1994) Expressing coons-gordon surfaces as NURBS. Computer-Aided Design, **26**, 2, 145-155.

Massen, R. (1994) Bridging the gap between optical 3D digitizers and CAD/CAD. Proceedings of International Conference on Rapid Product Development, IMS Intelligent Manufacturing Systems, 134-145.

Medland, A.J. Mullineux, G. Butler, C. and Jones, B.E. (1993) The integration of coordinate measuring machines within a design and manufacturing environment. Proceedings of the Institution of Mechanical Engineers, Part B: Journal of Engineering manufacture, **207**, NB3, 91-8.

Menq, C.H. Yau, H.T. and Lai, G.Y. (1992) Automated precision measurement of surface profile in CAD-directed inspection. IEEE Transactions on Robotics and Automation, **8**, 2, 268-78.

Merat, F.L. and Radack, G.M. (1992) Automatic inspection within a feature-based CAD system. Robotics and Computer-Integrated Manufacturing, **9**, 1, 61-9.

Motavalli, S. and Bidanda, B. (1994) Modular software development for digitizing systems data analysis in reverse engineering applications: case of concentric rotational parts. Computers & Industry Engineering, **26**, 2, 395-410.

Pahk, H.J. Kim, Y.H. Hong, Y.S. and Kim, S.G. (1993) Development of computer-aided inspection system with CMM for integrated mould manufacturing. CIRP Annals, **42**, 1,

557-560.

Rogers, D.F. and Fog, N.G. (1989) Constrained B-spline curve and surface fitting. Computer-Aided Design, **21**, 10, 641-8.

Sarkar, B. and Menq, C.H. (1991) Smooth-surface approximation and reverse engineering. Computer Aided Design, **23**, 9, 623-8.

Smith, K.B. and Zheng, Y.F. (1993) Multi-laser displacement sensor used in accurate digitizing technique. American Society of Mechanical Engineers, Production Engineering Division (Publication) PED: Manufacturing Science and Engineering, **64**, 23-31.

Smolky, R. and Vrana, J.J. (1992) CMM: in process inspection analysis on a CAD system. SAE Special Publications: Simultaneous Engineering in Automotive Development, **935**, 57-68.

Steger, W. Haller, T. and Roth-Koch, S. (1994) Efficiently digitizing of free-formed moulds with 3D laser-scanning. Proceedings of International Conference on Rapid Product Development, IMS Intelligent Manufacturing Systems, 362-74.

Stevensen, W.H. (1993) Use of laser triangulation probes in coordinate measuring machines for part tolerance inspection and reverse engineering. Proceedings of SPIE - The International Society for Optical Engineering, **1821**, 406-14.

Yau, H.T. and Menq, C.H. (1992) An automated dimensional inspection environment for manufactured parts using coordinate measuring machines. International Journal of Production Research, **30**, 7, 1517-36.

8   BIOGRAPHY

Suogui Zhang is an associate professor of mechatronics engineering in Taiyuan University of Technology, P.R.China. He received a MSc in Machinery Dynamics from Taiyuan University of Technology and now a doctoral candidate working in Integrated Manufacturing System in South Bank University, U.K. Current research interests include CNC, CMM, CAD/CAM, CIM, reverse engineering.

Dr. A. Ajmal obtained his PhD from University of Manchester (UMIST) in Computer-Aided Manufacturing and a MSc in Advanced Manufacturing Systems from UMIST. After completing his BSc (Hons) in Mechanical Engineering, he joined the Consulting Engineering Company. He worked in the University of Manchester & the University of Edinburgh as a lecturer and supervising research projects on GKS & CAM. Currently working as a senior lecturer in Computer-Integrated Manufacturing at the South Bank University. His research interest includes, development of integrated manufacturing systems, manufacturing planning and expert systems.

Shuzi Yang is a professor of mechanical engineering, president in Huazhong University of Science and Technology, academician of the China Academy of Sciences. Current research interests include automatic control system, mechanical default diagnostic, artificial intelligent, intelligence manufacturing system.

PART FIVE

Application of CAD Technology

19

The design and realizing technology of railway line urgent repair expert system

M. S. Liu   Y. H. Ma
Computer center, Shijiazhuang Railway Institute,
Shijiazhuang, P. R. China, 050043

Abstract

The building of RLURES (Railway Line Urgent Repair Expert System) is a beneficial experiment of artificial intelligence theory, expert system technology and CAD technology in the field of railway urgent repair. The paper has introduced the system's structure, design, realization technology, etc.. In the end, it has provided the examples of application. The design idea and its realization technology of the system have certain referential value in the application of expert system in the field of engineering urgent repair.

Keywords

Railway line urgent repair, urgent repair design, expert system, computer aided design

1 INTRODUCTION

Railway line urgent repair is an extremely complicated engineering urgent repair, which involves the making up and designing of the urgent repair plan, the drawing up of need's plan of urgent repair appliance and the construction progress arrangement, etc. under various damaging conditions. Traditional practice is to select and send some experienced engineers and technicians to make an on-the-spot investigation and to provide the urgent repair plan. This practice, on the one hand, need a longer time to be made up. On the other hand,it can't provide the necessary urgent repair design and the construction organization design, thus prolongs the whole period and lowers the working productivity. In order to make up the above weakness, we have developed the railway line urgent repair expert system, which is one with the feature of diagnosing and designing. The system can make up quickly the urgent repair plan according to the railway line damaging, the on-the-spot manpower and appliance available. When the urgent repair plan is confirmed, the system will make further design to urgent repair plan, and give plan design graphics and related construction progress map, and some tables and charts of measurement, etc.. The article first introduces the systematic structure of the system, the design and realization technology of the systems, and so on. In the end some practical examples of the system is provided.

2 RLURES'S SYSTEMATIC STRUCTURE

RLURES is an expert system that helps diagnosing and designing. At first, the system should have the function of diagnosis, that is the ability to make up quickly the urgent repair plan according to various damage conditions and on-the-spot manpower and appliance available. Secondly, the system should have design function, that is the function to make a engineering design, produce urgent repair design graphics, construction progress map and some tables and charts of measurement, etc., according to given urgent repair plan. Figure 1 shows our RLURES's systematic structure. The function of each model is explained as follows.

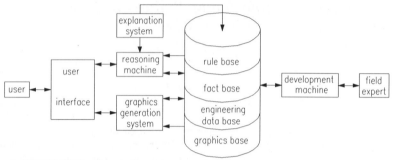

Figure 1 RLURES'S Systematic Structure.

2.1 Development machine

Development machine is used to gain the urgent repair knowledge by interacting with field experts. The knowledge is formatted and is stored in the rule base. Meanwhile the knowledge base can be maintained by development machine.

2.2 Knowledge base

The systematic knowledge base consists of four parts: rule base, fact base, engineering data base and graphics base. The rule base stores the field knowledge and graphics effectively by means of rule-frame description method. Fact base stores the source facts. While the data used in designing the urgent repair plan and the designing results are stored in the engineering data base. The graphics base is the collection of legends, which provided basic source material for the generation of design graphics of urgent repair plan.

2.3 Reasoning machine

According to the knowledge in the knowledge base, the damaging condition of railway line and the on-the-spot manpower and appliance available, the reasoning machine make out the urgent repair plan, prepare for the use of graphics generation system.

2.4 Graphics generation system

Under the control of graphics reasoning machine, the graphics generation system makes use of the graphics knowledge (graphics rules—generation rules and graphics putting together rules) in the knowledge base, the graphics block in the graphics base, and the property structure in the data base to form the design graphics for the urgent repair plan automatically. See Figure 2 for its structure.

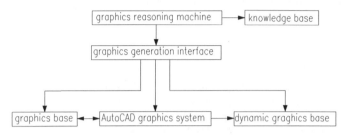

Figure 2 Sketch Map of the Graphics Generation System.

At first, the system gain the urgent repair plan according to reasoning, make up engineering design for it, and store the results of design into engineering data base. And use graphics reasoning machine to call the graphics rules in the knowledge base and graphics legends to produce graphics generation plan; then stores it in certain structural form into the dynamic graphics base. Finally, by means of graphics generation interface, system can call Auto CAD separately to form design graphics for urgent repair; call the construction progress figure generation model to form the construction progress figure; call the material list generation model to form major appliance consuming lists, etc..

2.5 Explanation system

The explanation system is mainly used to explain the urgent repair plan gained from reasoning. The users may ask such questions as: How is the urgent repair plan gained?, Why is such kind of urgent repair plan adopted?, etc.. You can get satisfied reply by using explanation system.

3 THE DESIGN AND REALIZATION OF RLURES

3.1 The designing ides of RLURES

RLURES acts as a knowledge base system, we'll describe it major designing idea in the following two ways.

Knowledge expression
In the railway line urgent repair expert system, rule-frame system is adopted to describe the knowledge involved in the system. The production rule is mainly used to describe the method in which the field experts to solve the problems. In the graphics system the production rule may also be used to describe the graphics generation rule(Liu Ma, 1994). In RLURES, the form of production rule has been dealt with in detail in(Liu, 1994), here it is omitted.

Much knowledge should be described in frame in the problem of engineering design(Li, 1991). Such as the attribute of graphics legends in graphics generation system; the structural plane in urgent repair, etc.. In RLURES, the frame is described as array, the name of array is corresponding to the name of frame, the elements of array are corresponding to each slot of frame. The type of element value can be described in *EVALUE* structure:

```
    typedef struct{
            char val_type;        /* the type of value */
            union es_val val;     /* store value field */
            }EVALUE;
```

where:

val_type take 0~6 to show the type of value;

val store value field, it's defined by union *es_val*;

union *es_val*{

```
            es_pos estring;        /* the value of series of character */
            es_byte elogical;      /* the value of logical type */
            short eint;            /* the value of short integers */
            long elong;            /* the value of long integers */
            es_real ereal;         /* the value of double precision */
            es_pos edim;           /* the value of array */
            es_pos ecomplex;       /* the value of complex */
            }
```

Because the series of character and arrays are under irregular scope, the occupation of their length in inner storage is uncertain, the data of uncertain length can't be described directly in the union. So, we must use the structure of certain length to express the data of uncertain length. *es_pos* is such a kind of data structure, and its definition as following:

```
    typedef struct{
            long epos;            /* record the position of data block in store medium */
            short elen;           /* record the size of block (the number of bytes) */
            }es_pos;
```

in order to save inner storage space, we store the data of uncertain length expressed by *es_pos* into outer storage. You may call it from outer storage into inner storage when needed. Thus, it can't restrict the size of series of character and arrays.

es_byte is a type of unsigned character:

typedef unsigned char *es_byte*;

es_real is a type of double precision:

typedef double *es_real*;

The element value may be one of the seven kinds of parameter type stipulated above, i.e. may be another frame. The whole of the frame is regarded as a value of parameter. It has the same treatment as the other kinds of parameter. The advantage of this treatment is to combine the normal frame theory with the rule theory naturally. Thus, it makes the frame consistent completely in form with the classic rule theory. The definition of slot is as follows:

```
    typedef struct{
            char *slot_name;      /* the name of slot */
            EVALUE slot_val;      /* the value of slot */
            }ESLOT;
```

If the slot is regarded as having order, its position can't be changed at will. The position of each slot in array is sole under such conditions. So, the name of slot can be omitted and the slot is corresponding to the element in array. Therefore, when the type of parameter is a structure, the parameter becomes a frame, the name of the parameters is the name of frame.

Reasoning machine

It is difficult for the system to use the only reasoning way to design the urgent repair plan, considering inconsistency of the urgent repair plan under the same urgent repair policy. In this case, the system provides different reasoning method: forward reasoning; backward reasoning; forward-backward hybrid reasoning and fuzzy reasoning, etc.. The use of different reasoning machines is under the control of meta knowledge.

In graphics generation system, there is a graphics reasoning machine used specially in the generation graphics. Graphics reasoning machine is a subsidiary reasoning machine in reasoning system. It adopts the backward reasoning method directly to the object(Guan, 1988). The algorithm is the same as the normal reasoning algorithm to the object. The way of joint match, in addition to rule match, use frame match in a large scale. In the graphics rule, there may appear graphics on the left side and right side of the rule. These graphics are described as attribute structure in knowledge base, i.e. frame. So, the graphics reasoning machine matches with the graphics rules is the way of the frame match.

Suppose graphics rules $R=\alpha \rightarrow \beta$, R contents with the grammar in context, on the left, α is corresponding to attribute frame A, on the right, β is corresponding to attribute frame B. So, R can be shown equal In value as: $R=A \rightarrow B$. The fact base written as $\{F\}$. F is a fact frame. If $x \in \{F\}$ exists, makes $A \sim x$, fact frame similarly matches with premise frame, and rule R is matched.

For matched rule R, if the concluding part (on the right) or attribute frame B is true, which becomes a new fact, can be added to fact base. The graphics reasoning makes use of graphics rule to do backward reasoning in graphics generation tree till reach the object (starting sign).

3.2 The realization technology of RLURES

The building method of system knowledge base and reasoning machine has been introduced above, now, the other realization technology is discussed as follows.

The decomposition of urgent repair object

To urgent repair object, the railway line urgent repair involves the urgent repairs about railbed, retaining wall, culvert and the building of middle station. In order to make the urgent repair plan, we decompose the railway line urgent repair to railbed urgent repair, retaining wall urgent repair, culvert urgent repair and middle station urgent repair, and build up corresponding knowledge base, without changing the premise of keeping the system structure and reasoning method of railway line urgent repair expert system. We adopt the multiple expert knowledge technology introduced in(Macher, 1989), separately draw up the urgent repair plan of railbed, retaining wall, culvert and middle station, according to the real damage condition, to solve the designing problem of entire railway line urgent repair plan.

The generation of design graphics in urgent repair plan

As the type of diagnostic designing expert system, it can draw up the urgent repair plan according to damage condition and urgent repair condition, and can generate the designing graphics of urgent repair as well. In railway line urgent repair expert system, the system working out urgent repair plan varies according to different damage condition. The system is required to generate corresponding urgent repair designing graphics. Obviously, the traditional CAD technology is difficult to satisfy these needs. The graphics generation system based on knowledge must be built up.

The graphics generation system in railway line urgent repair expert system builds up graphics legends based on the analysis of different urgent repair plan, to generate sentence pattern through graphics generation rule, and makes use of graphics putting together rules to put the generation sentence pattern together. Finally, it provides the construction graphics. Specifically, the realization technology of graphics generation system mainly include the following:

Step 1. Build up graphics legends

Build up the generation tree of different urgent repair plan, determine the graphics legends, and describe and express it in the inner of a computer.

Step 2. Determine the generation rule

Due to urgent repair plan generation tree is an and/or tree, it's easy to write the generation rule corresponding to urgent repair plan according to generation tree.

Step 3. Standard generation rule

Use production rule and frame to describe the generation rules and store them into knowledge base for graphics reasoning.

Step 4. Form generation sentence pattern

Gain the corresponding generation sentence pattern of urgent repair designing graphics according to the generation rule of urgent repair plan.

Step 5. Put graphics together

Sentence pattern is merely the graphics forming plan, i.e. the map collection of graphics and the order of figure element. Such as, generation sentence pattern:

$$\{t_{22}, b_{41}, d_{32}, t_{22}, l_{21}, t_{22}\}$$

it indicates the over bridge by using a double-line wooden platform, wooden cage base, double 3-line wooden pile pier base and double I-shaped beam. The sentence pattern can't directly combine and produce urgent repair construction map, for the lack of geometric size and building up size of all kinds of structural object for urgent repair, or the lack of the geometric data of graphics putting together and geometric size of graphics legends. So, sentence pattern must be put together.

Putting together means to determine the geometric relations between graphics legends. The realization of putting together is by means of the reasoning machine using the graphics putting together rules.

Putting together graphics includes two points: first, determine the geometric size of graphics legends itself, it's gained by binding transformation the graphics legends in graphics base. That is, transmit the value to the graphics legends which have formal parameters, and makes it an actual graphics. Secondly, determine the position size of graphics legends, i.e. the putting together size of the urgent repair structure.

Step 6. Graphics transformation

In graphics generation, actual graphics are considered to have gained by transformation primitive. In railway line urgent repair expert system, mainly involve the transformation of binding, translation, rotation, mirror image, scaling, etc..

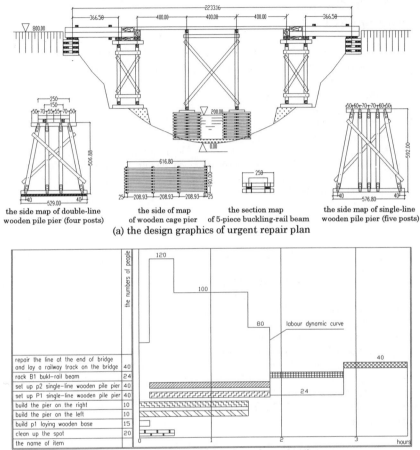

(a) the design graphics of urgent repair plan

(b) the progress map of urgent repair construction

Figure 3 The Urgent Repair Plan Design Graphics of the System Generation.

4 EXAMPLES IN PRACTICES

Railway line xxx was hit by flood at xxx-xxx DK200+198, the railbed was damaged by flood, the height of railbed is 4.5 meters, the length of gap is 24 meters, the depth of gap is 8 meters, the depth of water is 2 meters in the hole. It's required urgent repair within 12 hours.

According to the information input by the user, the system draws up the urgent repair plan through reasoning. That is: buckling-rail beam, wooden frame pier base, wooden cage base cross wooden pier base. After the plan is confirmed by the user, the system immediately generate the plan designing graphics and construction progress graphics, shown as figure 3(the material list omitted).

5 CONCLUSION

In designing the urgent repair plan, the system is tested and evaluated completely under the condition of practical operation and many imitating railway line damages. It includes the right test correct knowledge in the knowledge base; the right way of knowledge expression; system reasoning, the correctness of graphics generation, etc., and the practical test of the explanatory function of the system and the user interface, etc.. Many experts in urgent repair have been invited to test it, and the result is that the designing of the system plan is reliable and correct, the graphics generation is quick and practical.

RLURES is the beneficial experiment of artificial intelligence theory, expert system technology and CAD technology in the field of railway line urgent repair. The designing idea and realization method of the system has certain referential value in the application of expert system in the field of engineering urgent repair.

6 REFERENCES

Guan, J.W. and Liu, D.Y. (1988) Knowledge Engineering Principle. Jilin University Publishing House, Jilin.

Li, G.Q. and Luo, C.J. (1991) The Principle and Program Design Method in Engineering Design Expert System. Meteorological Press, Beijing.

Liu, M.S. (1994) Intelligent Computer Aided Design System ICADS and Its Applications: China Intelligent CAD'94 (ed. Y. H. Pan), 1st China Conference on Intelligent CAD/CAM, Tsinghua University Press, Beijing.

Liu, M.S. and Ma, Y.H. (1994) The Graphics Generation Technology in Railway Line Urgent Repair Expert System: Proceedings of the 4th International Conference on Computer-Aided Drawing, Design And Manufacturing Technology(ed. R. X. Tang), International Academic Publishers, Beijing.

Macher, j. (1989) A Review of Expert System Development Tools, Eng. Computer. Vol.6, Mach 17:13-4.

7 AUTHOR BIOGRAPHY

M. S. Liu, an engineer, is engaged in scientific research for a couple of years, whose research fields include artificial intelligence, export system, computer aided design, computer visualization, etc.. Many achievements in scientific research have been yielded and many papers have issued in above the fields.

Y. H. Ma is an engineer, whose research fields are DBMS, operation system, computer aided design, knowledge engineering, etc..

20

Computer Aided Design and analysis of Unidirectional Fibrous Composites

LOVI RAJ GUPTA

Lecturer, Mechanical Engineering Department,

M.I.T.S., Gwalior [MP], INDIA.

Abstract

A Unidirectional fibrous composite consists of parallel fibers embedded in a matrix. Several unidirectional layers can be stacked in a specified sequence to fabricate a laminate.

In the present work, a computer integrated approach is done for the design and analysis of Unidirectional fibrous composites. A computer software is written to predict the longitudinal strength and stiffness, transverse strength and stiffness, shear modulus, Poissons' ratio, thermal expansion coefficient and moisture expansion coefficient for the chosen volume fraction and material of matrix and fiber. The software incorporates an in built libraries of materials for both fibers and matrix. This gives an ease in selection and also provides vide variety of combinations. Materials can also be specified by the user, mentioning their engineering properties.

The package also deals with the theoretical aspects of Unidirectional fibrous composites in the precursor. In this county the basic definitions and nomenclature is explained through animated demos and graphics screens. This gives an on- line reference of the theory involved in the design of Unidirectional fibrous composites.

Keywords

Unidirectional fibrous composites, fiber, matrix, volume fraction, software, material libraries

1 INTRODUCTION

Composites dealt in the present study are formed by stacking several layers of parallel fibers embedded in matrix. Each of these layers can be termed as a lamina or a ply.

As composites are formed by union of two distinct phases (matrix and fiber) the properties are entirely different from physical properties of matrix material and fiber used. Behavior of

Unidirectional fibrous composite varies depending on the direction of stacking of parallel fibers. The direction parallel to fibers is longitudinal and the direction orthogonal to fiber is transverse. These composites have strength and stiffness different for both longitudinal and transverse i.e. the behavior is different depending on fiber orientation. This variation in properties makes the analysis of Unidirectional fibrous composites complicated. The intricacy is further enhanced by added properties of thermal expansion coefficient and moisture absorption coefficient.

To overcome complexities involved in numeration of physical properties of Unidirectional fibrous composites and to enable number of combinations of both fiber and matrix a computer based software is written in the present work. The software includes built in libraries of fiber and matrix materials.

2 SOFTWARE STRUCTURE

In the present work a computer software is written in Turbo-C to work on DOS based machines (PC ATs). The software developed is a menu driven package incorporating icons and dialogue boxes for data input and output. These elevate user interaction, thus making the software user friendly.

Precursor

The package in the foremost screen delineates the title along with the author's name and affiliation. The second screen defines fibers and matrix and also gives a gist about their influence in the composite material formed by their union. This screen also lists some of the commonly used fiber and matrix materials. On a keystroke an animated demonstration of longitudinal and transverse axes of unidirectional fibrous composite is presented. These screens add an advantage of incorporating on-line reference in the software.

Input

After explaining the theoretical concepts of unidirectional fibrous composites a slide for input procedure appears. This explains the input procedure involved in the software. Here, three options are available with the user :

• to go to material selection menus

• to go to lesson screens

• to exit.

On using the option of material selection menu the "*MATRIX MATERIAL MENU*" is shown in form of an icon. This menu enables the user to select a matrix material from the listed items and also provides a flexibility for user designated material. On selection of a particular matrix material the properties are rendered from the matrix material database. If the selection is done other than the library resident material the user has to specify following properties

° Density   [g/cu.cm.]

° Tensile Strength [MPa]

° Tensile Modulus [GPa]

° Thermal Expansion coefficient [10E-6/Deg.C]

° Water absorption coefficient [% in 24h]

Then the "*FIBER MATERIAL MENU*" is presented and the fiber material is selected in similar manner to that of matrix material selection. At this junction vide variety of materials for both matrix and fiber are available with the user in material libraries. This enables the user to form number of combinations of fiber and matrix to analyze the composite. A dialogue box then appears inquiring about the volume fractions of fiber and matrix to be used in designing the user specified unidirectional fibrous composite.

Numeration

Input procedure has rendered the user specified volume fractions of fiber (V_f) and matrix (V_m) to be used and the physical properties of user selected materials. In this section the calculation procedure used in the software is elaborated.

The numeration is based on a mathematical model with following assumptions :

• Fibers are parallel to each other (Unidirectional)

• Fibers are of uniform size

• Uniform material properties of both fibers and matrix

• No voids

• Fibers are elastic.

Step I : Calculation of composite density (ρ_c) by

$$\rho_c = \rho_f V_f + \rho_m V_m$$

ρ_f – Fiber density

ρ_m – Matrix density

Step II : Calculation of longitudinal modulus (E_L)

$$E_L = E_f V_f + E_m V_m$$

E_f – Fiber modulus

E_m – Matrix modulus

Step III : To ensure fiber controlled failure of composite formed minimum volume fraction of fiber (V_{min}) is evaluated and depending on this longitudinal strength is calculated

If $V_f > V_{min}$, $\sigma_c = \sigma_{fu} V_f + (1 - V_f)\, \sigma_{mu}$

If $V_f < V_{min}$, $\sigma_c = (1 - V_f)\, \sigma_{mu}$

σ_c – Composite strength in longitudinal direction

σ_{fu} – Ultimate fiber strength

σ_{mu} – Ultimate matrix strength

Step IV : Then the transverse modulus (E_T) is calculated as

$$E_T = \frac{1}{\dfrac{V_f}{E_f} + \dfrac{V_m}{E_m}}$$

The modulus is also evaluated by Halpin–Tsai equation (micro-mechanics approach)

$$\frac{E_T}{E_m} = \frac{1 + \zeta\eta V_f}{1 - \eta V_f}$$

$$\eta = \frac{\dfrac{E_f}{E_m} - 1}{\dfrac{E_f}{E_m} + \zeta}$$

$\zeta = 2$ for circular cross-section

$\zeta = \dfrac{2a}{b}$ for rectangular cross-section

Step V : The ultimate transverse strength of composite is then evaluated by

$$\sigma_{Tu} = \sigma_{mu} \left(\frac{E_T}{E_m}\right) \left[1 - (V_f)^{1/3}\right]$$

Step VI : The shear modulus (G_{LT}) is evaluated by

$$\frac{G_{LT}}{G_m} = \frac{1 + \eta\zeta V_f}{1 - \eta V_f}$$

Where $\eta = \dfrac{\dfrac{G_f}{G_m} - 1}{\dfrac{G_f}{G_m} + \zeta}$ and $\zeta = 1$

Step VII : The Poisson's ratio in LT direction (v_{LT}) is calculated

$$v_{LT} = V_f v_f + V_m v_m$$

v_f – Fiber Poisson's ratio

v_m – Matrix Poisson's ratio

and the Poisson's ratio in TL direction (v_{TL}) is numerated empirically by

$$\frac{v_{LT}}{E_L} = \frac{v_{TL}}{E_T}$$

Step VIII : The thermal expansion coefficients (α_L, α_T) are evaluated by

$$\alpha_L = \frac{1}{E_L} \left\{(\alpha_f E_f V_f) + (\alpha_m E_m V_m)\right\}$$

$$\alpha_T = (1 + \nu_f)\,\alpha_f V_f + (1 + \nu_m)\,\alpha_m V_m - \alpha_L \nu_{LT}$$

Step IX : The stress concentration factor (SCF) and stress magnification factor (SMF) are calculated

$$SCF = \frac{1 - V_f\,[1 - (E_m/E_f)]}{1 - \sqrt{\dfrac{4V_f}{\pi}}\,[1 - (E_m/E_f)]}$$

$$SMF = \frac{1}{1 - \sqrt{\dfrac{4V_f}{\pi}}\,[1 - (E_m/E_f)]}$$

Output

After numeration the output screen is presented in two halves. First half indicates the volume fractions specified by the user and also screens the physical properties of fibers and matrix selected by the user. The second half enlists the numerated properties of unidirectional fibrous composite formed by the union of user specified fiber and matrix. Along with these properties an icon appears with options

(i) to print properties

(ii) to write properties to a file

(iii) to re-enter materials and volume fractions

(iv) to exit.

If option (i) is selected both the properties of parent fiber and matrix along with the numerated properties of composite are printed on a printer. Option (ii) enables the user to save the current combination properties in a file which can be referred at any point of time. Option (iii) takes back to the input procedure screen and allows a fresh input.

3 CONCLUSIONS

In the present work a computer code is written to analyze the performance and to evaluate the physical properties of a user specified unidirectional fibrous composite. The software includes the in-built libraries of material for both fibers and matrix from which the user has to select the material. The materials can also be designated by the user defining their physical properties, this provides an added advantage of global usage of the software. This feature enables vide variety of combinations of fiber and matrix that can form unidirectional fibrous composite to be analysed. The software not only provides a hard copy (printout) of the numerated properties of the user specified composite but also saves it in a user specified file which can be referred to any point of time and can serve as a database for properties of various unidirectional fibrous composite in future. The package also gives a gist of theoretical aspects through graphic screens and animated demos providing an on line reference of theory involved to the user.

4 REFERENCES

Agarwal, B.D. and Broutman, L.J.(1990).Analysis and performance of fiber composites.John Wiely & Sons INC., New York.

Bootle and Kelly, S.(1992).Mastering Turbo-C.BPB Publications, India.

Tsai, S.W. and Hahn H.T.(1990).Introduction to composite materials.Technomic Publication Co.,

5 BIOGRAPHY

The author has graduated (**Bachelor of Engineering**) in Mechanical Engineering from Madhav Institute of Tech. & Science (M.I.T.S.), Gwalior (MP), INDIA, and has done his postgraduate (**Master of Technology**) from Indian Institute of Technology, Kanpur (UP), INDIA. He is currently working as a Lecturer in Mechanical Engineering Department, M.I.T.S., Gwalior (MP), INDIA since 1988. The field of working and interest incorporates Design, CAD, Composite materials and Fracture mechanics.

21

Development of a Computer Aided Design System for Milling Fixtures

M. Hua, Y.K.D.V. Prasad, K.P. Rao and H.K. Fai
Department of Manufacturing Engineering
City University of Hong Kong
Tat Chee Avenue, Kowloon, HONG KONG
Tel: (852) 2788-8420; Fax: (852) 2788-8423

Abstract

The traditional approach of designing good fixtures relies heavily on the experience of the tool designer. A good CAD system for fixture design helps in reducing the design time and also in producing quality designs by novice users. In this paper the development of a Computer Aided Design System for Milling Fixtures (CADSMF) has been presented. The system has been developed in a p/c environment under DOS operating system by using 'C' language and is interfaced with AutoCAD software. Additionally its necessary parametric programs for various standard components and assembly generation programs are coded in 'AutoLISP' and interfaced with the main system. The system is capable of generating part libraries for various standard fixture elements and also to guide a designer in an interactive mode towards the final design of fixtures. Its effectiveness has been tested with workpieces having different milling profiles.

Keywords

Fixture design, CAD/CAM, milling, CNC programming, tool path

1 INTRODUCTION

Fixtures are commonly used to hold components firmly in position during manufacturing processes such as machining, welding or assembly. They essentially consist of three basic types of elements: locators, clamps and supports. Designing of fixtures is a highly complex and intuitive process and it is also a critical activity since substantial amount of time and cost are involved in designing and subsequent manufacturing of a fixture for a part and a chosen operation. Although the latest CAD/CAM technology and the use of CNC machines decrease the time involved in manufacturing a product, the need for a fixture can not be eliminated. In fact the use of latest manufacturing techniques and mass production systems increased the necessity of the usage of cost effective and good quality fixtures. Due to the requirement of heuristic knowledge of experienced designers, development of computer aided fixture design systems has been attempted by few researchers in the past.   Recent developments in CAD/CAM techniques, especially in the field of graphic interfaces, knowledge representation, feature recognition and extraction techniques have provided good opportunities to develop automated fixture design systems.

An ideal milling fixture design system should reduce the time and cost in designing and fabricating special-purpose fixtures and should also reduce the time and cost associated with storing and retrieving the fixtures between different manufacturing operations.   Since the parameters to be considered are many, it may not be possible to study different alternatives manually. To overcome such difficulties in fixture design, many researchers have attempted to develop   automated fixture design systems by applying latest CAD/CAM and artificial intelligence techniques.   Nee et al. (1989) have introduced artificial intelligence (AI) and Computer Aided Design concepts. In a later work, Nee and Senthil (1991) and Senthil et al. (1992) developed a rule-based expert system for automatic fixture design (AFD) in a flexible manufacturing environment by using a solid modeller. Nee and Zhang (1993) developed design interfaces between process planning and fixture design for prismatic parts. Pham et al. (1989) developed a fully-automated knowledge based system for jig and fixture design, in which they incorporated heuristic and analytical principles in the form of explicitly declared facts and rules as well as numerical procedures. Cabadaj (1990) has incorporated the influence of clamping and cutting forces on the setting up of workpiece in his fixture design system. Trappey and Liu (1989) and Huang and Trappey (1992) developed an automated fixture design system by making use of the Projective Spatial Occupancy Enumeration (PSOE) approach to simplify the representation of a general shape workpiece either prismatic or non-prismatic. Whybrew and Ngoi (1992) have employed a spatial representation technique in their modular fixture assembly system. They used this spatial representation technique to describe the geometries of the various elements of the fixture, workpiece etc. and their relative and absolute positions. Jerry et. al. (1993) have developed an integrated fixture planning system with an emphasis on the integration of computer aided fixture planning (CAFP) with computer aided process planning (CAPP). Even though most of the systems that have been reviewed are very efficient in terms of knowledge representation, effective fixture design and interfaces, these systems need the support of expensive software and workstation capabilities.   In the present work a Computer Aided Design System for Milling Fixtures has been developed for use with a personnel computer under the most popular AutoCAD environment by incorporating features such as automatic location and positioning, automated assembly generation with an extensive database support for various standard elements.

2   DESIGN METHODOLOGY

The design of a fixture is dependent on a large number of inter-dependent factors such as shape and material of the workpiece, pre-machined surfaces and stipulated tolerances, machining operations and type of machine tools to be used, workpiece handling and safety considerations, and miscellaneous considerations such as chip removal, coolant flow, etc. It may be difficult to develop a fully automated system which can incorporate all the inter-linked factors mentioned above in order to generate a final fixture design. Safety for instance can not be incorporated in the system without due judgment from a skilled designer.

The present system is developed mainly for designing fixtures that are suitable for two types of workpieces, namely prismatic and cylindrical. There are a total of 12 (2 x 3 x 2) linear and rotational movements along the x, y and z - axes, including both positive and negative directions. By means of supporters and locators at least nine movements can be restricted, while the remaining three possible movements are constrained by clamps (Ruffley, 1989, Cogun, 1992, Noaker, 1992). For a prismatic workpiece, the 3-2-1 locating principle is used to configure the external locating points. In the first stage, the three-point supporting principle is used to assign three supporting points on the first datum plane; these points will be located as far apart as possible in order to increase the workpiece stability. Five movements will be restricted by following the three point supporting principle. In the second stage, two points are assigned on the second datum plane and can restrict further three possible movements. In the third stage, one point has to be assigned on the third datum plane which can restrict one more movement. For cylindrical parts, external locating principles have been applied to restrict the possible movements of the workpiece. Several components are incorporated in the system for properly supporting and locating the workpiece by following the external locating principles. Some of the components are V-blocks for locating or supporting externally cylindrical parts, adjustable supports, and locators for supporting and locating non planar or rough surfaces.

3   MODULAR STRUCTURE OF THE SYSTEM

The modular structure of the Computer Aided Design System for Milling Fixtures (CADSMF) is shown in Figure 1. This design system mainly consists of seven modules, namely: Input Module, Calculation Module, Database Module, Automatic Selection Module, Positioning Module, Interference Checking Module, Assembly and Bill of Materials Generation Module; each module being responsible for some independent operation in the process of designing fixtures for milling operations. The input to the system consists of a series of prompting statements requiring user information about workpiece material, feed per tooth, diameter of the tool, number of teeth, etc. The system then calculates the forces involved in milling and selects a range of appropriate clamping elements. After which, it searches for the corresponding locating, guiding and base elements, and a 3-D graphical fixture will be displayed. The designer can modify the drawing by selecting the elements interactively from the user menu or by using graphical commands. Important activities, processing and typical considerations involved in various modules are briefly mentioned below.

The *Input Module* is mainly responsible for the extraction of information about the geometry of the workpiece and also the milling feature for which the fixture is to be designed. DXF file format has been used to transfer drawing data between AutoCAD and fixture design system. This extracted information will be transferred to the *Force Calculation Module*, in order to estimate the cutting and clamping forces. The cutting force that acts on the tool during milling operation depends on the specific cutting energy of the work material and the material removal rate. Once the cutting force is evaluated based on the above criteria, the system estimates the clamping force based on the cutting force, the safety factor incorporated and other relevant parameters. Based on the estimated clamping force and clamping capacity of different clamps, the system selects an appropriate type of the clamp from the database, with a consideration to optimize the number of clamps.

In the *Database Module* parametric programmes are developed in AutoLISP language to build libraries of various standard elements used in the design of fixtures such as adjustable

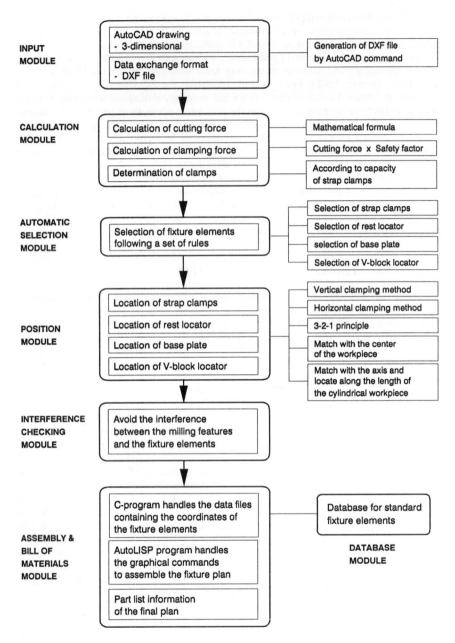

Figure 1   Modular structure of the system for designing milling fixture.

locating buttons, locating pins, strap clamps, rest locators, V-block locators , edge clamps and base plates. The libraries are created for Misumi standard components, Japan, which are widely used in Hong Kong manufacturing industries. Based on the estimation of forces the *Automatic Selection Module* determines the size and number of standard parts required for each function and selects the standard parts from the *Database Module* based on a set of sample rules that have been incorporated for the selection of the type of the clamp and number of clamps. The system makes use of a set of heuristics that will aid in the selection of various fixture elements in a semi-automated manner. Fixture set-up functions, such as positioning and fixing, are created as major constraint rules for the refinement of fixture structure. Various generic elements (with definitions of function, principle of action and nature of contact) are chosen by a set of rules which specify the generic positioning and fixing element classes. Finally, the standard elements (specific fixture components) are selected for fixture assembly that are appropriate for simple prismatic workpieces.

Once the various standard parts are accessed from the database the *Positioning Module* determines the position and location of each standard element depending on the dimensions of the workpiece and the force analysis. A workpiece must be accurately located and firmly clamped to ensure that there is no movement during machining and the cutting tools do their work quickly and efficiently. A suitable method of locating the workpiece is very important and will depend on many factors like geometrical shape of the workpiece, surface finish of the datum, features of the workpiece, and operation to be performed. After the DXF file is extracted, then the system identifies the information about the workpiece, profile to be machined from it. The rest locators will be positioned and oriented according to the size of the workpiece, following the 3-2-1 principle. Before finalizing the position of various standard elements, the *Interference Checking Module* checks for the interference between the various fixture elements and the milling area in order to ensure uninterrupted motion of the tooling elements during the entire milling operation for the chosen feature. If any interference is detected, then this module reorients the position of the fixture elements until no interference occurs. When a prior position of the fixture elements are blocked by the milling features, the system will reorient the position of each fixture element according to locating, clamping and supporting rules until the desired plan has been achieved. Figure 2 shows the typical approach used by the program to check and avoid the interference between the milling features and the fixture elements.

Finally the *Assembly and BOM Generation Module* automatically generates 3-D graphic display of the final fixture along with the bill of materials. The system is capable of generating four major views such as plan, elevation, side view and the Isometric view of the designed fixture simultaneously for better visualization, and a bill of materials will be shown only in the plan view of the assembly. The AutoLISP programs in the system extracts the coordinates of the fixture elements from several data files. Then, the AutoLISP programs handle the graphical commands to assemble the fixture plan in a three dimensional view. After the final fixture display, the user can click the 'BOM' button on the menu bar to extract the parts list information for display on screen. The system has been implemented in an interactive manner by designing a user menu as shown in Figure 3. By making use of this menu, user can interact with 'C' and AutoLISP functions under AutoCAD environment. Users can also edit the plan view of final fixture by inserting the fixture components from the menu bar. Finally, the user menu also provides a 'VIEW DISPLAY' for the user to modify the isometric view to a 4-view display simultaneously or independently. An on-line 'HELP' function has been incorporated so that the user can obtain more information about the system at any stage of the design.

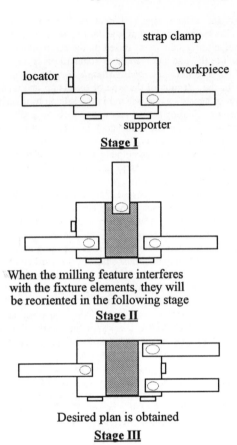

strap clamp

locator   workpiece

supporter

Stage I

When the milling feature interferes
with the fixture elements, they will
be reoriented in the following stage

Stage II

Desired plan is obtained

Stage III

Figure 2 Procedure for interference checking between fixture elements and milling feature.

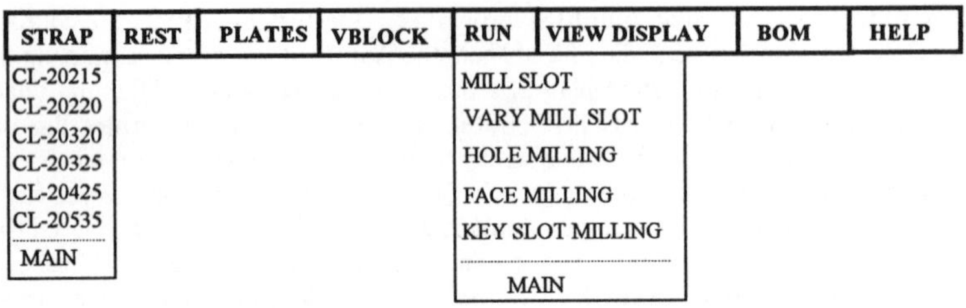

STRAP	REST	PLATES	VBLOCK	RUN	VIEW DISPLAY	BOM	HELP
CL-20215				MILL SLOT			
CL-20220				VARY MILL SLOT			
CL-20320							
CL-20325				HOLE MILLING			
CL-20425				FACE MILLING			
CL-20535				KEY SLOT MILLING			
MAIN							
				MAIN			

Figure 3 User menu of the CADSMF system.

4   CASE STUDY

The fixture design capabilities of the system are demonstrated with the help of producing a slot by milling. The part for fixturing together with its raw material form and machining profile is shown in the Figure 4. The 3-D drawing of the component and profile to be machined are created in AutoCAD environment and necessary drawing information is extracted as a DXF file, which formed a part of the input to the system. Other relevant information such as material of the workpiece, tooling details (diameter of the cutter and number of the teeth etc.) are provided by the designer interactively. After calculating the cutting force and clamping force, following the heuristics, the system selected suitable locating and clamping points as well as locating and clamping elements. Suitable standard parts such as locating pins, strap clamps, supporting elements and base plate from the data base are selected.

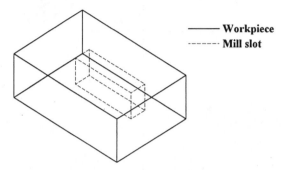

——— **Workpiece**
------- **Mill slot**

Figure 4   Isometric view of a workpiece with a mill slot.

The selected elements with reference to the workpiece are assembled automatically to give final fixture assembly. The isometric view of the fixture assembly is shown in Figure 5. With the help of the interactive user menu, four views of the fixture assembly can be generated as shown in Figure 6. It can be seen that the design solution generated by the system is similar to that of a design created by an experienced designer.

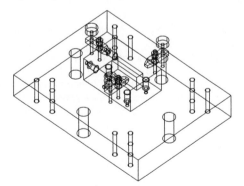

Figure 5   Isometric view of the fixture assembly

Top View   Isometric View

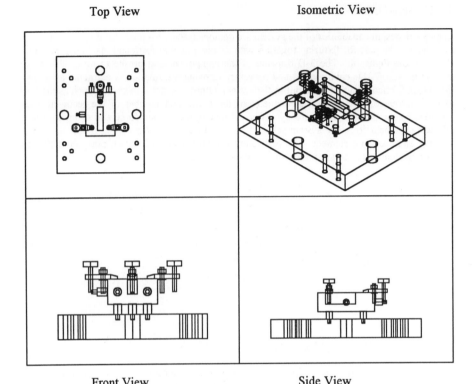

Front View   Side View

Figure 6 Display of the fixture in 4 different views.

5   CONCLUSION

Details pertinent to the attempt made by the authors in the development of a computer aided design system for milling fixtures has been presented. The system has been developed on a personal computer by using 'C' programming language. The parametric programmes for various standard elements used in the fixture design are developed in AutoLISP language. Rules implemented for calculation of cutting and clamping forces and selection of fixture elements are discussed. Finally the fixture assembly generated, along with four views of assembly and bill of materials for the selected slot milling, is presented.

6   ACKNOWLEDGMENT

The authors would like to acknowledge the financial support received from the City University of Hong Kong, through strategic research grant 700295, for conducting this work.

7 REFERENCES

Cabadaj, J. (1990) Theory of computer aided fixture design. *Computers in Industry*, **15**,141-147.

Cogun, C. (1992) The importance of the application sequence of clamping forces on workpiece accuracy. *Transactions of ASME, Journal of Engineering for Industry*, **114**, 539-542.

Huang Shih-huai and Trappey, J.C. (1992) Integration of modular fixture database- fixturing knowledge base and 3-d fixture planning interface. *Computers & Industrial Engineering*, **23**, 381-384.

Jerry, Y.H. Fuh, Chao-Hwa Chang and Melkanoff, Michel A. (1993) An integrated fixture planning and analysis system for machining processes. *Robotics and Computer Integrated Manufacturing*, **10**, 339-353.

Nee, A.Y.C., Bhattacharya, N.and Poo, A.N. (1985) A knowledge-based cad of jigs and fixtures. *SME Publication*, **TE85-902**, 19 - 26.

Nee, A.Y.C. and Senthil Kumar, A. (1991) A framework for an object/rule-based automated fixture design system. *Annals of CIRP*, **40**, 147-151.

Nee, A.Y.C. and Zhang, Y.F. (1993) Design interfaces for computer-aided process planning and fixture design. *Proceedings of IMCC93*, 217-223.

Noaker, Paula M. (1992) Commonsense clamping. *Manufacturing Engineering*, **109**, 43-47.

Pham, D.T., Nategh, M.J. and de Sam Lazaro, A. (1989) A knowledge-based jig-and-fixture designers assistant. *The International Journal of Advanced Manufacturing Technology*, **4**, 26-45.

Ruffley, Douglas W. (1989) Selection factors for clamps in fixtures. *SME Technical Publication TE89-352*, 27-31.

Senthil Kumar, A., Nee, A.Y.C. and Prombanpong, S. (1992) Expert fixture-design system for an automated manufacturing environment. *Computer Aided Design*, **24**, 316-326.

Trappey; J.C. and Liu, C.R. (1989) A structured design methodology and metadesigner, a system shell concept for computer aided fixture design. *Proceedings of ASME Design Automation Conference*, Montreal Canada, 309-313.

Whybrew, K. and Ngoi, B.K.A. (1992) Computer aided design of modular fixture assembly. *The International Journal of Advanced Manufacturing Technology*, **7**, 267-276.

8 BIOGRAPHY

M. HUA received B.Sc. and M. Phil. degrees in Mechanical Engineering from the Newcastle Upon Tyne Polytechnic, UK, and then worked as Research Assistant in the department. He received PhD from Aston University, Birmingham, UK, in 1986. Subsequently he worked as Research Officer in School of Mechanical Engineering, Bath University, UK, Chief Mechanical Engineer with Data products Components (HK) Ltd., Product Manager with Harting Electronik Ag, and Electro-Mechanical Product Engineering Manager with Data products Components (HK) Ltd. He joined the City University of Hong Kong in May 1993. His research interests are in metal forming, surface engineering, CAD/CAM, CAPP, heat transfer.

Y.K.D.V. Prasad received his B.Tech. and M.Eng. degrees in Mechanical Engineering in 1983 and 1986 respectively. He received his Ph.D. degree in CAD/CAM applications in sheet metal forming from IIT, Bombay, in 1993. After a brief industrial experience, Dr. Prasad worked as a Lecturer in the Department of Mechanical Engineering at Siddhartha Engineering College between 1985 and 1992. Since December 1992 he is associated with the Department of Manufacturing Engineering at the City University of Hong Kong. His research interests are in the area of CAD/CAM, tool design and sheet metal forming. He is a member of several professional organizations.

K.P. Rao graduated in engineering in 1976 and obtained a master's degree from in 1978. After a brief industrial experience, pursued research studies and obtained a Ph.D. in 1983. He worked as a Research Engineer until 1986 at the Central Metal Forming Institute, Hyderabad, India, and later pursued post-doctoral study at the universities of New Brunswick and British Columbia, Canada, before moving to the City University of Hong Kong in 1990 to engage in full-time teaching and research, and is the course leader for two degree courses. His research is mainly in the area of metal forming and material processing, and has extensively published in these areas.

H.K. Fai was a degree student during 1991-94 at the Department of Manufacturing Engineering, City University of Hong Kong, and has graduated in 1994.

22

Constraint-Based Elevator Design Support System

A. Ishida, Y. Arai, S. Akasaka, N. Haga
1st. Department
Production Engineering Research Laboratory, Hitachi, Ltd.
292 Yoshida-cho, Totsuka-ku, Yokohama-shi 244 Japan
Tel : 81-45-860-1651, FAX : 81-45-860-1621
E-mail : ishida@perl.hitachi.co.jp

K. Katsuta
Elevator Design Dept.
Mito Works, Hitachi, Ltd.
1070, Ichige, Hitachinaka-shi, Ibaraki-ken, 312 Japan
Tel : 81-292-73-3111, FAX : 81-292-73-4644

Abstract

We describe a design support system for configuration of elevators to customer specifications. The knowledge base of the system consists of various design attributes and constraints which form a constraint network. Based on this knowledge base values of attributes are inferred from a customer's requested specifications by using constraint propagation. After examining the elevator design process, the following functions were found to be indispensable for realizing a practical design support system: 1) modification of the constraint network during the design process when the model changes; 2) control of the direction of constraint propagation; 3) utilization of default constraints; 4) constraint relaxation; and, 5) countermeasure guidance stored in the know-how database by expert designers when conflict occurs during the inference process. The effectiveness of this system was evaluated by application to elevator design, and the design time was reduced from 4-8 hours to 30 minutes for a typical order.

Keywords

Design Engineering, Design Knowledge, Expert System, Intelligent CAD, Constraint Network, Constraint Propagation

1 INTRODUCTION

In the elevator sales engineering division, it is necessary to determine machinery and layout specifications quickly to meet a customer's requirements such as load capacity, speed and building specifications, and rapidly finalize orders to production. However, determining order specifications requires a sales engineer to consider carefully while referring to large amount of technical information in manuals. This requires much time and man power. Consequently, we have developed an elevator design support system that permits specifications to be constructed interactively on a workstation in which specialized knowledge such as design rules and know-how is stored. In this report, we describe the system outline and its functions.

2 THE DESIGN PROCESS

First, we examine the elevator design process. An example of an elevator design process is shown in Figure 1. The design process is as follows:

1) *Input of customer's request*: the customer has requested a change in the depth of the elevator shaft.

2) *Determination of product structure*: the hanging weight position must be changed from the rear to the side due to a necessary minimum gap between the weight and the inner wall. This in turn makes it necessary to change the traction machine orientation in the machine room.

3) *Calculation of specifications values*: the traction machine position(x,y) is calculated from the cage and weight positions while the sales engineer refers to manuals.

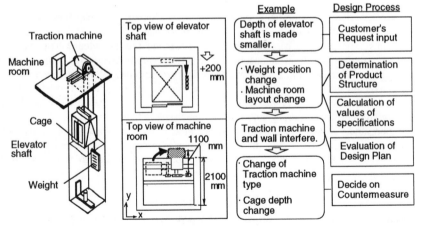

Figure 1 Example of the elevator design process.

4) *Evaluation of design*: the machine room wall interferes with the traction machine.

5) *Decide on countermeasure*: it becomes necessary to take a countermeasure such as a change of traction machine type, or a change of cage depth.

This process is repeated until all design specifications are met with no conflict. To develop a system for configuration of elevators efficiently in this process, methods for both representation of design knowledge and guiding sales engineers through the design process must be devised.

3 REPRESENTATION OF DESIGN KNOWLEDGE

The design knowledge is described as rules that govern the relationships among local attributes summarized as formulas, tables and flowchart in manuals. An expert designer possesses an image of the network of rules (constraints) and the related attributes, and thus is able to grasp the overall order of design steps needed based on this image, and then proceed with the design process efficiently. Consequently, we have adopted a knowledge representation which consists of attributes and constraints that allows development of design plans by propagation of constraints.

In the example in Figure 2, load, door width OP, and door position Xd are given, after which cage width and cage depth can be inferred using a table of feasible combinations of values (C1). Next, gap L is inferred from doorwidth OP by the relationship C2 and then cage position Xc and gap d are inferred by C4 and C3.

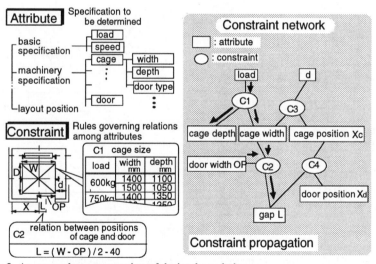

Figure 2 An example representation of design knowledge.

4 REQUIRED FUNCTIONS FOR GUIDING THE DESIGN PROCESS SUCCESSFULLY

After applying this approach to a elevator design, problems in current constraint propagation methods were observed. The following functions were found to be indispensable.

1) *Adaptability of the model*: in cases such as Figure 1, because of the change to the product structure, i.e. weight position type, new parts such as a rail bracket are necessary (Figure 3). As a result, the structure of the constraint network must be modified in the middle of design to cope with this change of product structure.

2) *Utilization of default constraints*: there are no instructions in manuals on how to systematically progress through the design process, although there are local design rules among attributes. Therefore, the know-how that experienced designers possess must be described in the knowledge base. For example, cage position Xc can be inferred from door position Xd. However, if door position hasn't been given, cage position can't be inferred.

To deal with this, experienced designers would position a cage in the shaft using the rule that a cage should be centered in the middle of a shaft, i.e. $Xc = Xs / 2$, if there is no customer request about door position (Figure 4). These types of default constraints are weak and ignored if a strong constraint exists, but they are necessary to progress through the design process.

3) *Control of the direction of constraint propagation*: in constraint propagation, values of attributes can be determined in both directions. However, sometimes only one direction make sense, so it is necessary to define direction of propagation in these cases.

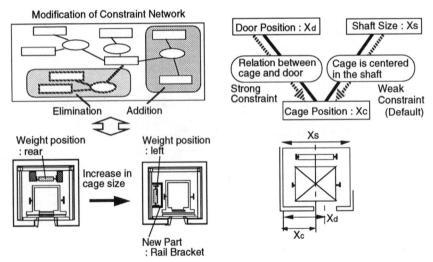

Figure 3 Change of product structure.   **Figure 4** Default constraint.

4) *Constraint relaxation*: specification of standardized parts is generally good on points of productivity and cost. Thus, in the beginning of the design process, order specifications should be inferred using recommended standard values. If the customer's request cannot be satisfied with standardized parts, the variety of values must be increased by constraint relaxation.

5) *Countermeasure guidance*: there are two countermeasure methods when design conflicts occurs. One is local conflict dissolution by changing values of attributes, or in other words, constraint relaxation within the considered constraint network. The other is global conflict dissolution by changing the product structure. Know-how of an experienced designer in this case is necessary.

To develop the above functions, we define the following items of design knowledge (Figure5):

a) *Conditions for activation of attributes and constraints*: various combinations of attribute and constraint values may require that other attributes and constraints be considered. When these combinations of values are entered in design process, related attributes and constraints are activated or inactivated to change the model dynamically.

b) *Definition of default constraints:* if all related attributes are given, this default constraint is ignored.

c) *Constraint weights*: if there are several constraints which can be inferred, constraint propagation is executed based on pre-defined weights.

d) *Definition of the direction of constraint propagation*: a flag is used to indicate which directions inferences can be made or not made. In the below example, inference of attribute Z is prohibited when attribute X and Y are already given.

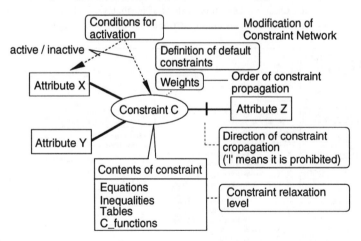

Figure 5 Description of attributes and constraints.

e) Definition of constraint relaxation levels: relation between contents of constraint and constraint relaxation levels is defined.

f) Countermeasure know-how of designer for global conflict dissolution is expressed by rules (IF..THEN..) in a separate know-how DB. A trace of the inference path is recorded for local conflict dissolution in another file.

5 PROCESSING MECHANISM

By using attributes and constraints described by designer in factory, trace of inference and countermeasure know-how, the system controls the design process as follows (Figures 6 and 7) :

1) *Input / modify of values of attributes*: input of the customer's required attributes by a sales engineer.

2) *Adapting the model*: referring to conditions that activate attributes and constraints, the structure of constraint network is modified when such conditions occur.

3) *Control of constraint propagation execution order*: referring to the type of constraint (table, equality, c_function, etc.), the default constraint, and direction of constraint propagation, constraints which can be activated are found (Figure 6). Consistency check is not executed for default constraint.

   The order of constraint execution is determined by selecting the constraint which has maximum weight among all activated constraints.

4) *Local constraint propagation*: the system executes local constraint propagation. There are two types of execution. One infers values of attributes which are not fixed and the other checks for inconsistencies among attributes. If only one suitable value exists for

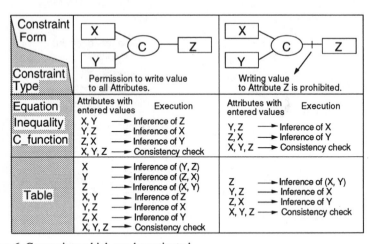

Figure 6 Constraints which can be activated.

an attribute the system enters the value into the attribute. If more than one suitable value exists, the system displays the candidates and asks the sales engineer to select one.

5) *Planning countermeasure*: if a design conflicts occur, the system can give countermeasure advice by either searching the inference record or referring to the know-how DB.

6) *Repeat of above process*: the above process is repeated until all values are calculated with no conflict.

7) *Display of design result, diagram drawing and document output*: after the system infers the design solution , it displays the design result, draws the diagram and outputs the document.

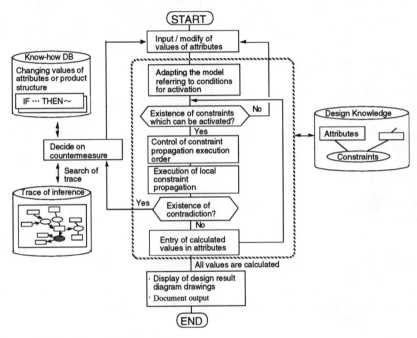

Figure 7 Flowchart of the processing.

6 OUTLINE OF DEVELOPED SYSTEM

Based on the functions described in Section 4, we developed an elevator design support system. This system was implemented as interactive type software on a Hitachi 2050G workstation. The design solution can be quickly calculated in about two minutes. The design solution is output in the form of an installation diagram and shown to the customer. Then, after the customer's approval is obtained, a production order is placed with the factory through a computer network.

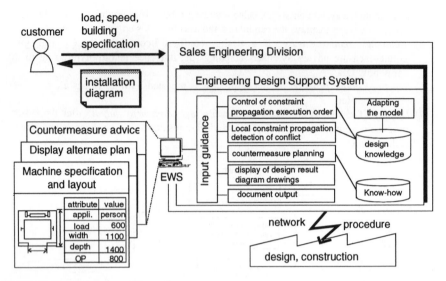

Figure 8 Total system configuration.

7 CONCLUSION

We have developed an elevator design support system which can efficiently guide the design process. The following functions were found to be indispensable for realizing a practical design support system: 1) modification of the constraint network during the design process when the model changes; 2) control of the direction of constraint propagation; 3) utilization of default constraints; 4) constraint relaxation; and, 5) countermeasure advice when conflicts occur during the inference process. When we used this system on a full-scale design problem(1600 attributes and 3300 constraints), we confirmed that the installation diagram could be obtained within a fairly short amount of time(about 30 minutes).

REFERENCES

Steele, G.L. Jr. (1978) The Definition and Implementation of a Computer Programming Language Based on Constraints, AI-TR-595, MIT.

Yoshikawa et al (1991) Intelligent CAD. Asakura Shoten, Japan

Arai, Y. and Akasaka, S. and Ishida, A. (1994) Expert System for Designing Elevators to Customer Specifications, The 12th Design Symposium of JSPE, Tokyo.

BIOGRAPHIES

Atsuhiro Ishida was born in 1965. He received a B.S. degree in 1986 in Mechanical Engineering from Tokyo University. He is a researcher at the Production Engineering Research Laboratory, Hitachi, Ltd. His present interests include knowledge-based design support systems. He is a member of JSPE and JSME.

Yoshinao Arai was born in 1953. He received a B.S. degree in 1975 and a M.S. degree in 1977 in Precision Mechanical Engineering from Tokyo University. He is a senior researcher at the Production Engineering Research Laboratory, Hitachi, Ltd. His present interests include production management and intelligent CAD systems. He is a member of JSPE, JSME and AAAI.

Shingo Akasaka was born in 1961. He received a B.S. degree in 1984 and a M.S. degree in 1986 in Industrial Engineering from Waseda University. He is a researcher at the Production Engineering Research Laboratory, Hitachi, Ltd. His present interests include knowledge acquisition for intelligent CAD systems. He is a member of JSPE and IPSJ.

Noriyuki Haga was born in 1965. He received a B.S. degree in 1991 in Mechanical Engineering from Tohoku University. He is a researcher at the Production Engineering Research Laboratory, Hitachi, Ltd. His present interests include artificial intelligence. He is a member of JSPE.

Katsuyoshi Katsuta was born in 1950. He received a B.S. degree in 1970 in Mechanical Engineering from Hakodate Technical College. He is a senior engineer in the Design Department at Mito Works. His present interests includes rationalization of elevator design.

23

Successful examples in CAD/CAE teaching

Kaminaga, K.*, Fukuda, Y.**, Sato, T.*
* Department of Metallurgy
** Computer Center
Shibaura Institute of Technology
3-9 Shibaura, Minato-Ku
Tokyo, Japan.
TEL: 81(03)3542-3201 Ext. 2316
FAX: 81(03)5476-3161

Abstract

Stimulating interest in the beginner is the most important aspect of any educational program for beginners. The same is true for introducing CAD/CAE to a beginner so that the beginner masters the principles of CAD/CAE and how to use the software in the manner of playing a computer game, without instilling a feeling that mastering CAD/CAE is 'difficult' or ' complicated.' A simple CAD/CAE introductory software program that enables the student to learn the principles of CAD/CAE is more effective for this purpose, rather than a multi-functional, high accuracy CAD or CAE program usually sold in the market, The results of implementation of introductory CAD/CAE education, based on the premise mentioned above, are reported in this paper.

Keywords

Introductoly CAD Education,Introductoly CAE Education,Examples

1. INTRODUCTION

An important point to be remembered when educating and rearing specialists or operators in CAD or CAE is to ensure that the student does not develop a dislike for CAD/CAE, or does not develop a feeling that its study is too difficult and complicated. The student must be offered education so that he/she enjoys the journey through the course, while picking up the principles of CAD/CAE naturally. However, educating the student with CAD/CAE as a 'black box,' where the student knows only the operating principles but not the fundamentals of CAD/CAE, results in a permanent feeling of insecurity in the student. Therefore, a method that simply and easily introduces the basics of CAD/CAE to the student is necessary. An introductory course on CAD/CAE was given to the students of the Department of Metallurgy,

Shibaura Institute of Technology, based on the concept mentioned above. The results were very satisfactory and are reported here.

2. A FOUR-STEP APPROACH TO INTRODUCING CAD USING CAD SOFTWARE

Several multi-functional and easy-to-use CAD software are currently available in the market. These software, however, are not suitable for teaching CAD to a beginner. This is because these CAD software cannot be used unless numerous CAD commands and operating procedures are memorized without understanding the basics of the CAD software. To find a way out of this difficulty, the authors have developed a three-level CAD software aimed at the beginner. After initial training in CAD using this software, the students were drilled in the use of commercial CAD software available in the market. The results of adopting this approach were extremely satisfactory.

2.1 Level 1 CAD software

The Level 1 Cad software consists of the drawing area in the upper part and the BASIC program area in the lower part, as shown in Figure 1. The program is written such that the LINE statement and the CIRCLE statement of the BASIC program are always displayed on the CRT, whenever a straight line or a circle is drawn.

The objectives of Level 1 CAD software are:
(1) to create an awareness in the student that a CAD program can be written using less than 10 lines of a BASIC program
(2) to emphasize to the student that the LINE statement and the CIRCLE statements of the BASIC program are the building blocks of the CAD program
(3) to create an awareness in the student that CAD is used by input of coordinates (that the CAD software is not a 'black box)
(4) to make the student understand that the Y coordinate axis used in this level is inverted.

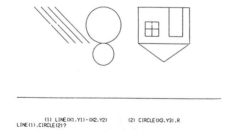

(1) LINE (X1,Y1)-(X2,Y2) (2) CIRCLE (X3,Y3),R
LINE (1), CIRCLE (2)?

Figure 1 Basic drawing using Level 1 CAD program.

2.2 Level 2 CAD program

The Level 2 CAD program was developed by slightly modifying the Level 1 CAD program.
The Level 2 CAD program consists of the drawing area, the menu area, and the data input area.
(See Figure 2.)

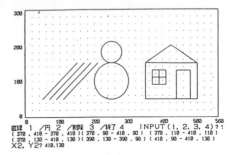

Figure 2 Basic drawing using Level 2 CAD program.

The educational objectives of the Level 2 CAD program are:

(1) to create an awareness in the student that figures can be drawn on the screen by methods
that are simpler than the method adopted in the Level 1 CAD program
(2) to stress the point that the LINE and the CIRCLE statements are fundamental parts of a
BASIC program, offering the same picture quality
(3) to emphasize to the students that "DEL" command superimposes a black-colored line on
the white line formed by the LINE statement or the CIRCLE statement, used to create a line or
a circle

Figures 3 to 5 show examples of figures drawn by students after learning how to draw simple
figures using the Level 2 CAD program.

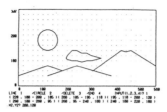

Figure 3 Example (1) of figure drawn by a student.

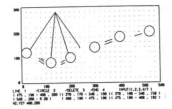

Figure 4 Example (2) of figure drawn by a student.

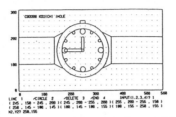

Figure 5 Example (3) of figure drawn by a student.

2.3 Level 3 CAD program

The ellipse command dimensional input commands were added to the Level 2 CAD program, and these CAD commands were arranged in a hierarchical structure. Figure 6 shows a typical example of a basic drawing using the Level 3 CAD program.

The CAD operations for input of dimensions consist of the following hierarchical commands:

Choose "Edit Figure" command by mouse
Choose "Dimensions" command
Choose "Horizontal Figure" command
Input start point coordinates x1, y1
Input end point coordinates x2, y2
Input dimension input position coordinates x3, y3

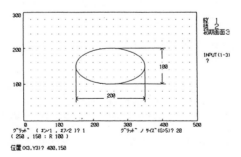

Figure 6 Basic drawing using Level 3 CAD program.

2.4 Level 4 CAD program

The commercial software Auto CAD GXIII available in the market was used as the Level 4 CAD program. (See Figure 7.)

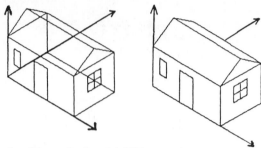

Figure 7 Basic drawing drawn using Level 4 CAD program.

Compared to the year when a commercial software was used from the start of the educational course, the four-step approach adopted this time for teaching the students gave very satisfactory results. The students had a sense of security and were more reassured when they used the commercial CAD software after mastering the first three basic levels, compared to the students who used the CAD software merely as a "black box." Also, the students who started out directly with the commercial software were confused and found the hierarchical structure of the CAD commands troublesome to master. However, when the students started using the commercial CAD software after using the Level 2 and Level 3 CAD software, they treated the commercial software as an extension of the CAD software developed by the authors, having more functions and convenient to use.

3. CAE EDUCATION FOR BEGINNERS USING A "GAMESOME ATTITUDE"

Several excellent CAE software are also available in the market but these are very expensive and have a large number of functions requiring considerable skill, therefore, they are not suitable for educating beginners. The CAE teaching program was implemented using the introductory CAE program developed by the authors and an education CAE software developed by Prof. Toshiro Miyoshi of the University of Tokyo. The teaching of CAE for Finite Element Method (FEM) is reported here. Similar concepts were adopted for Boundary Element Method (BEM) and Finite Differential Method (FDM).

A brief introduction about the FEM was given to the students before they started the actual study to stimulate their interest. The Finite Element Method was introduced to the students as a very useful and convenient simulation method developed by an engineer of the Boeing Company, USA, indispensable in the design of automobiles, planes and skyscrapers. Explanations about the usefulness of FEM were given with the help of examples, such as the cantilever beam shown in Figure 8(1). For this cantilever beam, the deflection and stress distribution can be easily calculated by merely substituting numerical values in the equations of strength of materials. However, the calculations for the cantilever beams shown in Figure 8(B) and Figure 8(C) are a bit more complex and solutions are sometimes not available by ordinary calculation methods. By using the Finite Element Method, however, the solutions to the beam problems in Figure 8(B) and Figure 8(C) can be obtained by input of data in a sequence similar to the problem in Figure 8(A). This is illustrated in Figure 9. The students were taught to use the FEM program developed by the authors. This program has been written so that simulation is performed by interactive data input on a section paper (graph paper) displayed on the screen.

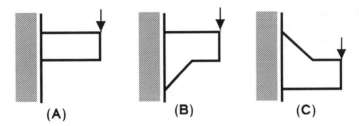

Figure 8 Elastic deformation of cantilever beams.

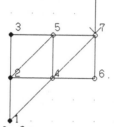

Figure 9 Data input screen of FEM software.

After becoming familiar with the use of software developed by the authors, the students were taught the CAE methods using CAE software developed by Prof. Toshiro Miyoshi of the University of Tokyo. The students were given the work of FEM simulation of any topic of their choice. The examples shown in Figures 10 to 14 are examples of FEM simulation performed by the students in a 'gamesome attitude.'

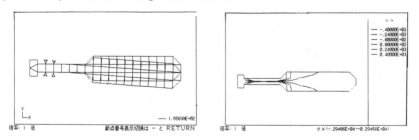

Figure 10 Deflection and stress on a baseball bat when a ball strikes it.

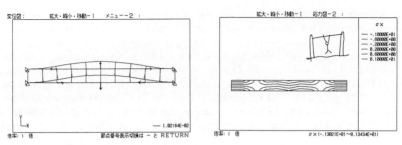

Figure 11 Deflection and stress on a horizontal bar (gymnastics) when a person suspends himself from the bar.

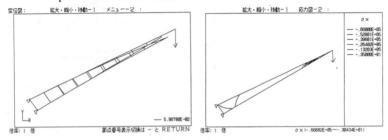

Figure 12 Deflection and stress on a fishing rod when hauling a fish.

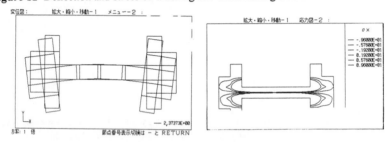

Figure 13 Deformation and stress of a bar bell (weightlifting).

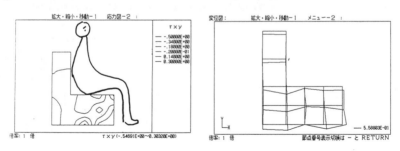

Figure 14 Deflection and stress on a chair when a person sits on it.

4. CONCLUSION

The authors have found that drilling the students first in the principles and usage methods of simple CAD/CAE programs written specifically for this purpose, and then exposing them to the commercial CAD/CAE software available in the market has better effects compared to directly exposing the students to commercially available CAD/CAE software.

5. BIOGRAPHY

Dr. Toshihiko Sato is a professor at the Department of Metallurgy, Shibaura Technical University, 3-9 Shibaura, Minato-ku, Tokyo, Japan. He earned three degrees - a bachelors in engineering (1961), a masters in engineering (1963), and a Ph.D. (1966) - from Tohoku University, Japan. He has worked at Shibaura Institute of Technology since 1966.
Ms. Kyoko Kaminaga is a research fellow at Shibaura Technical University, where she received BS degree in metallurgical engineering.
Mr. Yuichi Fukuda is an assistant professor at the Computer Center of Shibaura Technical University, where he received BS degree in electronic engineering.

Theory and Methodology of CAPM

24

Project Management Techniques Applied to Industrial Manufacturing

A. Rolstadås
Professor
University of Trondheim
Department of Production and Quality Engineering
N-7034 Trondheim, Norway
Phone: +47-73593785, Fax: +47-73597117, E-mail: arolst@protek.unit.no

Abstract

There are two types of planning tools: process oriented and project oriented. The project oriented tools are becoming widely applied, not only for one of a kind production, but also in batch and continuous manufacturing. To discuss project management, a project model must be defined. The project control baseline comprises a work breakdown structure, a schedule and a cost estimate. Planning techniques for each of these are discussed. Project control can be done according to six different principles: earned value, productivity development, resource availability, physical completion, tendency development and forecasting. These principles are discussed.

Keywords

Project management, Project planning, Project control, Project work, Project model

1 PROJECT ORIENTED WORK

Industrial production started based on projects. Each house constructed, each bridge erected, each automobile manufactured, etc. was tailored to the customer and performed as a one of a kind task. Henry Ford with his assembly line and mass production technology changed this. Products and parts were made in batches or even continuously.

The project approach to production may therefore be regarded as the original one. It is still the dominating one in most construction business and in research and development. It is gradually also penetrating into the manufacturing industry. In many ways project oriented work seems to be the future way of production. This paper shall therefore cover some of the most used techniques for planning and controlling projects. This will be referred to this as **project Management.**

Actually there exist two types of planning tools:
• Process oriented
• Project oriented.

The process oriented tools cover flow oriented (continuous), and batch oriented (periodic) production. The project oriented tools cover one of a kind production. As mentioned, the project oriented tools are gradually taking over for the classical process oriented tools.

Almost any individual in industry or business has in some context experienced project work. Roughly projects can be classified in two categories:
• Internal development projects
• Project production.

Internal development projects may cover any internal activity of an enterprise. Examples include development of a new product, implementation of new technology, construction of a new plant, study of productivity improvement, etc. Many of these projects are relatively small projects, and they may be executed by internal resources, external vendors or consultants or any mix of these two extremes.

Project production includes the situation where the company's end product to the customer is a project. An engineering company is typically an enterprise "selling" projects. The same is true for a shipyard. Another example is a research institute.

There are some important characteristics of a project which may help explain why projects will be the future mechanism of managing production:
• Projects are goal-oriented and thus help to keep focus on what is relevant.
• Projects are one of a kind and thus prevents the establishment of procedures, systems or habits that may be unnecessary
• Projects are by nature multidisciplinary and in this way promotes communication across the organization of the enterprise.

In addition to this, there are a number of well developed techniques and systems available on the market that can effectively help the planning and controlling of a project.

2 THE PROJECT MODEL

For further discussion of project management techniques on a general basis it is necessary to establish or define a project model. In this context a project model will distinguish between **owner** and **contractor**. The owner is the ultimate customer or user of the project results. The owner will define the work to be carried out by the contractor and will give all necessary frame conditions for it. Contractor is the party physically carrying out the work.

The project management techniques will at large be the same for both the owner and the contractor role, but they will of course be applied at different levels. The detailed plan of the owner will actually represent the top level plan of the contractor. Throughout this text, everything is viewed from the contractor's point of view.

The project model will further assume that some task is to be accomplished. This task may be the delivery of a product, providing of services or construction work, development of technical specifications, research, etc. The task involved may be of two types:
• Engineering
• Production.

Engineering includes all technical documentation. Production includes fabrication, assembly and commissioning. In this way the manufacturing of an enterprise can be regarded as consisting of three basic processes as indicated in figure 1. Engineering and production are the processes already discussed. Actually the principal difference between them is the type of flow. In engineering, the flow represents documents or technical information. In production, the flow represents materials. The third process represent the internal planning and control (management). This process controls the other two. The flow here represents information.

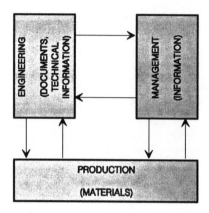

Figure 1 Basic manufacturing processes.

The two types of tasks involved can be regarded as a number of contracts to be accomplished. That is, the project comprises a number of contracts organized in a hierarchy. To carry out the work of the contracts will require resources, such as:
• Internal resources
• Contractors.

For the contractors it is convenient to distinguish between:
• Consultants
• Fabricators
• Suppliers

Consultants provide engineering and management services. Fabricators produce the physical installations. Suppliers delivers goods and services to be used in engineering or production.
    Finally it is assumed that the project is executed by a dedicated organization. This may either be a task or a matrix organization.
    In summary, the project model assumes that the project contains engineering and/or production and that it is comprised by a number of contracts, each carried out by a contractor that may be a consultant, a fabricator, a supplier, or carried out by internal resources. The project is executed by a project organization.

3 THE PROJECT DEFINITION

There exist many definitions of a project. Some prefer a vague definition such as "something with an ultimate goal". In this text a more precise definition will be used:

> *"A project is a one of kind task with a clearly defined goal to be reached within a given time and cost frame."*

This definition directly highlights the three main control variables of a project:
• Scope of work
• Time
• Cost.

A plan for these variables is referred to as the **project control baseline**.
   Project management was defined as planning and control of projects. **Planning** includes:
• Define objectives
• Prepare plan
• Order work.

Control will in this respect include:
• Monitor performance
• Take corrective actions.

In other words: Planning means establishing the project control baseline, while control means monitoring performance by comparing actual achievements against the control baseline and also taking necessary corrective actions.

4 STRUCTURING THE SCOPE OF WORK

Structuring the project scope of work involves establishing one of more hierarchical structures of the work, usually obtained by a top-down approach. Usually there are three different ways to structure the scope of work:
• Technical
• Organizational
• Economical.

This means that it is not only the physical tasks to be accomplished that are structured, but also the project's organization involved as well as the costs associated with accomplishing the tasks. **Technical structuring** means breaking the project down into managerial work assignments. **Organizational structuring** means breaking the project organization down into departments, groups, etc. to which work can be assigned. **Economical structuring** means breaking the total cost down in items suitable for exercising economical control.
   The basic methodology for structuring projects, stems from the US Department of Defence cost/schedule control system criteria from the late 1960ies. This basically operates

with a two dimensional structure:
WBS - Work Breakdown Structure
OBS - Organization Breakdown Structure.

These correspond to the earlier mentioned technical and organizational breakdown.

A **Work Breakdown Structure** (WBS) is a hierarchical breakdown of physical units (products/parts) and tasks (work activities and services) that are necessary to complete the project. A WBS is therefore an end-item oriented tree showing the sub-division of the project, products, deliverables, or items to be built, work tasks and services required to complete the project. It is very similar to a traditional bill of materials used in production planning. However, whereas the bill of materials always shows physical units of the product, the WBS may also include activities, services, etc. Few guidelines and almost no theory exist on how to do a WBS. This is in contrast to the impact the design of the WBS may have on success or failure of the project.

The **Organization Breakdown Structure** (OBS) is constructed in a similar way as the WBS. It shows the internal organization of the project. Level 1 represents the total organization. Level 2 represents the breakdown on the first division of the main organizational elements, etc.

The WBS breaks the project down until its elements at the lowest level are identified, and they are in turn divided into work assignments for individual groups. The OBS breaks the organization of the project down until the individual functional groups are identified. The contribution of each of these groups to the project is then made up of work assignments on the individual lowest level WBS elements. Thus the work assignments of individual groups on the lowest level WBS elements are common to both the work and organization breakdown structures and represents the foundation blocks of both structures. This two-dimensional structure is illustrated in figure 2.

The intersection between the two structures represents a work element with the following characteristics:
• A single person is responsible for it
• A formal specification of the work involved
• Its own estimates
• Its own plans for work schedule, resources and expenditure reports
• Its own analysis and reports.

Such an element is referred to as a **cost account**. This terminology is in line with the US Department of Defence's specification and its now being applied in many of the available software packages for project management. However, the term does not fully indicate the significance of the work element since it focuses too much on cost.

A cost account will include a scope of work and will identify the organizational unit responsible. By aggregation it is possible to obtain cost and performance at all levels of WBS and also per organizational unit at all levels of the OBS.

The cost account can be further split into work assignments. Each such assignments is denoted a **work package** (WP). A work package will represent a discrete task, activity, job or material item. In the majority of the cases it is represented by an activity used in planning. It will typically have a defined start and finish, an end product or goal of some kind, a relatively short duration and it will be the responsibility of a single organizational entity.

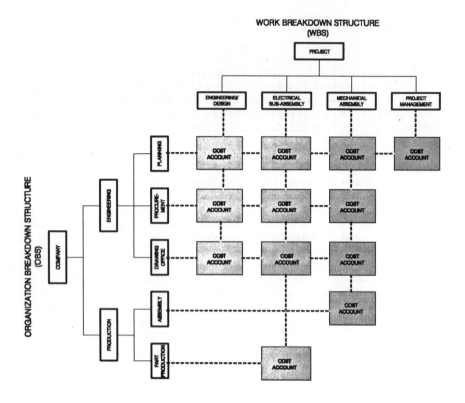

Figure 2 Integration of OBS and WBS into Cost Accounts.

In figure 3 it is shown how each cost account is split into work packages. For each work package the following data are planned:
• Cost
• Time
• Resources.

These are the crucial planning data. A common practice is to assemble these data on a form also including a description of the scope of work. Such a form is called a **CTR-sheet** and include all the data for that WP in the project control baseline. The collection of CTR-sheet are referred to as a **CTR-catalogue** and this contains the complete project control baseline.

In figure 3 the scope of work represented by work packages is shown and it is indicated a schedule and a cost estimate for each work package. For the cost estimate a new break-down structure is introduced: **Cost Breakdown Structure** (CBS). The CBS is usually a hierarchical breakdown of the resources involved. It is also referred to as code of account.

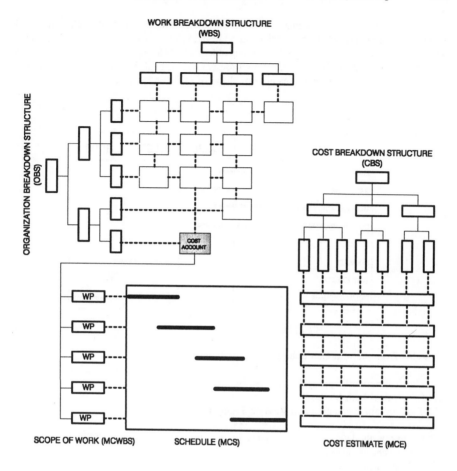

Figure 3 Project control baseline.

5 PROJECT SCHEDULING

Scheduling of projects is usually different from scheduling of batch production. The main difference is that projects usually requires that precedence relationship between activities are more important and therefore network type of techniques have to be applied. Of course, the most used scheduling techniques is the bar chart or the Gantt-chart. No diagram has ever conquered this in ability to consider dependant activities. Dependency is well handled by networks. The output of a network, may however be communicated by a Gantt-chart.

Figure 4 shows an example of a Gantt-chart. The duration of each activity is indicated by the length of the bar. Milestones are indicated by triangles. Such a Gantt-chart can be extended by linking the activities dependent on each other. It can also be used to monitor progress. A second bar can show progress. A vertical line indicating current date, can be used to compare if activities are on, ahead of or behind schedule.

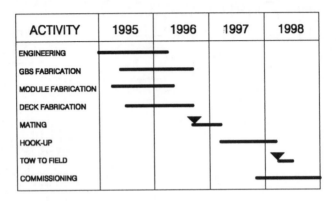

Figure 4 Example of a Gantt-chart with milestones.

The classical network approach comprise two techniques:
CPM - Critical Path Method
PERT - Program Evaluation and Review Technique.

CPM work with deterministic estimates of activity duration whilst PERT is capable of handling stochastic activity duration.
        The purpose of a network planning is:
• To determine the project duration
• To determine which activities that influence project duration.

Calculation is done in two steps: a forward and a backward one adding and subtracting duration respectively. In this way the following data for each activity is fixed:
ES - Early Start
EF - Early Finish
LS - Late Start
LF - Late Finish.

In addition float is determined. Float is the flexibility we have to move an activity in time without affecting the project duration. Activities with zero float are critical. A chain of critical activities from start to finish represent a critical path.

Activity duration is usually estimated using the trapezoidal method. This assumes that the level of resource consumption follows a trapezoid as depicted in figure 5. The area of the trapezoid corresponds to the scope of work. If the build up and run down periods are known the total activity duration can be expressed as:

$$T = \frac{S}{c \cdot B_{max}} + \frac{1}{2}(t_o - t_n)$$

where:

S - Scope of work
c - Effective manhours per time unit and per work position
B_{max} - Maximum manpower
t_o - Build up period
t_n - Run down period.

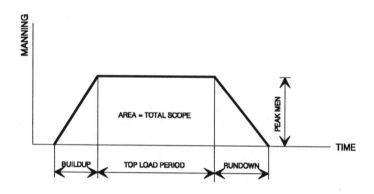

Figure 5 Example of a load profile.

The classical approach to network scheduling has later been extended to cover other types of problems. The most important extensions are:
• Overlapping activities
• Time/cost tradeoff
• Resource allocation.

Overlapping activities are possible by **precedence networks**. Such networks allow any combination of the four basic types of precedence relationships shown in Figure 6. The variable x indicates the overlap or time phasing involved. Such networks are useful since they allow a more realistic planning.

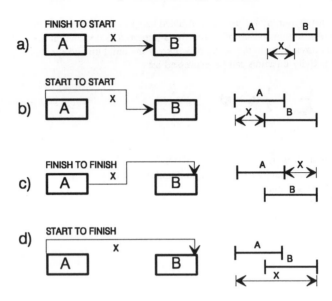

Figure 6 Activity dependence in a precedence network.

Time/cost tradeoff is based on the existence of a mathematical relationship between the cost of an activity and its duration. By associating a cost proportional to the total project time, an optimal solution to the duration of each activity may be found. The problem is combinatorial by nature, and heuristic techniques will have to be applied to obtain a practical solution.

Resource allocation is also a combinatorial problem and likewise calls for heuristic. The quality of many software packages for project planning, differs with the respect to the quality of the resource allocation algorithm applied. Two types of problems are usually considered:
• Resource smoothing
• Resource allocation.

The two main variables in such a resource problem are the level of available resources and the project time. Only one of these parameters may be kept fixed.

In the resource smoothing problem the project duration is fixed and the resource level is kept as low as possible. In the resource allocation problem, the resource level is given and the project duration is minimized.

In both cases the flexibility in planning is first of all in utilizing the available float in the network.

6 COST ESTIMATING

The main overall purpose of a cost estimate is to serve as the basis for cost control, i.e. to control that the amount of resources used is kept within the limit anticipated in project evaluation. The economics of a project is based on some assumption of total costs. Any serious deviation from this may affect the total project economics, and may turn a profitable project into a non-profitable one. It is a very important task for project management to control this situation.

Experience shows that when a cost estimate is set up, not all costs are identified. Therefore, any cost estimate will include a **contingency** to serve as a provision for those unidentified costs that are likely to occur. However, contingency does not include variations not likely to occur. Real life has shown that such changes do occur, and a different type of provision is therefore in addition necessary. This type of provision is called a **project reserve**.

The calculation of a project contingency is a matter of great concern. A theory has been developed based on the assumption that the estimate provided by an estimator is the most likely value. For control purposes, however, a fair estimate should have equal chances of overrun and underrun. In statistical terms this means the median. In this context the median

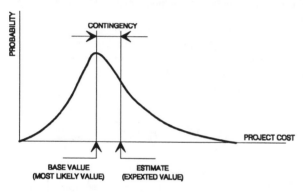

can be approximated by the expected value which is more convenient to compute. Assuming that the estimated costs follows a skewed distribution as depicted in Figure 7, the contingency can be defined as the amount that must be added to the base estimate to obtain a 50/50 probability. The base estimate is then defined as the most likely value provided by the estimator based on the existing drawings.

Figure 7 Base estimate and contingency.

For estimating two principally different methods apply:
• Synthetic methods
• Analytical methods.

The synthetic methods are applied in early stages and can with limited accuracy be applied to obtain a quick rough estimate. The basis is known cost of some similar facilities. The estimate is computed by applying scale factors and price indexes to this cost.

There are two types of synthetic techniques:
• Relational estimating
• Factor estimating.

The relational estimating is based on a previous similar project with a known cost. The assumption is that the new project will have the same cost. However, corrections are made to compensate for:
• Difference in capacity
• Different location
• Different time.

Factor estimating is based on the assumption that the relative share of various cost categories are invariant. Estimating one category, then gives the total cost split on the categories if the shares are known. The first to discover this was the chemist H.J. Lang. He estimated the equipment costs and developed the following formula:

$$C = E \cdot F$$

where C is the total costs, E the estimated equipment cost and F a factor (Lang-factor) defining the cost share of the equipment. For chemical plants F is approximately 3,6. The accuracy of factor estimating can be enhanced by splitting on several categories and considering only battery limits.

The analytical methods require more work as they are based on some breakdown of the total costs. This, however, requires some engineering to be done. The cost is estimated for each item in the WBS, and a total is summed up. The estimate is usually based on some computation of bill of quantities applying units rates and price and productivity indexes. Figure 8 illustrates the estimating process by an estimating chain.

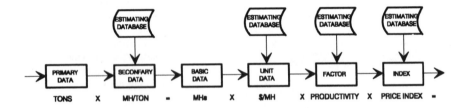

Figure 8 Example of estimating chain for labour cost.

7   PROJECT CONTROL PRINCIPLES

Project control shall secure that plans are followed with respect to scope of work, time and cost. In the succeeding some of the main principles applied in project control will be described and briefly discussed. These principles can be applied to control of work volume and to a certain extent to control of time. Control of volume, control of time and control of costs are the three different types of control applied. Control of costs is not discussed in this text. Here the main focus is on control of volume.

The main project control principles fall into six categories:
• Earned value
• Productivity development
• Resource availability
• Physical completition
• Tendency development
• Forecasting

The principle of **earned value** involves a measurement of what has physically been completed. This is compared to the planned value. Thus a real measure in volume of work of how far we have progresses compared to the plan is obtained. There are two options, either to measure manhours (earned hours) or costs (earned value).

Figure 9 shows an example. The solid curve (ECWS) represents the planned manhours (cumulative). The dashed curve (ECWP) shows the earned manhours. At the time of reporting (t), the difference between ECWS and ECWP represents a delay in volume of work, i. e. the number of manhours not completed according to the plan.

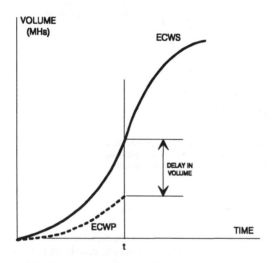

ECWS - Estimated Cost of Work Scheduled
ECWP - Estimated Cost of Work Performed

Figure 9 Progress Control of work volume.

The second control principle is to study **productivity development**. Productivity in this respect is defined as the ratio between planned and actual resource consumption. Productivity is therefore a measure of how close to the plan the real progress is.

Figure 10 shows a progress diagram based on costs. It shows the estimated costs of work scheduled and earned value (estimated costs of work performed). (In addition the actual cost of work performed is shown). The productivity is the ratio between the curves ECWP and ACWP. At the time t, the total cost deviation is the difference between ACWP and ECWP. This is the sum of a budget deviation (ACWP - ECWS) and a volume deviation (ECWS - ECWP).

A full project control must be based on three curves shown in Figure 10. However, this is seldom the case in practice. Many companies only consider the ACWP and ECWS curves, i.e. they focus only on the budget deviation.

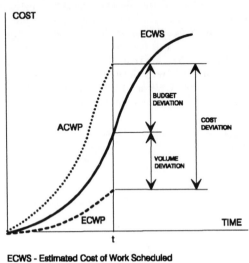

ECWS - Estimated Cost of Work Scheduled
ECWP - Estimated Cost of Work Performed
ACWP - Actual Cost of Work Performed

Figure 10 Example of a progress control diagram.

A poor productivity may be due to the fact that the work requires more resources then what was anticipated. The reason for a deviation in volume may be due to insufficient availability of resources. It is therefore important to monitor the **resource availability**. This is the third principle, and it is usually done by a histogram showing actual versus planned cumulative manning.

The progress diagram used so far gives no information on schedule progress. The best way to do this is by comparing with milestones. This is called monitoring the **physical completition**, and this represents the fourth control principle.

Most companies look to cumulative or aggregate figures when controlling a project. However, the first derivative, i.e. the tendency, gives better information. It opens a possibility for the management to react on a deviation at a much earlier stage. **Tendency development** is the fifth principle.

These tendencies may also be used to forecast where the project will end with respect to cost and schedule. This type of information is usually very valuable since it increases the awareness of reaching the project goal. **Forecasting** is the sixth and last principle. This is illustrated in figure 11 where expected delay and cost overrun at project completition are shown. The forecasted scope of work or cost is usually done by assuming that the productivity will remain the same for the rest of the project as up to the time of reporting (t).

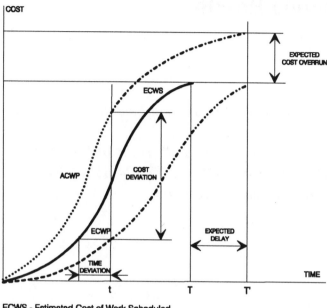

ECWS - Estimated Cost of Work Scheduled
ECWP - Estimated Cost of Work Performed
ACWP - Actual Cost of Work Performed

Figure 11   Progress control diagram with forecast to complete.

8   BIOGRAPHY

Asbjørn Rolstadås is professor of production and quality engineering at the University of Trondheim, The Norwegian Institute of Technology. His research covers topics like numerical control of machine tools, computeraided manufacturing systems, computeraided production planning and control systems and project management methods and systems. He is a member of the Norwegian Academy of Technical Sciences, the Royal Swedish Academy of Engineering Sciences and APICS. He serves on the editorial board of the journal Computers in Industry, and is the editor of the International Journal of Production Planning and Control. He is president of IFIP and past chairmen of IFIP TC5 and WG5.7. He is also past president of the Norwegian Computer Society and the Nordic Data Processing Union.

25

Standard Analysis of Share of Production Results

Wen Yuan Lin
Graduate Institute of Management Science, Providence
University
200 Chungchi road, Shalu 43309 Taichung Hsien, Taiwan
TEL:(04)632-8001 ext. 610   FAX:(04)632-2450
E-mail:wylin2 @simon.pu.edu.tw

Abstract

Standard Analysis of share of production results is a specialized application of the Standard Cost principal. It breaks down and reveals the sources of payoffs and deficits under the Share of Production Plan. These sources include: (1) physical productivity, (2) selling price, (3) raw material consumption, (4) indirect manufacturing expense, (5) administrative expense, (6) selling expense, and (7) hourly labor rate. Thus, through the operation of Standard Analysis it is possible to take the dollars of a payoff or deficit and to determine precisely how many dollars were obtained or lost by each of these seven sources being better or worse than Standard. Moreover, the results of Standard Analysis can be supported by statistics which compare actual details with their respective standards.

Keywords

Standard Analysis, Share of Production Result, Value Added, Rucker Plan, Standard Cost

1 INTRODUCTION

Although the means for developing a gain sharing have been around for many years, more have been studied about the employee involvement, communications, and other behavioral aspects of gain sharing than the measurement aspects. In this paper, the standard analysis of Share of Production Results which is based on the standard cost principle is developed. The study is to provide a framework for measurement of Share of Production Results. It breaks down and reveals the sources of (1) physical productivity, (2) selling price, (3) raw material consumption, (4) indirect manufacturing expense, (5) administrative expense, (6) selling expense, and (7) hourly labor rate. Thus, through the operation of Standard Analysis it is possible to take the dollars of a payoff or deficit and to determine precisely how many dollars were obtained or lost by each of these seven sources being better or worse than Standard. Moreover, the results of this Standard Analysis can be supported by statistics which compare actual details with their respective standards.

2 ESTABLISH STANDARDS

The first step in establishing the basic standards for this analysis is to take the Share of Production Base Period (1977-81) and organize all the essential figures. In the case of we shall work with annual data for the fiscal years ended Jan. 31, 1977 through 1981 inclusive. So the analysis involves 1982 operations compared to that base period.

a. Compute regression Equations - We regressed Sales Value of Production in turn on Material, Indirect Manufacturing Expense, Administrative Expense, Selling Expense, Wages, and Salaries. We then regressed Material on Sales Value of Production, Indirect Manufacturing Expense, Administrative Expense, Selling Expense, Wages, and Salaries. We went ahead, right through the list, regressing each account category on all the rest. This gave us all standards needed. Here are the equations for Sales Value of Production as a sample:

SV of P = -2.315 6062 E+4 + 3.338 3483E+0 * Material

  = 1.287 1094 E+5 + 9.101 4462 E+0 * Ind.

  = 2.064 5012 E+5 + 2.039 7615 E+0 * wages

  = -1.095 7931 E+5 + 1.506 5105 E+1 * Administrative

  = -2.599 4575 E+5 + 2.837 0581 E+2 * Selling Exp.

  = -3.462 2937 E+4 + 1.486 1968 E+1 * Salaries

We need one more equation which was not included in the original audit, namely, Wages regressed on man-hours.

3 COMPUTING STANDARD VALUES

Having all the tools to work with as described above, we now have a particular year (1982) under the plan to which we wish to apply Standard Analysis.

a. Preliminary Computations - Our first step is to write a program by using the SAS to solve the equations described above and average the results. We get:

Sales Value of Production estimated from:

Material	930,042.56
IME	1,136,484.77
Wages	944,682.64
Admnstv.	1,389,564.35
Selling	607,626.62
Salaries	1,204,671.99
1982 Actual	<u>1,095,989.00</u>
Average Including 1982 Actual	1,044,294.85

We repeat this process for the other components of the model and thus obtain the standards we need:

Sales Value of Production	1,044,294.85
Materials	319,459.23
Indirect Mfg. Expense	99,438.00
Administrative Expense	76,000.72
Selling Expense	<u>4,040.91</u>
Purchases of Intermediate Product	<u>498,938.86</u>
Gross Product	545,355.99

Standard Wages estimated from Standard Gross Product using the equation from the revised regression summary: 402,166.36

Standard Man-hours estimated from the regression of wages on man-hours:

68,681.28

4 PRODUCTION OVER OR UNDER STANDARD

In any given Share of Production Period, Actual Units produced will always be more or less than Standard units produced. In our analysis, "Units" means Gross Product Dollars. The number of boards produced by ' A' company is not really a useful category of output.

a. Production Over Standard - If actual units produced are more than standard, then part of the Actual Sales Value of Production and its difference from Standard will be the "normal" sales income expected from that excess production. Similarly, there will be normal material costs, and normal supply costs, administrative costs, and selling costs chargeable to this excess production, and by subtraction there will be "normal" gross product dollars due to it. Likewise, there is a "normal" labor cost input that is attributable to it, which is the normal Gross product value times the Share of Production per cent.

b. Production Under Standard - Conversely, when actual units produced are less than Standard, we have the same normal differences, only in reverse. That is, there will be a lack of sales income from the production that was not made, and similarly a saving in the cost inputs not expended, with a lack of gross product dollars from the net difference. And there will also be a saving in Wages which were not put into the production that was not made.

c. Segregation of Normal Differences - It is our purpose in Standard Analysis to separate out such differences as are attributable to normal income and normal costs (positive or negative) due to production being over or under Standard, because the remaining differences from Standard are the true economic factors contributing to a Share of Production result, whether payoff or deficit. These normal differences, however, will be used later to prove the arithmetic accuracy of our Standard Analysis computations.

5 BASIC PRINCIPLE OF STANDARD ANALYSIS

The basic principle of Standard Analysis is that a Share of Production Payoff or Deficit is (1) the Difference between Actual and Standard Production Values multiplied by the Share of Production Per Cent, plus or minus (2) the Difference between Actual and Standard Total Labor Cost.

  a. Production Value Difference - The Production Value Difference must now be broken down into its component parts of Price, Material, etc. The Share of Production Per Cent times each of these component parts gives us Labor's share of the gains or losses due to these components.

  b. Labor Cost Difference - The Labor Cost Difference must also be broken down into its component parts of Labor Rate and Physical Productivity.

  c. Combined Difference - When we combine algebraically Labor's Share of the Production Value components with the Labor components, we arrive at the figure which is the actual Share of Production payoff or deficit.

  d. Income and Cost Differences - In general, Sales Value of production and Gross Product are elements of income, and to show favorable differences, actual income figures must be higher than Standard. Material cost, supply cost, etc. are elements of cost and to show favorable differences actual cost figures must be lower than Standard. Unfavorable differences, of course, occur when income figures are lower than Standard and the cost figures are higher than Standard.

6 DETAILED STANDARD ANALYSIS OF THE PRODUCTION VALUE DIFFERENCE

First of all, we need the 1982 Actual results:

Sales Value of Production	1,096,989
Materials	285,530
Ind. Mfg. Exp.	110,727
Administrative Exp.	99,511
Selling	3,058
Pchs. of Intermediate Product	498,826
Gross Product	598,163

Now, let's assume that 'A' company has an operating Rucker Plan. Standard wages, or what Rucker calls Share of Production (SOP) Credit, can be calculated from the wages equation of the basic economic audit:

SOP Credit for GP = 598,163	448,176.44
Less: Actual 1982 Wages	361,921.00
Share of Production Balance	86,255.44

So, if the company had been operating during fiscal 1982 with a Rucker Plan, the manager would have had to shell out a bonus of $886,255.44. We need one more actual figure, man-hours. For 1982 it was 43,040.

a. Price Component - The basic 1977-81 standard for Sales Value of production is:

$$SV \text{ of } P = 1.564\ 2312\ E+4 + 1.885\ 1690\ E+0 * GP$$

If we substitute 598,163 in this equation and solve it we get standard SV of P = 1,143,280.66. Therefore, the Gross Difference due to price is 1,096,989 (actual) - 1,143,280.66 (standard) = ($46,291.66), which is unfavorable as designated by the company.

b. Material Consumption Component - Standard Material is given by:

$$Material = 1.235\ 85965\ E+4 + 5.628\ 04228\ e-1 * GP$$

Now, look at "Normal" as a volume variance and the others as budget variances. It is calculated as 598,163 (actual) - 545,355.99 (standard). A proof of arithmetic precision is had in the summation.

7 DETAILED STANDARD ANALYSIS OF THE LABOR DIFFERENCE

The following computations will show how the various components of the Labor Difference are determined.

a. Normal Labor Input for Production Over or Under Standard - If the Normal Gross Product Difference is multiplied by the Share of Production Per Cent, the resulting figure will show the normal amount of Labor Cost consumed in Production over standard, or saved in production under standard. So we have:

$$0.8712\ 8729\ E-1 * 52,807.01 = 46,010.09$$

This represents standard labor for production over standard, so as to the Labor Difference it is unfavorable.

b. Labor Rate Component - Standard wages for actual man-hours is calculated from:

Wages = -6.056 0876 E+4 + 6.737 3123 * 43.040 =   229,413.05

Less: 1982 Actual   361,921.00

Labor Rate Difference   (132,507.95)

c. Physical Productivity Component - Having determined the portion of the Total Labor Cost difference that is due to production over or under Standard, and the portion that is due to difference in average labor rate, we find that the balance is due to physical productivity better or worse than Standard. In computing the dollar value of physical productivity, we must determine a figure which when an added algebraically to the normal labor cost difference. For the company we used in this study, there are two basic elements involved in placing a dollar value on physical productivity above or below Standard: Man-hours and value.

We have:

Standard wages for Standard Man-hours(68,681.28)   402,166.36

  Less: Standard Wages for Actual Man-hours(43,040)   229,413.05

  Dollar savings from savings in man-hours   172,753.31

  Less: Normal Input   (46,010.08)

  Physical Productivity Difference   218,763.39

Then the Labor Difference is given by the algebraic sum:

   (46,010.08) + (132,507.95) + 218,763.39 = $40,245.36

which is favorable.

8 SUMMARY OF STANDARD ANALYSIS

With the completion of the above computations we are now ready to draw all the threads together into a summary.

a. Labor's Share of GP Differences - First, we must determine Labor's Share of the various components that make up the GP differences. We do this by multiplying these components by the Share of Production Per Cent. Doing this we have:

Labor's Share of GP Differences

[0.8712 8729 * Gross Differences]

Price	(40,333.33)
Material	55,306.96
IME	(661.93)
Admin.	(15,382.93)
Selling	1,071.24
Normal	<u>46,010.08</u>
Sum	46,010.09

b. Labor's Share of the Labor Differences - For these we take the actual dollar values of (1) the Labor Rate difference and (2) Physical Productivity. It will be found that the Share of Production Per Cent times the Normal Production Value Difference (7a above), will be offset by the figure for Normal Labor Input for production over or under standard. Thus our comparison of 1982 operations with the base period 1977-81 yields:

Physical Productivity	218,763.39
Price	(40,333.33)
Material	55,306.96
IME	(661.93)
Admin.	(15,382.93)
Selling	1,071.24
Labor Rate	<u>(132,507.95)</u>
SOP Balance	86,255.45

which exactly reconciles with the SOP balance noted in (6) above. We therefore may conclude as follows:

(i)The manager obviously made some huge reductions in his labor force in 1982, keeping the best, but at higher rates, relatively, than during the base period. Even though the force he kept cost more, they produced more.

(ii)The manager has a horrible Wages equation. Standard wages are 74% of the Gross Product Dollar, or 65% of the Value Added Dollar. That same figure for industry SIC 3679, all of the manager's competitors taken together, is 29%. So the manager's labor cost is 122.6% of the industry.

(iii)The way our model of the firm works, GP is a "derived" figure. If it is too small, the answer lies in increased price, which is bound to cause the manager some problems these days.

(iv)Since there is an unfavorable variance in Administrative Expense, the manager didn't take up the slack in the office that he did in the shop.

(v)If the manager had had a Rucker Plan in operation in fiscal 1982, and had had to disburse that $86,255.44 Share of Production Bonus, instead of showing a profit of $44,276, he would have shown a loss of $41,979.

Selected Data For Fiscal Year Ended 31 January 1982

	--------STANDARD---	--------ACTUAL----
Sales Value of Production	1,044,294.85	1,096,989
Materials	319,459.23	285,530
Indirect Mfg. Expense	99,438.00	110,727
Administrative Expense	76,000.72	99,511
Selling Expense	4,040.91	3,058
Purchases of Int. Product	498,938.99	498,826
Gross Business Product	545,355.99	598,163
	===========	========
Wages	402,166.36	361,921
Man Hours	68,681.28	43,040

GROSS DIFFERENCES FROM STANDARD
for the
Fiscal Year Ended 31 January 1982

Sales value of Production (Price)	(46,291.66)*
Materials	63,477.29
Indirect Mfg. Expense	(759.72)
Administrative Expense	(17,655.40)
Selling Expense	1,229.49
"Normal Difference"	52,807.01
Overall Difference	52,807.01

* ( ) indicates unfavorable variance

'A' company
SUMMARY OF FY 1982 PERFORMANCE
COMPARED TO 1977-81 BASE PERIOD
Labor Component

Physical Productivity	$218,763.39
Price	(40,333.33)
Material	55,306.96
Indirect Mfg. Expense	(661.93)
Administrative Expense	(15,382.93)
Selling Expense	1,071.24
Labor Rate	(132,507.95)
SOP Balance	86,255.45

9 REFERENCES

Graham-Moore, B. and Ross, T.L. (1990) Gain Sharing, Plans for Improving Performance, BNA Books, Washington, D.C.

Moore, B.E. and Ross, T.L. (1978) The Scanlon Way to Improved Productivity, Wiley-Interscience, New York.

Rucker, A. (1937) Labor's Road to Plenty, Page, Boston.

10 BIOGRAPHY

Wen Yuan Lin received the B.A. in Business Administration from Tatung Institute of Technology, Taipei, Taiwan . He received his M.S. in Operations Research from Case Western Reserve University, Cleveland, Ohio, and Ph.D. in Industrial and Management System Engineering from Arizona State University, Tempi, Arizona, in 1975 and 1993 respectively.

Dr. Lin is at present an associate professor of Graduate Institute of Management Science at Providence University. He is also the Director of Graduate Institute of Management Science. He has been active in professional and Management works for U.S. corporations for the past about twenty years.

26

Product and activity relationships in project deliveries

S. Törmä and M. Syrjänen
Helsinki University of Technology
Otakaari 1, FIN-02150 Espoo, Finland
tel.: +358-0-451 3250, fax: +358-0-451 3293
email: sto@cs.hut.fi or msy@cs.hut.fi

Abstract

The activities of a project where a complex product is created are mostly focused on the creation of the parts of the product or documents specifying the parts. We describe a way to represent a project so that the product and activities can be linked to each other. The linking facilitates the transfer of information between the technical side and the management side of a project.

Keywords

Project management, planning, scheduling, one-of-a-kind production

1 INTRODUCTION

Delivering a product such as a process plant to a customer is a long and complex task. It includes the engineering, manufacturing, procurement, shipment, installation and inspection of components, construction of buildings, preparation of manuals, and startup of the functional systems of the plant. An effort involving many disciplines and companies must be managed so that the delivery of the plant takes place at the due date of the project at latest.

This kind of tasks are nowadays typically managed as projects. The activities dealing with one particular product – but possibly belonging to different disciplines or even executed by different companies – are organized into one activity network. The scheduled activity network can be visualized and it provides the participants of a project a shared road map to achieve the goal of the project, the delivery of the product. Project management approach provides powerful scheduling and coordination methods (Pagnoni, 1990) as well as many commercial tools that support them.

The standard project management methods and tools focus on the management of activities and resources of projects. In project deliveries, however, the activities are mostly focused on the creation and manipulation of product related objects: parts, documents, and part groups. In standard methods the parts and activities are often related to each other only in an intuitive way: for example, the names of the activities may contain part codes or names of documents (see figure 1).

If the relationships between the parts and documents of a product would be explicitly represented in a project plan, information could be transferred between the product and the activity network in a more flexible way. This would benefit both the technical side and the management side of a project (figure 1).

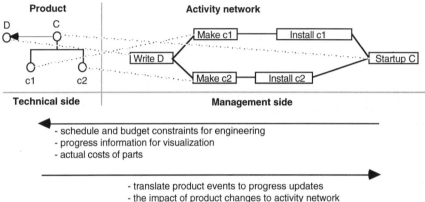

Figure 1 The linking between the technical and management sides of a project and possible interactions between the two sides

We have addressed this question in MUSYK-project* where a prototype of a special purpose project management tool called Process Plant Project Manager (3PM) has been developed. This paper describes how projects are represented in 3PM so that the technical and management sides of the project are linked to each other.

When a linking is established between product and activities, more accurate information about the schedule and budget constraints could be used in different technical tasks of the project. Engineering could take into account the schedule and budget constraints concerning the manufacturing of a part. Procurement management would have exact information of the schedule criticality of different purchases. Moreover, the progress of the product could be visualized from the product perspective. In addition, more accurate costing of the parts could be supported.

The advantages at the management side deal with the possibility to automatically translate information about product related events to progress updates in the project management

---

* MUSYK (Integrated <u>Mu</u>lti-level Planning and Control <u>Sy</u>stem for One-of-a-<u>K</u>ind Production) is the project number 6391 in ESPRIT III program.

system. Also the change management can be improved (Törmä, 1995). In addition, the correctness of a project plan with respect to the product can be automatically verified.

Many conceptual models for linking the product and activities in construction projects have been provided. Luiten (1994) contains a survey of different approaches and presents a way to create the linking through the states of a product. We extend the representation of the product in such a way that also the other phases of the project besides installation (or construction) can be covered with this approach. This makes it possible to apply this approach to project deliveries. In addition, we outline a way to link the activities of different product related objects by mapping the states of the objects to each other. We list some properties that the project plan should satisfy in its relation to the product and finally discuss about the benefits of the linking.

2   PRODUCT-ACTIVITY RELATIONSHIPS

This section discusses about the objects that are worked on in a project and their relationships to activities. In order to be able to represent the product-activity relationships we define concepts for all of the product related entities that are independently manipulated in a project. Then we show how the concept of state can be used to link parts, documents, and part groups to activities.

2.1 Project objects

Figure 2 contains a diagram of the classes that represent product related objects. The diagram has been drawn in the notation of Object Modeling Technique (OMT) (Rumbaugh 1990). Rectangles represent classes, triangles represent superclass/subclass relationships and lines between classes represent relationships. A black circle in the end of a relationship denotes a multivalued relationship.

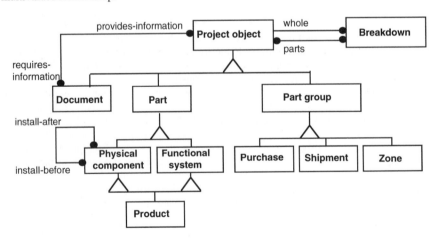

Figure 2   Project-object classes.

Project-object* is a class that contains all those objects towards which activities are directed in a project. It includes objects that are created during a project (parts and documents), as well as groupings of those objects.

A **part** represents any (aggregate or atomic) component of the final product. It is a superclass of both physical components and functional systems of the product. Examples of physical components are pipes, tanks, pumps, buildings, etc. Examples of functional systems are mass flow system, automatization system, etc.

Document refers to any information entity either on paper or in electronic form. Technical documents (drawings, process calculations, material lists) are mostly created in the engineering phase of a project and they provide information about the components for manufacturing, procurement and installation. There are also management documents such as budgets and schedules and commercial documents (inquiries, quotes, contracts).

Part-groups represent sets of parts that are processed together in some phase of a project. Examples are a purchase (a group of components that are procured from one vendor), a shipment (a group of components that are shipped together) and a zone (a group of components that are installed as one whole from the project management perspective).

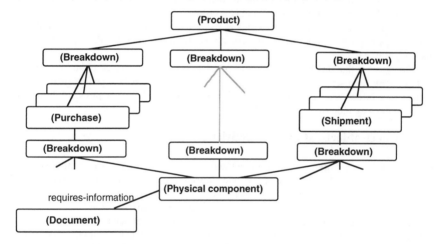

Figure 3 An instance diagram of a fragment of a product. In addition to the primary breakdown there are also breakdowns to purchases and shipments. One information requirement relationship is also presented.

Project objects may be decomposed into subobjects through the **breakdown** relation shown in figure 2. The primary breakdown is produced by engineers who design the product. During project planning new breakdowns will be developed to procurement and shipment purposes. The membership of an object in a part-group is represented with the breakdown relationship. Each part may participate to multiple breakdowns simultaneously. An example is shown in

---

* We use the term project-object in a more restricted meaning than it is used in (Luiten, 1994) where it covers also activities and resources. See Discussion.

figure 3 which contains an instance diagram in OMT-notation. The rounded rectangles represent instances and the label in parenthesis is the class of the instance.

A breakdown can be exclusive which means that its structure is a tree. New breakdowns may also be declared to be complete with respect to some other breakdown, most likely with respect to the primary breakdown.

Project-objects may have various other relationships between each other. We discuss here about two relationships that are interesting because they create precedence constraints between the activities of different project objects:

- **install-before** relationship between two physical components
- **requires-information** relationships between a project object and a document

Install-before means that a component has to be installed before another component. The reason for that relationship is usually a situation where one component is supported by another component (Darwiche, 1988) but may also arise as a consequence of safety regulations or from the need to provide weather protection to a component.

Requires-information relationship means that some project-object requires a document to be created. This can be another document (for instance, pump design document requires pressure loss calculations and layout drawings as its prerequisites) or a group (purchase that contains pumps requires pump design) or a component (to install a pump, the installation manuals are required). ,

3PM assumes that this kind of requirement relationships are provided either by some other system or by the user of the system. An example of the former is a geometrical reasoning system that infers what are the install-before relationships between physical components based on the way they support each other.

2.2 Project-objects, states and activities

Project objects are worked on in a project and they change as the result of the work. This means that a project-object can be in multiple **states** during a project. A state is represented as an entity that is associated with one project-object as shown in figure 4.

A state has a set of **state attributes.** They represent those aspects of the object that (1) may change during a project and (2) are interesting from the project management point-of-view. Location and owner satisfy both of these conditions while color, for instance, does not – in most cases – satisfy the latter one. Each attribute has a domain of allowable values.

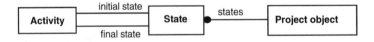

Figure 4   Relationships of activities and project-objects

The set of state attributes as well as the sets of their allowable values depend of the type of the project object. In 3PM the following set of state properties is used for a project object:

- physical: What is the physical status of the project-object (--, designed, created, in-use)?
- location: What is the location of the project-object (a geographical location)?

- ownership: Who owns the project-object (a company)?

States can be connected to each other through activities. An activity represents a transformation between two states. An example is shown in figure 5.

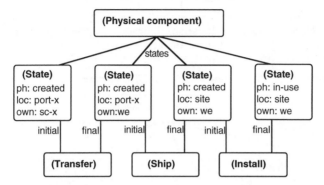

Figure 5   An instance diagram of a project-object, its states and the related activities.

An activity transforms the values of the state attributes between its initial and final states. For instance, the following is the definition for a transfer activity that changes the ownership of some project object while ensuring that the values of other attributes do not change:

```
(define-activity transfer project-object
  :initial (state :owner :?owner1
                  :physical :?physical
                  :location :?location)
  :final   (state :owner :?owner2
                  :physical :?physical
                  :location :?location)
  :test    (<> ?owner1 ?owner2))
```

2.3 Mapping the states of different objects

The states of different objects are not independent from each other. On the contrary, many of the states of an object are directly mapped to the states of some other object. An example of this is shown in figure 6.

The mapping can be established between the states of two objects that are related to each other through some the relationship discussed previously. Examples of how the mapping can be done are the following:

- Breakdowns: The initial state of independent phases of objects is mapped so that the similar state of a subobject must precede the state of a superobject (and in a corresponding way for the final state).

- Install-before: If B belongs to install-before(A) then the state where B is created must precede the state where A is installed (in-use).

- Requires-information: If D belongs to requires-information(A) then the state where D is created must precede the state where A is designed.

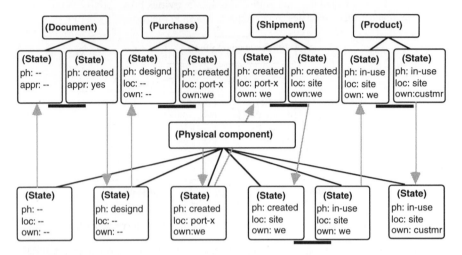

Figure 6   The mappings between the states of a physical component and the states of its design document, purchase, shipment and product. The gray arrows show the state mappings and the black bars show the independent phases of project-objects.

The gray links in the figure 6 show these mappings. For instance, when a document has been created, it means that the physical component that it describes has been designed.

Figure 6 also shows those phases – limited by the initial and final states – where a project object will be manipulated as an independent entity (the black bar below the states). Those are the phases that need to be planned by a project planner or an automated planning system. That is, activities that are able to transform the initial state into the final state must be created.

It should be noted that in the example situation the installation activities are the only activities that have to be planned to the physical component itself.

2.4 Activity network

The activity network for the project is composed of the network fragments of each project object. These fragments are connected to each other on the basis of state mapping relationships.

A mapping between two states S1 and S2 may either be translated into precedence constraints or hierarchical relationships between every activity ending in S1 and every activity starting from S2. This mapping takes care that, for instance, a transportation of a shipment may only be started after every purchase with which it shares common objects has been delivered.

Because the state mappings between groups go through the members of the group, we are able to verify that each member will meet the activities of those groups that it belongs to.

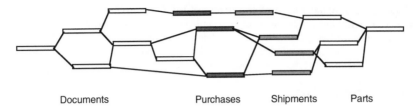

Documents   Purchases   Shipments   Parts

Figure 7   The overall structure of one level of an activity network.

2.5   Correctness of activity network

There are properties that a correct activity network should satisfy in its relation to the product.

- Completeness: The activity network is complete if the final state of each part can be achieved with an unbroken sequence of activities starting from its initial state within the project.
- Non-superfluity: The activity network is non-superfluous if each of the activities in the network contributes to the achievement of the final state of some of the parts.
- Accuracy: The activity network is accurate if all the parameters of activities (durations, budget values, etc.) are correct when derived from the product description.

The completeness of a project plan can be verified by checking that each of the states of each project object is achieved. Achieving means either that there is an activity that produces the state or that there is a state mapping that defines which other states must be achieved before this one and the state mapping is complete with respect to the relationships of the object.

The non-superfluity can be verified by checking each activity in the network and ensuring that both its initial and final states belong to some object of the current product description. Accuracy can be verified by going through all activities and ensuring that their values are same as computed values.

3   BENEFITS

The linking presented has advantages both at the technical side and at the management side of a project.

3.1   Technical side

The advantages gained at the technical side are that more accurate information about the **schedule and budget constraints** could be used. Engineering can take into account the schedule and budget constraints concerning the manufacturing of a part to be designed. Procurement management would have exact information of the schedule criticality of different purchases.

Moreover, the **visualization of the progress** from the product perspective could be supported. For instance, only the already installed components could be presented, or the components that are behind the schedule could be highlighted.

In addition, the linking could create a possibility to support more accurate **costing of the parts** of a product because the actual costs of activities could be directed to parts.

3.2 Management side

The linking provides the possibility to automatically use information of the **progress of documents and parts**. Product related events can be transferred to the project management system. For instance, the status information maintained by a document management system or a production management system could be transferred to 3PM. The approval of a document or the completion of a part could automatically be translated into an update of the status of the related creation activities in the project plan.

When engineering data management systems and identification methods for parts (e.g., bar coding) get into more widespread use, the product-activity relationships will provide possibilities to base the project coordination into much more reliable information than what can be provided by humans.

The linking can also support **change management**. When a product is changed, it would be easier to determine what changes to activity network must be done. This is described in more detail in (Törmä, 1995).

In addition, there are some **correctness properties** of the relationships between product and the activity network that could be checked by the project management system. An example is the completeness of the activity network with respect to the product. In practice this means that we will be able to verify that all parts of the product end up being designed, manufactured, shipped and installed. This removes a source of omission errors from a project plan.

4   DISCUSSION

We have described the way the relationships between product and activities are represented in a special purpose project management system, 3PM. The class diagrams presented are simplifications of the actual representation of 3PM. However, they show the main concepts and the nature of relationships.

The research to automatically generate the activity network for a project delivery is under way and will constitute the main contents of our further research work.

We have only discussed about the way to link product related objects (project objects) to activities of a project. This covers the most value-adding activities of a project delivery. However, there are also other kinds of interesting activities. Some of them deal with setup and maintenance of resources. Examples are construction of temporary resources such as scaffoldings, cleaning activities and activities where large resources such as cranes are moved. Furthermore, there are activities are focused on other activities. These could be called management activities and they include the planning and coordination activities of a project.

It is worth further research whether the representations discussed in this paper could be extended to cover also resource and management activities. We believe that it is possible if a proper way to represent the states of resources and activities can be devised.

5   REFERENCES

Luiten, G.T. (1994) Computer aided design for construction in the building industry, Ph.D. Thesis, Technical University of Delft.

Pagnoni, A. (1990) Project engineering: Computer-oriented planning and operational decision making, Springer-Verlag, Heidelberg.

Darwiche, A., Levitt, R.E. and Hayes-Roth, B. (1988) OARPLAN: Generating project plans in a blackboard system by reasoning about objects, actions, and resources. *Artificial Intelligence in Engineering Design, Analysis and Manufacturing*. Vol. 2, No. 3.

Rumbaugh J., M. Blaha, W. Premerlani, F. Eddy, W. Lorensen, (1990) Object-oriented modeling and design, Prentice-Hall, ISBN 0-13-630054-5, New York.

Törmä, S., Karvonen, S. and Syrjänen, M. (1995) Management of changes in one-of-a-kind production, Computer Applications in Production and Engineering: Proceedings of CAPE'95, (ed. Sun, G.H.), Chapman-Hall.

6   BIOGRAPHIES

Seppo Törmä (MSc.) is a research scientist in Institute of Industrial Automation in Helsinki University of Technology. His interests cover project and production management, object oriented methods, constraint satisfaction techniques and cooperative problem solving. He is member of American Association for Artificial Intelligence.

Markku Syrjänen (Ph.D Eng) is the professor of Knowledge Engineering in Helsinki University of Technology. His field of special interest is knowledge-based production management. He has been the vice-chairman of Finnish Society of Information Processing Science 1982-83, member of National Research Counsel for Technology (a subunit of Academy of Finland) 1986-88, and chairman of Finnish Artificial Intelligence Society 1990-92.

27

Constraint production management using Fuzzy Control

Dr. K. Mertins, R. Albrecht and F. Duttenhofer

Fraunhofer-Institute for Production Systems and Design Technology (IPK)
Department Planning Technology
Pascalstr. 8-9, D-10587 Berlin, Germany
Phone ++49-30-39006-259, fax ++49-30-3911037
E-mail <frank.duttenhofer@ipk.fhg.de>

Abstract

The functions of short-time planning and control of orders on the factory's shop floor can be supported by Production Management Systems. A scheduling system capable of reflecting the planning staff's experience has been implemented. Due to the complexity of the scheduling problem and the typical lack of precision in shop expert's planning strategy descriptions a system was designed based on Fuzzy Logic.

This article presents concepts, theory and implementation of a Fuzzy Logic scheduling system for an aluminium foundry and mill.

Keywords

Production management system, order scheduling, fuzzy logic.

1 INTRODUCTION

Planning and control functions for the production process in manufacturing industries are supported by Production Management Systems (PMS). The operational component of a PMS has to optimize the production schedule without altering the production system or product sequence. The main objective of order scheduling as a basic subfunction of a PMS is to ensure that all products will be manufactured at minimum costs, with appropriate quality and in time. Minimum costs can be achieved in pursuing substitutional objectives like minimum inventories, high machine utilisation rates and short order throughput times [4].

Applying a phasing approach the complexity of the scheduling process can be reduced. The process can be divided into subsequent phases: coarse planning, capacity planning, order release, and order fine-scheduling. Order release and order fine-scheduling should be accomplished on the spot, integrating a motivated shop staff's expert knowledge and experience. By this the shortest possible reaction time can be achieved in order to cope with interrupts and delays of the manufacturing process. Shop floor control systems support these functions [2, 3].

2 DECISION MAKING SUPPORTED BY FUZZY LOGIC

The main objectives of order scheduling are the minimization of set-up expenditure and pursuing shortest lead times. Those objectives are contrary because minimizing the set-up expenditure will increase the stock on hand and the lead time of the order. For an optimal solution both targets have to be taken into consideration. Conventional methods of order scheduling will not solve the conflict because they generally aim at only one target.

A reduction of the set-up expenditure can be achieved by installing quickly acting tool change systems and performing multiple manufacturing steps in one chucking. This has been realized in metal cutting and assembly industries, whereas in metal forming industries similar efforts were made which reduced the importance of building machining sequences and allowed the concentration on lead time reduction.

In an aluminium foundry and mill the process steps include melting of aluminium, founding into bars, annealing, milling, stretching of plates and coils, and cutting. Process times vary from a few minutes to hours or days. An annealing furnace only works efficiently processing full loads, so flexibility cannot be increased reducing the lot size. The processing and therefore scheduling of optimal charges is the main planning objective. It cannot be pursued by manual order scheduling because of the complexity of order dependencies. An information system has to support the planning staff sequencing orders and collecting them to charges and machining sequences.

Nowadays also in forming industries shortest lead times have to be a planning objective in order to decrease production expenditure and delivery times and to gain access to new markets.

The ratio of set-up time to application time in the process of rolling metal alloys becomes disproportionate. Therefore an order fine-scheduling module of a shop floor control system designed for an aluminium mill [1] has to perform the additional task of building machining sequences depending on set-up state sequences. The quality of the module shows in

1. the applicability of the resulting machining sequences and in
2. the throughput time of orders.

An empirical analysis of currently practised manual planning strategies revealed that they can only partially be modelled using binary logic. Experts at the shop floor apply definite rules for order sequencing regarding one single set-up state. Their strategy for sequencing the set-up states within a planning period could only be put down unprecisely. The reasons are the conflict of the planning objectives mentioned above and the unprecise way of human communication and interaction.

Fuzzy Logic is a means for modelling problem solution strategies based on human experience, knowledge, intuition and heuristics. Humans preferably handle qualitative and not so often quantitative values when making decisions. Qualitative values can be modelled directly with Fuzzy Logic.

Conventional mathematical methods are often too complex or not appropriate for solving sequencing problems in a given time frame. Even when dealing with minor problems a fuzzy strategy can be applicable if exact values are not necessary or cannot be collected or used.

Adapting human problem solving strategies and fuzzy values of the variables needed in an automated planning system simplifies the documentation of the planning process and the evaluation of its results.

3 OVERVIEW OF FUZZY LOGIC THEORY

The following chapter presents Fuzzy Logic theory insofar as it is necessary to understand the implementation of the planning system [4].

Fuzzy Sets

Defining a precise set of values is usually done giving the exact values or the range of values. For any value it can be specified whether it belongs to a certain set or not. Another approach is to define a membership function $\mu(x)$ which specifies in binary logic whether a value x belongs to the set or not [5].

In a Fuzzy Set the values of the membership function are floating point values in the range between 0 and a finite upper bound. In a normalised Fuzzy Set the upper bound equals to 1. Fuzzy Sets can be implemented in a computer using arrays of values or defining a membership function. A function consumes less memory and allows an adjustable granularity of input values.

Classes of functions are each defined by a set of parameters, e.g. a triangle-shaped function is defined by a value m with a maximal function value $\mu(m)$, by the minimum and maximum values μ_{min} and μ_{max} and by the distances of m to both corners (a and b, see Fig. 1). Trapezoids and generalised LR-(left-right)-Fuzzy Sets can be defined, the latter using non-linear functions.

To reflect terms of human communication in a computer, linguistic variables can be defined as collections of Fuzzy Sets. Linguistic variables contain common qualifying words and phrases in contrary to numerical variables.

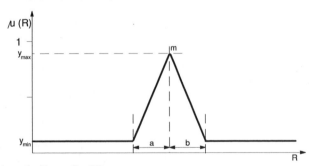

Figure 1 Triangular Fuzzy Set [5]

Operators on Fuzzy Sets

Elementary operators on Fuzzy Sets are intersection (minimum) and union (maximum) and the unary complement. The minimum and maximum operators satisfy the commutative and the associative law which enables a recursive combination of more than two sets in any processing order. Further operators can be defined by generalising the elementary operators. A combination of minimum and maximum behaviour of an operator is feasible using parameters. Criteria for the classification of Fuzzy Set operators are e.g. associativity, commutativity, monotony, continuity, and other mathematical properties. An important non-mathematical criterion is the adaptability of an operator by parameters in order to guarantee an optimal reflection of human implication. The method of modelling human terms in operators has not yet been put on a theoretical basis and is due to the user.

Fuzzy Logic

Fuzzy Logic provides methods for fuzzy implication using Fuzzy Sets. The expert's knowledge needed for automated inferencing is implemented in "IF condition THEN conclusion" rules. The linguistic variables in those rules are combined using Fuzzy Set operators. All facts and rules are assigned a special degree of certainty. Applying the rules should lead to useful and problem solving results. Therefore the definition of Fuzzy Sets of a linguistic variable should satisfy several conditions, e.g. limitations of the degree of overlapping regions. Consistency and completeness of the rules are a precondition.

4 SET-UP STATE SEQUENCING USING A FUZZY LOGIC APPROACH

In Fuzzy Logic calculations, functions replace simple numbers, which leads to an increased expenditure compared to binary logic. Therefore applying Fuzzy Logic is only recommended if a solution can not (or can only with a disproportionate effort) be found applying conventional methods, as in the example application described here. Fuzzy Logic is used for the selection of the next appropriate set-up state out of a set of possible solutions. All other planning tasks are done by numeric and heuristical strategies [4, 6].

The determination of the sequence of set-up states is initiated by an analysis of all operations in a certain time interval. For all possible set-up states the time interval until the next set-up is calculated. The number of operations in this time interval is counted. So the mixture of orders to plan is changing with the ongoing scheduling process. The determination of the sequence of set-up states and the scheduling of operations are performed interchangeably. Only that set-up state is searched which shall be realised at the current state of planning. After that step all operations belonging to the chosen set-up state are scheduled until one of the stopping criteria matches (Fig. 2).

A Fuzzy software tool has been implemented for the calculation of an optimal set-up sequence. The way of implementation is analogous to the development of a Fuzzy Controller (Fig. 3). For the evaluation of the result of a sequencing process both time- and machine-specific criteria have to be considered.

The slack time of an order is one of the time related values. The smaller the remaining processing time the smaller the slack time may be without increasing the likelihood of a delay of the completion time. In order to be able to determine the slack time of orders with different remaining processing times the slack time has to be put in relation to the remaining processing

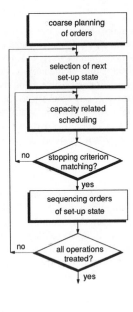

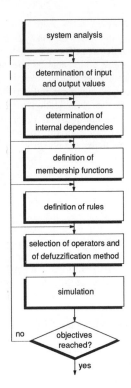

Figure 2 Process of planning [4]

Figure 3   Development of a Fuzzy Controller [6]

time. The result is called relative slack time. Assumptions are small transportation and laying-times in relation to processing times [7].

In case of the factory described here those conditions are not satisfied. Processing times sometimes are much shorter or even as long as the technically necessary laying-times. Therefore the relative slack time is calculated using the minimally needed remaining lead time. Transportation times are short in relation to transition times and have therefore been neglected.

The relative slack time is bound to an order, so a similar value related to set-up states has to be found which represents the relative slack time of orders in the circulation time interval. This value is calculated as follows: sum of all relative slack times of orders in the possible circulation time interval divided by the number of operations in that time interval. This value is called relative buffer.

For a roll of a mill the machine-specific value is the degree of utilization of the capacity dependent on its life time (number of planned orders). It is calculated using the number of operations with equal roll types in the circulation interval.

For the use in Fuzzy Logic all input values and the results are translated into linguistic variables. An example of Fuzzy Logic application describes the selection of the succeeding roll type of an aluminium rolling mill. The exact input parameters are transformed into linguistic

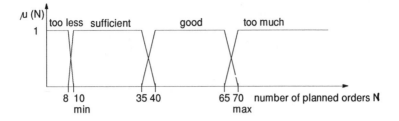

Figure 4 Membership functions of the linguistic variable `number of planned orders´ [4]

variables, as well as the output parameter which contains the decision for a specific set-up state. The membership functions of the linguistic variables are shown in Fig. 4, 5 and 6. The Fuzzy Sets are implemented in the shape of trapezoids. This shape should be used if the output of the rules shall not change throughout ranges of the input parameters, that means, a separation of linguistic values and their meanings is not possible or convenient.

The linguistic variable named number of planned orders (Fig. 4) is defined by four membership functions. The positions of the maximum values have been determined in correlation to the minimal (10) and maximal (70) number. Those values correspond to special roll types, so for each roll type separate membership functions have to be specified.

For the linguistic variable relative buffer five membership functions had to be specified (Fig. 5). The positions of the maximum values were chosen arbitrarily. Realistic values will be determined in the future along with the test of the module under production conditions. The width has been set to 1.5 shifts so far and has to be adapted to the maximum values.

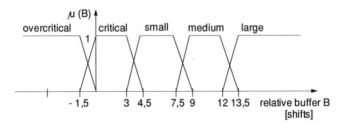

Figure 5 Membership functions of the linguistic variable `relative buffer´ [4]

The linguistic variable quality of set-up state is modelled by five membership functions (Fig. 6). The values shall indicate the quality of a set-up state in correlation to the input parameters (value in percent of quality). The positions of the maximum values are spread homogeneously, the width was set to 5 %.

The conclusion is based on rules of the type "IF number of planned orders is sufficient AND relative buffer is small THEN ... ". The aggregation operator determining the degree of matching the facts with the input parameters has to be realised by an AND operator. Because of its high efficiency the minimum operator was chosen.

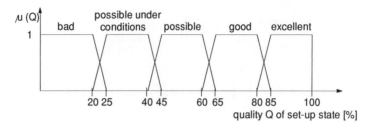

Figure 6 Membership functions of the linguistic variable `quality of set-up state´ [4]

To implement an implication operator determining the degree of validity of the conclusion the minimum operator was taken. As a consequence the minimal membership value of both input parameters (number of planned orders is sufficient and relative buffer is small) produces a value of the membership function "good" of the linguistic variable set-up state quality. No relevant factors were defined, i.e. all rules are regarded as being reliable without any restrictions. The factors equal to 1 for each rule. A necessity might arise in the future to introduce relevant factors in the model.

The single results of all rules are combined by accumulation choosing an OR operator. The result represents the final solution of the inference problem. The maximum operator was chosen because of its high efficiency.

The rules are based on the statements of experienced staff. The rules type is of the form "IF number of planned orders is sufficient AND relative buffer is small THEN quality of set-up state is good". In Table 1 all combinations are presented to form rules.

Table 1 Rules for valuation of a set-up state [4]

		number of planned orders			
		too less	sufficient	good	too much
	overcritical	good	excellent	excellent	excellent
	critical	possible	good	good	good
relative	small	possible under conditions	possible	possible	good
buffer	medium	bad	possible under conditions	possible	possible
	large	bad	bad	possible under conditions	possible under conditions.

Final result of the inference method is a Fuzzy Set used to evaluate a given set-up state. To be able to compare the result with other solutions it is necessary to retranslate the obtained Fuzzy Set into a concrete value (defuzzification). Methods for defuzzification are:
- division of the Fuzzy Set in two halves of the same area size and
- calculation of the centre of gravity of the resulting Fuzzy Set.

The resulting value will be the best representation of the information contained in the input Fuzzy Sets. In the current implementation both methods are applied in order to gain further experience.

5 INTEGRATION IN THE PRODUCTION MANAGEMENT SYSTEM

A production planning system performs the coarse planning of all orders. The results are transferred to the scheduling system to determine the final order sequence for all machines.

The sequence can then be changed manually using disposition tools with dynamic graphics. These tools also allow an automatic recalculation of the sequence and the creation of multiple scenarios in order to be able to try different sequencing methods or test the effects of scheduling parameter manipulations.

6 CONCLUSION

This paper presented a set-up scheduling system based on Fuzzy Logic. Benefits for system users are easy intelligibility of the system and intelligible documentation of a complex planning process. The planning system can easily be modified in order to adapt it to changing planning conditions and strategies. Due to its high performance the system allows quick reactions on unpredictable events in the manufacturing process. Costs will be reduced by choosing the right set-up plan for most cases.

Until today only a small part of Fuzzy Logic theory is used to implement planning and control systems, despite the good prospects of the technology in the field of decision making [8]. In the near future more efficient Fuzzy systems will be realised. Some scientists stated that Fuzzy Logic will be one of the key technologies of the next century, applicable in many fields which until today have shown resistance to automation attempts.

7 REFERENCES

[1]   Keller, A.; Mertins, K.; Albrecht, R.; Duttenhofer, F.: Auftragssteuerung für Gieß- und Walzprozesse. ZwF 87 (1992) 12, pp. 669-672
[2]   Mertins, K.; Albrecht, R.; Steinberger, V.: Werkstattsteuerung - Werkstattmanagement. Wegweiser zur Einführung. Carl Hanser Verlag, München, Wien 1992
[3]   Mertins, K.; Tonn, F.; Wegener, U.; Wilksch, S.: Entwicklung und industrieller Einsatz eines Fertigungsleitsystems. ZwF 87 (1992) 2, pp. 75-79
[4]   Mertins, K.; Albrecht, R.; Wegener, U.; Duttenhofer, F.: Set-Up Scheduling by Fuzzy Logic. Proceedings of the Fourth International Conference on Computer Integrated Manufacturing and Automation Technology. IEEE Computer Society Press, Los Alamitos CA 1994, pp. 345-350
[5]   Tilli, T.: Fuzzy-Logik: Grundlagen, Anwendungen, Hard- und Software. Franzis-Verlag, München 1991
[6]   Tilli, T.: Automatisierung mit Fuzzy-Logik. Franzis-Verlag, München 1993
[7]   Zimmermann, H.J.: Unscharfe Logik stößt auf großes Interesse. VDI-Nachrichten Nr. 20, 17. Mai 1991, pp. 14-15

[8]   Zimmermann, H.J.: Fuzzy-Technologien: Prinzipien, Werkzeuge, Potentiale. VDI-Verlag, Düsseldorf 1993

8 BIOGRAPHIES

Dr.-Ing. Kai Mertins, born in 1947, has studied Control Theory in Hamburg and Economy together with Production Technology at the Technical University of Berlin. He became member of the scientific staff of the University Institute for Machine Tool and Manufacturing Technology (IWF), Berlin/FRG. Since 1983 he had been head of the department „Production Control and Manufacturing Systems" at the Fraunhofer-Institute for Production Systems and Design Technology (IPK), Berlin/FRG, where he is Director for Planning Technology since 1988. Special field of interest: Manufacturing strategy development, planning for production systems, shop floor control and simulation.

Rolf Albrecht, born in 1950, has studied Engineering at the Technical University in Berlin and finished his studies with a diploma in 1978. Since this time he worked as a scientist for different institutions of the TU Berlin. From 1980 to 1985 he worked as an author for the scientific journal „ZwF-Zeitschrift für wissenschaftliche Fertigung". He took part at different scientific programs at IBM Heidelberg. Since 1990 he is head of the department „Production Control" of the Fraunhofer Institute in Berlin.

Frank Duttenhofer, born in 1964, has studied Computer Science at the University of Erlangen-Nürnberg. Since 1990 he is member of the scientific staff of the Fraunhofer Institute in Berlin. He is now head of the scientific group „Manufacturing Information".

28

Strategic issues in enterprise management information system development in China

Renchu GAN and Jueying YANG
College of Management, Beijing Institute of Technology
P. O. Box 327, Beijing 100081, *CHINA*
Tel. 86 1 842 8274, *Fax.* 86 1 841 2889

Abstract

Enterprises in China are facing new opportunities and challenges in developing their management information systems (MISs). The paper analyzes two crucial environmental issues, viz. social changes and enterprise performance, that are uniquely encountered in MIS development in China as are different from those encounterable in the West. In order to cope with these crucial environmental issues, strategies of MIS development are presented and described in four dimensions: supporting reforms, managing the changes, parallel development and development in stages.

Keywords

Management information system, development strategy, enterprise, China

1 INTRODUCTION

With the upsurge in the construction of Information Super-Highway in the international society (Hu, 1994) and the development of the national information infrastructure project series "Three Golden Projects" (Qu, 1994) in China, Chinese enterprises are facing new opportunities and challenges in the development of their management information systems (MISs). MIS systematically processes information from inside and outside of the organization to support managers of various levels in their management and decision activities in order to attain the overall goals of the organization (Davis, 1985). To adapt to the dramatic changes in the society and technologies and to meet with severe competition from both international and domestic markets, managers of enterprises often have to reengineer their business and explore new ways for their survival and successes. MISs have been considered as strategic tools to strengthen their competitive advantages. Country-wide efforts in enterprise informatisation in China started at late seventies when China began to open to the outside world and to reform her economic system. It provided the opportunity for Chinese enterprises to make use of experiences and results of the development of MIS in Western countries. Given more than a decade and half of efforts in developing national economy and spreading information technology in the industries, the financial and technical conditions for MIS development have been continuously improved. Billions of Yuans (RMB) were spent and are to be invested for MIS development in the country. However, problems and dissatisfaction have frequently puzzled the users and developers of the systems. Despite the achievements of some pilot MIS projects in a few leading companies, yet many massive projects in some enterprises are riding on the rocky roads. To explore their ways to success remains a hard task for the MIS researchers and practitioners in China.

2   CRUCIAL ENVIRONMENTAL ISSUES

MIS blends advanced information technologies with modernized management methodologies. China has an ambitious plan to apply such technologies and methods in order to speed up her industrial development and to meet the international standards for her world-wide economic activities. It should be emphasized that the effectiveness of the diffusion of advanced technologies and/or management methods in a socioeconomic system rely strongly on the adaptation of them to the application environments. The environmental issues that the enterprises encounter in the development of MIS in China are much different from those faced in the West. The following two environmental issues seem to be more crucial:

2.1   Social changes

The environments facing the enterprises in today's world are dynamic and often turbulent. Considerable changes are caused by the technology shift, economic growth and social transformation. The ability of adapting to such changes is critical to the success in MIS development. It is well known that China has been undergoing distinguished transformation from a planned economy system to the socialist market economy system. It causes radical changes in various aspects in the society. For instance, the structure, behavior and operational mechanism of domestic markets are gradually changing. The functions of the government in economic affairs are under reform and the relations between the government and the enterprises are undergoing regulation. The structure of industrial production of the society is under readjustment. Most enterprises in China are in their processes of reform. Especially, in the state-owned, large and medium sized enterprises that dominate China's economy, the organizational structure, management system and operational mechanism are radically changing. In the West, the changes in the interior and exterior environments of the enterprises were relatively stationary during the last three decades. However, for enterprises in their MIS development, never before have the social changes been so sharp and momentous as it is in China today. It do and should strongly affect the way of MIS development.

2. 2   Enterprise performance

itch the efforts of a decade and half in reforms, the vitality and the economic environments of the enterprises have progressively been improved. Generally speaking, however, the performances of the enterprises are still lagging much behind the international standards. According to a survey on enterprise management in the mainland of China (Li, 1993), the average per capita net value productivity in machinery industry in 1989 was only 1/26 of that in the United State, and 1/10 of that in Taiwan. The impediments toward higher performances came from both the technical and management sides. Technically, only 13 percent of the production facilities in China's industry were up to the advanced world standards. Nevertheless, management problems today may be more critical. A number of key enterprises received massive doses of new technologies and the technical conditions can compare with their Western competitors, yet the efficiency and effectiveness of their productive performances are still out of date. The main obstacles to management excellence in China's industry are lack of a solid and scientific base of enterprises management, lack of qualified managerial personnel and lack of steady and continuous efforts in management improvement. In this connection, MIS in an enterprise should be developed in parallel with the progress of surmounting the obstacles mentioned above.

3   DEVELOPMENT STRATEGIES

3. 1   Supporting the reforms

Nowadays, reform is the first priority in the social and economic development in China. At the enterprise level, the main purpose of reformation lies in the strengthening of the vitality and the competitiveness of the organization. It is important that the MIS must be developed in keeping with the processes of economic reformation. Unfortunately, the reformation and MIS development are often considered as two separate processes and little attention has been paid to the coordination between these two processes in an organization. Conflict problems may occur when the MIS is incompatible to the scheme of organizational reform, and/or the necessary arrangements for MIS development and operation are neglected during reformation. It may lead to the failure of MIS development in enterprises. In order to achieve success, one should adhere to the following principles:
- MIS must fully support the objectives and activities of reforms;
- MIS development ought to be located within the reform schedules and planned as one of the major strategic measures for attaining the overall goals of reform;
- The structure, operational mechanism and functions of the MIS should be in accord with the requirements of the organizational reform and business reengineering;
- The steering committees of reformation and MIS development need to be aligned in an enterprise and the chief executive is required to head these two committees.

3. 2   Managing the changes

As mentioned above, managing changes is a key factor to the success of the MIS. Conventionally, the structure of MIS is function-oriented and based on the managerial hierarchy of the organization in order to meet the information requirements of the managers in various functional departments and managerial levels. However, the conventional model of the MIS may fail to meet the demands of the enterprise while its organizational structure, managerial mechanism and information requirements are dynamic and uncertain. Referring to Chinese traditional philosophy (Zhang, 1990 and Zhou, 1992), we introduce two principles for dealing with changes as follows:
- Principle of Constancy: Coping with various changes by a constant manner;
- Principle of Contingency: Taking contingency measures to meet emergency events.
These two principles seem to be contradictory, yet they are complementary to each other in reality.

According to the principle of constancy in developing MIS, one should identify the fundamental and consistent factors behind the rapid and turbulent changes, and found a solid and robust base of the system to cope with these factors. For example, the basic attributes of the personnel, products, assets and debts, and the essential trends of the market and technology may be the key factors for some manufacturing companies to form their business strategies for a relative long period of time; On the other hand, following the principle of contingency, the MIS should be very agile, flexible and responsive to adapt to the changing phenomena, and be able to produce quick response in solving emergency problems at the tactical level. In order to meet those requirements, the system should be problem-oriented rather than function-oriented and some components of the system may be constructed in an ad hoc fashion. In order to comply with both the principles of constancy and contingency in a complementary manner, the MIS may consist of a solid and robust base and a number of agile and flexible front-ends.

3. 3   Parallel development

As noted in the previous section, the MIS is an integrated body of technology, management system and people. The issues to be delta with in MIS development are of three dimensions: technical, organizational and human-oriented. Conventionally, great efforts have been taken in developing hardware and software, yet the issues of the rest two dimensions are often ignored. Technology has often been oversold while little attention paid in developing the related organizational elements and personnel phases within the schedule of MIS development. It should also be noted that technical impediments to the success of MIS have continued to fall with the rapid development of information technology in today's world, and however, the impediments from the organization and human sides have been dramatically increasing along with radical social and technological changes and severe competition in the marketplace. Although advanced information technology can be quickly accepted at present, the improvement of management and personnel quality is rather a slow and gradual process. Naturally, the information system in an enterprise can function effectively and efficiently if, and only if, it runs with an advanced management system and qualified personnel. Unfortunately, however, the managerial and personnel situations in many enterprises in China, generally speaking, remain still far behind the world standards in making effective and efficient use of today's information technology. These problems are also the main obstacles to attaining high productivity and competitiveness of the enterprise. Facing this unpleasant reality, it is not the case that one should wait for the results by improving the management and educating the people first before the MIS development. On the contrary, MIS development should proceed as a strategic measure for the continuous improvement of the enterprise performance, and therefore, the development of the management system and personnel must go in parallel with the penetration of information technology in the organization. To implement the parallel development strategy, the following viewpoints must be emphasized:

- MIS development is an integration process rather than one of technology transformation. It integrates the information technology, management system and people and creates favorable working conditions for the people in acquiring and using information to support their managerial activities.

- Among the three major parts of the integral whole, the people is the core of the system. Greatest efforts must therefore focus on the development of the people during MIS development, including the staff members and workers, developers and users, and the top managers and their subordinates as well. Training and educating those people to understand the necessary elements of information technology, information system and contemporary management methodologies, and how to make effective use of them: ——This is the key to success according to the experiences and lessons learnt for a number of pilot MIS projects in China.

- The MIS is a human-centered system rather than a technology-centered one. The major indices of success lie in the improvement of the effectiveness of people's work. All the arrangements of the hardware, software and communication network should provide services, supports, conveniences and favorable working conditions to the people, and promote the understanding, harmonization and cooperation between individuals and departments of the organization. In no case should the people be driven and controlled by machines with some conventional MIS schemes. It should also be avoided that the system produces too little benefit, but too much bother to the people like in the case of some unsuccessful systems.

- The improvement of management system in keeping with the development of the people and technology is another heavy task. It should be noted that economic reforms at the enterprise level may concentrate on the improvement of the political and macro — economic environments, and reconstruction of the ownership, missions, goals and organizational structure of the enterprise, which all have strong impacts on the structure, performance and mechanism of the management system. As mentioned above, MIS development must match with the pace of the reform, meet its needs and fully support its objectives and activities. Besides, significant efforts should be taken in reengineering the management processes and renewing management methods in every managerial level with contemporary methodology in management science and information technology, and founding a solid and scientific base of rules, regulations and standards of managerial activities.

• It is of course not the case that technological issues are to be ignored in MIS development. Despite the fact that the advantages and potentials of contemporary information technology have been widely recognized. the issues of changeability, reusability, reliability and standardization of hardware and software are still the impediments to the effective and efficient use of it. Furthermore, lack of sophisticated and systematic techniques remain a main obstacle to the integration of various components of computer hardware, software, databases, communication network in a flexible and robust body. The success of MIS development may also depend upon the efforts in overcoming difficulties in their technical aspects.

3. 4 Development in stages

Since the development of MIS in an organization involve a process of continuous improvement, it should proceed gradually along with the progresses in business reengineering and organizational reform. In carrying out such a task through a long duration of time, it is necessary to identify the features and characteristics of distinct stages in the growth of an MIS. A well known model of MIS growth: Nolan stage model was developed in 1974, in which four stages of MIS growth were identified (Gibson, 1974). A revised version having six stages was introduced in 1979 (Nolan, 1979). The Nolan model has been widely recognized and used in the planning and diagnosis of MIS development. Basically, the Nolan model describes the features of the processes of technological diffusion together with relevant organizational learning in an organization. It explains the logic of the development with stationary environments. The Nolan stage model may not be applicable when the environmental changes are often abrupt and not technology—driven, as those encountered in today's world, especially in China. The author proposed a four-stage model to describe the features of growth of computer-aided production management (CAPM) (Gan, 1992) in China. The four stages identified in the CAPM growth model are: initiation, presystemization, systemization and integration. We might revise and extend the model for CAPM growth into a growth model for MIS. The new model describes the four stages of MIS growth as follows:

Initiation stage
Generally, development of MIS in an enterprise may start at a few pilot computer-based application projects in some departments or functional areas. Each project has the objective, schedule and task force of its own. At this stage of MIS growth, most of the application projects in enterprises are relatively small and deal only with man-

agerial problems at the levels of transaction processing and/or operational reporting. Only a small part of the personnel are involved in the development and use of these computer—based systems. The progresses in this initiation stage will significantly affect further development of the MIS. Careful selection and planning of the pilot projects is of vital importance in this stage. Analysis and assessment of the environments and conditions must proceed in three perspectives: technological, managerial and human-oriented. Emphasis should be placed on the later two perspectives since the effectiveness of the computer-based system greatly lies on the adoption of the management system and the accountability of the related personnel. If a department or division has completed a pilot project with success, it may be considered as a model for others to imitate, and the persons who joint the project may become the backbone of the contingent for MIS development in the enterprises. Therefore, the programs of training and educating the task force and users should be executed in parallel with the system development.

In order to manage the changes and meet the emergency needs of the organization, implementing a long—term program should be avoided at this stage. One may divided a relatively large project into some smaller projects and carry out each of them in sequence in a short term. The developers can attain quick results and the users receive early benefits from a short term project. It may create suitable climate and promote enthusiasm for the development. Furthermore, producing quick results is a key to success of MIS development under rapidly changing and turbulent environments. The initiation stage goes to the end if the pilot projects in the main functional areas of the enterprise are completed.

Normalization stage

The completion of several small, short—term projects has little effect to the strategic advances of the enterprise. However, it may embody many improvements and failures in the integration of the information technology, management methods and people in a certain area of the enterprise. Conflicts and contradictory problems may occur between the operation of the computer systems and managerial activities of the people involved. In the normalization stage, the major efforts devoted to the MIS growth are:

- Modification, revision and expansion of the existing computer-based information systems;
- Normalization of information processing either by computer systems or manually;
- The standardization and rationalization of the management system mainly at the operational and/or tactical level.
- Coordination of the activities in the above three dimensions to deal with the con-

flicts and contradictory problems.

The normalization stage is a transitional one. However, the successes in the efforts of these three dimensions as mentioned above may create favorable conditions for the growth of MIS in the coming stage.

Systemization stage

Based on the progression of MIS development in the previous stages, acomphrehensive and integrated information system for managerial purpose can then be planned and implemented in this stage. The system integrates all the subsystems in the various functional areas to meet the needs of strategic advances in an enterprise. As mentioned above, the system should fit the scheme of economic reform and consist of a solid and robust information base and a number of agile and flexible front-ends. It should be noted that the systemization or integration does not mean that every thing in information processing has to be done by the computer system. The main purpose of MIS development in this stage is to create favorable working conditions, produce quick responses to the environmental changes and provide effective supports. The computers and other equipment in the MIS are installed as service machines rather than controllers to the managers in every level. MIS development in the systemization stage involves quite a few heavy tasks to be fulfilled and may last a longer period of time. One may decompose the stage into several substages in accordance with the features of the problems facing him in the process of development.

Integration stage

Besides the MIS development, computer — aided manufacturing (CAM), computer-aided design (CAD), computer-aided process planning (CAPP), computer-aided quality control (CAQ) and other CAXs have each been rapidly finding their respective places along with the penetration of information technology into the enterprises. As a consequence, the MIS development in the last stage often has to merge with them towards a full integration of all such functions. The full integration has received much attention in the literature during the last two decades. In 1973, a model of computer-integrated manufacturing (CIM) was proposed as a means of corporate strategy in the manufacturing industry (Harrington, 1973). Since then, CIM has become a hot topic and a host of researchers and practitioners in the West as well as in China have been engaged in the development of CIM technology and systems. Unfortunately, the progresses in CIM development have still been much behind the promises of the developers and the expectations of users.

The experiences and lessons learnt in CIM development have shown that full integration of the technology, management and people in an enterprise may cause significant organizational changes, for instance, it may change the ways of management and production, the ways of doing business and ways of working and thinking of the individuals. In order to achieve success, it is necessary to reengineer the business, reconstruct the management system and reeducate the people involved. these have to bee long-term processes and should proceed step by step. It should be emphasized that the enterprise is a human—centered system and the integrators have to be the people, not the computers. The computer system and communication network are only aids to the people.

The stage model of MIS growth is for identifying the basic features in MIS development and enables the planning and implementation of the MIS to be made on a realistic and rational base. In China, most large and medium-sized enterprises in the manufacturing industry are at the initiation stage of the MIS development, 20—30% of them have reached the normalization stage, and less than 10% of them are now devoted to the development at the systemization stage. Some pilot projects of CIM development are underway in a few leading companies in China (Wu 1994), yet there is still a long way to go for them to achieve the full integration.

4   SUMMARY

Environmental issues encountered in MIS development in China are more crucial than that in the West. China has been undergoing a great transformation from a system of planned economy to the market economy system. The economic reform has caused radical changes in various aspects in the society that have significant impacts in MIS development.

Although the macro-economic situation and the vitality of the enterprises in China have been progressively improved, yet in the average, the performances of the enterprises remain much behind the international standards. Managerial problems are even more critical than the technical ones. This is also a main obstacle to the successful development of the MIS.

In order to cope with the crucial issues mentioned above and achieve success in developing MIS, a set of strategies are presented and described in four dimensions. Firstly, MIS development must fully support the economic reform, which dominates the social and economic developments in China. Secondly, manage the changes by complying with two opposite, but complementary principles: principle of constancy and principle of contingency. So that the MIS may consist of a solid and robust base and a number of agile and flexible front-ends. Thirdly, t develop an MIS in the enterprise, improvement of the management system and development of the personnel must go in parallel to the penetration of the information technology. Fourthly, MIS development is a continuously improvement process rather than a short term project. It should proceed progressively and go along in stages. A stage model of MIS growth is presented which consists of four stages as: initiation, normalization, systemization and integration.

REFERENCES

Davis, g. B. and Olson, M. H. (1985) Management information systems: conceptualfoundations, structure and development. McGraw.—Hill Book Company, New york.

Gan, Renchu and Zhang jianjun (1992) Computer—aided production management in China. *Computers in Industry*, Vol. 19, No. 1.

Gibson, C. F. and Nolan, R. L. (1974) Managing the four stage of EDP gowth. *Harvard Business Review*, January—February.

Harrington, J., Jr. (1973) Computer integrated manufacturing. Industrial Press.

Hu, Daoyuan (1994) Development of the information super—highmay: a survey (in Chinese). *CIE Spectrum*, No. 4.

Li, Dongjiang, et al. (1993) On the rationalization of enterprise management in China (in Chinese). *Management World*, No. 6.

Nolan, R. L. (1979) Managing the crises in data processing. *Harvard Business Review*, March—April.

Qu, Weizhi (1994) Speed up the development of the 'Three Golden Projects' to contribute to the informatisation of the national economy in China (in Chinese). *China Inforworld*, No. 8.

Wu, Cheng (1994) Advances of CIMS in China and the new features of research and application of CIM technology (in Chinese). in *Trends of Computer Integrated Manufacturing*, Science Press, Beijing.

Zhang, Xiaomei (ed.) (1990) Thirty—six ways out: a collection of Chinese ancient stratagems (in Chinese). Tongji University Press, Shanghai.

Zhou, Hengxiang (trans.) (1992) Art of war: a translation of Sun's work from ancient Chinese prose (in Chinese). Guizhou Renmin Publishing House, Guiyang.

29

A productivity study to support the future manufacturing concept

A. ROLSTADÅS
The Norwegian Institute of TechnologyUniversity of Trondheim Department of Production and Quality Engineering

B. MOSENG
SINTEF Production Engineering

SUMMARY

The paper describes different methods (self audit, extended audit and self assessment) for measuring of productivity and competitiveness. The methods are developed in the Norwegian Productivity Program - TOPP, and have been used in approximately 40 companies participating in the program. The self audit and external audit are developed mainly for mechanical industry (including shipyards, offshore and electromechanical industry), while the self assessment is a modular system consisting of a set of business processes and indicators which the companies can use to build their own self assessment system for continous improvement.

1   INTRODUCTION

The future enterprice will be based on global manufacturing. This means that the different functions of an enterprice such as product development, engineering, part production, assembly etc. may be geographically spread. This is referred to as the virtual enterprice.

The need for the virtual enterprice is based on the international market competition. The enterprice of today competes on an international market. To survive it must adopt to the best practice world wide. Productivity and competiveness then becomes a significant issue.

2   PRODUCTIVITY MEASUREMENT

Measurement of productivity and competitiveness in industry companies is difficult and raise several questions. What is productivity? Does productivity means the same for different people and organizations? Different viewpoints and positions give different answers.

The classic definition of productivity is produced goods pr. unit of production factors. This definition indicates a focus on the value adding and physical production process. The physical production process is well defined in most enterprices, and input and output can easily be counted, measured or calculated. Internal services, however, are not so easy to describe, and even more difficult to measure. Industry has a tradition of focusing on 'blue collar workers' rather than 'white collar workers'. This view have changed in the 1980s, but there are still people who defend the old way of thinking in respect of productivity.
However, it has no sence to produce a lot of products if there is not a market. Customer satisfaction is an extremely important factor in obtaining competitiveness.

Measuring of performance and productivity should be done from different viewpoints.
Figure 1 shows 3 dimensions.

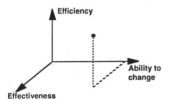

Figure 1 Performance measurement

In business terms the 3 dimensions could be described as follows:

Effectiveness   - to which extent are customer needs satisfied

Efficiency   - to which extent are the total resources in the company used in an effective and economic way

Ability to change   - to which extent are the company prepared to handle changes in surrounding conditions (strategic awareness)

When establishing a methodology for productivity measurement it is important to have in mind all 3 dimensions. A lot of surrounding factors will influence the productivity and competitiveness of a company. Some important factors to consider are shown in figure 2.

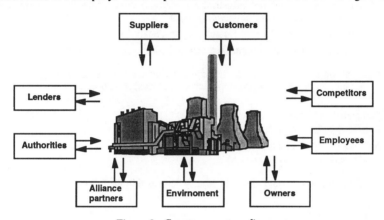

Figure 2   Company surroundings

3   THE NORWEGIAN PRODUCTIVITY PROGRAM - TOPP

Introduction: Norwegian economy is based on export of raw materials and semi finished products. Companies working on the international market are often subsuppliers to multinational customers. The contribution from Norwegian industry to the gross national product (BNP) is low (18%).
Those factors were important arguments to look at productivity and international competitiveness in Norwegian industry. To meet this challenges the productivity program TOPP has been launched in Norway.

The overall goal of the TOPP program is

> **to focus on total productivity for the whole enterprise and stimulate an industrial climate that improves international competitiveness**

Important objectives and key issues in TOPP are
- time to market
- quality
- flexibility
- total cost

The program involves cooperation between
- The Federation of Norwegian Engineering Industries
- The Norwegian Institute of Technology (NTH)
- Industrial companies (~ 40)

and is sponsored by
- The Research Council of Norway

The TOPP program is planned for the period 1992-95. Following subprograms exist:
1. Analysing company productivity and competitiveness (self audit, extended audit, self assessment, benchmarking etc.)
2. Implementing actions for industrial productivity improvements (industrial projects, seminars, courses, industrial networks, etc.)
3. Generating new knowledge (research projects, analysing productivity data, etc.)
4. Long term competence program (education , courses, dr.ing. programs, master degree)

Methods: Taking into consideration the different views and dimensions in productivity and performance measurement, following approaches (methods) have been developed and used

a) Self audit   (questionnaire)
b) Extended audit   (performed by experts)
c) Self assessment   (continuous improvement)
d) Benchmarking   (breakthrough)

The different methods a),b) and c) will be briefly described in the following.

4   SELF AUDIT

Methodology: The self audit is based on a questionnaire answered by the companies. The questionnaire consist of 3 parts asking for different types of information.

Part 1   Facta data

Part 2   General evaluation of functions and system variables (by individuals)

Part 3   Detailed evaluation of functions and system variables (by groups)

Part 1 consist of general information about the company, product data, cost distribution, maintenance cost, quality cost, customers, sub suppliers, financial and economic data etc. The data is facta or estimates.

In part 2 individuals (~20) in the company are asked to give their own evaluation of different functions (primary and support) and system variables (facilities, equipment, personnel, organization etc.). This evaluation is given confidentially and is later compared with information given in part 3 (see figure 4).

Part 3 is the largest and most detailed part of the questionnaire. The company is asked to build groups to evaluate a lot of detailed questions concerning all functions and system variables. Example of questions to evaluate the design function is shown in figure 3.

Functions and system variables are evaluated on a scale 1-7 where
  1 is far behind competitors
  4 is on same level as competitors
  7 is "best practice"

By design we mean the function responsible for the production support for order production or in connection to product development. Typical activities will be analysis, calculation, work out drawings etc.

6201 Evaluate the following areas/aspects as to current situation, and give a realistic possibility for improvement by estimating the status after 2 years. Judge how important the factors are for the company's competitiveness (N - No importance, M - Medium importance, G - Great importance).

Very bad			Medium			Very good	
1	2	3	4	5	6	7	
			Status today		Realistic status in 2 years	Importance for the company	

		Status today	Realistic status in 2 years	Importance for the company	
a)	Tools and technical facilities	1--2--3--4--5--6--7	1--2--3--4--5--6--7	N--M--G	
b)	Design competence in own company	1--2--3--4--5--6--7	1--2--3--4--5--6--7	N--M--G	
c)	Design investigation and work through	1--2--3--4--5--6--7	1--2--3--4--5--6--7	N--M--G	
d)	Laying down of product parameters (function, form, material, dimension etc.)	1--2--3--4--5--6--7	1--2--3--4--5--6--7	N--M--G	
e)	Presentation of production support (drawings, piece lists etc.)	1--2--3--4--5--6--7	1--2--3--4--5--6--7	N--M--G	
f)	Quality safeguarding system for design work	1--2--3--4--5--6--7	1--2--3--4--5--6--7	N--M--G	
g)	Procedures for treatment of change orders	1--2--3--4--5--6--7	1--2--3--4--5--6--7	N--M--G	
h)	Systematic registration and reuse of empirical dates	1--2--3--4--5--6--7	1--2--3--4--5--6--7	N--M--G	
i)	Co-operation with production/production planning	1--2--3--4--5--6--7	1--2--3--4--5--6--7	N--M--G	
j)	Market and customer contact	1--2--3--4--5--6--7	1--2--3--4--5--6--7	N--M--G	
k)	Co-operation with contractors	1--2--3--4--5--6--7	1--2--3--4--5--6--7	N--M--G	
l)	Co-operation with purchase	1--2--3--4--5--6--7	1--2--3--4--5--6--7	N--M--G	

Figure 3. Questions to evaluate the Design function (example)

Report - analysing of data: The data is collected and stored in a database. Different types of reports can be generated automatically. Individual reports based on data for one company or reports based on average data from a given number of companies can be produced. The report layout is the same which means that companies easily can compare own data with data from groups of other companies. The report consist of 3 parts

a) Main report with key indicators (10 - 15 pages)
b) Enclosure for comparison (~ 50 pages)
c) Enterprise specific enclosure (detailed economic and financial data) (~15 pages)

Figure 4 shows results from evaluation of primary functions.

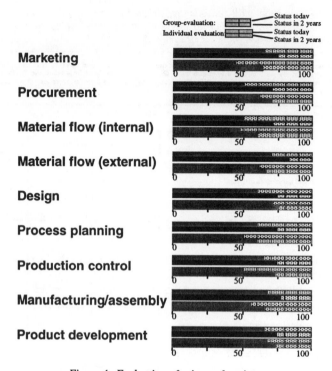

Figure 4. Evaluation of primary functions

5   EXTENDED AUDIT

Methodology: The extended audit is performed by experts analysing the company. Measuring of productivity and competitiveness is done on 2 levels.

- a) **Company level** - using indicators and key factors with focus and viewpoint on the whole enterprise
- b) **Company split up level** - using indicators and key factors with focus on parts and specified areas in the company

An overview of the different areas (37) in the model is shown in figure 5. Characteristics for the analysing areas are:

Company level

Economy - indicators and key factors which describes the economic conditions and potential. An important issue of this analysis is to evaluate the economic capacity for future investments and change.

External relations - indicators and key factors to measure the ability of the company to exploit and take advantage of surrounding conditions. This includes evaluation of customers, competitors, use of external resources, strategic alliances, etc.

Internal relations - the ability to use existing internal resources (products, machines, personnel etc.) in a time and cost effective way, and to take care of the internal milieu.

Ability to change - the ability to foresee and be prepared to meet new trends and quick changes in the environments.

Company split up level

Systems variables - primary resources and conditions necessary to produce products

Functions - Activities in the product life cycle. The functions are divided in primary and support functions.

Cycles - to follow and analyse flows between functions. Flows could be of different types (material , information etc.). The objective is to measure co-operation and infrastructure between functions.

Philosophies - to analyse overall production and management philosophies used in the company

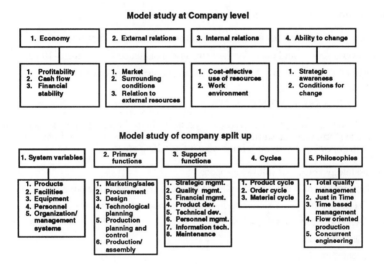

Model study at Company level

1. Economy	2. External relations	3. Internal relations	4. Ability to change
1. Profitability 2. Cash flow 3. Financial stability	1. Market 2. Surrounding conditions 3. Relation to external resources	1. Cost-effective use of resources 2. Work environment	1. Strategic awareness 2. Conditions for change

Model study of company split up

1. System variables	2. Primary functions	3. Support functions	4. Cycles	5. Philosophies
1. Products 2. Facilities 3. Equipment 4. Personnel 5. Organization/ management systems	1. Marketing/sales 2. Procurement 3. Design 4. Technological planning 5. Production planning and control 6. Production/ assembly	1. Strategic mgmt. 2. Quality mgmt. 3. Financial mgmt. 4. Product dev. 5. Technical dev. 6. Personnel mgmt. 7. Information tech. 8. Maintenance	1. Product cycle 2. Order cycle 3. Material cycle	1. Total quality management 2. Just in Time 3. Time based management 4. Flow oriented production 5. Concurrent engineering

Figure 5. Analysing areas - extended audit

The 37 analysing areas are broken down in a hierarchy of subindicators which are given points and weighted. Check list for all levels exist. Indicators for evaluation of the TQM process are given in figure 6.

No.	Description sub indicators	Points	Weight
1	Goals and strategy	3	10
2	Know-how and competence	5	10
3	Organisation and management	2	5
4	Information and analysis systems	2	5
5	Strategic quality management	2	15
6	Use of human resources	2	5
7	Quality control and qualtiy improvements	3	5
8	Results of quality work	4	20
9	Customer satisfaction	3	20
10	Cooperation and information	4	5
	Calculated	3,15	100%
	Total score	3	

Figure 6 Evaluation of TQM (example)

Report: The audit is performed by an expert team (normally 2 persons) asking questions and analysing all relevant areas in the company.

The results are presented in a report organized in 3 parts.

1) Analysis of main productivity key factors and indicators (~5 pages)

2) Proposed actions and plans for the company to obtain better results (5-10 pages)

3) Detailed description of subindicators for all analysing areas (~50 pages)

6   SELF ASSESSMENT

The TOPP Self assessment is a modular system to help the companies to build their own system for performance measuring. The objective is to follow trends and check if a business process is in control and continuos improvement is taking place.

The methodology consist of following steps:
   a) Identification of critical and important business processes in the company
   b) Selection of analysing areas and indicators to measure the business processes
   c) How to organize the self assessment
   d) Collection of data
   e) Presentation of results
   f) Evaluation of results, actions

An overview of the methodology is shown in figure 7. 22 different business processes with analysing areas, indicators and measuring techniques are described.

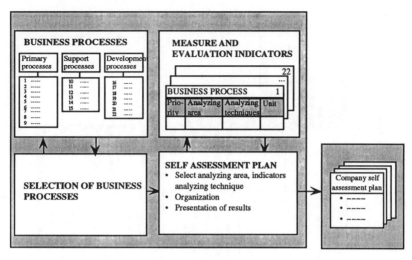

Figure 7   Self assessment

7   RESULTS

All the described methods have been used in the TOPP program. The number of companies involved are:

a)   Self audit   ~ 60 companies

b)   Extended audit   ~ 40 companies

c)   Self assessment   individual use

A lot of data about productivity and competitiveness are collected in the audits. A special research project has analysed those data, trying to find some general results and conclusions concerning the status of productivity and competitiveness of Norwegian industry. 8 different reports have been written focusing on different topics. The reports are:

1. General characteristics and possibilities

2. Analysing methodology

3. Analysing of procurement

4   Analysing of quality

5   Analysing of production planning and control

6   Analysing of product development

7   Analysing of marketing and sales

8   Analysing of strategy development

Figure 8 show a gap analysis from evaluation of different functions in the companies (self audit questionnaire).

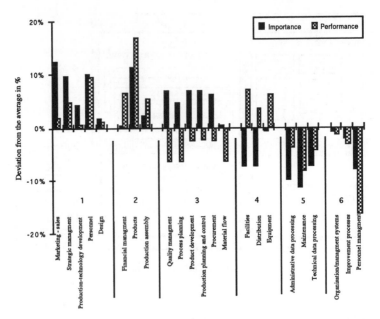

Figure 8 Gap analysis concerning performance and importance of different analysing areas (37 companies)

The analysis is showing gaps between performance and importance of the different functions. The data are calculated as deviations from the average value for all the functions. The functions are sorted in 6 groups. Group 1 shows functions where both performance and importance are better than the average. However the importance is greater than the performance, which indicates that the area is good, but there is still a lot to do.

A number of functions (group 3) show high importance and scores performance less than the average. These funcions are

- Quality Management
- Process planning
- Product development
- Production planning and control
- Procurement
- Material flow

This gap analysis is based on average values from 37 companies. However the companies are quite different (size, products, production methods etc.). Another grouping of the companies in more homogeneous branches would perhaps give other results.

A lot of detailed analysis on spesific areas have been performed (ref. mentioned reports and special reports for branches). Analysis of the design function indicates that improvements have to be done in

- design review
- definition of product parameters
- quality assurance in design
- systematic registration and reuse og knowledge and experience data

8   CONCLUSIONS

Measuring of productivity and competitiveness for a whole enterprise is difficult. To find indicators, key factors and analysis areas which are relevant to different types of companies, and to harmonize this into a common methodology is even more difficult.

The methodology described is based on different approaches. However in any case there is much individual judgment. It is difficult (or impossible) to measure total productivity by summarizing local measurements of sub areas. This is the reason for analysing the enterprises from different viewpoints. History has shown that economic success one year does not guarantee a competitive edge. Many perspectives that measure different factors have to be used to get a better understanding of competitiveness.

REFERENCES

(1)   TOPP Questionnaire for measuring of competitiveness (54 pages)

(2)   TOPP Self audit company report (~75 pages)

(3)   TOPP Extended audit - method handbook

(4)   TOPP Extended audit - operation handbook

(5)   TOPP Extended audit - company report (~70 pages)

(6)   TOPP Self assessment handbook

(7)   TOPP Benchmarking handbook

(8)   Moseng B., Bredrup H.: A methodology for industrial studies of productivity performance, Production Planning and Control, 1993 vol 4 no. 3

(9)   Moseng B.: Productivity Measurement, Methods and tools developed in TOPP. IFIP WG 5.7 Workshop Benchmarking - Theory and practice, june 94.Trondheim,Norway

(10)   Bredrup R.: PMS Factors influencing on Enterprice Profitability. 8th International Working Seminar on Productions Economics 1994. Innsbruck, Østerrike

(11)   Bredrup H.,Bredrup R., Rolstadås A.: PMS contribution to Improved Enterprice Productivity. IFIP WG 5.7 Working Conference on Evaluation of Production Management Methods 1994, Porto Alegre, Brasil

(12)   Bredrup H., Bredrup R., Estensen L.: Factors influencing Effectiveness of Investments in Modern Manufacturing Technology. Automation '94, Taipei, Taiwan.

(13)   Bredrup H., Bredrup R.: Performance Measurement to support Continous Improvement of Customer Satisfaction. EOQ '95, Lausanne, Switzerland

(13)   Sink D.S., Tuttle T.: Planning and measurement in your organization of the future. Industrial and Engineering Management Press, Norcross, GA. 1988

(14)   Porter M.E.: The Competive Advantage of Nations. Macmillan, London 1990.

The references 2, 3, 4, 5, 6 and 7 are written in Norwegian

BJØRN MOSENG obtained his MSc in Mechanical Engineering at The Norwegian Institute of Technology, University of Trondheim , in 1970. He is now Research Manager at SINTEF Production Engineering, section Integrated Manufacturing Systems. During the last eyars he has been involved in the Norwegian Productivity Program TOPP as project leader. His main fields of competance are technology planning, CAD/CAM/CIM and information technology. He has been involved as project manager in several international projects (ESPRIT, BRITE-EURAM etc.).

ASBJØRN ROLSTADÅS is professor of production and quality enginering at the University of Trondheim, The Norwegian Institute of Technology. His research covers topics like numerical control of machine tools, computeraided manufacturing systems, computeraided production planning and control systems and project management methods and systems. He is a member of the Norwegian Academy of Technical Sciences, the Royal Swedish Academy of Engineering Sciences and APICS. He serves on the editorial board of the journal Computers in Industry, and is the editor of the International Journal of Production Planning and Control. He is president of IFIP and past chairman of IFIP TC5 and WG5.7. He is also past president of the Norwegian Computer Society and the Nordic Data Processing Union.

30

Generate Parallel Behavior Structure for Automatic Workshop Problem with GP

Wenwei YU, Yukinori KAKAZU
Department of Precision Engineering
Hokkaido University
West-8, North-13, Sapporo, Hokkaido, Japan
Tel: +81-11-706-6443
Fax: +81-11-700-5070
Email: kakazu@hupe.hokudai.ac.jp

Dan JIN
Mechanical Engineering Department
Shanghai Jiaotong University
1954 Huashan Road, Shanghai, China

Abstract

Existing researches that apply hierarchical behavior for AI problem, but in a few instance, hierarchical behavior is abstracted for multi-agent problems solving. In this paper we introduce a scheme that creates a series of behavior structures which pursue parallel mechanism in acquiring the action plan of multi-agent in the case each agent should autonomously make decision, i.e., make decision independently according to the dynamic environment. GP(Genetic Programming) is used for evolution of the behavior structure. By solving the Automatic Workshop Problem wherein the working and moving schedule should be made by each robot considering obstacle avoidance, process order and working ability of itself, we show that the scheme is viable and promising.

Keywords

Multi-agent system, Parallel, Behavior, Genetic Algorithm, Genetic Programming, Job Process Problem, Automatic Workshop Problem, AGV Plan, Control Problem

1 INTRODUCTION

1.1 Background

Developing schedule for automatic workshop is a task more complex than ordinary Job Process Schedule Problem, because of the combination of variety of robots and moving trajectory conflict besides of the job process order. The space of feasible schedules grows exponentially as there are increases in the number of different jobs that must be processed, number of operations required by each job, number of kinds of robot and number of robot of each kind. This multi-variable changeability makes the traditional scheduling, i.e., collective and supervisory scheduling method lack of flexibility and even impossible, while finishing the works by several agents with different level of hierarchical behavior rules shows feasibility and effectiveness to be high level learning and searching scheme.

On the other hand, the searching for an optimal or even sub optimal behavior structure for each kind of robot also suffers from unrealistic search time and computational complexity while the condition and action grows. Recently, there has been wide research interest in applying model-less methods such as GA(Genetic Algorithm), GP(Genetic Programming), NN(Neuron Network) and RL(Reinforcement Learning) to the searching and learning problem[1][2]. These kind of methods can solve the problem with less dependence on the prior knowledge and model in less time in comparing with conventional algorithms. Among these model-less methods, GP is a promising one, which is an adaptive optimization algorithms based on principles of natural evolution[2]. In this paper, a series of hierarchical behavior rule sets represented by behavior tree are created using GP for each agent, it is the agents that autonomously search the path and coordinate to complete the tasks.

1.2   The Problem Description

The automatic working factory problem can be described like followings:
Definition of F, C and A

F=   {R, M, B}
        where
        R=   {S_i , i = 1~SPECIE_NUM}
                S=   {r_j , j = 1~robot_number(i)}
        M=   {W, N}
                W=   {w_k^l , k = 1~WORK_NUM, l = work_level}
                N=   {n_p , p = 1~NODE_NUM}
        B=   {W', E}
                W'=   {w_q^l , q = 1~REMAIN_WORK, l = work_level}
                E=   {e_q^t , t = 1~APPLIED_ROB_NUM}

here, F stands for Factory configuration, consisting robot set R and map set M;
the robot set R includes species S_i, each species S_i contains several robots r_j;
the map set M includes node set N and work set W, see Figure 1 for illustration;
B is the bulletin for communication and management, involving remain work set W'
and evaluation set E which is send by the robot applied for the work from bulletin,
the robot with minimum evaluation of a work will get the work, i.e.,

$w_q \Rightarrow r_{t'}$, $e_{q'}^{t'}$ = min{$e_{q'}^t$, t = 1~APPLIED_ROBOT_NUM}, see Figure 2 for illustration.

C=   {C(r) I r ∈ S}

the condition set C of robot behavior.

A=   {A(r) I r ∈ S}

the action set A of robot behavior.

• Problem:
        Generate the rule set as follows under the given condition of F, C and A;

        $Rule_i${C,A}(S_i) ⇒ min(W'),   i = 1~SPECIE_NUM

        in which $Rule_i$ is the hierarchical combination Tree of C and A, for robot species i.
        From the description of the problem, we can see that:
• more than one set of rule are necessary for more than one species of robot.

• each set of rule need evaluation.

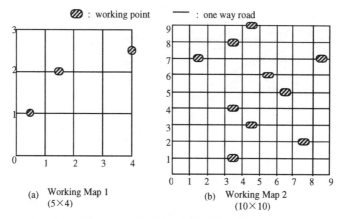

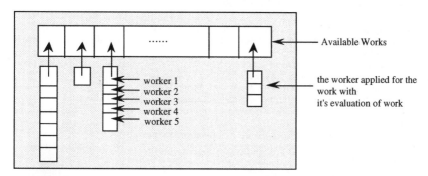

Figure 1 Examples of working maps of 5×4 and 10×10 problem.

Figure 2 The manage bulletin of the working factory problem.

2 APPROACHING TO THE PROBLEM BY GP

2.1 Introduction of Problem Oriented GP

According to the before working and papers on GP, GP's weak points, the large population and generation comparing to small problem and damage of good sub tree can be understood.

Trying to reduce the calculation and keep the good sub tree of behavior, we introduce the island model[5] to the method. Showing in Figure 3, for s species robot problem, we have s population pools, each has the pool size p. Then, for the evolution of s set of rule:

• According to the island model, we divide the each population pool by island number n, then we have each pool has n islands with island population of p/n.

• After each certain generations (migration interval), certain number individuals (migration rate), will migrate to other island. The migration happens like ring, i.e., the individuals from island i migrate to i+1, while those from island n go to island 1.

• The genetic operations like crossover, mutation happen just in the range of island.

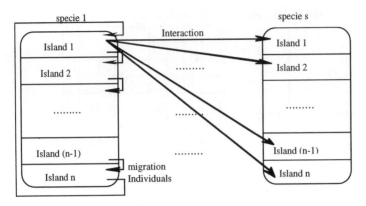

Figure 3 The island model introduced to the problem oriented GP.

2.2 The Condition Set C and Action Set A

After analyzing the problem presented in segment 1.2 and characteristics of GP, the domain dependent condition set C and action set A are decided and shown in Figure 4 in which each terms has the meaning as follows.
- *communication* makes the robot register to bulletin, evaluate the each work in bulletin, then bid the work with other registered worker.
- *working* do one work step if in fit work place with fit work ability.
- *waiting* makes the robot wait one time step on the current position.
- *random-step* randomly changes the direction of the robot, then makes the robot move one step in the new direction.
- *get-path* calculates the path between the current position and the object.
- *one-step* moves the robot one step along the gotten path or just randomly if without path.
- *back* moves the robot one step back.
- *exit* makes the robot exit from the working factory.

CONDITION		ACTION	
C0	IF-GET-PATH	A0	communication
C1	IF-REACH-DES	A1	working
C2	IF-GET-COLLIDE	A2	waiting
C3	IF-DIR-RATION	A3	random-step
C4	IF-OUT-PATH	A4	get-path
C5	IF-GET-WORK	A5	one-step
C6	IF-REACH-WORK	A6	back
C7	IF-WORK-FINISHED	A7	exit
C8	IF-FIT-WORK		

Figure 4 The domain dependent condition set and action set of the approach.

These conditions and actions will be used to generate the behavior rule set for the robot worker of the factory, and the behavior rule set should be evaluated according to the working result of the workers.

2.3 The Evaluation of the Behavior Rule

The evaluation of each behavior set of each species is realized according to the following steps:
• The evaluated rule set will form pair with the representatives of the island of rule set pool for other species of robot, i.e.,

$$\text{Pair}_j : \{\text{Rule}_1, \text{Rule}_2, ..., \text{Rule}_e, ..., \text{Rule}_i, ..., \text{Rule}_s, i=1 \sim s, e = \text{evaluated rule set}$$
$$\text{number} \}, j = 1 \sim s^I, s = \text{SPECIE\_NUM}, I = \text{ISLAND\_NUM}.$$

• The robot set R will enter the factory field M with the Pair$_j$ correspondingly, i.e.,

$$S_i \Leftrightarrow \text{Rule}_i, i = 1 \sim s.$$

• Within a preset time limit T, the robots will act according to their behavior rule in the factory.
• The remaining works RW and the used time UT, consumed energy CE will be used to evaluate the Rule$_e$,

$$F'(\text{Rule}_e) = w_{all} \times (w_{time} \times UT_{all} + w_{work} \times RW_{all} + w_{energy} \times CE_{all}) +$$
$$w_{species} \times (w_{time} \times UT_{species} + w_{work} \times RW_{species} + w_{energy} \times CE_{species})$$

here, e = evaluated rule set number, *all* means the result of all robot set R and *species* mean the species evaluated. And we have

$$w_{all} + w_{species} = 1.0$$

• One Rule set can form s^I pairs with the rule set of pool for the other species, the average value

$$\text{Fitness}(\text{Rule}_e) = \sum_i F_i'/s^I, \quad i = 1 \sim s^I,$$

the maximum value

$$\text{MaxFit}(\text{Rule}_e) = \max(F_i'), \quad i = 1 \sim s^I$$

here, we use the average value as the fitness because we believe that this can prevent the solve from convergence to rule set with too many random elements.

3 NUMERICAL EXPERIMENTS

3.1 Experiment Condition

The experiment parameters and condition are shown in Table 2.

Table 2 Experiment parameters and condition

Exp	Gen	Pop	Map	Time Limit	Island num/rate/inter	Remain works/steps	Weight all/specie	Worker specie/num
1	100	100	Map1	20	10/2/5	1/3	0.2/0.8	2/2
2	100	200	Map2	60	10/2/5	4/16	0.2/0.8	2/2
3	100	200	Map2	30	10/2/5	0/0	0.0/1.0	2/6
4	100	200	Map2	60	no model	4/16	0.2/0.8	2/2

Note:
• the Map1 is shown in Figure 1, (a), the Map2 is shown in Figure 1, (b), each work contains 4 steps;
• the genetic parameters: crossover rate = 0.85, mutation rate = 0.10, copy rate = 0.05 the mutation threshold = 0.10;
• maximum tree depth is set to 10.

3.2 Experiment Results and Discussion

• The behavior tree changes with the environment:
when in small map containing a little works, the rule set become convergent quickly with
the smallest condition and action element set involves *one step, working, if reach work,*
Figure 5, (a) is the tree of the experiment 1, the random search is enough for the task (see
Figure 4 for reference of condition and action meaning).

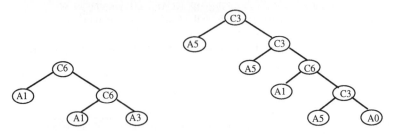

(a) The tree of experiment 1   (b) The tree of experiment 2
Figure 5 The tree of experiment 1 and 2.

When changed to the Map2 with more works and search field, the rule set extend to
some set with the action element of communication, one step and other necessary element,
Figure 5, (b) show the tree constructed by the condition *C3, C6,* action *A1, A0, A5* .
When the robot number increases, the possibility of collision and the need of
coordination between the different species of robot developed, so the more complex rule
set created for the adaption of the environment, see Figure 6, (a).

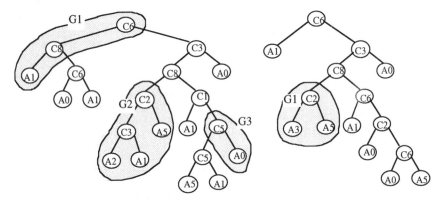

(a) the tree of species 1 in experiment 3   (b) the tree of species 2 in experiment 3
Figure 6 the result tree of experiment 3.

• The effective tree is usually made up with some effective sub trees which aim at the sub
goals of the whole task. In some experiment, the sub tree also can be understood, for
example, when we see the Figure 6, (a), the G1 is a branch for working, which can be
expressed as:
    IF REACH WORK
        IF FIT WORK
            working
and G3 is a branch for getting work and walking to work position, which also can be
expressed like

```
IF NOT FIT WORK
        IF REACH DES
                working
        ELSE
                IF GET WORK
                        one-step
                ELSE
                        communication
```

while G2 coordinates with Figure's (b) G1 when collide with each other

species 1's G2	species 2's G1
IF GET COLLIDE	IF GET COLLIDE
IF GET DIR RATION	waiting
waiting	ELSE
ELSE	random-step
working	

```
ELSE
        one-step
```

• Comparison between on island model and without island model

Figure 7 is the fitness comparison between experiment 4(Figure 7, a) and experiment 2(Figure 7, b), here Fitness = max($Fitness_i$), i = 1~POPULATION_NUM

$$MaxFit = max(MaxFit_i), i = 1\sim POPULATION\_NUM$$

$$Average = \sum_i (Fitness_i), i = 1\sim POPULATION\_NUM$$

$Fitness_i$ and $MaxFit_i$ have been explained in segment 2.4. In Figure 7, each migrating interval, there is a vertical line to show the migration happening. We can see that, with ordinary coevolution method, the solve can tend to convergence steadily, while on island model, the migration lead to the change to the spatial structure, and make the fitness curve especially average curve variating intensely, but not damaging the up-tendency of the curve.

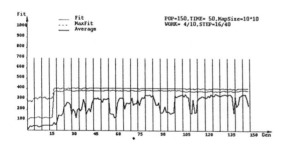

(a) fitness curve of experiment 4

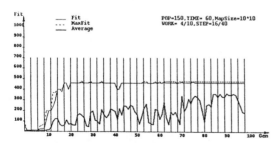

(b) fitness curve of experiment 2

Figure 7 Comparison between the fitness curves of experiment 4 and 2.

4 CONCLUSION

This paper has described the scheme using GP to generate and evolve the hierarchical behavior structure for the automatic working factory problem.

We can see that, when using distributed planning and control policy instead of traditional planning and control policy, the practical problem can be solved with flexibility on high level, i.e., be solved through the cooperation and adaption of the individuals.

The behavior rule set generated and evolved by GP can well fit the environment, makes the robot individual present some effective performance in the working environment. It seems that the GP can be used to explore more complex behavior structure of practical problem.

Using the Island Model leads to the reduction of the calculation, while not damaging the up-tendency of the fitness curves.

5 REFERENCES

[1]Jonalthan H. Connell, Sridhar Mahadevan et al (1993). Robot Learning. Kluwer Academic Publishers.

[2]Branco Soucek and the IRIS Group (1992). Dynamic, Genetic and Chaotic Programming, the Sixth Generation. A Wiley-Interscience Publication.

[3]Lydia Kronsjo, Dean Shumsheruddin (1991). Advances in Parallel Algorithms. Black Scientific Publications.

[4] Gregory J.E. Rawlins el al (1991). Foundation of Genetic Algorithms. Morgan Kaufmann Publishers.

[5] Lawrence Davis et al (1991). Handbook of Genetic Algorithms. Van Nostrand Reinhold.

PART SEVEN

One-of-a-Kind Production

31

Closed loop scheduling and control of One-of-a-Kind Production

Y.L. Tu, H. Holm & U.B. Rasmussen
Department of Production, Aalborg University
Figiberstræde 16, 9220 Aalborg Ø., Denmark
Phone: +45 98 15 8522, Fax: +45 98 15 3030
E-mail: i9yiliu@iprod.auc.dk

Abstract

In this paper, a closed loop cybernetic system is proposed for real-time scheduling and control of production in an automatic shop floor or on an automatic production line for One-of-a-Kind Production. By a concept of the so-called Product Production Structure, the designs and production processes of different types of products can be formulated into a set of data. These data can be 'read' by the control system and thereby the control system makes schedule of as well as real-time controls the production of the products. In this manner, the control system gains enough robustness to cope with the great diversity of product types (or kinds) in One-of-a-Kind Production. By the simulation module and experience time estimates of production processes, the control system is designed as an estimator to be able to make the production schedule before starting a production of a product or a batch of products with different types. By its closed loop, the control system is also designed to be able to real-time control the production and update the production schedule according to states of a shop floor or a production line during the production.

Keywords

CIM, cybernetic modelling, One-of-a-Kind Production, real-time scheduling and control

1 INTRODUCTION

The main contents in this paper report a production scheduling and control system developed in the ongoing research project under the title of "Cybernetic Modelling and Control in One-of-a-Kind Production (CMCOKP)". This research project is one of the 7 research projects in the main research program, namely IPS (Integrated Production Systems) II Research Program. The IPS II is a research program under the Danish Technical Research Council aimed at developing new approaches to industrial integration and strengthening research cooperation between several Danish academic research organisations and Danish industry.

According to the definitions made by Trostmann et al. (1993), a OKP (One-of-a-Kind Production) industry can be characterised as:

- the design of its product changes with almost every new order, and
- almost every one of its customer orders contains one and only one specimen.

The physical examples of OKP industries can be easily found in present heavy industries, e.g. a boiler manufacturing company or a shipyard.

In OKP industries, two main problems are identified in the following from the production control point of view:

- The structure (or design) of the product may change particularly much and frequently.
- Corresponding with changes of the product structure, the shop floor layout or production system configuration may change particularly much and frequently.

Because of these two problems, the control systems for automatic control of OKP (One-of-a-Kind Production) normally need much higher flexibility than the control systems for automatic control of mass production and even batch production. From a view point of control system structure, the higher flexibility of a control system is normally in conflict with its robustness. However, the reliability and feasibility of a control system depends very much on the robustness of the control system. This conflict has made the great difficulties to implement an automatic scheduling and control system in OKP industries, such as shipyards.

To resolve the problems for automatic control of shop floors in OKP industries is the main goal of the CMCOKP research project. The CMCOKP has been carried out co-operatively by the Department of Production of Aalborg University, the Control Engineering Institute of Technical University of Denmark and the Odense Steel Shipyard Ltd. in Denmark.

To develop and implement technologies and concepts for the design of automatic shop floor control systems in One-of-a-Kind Production industries by linking the industrial objectives with these technologies and concepts, a ship web welding assembly line at the Odense Steel Shipyard Ltd. is chosen as a pilot shop floor to running the CMCOKP research project. This ship web welding assembly line is named B13 line at the Odense Steel Shipyard Ltd.

2   PRODUCTION SCENARIO OF THE B13 LINE AND PRODUCT PRODUCTION STRUCTURE OF A CASE WEB

2.1   Descriptions of the B13 line and a case web

At the Odense Steel Shipyard Ltd., a ship body (or ship tank) is welded by hierarchically decomposing it into Blocks, Sub-blocks, Webs and metal plate cutting parts (i.e. base plates, brackets, flat bars, etc., see examples in Figure 1 (a)). The B13 line is one of the shop floors in the shipyard to weld different types of webs. One example of such webs to be welded on the B13 line is shown in Figure 1 (a).

As shown in Figure 1 (a), the case web consists of a base plate (BP2), three types of stiffeners, i.e. a large bracket '34STR', three large flat-bars 'S28', 'S32' and 'S37', and two small flat-bars 'S27' and 'S30'. There are a lot of different types of webs to be welded on the B13 line. Each of those webs generally consists of a base plate with different size and plan geometric model and a number of different stiffeners.

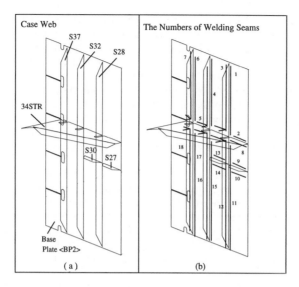

Figure 1   A case web.

At present time, the B13 line at the Odense Steel Shipyard Ltd. is a man-machine combined ship web welding assembly line. By the CMCOKP research project, a new automatic (or robotised) B13 line is going to be designed for the shipyard to remodel its present B13 line. The layout and working procedure of the automatic B13 line are quite different from the present B13 line at the shipyard.

The discussions made in this paper are focused on the new automatic B13 line. To simplify the description in the following, the term of B13 line will be simply used to refer to the automatic B13 line but not the present B13 line at the Odense Steel Shipyard Ltd.

According to the design made by Ørum-Hansen et al. (1994), the B13 line consists of an one-band conveyor and three types of robots. The space of the conveyor is subdivided into 5 even working stations. At every movement of the conveyor, the web panels (or base plates) placed on the conveyor will be transferred from one station to another station. The layout of the B13 line is scratched in Figure 2.

	Full-welding	Tack-welding (stiffeners)		Placing (base plates) Tack-welding (large flat bars)
	R8	R2		
	R9	R3		R0
	R10	R4		
	R11	R5		R1
	R12	R6		
	R13	R7		
Station 5	Station 4	Station 3	Station 2	Station 1

←

Figure 2   Layout of the B13 line.

In Station 1 of the B13 line (see Figure 2), two set-up robots are designed to place base plates of webs and then to tack-weld large flat bars (the length of the bar > 3 m). The robots work in Station 1 are shown as R0 (to place base plates, i.e. 'BP2' in Figure 1 (a)) and R1 (to tack-weld large flat bars, i.e. 'S28', 'S32' and 'S37' in Figure 1 (a)) in Figure 2. R0 and R1 always sequentially work in Station 1 by the sequence of R0→R1.

In Stations 2 and 3, maximum six tack-welding robots are designed to tack-weld stiffeners on base plates. These robots are shown as R2 through R7 in Figure 2. R2 through R7 can freely work either in Station 2 or Station 3. In the following description, this type of robots will be simply called a TR or TR's. In Station 2, the small flat-bars (i.e. 'S27' and 'S30' in Figure 1 (a)) are planned to be tack-welded, and in Station 3 the large bracket (i.e. '34STR' in Figure 1(a)) is planned to be tack-welded.

In Stations 4 and 5, maximum six full-welding robots are designed to completely weld stiffeners on base plates of webs. The welding seams for the case web shown in Figure 1 (a) are identified and coded in Figure 1 (b). These robots are shown as R8 through R13 in Figure 2. R8 through R13 can freely work either in Station 4 or Station 5. In the following description, this type of robots will be simply called an FR or FR's. In Station 4, welding seams '1', '2', '4', '6', '9', '10', '11', '13', '14', '15' and '17' for welding the case web are planned to be carried out. In Station 5, the rest of welding seams for welding the case web are planned to be completed.

The design and control of the three different types of robots, i.e. set-up robots R0 and R1, tack-welding robots R2 through R7 and full-welding robots R8 through R13 mentioned above are described by Ørum-Hansen et al. (1994).

2.2 Product Production Structure of the case web

The concept of Product Production Structure was firstly proposed by Drs Nielsen and Holm (1990). The Product Production Structure of a product is an joined logic illustration of the product design (or product decomposition) and production (or process planning). A Product Production Structure consists of three types of legends: circles, boxes and arrow lines. A circle on the Product Production Structure represents a product state. By the product states, particular the product states which mean the parts of the product, the design of the product is illustrated. A box on the Product Production Structure represents a process. The arrow lines on the Product Production Structure indicate the sequence of the processes by which the product is produced.

The Product Production Structure for the case web described in Section 2.1 is drawn in Figure 3. In Figure 3, the boxes are processes or operations and the circles are product states. The rectangles T1 through T5 mean six working stations of the B13 line.

To get the clear illustration in Figure 3, most of the intermediate product states are omitted. Only the final product state (circle 'xxxep'), the components (circles in rectangles T1, T2 and T3) of the case web and the welding seams (circles in rectangles T4 and T5) are shown in Figure 3.

To read the boxes in connection with their associated circles and the descriptions made in Section 2.1, the meanings of the operations shown in Figure 3 can be derived.

According to the Product Production Structure of the case web and the time estimate of each of the operations, a set of structured data can be formulated and stored in a data base.

For example, operation T11 in Figure 3 can be stored as 'T11=S1_10:0_bp2' in the data base. This data means T11 is carried out in station 1 (as 'S1'), normally takes 10 minutes (as

'10'), it is a set-up operation (as '0') and it operates on the base plate 'bp2'. By reading such data of all the operations for producing webs on the B13 line, the control system of the B13 line can make the production schedule for and control the production on the B13 line. The control system of the B13 line is described in the following section.

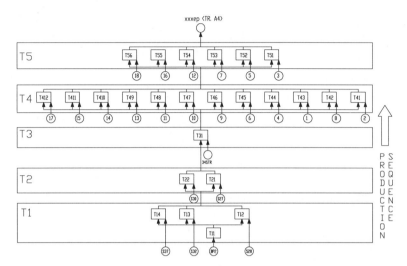

Figure 3   Product Production Structure of the case web.

3   CLOSED LOOP SCHEDULING AND CONTROL SYSTEM

At the Odense Steel Shipyard Ltd., a three-level (simply from the top to the bottom called A, B and C levels in the shipyard) hierarchical system has been used to schedule and control its production above the shop floor level. To simplify the description, this control system is named Factory Control System in this paper. The inputs to the B13 line control system are the outputs from the sub-control systems on the C level of the Factory Control System. An output from the Factory Control System to the B13 line control system is a batch of webs (normally with different types) to be welded on the B13 line in a certain time period T (e.g. a week or a month). It is simply called a C-schedule in the shipyard.

According to a C-schedule, the control system of the B13 line needs to further decompose webs into basic metal plate cutting parts and schedule as well as control the production of these webs on the B13 line.

The overall control structure of the B13 line control system is shown in Figure 4.

As shown in Figure 4, by equations (1), (2) and (3), the B13 line control system will firstly determine how many robots are needed to be configured on the B13 line according to a C-schedule, i.e. a batch of webs (as input path *'tasks'* shown in Figure 4) to be welded in a certain time period (viz. T in Figure 4).

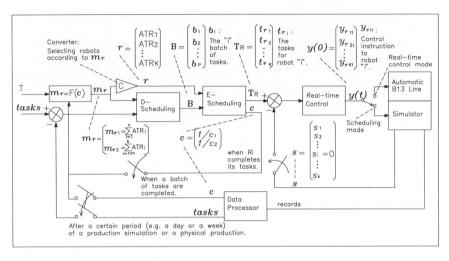

Figure 4   The control system of the B13 line.

$$
mr = \begin{bmatrix} mr_1 \\ mr_2 \end{bmatrix} = F(c) = \begin{bmatrix} \sum\limits_{i=1}^{h} T_{1i} \Big/ T & 0 \\ 0 & \sum\limits_{i=1}^{m} T_{2i} \Big/ T \end{bmatrix} \begin{bmatrix} \dfrac{1}{c_1} \\ \dfrac{1}{c_2} \end{bmatrix} \tag{1}
$$

T_{1i}: a time estimate of a tack-welding operation.
T_{2i}: a time estimate of a full-welding operation.

$$
c_1 = (\frac{1}{n} \sum_{i=1}^{n} \text{working time of } TR_i/(\text{working time of } TR_i + \text{ waiting time of } TR_i) \tag{2}
$$

$$
c_2 = (\frac{1}{k-n} \sum_{j=n+1}^{k} \text{working time of } FR_j/(\text{working time of } FR_j + \text{ waiting time of } FR_j) \tag{3}
$$

As stated in Section 2.1, robots R0 and R1 always sequentially work in Station 1. Hence by equations (1), (2) and (3), only the numbers of tack-welding robots (TR's) in Stations 2 and 3 and full-welding robots (FR's) in Stations 4 and 5 (see Figure 2) are calculated. It should point out here that m_{r1} (or m_{r2}) may be not equal to the number of TR's (or FR's) since ATR_i (see Figure 4) may be not equal to 1. This is because the robots (TR's or FR's) may work with different speeds. If a robot works with the average speed, its ATR equals 1. Otherwise, the ATR may be more than 1 (faster than the average speed) or less than 1 (slower than the average speed).

According to m_{r1} and m_{r2}, the Converter (as C triangle shown in Figure 4) will determine the number of TR's (viz. 'n') and the number of FR's (viz. 'k–n') by selecting proper TR's among R2 through R7 and FR's among R8 through R13. Then the D-scheduling module in Figure 4 will consequently determine the sequence and the lead time for producing each of the webs in the C-schedule.

The E-scheduling module in Figure 4 assigns and controls operations to each of the robots which are installed on the B13 in a certain production period. As mentioned in Section 2.1, the B13 line is designed as a five-station production line linked by an one-band conveyor. This design implies that each of the robots on the B13 line should be assigned the proper amount of tasks so that it has same working time as the others in order to gain a higher overall working efficiency. According to this criterion, the E-scheduling module assigns tasks for each of the robots by the following algorithms:

```
y=1; x=0; Number_of_robots=k;
Repeat
        Time_quota_of_robot_[y]=(∑ T₁ᵢ (or ∑ T₂ᵢ)/Mᵣ₁ (or Mᵣ₂))· ATRy ;
        Repeat
        x=x+1;
        to assign Operation_[x] to Robot_[y];
        Until ∑ Time_estimate_of_operation_[x] >= Time_quota_of_robot_[y];
y=y+1;
Until y>k;
```

After a schedule has been made by the E-scheduling module according to the above mentioned algorithms, waiting times between robots are irrevocably generated. By equations (2) and (3), c_1 and c_2 will be updated. To catch the production period (i.e. T) set by a C-schedule, m_{r1} and m_{r2} will be updated by equation (1). According to the new m_{r1} and m_{r2}, the D- and E-scheduling modules will generate new schedules and then the new c_1 and c_2 will be calculated by equations (2) and (3) again. In this manner, the control system will be recursively running until the desirable production schedules are approached. These production schedules include a time phased table for the production of webs on the B13 line, a time phased table for all kinds of operations associated with the carrying out robots, the number of the TR's and the number of the FR's in a certain production period. To start this closed loop, c_1 and c_2 are initially set as 1.

The Real-time Control module in Figure 4 is a dispatching and monitoring module. According to the schedule made by the E-scheduling module, it dispatches operation control instructions either to the Simulator in Figure 4 if the control system is working on the off-line scheduling mode or to the robots on the B13 line if the control system is working on a real-time control mode.

After a scheduling procedure (the system links with the Simulator) or a production procedure (the system links with the B13 line), according to the 'records' the Data Processor in Figure 4 will statistically estimate executing times for the same or the same kind of operations and feeds back these estimates to the scheduling modules (as *'tasks'* feedback path shown in Figure 4) as well as calculate c_1 and c_2. According to the new time estimates and new c_1 and c_2, the control system will recursively approach new schedules (if the system works on an off-line scheduling mode) or update the existing production schedule (if the system works on the real-time control mode).

4 CONCLUSIONS

According to the discussions made in this paper, it can be concluded that
• The control system proposed in this paper has system configuration flexibility.

- By applying the concept of Product Production Structure, the control system gains enough robustness to be able to cope with a wider product type domain in One-of-a-Kind Production.
- By the scheduling algorithms presented in this paper and a recursively approaching procedure conducted by the control system, the just in time production schedules can be achieved by the control system.

Although the control system structure, algorithms and concepts presented in this paper are developed particularly for shop floor control in One-of-a-Kind Production, they may also be references for the design and development of shop floor control systems in other industries.

5   REFERENCES

Nielsen, J.A. and Holm, H. (1990) Closed loop control of automatic shop floor systems in One-of-a-Kind Production. *Proceedings of the Nordic CACE Symposium, Technical University of Denmark*, 8.20-9.

Trostmann, E., Conrad, F., Holm, H. and Madsen, O. (1993) Cybernetic modelling and control in Integrated Production Systems. - A project overview. *Proceedings of the eighth IPS Research Seminar*, 213-25.

Ørum-Hansen, C. and Laursen, R.P. and Trostmann, E. (1994) Geometrical simulation of generic layout for robot assembly and welding tasks. *The Ninth IPS (Integrated Production Systems) Research Seminar, Fuglsø, Denmark*, 305-28.

6   AUTHORS' BIOGRAPHIES

Yiliu Tu is a post-doctoral research fellow at the Department of Production, Aalborg University, Denmark. He received a B.Sc. in Electronic Engineering and an M.Sc. in Mechanical Engineering both from Huazhong University of Science and Technology in China. In 1993, he received his Ph.D. from Aalborg University of Denmark. His present main research interest is real-time scheduling and control of automatic shop floors in One-of-a-Kind Production. He is a senior member of SME and CASA/SME.

Hans Holm received his B.Sc. in Production Engineering in 1970, Engineering Academy of Denmark, Aalborg, M.Sc. in Mechanical Engineering in 1975, the Technical University of Denmark, Copenhagen, and Ph.D. in Control Engineering in 1980, the Technical University of Denmark. He has been working in the fine mechanical industry as well as in heavy manufacturing industry with design, manufacture and control. In research he has primarily been working on control in manufacture. Since 1987 he has been an associate professor at University of Aalborg, where his main subject is control for manufacture.

Ulla Brandt Rasmussen received her M.Sc. in Production Engineering in 1992 at the Department of Production, Aalborg University, Denmark. Since August 1992 she is a Ph.D. candidate within the area of dynamics of shop-floor control. Her special research interests are real-time control of automatic shop-floor systems within the area of One-of-a-Kind Production, and the application of Coloured Petri Nets for shop-floor control systems.

32

Real-time shop floor control of OAK production system

R.P.Laursen, C.Ørum-Hansen, E.Trostmann
Control Engineering Institute
Technical University of Denmark, building 424, 2800 Lyngby,
Denmark. Phone +45 45934419, Fax +45 42884024, e-mail
bob@ifs.dtu.dk

Abstract

This paper presents a generic reference architecture for an automatic real-time production control system for one-of-a-kind production. The reference architecture is based on an off-line and a real-time part. The off-line part contains (re)Scheduling and Job decomposition modules. In the real-time part are the following modules active: Fine (re)scheduling, Dispatching, Shop floor, Observer/estimator and Simulator/monitoring. A production line from a ship yard is applied to the reference architecture as a case. The modules in the reference architecture are identified and described in the case.

Keywords

One-of-a-kind, real-time production control systems, automation, robot, generic architecture, shop floor control in ship production.

INTRODUCTION

In "One-of-a-kind" (in the rest of the paper abbreviated to OAK) production systems it is experienced that real-time shop floor control has great difficulties in staying robust towards disturbances and being able to meet the required through-put time, productivity and quality measures.

Disturbances are primarily caused by nonpredictable variations in materials supply, part tolerances, duration of the various processing cycle times, availability of tools/equipment and equipment break downs. These kinds of events often require a real-time rescheduling of the production plan of the shop floor activities. Such a rescheduling activity requires a real-time handling of a huge amount of data, hence it may often have a negative impact on productivity, quality and cycle time.

Todays production control systems (Browne, 1988) deals with automatic material handling and storage systems, automatic fabrication/assembly processes, computer aided testing and computer aided process planning.

In the literature real-time feedback from the shop-floor is mentioned, but the use of this information for a real-time rescheduling is not discussed in details, so there don't exists any reference architecture for a real-time shop floor control for a OAK production system.

The aim of this work is to introduce concepts of automatic control (feed back techniques) in order to control real time rescheduling processes and thereby handle production control activities for OAK production systems in a more robust manner.

GENERIC REFERENCE ARCHITECTURE

A generic reference architecture for an automatic real-time production control system for OAK production is proposed in figure 1. As indicated the reference system comprehends several different functional subsystems: Off-line (re)scheduling, job decomposition, real time (re)scheduling, real time dispatching, shop floor production, estimator/observer, simulator/monitoring. These subsystems are shortly described in general terms in the following sections.

Off-line (re)scheduling

The function of the off-line scheduling is to produce a master production plan (Tu, 1993). A master production plan is the time sequence of tasks for producing OAK products. The planned sequence will have great influence on the material logistic, because the planning must make sure that all materials must be in the right place at the right place and in the right quality. A rescheduling of the master plan is in focus if unforeseen events appears on the shop floor, which will change the production conditions so that the plan may not be held.

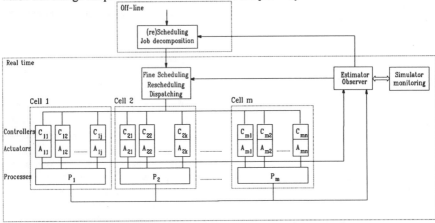

Figure 1 A generic reference architecture for an OAK production control system.

Job decomposition

Through a job decomposition process the master plan is broken down into a hierarchy of tasks and subtasks, called the product production structure (Tu, 1993), based on the processes needed for the production of the product. In figure 2 an example of a product production structure is shown. The product production structure ends at the lowest task level, called elementary tasks.

The product production structure illustrates the sequential hierarchy of tasks necessary to be carried out when the product is produced. The production starts with the elementary tasks, then the subtasks and further up in the hierarchy until the OAK product is finished.

An elementary task is the smallest task entity to be considered. A further decomposition will dissolve the task objective, e.g. by leading to an unacceptable level of quality, e.g. the welding of a seam must not be interrupted.

The job decomposition have, via the product production structure opened the possibility to intervention on several levels in the task hierarchy. E.g. where the lowest level (the elementary task) will concern only one robot, the second lowest level (subtasks) will concern task division between robots in a cell, and so on to the top level with the entire OAK product.

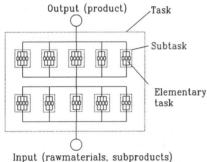

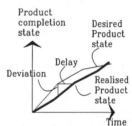

Figure 2 An example of a schematic production structure.

Figure 3 An example of a manufacturing state graph.

Fine (re)scheduling

Subtasks and hereafter elementary tasks are scheduled, in the fine scheduling module, with respect to various optimisation rules, e.g. physical connectivity, time consumption, quality. The rules are selected so a specified criteria for an optimal production is achieved. The current manufacturing state also influence the fine scheduling, because resulting deviations from the previous executed tasks (caused by unforeseen events) may change the schedule. Larger deviations at the sub task level may result in the need for a rescheduling of tasks that will make it possible to complete the tasks according to the master plan.

Examples of unforeseen events that cause deviations (from the schedule) during the execution in a cell, can be a too big gab between two parts being welded together, welding pistol runs out of wire, robots are unable to grasp materials. Such deviations in production parameters requires a rescheduling in real-time of the subtasks for correcting the error and for obtaining an updated optimal schedule under the new production state in the cell.

Dispatching

The scheduled elementary tasks are assigned to the actual machine (e.g. a robot) by the dispatcher. On basis of termination signals from the previous task, the dispatcher initiates the next elementary tasks e.g. by loading robot programs from the database and download it to the relevant controllers on the shop floor.

Shop floor system

The shop floor is here considered as a production system consisting of the three elements, controllers, actuators and processes. This three-component model can be considered in several levels in an OAK production system, generally leading to a recursive description of all levels in the shop floor. The terms are here defined as follows:

Controller   Unit which creates signals to the actuators so that the motion of the actuators follows the specified references as close as required.

Actuator   Device, which introduces some sort of mechanical, electrical, thermal power to activate or execute a process.

Process   Transformation of the input material to the desired output/state.

Observer/estimator

For being able to perform real-time control of an OAK production system, states from the production needs to be collected and fed back from the shop floor to the control unit. Decisions of external interference is supported by different visualisations of the real-time development on the shop floor.

Different manufacturing states are observed/estimated by the observer/estimator unit. The observed states are chosen so the remaining states can be estimated, and together create an adequate real-time description of the entire production.

The necessary states to be observed and the sample frequency have great impact on the stability and controllability of the entire production control system. A too high sampling frequency may result in rescheduling in a situation where the local controller is able to handle the situation without external interference. On the contrary, a too slow sampling frequency may not react fast enough and lead to an expansion of the problem. If the proper sample frequency is not chosen, the unforeseen event can make the control problem propagate a level up and the entire system may become unstable (Rasmussen, 1993). If the right manufacturing states are not fed back, the real-time production control system are not able of taking the correct decisions, and a non-optimal production sequence may be executed.

The observer/estimator unit observe/estimate manufacturing states for which reason a discrete model of the shop floor system must be formulated. Some methods for using discrete methods are described in (Franke, 1994) and (Baccelli, 1992).

The manufacturing state variables consist of product state variables, product producing state variables and configuration- and preparation state variables. An extended description can be found in (Holm, 1994).

Product state variables consist of geometric-, material/surface property state variables. Product producing state variables consist of material transformation- and assembly state variables. Preparation state variables consist of product configuration state variables, product preparation state variables and equipment preparation state variables. To use a set of these state variables, they must of course be observed and/or estimated.

The product states are often assumed to be as they are specified, e.g. the geometric states (e.g. dimensions) of a product are assumed to be as the drawings specify.

Product producing state variables can be (observed by sensors or) estimated based on other state variables. E.g. the product producing state variable "% finished" can be estimated based on positions, velocities, etc., of the current equipment. E.g. a welding is half (50%) finished if the attached welding robot is half through the welding trajectory.

The states of the production equipment (a subset of configuration- and preparation state variables) are used for local control and for monitoring.

Simulator/monitoring

Observed and estimated manufacturing states are used for real time simulation of the production system. Such simulations are used for creating real-time information to the operators so they can observe the state and the course of the production so they can intervene when necessary, e.g. when a machine breaks down.

The monitoring is based on real-time information such as a graph illustrating the product state as a function of time (figure 3), a list of the current task of the robots, etc.

A CASE APPLYING THE GENERIC REFERENCE ARCHITECTURE

A dedicated existing production line has been redesigned using a fully automated concept. The automated production line will then be used as a test bench for the testing of real-time production control procedures. The chosen line is a typical ship web-panel production line. Web-panels are frequently used elements in shipbuilding for obtaining sufficient stiffness in the ship structure.

Shipbuilding is today, like most production areas, confronted with continuing increased competition. This requires improvements in quality and productivity. One way to obtain this, is to introduce a new reference architecture for automatic real-time (feed back) production control systems whereby through-put time, delivery times are met, production delays, idling times are avoided and requirements on efficiency, productivity and quality are achieved.

The design of web-panels is most often one-of-a-kind, which means that the production of the web-panels can be regarded as an OAK production system. On basis of the requirements on an existing line today and the desired improvements, a new physical layout of the line has been developed, figure 4. A simulation model of the production line is also developed and implemented. The concept laid down in the reference architecture is being tested and evaluated by simulation of the modelled line.

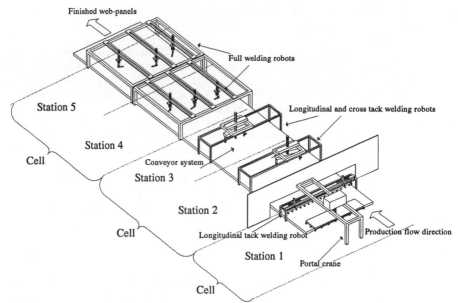

Figure 4 The one-of-a-kind production line for the shop floor control system of the web-panel line.

When receiving a task for full welding seam no. 1.

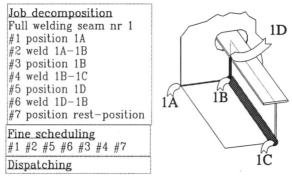

Job decomposition
Full welding seam nr 1
#1 position 1A
#2 weld 1A-1B
#3 position 1B
#4 weld 1B-1C
#5 position 1D
#6 weld 1D-1B
#7 position rest-position

Fine scheduling
#1 #2 #5 #6 #3 #4 #7

Dispatching

Figure 5 Example of the tasks executed in the module.

In figure 6 (element in level D) a product from the web-panel production line is shown. A web-panel consists fundamentally of two types of elements, stiffeners and web plates. The production sequence of a web-panel according to the layout in figure 4 as follows.

Station 1: Placing of web plates on the rollers with the portal crane. Tack welding of longitudinal stiffeners with the long tack robot. Station 2+3: Tack welding of the shortest longitudinal and all cross stiffeners with the long/cross robots. Station 4+5: Full welding of longitudinal and cross stiffeners with the full welding robots 4-9.

The generic reference architecture (figure 1) can be used on different levels of the production. Each time a lower level is examined, the modules of the shop floor (controllers, actuators and, maybe, processes) are expanded to contain the whole reference architecture (new objects are inspected but the same model is used). The case will be described according to the division of subsystems carried out in the theoretical section. The various levels used in the case will be described shortly in the following.

From the product production structure (a schematic hereof is shown in figure 2) is some information of the job decomposition obtained. The product production structure contains only the levels D, E and F (explained below). The D level contains the tasks, the E level contains the subtasks and the F level contains the elementary tasks

At the lowest level, F, the elementary tasks (e.g. a welding seam) are performed, see figure 6. At the level above F, is E, where the subtasks, consisting of a set of elementary tasks, are performed (e.g. a set of welding seams). The D level contains the performance of a set of subtasks (e.g. manufacturing a complete web panel). The level C contains the whole production line as shown in figure 4. The level B is the assembly of the web panels (coming from the shown line) to blocks. Finally the A level is the assembly of blocks to a ship.

LEVELS

A Ship

B Block

C Section

D Element (Task)
    (eg. a web panel)

E Subtasks
    (eg. a welding of a corner)

F Elementary tasks
    (one welding seam)

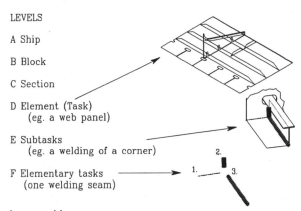

Figure 6 Task decomposition.

The following two subsections describe the activities in the case which fills in the various modules in the generic reference architecture (figure 1). First subsection (off-line activities) contains a description of the off-line modules: scheduling and job decomposition. These are only carried out once (or rare, depending on how the production elapses). Second subsection (real time activities) contains a description of the real-time modules: Fine (re)scheduling, dispatching, shop floor, estimator/observer and simulator/monitoring.

Off line activities

A master plan is generated, containing a description of which web panels to be produced. A re-schedule could take place if some unforeseen events occur. That could be a break down of production equipment, causing that the master plan should be completely remade. It could be lack of materials, causing a production of the next batch of panels to be advanced in the overall schedule. The master plan generation and rescheduling is mostly taken place in the top level of the production. Hence it covers a large time horizon, which gives the rescheduling a big impact on the production.

The master plan is broken down (job decomposition) in to a set of sub tasks. At level D the sub tasks are jobs for the different production cells (process: tack welding, full welding), e.g. full weld the panels a, b and c in cell x.

The shown production line in figure 4, contains 3 production cells: Station 1 (one cell, tack welding long elements), station 2 and 3 (one cell, tack welding short elements) and station 4 and (one cell full welding elements).

The E level is the job decomposing to various processes (processes: tack welding, handling, etc.), e.g. some robots are handling units and other are distinctly tack welding units. At the lowest level (F) (process: moving elements of a robot) decomposed jobs are trajectories for e.g. welding seams.

Real-time activities

Real-time fine (re)scheduling is first taking place at the E level. Because at the D level the master plan defines the order of web panels to be produced. At the E level, the fine scheduling determines which robots to do what, e.g. robot 2 perform subtask E5 (tack weld) or robot 7 perform subtask E41 (full weld). If an unforeseen event occur (e.g. a robot breaks down) the resisting resources (robots) have to share the extra burden. This requires a rescheduling of the current plan. One of the goals in the scheduling (and rescheduling) is to obtain equal time con-sumption of all the resoutces in one cell. The opposite situation could also occur: A resource (a robot) terminates earlier than expected. This requires also a rescheduling so the available re-source can assist the other resources (e.g. to terminate earlier).

At the lowest level F, the schedule is, again, predetermined, as these elementary tasks are collected in robot programs (performing E-level subtasks).

In figure 6 an example of the actions at the F level, described above (job decomposition, fine scheduling and dispatching) is shown. Here a task for full welding of the corner ABCD in a ship section is received from the scheduler. The optimisation in this case is carried out with respect to quality at the expense of time: The schedule could have been faster by reversing the direction of the welding seam #6, but then the welding quality would have decreased.

Dispatching at the D level is only to distribute the generated schedule to the cell controllers. At E level the robot programs are loaded from some off-line programming department and downloaded to the robot controllers. Moreover, the dispatcher determines when to start new processes (e.g. robot programs). This is based on termination feed backs from processes/actions which are previous to the current process. E.g. one welding is not started before another process (previous to the current) has terminated.

At the D level, the shop floor represents the production line, where each controller is a cell controller, each actuator is a production cell and each process is either tack welding or full welding.

At the E level, the shop floor represents the production cell, where each controller is a robot controller, each actuator is a robot and each process is either tack welding or handling (in the tack welding cell).

At the F level the shop floor represents the robot unit, where each controller is a joint controller, each actuator is a (robot) actuator and each process is the moving of the actuators.

In the D level, the estimator/observer receives states from the different production cells. These states describes how the production is running concerning, time, quality, etc. The states are delivered further on to the scheduling unit, which compare the observed states to the planned states.

In the E level, the observer/estimator observeres/estimates states from the production equipment e.g. robots (position, velocity, activity state, etc.), from the process, e.g. welding current and from the products, e.g. "%-finished". The states are delivered further on to the scheduling unit, where the original plans (desire states) are compared to the observed/estimated states.

On the F level, the estimator/observer is a part of the robot controller. It receives the feed back states from the actuators (and maybe from the process). If the states fed back are not sufficient, they can be used for estimating states such that the process is controllable. These states are used for evaluating the current process in the scheduling unit, where they are compared to the reference set.

A monitoring unit could, in this case, be a unit showing the state of various process parameters, e.g. welding current. The simulation unit could show the whole production line in 3D graphics to show the positions of the production equipment and the processed products and perhaps even record the fed back states such that the production could be shown again to for instance reveal errors in the production.

PERSPECTIVES

The proposed real-time control architecture, figure 1, is currently being evaluated and validated by simulation tests based upon the application case in figure 4. The achieved results will be discussed with respect to portability and applicability in other production domains and design rules for one-of-a-kind production systems will be derived.

REFERENCES

Baccelli F., Cohen G., Olsder G.J. and Quadrat J. (1992) Synchronisation and linearity: An algebra for discrete event systems. Wiley. ISBN 0-471-93609-X.

Browne J., Harhen J. and Shivnan J. (1988) Production management systems: A CIM perspective. Addison-Wesley Publishing Company. ISBN 0-201-17820-6.

Franke D. (1994) Sequentielle systeme: Binäre und fuzzy automatisierung mit arithmetischen polynomen. Studium Technik. ISBN 3-528-06527-3.

Holm, H. et al. (1994). An Outline for a State Space Based Manufacturing Theory: Control Part of Automatic OAK Production. Department of Production, University of Aalborg.

Rasmussen U., Holm H. and Smartt H.B. (1993) Dynamics of multi robot task sequence control: Introductory considerations. Methods. Expected results. Proceedings of the eight IPS research seminar held at Fuglsø, Denmark on 22-24 March 1993. ISBN 87-89867-26-2.

Tu Y. (1993) Real-time scheduling and control of material handling and equipment set-up in automatic shop floor for OAK production. Ph.D. thesis Department of Production, University of Aalborg, DK. ISBN 87-89867-30-0.

ACKNOWLEDGEMENT

The research presented in this paper has been sponsored by Statens Teknisk-Videnskabelige Forskningsråd (STVF) , the Danish Technical Research Council, under the research program for integrated Manufacturing systems, "Integrerede Produktionssystemer 2(IPS 2)". We gratefully acknowledge the support, which we have received.

Further the corporation with Odense Steel Shipyard (DK) and Aalborg University (DK) is highly appreciated.

BIOGRAPHY

M.Sc. Robert Peter Laursen

received his M.Sc. degree in mechanical engineering from Control Engineering Institute at the Technical University of Denmark in 1993. Current work is on his Ph.D. at Control Engineering Institute, with the topic in automation of OAK production.

M.Sc. Claus Ørum-Hansen

received his M.Sc. degree in mechanical engineering from Control Engineering Institute at the Technical University of Denmark in 1993. Current work is on his Ph.D. at Control Engineering Institute, with the topic in automation of OAK production.

Professor, Ph.D. Erik Trostmann

received his M.Sc. degree in mechanical engineering from the Technical University of Denmark in 1954 and his Ph.D. in control theory and engineering at Case Western Reserve University in Cleveland in 1963. Since 1970 he has been a full professor in Control Engineering at the Technical University of Denmark. He has participated in many design and development projects in the areas of automatic machines and plants such as CNC machine tools, foundry machines food production machinery. He is currently engaged in research within CIME, CAD, CAM, FMS, CNC and Robotics. He has been a project member of the ESPRIT CAD*I project, no. 322, and NIRO project no. 2614/5109. At present he is engaged in the ESPRIT InterRob project 6457.

33

Management of changes in one-of-a-kind production

S. Törmä, S. Karvonen and M. Syrjänen
Helsinki University of Technology
Otakaari 1, FIN-02150 Espoo, Finland
tel.: +358-0-451 3251, fax: +358-0-451 3293,
email: {sto, ska, msy}@cs.hut.fi

Abstract

Changes are inevitable in one-of-a-kind production where the uncertainty is typically high. However, changes have to be implemented in a way that does not change the delivery date of the product which is firmly fixed by the contract. This makes change management a critical area in one-of-a-kind production. This paper presents how a project planning tool can help a project planner to clarify and evaluate the impact of a change and to replan the remainder of the project. The planner can interactively try alternative corrective actions, compute the resulting schedules and their total costs.

Keywords

Project management, one-of-a-kind production, reactive planning

1 INTRODUCTION

A company that sells process plants, power plants, harbors, ships or large buildings is seldom able to deliver two similar products. Each of them is one-of-a-kind: it has different functional requirements, different constraints, and different priorities. In practice, one-of-a-kind production is project oriented. Each project consists of several phases of a different nature: engineering, procurement, manufacturing, transportation, installation and start-up.

The uncertainty is typically much higher in one-of-a-kind production than in volume production for many reasons: the strong influence of the customer on the product specification, the open environment and long duration of a project (Kuhlmann, 1991 and Nielsen, 1992).

In spite of the high uncertainty, deliveries in one-of-a-kind production are typically based on firm fixed price contracts. The delivery date of the product is fixed and enforced with high tardiness penalties. Furthermore, the contractor is typically interested in maintaining or developing its reputation as timely supplier. In practice, this leads to a situation where it is almost impossible for the contractor to let a project slip from its delivery date.

The high uncertainty means that there will inevitably be changes during a project. However, because the contract fixes the time frame in which the project can operate, there is a need for good ways to manage changes and replan the project in a way that leaves its delivery date untouched. A study documented in (Burati, 1992) suggests that the average costs of changes in construction projects exceed 10% of the total costs of a project.

The importance of change management is generally recognized. For instance, Kimmons (1990) notes that "a successful project manager sees a large part of his job as the control of changes". However, many tools do not support efficient change management and partly for that reason for instance schedules are not kept current if changes happen. This means that they can not fully serve as the basis of operations management.

We have studied change management methods in PROOMU-project* where a software prototype for change management has been developed as a part of Process Plant Project Manager (3PM) (Törmä, 1995). 3PM supports a project planner in the evaluation of the impact of a change and allows him to incrementally revise the project plan. Multiple plan revision alternatives based on different selections of corrective actions can be explored in a flexible way and their total profits can be computed. The project planner can clarify the influences of corrective actions before committing to them. In addition, the total (planned) costs of a change can be defined. After a commitment is made the affected project parties can be informed about the changes. Furthermore, the system maintains a database of changes to support later analyses and thus the continuous improvement of project management practices (Karvonen, 1995).

Change management in scheduling systems has been studied under the name reactive scheduling. A good example is OPIS scheduling system (Smith, 1994). However, the problem area is broader in one-of-a-kind production where the changes to the product or the documents create the biggest problems. We do not provided automated replanning but an interactive tool that supports the project planner in clarification of the impacts of a change, replanning, estimation of the costs and informing about the changes.

2.   PLANS AND CHANGES

2.1 Project plan

A **project plan** is a detailed formulation of program of action to achieve the goal of the project. A project plan defines the activities to be executed, their durations and resources, and their budgets.

In a project delivery the **goal** is to create a physical product within the time and budget constraints. Consequently most of the activities concern the creation of the product or its design documents (figure 1).

---

* PROOMU ("Project change management") belongs to the ProDeal technology program of FIMET and TEKES in Finland.

A project plan should be **complete** with respect to the product. This means that the final state of each part should be achieved with an unbroken sequence of activities starting from its initial state within the project. The initial and final states of parts can be defined in a straightforward manner. See (Törmä, 1995) for details.

A project plan should be **non-superfluous** in its relation to the product: each of the activities in the project plan must contribute to the achievement of the final state of some of the parts.

A plan is **accurate** with respect to the product if all the parameters of activities (durations, budget values, etc.) are correct when derived from the product description.

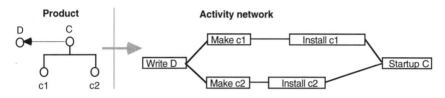

Figure 1 An example of a simple product C and its design document D and the corresponding activity network.

A plan must also be **feasible** in order to be viable as operational guidance (Smith, 1994). A feasible plan satisfies all the different constraints in the domain dealing, for example, with the capacities of resources and the availability of other required physical entities (such as parts) when activities start.

We call the plan of a project delivery **current** if it is feasible, and complete, non-superfluous, and accurate with respect to the product. The goal of change management is to keep the project plan current.

When a project plan is created, many assumptions are made about the structure of the product, durations of activities, the availability of resources, timely opening of activities, timely operation of suppliers, sufficient quality of work, etc. The probability that these assumptions will hold is lower in one-of-a-kind production than in volume production. The predictability is decreased by the amount of unique aspects present in each project: product design, site location and conditions, subcontractors, etc. In addition, the problems of insufficient or contradictory information are bigger in one-of-a-kind production due to the number and novelty of parties involved in a project.

2.2 Change requests

Any of the assumptions of a plan may fail and if the failing is large enough the plans that were based on it have to be changed. Typical change types are described below.

- A design change may originate from vague or inconsistent technical specification that can be interpreted in different ways. The customer is typically inclined to broader interpretation of the scope than the contractor. Moreover, it is quite possible that a customer wants to change his mind with regard to some technical aspect of the project. Design changes can be critical if they concern approved specifications that have already been sent to subcontractors or to manufacturing. Changes of this kind may cause major controllability problems during a project and result in many non-value adding activities that decrease the profit of the contract. Examples of non-value adding activities are redesign, remanufacturing and reinstallation of components.

- A quality problem means that some part of the product is unacceptable and can not be included into the final product as such. Either a new part must be acquired or the faulty part must be modified. Quality problems may lead to costly rework and difficult schedule problems especially if the rejected part is critical with respect to the schedule.

- Operational changes may arise from late deliveries or unavailable capacity (machine breakdown or sick-leaves of employees). Operational changes can usually be accommodated with schedule modifications. If there is not sufficient flexibility in schedules, scope changes must be considered.

The party that is **source of a change** is typically also responsible for the costs of implementing the change order. It can be any party: the customer, own company or some subcontractor. However, the project manager is typically responsible of the replanning and he makes the decisions about the actions that are required to implement the change.

The **timing of a change** has a strong influence on its impact. The later a (necessary) change is requested, the more costly and difficult it is to correct (Kimmons, 1990). In later phases much work may have been done that turns out to be useless (it has to be done again) or even harmful (the effects of the work have to be undone before the rework can be done).

2.3   Effects of changes

From the project management perspective changes have two kinds of consequences: schedule effects and cost effects.

Schedule effects can be either internal or external. Internal changes leave the delivery dates of the project untouched. They are not fatal but still undesirable for two reasons. If activities have to be moved later the risks of a schedule naturally increase because there will be less slack to correct subsequent changes. Second, it would be necessary to inform the resources whose schedules have changed.

An external schedule change means that the delivery of a project will be late. This can be very costly in fixed price project deliveries and should be avoided. There are often ways to arrange the corrective actions so that project can be completed in schedule. An example is fast subcontracting. However, this kind of corrective actions can be costly.

Because changes in project deliveries are mostly done in a way that preserves the delivery date of the project, their impact can usually be evaluated from their costs alone.

Cost effects of changes are often evaluated in an ad hoc manner. For instance, in the analysis in Burati (1992) only direct costs incurred from rework were included because total costs were not generally available. In multiproject environment some costs of a change may also be hidden because the fire-fighting actions to solve some problem may disturb the schedule of another project and cause additional costs for it.

The ability to compute the estimate of the total costs of a change is important both for the internal development work where the economic importance of the changes can be better understood and also for the negotiations of compensations when the source of the change is outside of the own company.

3   CHANGE MANAGEMENT FUNCTIONALITY OF 3PM

Process Plant Project Manager (3PM) is a special purpose project management system whose basic implementation was developed in MUSYK-project[*]. The main idea in 3PM is to integrate the scope, schedule and cost management in process plant projects.

3PM maintains a rich object-oriented model of the entities that are relevant for the management of a project delivery. In addition to the activities and resources it also contains a representation of the different objects manipulated in a project: parts, documents, purchases and shipments. This allows many product oriented activities (such as procurement and transportation) to be added to the activity network of a project and managed with the normal schedule and cost control methods (Törmä and Syrjänen, 1995).

With 3PM the project planner can interactively replan a project after a change to restore the feasibility of the plan and to ensure that the goals of the project will be met despite the change. The planner can clarify the impact of the change and explore alternative ways to correct the situation. The result is an incremental modification of the existing plan.

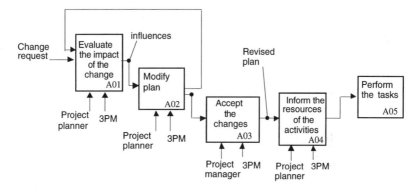

Figure 2   The change management process.

The goal of 3PM is to shorten the change processing time so that work that would turn out to be useless or even harmful because of changes can be discontinued as soon as possible. The reduction in processing time is caused by interactive planning capabilities of 3PM and by fixed and agreed on procedures that are obeyed in replanning.

Figure 2 shows the change management process that is carried out when a change occurs during the execution of a project. The first step is to evaluate the influences of the change. These influences may concern other objects (e.g., when a technical document changes, the parts or systems that it describes will also potentially change) or activities of the objects.

A project planner iteratively modifies the plan until an acceptable solution has been found.

---

[*] MUSYK (Integrated Multi-level Planning and Control System for One-of-a-Kind Production) is the project number 6391 in ESPRIT III program.

The project manager accepts the new plan and the resources that are responsible for the changed activities are informed about the change.

3.1 Events

3PM receives messages from outside containing progress information about activities and objects, and requests to change the model. The progress information contains actual values of time-points (start or end times of activities) and information that some state of an object has been reached. This information may come from the user interface or from some other system (e.g., from a document management system).

A message is interpreted which means that the model is updated with its information. It may additionally result in the creation of an event which – in the terminology of 3PM – describe a failed assumption.

The events are divided into the following groups:

- Object change   a document or part is added, removed or modified

- Object state change
  - Design change: document that was in state 'in-use' is moved to state 'draft'
  - Lost object: object that was in state 'created' turn out to be only 'designed'
  - Quality deviation: 'created' object is moved into state 'faulty-created'

- Activity change   the resources or the duration of an activity change

- Time-point change   activity is started or completed before or after its scheduled time

- Resource change   capacity reserved for an activity is not available

A change does not necessarily mean a conflict. It may also be an extension or reduction of the product description or to the activity network. However, it potentially results in the need of corrective actions.

3.2 Evaluation of impact

Changes are unique by their nature. It is impossible to determine the exact effects of a change automatically, at least not without an extensively detailed model of the domain. For this reason the responsibility for the evaluation of the impact of a change is left to a project planner. 3PM supports the planner in this task by keeping a register of the impact that has already been made and by visualizing the of the foreseeable effects that has to be checked or replanned by the project planner. There are two kinds of foreseeable effects:

- Possible foreseeable effects: things that may still be affected
- Necessary foreseeable effects: things that will necessarily be affected

The project planner must find out the real effects of a change from the set of possible effects.

The evaluation of the impact of a change proceeds in an alternating fashion with the modification actions. Many additional effects may be produced by planning decisions. E.g., the necessary effect of a change may be that some activity can start earliest at time point t1 that is later than its original schedule. However, due to the lack of capacity the activity is scheduled to start at time-point t2 that is later than t1. This will cause a larger impact on the succeeding activities. (The decision can either be made by the human planner, or an automatic planning or scheduling system.)

The computation of possibly effected entities (parts, documents, activities, resources) is based on their structural relationships. For example, a document has a link to the parts and other documents that use its information. The objects in the transitive closure of this relationship belong to the set of possibly effected objects if the contents of the document changes.

The objects are linked to activities through states. When an object changes, its activities are in the set of possibly affected objects.

The precedence constraints between activities can also be used to define the effects of a change in schedule. All activities linked through precedence constraint are possibly in the set of affected activities. The necessary changes can be determined with the computation of the time-bounds using the precedence constraints.

3.3 Modification actions

Events are processed and their impacts evaluated. The project planner uses the information about the impact of a change to try out different modification actions. The planner can interactively build alternative scenarios of how to solve the problems caused by the change.

The different type of events are processed in the following way:

- object change: planning
- object state change: planning
- time-point change: scheduling, planning
- resource change: scheduling, planning

Planning leads always also to scheduling. Planning means that activities are created, removed or their parameters are modified. In principle, planning can be done manually or automatically. (The automatic planning functionality of 3PM is under development).

Scheduling means that the resources and time-bounds of an activity are determined. If changes are too large, scheduling may not yield a satisfying result and activities have to be replanned. If that is not sufficient one possibility is to try a design change that would produce a product that is significantly faster to create.

3.4 Example

Figure 3 contains a simple example of how replanning proceeds when a design change is made to a product. The product C consists of two parts: c1 and c2. C as a whole is specified in the design document D. (The "product structure" is shown in the left side of the figure and the corresponding activity network at the right side of it).

In a point when both components c1 and c2 have been manufactured, a request for design change comes. (The bar shows the current time-point; all activities to the left of it have been executed). This means that the state of D is changed earlier and consequently a new activity is required in order to fill the gap between that state and the previous state of D.

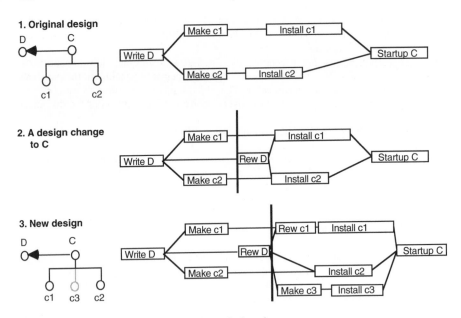

Figure 3   A simple example of replanning a design change.

The activities of D are replanned which leads to new version of the activity network. This version contains a rework activity "Rew D" where a new revision of D is produced. This activity has precedence constraints to all activities concerning the parts that D specifies. The effects of this change can be visualized.

The redesign leads to a new change request: the product structure has to been modified by adding part c3 and modifying part c1 (i.e., changing its state). The activities of both parts are replanned. As the result a rework activity for c1 is created as well as the manufacturing and installation activities for c3.

4   IMPLEMENTATION

This section describes the background mechanisms that contribute to the change management functionality of 3PM.

   Much of the change management functionality is based on the rich domain model that 3PM uses. The underlying database contains information about products and related projects, activities belonging to projects and internal and external resources that are used by the projects. 3PM can use the relationships between objects to determine possibly affected entities, persons that should be informed about changes, etc.

   A person is a subclass of a resource and the fact that a person is allocated to an activity is represented as normal resource reservation of an activity. When 3PM finds out that an activity is likely to replanned, those resources that are able receive messages (persons or companies) can be found out and a message can be sent to them to discontinue the work until a new plan has been made.

4.1 History maintenance

3PM has a history mechanism that makes it possible to maintain several different states for each object in the system. This mechanism is orthogonal with respect to the conceptual model of the system in a sense that all attributes of all object may have evolving values.

The history mechanism is based on an undo/redo mechanism described in Berlage (1993). Each time that user makes an operation that changes the state of a project a new context will be created. New values of each changed slot of each object are stored into the new context.

The history mechanism can support the **undo and redo operations** in the user interface. In any time user is able to cancel his or her operations or redo a canceled operation. The history may have multiple alternative branches. This can be used to support the **what-if analyses** in the system. User can create the project plan based on some set of decisions and then back up to the starting context and make an alternative scenario based on some new decisions.

The history mechanism can be used in change situations to support the evaluation of different ways to replan the remainder of the project. Alternative resourcing (such as using a subcontractor) may be considered. For each of the alternatives it is possible to define the final profit of the project after the new plan has been made. The difference of this profit when compared with the profit of the original plan can serve as the estimate of the costs of the change.

4.2 Visualization

The interactive support in the evaluation of the impacts of a change is based on the visualization capabilities of the user interface of 3PM. It has a large set of different views through which objects can be inspected: general trees and directed graphs, attribute inspector, activity network visualization, gantt diagram, three dimensional figure of a product, etc.

The history maintenance mechanism can be used to provide information about changes by (1) highlighting the affected activities in the gantt diagram of a project and (2) providing the possibility for the user to compare the new plan with the previous version.

5   DISCUSSION

3PM is a tool with which the project planner can interactively replan a project after a change to restore the feasibility of the plan and to ensure that the goals of the project will be met despite the change. The planner can clarify the impact of the change and explore alternative ways to correct the situation. The result is an incremental modification of the existing plan.

The goal of 3PM is to shorten the change processing time so that work that would turn out to be useless or even harmful because of changes can be discontinued as soon as possible. The reduction in processing time is caused by interactive planning capabilities of 3PM and by fixed and agreed on procedures that are obeyed in replanning.

First prototype of the change management functionality of 3PM has been built. The functionality of 3PM is being tested based on a model of a recaustization plant delivery project.

6   REFERENCES

Berlage, T. and Genau, A. (1993) From Undo to Multi-User Applications, in proc. of Vienna Conference on Human Computer Interaction 1993, Vienna, Austria

Burati, J.M. Farrington, J.J. and Ledbetter, W.B. (1992) Causes of Quality Deviations in Design and Construction. *Journal of Construction Engineering and Management,* 118.

Karvonen, S., (1995) Computer Supported Management of Changes, working paper, Institute of Industrial Automation, Helsinki University of Technology.

Kimmons, R.L. (1990) Project Management Basics: A Step by Step Approach, Marcel Decker, New York.

Kuhlmann, T. (1991) A Revolving Planning and Control System, *International Working Conference on New Approach Towards One-of-a-Kind Production* (eds. Hirsch, B.E. and Thoben, K.-D.), Bremen, Germany.

Nielsen, J.A. (1992) Concept for Real-Time Control of Automatic Shop Floor Systems in One-of-Kind Production, Ph.D. Thesis, Aalborg University, Denmark.

Smith, S.F. (1994) OPIS: A Methodology and Architecture for Reactive Scheduling, in *Intelligent Scheduling,* (eds. Fox, M. and Zweben, M.), Morgan Kaufman.

Törmä, S. and Syrjänen, M. (1995) Product and activity relationships in project deliveries, in *Computer Applications in Production and Engineering: Proceedings of CAPE'95,* (ed. Q.N. Sun), UK.

7   BIOGRAPHIES

Seppo Törmä (MSc.) is a research scientist in Institute of Industrial Automation in Helsinki University of Technology. His interests cover project and production management, object oriented methods, constraint satisfaction techniques, and cooperative problem solving. He is member of American Association for Artificial Intelligence.

Sauli Karvonen (MSc) is a research scientist in Institute of Industrial Automation at Helsinki University of Technology. His main interests are project management and manufacturing system design and corresponding applications of information technology. He is a member of the American Production and Inventory Control Society (APICS®).

Markku Syrjänen (Ph.D Eng) is the professor of Knowledge Engineering in Helsinki University of Technology. His field of special interest is knowledge-based production management. He has been the vice-chairman of Finnish Society of Information Processing Science 1982-83, member of National Research Counsel for Technology (a subunit of Academy of Finland) 1986-88, and chairman of Finnish Artificial Intelligence Society 1990-92.

PART EIGHT

Production Management Techniques

34

A Dynamic Scheduling System for Job Scheduling in low-volume/high-variety manufacturing

Zhang YaoXue Li GuangJie Di Shuo Cheng Hua Cheng KangFu
Dept. of Computer Science Technology, Tsinghua University, Beijing
Tel: (861)2561144 ext. 3596
E-mail: zyx@dcs.tsinghua.edu.cn

Kadota Toshiharu. Hayase Hiroto. Araki. Yoshida Osamu.
System Department of MEKTRON Company, Tokyo, Japan

Abstract: One of the most important issues in the process of development and usage of a manufacturing system is job scheduling. Though many scheduling criteria for job scheduling have been proposed, but most of them are impractical for use in the low-volume/high-variety manufacturing environment. This paper reports the development of a dynamic scheduling system for job scheduling in low-volume/high-variety manufacturing environment. The system provides us with a practical facility for job scheduling which takes into account of the influence of many factors such as machine setup times, cell changes, replacement machines and load balancing between machines. The system is based on a set of heuristic algorithms and client/server database technology. Numerical testing example are taken from a real-manufacturing factory of Japan. And high-quality results are efficiently generated.

Keywords: Client/server, scheduling, CPM

1 INTRODUCTION

This paper is concerned with the development of a dynamic scheduling system for job scheduling in a low-volume/high-variety manufacturing environment. At present, approximately 50~75% of manufactured parts falls into the low-volume/high-variety and mid-volume/mid-variety categories, and with the trend toward to increase in the variety of the products[2]. However, the commonly used scheduling algorithms or heuristics such as Shortest Processing Time(SPT), First Come First Serve(FCFS), and Shortest Due Date(SDD) are all too simplistic for use in scheduling of the low-volume/high-variety manufacturing. The reason is that these heuristics do not adequately address the influence of some other factors such as setup times, cell changes, replacement machines and over working time of personnels. Consequently, low-volume/high-variety manufacturing has always been plagued with difficulties and the scheduling task is left to the foreman of the manufacturing shop.

This paper reports the development of the scheduling system for job scheduling in low-volume/high-variety manufacturing environment. The objective of the system is to provide shop floor personnel with a practical scheduling tool which dynamically assigns the workers, the products to different types of cells(an assigned machines groups which can have different types) with satisfying due dates and maximizing machine utilization when the products data to be scheduled and the actual production progress has been given. The system also provides some analysis tools for analyzing the dally reasons, machine utilization and etc. The system uses the client/server architecture and different users can share the database in the server of network, so that the system becomes more powerful.

2 SCHEDULING PROBLEM IN LOW-VOLUME/HIGH-VARIETY MANUFACTURING

Because the low-volume/high-variety manufacturing is very plague with difficulties, the scheduling involves many factors to load a job onto a machine with the objective of satisfying due date and maximizing machine utilization. We first introduce these factors which influence the scheduling, and then summarize the scheduling problem. The factors used in our system can be listed as follows: order information, process design information, shop floor personnel information, cell and machine information, load balancing information, and actual production progress information.

We continually describe them in details:

1. Order information

Order information contains due dates, number of the products in an order, and the names of the products. We assume that the orders are all divided into different lots by a process designer of the manufacturing company. For specification convenience, we sometimes use the word "lot" instead of "product" in later discussion. An order from a customer may includes several lots or a lot may consists of several different orders for the same product. The due date is one of the most important factors which are usually considered in common scheduling algorithms. The difference in our system is that the due date is considered as assigned due date or basic due date.

2. Process design information

This information includes the parameters of the machines used in each process, the replacement machines, the process flows and the process priorities of each lot. Here, we assume that a process of any lot is performed by any machine belonging to an assigned machine group and the process designer gives the process flows for each lot.

3. Shop floor personnel information

Shop floor personnel information are personnel group names, numbers of the personnels, calendar of the personnel groups, calendar of every personnel, and the working time horizon of each personnel. To satisfy due dates of the orders, personnels are sometimes asked to work overtimes. Consequently, the working time horizons of the personnels are considered as variables in our system. However, the length of the working overtime is given by the foreman rather than the system, because the system can not automatically do such decision.

4. Cell and machine information

To manufacture low-volume/high-variety products, we compose the different machines into cells. A cell is a set of machines with different types or a line to complete one or more than one manufacturing process. There are three types of cells in our system. One of them consists of independent machines which can process different lots in different machines but only complete one process for every given lot(see Figure 1.a). Another one consists of machines which continually complete several processes but only for an assigned lot, and any machine of the cell does not do the same process with other machines in the same cell(see Figure 1.b). The last one is similar to the above second one, but a part of the machines of the cell can do the work on the same process(see Figure 1.c). These three types of cells are shown in Figure 1. The other types of the cells can be obtained by composing of the above three basic types.

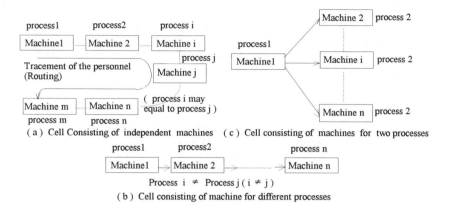

(a) Cell Consisting of independent machines   (c) Cell consisting of machines for two processes

Process i ≠ Process j (i ≠ j)

(b) Cell consisting of machine for different processes

Figure 1 Three types of basic Cells.

From the definition of the cells, we have that any machine can be the element of several different cells, but only one cell including such a machine is active at any assigned time. Moreover, we assume that any machine in our system must belong to a cell.

Another important parameters on the cells are the production time and the setup time. To schedule the low-volume/high-variety manufacturing, we give the capacity of every machine for processing one item and for processing one lot , respectively. Here, an item is a part of a lot, which cannot be divided further more. Consequently, the production time $P_{Ti}(m)$ of a lot in process i with machine m can be represented by:

$$P_{Ti}(m) = \begin{cases} N * t, \text{ or} \\ \\ t_{lot} \end{cases}$$

Here, N is the number of items of the lot, t is the process time of each item, t_{lot} is the processing time of the lot, respectively.

The setup time varies with the tooling, materials, and the products types. The setup time $S_{Tij}(m)$ of a lot in process i with machine m to next process j is given as follows:

$S_{Ti j}(m) = S_{tool} + S_{mat}$

Here, S_{tool} is the time to set up tooling for the lot, which depends on the tooling required and the current tooling on the machine; and S_{mat} is the time to change materials, e.g., change the back-up board in the manufacturing of flexible printed circuit.

Consequently, we have that the machine time T_i for the production of a lot in process i is given by:

$T_i = P_{Ti}(m) + S_{Tij}(m).$

According to this formula, we can calculate the time for the production of any lot in the above three basic types of cells.

5. Load balancing information

In low-volume/high-variety manufacturing, the load balancing between machines must be considered to obtain the maximizing machine utilization, specially, in the case of machine breakdowns or unsatisfying due dates, the replacement machines or cells should be used.

In our system, we assign a set of replacement machines or cells for each machine and each cell, then, if a machine breakdowns or the due date of the lot to be processed will not be satisfied, the replacement machine or cell will be selected according to our scheduling rules.

6. Actual production progress

The actual production progress information is important for dynamic scheduling. The system gets the actual data from shop floor by using data scanners, the information includes the status of every lot which has been scheduled or not, the status of the cells and the personnels in the shop floors, and the status of tools and materials.

3 SYSTEM DESIGN

We use the client/server model to construct our dynamic scheduling system. The reason why we use the client/server model is that the different shops in a factory always share the same database and the scheduling algorithm in the clients can be easily changed.

Figure 2 shows the data flow of our system.

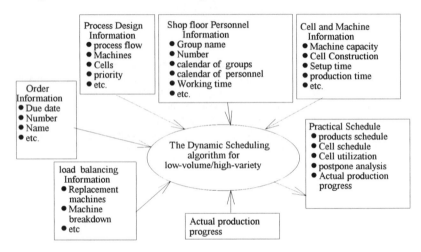

Figure 2 The data flow of the Scheduling System.

The System design includes five parts, i.e., the database design, the environment design, the heuristics design, the analysis algorithm design and the user interface design. We focus the discussion on the database, the environment and some heuristics of the system, rather than the analysis part and the user interface.

1. The environment

The network environment and the software environment our system required are shown in Figure 3. According to the Figure 3, the scheduling system obtains the order information, the process design information, the shop floor personnel information and other information except of the actual production progress information from the control center through a gateway. Moreover, the system gets the actual production progress information from shop floors through actual progress scanners which can be operated either automatically or in hand. The actual production progress data obtained from shop floors are sent to the database in the server at short intervals. The system performs rescheduling when it receives new actual production progress data and the user starts the rescheduling command. Here, we assume that the interval between two receptions of the actual data is smaller than the executive time of the scheduling system.

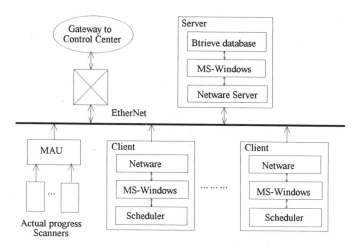

Figure 3 The construction of the environment of the scheduling system.

The database we used in our system is Btrieve developed by Novell Company. The Btrieve does not provide us with database management system, but only the format definitions, communication methods, and the development interfaces which can be invoked by using programming languages C, C++, PASCAL, BASIC and etc. The benefit of using Btrieve is that we can get a more flexible database system with high executive efficiency. Moreover, the strong network ability of the Btrieve is also what the scheduling system asks.

The MS-Windows is used to provide users with a user-friendly interface, and we omitted the discussion on this part in this paper.

2. Database design

The design of the database by using Btrieve includes the data file design and the schema design. Figure 4 shows the basic relations of the data files of the system. We construct 16 files for the system. These files are accessed by using some keywords such as Process Flow No., Machine Group No., Cell No., and etc.

The information introduced in above Section II are all put into these files of the database. For example, the order information, a part of machine balancing information, and the information on tools and materials are given in plan files. Figure 5. shows the design of the schema of the plan file.

3. The design of the scheduling algorithm

We propose a knowledge based step adjusting scheduling algorithm for our system. This algorithm firstly selects the scheduling objects, i.e., some lots, from the plan file according to the experimental maximum production ability of a shop; and then calculates the last starting time(LS) of every process for every selected lot according to the given due date or assigned due date of the lot. The LS of a selected in a process is the time that the production of the lot in the process must start. Otherwise, the due date of the lot can not be satisfied and the postpone will occurs. We can get the LS_i of a lot in the process i by using the following formula:

$$LS_i = LS_{i+1} - S_{Ti\,i+1} - P_{Ti}$$

Here, the LS_{i+1} is the last starting time of the lot in next process. If the next process is the last one, then LS_{i+1} is the due date. Moreover, the $S_{Ti\,i+1}$ is the setup time from process i to process $i+1$ and the P_{Ti} is

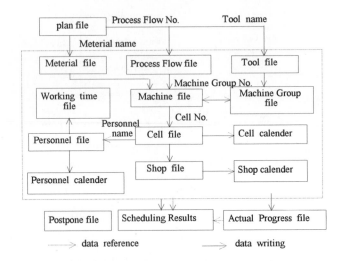

Figure 4 The relation of the files in the dynamic Scheduling System.

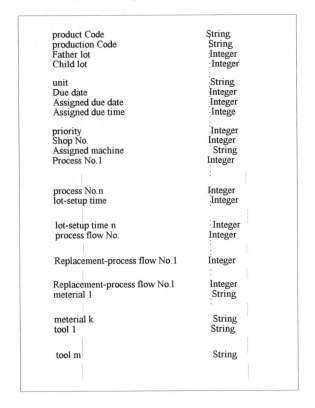

product Code	String
production Code	String
Father lot	Integer
Child lot	Integer
unit	String
Due date	Integer
Assigned due date	Integer
Assigned due time	Intege
priority	Integer
Shop No.	Integer
Assigned machine	String
Process No.1	Integer
process No.n	Integer
lot-setup time	Integer
lot-setup time n	Integer
process flow No.	Integer
Replacement-process flow No.1	Integer
Replacement-process flow No.1	Integer
meterial 1	String
meterial k	String
tool 1	String
tool m	String

Figure 5 The schema of the plan file.

the production time of the lot in process *i*. We must calculate the setup time and the production time to obtain the LS at first.

The next step of the algorithm is the calculation of the available machines and cells, including calculations of the personnels working time. This calculation is very complicate because there are too many constraint conditions on the machines, the cells, and the personnel working times. For example, the personnel include the formal employers and the part time. Moreover, the personnels should be divide into different classes according to their knowledge, experience and some other conditions. We use different data files to classify them.

When we have finished the above calculations, we assign all usable cells to the given lots process by process, until the final process of every given lot has been scheduled. Over 600 rules are used for this assignment in our system. However, if the overability occurred in any process at the same time, i.e., the usable cells are less than the quantity of the lots required; then, we use the following two methods to solve the problem:

1. Give a priority to each lot

The policy for deciding the priority is as followings:

    a. The lots with earlier due date have the higher priorities.

    b. If the user assigns a lot with a higher priority than others, then the lot has a higher priority.

    c. To those lots which have the same due dates and no user-assigned priorities, we give the following formula to calculate the priorities:

$P_i = a_1 * A + a_2 * B + a_3 * C + a_4$

Here, the a_1, a_2, a_3, a_4 are constants given by the user, and A represents the number of processes to be passed from the current process until the final process, B is the production time including the setup time from the current process until the final process, C is the waiting time until the current time. According to this formula, a lot with longer waiting time, longer production time and more processes will get higher priority.

2. Use replacement machines or cells

The selection of replacement machines and cells is based on two conditions, i.e., a set of candidates of replacement machines or cells to a given lot, and the utilization rates of these candidates. The system selects a candidate with lowest utilization rate and no loading in the current time as the replacement one. If there is no such a candidate, then the delay will occur.

Finally, the system shows the scheduling results to the user and enters the end state of the scheduling. The flow chart on the scheduling algorithm is shown on Figure 6., and we can briefly write the algorithm as follows:

Step1: Select the scheduling objects from the plan file and the postpone queues, which are not over the experienced maximum production ability of a shop, according to the due dates.

Step2: Check the materials and tools status required by the selected lots. If the required materials or tools are not ready, then the corresponding lots are added into the postpone queues and repeat to Step1.

Step3: Decide the process flows of the selected lots with materials and tools ready.

Step4: For each process of every selected lot, calculate the setup time and the LS.

Step5: For each machine calculate the cells it belongs to the active times of the cells, and the required personnels for operating these cells.

Step6: For each lot, assign the corresponding usable cells to it by the LS if each corresponding process based on the rules for scheduling. If overability occurred at any cell, then the system adjust the requirement to the cell according to the due dates and the scheduling rules.

Step7: If a lot can not satisfy its due date, then the system searches a replacement

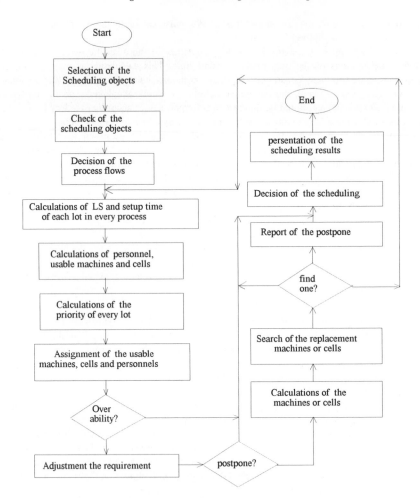

Figure 6 The flowchart of the scheduling system.

   machine or cell and go to Step4. If the replacement machine or cell can not be
   found, then this lot is added into the postpone queues.
Step8: Decide the scheduling and show the scheduling results.

4 IMPLEMENTATION OF THE SYSTEM

According to the design of the system outlined in the previous sections, we have implemented
the system in the BC++ language and the Btrieve database running on Netware 3.1.1 network
operating system and MS-Windows. The server is the PC486 with 200MB hard disk, and the
clients are the PC386 with 120MB hard disks. The system has been implemented for practical
usage with two objectives in mind: user-friendliness and high executive speed.

The system are mainly constructed with 6 modules as shown in Figure 7. These modules are: User interface module, Scheduling module, Analysis module, two Communication modules in the client and the server, and the database management module.

The scheduling module performs scheduling according to the heuristics given in previous sections, the analysis module analyzes the scheduling results and the postpone reasons. While the Communication modules provides us with the client/server architecture. Moreover, the database management module provides us with the operation methods to the Btrieve

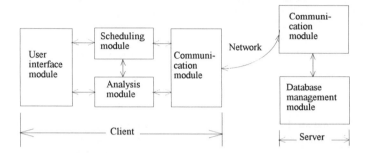

Figure 7 The module construction of the Scheduling System.

database. It is possible to install the Btrieve database and the management module into the client.

5 EVALUATION

The system has been used in a Japanese manufacturing factory which has over one thousand machines with different types, over one thousand personnels, and over ten shop floors. The factory which produces the Flexible printed Circuit manufacturing is a typical example in low-volume/high-variety. We have used our system in the scheduling of the exposure shop and the laminate shop of the factory. These two shops both have over one hundred machines and over one hundred workers. Using our system, the foreman can automatically schedule the shop production activities of 3~5 days with less than 1600 lots(a lot always includes 50~200 sheet parts). The executive time for scheduling in a 386/33 client is near to 2~5 minutes and the machine utilization can averagely arrive 90% in general cases. Moreover, comparing with the case of performing the schedule in hand, the system provides us with less postpone of the due date of the lots, because the machine utilization is increased and the personnels are used more reasonable by the system.

6 CONCLUSION

This paper has reported the development of a practical scheduling system for low-volume/high-variety manufacturing. This system has been implemented in client/server architecture and the influence of many factors on the scheduling process such as set up times, personnels overworking times, cell constructions, and replacement machines has been considered. The system has been used to the practical shop floor scheduling in a manufacturing factory of Japan and the scheduling results have provided with much better machine utilization and less postpone then the scheduling in hand.

REFERENCES

A.Kusiak and G.Finke (1988), "Selection of process plan in automated manufacturing systems", IEEE J.Robotics Automation, Vol.4 No.4, PP 397~402.

D.J.Hoitomt and etal. (1993), "A practical approach to job-shop scheduling problems", IEEE Trans. on Robotics and Automation. Vol.9, No.1, PP 1~11, Feb..

F.Vivers (1983), "A decision support system for job shop scheduling", European J.Operational Res., Vol.14 No.1, PP 95~103, Sept..

J.Ahn and etal. (1993), "Scheduling with alternative operations", IEEE Trans. on Robotics and Automation, Vol.9 No.3, PP 297~303, June.

Lee H.S.Luong (1994), "A database System for job scheduling in plastic injection moduling", IKDME, HongKong, PP 31~36.

BIOGRAPHY

Prof. Zhang Yao Xue obtained his master degree and doctor degree in computer science field from Tohuku University, in 1986 and 1989, respectively. In 1990, Prof. Zhang joined Department of Computer Science and Technology, Tsinghua University, and now is the Vice-chairman of the Department. He is currently heading groups that are researching network software, network design method, network interconnections and client/server computing.

35

The data management sub-system of the experimental push/pull production management system*

Assistant Y. Li and Professor D.W. Wang

The Research Institute of System Engineering,

Automatic Control Department, P.O.Box 135,

Northeastern University, Shen Yang (110006), P.R.China

Tel: (024)-3893000-3980(O), (024)-6859443(H)

Abstract

The paper introduces the data management sub-system of the experimental Push/Pull production management system, which is a mixture of the two popular manufacturing management methods - MRPII and JIT, in details, include the design of the database structure, the design of the data management function, and the design of the data management man-machine interface etc.. This satisfaction of the system provides a possible chance to applicating the hybrid Push/Pull strategy to Chinese manufacturing systems.

Keywords

Data management, MRP-II, JIT, Push/Pull, Manufacturing system.

*Supported by national found of nature science.

1 INTRODUCTION

Manufacturing Resource Planning (MRP-II) and Just-In-Time (JIT) are two kinds of production inventory management methods of batch   manufacturing   processes developed in western countries and Japan, separately, but they can't solve the real production problems independently and satisfactorily. More and more researchers and scientists have devoted themselves into how to combine the two methods, and have made a great progress. The hybrid Push/Pull production inventory control strategy is one of the most effective and satisfactive methods, which is just a result   of combining the MRP-II and the JIT. The control strategy enhances the function of MRP-II through the kanban control of JIT by using this through the production control. This kind of control strategy is successful in making minmum inventory and offer products to users in time.   Absorbing the advantages of the two methods, it provides an earliness/tardiness production planning method considering muiti-process capacity constraints for mass manufacturing. In order to simulate the real production process based on this theory and method, we design and develop an experimental Push/Pull production management system. In this system, we use the earliness/tardiness production plan methods to make the master production plan, schedule the production process, and use Push/Pull strategy to control the production. We design 26 data tables which cover every respect of the real manufactuing production process in this system. These tables are the basis and keys of the whole Push/Pull production management system. So how to manage and maintain the production datas becomes very necessary and important to the whole system.

The data management sub-system of the experimental Push/Pull production management system is designed as a powerful and independent system. It is a key for managers to manage the system resources and use the whole system. We use C as the main language to develop the system, and use Oracle Database as our database. The main menu of the system is programmed in C langurage, the other data management functions are developed by SQL*Form of Oracle. Following, we describe and introduce the data management sub-system from several defferent respects in details.

2 THE STRUCTURE OF THE WHOLE SYSTEM

The experimental push/pull production management system is powful. Its functions cover nearly every respect of managering the production process of an manufacturing system. In the system, we offer the following functions: basic data management, summing available-capacities, making master production plan, making rough-capacity balance and rough-capacity plan, making purchase plan and purchase simulation, monitoring the production processes, PUSH controlling of the materials, PULL controlling of the production processes, etc. The whole structure of the whole system is shown in Figure 1.

In Figure 1, those rectangles are function modules   and simulating processes. Those recangles with round tops are data tables. The lines with arrows stand for the direction of flow.

From the whole structure we can see that the data tables are very improtant to the whole system.

They are used to store systems' origal data resources, running informations and final results. Only through these data tables can the manager connect with the production management system.

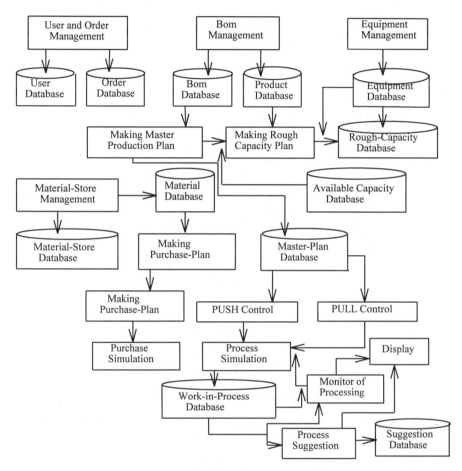

Figure 1 Structure of the whole system.

3 DESIGN OF DATA TABLE STRUCTURE

In order to ensure that the system run correctly and efficiently, we design the structure of every data table obeying the following rules:

- Simplify the structure of every data table, avoid using the diffuse data.
- The logic relation among data tables arrivals in 3-NF.

In this data management system, there consists 26 data tables which are shown in table 1.

Table 1 All data tables of the whole system

No.	Symbol	Description
1	USR	User records
2	ORD	Order records
3	ORDSUM	Summary order
4	HOF	Factory holiday
5	TIMP:L	Plan time
6	PBOM	Production bill
7	CBOM	Composition bill
8	PRTBOM	Part bill
9	SUBBOM	Sub-composition bill
10	PRD-CMP	Relation product and composition
11	CMP-PRT	Relation composition and part
12	PRT-SUB	Relationpart & sub-composition
13	PCS	Process bill
14	EQP	Equipment bill
15	MTRL	Material bill
16	MTSTC	Material storage
17	MTL-PCS	Relation process and material
18	MPS	Master production schedule
19	RCP	Rough capacity plan
20	PCH	Purchase plan
21	WIP	Work-in-process product
22	PDLV	Delivery-product record
23	EMG	Emergency information
24	RULE	Rule of process
25	SUG	Approve-process suggestion
26	PFM	Month production target

For example, we show structures of master-production-schedule table (MPS) and rough-capacity-plan (RCP) table in table 2 and table 3.

Table 2 Structure of master production plan (MPS)

Field	Description	Type	Length	Point	Remark
LY	Year	Char	4		Key
PLM	Month	Char	4		Key
WKDT	Work day	Char	4		Key
PRD#	No. of product	Char	4		Key
TYPE	Type of product	Char	1		Key
PLNQ	Plan quantity	Number	5	2	

Table 3 Structures of rough capacity plan (RCP)

Field	Description	Type	Length	Point	Remark
PLY	Year	Char	4		Key
PLM	Month	Char	2		Key
WKDT	Work day	Char	2		Key
PCS#	No. of process	Char	7		Key
VCQ	Available capacity	Number	4	1	
PCQ	Plan capacity	Number	4	1	

4 DESIGN OF THE DATA MANGEMENT FUNCTION

The function structure of the data management is shown in Figure 2.

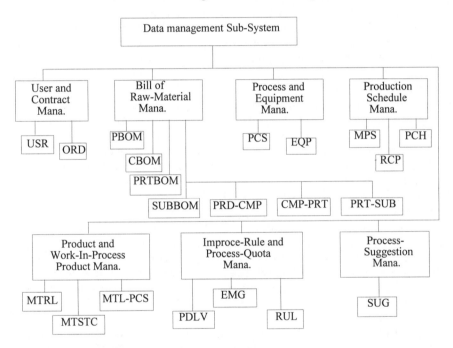

Figure 2 Function structures of data management.

Every data management function includes sub-functions: append/insert, browse, inquire and delete, which are shown in Figure 3.

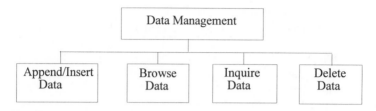

Figure 3 Function structure of management sub-system.

The data management provides plenty functions to manage the data, including:

① Append/insert the original, external data. The system provides an interface to append/insert all original, external data of the Push/Pull production management system for managers. While appending/inserting data, only the necessary data are required. Some information can be automatically obtained from other related data tables by the system itself. The system can operate several different data tables together at the same time. In order to avoid entering wrong data, system provides functions to distinguish data type, check valid data etc.. In the meantime, the system provides prompts in time. These functions can keep wrong data from being entered to the system, thus maintain the safety of the system.

② Data inquiring. The system provides an interface of inquiring data information from those authorized data tables. Through selecting various different data tables, the manager can inquire any data table with no-condition and multi-conditions. Among 26 data tables, there are only a few data tables can be operated by the manager, such as MPS, RCP. In other hands, some data tables are not allowed to be operated.The manager can't, yet needn't operates these data tables. They can be operated and maintained by the system itself, such as ORDHZ. The system will maintain the data consistences.

③ Data updating. The system provides an interface of updating data of those authorized data tables by managers. Through selecting various different data tables, the manager can update any authorized data tables. In order to maintain data consistences, the system operates several different data tables in the same time while one or more tables are updated by the manager, That means when the manager updates a data table, those related data table can be updated in the same time automatically. In this way, the system keeps the data consistences.

④ Data deleting. The system provides an interface of deleting data of those authorized data tables for managers. Through selecting various different data tables, the manager can delete any data in those authorized tables. In order to prevent from losing useful data, the system provides delete protection through COMMIT and ROLLBACK mechanisms of Oracle. Users can confirm aborting or committing the operation.

5 DESIGN OF THE SYSTEM'S MAN-MACHINE INTERFACE

When designing the system,we use pull-down menus, pop-up windows, multi-window and on-line help functions, etc., thus providing a very friendly man-machine interface to managers.

The main window of the system is shown in Figure 4.

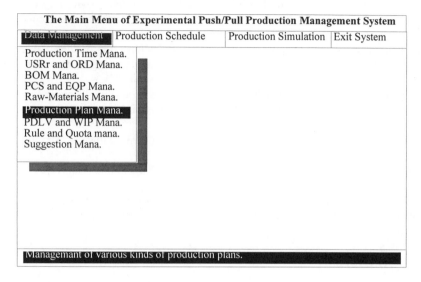

Figure 4 The main window of the system.

The system has following features in the design of the man-machine interface:

① Providing an integrated environment of data management for the manager. In this environment, the manager can accomplish all the works of data management. We use pull-down menus in main menu, and use pop-up windows or menus in every sub-menu.

② By using the multi-window technology, the data can be rolled up, down, right and left in the windows. Without the restriction of the window size, managers can append or insert data freely.

③ Using run-on-load on-line help, providing system information to manager. In any time, manager can get help by pressing a key.

④ Using tight anti-error techology to against the manager's wrong operation to prevent from losing useful data. The system does its best to provide data in maximum range.The system can catch Oracle error and give managers approciate prompt informations and wait for manager's responds.

6 REFERENCES

D.W. Fogarty and T.R. Hoffman, Production and Inventory Management, *South-Western Publishing Co.*, Cincinnati.

Hodgson, T.J. and Wang Dingwei, (1991) Optimal Hybrid Push/Pull Control Strategies for Parallel Multistage System: Part 1, *Int.J.Production Res.*, 29, (6), 1279-1287.

Hodgson, T.J. and Wang Dingwei, (1991) Optimal Hybrid Push/Pull Control Strategies for Parallel Multistage System: Part 2, *Int.J.Production Res.*, 29, (6), 1453-1460.

Li Ying and Wang Dingwei, (1993) The Determining Method About Key-process of Earliness/Tardiness Production Planning for Mass Manufacturing System, *The Symposium on Automatic and Computer Application of Liao Ning Province*, 81-84.

Li Ying and Wang Dingwei, (1994) The Relaxtion Approaching Method of Earliness/Tardiness Production Planning for Mass Manufacturing System, *The Second Symposium on Computer Integrated ManufacturingSystem (CIMS)*.

Wang Dingwei and Li Ying, (1993) Earliness/Tardiness Production Planning Method for Single-Product Manufacturing System, *The Symposium on Control and Decision of P.R. China*, 717-720.

Wang Dingwei, Chen Xianzhang and Li Ying, (1993) Structure Design of an Experimental
 Push/Pull Production Management System, *Proceedings of Int. Conference on CIM*, 138-141.

Y. Sugimori et al, (1977) Toyota Production Ststem and Kanban System Materiallization of Just-In-Time and Respect-For-Human System, Int. J. Production Research, Vol. 15, 553-564.

7 BIOGRAPHIES

Assistant Y . Li, working in the research institute of system engineering of Automatic Control Department, Northeastern University. Bachelor: Mathematicl Department of Liao Ning University, 1987-1991; Master: Automatic Control Department of Northeastern University, 1991-1993.

Professor D.W. Wang, working in the research institute of system engineering of Automatic Control Department, Northeastern University. Bachelor: Automatic Control Department of Northeastern University, 1978-1982; Master: Automatic Control Department of Hua Zhong College of Technology, 1981-1984; Doctor: Automatic Control Department of Northeastern University, 1993-.

36

PROCESSING OF SUPPLIER INFORMATION IN OFFSHORE PROJECTS

Professor Ola Westby
Department of Marine Systems Design, Norwegian Institute of Technology, University of Trondheim, Norway

and

Dag Runar Elvekrok, Dr.Ing. Student
Program on Applied Coordination Technology, Norwegian Institute of Technology, University of Trondheim, Norway
E-mail:dre@pakt.unit.no

ABSTRACT

The management of the engineering and fabrication of platforms for petroleum production in the North Sea is generally fairly advanced, however, there are considerable possibilities for improvements. If new types of procedures for information processing are implemented, the lead time of development projects may be reduced. Downstream applications of information in the projects can be made more productive as a result of improvements in the quality of information generated during detail engineering. This paper is concerned with project management, especially the processing of supplier information in the detail engineering phase of development projects. This issue will be of increasing importance for oil companies and engineering companies in the years to come.

The first part of the paper describes the normal processing of paper-based documents related to deliveries of packages from suppliers in the detail engineering phase. Drawings, specifications and communication documents are being repeatedly processed in numerous versions and revisions by suppliers, engineering companies, fabricators and oil companies. This is expensive and time consuming. The paper also presents an analysis of the management and control of computerised information processing in engineering.

The last parts of the paper outline suggestions which could reduce lead time in projects and at the same time improve the of quality of information in offshore development projects. One of the basic ideas to improve productivity and quality is to process information as a set of small information elements or entities independent of traditional objects like paper-based documents. Information models may consist of a number of documents and databases, but they may nevertheless be regarded as a set of integrated information entities. The terms and definitions for establishing procedures for processing information elements are also presented. This paper is partly based on research work initiated by Helge Moen, a senior specialist at Kvaerner Engineering a.s., Oslo.

1 BACKGROUND

The information communication between an engineering company and suppliers is critical for the successful accomplishment of any development project. Supplier information is critical regarding the equipment in the platform design. This equipment is divided into two groups: bulk equipment such as pipe parts, instruments etc., which usually is possible to purchase directly from suppliers and tagged equipment such as pumps, heat generators etc., with a complex structure that is engineered by the supplier.

Information processing for tagged equipment is complicated and critical for the progress of a project. The main purpose of this processing is the generation of specifications, purchasing and coordination and implementation of supplier information for engineering activities. In this paper engineering activities for the coordination and implementation of supplier and engineering company information are called downstream activities. The paper presents the information process involved in the supplier packages for tagged equipment.

It is not uncommen that there is a huge amount of supplier documents. Table 1 gives an example for a typical offshore project.

	Number	Number of revisions	Revisions (average)
Drawings	7440	11160	2.5
Documents	11600	16740	2.5

Table 1 Number of supplier documents in a typical offshore project, Gyda, for the North Sea (Kvaerner Engineering).[1]

A document consists of a set of information elements. The status and accept coding is related to the whole document and is stated by the information elements with the lowest quality in the document. One unimportant element may therefore delay acceptance of a document and the use of critical information elements of good quality. This is a significant factor in the processing of paper-based documents and a common cause that delay the completion of engineering. Time for new revisions of supplier documents that are not accepted by the engineering company is usual 2-6 weeks.

This paper describes the idea of information elements, the terms and definitions necessary for establishing procedures for processing information elements are also covered.

2 NORMAL PROCESSING OF PAPER-BASED DOCUMENTS FOR SUPPLIER PACKAGES

The processing of information from and to suppliers may vary from one engineering company to another. It may also vary between oil companies and depend on the size of development project. In order to quantify the principles of problems and solutions this paper considers a characteristic example in detail.

---

[1] Apart from this table the term "document" is used to mean both drawings and other documents in this paper.

2.1 Documents and status

Coding is used by both engineering company, oil company and suppliers. A code releases the information in a document for a particular purpose. The code is an alphanumeric combination according to a special system.

Revision number is a continuous number that shows the revision of the document.

Status code shows the action the document is issued for. Examples of package specifications are "Issued for Inquiry" (IFI) and "Issued for Order" (IFO). Revision number and status code are grouped together in one code on the document.

Accept code is a code the engineering company gives the supplier document after a review. This code is usually based on the quality of the supplier document and the change that the supplier document must go through before it complies with the engineering information. Examples are "Accepted" and "Not Accepted".

Class code is a number code that identifies the type of information of supplier documents. These numbers refer to (and are usual identical to) a paragraph in Norwegian Standard "NS5820-Supplier's Documentation of Equipment". The coding is explained when it occurs in the text and on the last page of this paper.

2.2 Main activities in processing information for supplier packages

There are three main activities in processing information for supplier packages. Figure 1 shows these activities as they usually appear as a part of the detail engineering phase.

The first is "Generating specifications" which specifies the equipment. The second is "Purchasing". The purpose of this activity is to select the supplier, place the order and follow up the delivery. The third is "Downstream activities" which are engineering activities that are directly dependent on the use of supplier information. This activity coordinates the engineering and supplier information.

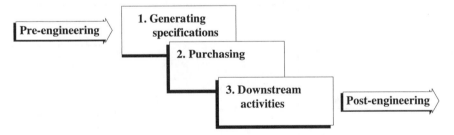

Figure 1 Main activities in processing information for supplier packages.

It is vital to start these activities as early as possible. However, a minimum of engineering information must exist prior to any of the main activities. This minimal information is generated in the pre-engineering phase or early in the detail engineering phase. Engineering activities may continue in parallel with the main activities. Figure 2 shows the document flow for the main activities.

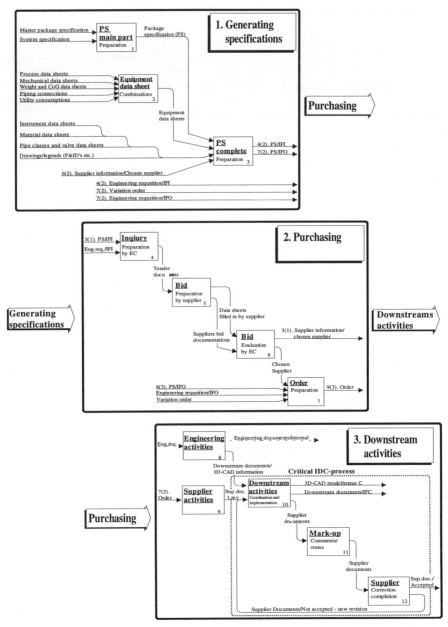

Figure 2 Detail description of the document flow for the main activities[2].

---

[2] Symbol description: Rectangles - activity, Arrows - document flow. A number on an arrow shows the preceding or the succeeding activity for the document flow. Example: 4(2) means activity box 4 on main activity 2 ("Purchasing").

2.2.1 "Generating specifications" - first main activity

A package specification (PS) is generated for each package that has to be purchased. A package engineer (PE) is responsible for preparing the PS which is generated in several steps. First, the main part of the PS is prepared. This consists of a master package specification (MPS) and a system specification. MPS consists of common descriptions and guidelines for all packages and is generated early in the engineering phase or by the client/oil company prior to the engineering. Only relevant parts of MPS are included in PS at this stage.

Thereafter, different data sheets are normally entered into PS as appendices. Equipment data sheets consist of a combination of different data sheets. Other data sheets, such as instrument and material data sheets etc., are directly entered into PS as appendices. There are additional data sheets to the ones illustrated in Figure 2. The last step is to complete the main part of PS and the data sheets.

Prior to "Generating specifications", all documents have a status. Depending on the status the documents may be released for implementation in PS. After entering the PS, the status of these documents is subordinate to the status of the PS.

The first PS status is "Issued for IDC" (IDC: Inter Discipline Check in the engineering company; not shown in the figure). After this the next status for PS is "Issued for Inquiry" (PS/IFI).

2.2.2 "Purchasing" - second main activity

The first purchasing activity is an engineering activity which prepares documents for inquiry. Documents for inquiry normally consist of PS/IFI and an engineering requisition. This is prepared by the PE and a purchasing agent at the engineering company (EC).

After that, a couple of selected suppliers receive the tender documents in order to prepare a bid. The data sheets that are included in the PS may have some open sections that have to be filled in by the supplier. The supplier fills in the data sheets and returns them with the bid documentation to the engineering company.

After receiving the bid documentation, the engineering company evaluates the bid and chooses the supplier. The engineering company prepares for order. Relevant information from the chosen supplier that is received with the bid is implemented in the PS before the PS is "Issued for Order" (PS/IFO). Documentation for order is almost the same as for inquiry, but may be subject to changes made by the engineering company. These changes are taken care of by means of variation orders.

2.2.3 "Downstream activities" - third main activity

This main activity occurs when the engineering company begins to receive supplier information that describes the engineering by the supplier. The purpose of these activities is to coordinate information from the supplier and engineering company.

The first activities show engineering and supplier activities in parallel. The results of the engineering activities are downstream documents or 3D-CAD information which is used in the downstream activities at the engineering company. The results of the supplier activities are supplier documents which are sent to the engineering company.

Downstream activities influence the different disciplines in the engineering company that are dependent on supplier information. These activities comment on the supplier documents and implement the supplier information in the downstream activities. The implementation of this information either refers to information in the 3D-CAD model or to downstream documents. These implementations cannot occur before the supplier documents are accepted by the PE in the "mark-up" activity.

After "mark-up", supplier documents are sent back to the supplier. If they are not accepted the supplier makes a new revision of the documents and returns them to the engineering company for a new review. If this is accepted the supplier starts to produce the equipment.

The activities inside the dotted line are the IDC-process of supplier documents performed by the engineering company. This process will now be considered in more detail.

2.3 IDC-process of supplier documents

The IDC-process is rather complicated with several participants and numerous paper-based documents flowing around. The IDC-process should cover every discipline or party that is dependent on supplier information. Figure 3 shows the IDC-process in detail. The document flow schema is vertically divided in three parts, one for each of the participants in the process.

2.3.1 Participants of the IDC-process

Supplier document control (SDC) is responsible for the administration of the supplier documents. This consists of external and internal distribution, registration, quality checking, copying, filing etc. of the documents. SDC has nothing to do with the information in the documents.

Downstream disciplines are the disciplines that receive the supplier documents on IDC and have activities that are directly dependent on the information in the supplier documents. They check information in the supplier documents for consistency with the downstream information and incorporate information on documents that are accepted by the PE at "mark-up". Dependent on the number of disciplines in the IDC-route it can be organised in parallel or sequentially.

Package engineer has the responsibility for the treatment of the information contents of the documents. This activity, called "mark-up", include implementation of the comments from the downstream disciplines and the PE on an original "mark-up", checking for consistency and status coding. An original "mark-up" is an IDC-copy that is used for the final comments by the PE. The PE for tagged equipment usually belongs to the mechanical discipline.

2.3.2 Document flow for the IDC-process

Document flow - detailed description. When SDC receives the supplier documents from the supplier, SDC registers the documents in a database and submits the documents to a quality check. A project review request (PRR), which includes a distribution matrix (DM) for the IDC-route is printed from the administration database.

The supplier documents are copied into several sets of IDC-copies. One of these IDC-copies will appear as the original "mark-up". After copying, the supplier's original and the original "mark-up" are filed at the SDC and the other IDC-copies are distributed for IDC. These SDC activities usually requires 1 day.

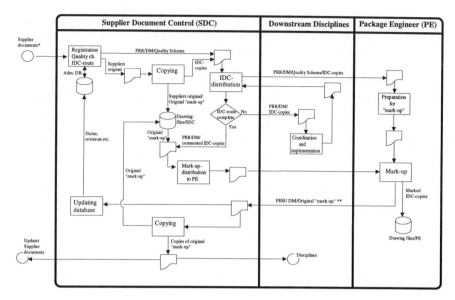

Figure 3 IDC-process of supplier documents inside an engineering company[3] .

The IDC-distribution to the PE for "mark-up" includes the PRR, DM, quality forms and one set of IDC-copies. When these documents are received, the PE will prepare for "mark-up" and make his own comments on the supplier documents.

The IDC-distribution to the downstream disciplines includes a PRR, DM and a set of IDC-copies. No documents pass directly between the disciplines. All distribution is through the SDC. If more than 3-4 disciplines are included on the IDC-route it is usual to organise the IDC-route so that it is parallel with several sets of PRR, DM and IDC-copies. The activity inside the downstream disciplines is coordination and implementation of supplier and engineering information. The downstream disciplines also usually make comments on the IDC-copies for incorporation by the PE at "mark-up". The time for IDC is usually 1 day per discipline.

When the IDC-route is complete, the different sets of PRR, DM and discipline commented IDC-copies are sent with the original "mark-up", to the PE for "mark-up". The PE makes the "mark-up" and returns the original "mark-up", PRR and DM to SDC. Marked IDC-copies from the downstream disciplines are filed by the PE in a drawing file. The time for "mark-up" is usually 2 days.

The engineering company's official revisions of the supplier documents cannot take place until "mark-up" is made by the PE (marked with **). Downstream documents with supplier information on the IDC-route will not be released in a new revision until the original "mark-up" is finished and released as an official revision by PE. Suppliers issue of the official revision of documents means that the documents will leave the supplier and depart at SDC (marked with *).

---

[3] Symbol description: Close circle - Start, Open circle -finish, Rectangle - activity, Cylinder - storage, filing, database etc., Diamond - decision, Cut rectangle - document.

When receiving the PRR, DM and the original "mark-up", SDC updates the administration database with status code, revision number etc. and copies the original "mark-up" in several copies. One of the copies is sent back to supplier for upgrading of the information and the status of the document. Disciplines which have marked for copies of original "mark-up" at the PRR also receive these copies. Finally the original "mark-up" is filed in a drawing file at SDC. The time for this activity is usually 2 days.

If the documents are accepted by the engineering company the supplier starts the production of the equipment. If the documents are not accepted the supplier upgrades the information in the documents and sends it back to the engineering company for an additional review. Normally, it is not necessary with a complete IDC-process as described in Figure 3 for the additional review. Sometimes this review is taken care of by the PE alone or the PE arranges an IDC-meeting with the involved disciplines. The time for the "mark-up" and IDC-meeting is usually 2 days.

2.4 Status dependencies and project progress

Project progress is dependent on the status of the downstream documents and the 3D-CAD model. These are again dependent on the status given by the engineering company's official revision of the original "mark-up". Figure 4 shows these status dependencies.

First the engineering requisition and package specification passes through IDC, IFI and IFO. This is in parallel with other engineering activities generating 3D-CAD model and downstream documents. The supplier will normally start the preparation of the bid when the inquiry is received, but no status dependencies occur at this stage.

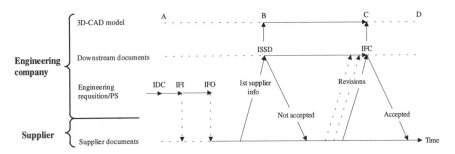

Figure 4 Status dependencies and project progress.

After order (IFO), the supplier starts the engineering of the equipment and after some time, normally 7-18 weeks, the engineering company starts to receive supplier documents for IDC and review. At the first receipt, called "1st supplier info", the document is normally not accepted by the engineering company.

However, usable information is normally transferred to the engineering activities at this moment. This gives the downstream documents status "Issued for Start Shop Drawing" (ISSD) and the 3D-CAD model status B. This status releases the information in the 3D-CAD model and downstream documents for limited use.

After correcting the documents the supplier sends a new revision to the engineering company. If the documents are accepted all the supplier information could be included in the downstream activities. This will give the downstream documents status "Issued for Construction" (IFC) and the 3D-CAD model status C. This releases all the supplier information included in the downstream activities for unlimited use. The dotted lines in the figure show that more than one revision may be necessary before the supplier document can be accepted.

As illustrated, it is of utmost importance for the project progress that supplier documents obtain a quality level that correspond to acceptance of the supplier documents as soon as possible. Otherwise, this will delay the completion of other engineering activities in the project. New revisions of supplier documents usually take 2-6 weeks.

A problem with the normal processing of paper-based documents is that the status refers to the whole document. This means that the status of a supplier document is based on the information of the lowest quality in the document. Acceptance of the document may be delayed because of unimportant errors or lack of information of minor importance. This is critical for the progress of the project.

The following will outline a theoretical means to avoid this problem .

3 THEORY OF INFORMATION ELEMENTS

3.1 Levels of information

A document is a collection of separate information elements. Though each of these elements has a unique meaning, they do not have the same importance for engineering. It is reasonable to expect that a traditional document consists of both important and unimportant information elements. However, there is no normal procedure that takes this into account. Figure 5 shows how information can be divided into different levels.

Document level shows how all the information in a document is related to the processing of a document independent of the importance or quality of the information elements. This level does not accept the use of good quality information elements, which may be critical for the project progress, unless all other elements are brought up to the same quality. This level corresponds to the normal processing of paper-based documents.

Element level of the documents shows how the information in a document can be split into separate information elements. One element can appear in different documents with different meanings regarding information. Example: A valve can appear as a physical object on an arrangement drawing and as a symbol on a piping and instrument diagram (P&ID). The intention of using this level is to make use of good quality information elements without bringing the rest of the elements up to the same quality. This might be done later in the process. As paper-based documents do not facilitate this level of information, it is necessary to establish new or revised procedures corresponding to this kind of information processing. Some ideas concerning such procedures are briefly described towards the end of this paper.

Element level of the product refers to a product model and is therefore entirely dependent on computer-based information models. Each element in a product model needs a description and a relation to other elements of the product model. If some of the minimum requirements for an element in the product model are not met, the intention with the elements in the product model will not be fulfilled. In object-oriented theory and information models this condition can be explained as follows: An object-oriented model is a system that consists of several separate information elements. Each of these elements needs a description of "what it is" which includes attributes and properties. The existence of the model as a system is represented by relations between the elements.

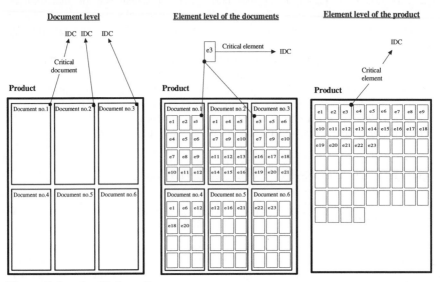

Figure 5 Levels of information.

The principle of processing information elements is similar for the element level of the documents and the element level of the product. The idea is based on the need to process critical elements separately in the engineering information model and the procedures for information checking.

Critical elements have to be identified at an early stage in the engineering process and be subjected to special attention by the supplier and engineering company. For instance, if these critical elements are a part of the inquiry, the engineering company may use the supplier's ability to comply with these special forms for information as a criteria in evaluating a potential supplier. The processing of critical elements at an early stage of engineering depends on the supplier's ability to produce these information elements at an earlier time than is usual. If the production of information is considered over time it is reasonable to assume that the quantity of information increases as shown in Figure 6.

The figure shows that a lot of information is produced by the supplier. However the information is not available for downstream activities at the engineering company until the information is completed on paper-based documents by the supplier and send to the engineering company. The figure also illustrates that new information procedures for exchange of supplier information between supplier and engineering company are interesting.

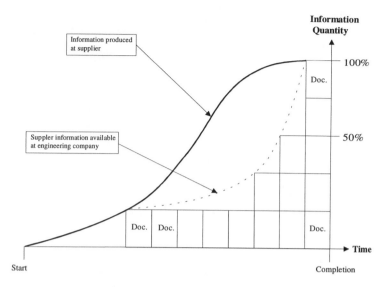

Figure 6 Information produced at the supplier and available information at the engineering company.

3.2 Advantages of using the element level of documents

There are several advantages in establishing procedures that can be used for the element level of documents. This type of processing is based on procedures that define critical elements instead of critical documents.

Critical elements have to be identified and defined. This differs between disciplines, for example:

- For logical information the process discipline may identify pipe size, temperature, pressure and volume as attributes to critical elements.
- The steel structure discipline may identify information about foundations, weight and centre of gravity as attributes to critical elements.
- The piping arrangement discipline may define piping connections, dimensions and volume required around the equipment as attributes to critical elements.

A way to process critical information elements is to represent them in separate documents that are especially tailored to need for critical elements in a discipline. It might also be useful to make pre-revision of documents that consist of critical elements and finish the documents afterwards.

Figure 7 illustrates the time reduced by splitting information into critical information elements and incorporate these elements in the engineering activities at an early stage.

The document level in Figure 7 illustrates how all information in a document has to be finished by the supplier before the engineering company can utilise it. If the IDC lasts for one week the information in the document will be accepted 8 weeks after order (WAO). This means that further downstream activities can not implement the supplier information for unlimited use until 8 weeks after the order.

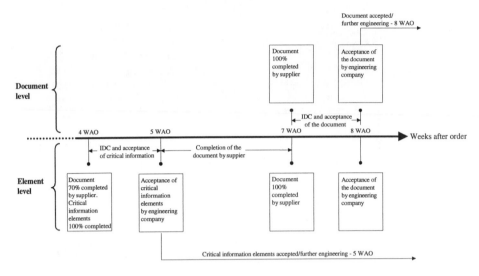

Figure 7 The effect of receiving critical information elements at an earlier stage.

The element level of the documents in Figure 7 shows how critical information elements are completed by supplier and sent to the engineering company at an earlier stage, for instance in a pre-revision of a supplier document. The IDC and acceptance of critical information elements will be finished 5 weeks after order and the supplier information will at this moment be released for unlimited use. The times for document level in Figure 7 are typically for a large development project in the North Sea. The times for element level are just examples that illustrates the effect of receiving critical information elements at an earlier stage. This is possible by using the theory of information elements.

The next session in this paper presents a concept for establishing procedures to check the information elements.

4 ESTABLISHING OF PROCEDURES FOR CHECK OF INFORMATION

This section is a systematic approach to the considerations to be implemented in new procedures for quality assurance based on small information elements.

For paper-based documents it is normally suitable to choose each piece of paper as an entity for information processing. For computerised information models, however, like 3D-CAD models, engineering databases and word processing archives, different formats and content of documents may be defined during print out. For information computer aided processing a "document" or entity will normally be chosen with different boundaries than for a printed document. The complete quantity of information that is available has to be structured in entities of information both for paper-based information and for computerised information.

An information entity will often correspond to some properties of a physical object, but could be something else. Dependent on the application to be performed, entities may for instance be chosen corresponding to design systems or geometric locality (ref. area plot plans) or fabrication tasks. For preliminary design purposes it is normally suitable to define the scope of entities based on functional systems, ref. flow diagrams etc. For fabrication planning it might be more convenient to use the task-based structure of entities, ref. work break down structure of networks, or even geometric structure for the evaluation of operability of a platform.

As explained, the scopes of entities will be fairly large if each entity corresponds to a normal paper-based document. For manual information processing this gives a fair, but limited overview. If the scopes of the entities are very limited, the number of entities in an information model will be large, but each entity will be relatively simple. The scopes of the entities may vary from one application to another.

Information related to packages from suppliers will normally be changed and checked numerous times by the suppliers themselves, the engineering company, the fabricator and the oil company. The changing and checking of information may be categorised in procedures according to the following criteria:
- integrity of each information entity
- compatibility between information entities
- consistency of information model
- completeness of the integrated information according to the needs of the project.

4.1 Integrity

The integrity check will include checking the information of the entity relative to technical requirements and specifications of the entity. The entity has to be large enough to make technical analysis meaningful or alternatively the analysis will have to be applied to a set of entities.

The status of an aggregated document will normally be set to the lowest status of all the information entities that are included. Too large entities may delay the upgrading of the status of entities caused by inaccuracies or errors in some unimportant data. In other words, in order to tag as much information as possible with a high status the information entities should be chosen small, but in order to make production of information meaningful this cannot be too small.

A normal P&ID is an example of a document which for some applications may be chosen as an information entity. But in a P&ID there are many objects which could be checked stand alone for integrity. This indicates that an entity should preferably be chosen corresponding to each instance of equipment in the P&ID. Such instances may be rotating equipment, a pipe or just a label in a diagram. For other applications the P&ID may be processed as a compound entity.

A formal error in the labels on some drawings is an example from a development project where the drawings could not be released for downstream applications despite the fact that the drawings were technically correct. In order to change the labels in this project the drawings had to be processed by more departments. This ended in unacceptable delays and a dispute in court between an oil company and an engineering company.

Most of the integrity checking is performed by discipline engineers. If the size of information entities is reduced, the number of disciplines involved in each entity may also be reduced. Procedures for integrity checking are normally not complicated.

4.2 Compatibility

Compatibility is defined here as the suitability of integration of more information entities. This is vital to generate compound entities. The check of compatibility is normally linked to physical interfaces between physical entities. A check of compatibility may often be performed as a check of whether the information about the entities complies with the specification for the whole model. As the size of information entities is reduced the number of compatibility checks tends to increase, but each of the checks will be simpler.

4.3 Consistency

Consistency checking is composed of a set of procedures for searching for specific types of errors. An important consistency error is collision between physical objects for instance pipes, equipment, cables and steel structure. Another example of inconsistency is excess of specification of platform caused by aggregation of a number of physical objects, for instance excess of budgeted weight. A normal consistency check is the so-called yellow line check.

Some consistency checks may be fairly complicated. But when an engineer or a department wants to introduce some information to a comprehensive 3D-CAD model or engineering database it is absolutely essential to check for consistency. Therefore strict definitions of information entities should be approved and procedures for consistency checks formalised. There is no easy way around this. However, it will help to introduce simple information entities with specified interfaces and relations to the rest of the information model.

4.4 COMPLETENESS

The check of completeness is a search for missing information. In the 1980s this type of check was concentrated on information which was essential to prefabrication and fabrication. In the 1990s more attention has been paid to information needed by an operator of petroleum production (ref. HTID and STID specifications of STATOIL). It is an essential and comprehensive task to run completeness checks for information from the very beginning of projects starting with the tendering procedures. Additional requirements for information at late stages in projects are expensive and sometimes impossible even for simple documents like maintenance manuals.

For practical purposes it is not too difficult to determine suitable scope for the information entities mainly based on the simplicity of interfaces and relations between entities. Optimal scopes for information entities seem to be considerably less than traditional drawings and bills of materials.

In the years to come, engineering will be more based on integrated information models like 3D-CAD models, engineering databases and word processing archives than paper- based documents. Therefore it will be necessary for each application to define suitable submodels for information processing, which will consist of a set of information entities.

5 CONCLUSIONS

In offshore development projects the information exchange between suppliers of physical packages and the engineering is costly in terms of time and resources. One of the reasons for this is that in order to avoid complications in downstream applications of information, comprehensive quality assurance procedures have to be run. One of these is the IDC.

A paper-based document consists of a set of information elements. All elements in a document have to be upgraded to a high quality before the document can be released for downstream applications. This is a bottleneck.

Information processing may be applied to information elements instead of paper documents. This makes it possible to reduce the lead time for critical information. There are two different possibilities for the implementation of this. One is the introduction of computerised models consisting of information elements. The other is to split paper-based documents into information elements. Both may result in considerable reductions in lead time in engineering.

As procedures for information processing of a set of information elements are established, consideration should be taken to check the integrity of each element and the compatibility and consistency of the total model and its completeness.

The information elements can be identified differently from one application to another. In general, the choice of more limited scope of each element leads to simpler checks of the integrity of the elements and the checks of consistency. However, in order to make the quality checks surveyable the information elements should not be too small and preferably be of a suitable size for normal evaluation by a specialist engineer.

The ideas presented in this paper may be applied to paper-based information processing, but will probably be more profitable as CAD/CAM-models, engineering databases and archives of word processing documents will represent the original data in a development project.

ABBREVIATIONS

Status

IDC	Inter Discipline Check
IFI	Issued for Inquiry
IFO	Issued for Order
ISSD	Issued for Start Shop Drawing
IFC	Issued for Construction
A,D	Start and finish of 3D-CAD model
B	ISSD for 3D-CAD model
C	IFC for 3D-CAD model

Documents

PS	Package Specification
MPS	Master Package Specification

PRR	Project Review Request
DM	Distribution Matrix
P&ID	Piping and Instrument Diagram

Miscellaneous

HTID	Handover of Technical Information and Documentation
STID	STATOIL Technical Information system in Operations
SDC	Supplier Document Control
PE	Package Engineer
EC	Engineering Company
WAO	Weeks After Order

REFERENCES

Kvaerner Engineering a.s., Brage Project:
- Package Specification: Shell and Tube Heat Exchangers
  Doc.no. 30-1A-NH-M49-06034 (include drawings).
- Supplier Document Control. Proc.no. 2500-C19.
- Supplier Data Requirements. Proc.no. 2500-C20.
- Supplier's Documentation of Equipment. Doc.no. 30-1A-NH-M44-00001.
- Project Coding Manual. Section VIII: Document codes, Section XI: Supplier Data List.
- Master Package Specification. Doc.no. 30-1A-NH-M49-06000. Extracts.
- Miscellaneous statistical data for the Gyda Project.
Elvekrok, Dag Runar, (1992) Vendor Information in Offshore Projects : Information control. Department of Marine Systems Design, Norwegian Institute of Technology, University of Trondheim.
Elvekrok, Dag Runar, (1993) Vendor Information in Offshore Projects : Selections and control of critical Information Elements. Department of Marine Systems Design. University of Trondheim.
Magnus, Bjørn Henrik, (1992) Requirements for Information to be handed over to Operations. Offshore Information Conference, Oslo, Norway.
Norwegian Engineering Industries Standardisation Centre, NVS, NS5820 - Supplier's Documentation of Equipment.
Marca, Davis and McGowan, Clement, (1987) SADT-Structure Analysis and Design Technique, McGraw-Hill.
Rumbaugh, James et al, (1991) Object-oriented modeling and design, Prentice-Hall.

BIOGRAPHY

Dag Runar Elvekrok will present the paper at the conference. Dag Runar Elvekrok was educated from the Department of Marine Systems Design, Norwegian Institute of Technology, University of Trondheim, Norway in 1993. He is now working with a Ph.D. which the subject of the thesis is supplier information in offshore projects.

37

Development of MIS For Automobile Repair Industry

D.M.Zhou , L.Zhang , B.C.Fu , F.Y.Wen
The Institute of Computer Applied Technology of HENAN University
Minglun Street, Kaifeng, Henan Province, P.R.China 475001
Tele:(0378)558833-506

Abstract

This paper is intended to give a general analysis on the production model and the management model applied to the automobile repair industry. Meanwhile, an auto–repair MIS system model and an implementation model is further described. An application instance of this auto–repair MIS model has been implemented in an auto–repair company and the expected results have been obtained.

Keywords

Auto–repair, MIS, Production Model, Management Model, System Model, Implementation Model.

1. PRODUCTION MODEL AND THE GOAL OF MIS FOR AUTO–REPAIR INDUSTRY

1.1 The Production model for auto–repair industry

The production model for the anto–repair industry can be summarized as a "problem–solving finite cyclic iteration". The term "problem" means to find out the failure or breakdown of the vehicle and then proceed to determine how many locations need to be repaired, while the term "solving" means to get the failure or breakdown repaired so as to remove the "problem". However, the correct determination of the failure or breakdown locations has not happened before the actual service unless it is an extremly simple one because it needs repeated modification through many actual services. It is therefore considered that the automobile repair process is the process of a "problem–solving finite cyclic iteration". The entire life cycle for auto–repair can be divided into following steps:

(a) Making a failure or break-down report;
(b) Dismantling for inspection and analysis;
(c) Preparing material or parts;
(d) Repairing the failure or break-down;
(e) Unit test;
(f) System test;
(g) Running test.

The first two steps are regarded as the problem space, while the remaining five steps as the solving space. The entire service process formed by the seven steps can be expressed by a fountain model as shown in Figure 1.

With reference to the fountain model, we may see that only the last two steps must be done on the basis of the completion of the pre-

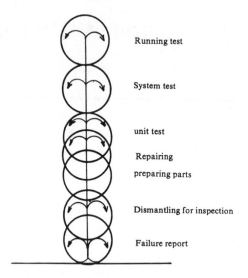

Running test

System test

unit test

Repairing

preparing parts

Dismantling for inspection

Failure report

Figure 1. FOUNTAIN MODEL OF AUTO –REPAIR PROCESS.

vious step while the remaining steps have two or three steps overlay with each other. Meanwhile all steps except the step of the material or parts preparation have the tendency to return to any of their previous steps (as a waterfall case). The fountain model has given a detailed description of the entire auto–repair process. which is also known as "problem–solving finite cyclic iteration".

1.2 Production management model for auto–repair industry

Based on the fountain model, the production management model for auto–repair industry can be described as a "finite plan–dispatching cycle" as shown in Figure 2. The service level, the repair items, the required man–hour and material or parts shall be determined by "plan" while the actual service work shall be assigned to the workshop by "dispatching" so as to get the breakdowns repaired, which is mentioned previously as "solving". The water-fall case in the fountain model makes the "plan–dispatching" become finite cycle.

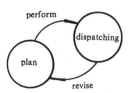

perform

dispatching

plan

revise

Figure 2. THE FINITE PLAN–DISPASTCHING   CYCLE PRODUCTION MANAGEMENT MODEL.

1.3 The goal of MIS for the auto–repair industry

The final goal of the MIS for auto–repair is to ensure all the service work done effectively. The details concerned are listed as the followings:

(a) To work out or revise the material or parts and man–hour plan based on the given service level and repair items and if necessary, to regulate the service level and repair items so as to achieve the dynamic evolution for production planning.

(b) To improve effectively the stock management so as to avoid parts supply hold–up caused by the dynamic evolution of production plan and unspecified vehicle types.

(c) To ensure a quick searching of the unique parts identification code. Being the bottle neck of the manual MIS, this will ensure the correct supply and eliminate errors.

(d) To perform the real–time job dispatching and effective job progress supervising, so as to optimizing a best way to make use of the man–hour resource and bring down the number of vehicle–day in the workshop.

(e) To carry out the real–time cost accounting in the unit of the single vehicle and labor accounting in the unit of each individual laborer so that it is possible to handle correctly the situation about the on–going production and eliminate the contradiction between production and profit or loss owing to the long cycle of the production, which is often encountered in the manual MIS.

(f) To be featured with additional functions of the MIS such as: finance management, equipment management, sales management, personnel management and labour management etc.

2. SYSTEM MODEL OF MIS FOR AUTO–REPAIR

We take the interactive workgroup model as the system model of the MIS for auto–repair so as to meet the goal of application and the required function of MIS as fully as possible. The cyclic iteration characteristics shown in the auto–repair production and management process enable the follwings to be dynamic evolution together with the production process such as: production planning (including service items, materials or parts and man–hour needed), spare parts supply (including stock supply and vendor supply), man–hour dispatching, quality assurance, cost accounting and fund employment etc. The MIS, therefore, should have the function of the real–time forecast in the production process (such as: the latest demand for service items, materials or parts , man–hour and fund etc) and the dynamic tracing capability in production state (such as vehicles under–repairing, vehicles pending for service, vehicles interrupted for the time being, vehi-

cle pending to be reworked, vehicles requesting an external assistance for repairing, stockage, and fund employment etc.) and production accounting (such
as: consumed man–hour and material or parts, the number of vehicle–day in the
workshop etc). All above funtions have indicated that the interactivity on the
nodes must be regarded as an important characteristic in the system model.

The hardware structure, data and function allocation as shown in Figure 3, the
data exchange and synchronization among nodes as shown in Figure 4 have reflected the work cooperation of all nodes in the system resulted from the characteristics about the workgroup in the model. On each node, the data are further divided into private data and sharing data for the purpose of reducing the data
transmission burden in the network and achieving the client / server structure.
The system model is implemented as followings: Firstly, we choose the Windows
For Workgroups (Microsoft WFW) as our platform. Secondly, we
comprehensively use the Database System Foxpor, Access and Excel supported
by the WFW and the techniques on OLE, DDE, e–Mail and the sharing resource
etc.

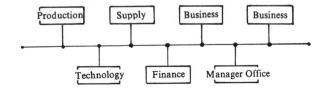

Figure 3.   SYSTEM STRUCTURE.

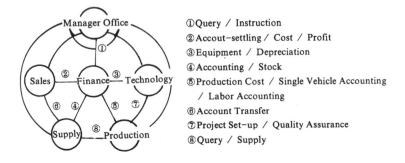

① Query / Instruction
② Accout–settling / Cost / Profit
③ Equipment / Depreciation
④ Accounting / Stock
⑤ Production Cost / Single Vehicle Accounting
   / Labor Accounting
⑥ Account Transfer
⑦ Project Set–up / Quality Assurance
⑧ Query / Supply

Fig 4   RELATIONSHIP OF NODES COOPERATION.

3. IMPLEMENTATION MODEL OF MIS FOR AUTO–REPAIR

3.1 Finite Plan Sequence

The plan to get a vehicle repaired is a 3–tuple $S = <J, P, T>$, where J is a list of service items, P is a list of ordered pair $<p,n>$ which specify the parts and the related quantity needed by items in J and T is the total man–hour required by items in J. In the light of the above 3–tuple plan we may see that J is initiative while P and T are passive. That is to say J determines P and T.

Suppose the finite plan sequence formed in the finite plan–dispatching cycle is :

$$S_0, S_1, S_2, \cdots, S_n$$

where the S_0 is called initial planning and $S_i(i>0)$ as intermediate planning resulted from the modification to S_{i-1}. The final intermedinte planning S_n is called terminal planning, representing the actual service items, consumed parts and their quantity together with actual consumed man–hours. As a result, the implementation of the finite plan sequence is composed of the initial implementation and amending implementation. In auto–repair MIS, both initial and amending implementation are achieved by man–machine interactive operations.

3.1.1 The implementation of initial plan $S_0 = <J_0, P_0, T_0>$

① $J_0 := <e_1^{(0)}, e_2^{(0)}, \cdots, e_{l_0}^{(0)}>$, where $e_i^{(0)}$ is the required service items specified by service expert after examing the failure or breakdown of the vehicle.

② As for $e_i^{(0)} \in J_0$, the initial value of the parts and related quantity required by the service items $e_i^{(0)}$ can be determined based on the material or parts quota f. Suppose E is the set of all service items and P is the set of all parts , then the material or parts quota f is the function $f : E \to \rho(P \times N)$, where N is the set of positive integer. The list of $<f(e_1^{(0)}), \cdots, f(e_{l_0}^{(0)})>$ makes up the initial value $\vec{P}_0$ of P_0. Attention that each $f(e_i^{(0)})$ will be a list composed by ordered pairs $<P_j, n_j>$; $\vec{P}_0$, therefore, is also a list of ordered pairs.

③ The service experts can obtain the verified initial material or parts plan P_0 confirmed by operating IN / OUT which move the ordered pairs from or to the list $\vec{P}_0 = <f(e_1^{(0)}) \cdots, f(e_{l_0}^{(0)})>$.

④ As for $e_i^{(0)} \in J_0$, the man–hour $g(e_i^{(0)})$ required by the service item $e_i^{(0)}$ can be determined by means of man–hour quota $g : E \to N$, then, $T_0 = \sum_{e \in J_0} g(e)$ is the initial man–hour plan.

3.1.2 The implementation of amending plan $S_k = \ <J_k, P_k, T_k> \ (k>0)$

① $J_k = \ <e_1^{(k)}, \cdots , e_{l_k}^{(k)}>$ is obtained from the list $J_{k-1} = \ <e_1^{(k-1)}, \cdots , e_{l_{k-1}}^{(k-1)}>$
through operating the IN / OUT by the sevice experts.

② Suppose $\triangle_+^{(k)} = J_k - J_{k-1}$, $\triangle_-^{(k)} = J_{k-1} - J_k$, calculate $f(\triangle_+^{(k)})$ and $f(\triangle_-^{(k)})$, then
$\vec{P}_k = P_{k-1} + f(\triangle_+^{(k)}) - f(\triangle_-^{(k)})$, where $"+"$ means the moving-in operation while $"-"$
moving-out operation.

③ Amend $\vec{P}_k$ by the service expert to obtain P_k.

④ $T_k = T_{k-1} + \sum_{e \in \triangle_+^{(k)}} g(e) - \sum_{e \in \triangle_-^{(k)}} g(e)$

3.2 Task dispatching

A group of service items which must be performed in sequence is called a job.
The task dispatching in an auto-repair MIS can be divided into two levels. The
first level dispatching designates the priority of the job and assignes the job to the
workshop or amends the priority of the job which has been already assigned to a
workshop. The second level dispatching is the service item dispatching which is
carried out in the workshop and used to preserve the order standing in the job.

The first level dispatching is a event dispatching, which will take place once there
is a job or the priority of a job to be necessarily amended. The rules to this function are: ① To keep the homogeneity on the same vehicle with different jobs for
the purpose of reducing production cost. For instance, The different jobs on the
same vehicle can be assigned to different workshops with the equal priority level
so as to make the different jobs on the same vehicle be done parallelly, aiming at
bringing down the number of vehicle-day repairing at the shop ;or all the job
priorties given to the same vehicle can be rearranged at one time, either to move
ahead or to put off. ② To balance the job load in different workshops so as to
make the most effective use of man-hour resource. For instance, the job will be
assigned to the workshop with minimum man-hour coefficient. The man-hour
coefficient is defined as (assigned man-hour / quota man-hour) to indicate the
current work intensity in the workshop. The performance result of the first level
dispatching makes all the workshops have a job queue in the order of the
priority.

The second level dispatching is also the event priority dispatching. Each service
item in a job will tend to go through many changes in state from entering into the
waiting queue to the end of the service. There are totally six possible states as: wa
iting for service, being ready, repairing suspending, waiting for material and waiting for the job end .Apart from the state of repairing, all the other states possess

a job queue in their own priority sequence. When touching off the dispatching, the second level dispatching will choose the first item from the item queue at being-ready state and send it into the repairing state. The event dispatching and states transmitting are shown in Fig. 5.

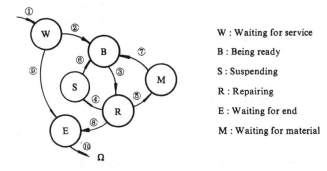

W : Waiting for service

B : Being ready

S : Suspending

R : Repairing

E : Waiting for end

M : Waiting for material

① To enter into the queue (from the first level dispatching). Waiting for service ② Materials ready ③ Dispatching ④ Forced stop ⑤ Short of material ⑥ Stoping relief & Being ready ⑦ Materials coming ⑧ Repair done ⑨ Job end ⑩ Job completion

Figure 5   THE STATES TRANSMISION IN THE SECOND LEVEL DISPATCHING.

3.3 The Search of Unique Part Identification Code

In the parts planning sequence $P_0, P_1, P_2, \cdots, P_n$, any P_k in the ordered pair $<p_k, n_k>$ contained in P_i is only the general name of the parts. The general name is considered to be the abstract name of the parts, in other words, what it represents is only the name of parts category, not the name of an actual part in this category. The name of an actual part is specified by the parts manufacturer known as the unique part identification code and written as IDp. In actual service, the required part can only be obtained when

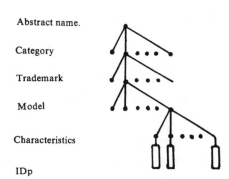

Abstract name.

Category

Trademark

Model

Characteristics

IDp

Fig.6   THE SEARCH TREE OF THE PARTS.

both the abstract name p_k and the IDp_k have been established together.

The abstract name of the parts can also be defined as a search tree as shown in Fig. 6. The root node represents the abstract name of the parts, while the first level intermediate nodes represent the vehicle category (e.g. truck or car etc.), the second level intermediate nodes represent trademark of different automobiles (e.g. Toyota, Santana etc.), the third level intermediate nodes represent the model of the vehicle (e.g. YH50, YH60 in Toyota Series and so on), the fourth level intermediate nodes represent the characteristics of the vehicle (e.g. the left seat steering wheel, mountain region disign etc), the leaf of the tree stands for IDp. We call a PATH the sequence of nodes which starts from the root node passing through all intermediate nodes to the leaf. Then the IDp is determined by the function $\psi(p_k, PATH)$ while the PATH is specified by the foundamental data of the vehicles underrepairing. The mapping method of the function ψ is the search tree as shown in Fig.6.

3.4 Direct Production Cost Accounting and Prediction

The direct production cost is composed of the material cost and labor cost. As a result of the dynamic evolution of the material planning in the whole production process from the material planning before serice to material consumption list after service, it is possible to account the in−use material cost at any time and it makes the prediction of the future material requirement possible. Meanwhile, the dynamic evolution of the man−hour planning sequence makes the prediction and account of the required man−hour possible.

4. DEVELOPMENT INSTANCE

This auto−repair MIS has been applied to DRAGEN Auto−Service Company. As this MIS system is based on the foundation of careful analysis about the production and management model for the auto−repair industry and paid special attention to the key issue of the finite plan−dispatching cycle in production management, it has greatly improved the management in the fields of material supply, sales, stock, labor, technology and financial affairs etc. On one hand, this auto−repair MIS system has made the production process show clearly by the detailed illustration of dynamic data and related figures to provide with the dynamic tracing function to the production states, thus the direct production cost has been dramatically reduced (including labor cost and the shop−vehicle−day). On the other hand, it has realized the real−time forecast on the requirement of the material supply and finance, thus, the management cost has been greatly reduced and the handling capacity greatly increased. In a word, the design of MIS system has achieved the expected results.

References

Fu.B.C. and other Co—authors, Application of Multimedia Database in the industry of Automobile Reparation, A Collection of Theses on CD—ROM and Multimedia CD & M '94 / China International Symposium.(1994)

Biography

Demin Zhou, Professor, Computer Department of Henan University. Birth Date: 1938, graduated in 1961 from Nankai University, major in computing mathematics. Specialized in research and development of software engineering and object—oriented technology.

PART NINE

Theory and Methodology of CIM

38

From CIM to Global Manufacturing

G. DOUMEINGTS[a] , Y. DUCQ[a] , F. CLAVE[a] , N. MALHENE[a]

[a] Université Bordeaux 1 - Laboratoire d'Automatique et de
Productique - GRAI - 351 Cours de la Libération 33405 Talence cedex
- France -
Tel : 56 84 65 30 - Fax : 56 84 66 44 - Email : doumeingts@lap.u-
bordeaux.fr

Abstract

The objectives of this paper are to show the evolution of CIM Systems, mainly in the Western Countries from the end of the seventies till now and the influence of the economic situation.

At the end of the seventies, the development of CIM was push by the technology. The result was a high level of investment and a low level of return on investment. The reason was mainly the inadequacy of the proposed solutions in comparison with the requirements. At the end of the eighties, the new economic situation changed the attitude : a contrario from the previous one, the problems pull the technical solutions in the factory combined with a reorganisation of the physical flow in order to introduce the Just In Time philosophy. Today, again due to the world wide facilities, we see a new paradigm : the Global Manufacturing in terms of market and of production of goods.

Key Word

CIM, Integration, Global Manufacturing.

1 INTRODUCTION

The objectives of this paper are to show the evolution of the CIM concept from the end of the seventies till now taking in account the strong pressure of the economic situation. This analysis of the evolution is more oriented towards the western countries but now, it seems the difference with the far eastern countries is becoming smaller.

In a first part, we will describe the situation at the end of the seventies when the evolution of the market influenced strongly the evolution of the industrial enterprises.

Since the end of the seventies, the market and the Industrial Enterprises have evolved. First, the level of competitiveness increased with the appearance of new competitors coming from new geographical area. Then, customer has become more and more exigent, searching cheapest products with appropriate quality, in shortest lead time. This evolution implied a lot of changes concerning the manufacturing systems : level of automation, advanced production management, quality, level of integration....

In this time, the development of information technology in industries led to CIM concepts implementation. This was considered as key element for competitiveness improvement, allowing shorter product life cycles, reduced costs.... At the end of the eighties, the new level of competitiveness and the economic difficulties led the Industrial Enterprises to search new techniques abd new methods more compatible with economic and human aspects. This new situation changed the CIM concepts. At the end, we will described the present situation with a new evolution based on a world-wide extension.

2 COMPUTER INTEGRATED MANUFACTURING (CIM) AT THE END OF THE SEVENTIES

At the end of the seventies, the increasing of the competitors number and of the customers wishes led to a complex market. On one side, the customer wanted customised products especially adapted to its needs. On the other side, due to the fact that the innovation transfer time decreased, the result was a great diversification in term of products and a stronger competition due to the saturation of the market.

To face to the new situation, the industrial world invested massively in the automation and the information technology. The Computer Integrated Techniques were created. The only problem was that in this time, the users, based on the assertion of the specialists, believed that the word COMPUTER can solve all the problems.

E. MERCHANT [Merchant 88], who was certainly the first researcher presenting the "Manufacturing System" in the sixties, defines CIM on this way : "CIM is a term coined to represent the full range of the capability potential which the digital computers holds for manufacturing". Figures 1 illustrates this definition.

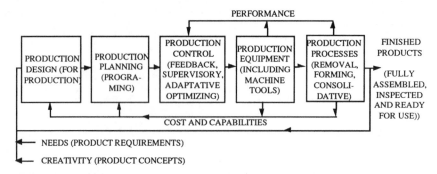

Figure 1 Computer Integrated Manufacturing System on MERCHANT point of view

Certainly this view of CIM was the earliest view. In 1984, G. DOUMEINGTS [Doumeingts 84] proposed an equivalent definition. He looked on the integration of all functions from the design of the product to its delivery with six major functions (figure 2) : Computer Aided Design, Computer Aided Manufacturing, Computer Aided Equipment Control, Computer Aided Production Management, Computer Aided Quality, Computer Aided Maintenance.

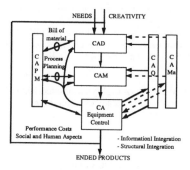

Figure 2 CIM on DOUMEINGTS point of view.

The main features of this system is the double integration :

❑   Integration by information : it exists a massive exchange of information between the various functions. For example the process planning produced by CAM is one of the basic information used by CAPM, Bill of Material produced by CAD is also used by CAPM for MRP function, etc... But the exchange of information between functions is not direct. This information must be transformed : it is impossible to use Process Planning in CAPM, as it is created in CAM function.

❑   Integration by structure : the structure of one function interferes with the structure of one another. It is typically the case with the planning of maintenance and the scheduling : the first must be taken into account by the second and vice versa.

The first CIM framework was proposed but despite the enthusiasm of the people for these CIM techniques, the integration was not a success. Indeed, the main preoccupation of the engineers and researchers was the development of Automation Systems. Blinded by technical challenges, they gave priority to Information Technologies and Automation. They did not take into account essential views such as social and human aspects, cost considerations, integration, etc... It was the great period of the FMS (Flexible Manufacturing Systems) and the UNMANNED Factory.

Led by the challenge of the full automation, the engineers forgotten that if we have to spend 20% of the investment to reach 80% of automation, it costs 80% of investment for the last 20% of automation. They were thinking only in term of investment (often very heavy), instead of thinking in term of return on investment. They forgot that the functioning of any system depends on the performances of its worse part.

Major other enterprises developed their own CIM project during the eighties : CATERPILLAR, John DEERE in US, Citroën in Meudon, Renault in Boutheon in France, ... If these projects were technical success, the resulting systems were not as performant as industrialists enjoyed. These first projects missed experience. There was a lack of global view and studies were not complete. For example, one failed because the ground couldn't support the AGV (Automated Guided Vehicle) implementing by the company. An other company didn't take into account the electric contacts robustness of the automatized mechanism it developed.
We will recall the excellent paper published in 1983 by J. HATVANY [Hatvany 1983] "Dreams, Nightmare and reality". We think this paper helped seriously the industries to

understand that the most important in such domain is to analyse and to understand the problems.

We want to mention an excellent project in France in 1985 : SNECMA Le Creusot, in which the human and economic point of views were taken in account beside the technical one. In this big project (70 millions of dollars), the GRAI method allowed to define the specifications of the factory control.

Anyway, the eighties saw the birth of highly automated Manufacturing Systems but also the inadequacy between objectives and the means to reach them.

3 CIM AT THE END OF EIGHTIES

At the end of the eighties, a new economic situation appeared. This was characterised by the growth of economic difficulties. So, it implied to take care seriously to the amount of money spent in investment. About all, the exigence coming from the market was stronger : more diversify products in less time !!!.

The economic pressure obliged also the company to consider the performance of the enterprise as the combination of three major criteria : cost, quality and lead time. Then, the improvement of the manufacturing system performance led to optimise the triplet [cost, quality, lead time].

In order to manage these changes, several sciutions were proposed.

The first solution was to increase the anticipation capacity. This led the enterprises to have a strong policy oriented toward market and products, to have a high modularity of products with common components and a very homogeneous sector of activities. It is why the enterprises evolved from an economy of scale to an economy of scope with coherence in the investments, in technologies and in knowledge, and with a good adaptation to the market : the companies focused on their core business.

The second one was to increase the adaptation capacity. This led the companies to have either an over-capacity or to develop a flexible manufacturing system.

The third solution was the improvement of product flow control. This led the enterprises to have a complete integrated manufacturing control system. This integration had to be implement from the design to the delivery of the product (Figure 3) and through the various decision levels from the strategic to the operational level.

These solutions imply to reorganize the manufacturing structure, to simplify the procedures and to improve the level of integration. Unfortunately, in this time, the integration was only at the technical level. It was necessary to develop new architecture at the conceptual in order to reach this situation. Due to the fact it is necessary to find, in a short time, the right solution to solve the identified problem, it is necessary to use not only architecture but also methodology to design and specify the solutions and to elaborate the specifications book in order to limit the risk of inadequacy between solution and problems. GIM (Grai Integrated Methodology) was one Methodology developed through ESPRIT Projects to answer to the industrial situation.

Moreover, in order to verify the achievement of the new objectives of these manufacturing systems, there was a need for performance measurement. This measurement was realised by the definition and the implementation of Performance Indicators Systems.

So, the new definition of CIM became:

"CIM stands for global methodological approach in the enterprise in order to improve the competitiveness of industrial enterprises.

This methodological approach is applied to all activities from the customer (product idea) to the customer (delivery of product), in an integrated way, using various methods and means (techniques, computers, automatic control)... in order to ensure simultaneously :

- [ ] productivity improvement,
- [ ] cost reduction, , increasing, ,
- [ ] fulfilment of lead times,
- [ ] quality of product,
- [ ] global or local flexibility of manufacturing system

In such an approach, the economic, social and human aspects have a place as important as the technical aspects.

So, according to this definition, the new manufacturing systems took in account not only the technical side but also the environment, the economic and the human aspects. :

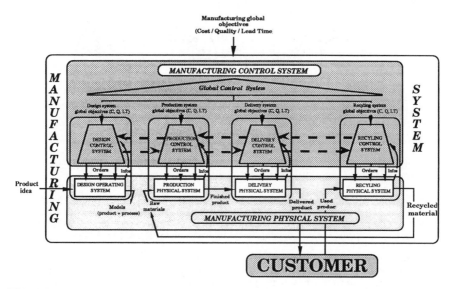

Figure 3 The integrated manufacturing system.

4 TOWARDS THE GLOBAL MANUFACTURING

During the last years, the market features evolved towards a internationalisation of exchanges. If the industrial enterprises can sell their products simultaneously in several regions around the world, the competitiveness level becomes more and more high. The industries are obliged to improve continuously their performances towards the optimisation of the triplet (Cost, Quality, Lead Time) for products and services. This new situation implies that the industrial enterprises design, develop, manufacture and distribute their products with a global point of view, searching appropriate partners in order to reach the performance level.

This manufacturing globalisation has two mains requirements for the considered enterprises. The first is linked to the building of a set of enterprises which create a chain from the design to the delivery of the products. These enterprises allow, by their abilities and their performances, to fulfil the customer needs at a minimum cost, with a appropriate quality and a short lead-time

[Doumeingts 1995]. This set forms a sort of "Virtual Enterprise" existing temporarily and in which it is necessary to develop the inter-enterprise relationships.

The second requirement is linked to the management of this "virtual enterprise". It consists to extend the concepts of integration, appeared in CIM approaches, to this set of enterprises located around the world (Figure 4). This integration is difficult because it must be realised in a distributed environment.

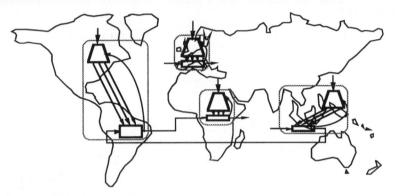

Figure 4   The Global Manufacturing System.

Regarding to this evolution towards globalisation of Manufacturing, several research projects were launched in which the GRAI/LAP is or was involved.

One of those, called "GLOBEMAN 21 : Enterprise integration for GLOBal MANufacturing towards the 21st Century" was developed in the frame of an international programme : IMS (Intelligent Manufacturing Systems). This project involving more than twenty partners (Europe, United States, Japan, Canada, Australia) was launched in March 1993.

During its feasibility study of one year, the objective was twofold:
        - to perform a comparative analysis of current practices in order to determine the "good practices" related to four kinds of production (one off, small and large batch, continuous / semi-continuous) and simultaneously for six regions around the world (Japan, Europe, EFTA, Canada, USA, Australia). This study had to bring to the fore the cultural, technological and economical influences on the performances of the production systems,
        - to determine, with the help of this comparative study, the research topics to promote in global production for the next ten years (horizon of the IMS programme) and which will support the future objectives of this project.

Based on the GRAI model [Doumeingts 84] and on the ECOGRAI method [Bitton 1990], a methodology to analyse and to compare the set of enterprises was developed. It includes three parts : a model and a formalism to describe the industrial practices, a structured approach to collect the data, several methods to analyse these practices and to derive the "Good Industrial Practices" [Doumeingts 1994].

The model is composed of three axes (figure 5): the managerial axis, the production axis, the evolution axis.

The managerial axis is decomposed in three levels : strategic, tactical and operational. The production axis is decomposed in various functions of product life cycle. The evolution axis contains two status:
- the current practice matrix (AS IS)
- the future practice matrix (TO BE) wished by the industrial partners.

So, for each function of the production axis and for each level of the managerial axis one identifies the objective, the drivers, the performance indicators and the boundary conditions (figure 5). This form the formalism : the Matrix (Figure 5)

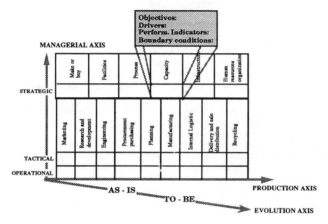

Figure 5 The Matrix.

In order to analysed the collected data and to derive the "Good Industrial Practices", four research centres have developed processing methods: "Coherence and Practice Analysis" developed by the GRAI/LAP and based on the coherence analysis of the various industrial practices inside the Matrix, "Cluster/Factor analysis" developed by Virginia University and based on the practices similarities study, "Profile analysis" developed by HUT and based on the QFD method, "Vizualisation" developed by CSIRO and allowing to vizualise graphically the Good Practices identified by GRAI/LAP. Finally, in order to derive the future research topics in Information Technology, Tokyo University developed an analysis method based on a deeper description of the Drivers.

This methodology was applied to 21 existing production systems related to 16 industrial partners. These 21 matrices (AS IS) were analysed with the five processing methods developed in the project. The synthesis of the results highlighted a set of "Good Industrial Practices".

In parallel, for each of the four domains, the future practices wished by the industrial partners were described with the TO BE matrix. The comparison between the AS IS and the TO BE matrices allowed to bring to the fore the future needs and also the future research topics.

Moreover, for each domain, the industrial partners identified the future research areas related to their type of production according the problematic of the globalisation.

One can also mention the Eureka TIME project (Tools and methods for the Integration and for the Management of Evolution of industrials firms) [TIME 93] in which the GRAI/LAP is also involved and which associates, at the European level, industrial enterprises and Research Centres of six countries (France, Portugal, Italy, Norway, Sweden, Finland). The objectives of this project are to develop methods and associated computer tools, to help the enterprises to place in the global market and to manage their evolution in order to adapt continuously. They can be classified into four categories.

The first one consists in providing a vision and an understanding on the new manufacturing management taking into account the continuous change of the economic environment. This new manufacturing management must allow the managers to elaborate and to implement a continuous strategic evolution.

The second one is to elaborate a methodology and supporting tools to implement this strategic evolution in the company.

The methodology must help the company :
- to define its Manufacturing Strategy,
- to elaborate the Evolution Procedures,
- to implement them.

The third one is to test, to improve and to finalise this methodology and the supporting tools, for relevant industrial situations, for various classes of manufacturing systems, based on the industrial partners involved in the project.

This experimentation will allow to provide several reference models, structured by type of problematic encountered.

The fourth one is to define a structure allowing to collect the evaluation reports of industrial companies, to determine the "Industrial Best Practices" at a European level and to provide anonymously the results by category of Reference Models.

Figure 6 shows all these conceptual elements.

The main phases of the evolution process appear :
- the system "as is", description of the present situation of the industrial system as well as of its environment, and the "mastered as is", obtained by eliminating deviations from the standard in the functioning of the system,
- the "should be" system, set of Objectives and Key Success Factors, which indicates the direction to follow and channels the evolution,
- the "next step" phase, which is the next steady stage reached from the initial situation ("as is") in the continuous improvement process towards the "should be",
- the action plan, set of actions to undergo and of guidelines to follow along the evolution path, allowing to evolve from one steady state to another.

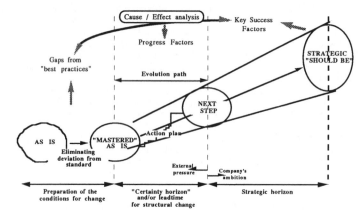

Figure 6 The TIME GUIDE model : Guiding the development of industrial enterprises.

The evolution process of industrial companies is undertaken as a project including five main phases (Figure 7).

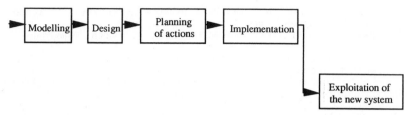

Figure 7 Main steps of a project

GIM (GRAI Integrated Methodology) is a method which allows to cover modelling and design steps. The term method means that GIM is composed of :
- conceptual models,
- structured approach,
- formalisms.

The modelling part of GIM allows to identify the elements of the "as is" system and their connections. It has to take into account both internal components and market-oriented issues (the company's customers, suppliers and competitors). GIM design step allows to design the model of the future system. Thanks to an user oriented approach, GIM allows to perform a model which is in total coherence with the user's needs. Moreover, it allows to take into account other essential elements such as the objectives of the enterprise, the present position of the enterprise on its market, the position of its competitors, some unconsistencies in the system, etc... To ensure the good appropriateness between objectives and means, GIM supports during modelling and design steps data providing by several additional tools such as self-audit tools or benchmarking tools.

5 CONCLUSION

If the CIM concept was strongly developed during the eighties in order to answer to the evolution of the market, it highlighted some limits. In particular, it led to build automatized systems without a real integration, to spend lot of money without thinking about the return on investment and to have a too important level of automation in industry without taking into account the human factors. It is why the end of the eighties, due to new economical difficulties, led the companies to answer exactly to the problems to solve with using in particular new methodologies and architectures to specify, to design and to implement manufacturing systems. With the use of these methods, the need for performance measurement highlighted to verify the achievement of the manufacturing systems objectives. Today, the manufacturing systems must be global, with several entities distributed in different environments, and completely integrated, from the design of the product to its delivery. The new problem which must be solved is the management of such global systems, how to control them, how to measure their global performances and how to make them evolve. The development of elaborated methods with associated computerized tools will allow to answer to this situation.

BIBLIOGRAPHY

BITTON M.(1990) "ECOGRAI : méthode de conception et d'implantation de systèmes de mesures de performances pour organisations industrielles",*Thèse d'automatique*, Université Bordeaux I.

DOUMEINGTS G. (1984) "Méthode GRAI, méthode de conception des systèmes en productique",*Thèse d'état es-Sciences,* Université Bordeaux I.
DOUMEINGTS G. (1984) Conference on Manufacturing Automation " - Sao Paulo - Brasil - July 1984
DOUMEINGTS G. , CLAVE F. , DUCQ Y., MARCOTTE F. (1995) "ECOGRAI, méthode de conception et d'implantation de systèmes d'indicateurs de performances pour organisation industrielles" , The third international conference ILCE '95 - Concurrent Engineering & Technical Information Processing - Paris 30/1-3/2/95.
DOUMEINGTS G. , MARCOTTE F. , CLAVE F. , DUCQ Y. (1994) " IMS Programme : GLOBEMAN 21 Methodology to evaluate the Best Industrial Practices worldwide " The third International Conference on Automation Technology - AUTOMATION'94 - Tapei - TAIWAN - July 1994.
MERCHANT (1988), Promalat 88, Dresden, GERMANY
TIME (1993) Deliverable D31: "Definition of the deliverables" Last version - 4/10/93
HATVANY J. (1983), "Dreams, nightmares and reality", CAPE Conference 83, 25-28 April, Amsterdam, The Netherlands.

39

Integration of information model(IDEF1) with function model(IDEF0) for CIM system design

Lu LingZhi, Ph.D.
Ang Cheng Leong, Ph.D.
Prof. Robert K L Gay
Gintic Institute of Manufacturing Technology
Nanyang Technological University, Singapore 2263
Tel:7991129 Fax:7916377
Email:gllu@ntuvax.ntu.ac.sg

Abstract

In this paper a methodology is proposed for the integration of IDEF1 with IDEF0, allowing an IDEF1 model to be generated easily from the corresponding IDEF0 model. The methodology involves: 1)the principle for the integration of IDEF1 with IDEF0 based on the concepts of IDEF methods; 2)the new requirements for IDEF0 diagrams at the relative bottom levels to meet the prerequisites for the integration of IDEF models; 3)the development of knowledge-based systems(KBIDEF) for the integration of the IDEF1 model with its corresponding IDEF0 model; 4)the design of two databases: object-oriented IDEF0 and IDEF1 databases, three libraries: Entity Class Library, Relation Class Library and Domain Relation Class Library, and two knowledge bases: Relation Analysis Knowledge Base and Domain Knowledge Base for CIM information system design. Finally, the paper suggests some areas for future work.

Keywords

CIM, IDEF, system design, information modelling

1 INTRODUCTION

IDEF(**ICAM DEF**inition) methodology was developed by the US Air Force to describe manufacturing organizations in a structured graphical form. Now many IDEF methods (IDEF0~IDEF14) are available or being developed. So far a lot of people have used, developed and studied on them all over the world. There are mainly six kinds of research and applications related to IDEF methods as follows:

(1) Most of the research related to IDEF methods deals with how to use IDEF methods in various application areas, such as modelling the structure of an organization[3], modelling a FMS[4], developing production schedule generation systems and so on;

(2) Some researchers develop generic or reference IDEF models in various domains;

(3) Some enhanced IDEF methods are developed for various applications. IDEFc is developed for mapping quality management systems. An enhanced IDEF0 is developed for project risk assessment[5];

(4) Some companies and organizations develop CASE tools to support IDEF methods[6] ~[7]. These tools make it easy for customers to use IDEF methods to model a system. Knowledge Based System Inc has developed AI0, Smarter, ProCap softwares for IDEF0, IDEF1, IDEF3 respectively[1]. Meta Software Corporation has developed Design/IDEF tool for IDEF0, IDEF1 and IDEF1X. Tsinghua University developed CASE tools for supporting IDEF0 and IDEF1X in Chinese language[6];

(5) Some of the research deals with the assessment of IDEF methods[9][11] or compare them with other methods(SSADM, DFD and etc) [10];

(6) The integration of IDEF methods is studied and its CASE tools are developed by the other researchers[8][12]. Godwin and etc studied the integration concept of IDEF descriptions[8]. System Modelling Corp tried to integrate SIMAN simulation tools with the IDEF0 model. But most of them only develop tools for the integration of IDEF0 model with IDEF2 model or other simulation model. Very little research has dealt with the methodology and CASE tools for the integration of IDEF1 models with IDEF0 model.

IDEF0, IDEF1, and IDEF2 are three separate methods, which are available for modelling functionality, information relationship, and dynamic procedure respectively. Because the three methods are independent, the complete IDEF modelling process can be very time-consuming and involves a tremendous amount of wasteful efforts in repeated capturing of the same data. In addition, there is also the possible incompatibility problem of the three IDEF models when they are built independently. These problems can only be solved by integrating the IDEF0, IDEF1 and IDEF2 models with CASE tool. Additionally, the CASE tool with the integration of IDEF methods makes it easier, faster and more convenient for users to model a system.

As decribed above, there is almost no research on the integration of IDEF1 models with IDEF0 models. In this paper, a methodology will be proposed for the integration of IDEF1 with IDEF0, allowing an IDEF1 model to be generated easily from the corresponding IDEF0 model. A CASE tool, **KBIDEF**(Knowledge-Based system for the integration of IDEF1 with IDEF0), will be developed.

2 BASIS FOR THE INTEGRATION OF IDEF1 WITH IDEF0

2.1 Basic Concepts of IDEF0 and IDEF1

IDEF0 is a method used to produce a function model which is a structural representation of the functions of a manufacturing system or enviroment, and of the information and objects which interrelate those functions. As a communication tool, IDEF0 enhances domain expert involvement and consensus decision-making through simplified graphical devices. As an analysis tool, IDEF0 assists the modeler in identifying what functions are performed, what is needed to perform those functions, what the current system does right, and what the current system does wrong. Thus, IDEF0 models are often created as one of the first tasks of a system development effort.

The basic concepts of IDEF0 is shown in Fig.1. A box represents a function. Arrows are constraints (input, output, control and mechanism) that define the boxes. "Data"(arrows) may be information, objects, or anything that can be described with a noun phrase.The input data(on the left) are transfered into output data(on the right). Controls(on the top) govern the way the function is done. Mechanism(on the bottom) indicate the means by which the function is performed.

The primary strength of IDEF0 is that the concise method has proven effective in detailing the system activities for function modelling. Additionally, the description of the activities of a system can be easily refined into greater and greater detail until the model is as descriptive as necessary for the decision making task at hand.

IDEF1 is a method used to establish the requirements for what information is or should be managed by the enterprise. It is designed to assist in discovering, organizing, and documenting the infomation image of physical and conceptual objects(e.g., people, places, things, ideas, etc.) found in the real world. IDEF1 provides a set of rules and procedures for guiding the development of information models.

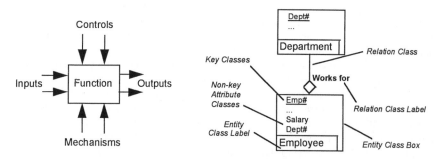

Fig.1. IDEF0 Function   Fig.2. An IDEF1 Diagram

The basic concepts of IDEF1 is shown in Fig.2. An IDEF1 entity represents information maintained in a specific organization about physical or conceptual objects. An IDEF1 entity class refers to a collection of entities or the class of information kept about objects in the real-world. Entities have characteristic attributes associated with them. The term attribute class refers to the set of attribute-value pairs formed by grouping the name of the attribute and the values of that attribute for individual entity class members(entities). A key class is a collection of one or more attribute classes which distinguishes one member of an entity class from another. A relation class can be thought of as the template for associations that exist between entity classes. The existence of a relation class is discovered or verified by noting that the attribute classes of one entity class contain the attribute classes of the key class of the referenced entity class member.

IDEF1 enforces a modularity that eliminates the incompleteness, imprecision, inconsistencies, and inaccuracies found in the modelling process. The IDEF1 modelling exercise provides a foundation for database design, gives a definition of the information structure, and provides a requirement statement reflecting the basic information needs.

2.2 Principle for the Integration of IDEF1 with IDEF0

IDEF family has about 14 methods, each of which is used to describe or model a system from its own perspective. Each type of model or description focuses on a relatively narrow set of relationships and system characteristics comprising a particular viewpoint of the overall system. They describe different information and knowledge of the same system. Therefore, each type of model cannot be converted or generated from another type of model directly and automatically.

However, as these models are the models of the same system they should have consistency to a certain degree with each other. Some rules can be found to obtain consistency among various types of models.

Secondly, these models share some common information or data sources although the models describe different charateristics or relationships of common data.

Therefore, the integration of IDEF methods, in which various types of IDEF models are created, is necessary and feasible on the basis of the two points mentioned above.

Specially for the integration of IDEF0 and IDEF1, the following consistency rules exist between two methods.

(1) The relation classes in IDEF1 models exist between output and input entity classes, and between output and control entity classes of the same function in IDEF0 models if the outputs are directly produced from the inputs and controls;

(2) The relation class labels in IDEF1 model are associated with the corresponding function labels in IDEF0 model to a certain degree;

(3) The global IDEF1 model is based on the hierarchy structure of the IDEF0 model.

IDEF1 and IDEF0 share the common data: the labels of entity classes. All the entity classes identified in the IDEF1 model must map to arrows at the relative bottom level of IDEF0 diagram. In other words, the entity class labels of IDEF1 models can be extracted from the relative bottom level of IDEF0 models.

Therefore, the integration of IDEF1 with IDEF0 is necessary and feasible.

3 KNOWLEDGE-BASED SYSTEM FOR THE INTEGRATION OF IDEF1 WITH IDEF0 (KBIDEF)

The flow diagram of knowledge-based system for the integration of IDEF1 with IDEF0 is shown in Fig.3. The flow is theoretically based on the principle mentioned in the section above and is supported by the technologies of knowledge base, object-oriented database and window programming. The system consists of two databases: object-oriented IDEF0 and IDEF1 databases, three libraries: Entity Class Library, Relation Class Library and Domain Relation Class Library, and two knowledge bases: Relation Analysis Knowledge Base and Domain Knowledge Base in addition to the main flow.

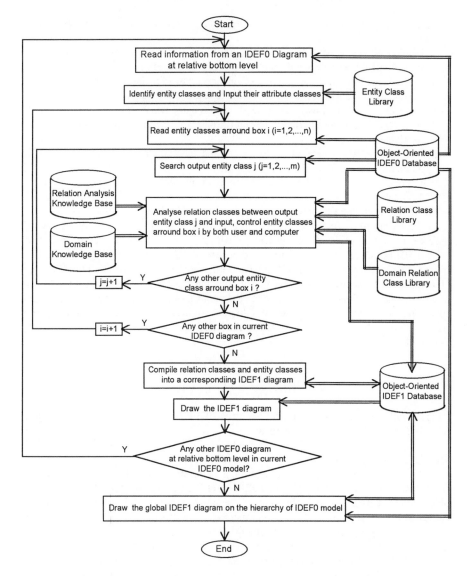

Fig.3. Knowledge-Based System for the Integration of IDEF1 with IDEF0(KBIDEF)

The logic for the integration of IDEF1 with IDEF0 is feasible with the following seven premises:

(1)The information of IDEF0 model in current IDEF0 CASE tool should be in object-oriented data format and easy to be read;

(2)Input, control and output arrows of IDEF0 diagrams at relative bottom levels should be entities;

(3)Relation classes exist only between output and input entity classes, and between output and control entity classes for the box currently analysed;

(4)It is possible to collect the glossary of relation classes for the Relation Class Library. For each application domain, the glossary should be gradually standardized;

(5)There are rules or knowledge(general and domain) which can be used to aid the analysis of relation classes and their types on the basis of function names and the structure of entity classes;

(6)It is possible to compile all entity classes and relation classes into a whole IDEF1 diagram on the basis of IDEF0 hierarchy structure;

(7)It is possible to draw IDEF1 diagram automatically from object-oriented database of IDEF1 model.

The above seven premises will be investigated in detail in the next section.

As shown in Fig.3., the Entity Class Library provides many generic entity classes and their key and attribute classes, on the basis of which the entity classes corresponding to the current IDEF0 model will be identified and generated. The current IDEF0 model and the related IDEF1 model are stored in the Object-Oriented IDEF0 Database and the Objected-Oriented IDEF1 Database respectively. The Relation Class Library provides the glossary of generic relation classes, and the Domain Relation Class Library provides the glossary of special relation classes for the system design in the domain. The existence and the type of relation classes are analysed on the basis of the Relation Analysis Knowledge Base and the Domain Knowledge Base.

4 STUDIES ON SEVEN PREMISES FOR THE KBIDEF

(1)The information of IDEF0 model in current IDEF0 CASE tools should be in object-oriented data format and easy to be read;

Now we have an IDEF CASE tool. It can be used to aid the drawing of IDEF0, IDEF1 and IDEF1X models. But it hasn't integrated three models. We have original C/C++ source code of IDEF CASE software and use it as the starting point of the new integrated IDEF CASE tool. The information of IDEF0 model in the current IDEF CASE tool is in object-oriented data format. It has data dictionaries, can export IDEF0 information in database format, and can creat five reports: Reference Report, Activity Report, Arrow Report, Consistency Report and IDEF Report. Therefore the information of IDEF0 model in current CASE tool is easy to be read. Based on the object-oriented data format of current IDEF0 and IDEF1 models, the object-oriented IDEF0 and IDEF1 databases are built for KBIDEF.

(2)Input, control and output arrows of IDEF0 diagrams at relative bottom levels should be entities;

According to original IDEF definitions, arrows are constraints that define the boxes. Arrows may be information, objects, or anything that can be described with a noun phrase. Therefore arrows of IDEF diagrams at the relative bottom level may not be entities. In order to integrate IDEF1 model with IDEF0 model, the following reqirements should be met in the IDEF0 model.

. The system currently analysed should be CIM information system.

. Its Inputs, Controls and Outputs(ICO) can only be data elements.

. It should be decomposed to the relative bottom level at which every ICO is not bigger than an entity class.

. All entity classes should be identified and then their key and attribute classes should be defined by the human aid and system.

(3)Relation classes exist only between output and input entity classes, and between output and control entity classes for the box currently analysed;

According to the original IDEF1 definition, a relation class is a meaningful association between two entity classes. In IDEF0 models, there are inputs, controls, outputs and mechanisms arround a box(function). Mechanism is the person or device which carries out the function. Therefore for the function currently supported by mechanisms, there should be no relation classes between the mechanisms and others arround the same box. Inputs and controls are only related to the corresponding outputs respectively and thus there are no relation classes between inputs and controls for the same box. Therefore relation classes exist only between output and input entity classes, and between output and control entity classes for the box currently analysed.

(4)It is possible to collect the glossary of relation classes for the Relation Class Library. For each application domain, the glossary should be gradually standardized;

After investigating many cases of IDEF0 and IDEF1 models, it is found that similiar labels of relation classes are used in the same application domain. For example, in the IDEF1 models of production scheduling, SATISFIES/SATISFIED, REQUIRES/ REQUIRED BY, BASED ON/BASES are always used as the labels of relation classes. Some words, like GENERATES/GENERATED BY, OUTPUTS/OUTPUTED FROM, CONTROLS/CONTROLED BY are common for the IDEF1 models of various application domains. Therefore it is possible to collect the glossary of relation classes for the Relation Class Library. For each application domain, the glossary should be gradually standardized and stored in the Domain Relation Class Library.

(5)There are rules or knowledge(general and domain) which can be used to aid the analysis of relation classes and their types on the basis of function names and the structure of entity classes;

After studying the IDEF models of several systems and basic concepts of IDEF0 and IDEF1, it is found that there are the following rules, for example, which are very useful to aid the inference of relation classes.

Rule 101: If the control entity class at IDEF0 box is something like "Requirements", the relation class between its output and the control may be "Satisfies/ Satisfied By";

Rule 201: If the function name at IDEF0 box is something like "Produce ... " or "Generate ... ", the relation class between its outputs and inputs may be "Based On/Bases" or "Generates/Generated From";

Rule 301: If the attribute classes of one entity class contain the attribute classes of the key class of the referenced entity class, a relation class exists between the two entity classes;

Rule 302: If the primary key class of child entity class is the same as the one
of parent entity class, the relation class between the two entity classes
is "Categorization";

Frequently the relation classes and their types are more dependent on domain knowledge. For example, the type of relation class between entity classes "Operator" and "Machine" is determined by the factory regulations and practical situation. The type of relation classes may be "1:1", "1:M" or "M:N". Because of the complexity of information models and the differentiation of systems, some relation classes and their types should be determined by human interaction.

(6)It is possible to compile all entity classes and relation classes into a whole IDEF1 diagram on the basis of IDEF0 hierarchy structure;

This can be done according to the following procedure:

a. Common entity classes are searched from various IDEF1 diagrams corresponding to the IDEF0 diagrams at relative bottom levels;

b. Various IDEF1 diagrams are integrated into one by identifing the common entity classes;

c. The whole IDEF1 diagram is drawn on the same page.

It is difficult to draw the whole IDEF1 diagram automatically on one page by computer. This should be aided by some human interaction.

(7)It is possible to draw IDEF1 diagrams automatically from the object-oriented database of the IDEF1 model.

Our current IDEF CASE software is coded in C/C++ language. Some drawing functions have been ready in the software. These functions in C/C++ language can be called to draw IDEF1 diagrams by KBIDEF.

6 CONCLUSIONS

In this paper, a conceptual framework of KBIDEF for the integration of information model (IDEF1) with function model (IDEF0) is presented. On the basis of IDEF0 and IDEF1 concepts, the principle for the integration of IDEF1 with IDEF0 is analysed. The new requirements for IDEF0 diagrams at the relative bottom levels are investigated to meet the prerequisites for the integration of IDEF models. Some knowledge and rules used for the integration are acquired for examples. The KBIDEF system consists of two databases: object-oriented IDEF0 and IDEF1 databases, three libraries: Entity Class Library, Relation Class Library and Domain Relation Class Library, and two knowledge bases: Relation Analysis Knowledge Base and Domain Knowledge Base in addition to the main flow.

Such a KBIDEF will overcome the major problems of manual generation of of IDEF1 models. The IDEF1 models generated by KBIDEF are compatible and consistent with their corresponding IDEF0 models. KBIDEF makes it easier, faster and more convenient for users to build the information models of CIM systems. KBIDEF will help to further enhance the popularity of IDEF1 models in industrial community and therefore accelerate the process of CIM design and implementation. A prototype KBIDEF for CIM information system design has been built at the GINTIC Institute of Manufactuing

Technology, Singapore. The development of a prototype system and the application test of the system are being carried out.

In the future the methodology of CIM information system design and implementation will be supported by the object-oriented technology. Hence, the authors hope to study further the integration of function model (IDEF0), information model (IDEF1), dynamics model (IDEF2) and object-oriented model (IDEF4). The process of CIM information system design will be further automated and accelerated by the future KBIDEF with the integration of four IDEF models.

7 REFERENCES

[1] Richard J. Mayer, Michael K. Painter, Paula S. deWitte, "IDEF Family of Methods for Concurrent Engineering and Business Re-engineering Applications", KBSI Technical Report, 1993

[2] SoftTech Inc, "Volume V - Information Modelling Manual (IDEF1)", 1981

[3] Ralph R. Bravooco and Surya B. Yadav, "A Methodology to Model the Functional Structure of an Organization" Computers in Industry 6 (1985) 345-361

[4] Robert K L Gay, Roland Lim and Lau Wai Shing, "Using IDEF Methodology for Functional, Information and Dynamic Modelling of a FMS", GIMT

[5] Cheng Leong Ang and Robert K.L.Gay, "IDEF0 modelling for project risk assessment", Computer in Industry 22 (1993) pp31-45

[6] Zhou Zhiying, Yang Jin and Hua Wei, "The CASE Tool for Supporting IDEF1X", Proceedings of ICCIM'93 Beijing, pp227-230

[7] Thomas C. Hartrum, Ted D. Connally, and Steven E. Johnson, "An Interactive Graphics Editor with Integrated Data Dictionary for IDEF0 Structured Analysis Diagrams", School of Engineering, Air Force Institute of Technology, OH 45433

[8] A.N.Godwin, J.W.Gleeson and D.Gwilliam, "The Integration of IDEF Descriptions", IMMS Group, Coventry Polytecnic

[9] Anthony N. Godwin, Joseph W. Gleeson and Dean Gwillian, "An Assessment of the IDEF Notations as Descriptive Tools", Journal of Information Systems Vol. 14, No. 1, pp13-28, 1989

[10] R. K. Maji, "Tools for development of information systems in CIM", Adv. Manuf. Eng. Vol 1 October 1988

[11] Francois Vernadat, "High-Level Modes for Organization and Information System Design of Manufacturing Enviroments. The New M* Approach", INRIA-Lorraine, Campus Scientifique, BP 239, Bd des Aiguillettes, 54506 Vandoeuvre-les-Nancy Cedex, France

[12] Victor E. Sanvido, Soundar Kumara and Inyong Ham,"A Top-Down Approach to Integrating the Building Process", Engineering with Computers 5, 91-103 (1989)

[13] U.S.Air Force, Integrated Computer-Aided Manufacturing (ICAM) Architecture Part II, Vol. IV - Function Modelling Manual (IDEF0), Air Force Materials Laboratory, Wright-Patterson AFB, Ohio 45433, AFWAL-TR-81-4023, June 1981a.

[14] U.S.Air Force, Integrated Computer-Aided Manufacturing (ICAM) Architecture Part II, Vol. V - Information Modelling Manual (IDEF1), Air Force Materials Laboratory,

Wright -Patterson AFB, Ohio 45433, AFWAL-TR-81-4023, June 1981b.

[15] T.J. Williams and etc., Architectures for integrating manufacturing activities and enterprises, Computers in Industry 24 (1994) 111-139

[16] An evaluation of CIM modelling constructs -- Evaluation report of constructs for views according to ENV 40 003, Computers in Industry 24 (1994) 159-236

8 BIOGRAPHY

Dr Lu LingZhi is a Research Fellow at the Gintic Institute of Manufacturing Technology, Singapore. His main research interests are CIM system design, IDEF methods, CAD/CAPP/CAM integration and manufacturing database. Before he joined Gintic he had ever held a few technology management positions (technical manager and directors) in companies of Singapore and China. He has directed CIM systems design for two companies. He has published over 20 papers on CIM, manufacturing database and expert system. He received his BSc(1983) and MEng(1986) from Beijing University of Aeronautics & Astronautics, and got his Ph.D.(1989) from Nanjing University of Aeronautics & Astronautics.

Dr Ang Cheng Leong is a senior research fellow at the Gintic Institute of Manufacturing Technology of the Nanyang Technological University in Singapopre. His qualifications include a BSc from the University of Singapore, an MSc from the University of Wales Institute of Science and Technology, and a PhD from the University of Nottingham. His current research interests include CIM planning, AMT justification, group technology, and modelling, design and analysis of manufacturing systems.

Prof. Robert K L Gay received his BEng, MEng, and PhD degrees from the University of Sheffield. He has many years of teaching and research experience. He joined Nanyang Technological University in Singapore as an Associate Professor in 1982 and is now the Research Director of CIM Division in Gintic Institute of Manufacturing Technology of the university. His current research interests include factory automation, AI applications and computer graphics.

40

A methodology for managing integration problems with CIMOSA

A.J.R. Zwegers, Eindhoven University of Technology
T.A.G. Gransier, TNO-TPD and Eindhoven University of Technology
PO Box 513, 5600 MB Eindhoven, The Netherlands
Tel. +31 40 474370, Fax +31 40 436492, Email azw@bdk.tue.nl

Abstract

The objective of this paper is to present a methodology for re-engineering processes in which the architectural framework CIMOSA supports an organisation in managing its integration difficulties. New market requirements demand new production strategies. Therefore, enterprises follow re-engineering processes and are faced with consequent integration problems. We investigated the re-engineering processes in three European enterprises that have applied CIMOSA in changing their manufacturing systems. Important factors in these re-engineering processes are the CIMOSA framework, the logical architecture, and the physical architecture. Appropriate usage of the CIMOSA modelling framework and the integrating infrastructure contributes to business integration and application integration. Although the CIMOSA framework has some gaps, the concepts behind it and the experiences gained show that it enables companies to manage integration difficulties.

Keywords

Enterprise integration, CIMOSA, Computer Integrating Manufacturing, Industrial automation

1 INTRODUCTION

Enterprises face integration problems due to necessary internal adaptations to cope with severe competition in the global marketplace. In order to survive at the global market, efficient operation and innovative management of change are essential. Heterogeneous manufacturing and information systems are of serious concern in improving the current enterprise operation. In addition, since companies try to meet competition and to exploit new opportunities, enterprise operations have to evolve continuously. Evolving enterprise operations create new heterogeneities of legacy and new technology systems, so that integration problems are amplified.

The objective of this paper is to present a methodology for re-engineering processes in which the architectural framework CIMOSA – Computer Integrated Manufacturing - Open System Architecture – supports an organisation in managing integration problems. CIMOSA claims to solve the above-mentioned problems (AMICE, 1993); it aims to offer support on two kinds of integration, namely 'business integration' and 'application integration'. Business integration is concerned with the coordination of operational and control processes in order to satisfy business objectives. Application integration, which affects the control of applications, means that cooperation between humans, machines and software programs has to be established. This paper shows how the use of CIMOSA supports an organisation in managing its integration difficulties.

Research was conducted in three industrial organisations throughout Europe, namely in the Greek aluminium foundry Elval, in the French car plant Renault, and in the German machine tool factory Traub. Results were obtained by document analyses and interviews with people from the three companies that applied CIMOSA from 1991 to 1994 in the context of the VOICE projects (Esprit project numbers 5510 and 6682). Detailed information about Elval, Renault and Traub is given by Arabatzis, Piérard and Schlotz respectively (VOICE, 1995).

This paper is organised as follows. In the next section, we present the concepts that play a role in an organisation's approach to the migration path from a current Computer Integrated Manufacturing (CIM) system to a future system. These concepts, namely a CIM system's logical and physical architecture, and CIMOSA, serve as a framework for the description of the system engineering activities. After this, a methodology for system engineering supported by the architectural framework CIMOSA is presented. Furthermore, we show how to accomplish enterprise integration during system engineering by means of CIMOSA. We conclude this paper with a discussion on our findings.

2   ENGINEERING APPROACH

When necessary, enterprises decide to change their manufacturing systems. Modifications to current CIM systems are required because of changing market requirements, business objectives, business requirements, manufacturing concepts or available technology. The decision to manufacture new products, might for example result in changes in the existing CIM system. The manufacturing systems that are in operation today will be re-engineered at some point in time to turn into next-generation manufacturing systems, so that current integrated systems become future heritage systems.

We look at a manufacturing system from a logical and physical point of view, i.e. we distinguish logical components and physical components. During the re-engineering process these components and their interfaces are defined. In other words, the re-engineering process involves the design of a logical and a physical architecture.

Elval, Renault and Traub applied the CIMOSA architectural framework in their re-engineering processes. Figure 1 visualises the approach of the three companies. They defined logical architectures by means of the CIMOSA modelling framework. The physical systems that provide enterprise operation according to the defined models, were designed with the help of the CIMOSA integrating infrastructure. Finally, the CIM systems were implemented and transferred into operation.

In the remainder of this section, the three concepts, namely the CIMOSA architectural framework, logical architectures and physical architectures, are presented in more detail. In the next section, we illustrate the roles of these concepts during a re-engineering process.

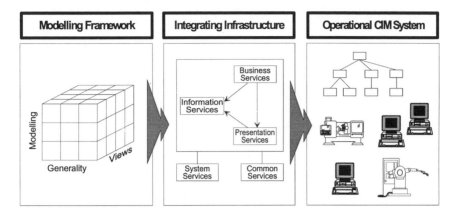

Figure 1 Engineering approach.

2.1 CIMOSA

The first concept is the CIMOSA architectural framework, which consists of the CIMOSA modelling framework and the integrating infrastructure.

CIMOSA modelling framework
In Figure 1, the modelling framework is represented by the CIMOSA cube. The cube allows to model different aspects and views of an enterprise. This three-dimensional framework has a dimension of generality, a dimension of enterprise models, and a dimension of views:
- The generality dimension is concerned with the degree of particularisation. It goes from generic building blocks to their aggregation into a model of a specific enterprise domain.
- The modelling dimension provides the modelling support for the system life cycle, starting from statements of requirements to a description of the system implementation.
- The view dimension offers the possibility to work with sub-models representing different aspects of the enterprise.

CIMOSA integrating infrastructure
The CIMOSA integrating infrastructure enables CIMOSA models to be executed; it allows the control and monitoring of enterprise operations as described in the models. Furthermore, it provides a unifying software platform to achieve integration of heterogeneous hardware and software components of the CIM system.

The integrating infrastructure is made of a number of system-wide, generic services. The business services control the enterprise operations according to the model. The information services provide for data access, data integration and data manipulation, and the presentation services act as a standardised interface to humans, machines, and software applications. A product that is connected to the presentation services, can be attached and removed without changing any other part of the information technology environment. Other services are the common services and the system management services. In Figure 1 the integrating infrastructure and the mentioned services are shown.

2.2 Logical architecture

The purpose of an architecture is to structure a complex system. It structures a system into modules and it determines the modules' interfaces. The logical architecture defines the functions of system components and their interfaces. Functions are assigned to components, allowed inputs and outputs are defined, and the relations between components are specified.

The importance of a logical architecture is derived from its relationship with the manufacturing system. The architecture determines how functions are distributed and coordinated. By fixing the functionality of the CIM system, a logical architecture determines the effectiveness of the system. However, changing market requirements make new, higher demands upon enterprises, which have to adjust their manufacturing systems according to the new objectives. Besides determining the system's effectiveness, the logical architecture is perhaps even more important because it establishes the limitations or possibilities for changing the system's functionality and effectiveness in the future (Dilts, 1991).

2.3 Physical architecture

The third concept is the physical architecture of a manufacturing system. Whereas the logical architecture defines component functions and their interfaces, the physical architecture defines physical components and their interactions. Examples of physical components are concrete objects such as computers, programmable logic controllers, networks, and manufacturing resources, but also nontangible objects such as software and NC-programs.

The design of a physical architecture is influenced by the evolving technology. On one hand, the technology opens opportunities for system development; on the other hand, system architectures have to be able to take the future opportunities of evolving technology into account. Future information and manufacturing technology components have to be integrated into the current system, since new objectives might result in the application of new technology. Both the logical and the physical architecture must provide flexibility to the system, so that the system can be adjusted to the new technology components.

3   SYSTEM ENGINEERING

In this section, a methodology for re-engineering processes is presented, based on the experiences gained by the three industrial organisations Elval, Renault and Traub during their re-engineering processes. For a more extended description the reader is referred to Zwegers (VOICE, 1995). In the next section, we focus on enterprise integration during system engineering.

The phases of the re-engineering process are schematically shown in Figure 2 and are described in more detail in the following subsections.

3.1 Determination of objectives

Enterprises change their production systems driven by certain objectives. At a specific moment, an organisation is dissatisfied with its current manufacturing system. Then, new objectives are determined for (part of) the system. These objectives are derived from the company's business strategy and are the starting points for system development.

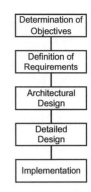

3.2 Definition of requirements

In the previous phase, the decision is taken to change the current CIM system. This phase is concerned with the definition of requirements for the future system.

Requirements are defined in a logical sense for the application to be developed, and in a technical sense for the application's underlying infrastructure. In other words, requirements are defined for both the logical and physical components of the manufacturing system. For the logical components, requirements are described in terms of a scenario. In this scenario, the needed functions and

Figure 2 System engineering phases.

information flows, and their place in the total manufacturing system are outlined. For the physical components, technical requirements are given. Besides the users' current needs, requirements take into consideration the existing manufacturing system, strategic aspects like standards, the factory environment and production process constraints.

3.3 Architectural design

After requirements are defined, the design of the CIM system commences. We have distributed the design activities over two phases: architectural design and detailed design. In the architectural design phase, the system's logical and physical architecture are defined, supported by the CIMOSA framework. In the detailed design phase, the system is worked out in more detail, based on the specified architectures.

Before the logical and physical architecture of the future system are specified, the existing manufacturing system is analysed. Knowledge acquisition can be carried out by modelling the current manufacturing system using the CIMOSA modelling framework. The requirements definition level and, if more detailed information is required, the design specification level allow to obtain the necessary information for the analysis of the current CIM system.

The CIMOSA modelling framework supports a user in defining his CIM system's logical architecture. The (particular) requirements definition level of the modelling framework allows the user to model the functional scenario description, that was defined previously. By modelling, the user structures the system component functions, defines the components' allowed inputs and outputs, and specifies the relations between these components. In other words, by modelling at requirements definition level a logical architecture is designed.

Along with defining a logical architecture, the designer also outlines a physical architecture, whether or not influenced by the existing infrastructure. When defining logical components, one immediately maps these components on physical ones; the logical architecture is depicted on the physical architecture. For instance, when the designer defines a control function, he might decide to execute this function on a workstation. In addition, the designer determines the interaction of this component with other physical components.

3.4 Detailed design

In the second design phase, the CIM system is specified in more detail, elaborating on the architectures that were defined in the architectural design phase. Both architecture specifications are further decomposed and worked out into detailed system designs.

By means of the CIMOSA design specification modelling level, a designer is able to detail a manufacturing system's functionality, taking the model at requirements level as starting point. The requirements definition level supports a user in the definition of his CIM system's logical architecture, whereas the role of the design level is to restructure, detail and optimise the required functionality in a consistent model. System optimisation can be supported by simulation, taking all business and technical constraints into account. By the specification of a model at design level, a designer describes the full functionality of a CIM system, while staying within the specification of the logical architecture. If the model at design specification level reveals inconsistencies, or the model lacks optimisation, it might be necessary to adjust the model at requirements definition level.

In addition to detailing the functionality of the system, the designer specifies the technology to be employed in order to achieve the required system functionality. Simply stated, CIMOSA prescribes that for each of the most detailed specified functions, called 'functional operations', a resource is assigned that provides the required capabilities. The prime task of the design specification modelling level is to establish a set of (logical) resources that together provide the total set of required capabilities. Some of these required capabilities are offered by the generic services of the integrating infrastructure.

A refinement of specified functions is accompanied by a refinement of the physical resources that provide the desired functions. An enterprise looks for adequate products, either commercial ones or user developed. The physical components, which are defined in the architectural design phase, are specified completely. For this specification, the definitions of the CIMOSA integrating infrastructure are used. Then, the products that fulfil the design specifications are selected, considering enterprise policies and constraints. CIMOSA states that the final build/buy decisions should result in a model that describes the implemented, physical system. Appropriately, CIMOSA calls this model the implementation description model. However, since this part of the modelling framework has been defined recently, we do not have any experience with it.

3.5 Implementation

The final phase constitutes of the implementation and release for operation of the specified CIM system. Implementation is based upon the results and decisions of the previous phases.

Implementation activities concern those tasks needed to bring the system into operation. During the implementation phase, the necessary new physical components are procured or built. These components are tested on a testbed and the correctness of the underlying logical model is verified. When the system passes the tests with satisfactory results, the system is prepared for transfer to production. Essential activities such as operator training and acceptance testing are carried out before transfer to production. Finally, the accepted CIM system is released for operation, after which it manufactures products until it is replaced by the next CIM system.

4   ENTERPRISE INTEGRATION

In the previous section, a methodology for system engineering is given, in which the role of the CIMOSA architectural framework is illustrated. Based on the experiences gained, we indicate in this section how enterprise integration can be achieved by means of CIMOSA. We proceed with the distinction in enterprise integration as presented in the introduction, particularly business integration and application integration.

4.1   Business integration

The CIMOSA models allow the user to design logical architectures. During the architectural design phase, the CIMOSA user is able to acquire knowledge of his current CIM system by means of modelling at the requirements definition level. This knowledge is needed in order to analyse the existing system. Then, the model is changed to take required modifications into account. By means of the specification of components that represent operational and control functions, the inputs and outputs of components, and interfaces between components, a logical architecture is defined.

The definition of a sound logical architecture, upon which the further design of the system is based, is the key to business integration. CIMOSA takes a modular approach to integration, i.e. enterprise operation is modelled as a set of cooperating processes which exchange results and requests. Only this exchanged data needs to have a representation that is common between the cooperating processes. In addition, CIMOSA provides the means to manage processes that are inherited from the past. By employing models for identification, analysis and coordination of existing and new functions, required logical architectures can be designed that define modular components and their interfaces. When the design and implementation of the CIM system follows the architecture specification, the requested integration of business processes will be obtained. In this way, the required level of business integration is achieved, which is reflected by the coordination of operational and control processes that fulfil the business objectives.

4.2   Application integration

The other aspect of enterprise integration, application integration, is concerned with the usage of information technology to provide interoperation between manufacturing resources. Whereas business integration can be regarded as the integration of logical components, application integration affects the integration of physical components.

Application integration is supported by the CIMOSA integrating infrastructure. Cooperation between humans, software programs, and machines has to be established. The integrating infrastructure provides services to integrate the enterprise's heterogeneous manufacturing and information systems in order to satisfy the business needs as identified with the help of the modelling framework. The CIMOSA specification of the integrating infrastructure can be seen as a reference model. Physical components that are designed and implemented according to this reference model provide desired features such as interoperability, portability, connectivity, and transparency.

5   DISCUSSION

Although the engineering process is presented as a sequence of phases, this does not imply that it follows a waterfall model. In contrary, designing CIM systems is an iterative process, with frequent jumps between the phases. Iterations are necessary because of increased insight, unexpected events, etc. Therefore, the presented methodology should be considered as a presentation of advisable activities, and not as a prescription of mandatory phases.

In addition, jumps between the logical (or functional) domain and the physical domain occur. The specification of both logical and physical components during an engineering process go hand in hand. A refinement of logical components is accompanied by a refinement of the physical components. Figure 3 presents design as the refinement process that interlinks the logical and the physical domain. Note that this observation is confirmed by Suh's theory on designing products (Suh, 1990).

Although the CIMOSA framework has some gaps, it enables companies to manage integration difficulties. The CIMOSA modelling framework is not complete; for instance, the three companies did not have the definition of the implementation description modelling level at their disposal. Furthermore, the integrating infrastructure was not specified completely. However, CIMOSA suffers most from the lack of commercial products that fulfil the CIMOSA (integrating infrastructure) specifications. Nevertheless, the concepts behind it and the positive experiences gained by the three industrial organisations show that CIMOSA provides valuable contributions to managing integration problems.

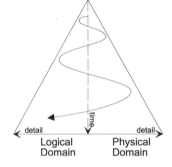

Figure 3 The design process.

6   REFERENCES

AMICE Consortium (1993) CIMOSA: Open System Architecture for CIM. Springer-Verlag, Berlin.

Dilts, D.M., Boyd, N.P. and Whorms, H.H. (1991) The evolution of control architectures for automated manufacturing systems. *Journal of Manufacturing Systems*, vol. 10, no. 1, 79–93.

Suh, N.P. (1990) The principles of design. Oxford University Press, New York.

VOICE Consortium (1995), Special issue on application and validation of CIMOSA. *Computers in Industry*, to be published.

7   BIOGRAPHY

Arian Zwegers has been working on the Esprit project VOICE II as a Ph.D. student from the Eindhoven University of Technology. His research interests include architectures in general, and control architectures in particular.

Theo Gransier has been working on both VOICE projects. First, he was responsible for the technical coordination of the modelling part. From 1993, he was project manager of the VOICE II project. His main research interests are distributed control architectures.

41

Modeling of virtual manufacturing devices for machining data generation

M. Matsuda [*] *and F. Kimura* [**]

[*]*Faculty of Management and Informatics, Sanno Institute of Management*
1573 Kamikasuya, Isehara, Kanagawa 259-11, Japan,
Telephone:81-463-92-2211, Fax:81-463-93-0554,
E-mail:matsuda@mi.sanno.ac.jp

[**]*Department of Precision Machinery Engineering, Faculty of Engineering, The University of Tokyo*
7-3-1 Hongo, Bunkyo-ku, Tokyo 113, Japan,
Telephone:81-3-3812-2111, Fax:81-3-3812-8849,
E-mail:kimura@cim.pe.u-tokyo.ac.jp

Abstract

The process of generating machining data is based on the translation of product description data into machine control commands. In other words, computer-manufactured machining data is produced from input product data using manufacturing environment data; a process known as virtual manufacturing. The use of manufacturing environment data is described here as virtual manufacturing devices (VMD). In this paper, in order to achieve accurate automatic machining preparation, a machining data generation system based on the above virtual manufacturing concept is proposed. VMD is the most important component in this machining data generation system. The general framework of VMD consists of a component model and a process model. As an example of VMD, modeling of milling machine is explained, and milling data generation system using this virtual milling machine is introduced.

Keywords

virtual manufacturing, modeling of manufacturing environment, process/operation planning, machining preparation, NC data generation

1 INTRODUCTION

Traditionally, the CAM system generates NC data, while the simulation system calculates cutter paths to find tool interference and to inform cutting time. The NC programming system also supports machining preparation and development of NC data. Recently, the demand for a total integrated manufacturing system based on product model that encompasses all stages, from design to production, is increasing in order to accurately produce a high quality product.

A highly accurate product machining system will be realized through the development of a consistent automatic machining system based on a given product model. This system will operate through all stages, from machining preparation to the commencement of actual machining. Recently, there have been two lines of research which are leading to the basic concept of the total machining system described above. One is the automated milling data generation system with an extraction function for machining features that is based on a product description with design features (Matsuda, 1991). This system revealed that the machining data generation process is a translation process — translating product description with design feature into an NC machine program that corresponds with the machining features. These machining features are extracted by applying machining constraints, such as tool diameter, to given design features. The other line of research culminated in an operation planning system, in which tool paths are modified in order to keep cutting force constant and to attain consistent machining accuracy based on a cutting process model (Tsai, 1991). This system showed that increasing machining constraints give rise to more optimum and precise machining data.

In this paper, virtual manufacturing is discussed as a generally applied field of machining constraints for translating product data to machining data. Following on from this,*Virtual Manufacturing Device (VMD)* is introduced as primary component of the virtual manufacturing system. Finally, the structure of a milling machine model and milling data generation system using this milling machine model are discussed as one application of VMD for virtual manufacturing systems.

2 CONCEPT OF VIRTUAL MANUFACTURING

2.1 Machining data generation process based on a product model

In order to produce actual mechanical parts, it is necessary to generate an NC machine program based on the product model created at the design stage. The product designer usually describes the product using design features. An NC data generation system usually calculates the cutter paths based on such input data as machining features and tools. The product model created at the design stage should be provided as input data to the machining system. The machining system should extract machining features and plan operations such as the selection of tools, determination of cutter speed, and calculation of cutter paths. In a machining system, this means product data with design features is translated into product data with machining features, and then into such machining data as machine control programs.

These translation processes are facilitated by applying constraints on machining factors such as machining methods, machine specifications, and machine functions. For example, the size of the tool diameter is the primary constraint in the above mentioned milling data generation

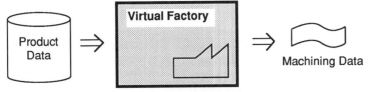

Figure 1 Concept of virtual manufacturing.

system (Matsuda, 1991), and the cutting force is main constraint in the above mentioned operation planning system (Tsai, 1991). These examples also show that the level of optimization and accuracy of machining data are proportionate to the number of machining constraints imposed and the accuracy of these constraints. The development of a translation system involves the implementation of mechanisms that apply machining constraints to product data. In other words, to realize this system it is necessary to build a virtual manufacturing factory within a computer and carry out machining in this virtual factory as shown in Figure 1.

2.2 Product data and machining data

Table 1 shows the relationship between product data, generated machining data and machining

Table 1 Product data, machining data and machining environment data

Machining Data	Product Data	Machining Environment Data	
		Machine Spec.	Simulation
process plan, method of fixing, type of tools	overall shape, form feature: (shape, dimension, location, accuracy), relationship between form features work-piece: shape	freedom of machine axes, jig/fixture, available tools	machining order, cutting time (rough)
tool size, tool paths, pick feed, depth of cut, cutting speed, feed rate	machining feature: (shape, dimension, location, tolerance) finishing accuracy, work-piece: material	tool diameter, tool shape, tool material, tool life, chuck, max. of cutting torque, max. of cutting depth	cutting time, tool interference, cutting accuracy (cutting force, chatter, etc.)
NC data		NC format, tool number, tool offset	trial/test

environment data. Machining environment data is divided into static data, such as machine specification, and dynamic condition data, which is obtained as a result of simulation. In Table 1, machining data has three levels. The generated machine data becomes more precise as the data moves down through the three levels. The top level is the process plan. The middle level is the operation plan, and the bottom level is NC data. Each machining data is decided based on the contents of the product data and environment data. Table 1 shows the exact relation between the generated items and their corresponding decision factors.

3   VIRTUAL MANUFACTURING SYSTEM FOR MACHINING DATA GENERATION

A product model is input in the virtual manufacturing factory, which results in the output of machine control programs. Figure 2 shows the structure of the machining system in the virtual factory. The product model description is translated into physical machine control commands by means of the control VMD in the virtual factory. Each VMD is linked to a physical manufacturing device (PMD) in the actual factory through a one-to-one mapping system.

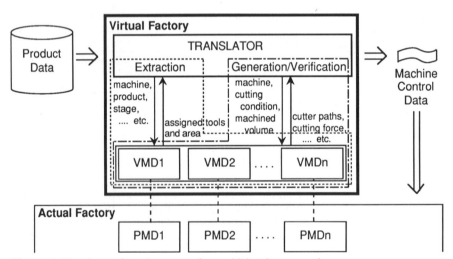

Figure 2 Virtual manufacturing system for machining data generation.

The translation procedure consists of four steps. The first step is the extraction of machining features through preliminary operation of the appropriate VMD. The second step is the simulation of machining by precise operation of VMD. The final steps are the evaluation and verification of generated machining data using the results of simulation.

4 STRUCTURE OF VMD

4.1 Concept of VMD

VMD has been used as the conceptual control object in ISO standard: Manufacturing Message Specification (MMS). VMD also has been discussed in the context of new programming environments for automated factories. These factories will utilize VMD as controlled objects which represent abstracted devices, such as robots, NC machines, automatic guided vehicles, and storage systems (Takata, 1993). In this programming environment, manufacturing programs would usually be developed to control VMD. Their abstracted operation commands for VMD would be translated into appropriate commands according to the protocol of each device through the class library system of PMD.

In this paper, the concept of VMD is extended so that VMD becomes the component machine which carries out manufacturing in a virtual factory. VMD is a computer model which represents the specification, function and behavior of PMD. VMD is the controlled object in the virtual factory.

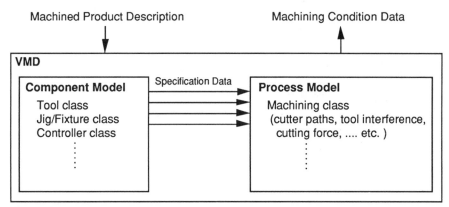

Figure 3 Conceptual structure of VMD.

4.2 Modeling of VMD

Generally, VMD is composed of two models: a component model and a process model, as shown in Figure 3. The component model is the description of the static elements of the machine, such as machine specifications, tools, jigs or fixtures, and controller. The process model is the description of the actual operation by means of simulation, such as cutter location, cutting force, cutting accuracy and error sensing. Each model has several variable classes, and through the combination of these classes, flexible VMD modeling can be realized.

5 MILLING DATA GENERATION SYSTEM

5.1   Modeling of milling machine

A milling machine is one type of NC machine. Pockets or cavities, including free surfaces, are machined by a milling machine. Therefore, a milling machine is an appropriate example of the application of VMD. In Figure 4, the structure of a virtual milling machine is simplified, such that its component is limited to tool class only and its process is limited to the milling class. This model is implemented using object oriented programming language C++.

The milling process object is organized into a hierarchy of process objects from the process for one tool at respective depths to the process for a product. The function of the lower level process is inherited by the upper level process. Each level process can be accessed depending on user requests.

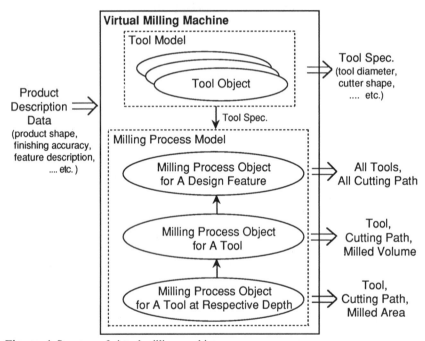

Figure 4 Structure of virtual milling machine.

5.2   Milling data generation system using virtual milling machine

Trial system for milling data generation using the above mentioned virtual milling machine is being developed as shown in Figure 5. This system is structured on the solid modeling system DESIGNBASE which is supplied by Ricoh Co., and is being developed using the object oriented programming language C++.

A pocket/cavity which is designed as a design feature at the product design stage is indicated by operator. The milling data generation system selects the tools to be used, extracts the volume to be milled by each tool, calculates the cutter paths for each volume, and generates G-code data. At every process the milling data generation system operates the virtual milling machine by means of the milling machine model.

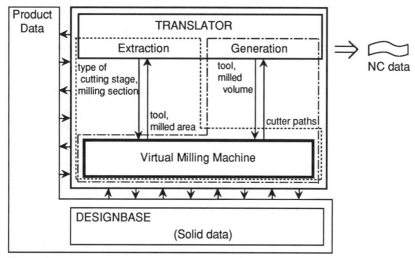

Figure 5   Structure of milling data generation system.

6 CONCLUSION

In this paper, the organization of a virtual manufacturing factory in a computer is proposed as the basis for a consistent machining system. The virtual manufacturing factory is a mechanism for generation of optimum machining data. VMD is the most important component of a virtual manufacturing factory. Virtual milling machine modeling by object oriented programming is given as an example of VMD. This virtual milling machine is effectively used as the main component of a milling data generation system.

7 REFERENCES

Matsuda, M. and Kimura, F. (1991) Extraction of machining features for milling data generation: Computer Application in production and Engineering CAPE'91. IFIP, 353-60.

Takata, M. (1993) A Programming Environment for Factory Automation: Proc. of Symposium on Manufacturing Application Programming Language Environments MAPLE'93, 215-24.

Tsai, M. D., Takata, S., Inui, M., Kimura, F., and Sata, T. (1991) Operation Planning Based on Cutting Process Model: Annals of CIRP, Vol.40/1.

8 BIOGRAPHY

M. Matsuda is an associate professor in the Department of Computer Science of Sanno Institute of Management. She received a Dr. Eng. degree in precision machinery engineering from the University of Tokyo in 1989. She has been active in the field of CAD/CAM, process/operation planning, NC programming and numerical control. Her recent research interests include product modeling for manufacturing and virtual manufacturing. She is a member of Japan Society of Precision Engineering.

F. Kimura is a professor in the Department of Precision Machinery Engineering of the University of Tokyo. He received a Dr. Eng. Sci. degree in aeronautics from the University of Tokyo in 1974. He has been active in the field of solid modeling, free-form surface modeling and product modeling. His research interests now include the basic theory of CAD/CAM and CIM, concurrent engineering, engineering simulation and virtual manufacturing. He is a member of IFIP WG5.2 and 5.3, and a corresponding member of CIRP.

42

Services and Tools for Group Editing
(An Extended Message Handling System)

B. Buksh,
Manager(Programming),
Institute of Drilling Tech.,
Oil & Natural Gas Corpn. Ltd,
Dehra Dun- 248001,
INDIA.
Ph. +91 135 27102 X5394,
Fax +91 135 28926.

B.S Doherty,
Lecturer, CSAM Department,
Aston University,
Birmingham, B4 7ET
United Kingdom.

+44 21 3593611
+44 21 3336215

A.P.H Jordan
Head of Computing Sevices,
University of Huddersfield,
Queensgate, Huddersfield,
HD1 3DH, U.K.
United Kingdom.
+44 484 420246
+44 484 516151

Abstract

A variety of group communications has been developed, and over recent years there has been increasing standardisation in communications to improve interconnectivity and provide a common flexible infrastructure in a wide spread communication network. This infrastructure is provided within the Open System Interconnection (OSI) environment. Within OSI the X.400 series of recommendations provide standard protocols and message formats that allow exchange of messages on a global basis.

Modelling of group activities within three application groups provide three sets of services and tools for a wide range of group communication activitities. The services and tools identified for group editing activities may be viewed in an OSI environment by defining a separate application type.

Keywords

Group communication, group editing, Informative, Supportive, Objective application group, computer supported cooperative work, SAGE project, information model, message handling system and directory services.

1   INTRODUCTION

Information technology has in recent years seen a move towards individual work-stations linked by networks. These networks have spread to allow global communications. The improved communication capability has been exploited in many ways. But recent development has been Computer Supported Co-operative Working (Taylor, 1990).

CSCW provides the computing environment that allows several people to work co-operatively on a common task (Wilson, 1991, and Cross, 1989). The growth of CSCW has led to development of tools for co-ordinating group activities. Some tools are available but do not fully address the need of particular group of people working cooperatively.

This paper gives a brief overview of the work of the SAGE project (Services for Activities in Group Editing), which sought to investigate the needs of a group of people working co-operatively and communicating in an OSI network environment, and to recommend services and tools to meet these needs. The work focussed specifically on the services and tools required for group editing activities. It has resulted in the proposal for a functionally based classification of group activities and recommendations to provide the OSI based services and tools by extending X.400 protocol.

The paper aims at defining a common messaging protocol on a specific area of group working (i.e. Group editing). The protocol is based on the X.400 series of recommendations, in particular the X.420 recommendation. The project sought to identify services and tools for management, storage and distribution in support of group editing activities. The work follows the direction proposed by the CCITT draft working document on group communication X.gc (Joint ISO/IEC/CCITT, 1991) and AMIGO report (Smith et.al. 1989).

The case study approach is used to identify user requirements and to determine common functionalities for a variety of group editing activities. To support the study, a prototype tool set has been implemented to provide the functions and services identified as needed to support group editing. This prototype has been implemented in the X.400 environment and was used both to confirm that the service definitions of the tools are workable and to provide a testbed for user feedback which improves the information obtained from interacting users. A prototype on a X.400 environment helped to refine user requirements as a source of new ideas and to test the proposed functionalities.

The literature shows that a set of basic requirements have been identified for a framework for a group communication base model. The 'SAGE' project uses the X.400 services and the definition of a basic group communication information model defined in the draft X.gc recommendations (Joint ISO/IEC/CCITT, 1991). The project adopts the same definition of terms as the X.gc working document. Terms not defined in the X.gc recommendations are defined for the this project.

2   THE 'SAGE' PROJECT AND THE OSI MODEL

The Application Layer is the highest layer in the OSI reference model and it can be considered as the layer in which all application processes reside. The work of the 'SAGE' project is in the application layer, and is particularly concerned with the X.400 (i.e. message handling system) recommendations. The position of the 'SAGE' project within the application layer is shown in figure 1.

X.400 and X.500 recommendations enable interpersonal electronic mail and distribution list in a geographical distributed environment. The services of X.400 recommendations from a message handling environment are exist within the application layer. The X.500 (Directory Services) are also a part of the services provided the application layer as a standard. Both X.400 and X.500 services do not support group activities such as co-operative writing, editing of periodical newletters, or meeting scheduling, which are based on the messaging systems.

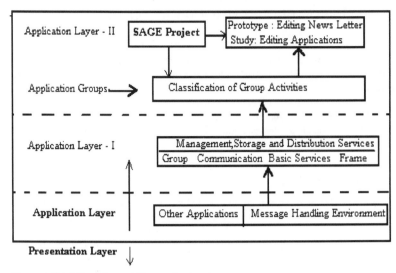

Figure 1 'SAGE' project within application layer

3 THE GROUP COMMUNICATION SYSTEM

The 'SAGE' project follows the group communication architecture model described by Weiss & Bogen (Weiss & Bogen, 1989) in which the arctitecture model consists of information objects (e.g. a contribution) and the communication entities (e.g. group user) which form an *Information Model*. The functionality and services provided by the group communication system are defined in terms of the operations (e.g. create, delete and submit) in order to provide the underlying functionality to support a particular kind of group communication process. These services are followed by the related support services (e.g. directory and message handling services).

The configuration of the group communication system is dynamic (Weiss & Bogen, 1989). For example a service agent is not permanently bound to one group or one activity. The information model comprises the group communication environment (i.e. entities, domains and information objects). The information model represents both information and members of the group as objects (Joint ISO/IEC/CCITT, 1991). The object classes in the model arranged into a clsss hierarchy. Each group communication activity may define its own specific object classes, derived from the standard set of base classes. The information model identifies three base classes: group communication item, group communication entities and group communication domain X.gc (Joint ISO/IEC/CCITT, 1991).

The services offered by group communication system are categorised into basic services and advanced services: basic services realized within group communication system, and may be viewed as common services. Advanced services are intended for handling some more detailed activity procedures for examle: voting and group editing. The basic services are made-up of a distribution service, an archive service and a co-ordination service for which access is made by special communication support system (Weiss & Bogen, 1989).

4   MODELLING OF GROUP COMMUNICATION ACTIVITIES

There are two main approaches described in the literarture. The models by Klehn & Kuna (Klehn & Kuna, 1991) are based on co-operation, organisation and environment. Jakobs (Jakobs, 1991) introduces Static, Semi-Static and Dynamic group classes for group communication. These classes or models are based on the nature of a group (co-operative group, organisational group, static group and dynamic group etc.), but these groups are not discussed in the light of group communication services and tools under the sub-group or class. Therefore, there is a need to describe the classes of group activities based on their functionalities to identify the services and tools for a set of class.

4.1   New Modelling Approach

The 'SAGE' project takes a different view on classifying the groups. It does not classify the group users. Instead, it classifies the group activities depending on their functinalities. This means a group has to be formed by the activities of similar nature: for example a set of activities witch distributes information (bulletin board). These classes would then be able to provide the services and tools needed by each class of activities in the group communication environment.

The concept of dividing the group communication system into a base model and a set of activities (Benford & Palme, 1993) has been used in this project. The classification of group activities proposed by the 'SAGE' project takes a similar approach to Benford & Palme to define services and tools for a set of activities.

The 'SAGE' project has introduced three classes of application group, for each of which there is a set of services and tools which meets the common functional requirement of the class. The classes are:
- '**Informative**' application group,
- '**Objective**' application group and
- '**Supportive**' application group.

Each category of the application group identifies the activities with similar functionalities. The Information application group category identifies activities which distribute information and sometime need replies (e.g. Bulletin board, Distance learning and Distrbution list). The Objective application group identifies those activities which need some kind of group decision making procedures such as voting and executive decisions. The Supportive application group covers cases where group of people work together on a single pool to complete a certain task and need supporting work within the group (e.g. cooperative writing and software development team).

The conceptual modelling follows current CCITT proposals, and each of these application groups have their own service agent. For example the services for supportive application group has to be provide by 'Supportive Application Group Agent' (SAGA). The agent should be able to meet group communication requirement of each application group in-terms of basic, advanced and information handling services.

Use of this classification allows the possibility of developing three sets of tools which would cover a wide range of group activities, rather than developing tools for individual activities. Group editing is concerned to be in the supportive application group.   The Group

Communication Services in terms of basic services, advanced services and rule based information handling services would be met by the supportive application group agent. The services of a group agent is not bound to any group or activity concerned. It is for the entire range of the class.

The proposed application groups that is Informative, Objective and Supportive application groups need to be implemented as independent application types (within X.420). These applications has to be identified through separate application identifiers for each group like inter-personal messaging and electronic document interchange system in the OSI environment.

5   PROPOSED MODEL FOR GROUP EDITING ACTIVITIES

This section describes a common model for group editing activities considering the results of three case studies (a newsletter, a technical paper and a software development team), prototype and user comments.

An editing environment can be modelled as a group of functional object. These objects may be identified as contributions, group editing system and group members. The group editing services would be met by a supportive application group agent. The group editing environment consiting of all its components is shown in figure 2, witch is a part of the supportive application group.

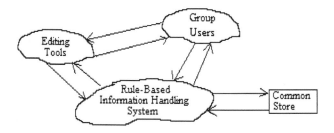

Figure 2 Group Editing Environment

The editing services and tool set is consists of about seventeen operastions on which group members (author, editor and reader etc.) have different type of access (read, write and modify) in relation to the nature of the member. The access control module grants permission to determine which user can perform which functional operations. The reason being, there is a requirement of privacy within the environment where information is shared between users and a requirement of integraty of information.

Every event of the activity has to pass through a Rule-Based Information Handling system which is capable of eximine the operation and take necessary action accordingly. Common Store is a place where group members have access of permission, granted on the contributions in the common store.

The operations provided by the 'SAGE' project is an extention from CCITT draft X.gc and the AMIGO report (Smith et.al., 1989) which provides a set of group editing services and tools. The group members would invoke these operations by means of three ports as discussed in the AMIGO report (Administration port, Retrieve port and Store port).

5.1   The Group Editing Services

A set of additional services and tools to support group editing are proposed in the context of the CCITT draft on group communication X.gc. The proposed services and tools are mapped onto the X.400 series of recommendations, with the Abstract Service Definition of the operational objects defined, along with their associated component files, by extending the X.420 protocol functionality. The work defines a set Abstract Service Notation One (ASN.1) for seventeen group editing services. The services inlude register activity and member, updatedocument, editrequest, editsuggestion, editreminder, acceptcontribution and editnotification etc. (Buksh, 1993).

The recommendations are proposed in the style of CCITT and are comprised of Abstract Service Definitions mapping onto the message handling system, and functionality extensions to the X.420 protocol.

The recommendations style are discussed in the light of global naming structure and represented by ASN.1 (abstract service notation one) object identifiers (Joint ISO/IEC/CCITT,1991). The abstract service definitions are given for the operations which are additional to the CCITT draft X.gc (Joint ISO/IEC/CCIT, 1991) for group editing activities. The mapping and X.420 protocol functionality extension is proposed in terms of component files (header components) readily available for editing operations. The support requirement for a group edititng environment is split into two parts: the operational part of the functions and the rule based information handling part. The set of abstract operations which access and manipulate group communication editing objects are defined in turn. The conceptual group communication operations are specified in terms of outline functions in the form (Benford et.al, 1990):

function name (parameters) --> Results

The operations are specified using OSI Abstract Service Definition conventions. The abstract operations should return status information indicating their success or failure as a part of the operation results. The operations must be given distinguished names to identify objects. The naming of objects is a critical for large distributed systems (Benford et.al., 1990). All objects are accessed by names provided them. Therefore each object must have at least one globally unique name. Group Communication Distinguished Names (DNs) provide the handle by which all objects accessed (Benford & Palme, 1993). An object may be identified by more then one name (e.g. aliases).

The examples of activity definitions are as given below:

Create-Activity (DN, domainDN, attributelist) --> attributelist, status

Creates editing activity with the given Distringuished Name and attribute in the Distringuished domain. The object should be created and the name of the common store is to be determined from its name. The object type identifies the class of object. The object can subsequently be removed from its distinguished domain.

Register-Group Member (DN, clusterDN, attributelist) --> status

The Abstract Service Notation (ASN.1) form of the above operation may be given as:

```
--PORTS        (within admin, retrieve and store)
--port types
       sage-admin PORT
              CONSUMER INVOKES       {
              CreateActivity,        }
       ::= id-pt-sage-admin

--Abstract Unbind Operation
CreateActivity ABSTRACT-OPERATION
              ARGUMENT        Craete-Argument
              RESULT          NULL
              ERRORS          {accessControlError,objectError,schemaViolationError,
                              attributeError}
       ::=2

Create-argument::= SET {
       name          [0]    DistingusedName,
       domainName    [1]    ObjectName,
       attributes    [2]    Attributes }

--AbstractErrors
accessControlError ABSTRACT-ERROR
       PARAMETER                   AccessControlError
       ::=0
--errors
AccessControlError ::={readOnly,readWriteOnly,writeOnly,notAccessible,
                       permissionDenied }
```

The other parameters and definitions for results and errors may also be defined likewise.

6   CONCLUSION

The OSI environment combines the multi-vendor systems for different users. The distrbuted application i.e. message handling system integrates multi-media user applications and is common carrier for electronic messaging systems.

The proposed application groups that is Informative, Objective and Supportive application groups should be implemented as independent application types (within X.420). These applications has to be identified through separate application identifiers for each group like inter-personal messaging and electronic document interchange systems in the OSI environment.

The three classes of application groups perhaps provides the three base models for Informative, Objective and Supportive application groups to develop three sets of services and tools for a wide range of group communication activities. Such services should contain the basic, advanced and rule-based information handling services (Buksh, 1993). There are still some issues to be discussed in detail, such as; complexity of the group service agent, categorisation of group activities into three application groups, investigation of common functionality within each group, definition of services and tools for each application group, extention of the CCITT framework model for the application groups and modelling of activities onto the framework model. There are many other issues. But this is the right time to start with standardising the group activities considering message handling system as a common carrier.

7   REFERENCES

Benford, S., Howidy, H., Shepherd, A. and Smith, H. (1990) Communication in Open System Environment: The GRACE Project Phase I Report, Nottingham University, U.K.

Benford, S. and Palme, J. (1993) A Standard for OSI Group Communication: Computers Networks and ISDN Systems, Vol./Part 14(5-6) 933-46.

Buksh, B., (1993), Services for Activities in Group Editing, Ph.D. Thesis, Computer Science and Applied Mathematics Dept., The University of Aston in Birmingham, U.K.

Cross, J.A. (1989) Technology Transfer in the 1990's:Evaluating Computer Support for Co-operative Work, Computer Trends in 1990's 17th Annual ACM Computer Science Conference, New York, ACM Press 474.

Jakobs, K.(1991) , Beyond the interface Group Communication Services Supporting CSCW, Support functionality in office Environment (ed. A.A. Venijan-Stuart, H.G. Sol & P. Hammersley), Elsevier Science Publishers B.V. North Holland.

Joint ISO/IEC/CCITT, (1991), Group Communications: A Working Document X.gc, Version 7 CCITT Study Group I/Q.15, VII/Q.18 and ISO/IEC Jtc 1/SC 18/WG 4 period 1989-1992 MH-534.

Klehn, N. and Kuna, M.(1991) A three Level Approach to Model Systems for Computer Supported Co-operative Work, Informatik, Information Report Vol/ Part-4 222-31.

Smith, H., Onions, J. and Benford, S. (ed.) (1989) Distributed Group Communication the AMIGO Information Model, Ellis Horwood Ltd. Publishers Chicherter.

Taylor, J.M.(1990), Co-operative Computing and Control, IEE Proceedings, Vol. 137 Pt.E. No. 1 U.K. 1-16.

Weiss, K-H and Bogen, M. (1989), A Computer Communication Service Architecture (ed. Smith, H., Onions, J. and Benford, S.) Distributed Group Communication the AMIGO Information Model, Ellis Horwood Ltd. Publishers Chicherter.

Wilson, P. (1991), Computer Supported Co-operative Work, Great Britain, Intellect Books Publishers.

8   BIBLIOGRAPHY

CCITT/ISO Recommendations, (1988), X.400: Data Communication Networks- Massage Handling Systems X.400-X.420, Volume VIII (Blue Book).

Palme, J. (1987), "Distributed Computer Conferencing", Computer Networks and ISDN systems, Vol./part 14(2-5) North Holland.

Sherif, M.H. & Sparrell, D. (1992), Standards and Innovations in Telecommunications, IEEE Communication Magazine, July 1992.

Strokes, A.V. (ed.), (1986), The OSI Model of Open System Interconnection, State of the Art Report, England:Pergomon Infotech.

PART TEN

CIM Systems

43

COMPUTER INTEGRATED MANUFACTURING IN CHINA

B. H. Li, C.Wu, J.T. Deng, Y .C. Gao, J. S. Xue, G. Q. Gu, S. S. Zhang
China National 863/CIMS Subject Expert Group
Tsinghua University, Beijing 100084, CHINA, Phone:01-2568184,
Fax:2568184

Abstract
863/CIMS is one of the major subjects of automation field in CHINA NATIONAL 863 HIGH-TECH PROGRAM. This paper introduces the strategic target, strategic deployment, working principle, research environment, research on basic and key technology, product pre-research and development, applied engineering and international cooperation of the 863/CIMS.

Keywords
Computer Integrated Manufacturing System

1 INTRODUCTION

It's well known that Computer Integrated Manufacturing (CIM) has become a focus of manufacturing industrial technology since 1980's, due to the stimulus of intense competition in the world markets and the support of the rapid development of some relative techniques. CIM philosophy has been widely accepted as the next generation philosophy to organize and operate enterprises. The development of CIM shows up in:

- The development strategies have been drawn up and executed in many developed countries (e.g. CALS and AMRF of U.S.A, CIME/ESPRIT of EEC, IMS of Japan), moreover a lot of research bases, a group of development teams, and some CIMS enterprises have been built up in different areas, with various scales.
- The CIM industry, including some relative unit technologies and system integration technologies, has been developed with an estimated output value of five billion U.S. dollars in 1995.

An open China must face the entire world, and build up the socialistic market economy. This means that the manufacturing industry of China also need utilize CIM to meet the intense competition inside and outside the country. On the other hand, a country like China with over one billion people must have its own CIM technology and industry. These are the two main reasons why we select 863/CIMS as one of the sixteen major subjects in the CHINA NATIONAL 863 HIGH-TECH PROGRAM.

In 1973, Dr. Harrington first put forward the concept of CIM in his book 'Computer Integrated Manufacturing'. During the past two decades, especially the latest decade, the concept of CIM has been enriched and enlarged without discontinuation. But till now, there is no an authoritative definition. This paper gives out a definition of CIM and CIMS.

'CIM is a new kind of philosophic theory used in organizing, managing and running the enterprise's production, it takes advantage of computer software and hardware, synthetically uses modern managing technology, manufacturing technology, information technology, automatic technology, system engineering technology, and it integrates organically the three relative factors of person, technology, running management in the whole process of enterprise's production, as well as information flow and material flow, and runs them optimally, to make service excellent, bring products to market timely, and realize product's high quality, low cost, so that enterprises will win the market competition.'

'CIMS is a complex system composed and run according to CIM philosophy'.

Since 1987, 863/CIMS has made striking achievements. CIMS is being paid more and more attention, and attracts more interest and investment. When 863/CIMS just began in 1986 and 1987, an opinion was that 'CIMS is far from us'. In 1989 and 1990, some medium or large scale key enterprises asked for carrying out CIMS engineering for the sake of development, everybody felt that 'CIMS steps for us'. The trend in 1993 and 1994 is that more and more enterprises, including quite a few medium and small size enterprises, need CIMS, so a very important thing now is that 'how to spread CIMS technology in China'. These three views typically reflect the trace of CIMS' development in recent seven or eight years in China.

In the next paragraphs, the 863/CIMS will be introduced in the following six aspects:

- plan outline (strategy target, strategy deploy, organization and management system, work's instructive principle);
- research environment;
- the research on basic and key technology;
- product pre-research and development;
- applied engineering (environment);
- international cooperation.

2 PLAN OUTLINE

2.1 Strategic Target

Target by 2000 year
- setting up 15~20 model enterprises and 100 general applied enterprises to use CIM successfully for demonstration in different areas (aeronautics, machinery, electronics, electrical equipment, costume, metallurgy, chemical engineering et al.), of different production characteristics (discrete, semidiscrete, continuous), under different basic conditions (whole CIMS, local CIMS, CIMS on PC), by different implemented schemes (based on imported unit technical equipment or technical equipment made in China);
- setting up a series of advanced CIM research centres/opened laboratories;
- organizing a CIM professional team of thousands of personnel hierarchically;
- achieving a series of advanced research progress;

- promoting the establishment of CIM high-tech business.

2.2 Strategy Deployment

For achieving the target mentioned above, 863/CIMS adopts the following deployment:

Two environments
- research environment (see (3));
- engineering environment(see (6)).

Four levels
- 'applied engineering' which is divided into two parts: 'model application enterprises' and 'general application enterprises', to use CIM technology successfully;
- 'software product development' ;
- 'key technology research and development';
- 'application basic research'.

2.3 Working Principle

- guiding by application, driving by technology, adopting finite targets, stressing the main points, combining with the situation in China, paying attention to practical results and forming business;
- advocating '863' spirit of 'Justice, Dedication, Creation, Reality, Cooperation';
- adopting expert leading mechanism under the leadership of National Science and Technology Commission and taking turns of one third of all experts in regular time;
- emphasizing the working principle of 'Team Work', which cooperates multi-disciplines to work, and system integration;
- introducing rolling and competitive mechanism to determine the projects;
- paying attention to CIM group building, especially the training of young strength;
- strengthening international cooperation;
- strengthening the cooperation with other plans and the related departments of China closely.

3 RESEARCH ENVIRONMENT

To provide a good environment for research on application basic technology and key technology , to provide an experimental base for the development of products, to provide the technology support for applied engineering, to establish a good condition for training personnel, as well as to provide a window for international cooperation and exchange, we concentrated our human power and financial resources in 863/CIMS on building an engineering research center and seven open laboratories with various speciality characteristics.

3.1   The Engineering Research Center of CIMS (CIMS-ERC)

This is one of the most important construction projects in 863/CIMS, constructed at Tsinghua University in five years from 1988 to 1992, and was verified and accepted by government at the early of 1993. It got the 'University Lead Award' from Society of Manufacturing Engineering of the United States.

CIMS-ERC tasks
- CIMS research center of integration technology;
- CIMS testing center for various individual techniques integration;
- transferring center of domestic and foreign advanced technology;
- training personnel center with special skill of various fields.

The technology outline
Supported by distributed database and factory network, CIMS-ERC realizes an experimental integration manufacturing system from engineering design, production scheduling and control to manufacturing, which can be depicted in the Figure 1.

CIMS-ERC has implemented the information integration of unit assembly design, part design, process planning and NC-program generation. In this way, the periods of product engineering design can be shorten greatly.

A flexible manufacturing cell has been built, consisting of eight working components: material store and transport, fixture and tool management, measuring tool management, etc. Mean while, the integration with CAD/CAM is implemented. The structure of its manufacturing control is the four-level hierarchical control of shop floor-cell-workstation-equipment. The shop floor level can produce a weekly job list. The cell level produces the bidaily rolling plan according to the weekly production schedule, production capability and the worksite information of shop. The workstations call the correspondent NC program according to the job instruction from cell controllers.

By design and realization of these tasks mentioned above, we obtain lots of experiences and technologies, which have been applied to other projects in 863/CIMS, especially in many model application enterprises.

3.2   Open Laboratories

A brief description about other seven open laboratories of CIM unit technology is given in table 1.

4   THE RESEARCH ON APPLICATION BASIC TECHNOLOGY AND KEY TECHNOLOGY

4.1   Application Basic Research

The research task on application basic technology is a kind of technology-driven research under certain application background. This task develops necessary explorations, verifications and new

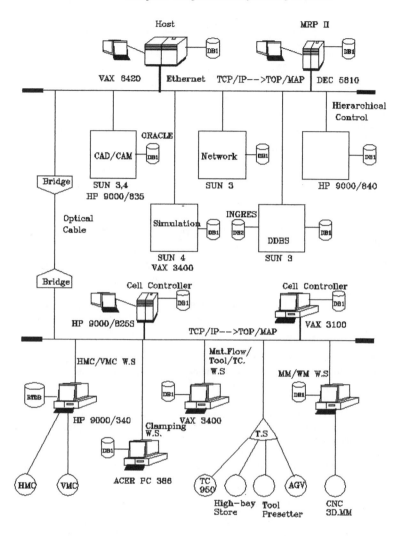

Figure 1   CIMS–ERC System Architecture

Table 1 Seven open laboratories of CIM.

Name	Commencement	Complete on	Location	Main Tasks
Product Design Automation Lab.	1987	1993.12	Beijing U. of Aeronautics & Astronautics	R & D of technology and products for CAD/CAM /CAE in CIMS
Process Planning Design Automation Lab.	1987	1993.12	Shanghai Jiao Tong U.	R & D of technology and products for CAPP in CIMS
Integrated Management Decision Information Lab.	1987	1993.12	Tsinghua U.	R & D of technology and products for computer aided management and decision in CIMS
Flexible Manufacturing Engineering Lab.	1987	1993.12	Beijing Research Inst. of Machine Tools	R & D of technology and products for FMC/FMS/FME in CIMS
Quality Control Technology Lab.	1987	1993.12	Xian Jiao Tong U.	R & D of technology and products for CAQ in CIMS
Database and Network Lab.	1987	1993.12	South-East U.	R & D of technology and products for Network and Database in CIMS
System Theory and Technology Lab.	1987	1993.12	Shenyang Inst. of Automation	R & D of technology and products for theory, simulation and AI in CIMS

ideas in the concepts, principles and methodology suitable for the forward development of world-wide CIM science and technology. This task can be divided into the following branch-subjects:

Management information systems

Emphasis on streaming production resources management, and scheduling at two levels of enterprise and shopfloor. A framework for toolkit integration according to the problem-solving conception has been developed to enable users to adopt, integrate various available tools, platforms and softwares for their own needs. For examples, within the framework, a tool of rapid prototype development support system named RADISS has been developed and has been validated by enterprises.

Design automation and CAD/CAM integration

For exploring the fundamental principles and technical way to implement product/process engineering design automation in order to shorten the time-to market (T), ensure the quality (Q) which can meet the requirements defined by market, and make the cost (C) competitiveness, Product/Process engineering is purposed as an unified, integrated and controllable automated engineering process.

   Three basic aspects should be clarified before a technical system constructed:
- an unified description for product/process engineering process, which should be available for computer processing also;

- a mechanism for mapping various views of product/process, which are defined in the course of product life cycle according to the different department's working areas with their specific views of the product;
- the technical way, how to implement the unified description of product/process engineering process, the mechanism of mapping, and how to specify the controllable product/process engineering system.

Base on the above, STEP technology, 'O-O' technology are used for describing unified product/ process engineering process, and resulted into a knowledge architecture model for further implementation. Also a prototype of CAD/CAM integration system was developed and named Computer Support Mechanical Development CSMcD which is based principles of the aspects mentioned previously. And a new concept called COMposition AP (COMAP) is proposed and studied for integrating different APs to support lifetime cycle of a product. It is based upon the existing STEP APs and focused on the horizontal integration and the interpretability of those APs.

By the way, using intelligence to enable product/process automation system to work properly in several domains is another important task. For example, design with tolerance analysis, assembly etc.

Integration, and support environment
Networking and enterprises data management in manufacturing automation environment is focused, especially to exploring the ability of enterprises data integration and sharing through advanced network, protocol engineering and database technology.

Main achievements are:
- communication controller for interconnection of different type machines;
- wideband Ethernet, wideband 1Mbps main bus network and baseband 3Anet;
- TOP/FTAM protocol for different type machines;
- data secrecy keeper with changeable key code;
- interconnection system of field-bus and Ethernet;
- different structural distributed database management system with object-oriented and semantic associated data model, relevant data language, host language and assimilator.

Some developed technology of Network/database R&D are used in industrials, says: Beijing No.1 Machine tool Plant (BYJC), Huabao Air Condition Factory (HB).

Shopfloor automation
Cell controller development platform has been developed and based on it, several controller with intelligent ability to adopt various control model can be developed. Many FMC has been developed, installed and operated by Chinese experts.

According to the investigation, that in Chinese enterprises there are needs of requiring information system to help shopfloor to manage production running, that is different from MRPII view, several projects are focusing to solve this problem.

Quality
Quality is very important for improving Chinese enterprises competitiveness. Quality is a problem of system management combined with system technology rather than just a SPC, or TQC. A

framework of conception for defining the problem of quality within CIMS has been studied, and also several practical quality management systems is developed for industrials, like QFD studies in JianHan Mechanical Factory, quality management system for Zhengzhou Textile Machinery Works, for Chengdu Aircraft company.

Others: CIM system open architecture, methodologies, and standardization
Many research efforts focusing on the basic framework, system architectures, methodologies and standardization have been studied in order to define relevant conceptions, principles, key problems and support methodologies, tools for applications.

Some ongoing application basic research topics
- CIMS software engineering and standardization, including the research of $IDEF_4$ technology and the realization of implementation prototype;
- CIMS architecture and deploying strategies, including the research of CIMS architecture on the concept of agile manufacturing, lean production and others, and the research of instances of application engineering of CIM integrated infrastructure with OSA;
- CIMS system and integrated technology, including the system technology and experimental system design of concurrent engineering and the management of product developing process of concurrent engineering;
- product design automation on the environment of CIMS, including the construction of product model using EXPRESS language in the STEP standard and product design process in concurrent engineering;
- integrated process planning design automation, including the automation of assembly process planning design and the automation technique of process planning in concurrent engineering;
- CIMS flexible manufacturing technology, including integrated mode of shop-floor management and control, and the prototype system of shop-floor management and control based on client/server system architecture;
- integrated management and decision information system, including the research of manufacturing oriented, distributed information model and physical model of management and decision supported system based on client/server distributed structure and O-O technology, also and the research of advanced management and administration and conceptual model and system architecture of relative management and administration decision information system;
- integrated quality system of CIMS, including the research of Quality Function Deployment (QFD) and the research of implementation of ISO 9000 in the environment of CIMS;
- computer network of CIMS, including the MMS realization of heterogeneous system, the modulator/demodulator of carrier band MAP network, and the new network technology supporting concurrent engineering;
- database technology of CIMS, including information integrated platform in the environment of CIMS, and the distributed real-time database management system;
- system theory and methodology of CIMS, including the optimization and intelligence of production planning and scheduling.

4.2 Key Technology

Serving the needs of application, the research of key technology emphasizes on solving the problems of technology which , once made a breakthrough, can bring enormous benefits in the application of next step.

Two kinds of key technologies has achieved great successes, which are listed as below:

1. It is the basic character of the first stage of CIMS that solving the problem of information integration under the different-structure environments. Through the hard efforts of years exerted by CIMS-ERC, which is the main representation, and all other special projects and a group of open laboratories, the key technologies of information integration have been grasped basically. This is very important for successful execution of CIMS in our enterprises.
2. The breakthrough of another key technology is the great improvement of STEP-based CAD/ CAM prototype system mentioned previously. It make the technology of our country in this aspect stay international level and receive acknowledgment and attention in the world.

The important projects faced by us in tackling the key technology at present are the following:

* concurrent engineering, for example, a product oriented (instead of parts oriented) assembly modeling technique is studied, that focuses on product data representation, assembly relationship extraction and design information propagation among different design components in different assembly hierarchies;
* the research of integrated application framework.

With the deep development of CIMS in our country, new key technology breakthrough projects will be formed incessantly.

5 CIMS PRODUCT PRE-RESEARCH AND PRODUCT DEVELOPMENT

CIMS product pre-research and product development level is an important work for hi-tech industrialization. The objective of this level is to transfer the research achievements into commodities on the base of the market requirements.

Since recent years, 863/CIM has supported nearly 20 projects, including CAD, CAPP, management and decision information system of manufacturing, schedule and control of production process, simulation and other computer products relative to the automation of factory. These products can be divided into three kinds: super-short-term product, short-term product, pre-research product.

5.1 Super-short-term Products

These are "short, flat, fast" products; their general development period is about one year. Typically ones are the following:

* Micro Computer Aided Drafting and Designing System;
* MIS Rapid Developer Based on Microcomputer Network;
* THDA-2000 Laser-disk Archives System;
* Engineering Database Management System on Microcomputer;

- Scheme Design Expert System Toolkit;
- Real-time Monitoring-Displaying System;
- Cast CAPP and 3D Simulation Analysis for Soliditing Process.

5.2 Short-term Product

It takes about two years to finish its development and marketing, on the base of its mature technology or prototype product. Typically ones are the following.

Computer-Aided Quality Information System (CAQIS)

Its application functions include quality planning, quality measuring and checking, quality analysis and estimation, and quality information management.

Its service functions include application creation and execution, datafile and database management, software component management, communication service and data transformation, and interface management.

$IDEF_0$, $IDEF_{1x}$-aided software design

This is a tool package used for IDEF modelling. In addition to IDEF Modelling, the package has store, modification, output and management functions of IDEF model. To support these functions, a graphic library, a Chinese character system and its editor have been provided.

Rapid Application Development and Integration Support System (RADISS)

RADISS provides CIMS/MRPII designers and developers with an open integrated support environment for MRPII application system creation, built on UNIX with SQL and C languages. It is used for describing structure and functions of CIMS/MRPII with formal languages, creation and maintenance with tools.

MRPII system in CIMS

The MRPII applies to enterprises that have multiple variety, medium/small batch, even large batch production.

The MRPII includes functions of production planning, materials requirement planning, capacity requirement planning, shop-floor job scheduling, inventory management, materials supplement management, sales management, accounting, cost management, facility and tool management, data management for MRPII in CIMS.

The MRPII has an open architecture, SQL language and interface with CAD, CAPP, and FMC/ FMS.

Comprehensive Information Acquisition System (CIAS)

CIAS is a tool for top-level managers to acquire comprehensive information about their enterprises, and an environment for the tool and application integration.

CIAS' functions include system control for searching, operation history and analysis, backup, inquiry/analysis report and graphics output , Modelling language, SQL for database, report database, report generating, communication and security control, model dictionary, man-machine interface, text editor, electronic memorandum, automatic trigger, window editor, industrial statistics, market forecasting, optimizing production plan model and aided decision model.

Integrated Manufacturing System Simulation Software (IMSS)

This is an integrated, manufacturing enterprise-oriented, and general purpose simulation software for scheme estimation and validation, production capacity analysis of manufacturing system, and optimization of short-term planning and scheduling of shop-floor.

IMSS has Modelling environment of non-language, animation creation, structured and object-oriented Modelling function.

The control and management of simulation process provides multi-window, menu driven, graphics input for friend interface, managing simulation resources in uniform, real time monitoring and animation functions.

IMSS provides variety of statistics analysis and automatic report generator.

CIMS control workstation developing platform

This is an open software framework, on which several special purpose modules can be assembled to construct machine workstation or cutting tool workstation.

It provides CNC and PLC communication interfaces with FANUC 6M, SIEMENS 850 and FAGOR 8025 at the workstation level.

It provides interactive user operation interface, user's defined communication control module, and fault-detection function.

Factory Computer Scheduler (FCS)

It generates achievable shop production schedules based on finite resource capacity by discrete event system simulation in real-time. It can describe the attributes of the shop's elements and events in managers' words and integrate with other management systems in CIM environment which make it react to any changes in the factories within minutes. The system is easy to use through the modeler's and scheduler's multi-menu-driven interfaces. An expert system is embedded for result analysis, appreciation and scheduling optimization and reports.

Cell Control System

The System is used for management and control system of small/medium enterprises FMS (FMC) in CIM environment or not. It includes Cell controller, Machine workstation, Material flow work-station, Tool workstation, and DNC interface.

Integrated CAPP System for case-type component

The system applies to designing, process planning and NC programming for general case-type components. It has the following functions:
- feature Modelling based on Euclid CAD system, creating information model in neutral file needed by CAPP and NCP;
- process planning in automatic creation and interactive made with GT technique;
- user-interface for describing manufacturing resources and effectiveness verification of process planning;
- generating 2D, 3D NC program for case-type component and post-processing;

- information sharing among part designing, process planning, manufacturing resources, and NC programming.

5.3 Pre-research Product

It is a kind of product pre-research with major study and development for the market requirement after 3 to 5 years, on the base of product and achievement technology combining with new ideas, concepts and principles of product developed in the world.

Integrated CAD/CAPP/CAM system

The system utilizes STEP as product model describing standard and integrates mechanical part designing, process planning and NC program controlling information in CIMS environment. The system is planned to be finished in 1995.

Manufacture management information system for large, medium enterprises

It is an advanced MRPII-system based on the Chinese environment.

6   APPLIED ENGINEERING (ENVIRONMENT)

Applied engineering is one of the important parts in 863/CIMS. The reasons for setting up applied engineering are the following:

- CIMS is an integration or optimization system of people, organization, technology, management and administration, thus it is necessary to master CIM technology completely by typical enterprises practicing.
- The works of typical enterprises practicing can guide the carrying out of CIMS in the enterprises of China.
- The practice of applied enterprises will check and expedite the research of 863/CIMS technology.

From 1989 to now, eleven enterprises, as listed below, have been chosen as 863/CIMS model enterprises: Chengdu Aircraft Industrial Corporation (CAC), Shenyang Blower Works (SB), Jinan First Machine Tool Works (JFMT), Shanghai NO.2 Textile Machinery Ltd. (STTM), Beijing NO.1 Machine Tool Plant (BYJV), Zhengzhou Textile Machinery Plant (ZTM), Dongfeng Motor Corporation (DF), Huabao Air Conditioner Factory (HB), China National Garments Research & Design Center (CNGC), Hangzhou Sanlian Electronic Compare (HSEC), Jingwei Textile Machinery Plant (JTMP). Those covers multiple types of machine-building industries (including single piece, multivarieties, small batch, large batch and other types), electronics industry, home electric appliance industry, clothing industry and so on. In addition, 863/CIMS is going to spread CIMS to metallurgical industry and chemical industry etc.

These enterprises realize CIM quite comprehensively. Shenyang Blower Works CIMS (SB-CIMS) will be introduced briefly as an instance in what follows.

Aimed at global benefit, SB-CIMS is an integrated automatic system with wide-ranging scale, complex technology. Its architecture, viewed from application, includes enterprise layer, workshop layer, cell layer, workstation layer and equipment layer; while from information it has computer

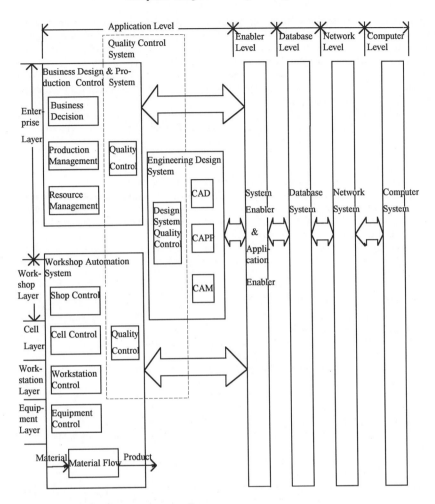

Figure 2 SB-CIMS system structure diagram.

network layer, layer, database layer, enabler layer and application layer. Thus, SB-CIMS's architecture is a two-dimensional hierarchical structure, and also an open structure. See Figure 2.

During recent two years, critical developments have been made in model applied enterprises. Among these, the breakthrough projects of Beijing No.1 Machine Tool Plant, Shenyang Blower Works and Chengdu Aircraft Industrial Corporation have passed verification and made great progress, which greatly benefited these enterprises and got high appraisements. For example, Shenyang Blower Works is discrete manufacturing enterprise with single piece production. Its quotation system developed by itself makes quotation period shorten from 6 weeks to 2 weeks, which reaches the world top level. This system can provide technique quotation, financial quotation and business quotation that accord with the international standard. The success in the quotation system severely enhances the competitive power of enterprise. Moreover, the period of design, preparation and product delivery has been shorten from 18 months to 10 ~ 12 months due to CIMS, and this greatly increases the global economic benefit of enterprise. Some similar instances lie in other CIMS enterprises.

863/CIMS promotes CIMS applied engineering directed by the principle that 'Benefit driven, Globally planned, Key points overcome, Stages implemented, Application spread'. The strategies of implementation of applied engineering are the following.

Organization mechanism

Under the leading of chief scientists and experts of 863/CIMS, we establish the leading group of applied engineering. This leading group consists of some experts of 863/CIMS expert group, and it is responsible for the selecting, designing, carrying out, checking and verification of application enterprises, moreover , we found the office of application engineering to be responsible for the management of plan and quality of applied engineering.

Every enterprise organized a leading group and union team of designing and carrying out, which consist of people from enterprise and 863/CIMS. Their specific duty is to implement their own objective of CIM.

The principle of implementation

- first, taking the machine-building enterprises (discrete type) as key link, and then expanding these technologies into other type/industry enterprises step by step;
- paying attention to the low cost CIMS applied in medium and small enterprises;
- combining the condition of our country, and set limited target;
- thinking highly of the integration and optimization of people/organization/technology /management and administration;
- making program from top to bottom, but carry out from bottom to top;
- taking benefit as the central task, making success while carrying out;
- breaking through with emphasis, by stages and in batches, carrying forward overall;
- strengthening the cooperation in many ways, such as the other works of 863/CIMS, main ministry of plants, foreign traders;
- making the standardizing of CIMS engineering, software engineering and quality assurance.

With efforts of many technologists in recent seven or eight years, key techniques in the first stage of CIMS, applied with a feature of information integration, have been solved basically. So 863/CIMS applied engineering begins to spread and apply in larger scale. Since 1994, 'general applied enterprises' level has been formed, i.e., in more enterprises, especially in medium or small size enterprises (or the parts cut off from large or medium enterprises), spreading CIMS application, whose scale is to be reduced to small size and period limited in two years. We will more often rely on national ministries, councils and the local departments to realize the comprehensive management of project application, setting, development, implementation and check and verification. In 1994 and 1995, 863/CIMS prepares to spread the CIMS application in about forty medium and small enterprises, and will further spread it in the Ninth Five-year Plan.

In CIMS application spread plan, we will spread a variety of 863/CIMS products developed in late years. It is expected that the application spread of CIMS engineering will drive the spread of CIMS products.

7 INTERNATIONAL EXCHANGE AND COOPERATION

Since the establishment of 863/CIMS subject in June 1987, different modes of international technical interchange and cooperation on various levels have been developed actively in order to keep abreast of the new developments of international advanced technology and to foster our own talented personnel of high technology. Now a new prospect was opened up and several channels were established and good results have been achieved.

Several forms of activity in international technical cooperation have been used in these years:

Important international conferences
Since 1987 863/CIMS subject has tracked the following conferences continuously: AUTOFACT, ESPRIT, IPO/ISO, CAPE, INCOM.

International technical cooperation
Since 1990, various technical cooperation between 863/CIMS subject and EEC, Germany, and other countries and some corporations on specialized trades, such as, DEC and HP, have been established. Both sides of the Taiwan Straits came into friendly contact too.

Holding international conferences in China
The international conference on CIMS technology has been held by CIMS-ERC in May 1992. About 80 persons, including 20 or more from the foreign lands, attended the meeting for academic interchanges, meanwhile, the development and achievements of CIMS-ERC have been discussed and evaluated, and good impression has been made. Delegations of Taiwan and Hongkong have attended actively the second 863/CIMS annual conference held at Shenzhen. The 1st. international Beijing-ICCIM'93 conference held by CIMS subject has got support from the well known international organizations SME, more than 80 representatives attended the meeting.

Inviting foreign scholars to give lectures in China from U.S.A., Europe, Australia and so on
In the past several years, the international 863/CIMS technical cooperation of CIMS subject has been steadily developed. We will strive to achieve great successes in the main cooperative topics and to open up actively the new prospect in order that the international academic exchanges become one of the important means for promoting the development of our CIMS technology.

8 CONCLUSION

Through seven years' exertion, 863/CIMS has made the following striking achievements:

- It has drawn up 2000 year strategic target and the Seventh, Eighth, Ninth FYP subtargets consummately.
- It has established and consummated an expert responsible system and organized a CIM professional team of 1500 personnel's.
- It has formed the strategy deployment of two environments and four levels.
- It has well constructed one engineering research center and seven open laboratories.
- It has got hundreds of progress on application basic technology and key technology researches.
- It has realizes some CIM beneficial systems of different scale in eleven 'model applied enterprises' with some breakthrough advancements, and started to spread the CIM application in 'general applied enterprises'.
- It has developed a variety of software products on CIM and started to set up several CIM hightech companies.
- It has launched many international cooperation projects with foreign countries and established good relations with relative domestic ministries, councils and departments.

We confirmly believe, through our hard work, the strategic target of 863/CIMS can be realized.

9 REFERENCE

Automation Expert Committee of the National 863 High-Tech R&D Programme of China. (1990) Proc. of CIMS-China 90, The National 863 High-Tech R&D Programme of China, Beijing.
Automation Expert Committee of the National 863 High-Tech R&D Programme of China. (1992) Proc. of CIMS-China 92, The National 863 High-Tech R&D Programme of China, Shenzhen.
China 863/CIMS Subject Expert Group. (1993) Proc. of International Conference on CIM (ICCIM'93), The National 863 High-Tech R&D Programme of China, Beijing.
China 863/CIMS Subject Expert Group. (1994) Proc. of CIMS-China 94, The National 863 High-Tech R&D Programme of China, Wuhan.
China 863/CIMS Subject Expert Group. (1994) Developing tendency of Computer Integrated Manufacturing technology and System. Publishing house of Science, Beijing.
China 863/CIMS Office. (1994) The Progress and Prospects of National CIMS Subject. The National 863 High-Tech R&D Programme of China, Beijing.

Acknowledgement
Thanks to Mr. Xiangyang Li and Ms. Qing Liu for their much help in the work.

44

Decentralized and collaborative production management - A prerequisite for globally distributed manufacturing

Hirsch, B.E.; Kuhlmann, T.; Maßow, C.
Bremen Institute of Industrial Technology and Applied Work
Science at the University of Bremen
Hochschulring 20, 28359 Bremen, Germany,
Tel: + 49-421-2185540, Fax: +49-421-2185510
e-mail: km@biba.uni-bremen.de

Abstract

Due to current market trends the necessity of decentralized and collaborative production management is increasing and will become more and more complex. In this situation the ESPRIT project DECOR specifies a framework for distributed intra- and interorganizational production management. Additionally, a general information system concept which designs, supports and co-ordinates multi-level, distributed and autonomous decision-making systems is being developed.

Keywords

Production Management Techniques; Decision Support Systems

1   INTRODUCTION

Traditionally, the challenge of production management has been to optimize and control processes and flows within a factory. Most existing production management systems are designed from this viewpoint and restricted to the management of in-house processes. The JIT-philosophy then broadens this perspective to include the co-ordination of in-house production with chains of suppliers and subsuppliers. This has required a changed production management concept towards a hierarchical inter-site organization.

Today three trends which stimulate a further transformation of production management concepts (Hirsch, 1994) are obviously:

Globalization of markets
Progresses in manufacturing, transport and information technologies and major political events are leading to globally open markets. This means most companies must operate in worldwide markets and with worldwide supply chains.

- Customer orientation
  As customer satisfaction is becoming the crucial factor for competitiveness for all types and phases of industrial manufacturing, companies have to develop effective strategies for orienting enterprises toward the customer. Customer orientation has to embrace the "time-to-market" and the "time-in-market" phases.
- Environmental awareness
  The social pressure increases to create environmentally friendly product and production systems. Therefore, manufacturing enterprises have to develop total life cycle concepts for their products.

These trends require dynamic and holistic production management concepts integrating globally distributed production processes of value-adding logistic chains. These chains have to be interpreted as a customer-order-oriented network of co-operating and autonomous enterprises. Therefore, the necessity of decentralized production management and decentralized decision-making will increase dramatically in the next few years and will become more and more complex.

The purpose of the ESPRIT project 8486 DECOR "Decentralized and Collaborative Production Management via Enterprise Modelling and Method Reusage" is to provide the European industry with the enabling technologies for decentralized production management to meet the competitive requirements for the rest of the decade.

The aim of the DECOR project is the development and application of a framework for managerial tools which is based on advanced information technologies and which is introduced for the purpose of distributed decision-making in existing and future heterarchical organization structures. The framework allows cost-effective integration within an open system architecture and will close the gap between existing production management applications and existing communication services.

2 TRENDS OF ORGANIZATION STRUCTURES

The present organizational structure of many European industrial organizations is characterized by the influence of Taylorist principles. This is reflected in the highly organized, functionally- oriented hierarchy. The weaknesses of such a system of organization are recognized by many enterprises. First attempts were the introduction of cost and profit centers as the basic organization structure (see Figure 1). These approaches promote the decentralization of production management but have an intra-factory orientation.

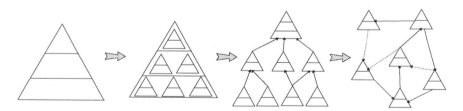

Figure 1 Trends of organization structures.

Due to the globalization of markets and production European industry operates necessarily in a highly distributed business environment. Also, the trend towards lean production and concurrent engineering increases the need for decentralization. European companies are also starting to establish manufacturing units in several different countries to provide better and more responsive customer service. This tends to lead to organizations with a weaker structure, i.e., a reduced degree of hierarchy and autonomous decision-making, thereby enabling faster, more flexible decisions.

This means European manufacturers have to work within globally distributed temporal and logistics networks of independent enterprises and production units. The challenge of the future is, therefore, to define production management concepts and enabling technologies allowing distributed production in a decentralized and collaborative manner.

However current information systems are designed to support the Taylorist organization structure: they were developed to optimize individual, defined tasks. In other words, the development of these systems has limited the organization to a functional orientation and has either ignored or neglected to take into account the incorporation of any one task into a chain of processes divided into different organizational units. This means that these decision making systems do not support the communication and information exchange between themselves. But the missing link between decision making systems and existing communication services prevent also a process oriented integration.

In order to achieve a process-oriented structure, the focus of the DECOR effort is aimed at creating a technical framework/system for supporting the integration of functions within a process chain. The solution of management problems must not be allowed to occur through increasing sequentialisation and building up the hierarchy for individual tasks, but should utilize infrastructural support for the single process as well as the overlapping process chains. This means on the one hand that the right information should be in the right place at the right time. On the other hand, organizational units should be able to draw upon decisions from other units, so that a type of PULL principle is created.

3 REQUIREMENTS TO FUTURE PRODUCTION MANAGEMENT SYSTEMS

Especially production management problems require a series of decisions with an increasing level (multilevel) of detail. Each level has a specific level with respect to the planning horizon, the aggregation and certainty of information. The contradictory objectives resulting from these differences are showing the increasingly importance of concepts for multi-level decision problems, also with respect to autonomous organizational structures which require decentralized decision making sub-systems.

Due to the business strategy to produce goods within a multi-site and multi-supplier environment - additionally enterprises will not be integrated in stable, long-term relationships but will have to work in changing consortia been constituted customer order oriented - decisions will have to be harmonized between different sites.

In addition to these rather strategic motivations for decentralizing control in manufacturing systems, there are several specific reasons at the operational level of production management supporting a decomposition of the overall problem and distributing control in solving it (DECOR, 1994).

Opposite to many previous approaches, production management in a distributed manufacturing environment should be based on both vertical and horizontal interaction of relatively independent, i.e., loosely coupled, and opportunistically behaving decision-making subsystems.

The thrust in the proposed approach is that a distributed production management system exhibiting the characteristics listed above can be constructed by having the individual decision-making subsystems within the production organization (e.g., work areas, departments, product factories, autonomous sites etc.) to exchange between themselves information regarding local operational constraints. If this constraint information (e.g., release date and due date of a particular operation sequence, delay of the release date of a particular order, etc.) is being passed both vertically, i.e., between the superior and the subordinate organizational unit, and horizontally, i.e., from peer to peer within the organization, and if the incoming information is being analyzed based on local circumstances before any action is necessarily taken, efficient co-ordination can be facilitated.

The production management system has itself to provide a solution along both the computational, i.e. the processing level of distributed management, and along the behavioral, i.e. the problem solving level of distributed management. Categorically, a system for distributed production management has to support the implementation of three classes of management knowledge:

1. Domain knowledge covering the representation of all the physical and conceptual entities in the production organization,
2. Problem-solving knowledge defining the actual production management procedures and problem-solving rules, and
3. Communication knowledge establishing co-ordination links within the distributed set of loosely coupled decision-making subsystems.

4 DECOR ARCHITECTURE

The DECOR architecture decomposes a production management system into individual Decision-making Subsystems (DMSs) (DECOR, 1993). No matter, what level of abstraction or aggregation, a single DMS always covers the production management problem in a two-level planning hierarchy as depicted in Figure 2. An individual DMS can therefore be responsible for the management of an production unit e.g. an entire factory and the associated sub-unit level e.g. production departments. Each sub-unit again, can be modelled to be covered by a "more detailed" DMS going down to the individual work area level with regard to its scope of problem solving.

The justification for the architecture is to distribute decision making in a production environment into fairly independent nodes, and co-ordinate amongst nodes to keep the overall system consistent. In the DECOR architecture, a DMS itself consists of a Problem-solving Module (PSM), a Message Management Module (MMM), and a Domain-modelling Module (DMM). These modules will be described in the following chapters.

Since the functional architecture (but not the contents) of any DMS is similar from one DMS to another, the DECOR architecture is capable of modelling a distributed production management system no matter what the level of abstraction is, i.e., if it is distribution amongst work cells in a factory or distribution amongst entire factories or both.

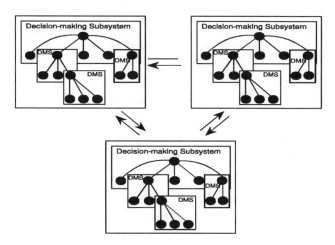

Figure 2 The DECOR architecture

4.1 Problem Solving Module

As a problem-solving and co-ordination environment, the DECOR target domain, i.e., the manufacturing organization may be characterized as semibenevolent: different subsystems in the hierarchy are self-interested and "greedy". They usually have conflicting goals and crossing local preferences. Sub optimal solutions may well be in conflict with the more global interests and preferences of the organization. This means that the competing subsystems have to compromise in favour of global concerns. A basis for making the compromises is the global set of system objectives which is decomposed and distributed amongst the subsystems. The crux of co-ordination in a dynamic manufacturing environment is hence adaptation of the local plans in response to various disturbances so that the well being of the overall system is guarantied.

Because of its inherent nature, distributed production management requires a co-ordination framework which combines goal-directed and opportunistic local reasoning with the need to ensure coherent network behaviour. Achieving coherence refers to the problem of defining co-operation strategies to assure that a community of agents will work together in order to optimize some global criteria. These criteria may involve balancing the work load between the agents, minimizing the problem-solving time, maximizing the quality of the solution, etc. It is precisely the competing yet co-operative nature of the environment which distinguishes a distributed production management system from many other decentralized settings.

The task of a specific decision-making subsystem in the DECOR architecture is to manage the manufacturing subproblem within the scope of the particular DMS in question. In our distributed model of production management, this task does not only entail the capability to solve a static management subproblem. Also, there must be a companion reactive capability for dynamically responding to both relevant changes in the local portion of the manufacturing system due to unanticipated events as well as changes communicated from external nodes. The local commitments made must be reflected in the local domain-model and also communicated to external nodes.

Our three-fold architecture (PSM,MMM,DMM) for local decision-making takes care of these requirements. In this context, PSM implements knowledge and methods for solving the local production management problem. The basic nature of local management problems is quite similar and suggests a uniform control architecture which will allow the application of an extensive set of tools for integration of problem solving software. Although basically similar, characteristics of local management problems will vary between decision-making subsystems within the entire manufacturing system, also within one factory. Hence, our PSM control architecture is able to support a variety of problem-solving methodologies for production management, and also problem solving strategies based on a combination of existing methods.

Moreover, our requirement was that the architecture should be able to exploit existing software for production management without having to resort to a costly software re-engineering process. Therefore, the architecture supports the integration of different methodologies and existing, heterogeneous systems. Lastly, the PSM architecture facilitates intra-node communication, i.e., communication with DMM and MMM.

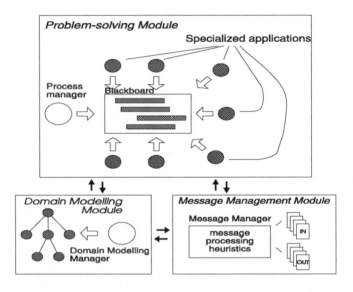

Figure 3: Structure of the Problem-solving Module and its relation to the other modules

In order to fulfill these requirement we have decided to use the Blackboard framework as a control metaphor for the PSM (DECOR, 1994). The Blackboard co-ordination metaphor casts problem solving as an incremental process wherein individual problem solving "experts" with well-defined areas of competence (knowledge sources) opportunistically contribute to the development of the overall solution (see Figure 3). The Blackboard approach has its power in organizing problem-solving by means of multiple independent knowledge modules, possibly based on different problem solving methods, and knowledge representation techniques. The Blackboard model is general in the sense that it supports state-of-the-art, opportunistic production management methodologies based on multiple knowledge sources as well as more monolithic methodologies. The blackboard model constitutes a framework which supports integration of heterogeneous production management software and offers large-grained modularity for distributed design, implementation, and maintenance.

4.2 Message Management Module

A fundamental approach of DECOR is to enable and support loosely coupled, hierarchically organized decision-making systems. Within such a distributed environment the decision making subsystems need to be able to interact, that is, to exchange different decisions between each other. These decisions may include task and responsibility assignments, management decisions, information about sudden changes in plans, etc. The loose coupling implies restricted communication between the nodes of the hierarchy, thus requiring intelligent communication control.

The communication in DECOR consists of information passed between nodes in the form of structured messages. Each node in the hierarchy uses a set of so-called heuristics to interpret the incoming messages and to guide and constrain the communication traffic. The interpretation of the messages is based on a number of different criteria, some of which are local and specific to the recipient, some of which are global and general in the hierarchy, and some are case-specific to the message.

The problem-solving message mechanism has to be able to answer the following questions:
1. When to send messages to other nodes? The system should be able to decide in which cases to actually communicate with which other nodes.
2. What are the contents of the messages? What data to exchange with the other nodes?
3. How to react to incoming messages? The system should trigger and perform the appropriate and necessary procedures.

In order to fulfill these requirements the MMM supports the administration and interpretation of different types of process models (Kuhlmann, 1994). In the following these different types and their management are described.

Global process models
With global process models the information flows between a network of decision subsystems are specified. The descriptions created by the subsystem responsible for the whole process can be product- or project-related. They contain the main tasks defined as activities, the related subsystems, e.g. those which perform the tasks, and the data flow represented as generalized documented classes.

Local process models
This model type enables the configuration of message management modules according to the specific environment of a subsystem. Local process models are installed expressly for each message management module and specify the standard processing of incoming and outgoing messages. These defined processing procedures enable both an heuristic and rule-based description of subsystem-related handling of messages.

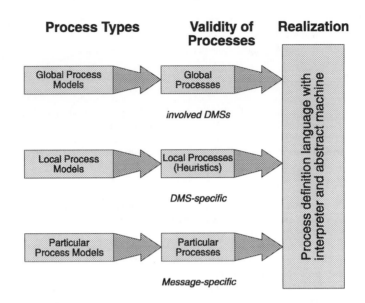

Figure 4 Process models

Particular process models

A particular process model enables the definition of a procedure describing the processing steps for a specific message during its life cycle. Since this kind of model enables that a particular sequence of a message be performed, a particular process model is directly connected to a message and is transferred with it (i.e. the message includes it).

Model management

For the management of the three different types of models, an MMM contains a process model administrator and a process model interpreter. The underlying mechanism is based on a process description language which allows the definition of processes and the control of the related information flow.

Based on the valid process model and the analysis of the included control data, the process model interpreter can trigger the necessary processing steps. During the model definition phase it also supports the testing of process models regarding their syntactical correctness.

4.3 **Domain Modelling Module**

The DMM models the underlying production data and knowledge with respect to the particular DMS in question and also acts as an interface to a distributed database which stores this information. The Domain Model captures all the production entities relevant to a particular decision-making subsystem. The entities may be either physical objects (products, resources, operations, etc.) or conceptual objects (customer orders, process plans, static or temporal relations, etc.). This means the DMM provides a comprehensive, "deep" model of the underlying manufacturing facilities and processes. It combines a multilevel representation of the basic entities and their relations in the decision domain with an active representation of the various constraints on these basic entities. The DMM provides both the MMM and the PSM with the required domain knowledge.

The DECOR distributed database accessed by the individual DMMs provides a single and unique repository for storage, access, distribution, and maintenance of all information and data utilized for and generated during the use of the associated application.

Another important feature of the DMM is the possibility to diagnose the performance of the related production sub-unit in order to evaluate the major bottlenecks and to perform deviation analysis.

5 SUMMARY

In order to support the current trend of production management approaches towards collaboration and distribution, the ESPRIT project DECOR has developed a conceptual framework and a set of software tools for the design and coordination of decentralized production management applications in distributed manufacturing environments. The first prototypes of the above-mentioned modules are available for MS-Windows. At the moment the implementation of these prototypes is being performed at two industrial sites. With these test cases the DECOR project demonstrates its solutions within:
- two contrary production types: one-of-a-kind and repetitive production.
- the three levels of production management: strategic, tactical and operational.
- different product life-cycle phases: production and after-sales.

6 REFERENCES

DECOR Consortium (1993) DECOR Project Description; ESPRIT III Project 8486, Bremen
DECOR Consortium (1994) DECOR Toolbox Architecture Document, Deliverable D5.1; ESPRIT III Project 8486, Berlin
Kuhlmann, T. and Marciniak, Z. (1994) Specification of the Message Management Module, Deliverable D2.1; ESPRIT III Project 8486, Bremen
Hirsch, B.; Crom, S.; Thoben, K.-D.; and Kuhlmann, T. (1994) New Manufacturing Paradigms - Their Contribution to Improve Customer Satisfaction, Shorten Time-to-Market and Lengthen Time-in-Market

45

Creating The Learning Company: A New Manufacturing Challenge

Jens O. Riis and Claus Neergaard
Dept. of Production, Aalborg University
Fibigerstraede 16, 9220 Aalborg, Denmark. Fax +45 9815 3030

Abstract

This paper discusses how adoption of a learning approach to create a learning company constitutes a new challenge to manufacturing companies. However, we need to give substance to this paradigm. First, we shall present a comprehensive model of learning in organizations which captures four important perspectives of learning. Then, implications will be delineated for decision support, empowerment of individual employees, the definition of a new role for management systems, and for providing substance to new organizational forms proposed.

Keywords

Learning in organizations, individual learning, collective learning, decision support, management systems, organizational structure, manufacturing paradigm.

1. THE NEED TO MAKE USE OF PREVIOUS EXPERIENCE

The need for industrial enterprises to be both highly adaptive and productive (agile and lean) has shifted the focus to also include the technical and administrative activities surrounding the production processes themselves, such as purchase, quality assurance, production planning and control, engineering design, invoicing, etc. In recent years, several methods have been developed which address these issues, such as Time Based Management, Activity Based Costing, Activity Chains, Business Process Re-engineering. However, we need a more comprehensive approach to deal with the continuous development of knowledge and an adaptive design of systems and organizations.

In the face of market requirements to provide customized products and services, learning from past experiences as well as the adoption of new knowledge from outside has become an important challenge to industrial enterprises, in particular one-of-a-kind producing companies, such as enterprises producing engineered facilities. Especially collective learning processes across organizational boundaries and disciplines are important.

Empirical studies which we have conducted show that management systems, e.g. production planning and control systems, in general do not support individual and collective learning; and often they constitute a barrier for learning processes, cf. Neergaard (1994). Many management systems are not designed to deal with learning aspects; rather they concentrate on planning and controlling the direct execution of activities.

Learning in organizations may be derived from the strategic issues of identifying and developing core competences by asking where it is important for the enterprise to develop its core competences, i.e. to learn. In this way, the issue of core competences is perceived as more than just technology in the sense of hardware and software, to also include "org-ware" and human aspects.

Developing core competences in an industrial enterprise calls for an integrative approach which combines various aspects of organizational learning processes.

We shall first present a comprehensive model of learning in organizations which captures four important perspectives of learning. Then, implications of using the learning approach will be delineated for elements of the emerging paradigm, such as decision support, individual motivation and qualifications, management systems, and organizational forms.

2. A FOUR-PERSPECTIVE MODEL OF LEARNING IN ORGANIZATIONS

A review of the literature on learning has resulted in a classification into four groups:

The individual behaviour perspective deals with informal learning processes of an individual. It captures information about human behaviour, for instance how individuals react in given situations and under specific conditions, as well as the personal interactions among people. Attention is focused on the informal, unconscious behaviour of a single organizational member and the interpersonal interactions among a number of members of an organization, cf. Argyris (1978, 1993).

The decision support perspective focuses on formal, individual learning processes in organizations. The main interest is how an individual decision maker learns in connection with problem solving situations. This includes the use of information technology and decision models to support decision making. The perspective is mainly used to study and understand how individual learning is influenced by available information technology and its institutionalized knowledge, cf. Nonaka (1991), Duncan & Weiss (1979), Riis (1992).

The management systems and organizational structure perspective concentrates on collective learning processes as guided by formal organizational structure and by management systems through formal planning and control processes, operating procedures and reward systems, cf. Riis (1978), Cyert & March (1963), Jelinek (1979). The allocation of responsibility and authority and the structure of divisions, departments and sections also regulate organizational learning processes, cf. Hall & Fukami (1979), Mintzberg (1983).

The corporate culture perspective represents what an organization knows, which is neither codified nor formalized in systems, cf. Schein (1985, 1988, 1990). The focus is on social, informal relations, collective habits, behavioural patterns and attitudes existing in an organization. Corporate culture is seen as emerging from collective learning processes and guides and shapes collective and individual behaviour.

The four groups represent different aspects of learning and may be ordered along two axes: (1) formal versus informal learning, and (2) individual versus collective learning, i.e.

- individual behaviour perspective being informal and individual,

- decision support perspective being formal and individual,

- management systems and organizational structure perspective being formal and collective, and

- corporate culture perspective being informal and collective.

The benefit of introducing four perspectives of learning lies not only in its capturing of four essential dimensions of learning in organizations, but primarily in its focus on the mutual interactions between elements of the four perspectives, as shown in figure 1. For example, to which extent do management systems support individual behaviour; are individual informal learning processes in tune with the intentions of the organizational structure and management systems; or do management systems provide information which may stimulate cross-functional learning?

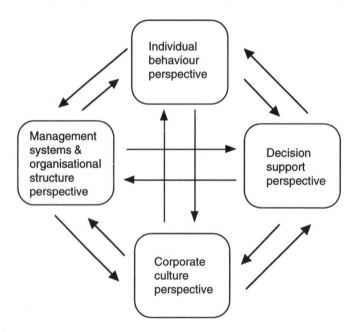

Figure 1   A Four-Perspective Model of Learning in Organizations.

The four-perspective model of learning may help relate a given technology to corporate strategy by identifying core competences which the technology may support. Furthermore, the necessary links may be spelled out between the technology and the organization which is to adopt and use the technology.

The model has been applied to four phases of a development project in which a new technology may be introduced: (i) definition of scope and objectives of the development project, including establishing the need for learning in view of strategic challenges; (ii) analysis and diagnosis of the nature and direction of current learning taken place in the enterprise - to be compared with the need; (iii) development of solutions, including appropriate technological means; and (iv) managing the implementation process.

3. IMPLICATIONS OF THE LEARNING PARADIGM

Adopting a learning approach, e.g. by working with the four-perspective learning model presented in the preceding section, has several significant implications for the industrial enterprise and the way in which it operates. In this section we shall discuss some of these implications which point to and give substance to a new paradigm: The Learning Company.

In addition to be instrumental for developing core competences in industrial enterprises, the learning approach suggests

o   A re-orientation of focus for the development of decision support,

o   Identification of individual roles and means for empowerment and self-governing,

o   A new role of management systems which support continuous learning processes on the basis of past experience and the creation of shared perceptions,

o   New organizational forms centered around learning networks.

3.1. Decision support for planning requires new systems

Traditionally, the development of information technology has been oriented towards providing a decision maker with the capability to quickly and accurately retrieve and duplicate relevant data and information. Although this has led to an increased productivity, individual learning has not been supported. This will require systems which can support a decision maker in combining different sets of data, e.g. by use of graphical presentations or statistical analyses. Furthermore, the decision maker should be able to identify patterns for instance by combining characteristics of the current situation, the action considered and chosen, and the observed effect. Such a set of information constitutes the nucleus of experience.

In recent years, decision support systems have been developed in several engineering disciplines; in the area of management systems for production planning and control primarily to scheduling. The focus in production management systems has been directed towards increasing the accuracy and concurrence of data which has resulted in marked improvements in industry. However, the decision maker is not supported in his/her effort to learn from past experience.

Many decision makers in different departments and sections of an industrial enterprise are involved in tasks pertaining to the overall delivery, quality and cost. Hence, a decision maker is part of a collective learning process. Decision support systems therefore should reflect the cross-functional nature of learning.

It is our experience that it is easy to point to means for improving individual decision making by using the four-perspective model to provide decision support which may stimulate individual learning and yet be in tune with the cross-functional relationships dominating the modern industrial enterprise.

Case example. In a Danish enterprise a so-called "intelligent" CAD system was introduced, incorporating a previously developed simplified and modularized product program. In contrast to the CAD system used in the past as a drawing program and for retrieving drawings, the new system included a 10-step procedure, whereby each designer was supported and directed through the design process. The designer was asked to select between pre-defined and previously used modules and components to configure the new product. Rules and knowledge were attached to each component in the system, thus allowing only certain connections between components. The system was interactive so that the designer, through the different choices he made in each of the 10 steps, was kept informed with regard to specific conditions he should be aware of, helping him to learn from past mistakes.

The case shows how computer applications such as a CAD system are able to stimulated and support individuals in their jobs by placing considerable emphasis on visualization and interaction.

3.2   Self-directed and empowered employees

A general characteristics of the literature on Continuous Improvements, Total Quality Management, Total Productive Maintenance, etc. is the proactive role expected of employees. They are assumed to accept a greater responsibility, to take initiative in promoting improvements, and to be more self-governed. However, this presupposes that employees are empowered to play this new role. In turn, this requires not only the necessary qualifications and skills, but a comprehensive understanding of the objectives of own task as well as their relationships to the objectives of the enterprise.

Although there is a general agreement about the future role of employees, no well-defined route exists for carrying out the needed transformation process. This will require that employees not only achieve a thorough understanding, but also change attitude and learn new behavioural patterns.

A learning approach will focus attention on the creation of a positive learning environment in which experiments may be carried out allowing employees to exceed known barriers. New working habits must be developed including the development of new roles. By definition, a role emerges as the result of a mutual adaptation process.

Thus, we believe that a learning approach may inspire an enterprise to adopt appropriate means for developing self-directed and empowered employees. This will include working with several perspectives of our learning model, such as personal training and coaching, redesign of management systems (e.g. wage system) and organizational structure.

3.3   A new role of management systems

Many management systems in operation have been designed to plan and control the execution of actions related to production, distribution, quality, etc. representing what may be called the first order of control. Such management systems are indeed very useful for the daily operation. However, they do not provide support for any effort to improve performance.

Empirical studies in order producing enterprises indicate that most management systems do not encourage individual learning and do non-intentionally constitute a barrier to any cross-functional learning.

A learning approach suggests that the role of management systems be seen as a means for guiding the attention of employees and for providing decision support for cross-functional learning, and not merely as an instrument for controlling the execution of activities. This assumes that employees are acting responsibly and are educated and trained to take initiative. In this way management systems may support the development of self-directed employees and of new flat and flexible organizational forms. Riis and Frick (1991) define the new role of management systems in this way:

o   to stimulate individual learning processes by motivating individual employees, so that experiences from other areas of the organization may be included in the performance of their own task.

o   to stimulate organizational learning processes by the development of shared perceptions (common experiences) related to key issues essential to the corporate strategy and the development of competitive strengths.

o   to use organizational learning processes in adapting to external and internal changes.

Instead of seeing the design of management systems as a "one shot" event with experts determining what is the best system and leaving employees passively to adopt it with no active participation, we suggest that learning processes be incorporated into the design process. This will also encourage improvements derived from individual contributions of employees in their everyday work.

Case example. An order producing enterprise wanted to improve its planning and execution of custom orders which cut across functional and organizational borders. Only a very poor basis for planning existed with no record of previous duration of activities. Instead of developing and introducing a large project management system for planning and control of activities, a standard PC project management software was introduced as a means for - through a dialogue with the involved employees - establishing rough estimates of the duration of key activities and a shared understanding of their interrelationships. In this way, the PC software was used as a learning tool to stimulate the creation of a shared perception of the issue and to develop rough estimates commonly agreed on. It was assessed that the building of mutual trust and respect across functions in the organization was as important as obtaining more precise estimates of the activities, and was seen as a means for creating ownership and commitment to plans.

3.4   Flexible and flat organizational structure

Current literature on organization theory and industrial practice offer strong arguments for seeking new organizational forms to cope with the demand for greater flexibility and at the same time being able to operate with precision, high quality and productivity.

Ideas such as organizational networks to replace the traditional one-dimensional hierarchical structure, lean management, and the use of multi-disciplinary teams are being developed and tested in practice. However, no solid theoretical basis with practical implications has yet been developed to provide substance to the many headings associated with world class manufacturing and lean production, to mentioned a few of the ideals proposed.

The development of methods for analyzing cross-functional activities represent an interesting new point of departure for obtaining a comprehensive understanding of the complex interrelationships between sections and departments. The methods are central for programs such as Activity Chains, Time Based Management, and Business Process Re-engineering. However, they represent only the beginning. The development of cross-functinal cooperation should be seen in a learning perspective.

A learning approach will seek answers to questions like: Where does learning take place? Who in the organization is learning? How may learning be stimulated along an activity chain?

Case example. An order producing enterprise decided to aim for a reduction of the lead time in the technical administrative order flow by 50 per cent. One of the parts of the order involved the development of individual customer solutions requiring close contact to the customer for discussion and approval, and the construction of prototypes in the workshops.

The enterprise decided to form a team of engineers and skilled workers, supported by a planner. This turned out to reduce the development time and to increase the quality. This new working mode is an example of using a means known from production as production groups to support and stimulate cross-functional learning in the technical administrative area.

4. TOWARDS THE LEARNING COMPANY

Based on the implications discussed in the preceding section we believe that the four-perspective model of learning may help provide more substance to the discussion which eventually will lead to a clear definition of operational terms and their interrelationships to represent the content of the learning company as a new paradigm.

We propose to define a learning company as

"An enterprise having an appropriate combination of the four types of learning capable of mutually supporting one another enabling the organization to continuously develop its preparedness for external and internal changes."

As discussed in the preceding section, the four-perspective model of learning has significant implications for key elements of an industrial enterprise. Moreover, the discussion and case examples revealed a very close interrelationship between the elements. They constitute a framework for learning by guiding the learning processes, and at the same time they are themselves changed as a result of learning processes.

A deeper understanding of the mutual interplay between these elements of an industrial organization is essential for the extent to which the new paradigm will have a positive impact on industrial enterprises. Hence, further research is needed along these lines.

5. CONCLUSION

The importance of being adaptive in a dynamic and turbulent environment has called for an increased focus on learning from past experience and from new knowledge obtained outside an industrial enterprise. A newly developed multi-perspective model of learning in organizations was presented as a means for providing more substance to the emerging paradigm of the learning company. In continuation, we outlined implications of adopting aj learning approach for elements of the learning company. For example, it was shown that learning may play a key role for empowering employees; that the development of decision support needs a re-orientation of focus to address learning processes; that a new role of management systems may be derived which support continuous learning processes on the basis of past experience and the creation of shared perceptions; and that new emerging organizational forms rest on learning networks and learning along activity chains.

Case examples showed that the new paradigm of the learning company includes a complex and not yet fully understood mutual interplay between the elements. Further research along these lines are needed.

6. REFERENCES

Argyris, Chris (1993): On Organizational Learning, Blackwell, 1993.

Argyris, Chris and Donald Schon (1978): Organizational Learning: A Theory of Action Perspective, Addison-Wesley Publishing Company.

Cyert, Richard and James G. March (1963): A Behavioural Theory of the Firm, Blackwell, Second Edition.

Duncan, Robert and Andrew Weiss (1979): Organizational Learning: Implication for Organizational Design, in Barry M. Staw, Research in Organizational Behaviour, Vol. 1., Greenwick JAI Press INC.

Hall, Douglas T. and Cynthia Fukami (1979): Organizational Design and Adult Learning, in Barry M. Staw, Research in Organizational Behaviour, Vol. 1., Greenwick JAI Press INC.

Jelinek, Marian (1979): Institutionalizing Innovation: A Study of Organizational Learning Systems, Praeger Publishers.

Mintzberg, Henry (1983): Structure in Fives: Designing Effective Organizations, Prentice-Hall.

Neergaard, Claus (1994): Creating a Learning Organisation: A Comprehensive Framework, Ph.D. Dissertation, Dept. of Production, Aalborg University.

Nonaka, Ikujiro (1991): The Knowledge-Creating Company, Harvard Business Review, November-December.

Riis, Jens O. (1978): Design of Management Systems - An Analytical Approach, Akademisk Forlag, Copenhagen.

Riis, Jens O. and Jan Frick (1991): Organizational Learning: A Neglected Dimension of Production Management Systems Design, p. 141- 150 in Advances in Production Management Systems by E. Eloranta (Ed.), North-Holland.

Riis, Jens O. (1992): Gathering Cross-Functional Experience (in Danish), Department of Production, University of Aalborg, Denmark.

Schein, Edgar H. (1985): Organizational Culture and Leadership, Jossey-Bass.

Schein, Edgar H. (1988): Organizational Psychology, Prentice-Hall.

Schein, Edgar H. (1990): Organizational Culture, American Psychologist, Feb.

Jens O. Riis is a Professor of Industrial Management Systems at the Aalborg University. He holds an M.Sc. in Engineering and a Ph.D. in Operations Research. His research covers areas such as design of production management systems, project management, and management of technology.

Claus Neergaard holds an M.Sc. in Engineering and a Ph.D. from the Aalborg University

PART ELEVEN

Modeling and Simulation in CAM

46

Modelling PCB facilities in ARENA - A generic data-driven simulation approach

R. de Souza

School of Mechanical and Production Engineering
Nanyang Technological University, Nanyang Avenue,
Singapore 2263. Fax (65) 7911859
e-mail: internet%"mrdesouza@ntuvax.ntu.ac.sg"

Abstract

This paper will present, from a simulation perspective, the building of a generic simulation model for printed circuit board (PCB) assembly lines. The stages in the evolution of the modelling over the project are discussed as are the experiences in the selection of an appropriate modelling platform. The paper will offer some poignant conclusions as to the potential of these models in the light of the recent leaps in the development of a new breed of modelling tools.

Keywords

ARENA, PCB, simulation, generic, data-driven

1 INTRODUCTION

Simulation as a technology has been around a great many years, however only relatively recently has the advancements in computing technology provided the impetus for the development of simulation tools that are far more user-friendly and powerful than was ever possible before. However, even with these significant advances, simulation still remains a much under-utilised tool in decision-making even in the relatively simple case of a PCB facility where immediate benefits may be realised.

Management of a PCB facility can be complex, particularly in the subcontractor sector of the industry where there is uncertainty of incoming orders, quantities and mix. Even where certainty in production does exist, it is often necessary to compare alternatives for future equipment and layouts and to compare work flow alternatives such as changing batch sizes.

Discrete event simulation is rapidly becoming a popular tool to address these manufacturing issues in electronics manufacturing assembly. Design, planning and experimentation through simulation models is ideal as one can plan or anticipate production requirements without undue material handling of the PCBs and downtime of expensive equipment. Presented in this paper is such a model that can rapidly simulate the range of manufacturing lines by just populating logic or data modules. The model has the capability to represent single, multi or mixed model production; single or double-sided boards; machine breakdowns, shift and break policies and of course, produce necessary output such as capability of meeting due dates and throughput.

The work reported in this paper is a development of the work in de Souza (1993) on the use of simulation for representation of a SMT lines, with SLAMSYSTEM from Pritsker Associates, which can be used to model manufacturing detail such as breakdowns, shifts, product mix etc. In this paper, the latest developments are presented. The model described is ported to a very new flexible simulator called ARENA from Systems Modelling. The graphical model is now completely data driven which is a distinct advantage. ARENA is used as the basis for the discussion.

2 WHY PCB FACILITIES?

PCB facilities were targeted for several reasons. First and foremost they are relatively simple paced serial lines capable of being modelled in almost all packages (although a significant amount of materials handling capability needs to be present to cope with the material movement and the sheer volume of production of printed circuit boards. Although these lines are simple in flow, there are many variations (centred around a basic three types) possible. Further the routing of products and their respective launching in terms of either single, multi- or mixed models onto the line add a new dimension of complexity. This is further compounded where the particular line or lines are set up for contract manufacture. It is the latter, and its significant variability, that makes discrete event simulation a natural choice. Add to the previous, the fact that electronics manufacturing is one of the biggest manufacturing sectors, with PCB assembly as a major supporting industry, in Singapore and it immediately becomes clear that there are significant potential benefits to be gained by providing a rapid development tool adaptable to a particular line for deployment by manufacturing management.

3 SELECTING A SIMULATION SYSTEM

The next stage was to select an off-the-shelf modelling package that was available locally, matched the task, flexible, available on a range of platforms (particularly on DOS personal computer and UNIX workstations) and most importantly, to the local collaborators, of moderate cost. The latter could be played down if the package could also be justified for other uses. The range of discrete event simulation software available at the time was Witness, Slamsystem, Taylor II, Siman Cinema and Automod.

Attention turned to evaluating all the aforementioned packages. A survey was carried out through building simple models and then progressively more complex models to the point that an obvious candidate package emerged. (de Souza and Loh, 1993). It emerged that Siman Cinema offered the greatest potential with its capabilities in materials handling, data processing, interfaces to spreadsheets and files and animation. The modelling perspective of separate model and experiment and permutations of the two was particularly appealing.

However, the survey was short-lived, as a new breed of flexible simulator was introduced in the form of ARENA. The latter being a "natural" progression by the company on Siman Cinema. This questioned the original survey on classifying our results by languages and simulators. This new product now bridged the gap between simulators (easier to use but restricted domain) and languages (flexible but with a greater learning span).

This new hierarchical flexible simulator in the guise of ARENA offered the capability to rapidly develop all that we had done but in a fraction of the time. Was a PCB manufacturing simulator now easier to build? The conclusion reached was positive in that no domain model was offered to a novice user and thus learning was still significant. Visual mass customised models had a role to play! Building the model in ARENA is now reported.

4 ARENA

ARENA is a complete simulation environment that supports all basic steps in a simulation study. The ARENA system includes integrated support for input data analysis, model building, interactive execution, animation, execution tracing and verification and output analysis.

The key idea behind ARENA is the concept of tailorability to a specific application area. The ARENA system is based on the unique concept of application templates. This is fundamental to flexibility and ease-of-use provided by ARENA. Templates provide the user with a modelling interface of modules that is familiar to the user. For example, at the highest level the user can pick and place a server module that can define a processing station. This module with animation constructs built-in can then be defined in a manner familiar to the user. The real power lies in using a domain specific template. In this instance the user now has a simulation tool that closely matches the real system being modelled - hence the user is presented with concepts and terminology that are focused on his or her problem. This dramatically reduces the level of modelling abstraction required by the user. However, the user is not limited to the constructs provided by the single domain. The user can combine domain restricted templates from one or more templates with the full modelling power of the SIMAN language and thereby avoid the modelling "brick wall" encountered with traditional hard-coded domain-restricted packages. Thus, ARENA is a package that can behave as a "simulator" for a restricted domain or migrate to a full-blown simulation language. Thus this environment could be properly described as a flexible simulator.

With ARENA the investment in simulation is not only protected but able to grow with the needs of the user. Third party provision of templates, at reasonable prices, is rapidly becoming available.

5 A GENERIC MODEL?

The word generic conjures up an image of a model capable of representing any line. In addressing the feasibility of this issueit was necessary to study the core process flow as aforementioned. Though, three main flows were evident, our practical investigations, on the other hand, showed that a basic linear flow persisted and that Types I and II assembly were most common. There was no significant variaition in the number of machines, types of board handled and hence their routing essentially because of technological series processing. The challenge thus lies in building a model that can easily accommodate product mix, schedule variations/disturbances and machine operating parameters. The model is generic in that it involved mass customisation for the particular target sector. This was successfully achieved and put to the test through a number of case studies.

6 THE MODEL

The animated model is shown in Figure 1. The process of building the simulation model consists of choosing the modules from a template that best represent the system processes, combining them in a directed network that represents the entity flow, and parameterising the modules with model-specific data. The modules are then substituted by realistic animated machine icons. The output from the model would automatically report statistics on job waiting time, queue length, server utilisation, time in system and throughput.

"Created" boards are held in a stacked queue to await launch into the line. Each board travels on an accumulating conveyor to a series of stations (boards may bypass some of these stations based on their respective attributes). The process flow, see figure 1, are screen printing, glue dispensing, chip mounting, second glue dispensing, second chip mounting, IC placement and depending upon board type the boards are either channelled to wave or reflow soldering. Finally board production statistics are gathered at the final module. Various screens depiciting this process are shown in the series of figures 2-8.

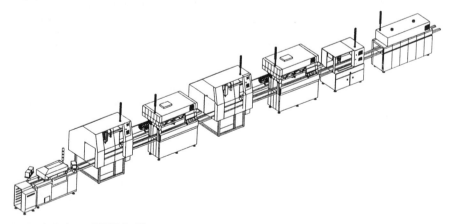

Figure 1 Animated PCB facility.

The model is data-driven in that most of the model parameters are predefined. For example, shown in figure 2 is the module definition of attributes of the product mix and the schedule. This essentially defines the routing and sets the model termination criteria. Figure 3 shows the module that permits the user to define operating parameters of each machine as process steps and operating profile. Figure 4 shows the module definition of the conveyor characteristics. Figure 5 sets the module's statistics for collection eg. time-in-system. The collected statistics are displayed in a report as shown in figure 6.

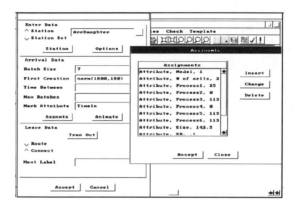

Figure 2 Module displaying attributes of the PCB that define routes.

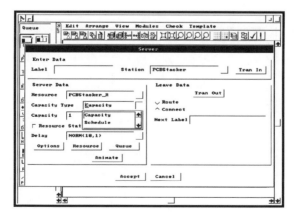

Figure 3 Module displaying operating parameters.

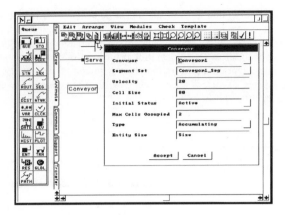

Figure 4 Module displaying conveyor parameters.

One objective in modelling the PCB facility is to assess the time in the system for each board and the throughput. This is achieved by setting tallies and counters in the statistics module. These results are then reflected in the report. The model shown in this example had a schedule time of 2000 time units and 8 replications.

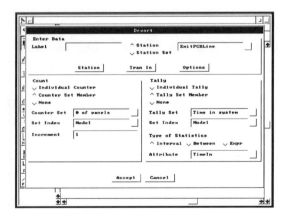

Figure 5 Statistics collection module

```
                     SIMAN V - License #       1
                     YOUR COMPANY NAME HERE
                   Summary for Replication 1 of 1

Replication ended at time      : 28800.0
                          TALLY VARIABLES
Identifier               Average   Variation  Minimum    Maximum    Observati
Time in system Dauhter   2059.3    .14079     1662.2     2474.2        7
Time in system MClock    2569.1    .08472     2275.4     2879.4        7
Time in system Base      10228.    .09590     8566.6     11882.        63
                      DISCRETE-CHANGE VARIABLES
Identifier               Average   Variation  Minimum    Maximum    Final Val
StQ1_R Busy              .00000       --      .00000     1.0000     .00000
StQ1_R Available         1000.0    .00000     1000.0     1000.0     1000.0
# in QQ1                 .84257    7.2976     .00000     69.000     .00000
# in Conveyor3_Q         .00000       --      .00000     1.0000     .00000
# in Conveyor2_Q         .00727    11.688     .00000     1.0000     .00000
# in Conveyor1_Q         3.7398    2.3136     .00000     49.000     .00000
                              COUNTERS
               Identifier                Count   Limit
               # of Daughter panels         7    Infinite
               # of MClock panels           7    Infinite
               # of Base boards            63    Infinite
                             FREQUENCIES
                                       --Occurrences--   Standard   Restric
Identifier               Category      Number  AvgTime    Percent   Percent
STATE(PCBStacker_R)      STACKER       65      21.743        4.91      4.91
                         IDLE          66      414.94       95.09     95.09
STATE(ScreenPrinter_R)   SP            77      33.323        8.91      8.91
                         IDLE          78      336.33       91.09     91.09
STATE(Glue1_R)           G1            63      36.939        8.08      8.08
                         IDLE          64      413.63       91.92     91.92
STATE(ChipMounter1_R)    CM1           140     65.264       31.73     31.73
                         IDLE          141     139.45       68.27     68.27
STATE(Glue2_R)           G2            63      43.266        9.46      9.46
                         IDLE          64      407.41       90.54     90.54
STATE(ChipMounter2_R)    CM2           140     66.577       32.36     32.36
                         IDLE          141     138.15       67.64     67.64
STATE(ICPlacement_R)     ICP           140     66.169       32.17     32.17
                         IDLE          141     138.55       67.83     67.83
Execution time: 0.07 minutes.
```

Figure 6 Report module.

7 CONCLUDING REMARKS

Simulation technology is reaching the point where it can have a significant and widespread impact on the design and operation of nearly any complex system. It is rapidly being being viewed as a natural and indispensable tool for making significant design and operational decisions. This paper has presented an exploratory view of a model targetted at simulation practitioners in this part of the globe and described one tool that provides the greatest possibility for addressing these issues.

As discussed, modelling has been made so much easier with the advent of the new breed of modelling tools. However, selecting, creating and editing large connected visual models still remains a cumbersome task and requires a definite skill level in the tool of choice. Building visual models provides a base from where more sophisticated models can be developed. The model presented can be considered as mass customisation for the particular target sector.

Whereas in the past model building would address the requirements of a facility either in particular or generically, with the new tools it is becoming increasingly important to parameterise and modularise. This gives added flexibility, power and productivity to the subsequent modeller.

8 ACKNOWLEDGEMENTS

The simulation group are gratefully acknowledged as is the support given by ER Mechatronics (S) Pte Ltd. and the Applied Research Fund.

9 REFERENCES

de Souza, R. (1993) Design and development of a generic PCBA line simulator", *Journal of Electronics Manufacturing*. Vol. 3, No. 3, 133-44.

de Souza, R., and Loh, C. L. (1993) A review of some simulation approaches for the modelling of manufacturing systems. *Proc. 2nd Int.. Conf. on CIM*, Singapore, 469-75.

10 BIOGRAPHY

The author is a Senior Lecturer in the School of Mechanical and Production Engineering. Research interests include flexible manufacturing systems and the application of discrete event simulation to industrial problems. The author has consulted for many organisations in Singapore, presented many technical seminars and has chaired or organised technical conferences in many countries. Over 50 articles have been published and a book co-authored. The author is also an editor of the Journal of Electronics Manufacturing.

47

Using Discrete Event Simulation for the Improvement of a Traditional Mass Production System

P.F. Cunha [1] and *R.M. Mesquita* [2]

[1] EST/IPS - Escola Superior de Tecnologia do Instituto Politécnico de Setúbal
  Rua Vale de Chaves, Estefanilha, 2900 Setúbal; Portugal; Tel: 65.761621, Fax: 65.721869
[2] Associate Professor - Institute of Materials and Production Tecnologies
  INETI - Instituto Nacional de Engenharia e Tecnologia Industrial
  Estrada do Paço do Lumiar; 1699 Lisboa; Portugal; Tel:1.7164211, Fax:1.7160901

Abstract

Many different studies have been presented showing the successful use of simulation to support the implementation of Advance Manufacturing Technologies, specially in batch production systems. The design and analysis of a Flexible Manufacturing System is a typical example refereed in the literature to show the advantages of simulation. However due to the potential benefits that exist in manufacturing systems, where the products are manufactured in high volumes using dedicated lines, a typical mass production system was chosen to demonstrate the broad interest of using Discrete Event Simulation (DES). In this paper the experience of developing a simulation project is presented and the benefits obtained from using DES to the analysis and evaluation of proposals to be implemented in a manufacturing system will be discussed.

Keywords

Advance manufacturing technologies, flexible manufacturing systems, discrete event simulation, MRP, job shop production, batch production and mass production.

1. INTRODUCTION

In today´s economy the evolution of manufacturing industries is rapidly changing due to many different factors. The increased labour costs, the rising overall manufacturing costs, the high cost of capital and also the increased competition from foreigner and inside companies are some characteristics of competitive environment which involve the manufacturing industry. To keep the competitiveness, the companies are challenging to search for productivity improvements, adopting new organisational concepts to manage the manufacturing system and to increase the flexibility of designing and manufacturing new products. In this new environment the chosen technology and the used management approach are relevant requirements to run the manufacturing system and to reach a competitive position. The flexibility has been a keyword to take into account by top managers.
Many studies have been presented showing that the low performance of a Flexible Manufacturing System (FMS) is mainly concerned with the way the overall system is managed (Kovács, 1992). A modern management approach is usually connected with more competitive companies where a continuous search for improvements in the manufacturing system is a target. One of the of these

manufacturing strategies is the manufacturing process simplification and the weigh reduction of non-value added operations in the system, using a high level of staff involvement. The reduction of lot sizes, stocks, lead times and the balance improvement between production rates and market needs, are standard criteria in the evaluation of the system performance. In opposition the maximisation of machines use, increment of lot size and inefficient information flow inside the company are few characteristics of traditional organisations, sometimes with different hierarchical levels. In this organisational structure it is possible to observe an increment of specialisation in terms of humans resources and manufacturing capabilities in opposition with the flexibility improvement desired in modern manufacturing systems. In any type of approach used to manage the manufacturing system, the lack of confidence in the expected results by implementing of a specific strategy has been a restriction for an effective system management. The possibility to know in advance the performances of the system, after the proposed changes have been implemented, will help the managers in decision making. A wrong decision could have a major impact in the company's competitive position.

Due to the high number of variables and parameters, involved in the characterisation of a manufacturing system, the Discrete Event Simulation (DES) has being proved to be a very useful technique for the design and analysis of a manufacturing system. The simulation allow to check the results of the proposed changes and allow to test the effectiveness of new ideas.

2. APPLICATION OF SIMULATION

Simulation, as a technique, has being used more often for different type of manufacturing systems. Numerous studies have being publishing showing different applications of simulation which, as a technique allow to enhance the manufacturer ability to make tactical and operational decisions concerning new technologies or as a decision support tool on an everyday basis. Chase and Aquilano (1989) presents that in general the Discrete Event Simulation can be used in the analysis and design of a manufacturing system or in the scheduling of a production system. They present also DES as a useful technique for training managers and workers in how the real system operates, demonstrating the effects of changes in system variables and developing new approaches about organisational relationships for a better functionality of the system.

In the analysis and evaluation of an existing or proposed manufacturing system the simulation has proved to be an effective technique. During the Design phase for the implementation of a manufacturing system, simulation tend to focus on equipment and material handling requirements. The output of the simulation studies aids the design of the system specification and the evaluation of system performance (Adiga,1991). Simulation can be used also to test different control and dispatching strategies. In this context, there is a need for system performance prediction rather than system failure or malfunction assessment after the problems occurrence. The more problems can be anticipated, the better the chance of enacting timely and effective solutions (Castek, 1986). This can be done for example through the integration of simulation capabilities with the existent manufacturing planning and control system. Simulation can highlight which requirements due to resources constraints will be late or if capacity is available to complete the production schedules. Thus the simulation model contribute to decisions concerning an effective strategy to reach the defined target for the manufacturing system.

The manufacturing layout and the technology used can be conditioned by the variety of products and the volume produced in each manufacturing system. Products manufactured in a very high volume are effectively handled with mass production systems. The high volume justifies large capital expenditure for equipment designed as product specific. Product variation and even engineering design changes are often difficult and expensive to incorporate. At the opposite end, the job shop production systems have to handle frequent variation in the product design. Thus the batch sizes are small and doesn't exist commonality of routing for the work that constantly arrive. Those job shop systems are designed to accommodate high product variability, paying the price in low efficiency. In most industrialised countries the majority of discrete manufacturing systems fall between the process-oriented job shop and product oriented dedicated line, in an area called "mid-volume, mid-variety" (Kochan, 1986). The discrete manufacturing systems classified in this area are also known as a batch production systems.

All types of manufacturing systems offer ample opportunity for the useful application of simulation. In spite of the investments that have being observed in the implementation of Advanced Manufacturing Technologies, specially in batch production, the interest in developing a simulation project also increase when it is observed an increment in the complexity of the system or where changes in the process which involve high capital investment are proposed. The production facilities in mass production systems consist of highly specialised and dedicated machines and tools capital intensive. As in FMS the call for high capital investments and the increased complexity of the system highlights the need for cost effective methods for the design and analysis of a manufacturing system. Thus the potential benefits obtained from a simulation project depends on the complexity of the system together with the overall amount of required investment.

3. CASE STUDY

The general discussion presented above allowed to identify the interest of simulation for the analysis and the improvement of production systems and to define levels of application where DES can be used. simulation has being presented as a tool to support the manufacturing management and operations. This section is concerned with the use of DES as a tool to support the company to improve its manufacturing system. Our approach is to look at an existing traditional manufacturing system, where an high number of shortages, unreliable deliveries, high level of work-in-process (WIP) was observed, and to see through the developed study how a simulation model can be used to support the manufacturing engineers and managers to identify the existent problems and suggest a strategy to solve them (Cunha, 1993).

3.1. Company profile

Fábrica Nacional de Moscavide (FNM) is one of the national defence industries. It is considered to be a medium sized enterprise which essentially manufactures light ammunition for small arms of 7.26mm, 5.56mm and 9mm calibre. The current manufacturing system that exist in FNM can be considered to be mature and consist mainly of old equipment. Most of the operations which characterise the manufacturing system are executed by unsophisticated machines in a similar sequence to each calibre. The organisation of the manufacturing system at FNM can be termed as a mass production system. The existent complexity in the manufacturing process management at FNM is connected with the high number of machines involved in the process and due to some restrictions that exist in the material flow throughout the manufacturing process. The physical distance between sections where each component is manufactured and the specific characteristics of the market for this products impose some relevant rules to manage the manufacturing system. Figure 1 presents a sequence of basic manufacturing routes involved in cartridge manufacture.

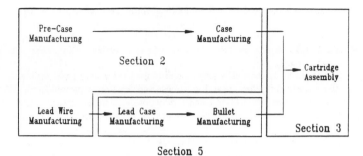

Section 5

Figure 1 Manufacturing sequence installed in FNM plant for both cartridge manufacture.

In general the highly dedicated machine, the defective organisation and the long lead times of the manufacturing system means that the profitability of the company is very susceptible to variations in the master plan.

3.2. Project Objectives and Methodologies

The aim of the project was to investigate how the performances of the manufacturing system at FNM can be improved. The use of Discrete Event Simulation allow an understanding and definition of the problems and opportunities existing in the present manufacturing system. To understand and define the existing manufacturing system at FNM, a set of interviews were established with the top managers and their operational staff. Hence, different types of problems were identified and a list of issues which have influence in the manufacturing system was obtained. Some of the issues were the following:

- Instability in the master production plan;
- Low flexibility of the process to deal with changes imposed by new orders;
- High manufacturing lead times;
- Existence of large WIP as a way to solve breakdown problems;
- Different transfer lot sizes for different sub-components of the same product;
- Extensive paper work required between different areas of management;
- Lack of information between departments connected with manufacturing;

From this preliminary analysis of the overall manufacturing system it was identified that the main problem is the instability of production flow. It was usual to observe delays in production plans or disruptions in the production flow due to the unavailability of raw materials or consumable products. The high level of work-in-process (WIP) is used as a management approach to compensate the effect of the low reliability of the manufacturing system and the instability of the master plan. The existence of high manufacturing lead times also helps to increase the level of WIP. Improvements of the manufacturing system performances at FNM, i.e. mainly reducing the WIP and increasing the accuracy and realism of the manufacturing plan and control system, will allow to increase the competitiveness of the company.

To use DES techniques a well defined methodology was followed. First, a set of specific data related to the real manufacturing system was collected. To built the simulation model a few assumptions were made. After the model had been built, its validity was checked to assess the reliability with which the simulation model reflected the behaviour of the real system. To identify the existing problems in the manufacturing system the simulation was run with real data. From the identified problems further developments and changes in the simulation model were done to test the different solutions.

3.3. Model Development

To identify the characteristics of the manufacturing system and to allow a valid representation of the system to be constructed in the form of a simulation model, a commercial simulation software was used. The simulation system WITNESS was specially designed for building simulation models of manufacturing systems. The animation and the graphical facilities available in WITNESS allow a easy visualisation of a complex manufacturing system operation. The development of a simulation project has unique aspects which must be carefully managed to ensure their effectiveness. Familiarity with the manufacturing system and its operation is a basic requirement for accurate modelling. Different factors which limit the performance of the system must be investigated. Bearing in mind that the simulation results are only as real as the input data, it was essential to collect and use the most correct modelling information as possible.

During the construction of the simulation model a variety of assumptions were made. These assumptions allowed the model content and structure to be simplified without losing an acceptable level of accuracy. Essentially, the assumptions were made to remove factors which would have no influence on the performance measures under consideration. Previously to model development, the level of detail which would allow the implementation of each operation in the manufacturing system was defined as part of the simulation strategy. To assess its validity, the results of the simulation model were compared with the data retrieved from previous manufacturing runs under the same conditions and/or against the empirical knowledge of the manufacturing process. The accuracy of the model was also assessed using the available information about executed production plans which could confirm the predicted results of simulation.

3.4. Analysis of Manufacturing System Behaviour

After the simulation model had been validated, different approaches to analyse and assess the behaviour of the manufacturing system at FNM were defined. To test the manufacturing system, the production of a small amount of cartridges was used to simulate the manufacturing system, with no constraints in the production flow. Then, a production plan was prepared by FNM staff, to satisfy a dummy order for both calibre, and was used as input data to drive the model. This production plan was developed using the company manufacturing planning and control system (MPC). Through this analysis it was possible to observe the way the system changes and to identify the following main important constraints of the system:

- The level of WIP observed during the manufacturing run is significant. Through the simulation it was shown that by running the model with the developed production plan, high levels of WIP would be produced throughout the manufacturing process.

- Existence of bottlenecks in the process due to operations shared between different products and differences in cycle times. The influence of degreasing operation, shared in the manufacturing process for both calibres, can be observed in Figure 2.

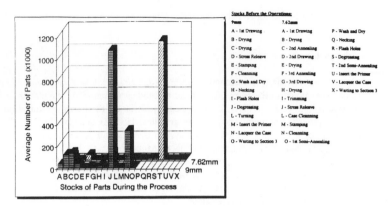

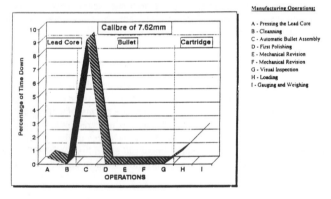

Figure 2 Average number of parts in stock throughout the pre-case and case manufacturing process.

- The rate of breakdowns relative to the system capacity. Throughout the manufacturing system, the machines which have a higher rate of breakdowns were identified through the simulation. Figure 3 shows the breakdown analysis for lead core, bullet and cartridge manufacturing process for 7.62mm calibre.

Figure 3 Percentage of machine down time in lead core, bullet and cartridge manufacturing process.

3.5. Manufacturing System Improvements

By using the production plan generated by the MPC system available in the company for driving the simulation model, the results suggests that different approaches be used to release the observed constraints. The main possible approaches were:

i) The use of additional equipment in the pre-case manufacturing process to avoid bottlenecks in the operations carried out at the same machine for both calibres.

ii) The implementation of a new Manufacturing Planing and Control system in order to improve the production plan.

iii) The reduction of the breakdown rate, considering that this could be done by implementing a preventive maintenance for the automatic bullet assembly operation. This operation, being a critical operation, was rated with the highest breakdown rate.

iv) The enhancement or adaptation of existing equipment in order to increase the capacity of the weighing operation necessary for the lead core manufacturing, where the build-up of WIP occur.

Both the implementation of a maintenance plan for automatic bullet assembly operation (iii) and the increase of capacity in the weighing operation in the lead core manufacturing process (iv) are approaches from which an increase in the capacity for each operation and the reduction of WIP could be expected. However, due to the higher throughput time and cost per unit in the pre-case manufacture, this process was considered critical in the overall manufacturing system. Thus pre-case manufacture was defined as a priority for the simulation study. The improvement of this set of operations can be done either by releasing or removing the bottlenecks or by developing a new production plan. The former solution involves low capital investment and the later requires a new MPC system based on the MRP concept. To demonstrate the effect of the MPC system on performance a prototype was developed. Figure 4 present the information flow when the new MPC system was integrated with the simulation system.

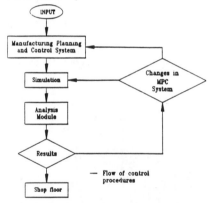

Figure 4 Integration of a MPC system with the simulation system.

3.6. Results and Discussion

The results obtained from the simulation model when it was tested with the data, reflecting the above mentioned improvements, allow to identify different performances. The main differences identified through the analysis of the simulation results can be summarised as follow:

- *Throughput time* - The proposal to add a second degreasing machine showed that the overall throughput time could be reduced 29%. Using the new production plan to drive the manufacturing system, the average time that the parts spend in the system could be reduced by 33%. The proposal to use the new MPC system to run the manufacturing system was more concerned with the improvement in the overall system. For this reason it is expected that higher reductions in the throughput would be reached than when a specific change in the pre-case manufacturing system, like a new degreasing machine, was implemented.

- *Bottlenecks* - The highest build-up of stock occurred ahead of the degreasing operation, specially when producing both calibres at the same time. Table 1 shows the average amount of parts stocked up during each simulation run, as a percentage of the produced quantities.

Table 1 Analysis of stock of parts in Degreasing operation Buffer.

	FNMs Production Plan		Using a 2nd Degreasing Machine and FNM's Plan		New MPC system without 2nd Degreasing Machine	
Calibres	9mm	7.62mm	9mm	7.62mm	9mm	7.62mm
Cartridges Produced	9.8%	17%	0.1%	0.1%	5%	9.5%

The observed decrease of stock in the simulated experiment, using a second degreasing machine and a new production plan was approximately 99% and 45% of the initial simulation run when a company production plan was simulated.

- *Level of Work-in-process* - The average reduction in the amount of WIP throughout the manufacturing process was 43% and 35% when proposals of a new degreasing machine and the use of a new approach to develop the production plan were tested in the simulation model. Figure 5 shows the obtained average levels of WIP during the simulation time.

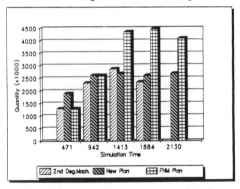

Figure 5 Average of work-in-process for pre-case manufacturing process.

- *Balance throughout the manufacturing system* - Introducing a new machine in the manufacturing process or using a new production plan, was observed a better balance between material flow and each different stages of the manufacturing process. Thus in each experiment was obtained a reduction of 39% and 61% in the average time that the purchased raw materials or components took to be transformed or assembled. Those improvements can be obtained reducing the bottlenecks or taking into account the real constraints in the production plan development.

4. CONCLUSIONS

Discrete Event Simulation proved to have a good capabilities to be used in modelling typical mass production systems. As an engineering tool it provides relevant information about the manufacturing facilities and brings a significant flexibility into the system analysis. The ability to change the model easily and to manipulate the input and output data of the simulation system increase the interest to use DES in an industrial environment. After the constraints to be identified in the simulation model different experiments can be plan to test and evaluate the improvements obtained in the manufacturing system performance.

The possibility to integrate a developed Manufacturing Planning and Control system with the simulation system allows to use the output data of MPC system to drive the simulation model. Thus the simulation system demonstrated to be very effective to show the potential benefits for the manufacturing system when a different approach to generate the production plan was used.

Inside the company, the development of simulation project create a general attitude of enquiry and deeper thought about the system. The results of simulation were used to support some strategic decisions that were taken after this study.

5. REFERENCES

Adiga, S. and Dessouky,M.; Flexible Manufacturing Systems Simulation; Handbook of Flexible Manufacturing Systems; (K.Jha Nand eds.); pp 89-109; Academic Press

Castek, D.M. and Lichtefeld Jr., R.A.; How Simulation Technique can Support Tactical and Operational Decisions; 2nd International Conference on Simulation in Manufacturing; pp 169-175; June (1986)

Chase, R.B. and Aquilano, N.J.; Production and Operations Management - A Life Cycle Approach; Irwin; Boston (1989)

Cunha, P.; Discrete Event Simulation for the Improvement of a Manufacturing System; MSc Thesis; Staffordshire University (1993)

Kochan, D.; CAM Developments in Computer Integrated Manufacturing; Spring Verlag; Berlin (1986)

Kovács, I. ; Novas Tecnologias, Organização e Competitividade; Sistemas Flexíveis de Produção e Reorganização do Trabalho; CESO I&D; pp 17-67; Lisboa (1992)

6. BIOGRAPHY

Pedro Filipe do Carmo Cunha
He is graduated in Materials and Physical engineering in 1988 from Faculdade de Ciências e Tecnologia of Universidade Nova de Lisboa. Since 1991, he is Senior Professor at Escola Superior de Tecnologia (EST-Higher School of Technology) of Setúbal. In 1993 he obtained the Master degree in Computer Aided Engineering at the Staffordshire University in U.K. During is activity at EST he has been involved in different research projects and has developed some actions to the industry.

Ruy Manuel Dias Mesquita
A graduate in Material Engineering in 1977 from the Instituto Superior Técnico (IST-Higher Technical Institute), he obtained his doctorate in Mechanical Engineering at the Technical University of Lisbon in 1988. Associate professor in mechanical Technology at IST since 1991. He has being developing a research activity in the area of Production Engineering, with results published in conference proceedings and in Portuguese and international periodics. At present, he is the director of the Institute of Materials and Production Technologies at INETI-National Institute of Engineering and Industrial Technology.

48

Heracles - Automation Design And Control Made Easy

Authors:
V. Rossetti, G.C. Rizzetti - Pirelli Informatica
v.F.Testi, 250 MILANO (Italy)
Tel +39 2 6442 5661, Fax +39 6442 4999, Email 100136.1036@compuserve.com

W. Wohlgemuth - Siemens
S. Wheeler - BMT
G.C. Citron - SNIA
S. Tresse - Kade-Tech Recherche

Abstract

This paper describes the methodological and technological approaches followed in the ESPRIT project 6464 HERACLES, aiming at developing a set of software tools to support engineers in the activities of designing automation plants, providing means to automatically produce schematics, PLC code, diagnostic and simulation models using an object-oriented design approach which foresees re-use of existing plant components.

Keywords

PLC, automation, design, OOD, CAD, diagnostics, simulation

1. OBJECTIVES OF THE PROJECT

HERACLES addresses the creation of a software tool for the high-level modelling of automated systems, providing the knowledge-based support tools to give practical assistance in the system engineering activity, down to maintenance and support at application level.

The design of automation systems is a complex engineering activity, which aims at producing the Functional Specifications, Input/Output (I/O) List, Electrical Schematics, etc. for a whole automated system. This activity, benefits only very superficially from CAD tools, which generally do not capture the functions of components. Engineering is the starting-point for the production of low-level Programmable Logic Controllers (PLC) software. Without formal methods at the engineering level there is no uniformity at the code level: this hampers maintenance, diagnosis and support.

HERACLES' primary objective is to increase the productivity by:
- computerising routine functions, allowing him to concentrate on high-level design
- providing the tools to help the engineer create, modify and analyse outputs in a fraction of the time taken with conventional approaches
- reducing redundancy and inconsistency of the information produced during the whole development process
- re-using knowledge from previously produced automated systems; users' experience in developing automated handling plants shows that 80% of the basic components and devices could be re-used in a new plant
- defining tools for those design sub-processes which were not yet tool assisted.

The multiple functionalities offered by HERACLES support a wide range of users in their work: high-level plant designers, PLC software specialists, electrical engineers, and commissioning and operating staff (down to shopfloor operators). Each of these users will find a valuable support to their activities.

The system developed consists of modules which assist design, encapsulate low-level coding rules, compile high-level language into PLC software, interpret plant structure, and create an integrated overall plant automation model.

2. THE UNDERLYING CONCEPTS

The development of an automated system consists of several processes. These can be summarised as:
- Planning the automated system extension and the required performance.
- Designing the automated system topology and its components (objects).
- Designing the electrical schematics.
- Designing the Control Software
- Testing the automated plant and related control software. If a Simulation Tool is used in this phase, the simulation model has to be designed first.
- Designing the information for other Runtime Systems (e.g. Monitoring and Diagnosis systems).

The basis for the project's requirements and specifications work has been the development of automated handling plants. This is the project user domain, and it has proved so far to be a sufficiently general process, which does not hamper in any way the generic validity of the results obtained.

The development of an automated plant can be described in 3 phases: design, testing and running.

The **design phase** (high-level design is considered here) produces a series of documents and specifications, which serve either as input to a next design step or development phase, or as plant documentation. Outputs include:
- Specifications analysis, such as civil drawings, layouts, plant behaviour descriptions, lists of machinery;
- High-level design, including topographic drawings, functional and behavioural specifications of the plant components;

- Detailed design, including PLC software code, electrical schematics, Control hardware, Bill of materials.

The **testing phase** involves machinery and electrical panel tests, as well as tests on the PLC software.

Finally, the **running phase** includes:
- Plant components assembly
- Plant start-up
- Customer commissioning and training
- Plant management and maintenance

Traditionally, the development of the topology, the electrical parts and the control software are separate processes in the sense that the interactions between the processes and the management of common data are not tool assisted, the re-use of predefined data (e.g. I/O-Signals) is not planned and consistency checks are not automated. In many development processes the usage of simulation tools for planning and testing the system is not provided, so many problems occur for the first time during the start-up/runtime phases of the system and produce very high costs.

The requirements to use a monitoring system and/or a diagnosis system in the runtime environment to display to the user the actual plant state often arise during the first usage of the automated handling plant when problems occur. In this case the effort for acquiring the necessary data/information for these systems is very high as they are also separate processes and were not planned in the first design phase.

Moreover, in the traditional development process the re-use of existing knowledge about a predefined automated handling plant is difficult, as no methodology and no tools exist to help the user do this, so a lot of effort is spent for reproduction of data that already exists in similar projects.

In summary, the following weaknesses characterise traditional development processes:
- The development process is split into separate processes. This leads to increased efforts for interchanging the produced development information between the different development processes.
- Data produced in one process cannot be directly re-used in other processes.
- Consistency checks can not be performed automatically.
- Problems and faults often occur for the first time in the runtime phase as no simulation tool was used for planning and testing the automated system. This brings higher costs for putting an automated system into operation as it is not completely tested before the installation.
- The integration of monitoring and diagnosis systems is very expensive as they are usually not planned in the first design phase, and the data needed must be prepared (usually manually) as an afterthought.
- The re-use of knowledge is not provided for.
- The standardisation and certification of the automated plants and related control system is not supported by high-level design tools.

3. THE HERACLES APPROACH

HERACLES proposes the use of an integrated set of tools to support the user in all three phases, from design, to testing and running the newly developed automated plant.

The development process is radically modified, as some of the operations that were traditionally performed manually are now performed automatically by HERACLES. Moreover, as the user is assisted in the remaining operations, he must also in some way adapt himself to a new development methodology, where he is guided by the HERACLES system.

The introduction of HERACLES thus effectively supports all types of users involved, eliminating repetitive tasks for plant designers while enforcing the use of company standards, ensuring fast and automatic generation of PLC software, and delivering error-free and consistent testing and maintenance support systems for commissioners and shopfloor operators.

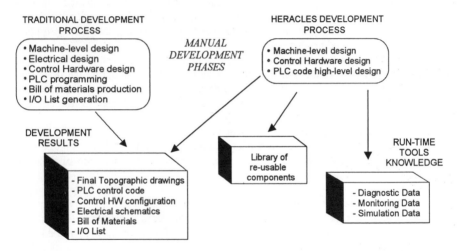

Figure 1 Traditional vs. HERACLES development processes.

Figure 1 compares the traditional versus the. HERACLES development processes. The main advantages can be seen as:
- the number of manual phases required by the improved process are dramatically reduced;
- only the new context produces directly the knowledge for the run-time tools;
- run-time tools data and development results are consistent with one another because jointly derived from the same common source;
- the run-time knowledge can be easily re-generated and updated to respond to changes at the design level.

HERACLES objectives are not to simply automate existing user processes, but to review and improve them. The questions asked were not only *how are we going to achieve this?*, but also *how could the plant designer or engineer do this in a better way ?*

Answers to these questions were on one hand adopting an object-oriented approach to plant design, and on the other hand to support the user at best by automatically generating plant design and engineering outputs.

In HERACLES, the traditional design and engineering process has been broken up into four major phases, from design to run-time management, with intermediate knowledge generation and testing phases (Figure 2). Software modules to support each phase have been developed or integrated in the system from third-party vendors.

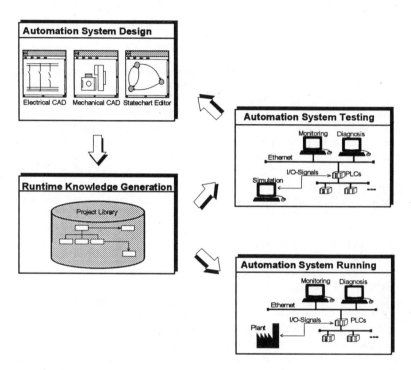

Figure 2 The HERACLES phases.

Phase 1: Automation System Design
Design tools provided by HERACLES include Mechanical and Electrical CAD systems, and High-level PLC Software Design tools.

Phase 2: Runtime Knowledge Generation
The design tools provide the design data and knowledge to a central repository (the Library). Knowledge compilers then take this data and build the knowledge necessary for the testing and run-time tools.

Phase 3: Automation System Testing
A Simulation tool is provided to test the functionality of the designed automation system, before it is actually implemented. It uses simulation models built automatically by the knowledge compilers.

Phase 4: Automation System Running
The PLC software necessary to run and control the plant is generated automatically, starting from its high-level design.
Monitoring and Diagnosis tools are automatically configured to provide run-time control of the automation system, once it is put into operation.

5. IMPACT OF THE PROJECT

5.1 Technological impact

HERACLES is a novel system in four ways. Firstly it is the first design system to recognise that many objects (*e.g.* grabbers, valves, sensors) within an automated plant have both hardware and software aspects which should be handled together, giving advantages in maintainability of the design and re-use of high-level design components. It brings many of the advantages of the proven software technique of object-oriented design to the engineering community, without forcing its use in inappropriate situations. Object-oriented design tools are not entirely novel *per se*, but most existing tools have only covered the software side of the problem, and do not cover the hardware components associated with the software behaviour.

Secondly, where in the past tools for designing, testing and running the plant have normally been separate, with no re-use of information between them, HERACLES makes these tools into modules of a system. All information passes through a library, and it is never necessary to enter the same information twice, so that for instance information originally entered in the CAD system will be available for use by the diagnostic system. Apart from a considerable saving in effort, this ensures the consistency of the design used by the different components.

Thirdly, the modular system has been designed to have an open architecture: it is entirely feasible to interface third-party modules to the system, or to substitute alternative components for those provided. Illustration of this is given by the integration of a third-party simulator which was never designed for this use.

Fourthly, the diagnostic system is unusually complete. With respect to the weaknesses of traditional systems (cf. 2.1), the HERACLES diagnostic system is particularly strong in diagnosing "low-level" problems because most of the data corresponding to the "rules" of the expert system is generated automatically from the same information as the PLC source code This saves in set-up time, and allows the diagnosis of many fault conditions which have not been explicitly anticipated by the set-up engineer. The data is used to build a model of the hardware and PLC software forming the design, which the run-time system used to reasons from symptom to cause even for novel faults.

5.2 Commercial impact

Today, three key requirements are essential to gain or maintain competitive advantage in the field of automation engineering:
- reactivity, i.e. respond quickly to new necessities - in our case, have a new plant in operation as soon as possible;

- quality, i.e. build a plant which will respond to all requirements of effectiveness and efficacy;
- price, i.e. deliver a plant at reasonable costs.

HERACLES provides considerable support in satisfying these requirements:

All the HERACLES modules aim at reducing the effort needed to design, test and automate a new plant: it provides re-use of design components, automatic delivery of testing facilities, reduction of the knowledge acquisition efforts for diagnostic and monitoring facilities, automatic generation of the PLC software. All these features contribute to a considerable reduction in the time needed to have a plant ready and working.

The support tools for the design activities enforce the use of standards and rules for "good engineering practice", which are embedded in the tools themselves. The automatic generation of testing data provides a means to the user to continually and incrementally test what is being designed in a thorough way, providing the facility to re-modify design and rapidly test different solutions, until the optimal one is found. When the plant is in its working life, adequate diagnostic facilities guarantee the supervision of its functionality.

The reduction of effort needed to deliver a new plant consequently reduces the costs of such a task, and thus also the price for the final customer.

6. RESULTS ACHIEVED

Reduced efforts for designing, configuring and programming the automated system

The central Library, accessed by all software modules, avoids repetitive and redundant data input; the plant layout data specified on the CAD system, for example, will be used for the synoptic diagrams on the simulation tool and on the diagnostic system interface. This reduced effort guarantees in addition an error-free configuration of the support systems (i.e. the testing and run-time tools) with respect to the original plant conceived on the CAD.

Re-usable components retrieved from the Library also help in reducing design and engineering time.

The early prototype compilers generate between 60% and 90% of the data for the testing and run-time modules. An exception is the monitoring compiler, which at the generates only 10% of the needed information for the monitoring module. This is because the monitoring tool used is currently not sufficiently "open" in its interfaces, to import data in an automatic way.

Maintain company know-how

The knowledge about the structure and the behaviour of an automated system and its components can be stored in the component library of the early prototype. This allows the company to re-use the knowledge of a former project in subsequent projects.

In the design phase we are now only using tools (Intelligent CAD and State Charts) which allow to describe the structure and the behaviour of the automated system independently of any run-time system. This guarantees that also by changing one or more run-time systems (e.g. a new generation of PLCs) the design knowledge can be re-used.

Reduce time and costs for commissioning

The software describing the behaviour of the components will be tested by the simulation tool of the early prototype. It is known also from other projects that the off-line testing of the software reduces the time and the costs putting an automated system into operation as most of the faults can be found before the developed software will be used on the real plant.

During the commissioning the available diagnosis system helps to find quickly the cause of faults. Normally, diagnosis would not be available in this initial phase because manually building the knowledge necessary is a task which is often tackled only once the plant is running. With HERACLES, the diagnosis data is generated together with the PLC data, providing the needed knowledge bases. At the moment about 90% of the diagnosis data can be generated automatically.

Improve the monitoring of an automated system

The actual state of the process will be displayed on the monitoring system. Parts of the monitoring data can be generated automatically by the HERACLES prototype (10%). The object-oriented view of the monitoring system now fits to object-oriented specification of the automated system and the generated PLC code.

Increase the quality of the run-time data

The prototype allows to re-use completely tested standard components (objects). No chance to make a fault except to choose the wrong object type.

The automatic generation of run-time data on the base of the design data improves the quality as faults based on misinterpretation or misspelling can be avoided.

Using graphical and structured tools to describe the design of the automated system led to better specifications, which are easier to understand. So the number of conceptual faults or misunderstandings decreases.

Improve the documentation

Use of the design/specification documents (data) as the base for the generation of the run-time data guarantees that the documentation fits with the data used by the run-time systems. This helps to resolve the problem of today's automated systems, where the gap between documentation and run-time data becomes broader over time. The gap has its origin in the fact that people make changes and corrections of the automated systems with tools which are on lower levels than the design tools, and have no interfaces to generate automatically an updated version of the documentation.

In HERACLES the behaviour will be described by the State Charts, a graphical description language. This improves the recognition of the main parts of the behaviour, the states of the object and the conditions (transitions) between the states.

49

Simulation-based evaluation of assemblability for machine parts

H. Hiraoka, Y. Takahashi and M. Ito
Chuo University
13-27 Kasuga 1-Chome, Bunkyo-Ku, Tokyo 112, JAPAN
telephone: +81-3-3817-1841, fax: +81-3-3817-1820
e-mail: hiraoka@mech.chuo-u.ac.jp

Abstract

A framework of evaluating assemblability using assembly simulation is proposed. Current assemblability evaluation methods perform evaluation by looking up the evaluation table. The table usually contains typical assembly features, to which assemblability scores are specified. Although the method is quick and is tuned for the shop floor based on the factory engineer's experience, it may not be applicable to new situations or other sites. This paper describes a simulation-based method for a rational foundation of the evaluation that can be easily applied to different situations. Evaluation is performed in two ways; One is based on the simulated motion of each part in assembly procedure. Possibility of deviation of the motion from nominal assembly motion is evaluated through the simulation. Another is based on the simulation of robot motion for the assembly procedure. Scores for assemblability are generated by evaluating those robot motions.

Keywords

Assemblability, assembly simulation, contact state, configuration space, assembly robot, operability

1 INTRODUCTION

Manufacturers are facing harsh situation these days. Lot size of production becomes smaller to comply with various consumers' needs more correctly. Changes or innovations of product design become frequent to catch their hearts quickly. Shorter term to the distribution of products is a requisite to overcome the competitors. These pressures in time make it difficult to do 'learning by practice.' This is why companies consider it important to evaluate the assemblability of product that greatly affects productivity of the shop floor, preferably in computerized way.

Current approach of evaluation of assemblability is based on experience and expertise of production engineers in the shop floor (Boothroid, 1983). Though the experience should not be neglected, we are in doubt of its effectivity as the products and environments are changing very rapidly these days.

To make the evaluation robust for changing situations, we consider rational and objective evaluation method is necessary. Measures generated using computer simulation based on physical and mathematical basis would be helpful for the purpose. Of course there exist measures or methods for assemblability that are acquired only by expertise and are indispensable, but we should clearly distinguish these two types of measures.

In this paper, we propose a scheme to evaluate assemblability of product using simulation of its assembly process. We describe two approaches; one is based on the behavior of parts in assembly process and the other is based on analysis of the motion of assembly robots.

2 SIMULATION-BASED EVALUATION OF ASSEMBLABILITY

Common way to evaluate the assemblability is to use looking-up tables as shown in figure 1. Features of product that may affect assemblability such as symmetricity of parts, and number and size of holes in it, are enumerated. Values of estimation are shown in the table corresponding to each of those features. Engineers can easily estimate the assemblability as a sum of those values. This looking-up table is arranged by production engineers of the shop floor. They decide and maintain the values in the table based on their experiences. As the method is mainly based on the experience or expertise of those production engineers, it is only applicable to their shop floor.

This 'know-how' based evaluation is quick and accurate for the restricted type of product and the particular shop floor, as it is best tuned for the manufacturing environment and the product family. However, the method has no explicit rational or logical basis. This makes it difficult to show the designer of the product what is wrong with the design. It may also be difficult to apply the method to different sites or changing situations as the manufacturing process or environment is not considered explicitly in the method.

As stated above, development of new products or new manufacturing process is becoming rapid these days. It is ambiguous that the experience can cover new situations. Range of applicability of the method should be clear and the result of evaluation should provide designers with sufficient information for the change of the design. A new method is considered necessary to suffice these needs.

We propose a scheme shown in figure 2 that evaluate the assemblability based on the computerized simulation of assembly process. Possible behaviors of parts and tools in assembly process will be predicted through the simulation. Factors on the assemblability are estimated by evaluating the obtained behaviors. As the method is based on rational basis with manufacturing process and environment explicitly considered, it will be applicable to any different situations if data about the product, process and environment is provided.

We notice that this method may be slow as it needs a lot of calculations for the simulation. It requires the precise information or the model of the product, tools and environment to get the accurate results. We also understand the utilization of experience or expertise of engineers is requisite to summarize the evaluation results and get the final assemblability value. Still we consider the new method provides the general evaluation that is clear for the designers of the product as well as the engineers of other process or sites.

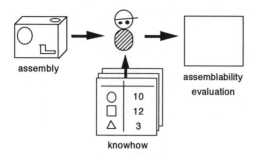

Figure 1 Current method of the evaluation of assemblability.

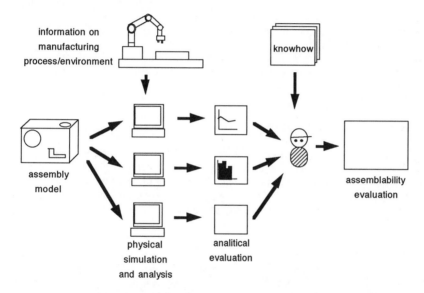

Figure 2 Proposed simulation-based evaluation of assemblability.

In this paper we discuss two approaches for this method. One is to evaluate the predicted behavior of parts and the other is to evaluate the simulated behavior of the assembly robot. The former is described in 3 and the latter in 4.

3 EVALUATION BASED ON SIMULATED BEHAVIOR OF ASSEMBLED PARTS

3.1 Basic concepts and assumptions

We simulate the possible behavior of parts in assembly process to get a factor of assemblability. Possibility to deviate from nominal states of the assembly procedure is estimated. If the possibility is high, it means that the product is difficult to assemble. Details for the evaluation are described in the following sections.

We gave the following two assumptions for this evaluation;
· Assembly procedure is given beforehand. Here we do not argue how to generate appropriate assembly procedures for the product. In other words, we evaluate an assembly product and its assembly procedure as a pair.
· Assembly procedure is carried out such that pairs of faces are in contact. The restriction gives reliable assembly procedures as well as making the problem simpler. For this type of assembly procedure, operations are performed stably by pushing the part to the contact face and sliding it along the face.

Besides these assumptions, our current system handles only polyhedral shapes. Analysis is made on kinematics and statics aspects. Examples shown here is two dimensional though we think its extension to three dimensional case is not difficult. See 3.6. We also limit the analysis to include one-point contact only.

3.2 Representation of state in assembly procedure

The state of two parts in contact can be described with pairs of geometric elements of both parts. We call this state "contact state" and use it as basic representation of state of parts in assembly procedure. Figure 3 shows possible contact pairs of geometric elements. Note that some special or rare cases are omitted such as a pair of vertexes in contact. Assembly procedure is represented by the sequence of face-to-face contact states.

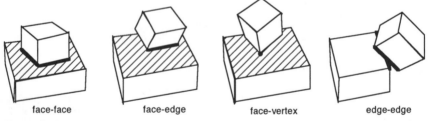

face-face　　　　face-edge　　　　face-vertex　　　　edge-edge

Figure 3 Contact state.

For two dimensional case as shown in figure 4, assembly procedure is assumed to contain at least one edge-to-edge contact state and slide the part keeping the contact. Possible degree of freedom (DOF) of the part to be mated is a translational motion along the contact edge. Tumbling motion that breaks the contact state has a rotational DOF around each of the two contact vertexes.

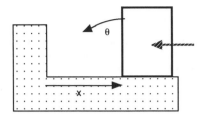

Figure 4 Possible motion of the part in contact in two dimensional case.

3.3 Prediction of the behavior of parts in assembly procedure

At each state of assembly procedure, some DOFs exist as described above. For assessing the assemblability by estimating behavior of a part in an assembly procedure, we enumerate changes of the state possible in the nominal states in the assembly procedure.

We introduce the concept of contact configuration space or cc-space for this purpose. It is an extension of configuration space (Lozano-Perez, 1986). A combination of position and orientation of the part in contact is represented as a point in the space that has every possible DOF as its axis. If a part is free, it has three DOFs in two dimensional case, but as the part is in contact with other part the number of DOFs i.e. the number of axis of the cc-space is reduced. For example, the part in figure 4 has two DOFs; one is translational motion along the contact edge and the other is rotational motion around the contact point. Cc-space corresponding to this state has translation distance x and rotation angle θ for each axis, as shown in figure 5.

Simulation is made based on this cc-space to get the possible changes of state. Cc-space is divided into mesh and every point at the crossing is checked if the part has any other interference with parts. Thus the space is segmented into two types of area, those with interferences and those without interferences. The border between two areas shows the contact state. We can get the possible change of contact state using this map in cc-space. In figure 5 nominal assembly procedure is represented by x-axis that is translational motion along the contact edge. The part may leave this nominal assembly procedure by tumbling around the contact vertex. In cc-space it is represented by leaving from x-axis into upper area. When some element of the part collides with other part, the point in cc-space reaches the border of the interference area.

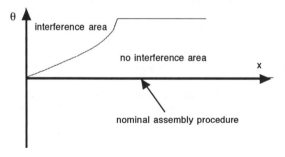

Figure 5 Contact configuration space for the part in figure 4.

3.4 State transfer graph

We get cc-space for each contact state of assembly procedure. Simulation on cc-space gives possible change of contact states from the nominal state. This is summarized in state transfer graph as shown figure 6. It represents possible changes of state around a nominal state of the assembly procedure. In the figure, solid arrows show the nominal assembly procedure.

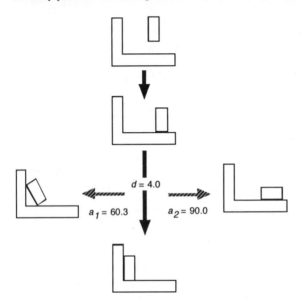

Figure 6 State transfer graph.

3.5 Evaluation of state transfer

We use the ease of the state transfer in assembly procedures as a measure of assemblability. If states easily transfer to the nominal state, we consider assemblability of the part is high. If states are apt to transfer to states other than the nominal assembly procedure, assemblability of the part is considered to be low.

As a way to estimate the ease of state transfer we propose to evaluate the average angle a from nominal state to other contact state in cc-space. If the average angle is large, it is considered that the collision states are remote and the part has high assemblability. Another measure we propose is the length of nominal state d. If it is short, the part is considered to have higher assemblability and vice versa.

These values are also shown in figure 6. Note that the tumbling direction is classified and the corresponding average angles before the collision are calculated. As another example, we evaluate the case of an L-shaped part. In this case the assembly procedure has two steps as shown in figure 7.

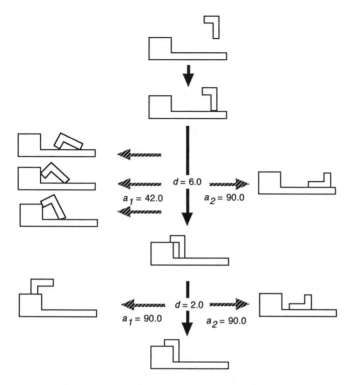

Figure 7 State transfer graph for an L-shaped part.

3.6 Possible extensions

We are now extending the method to have a three dimensional capability. The degrees of freedom of the part in motion for three dimensional case are analyzed as follows;

· A pair of faces in contact: When only one pair of faces is in contact as shown in the left side of figure 8, the part to be mated has two translational DOFs plus one rotational DOF along the contact face. If we think the motion that removes this contact pair is made only by tumbling i.e., rotating the part along the contact edge, there exists one additional rotational DOF for each contact edge.

· Two pairs of faces in contact: When two pairs of faces are in contact as shown in the right side of figure 8, motion of the part is restricted to one translational DOF along the common edge of two contact faces. Tumbling motion to break the contact may have one rotational DOF along the contact edge on each contact face provided that such motion is possible.

We think the extension is rather straightforward but we need a faster algorithm for collision check to lessen the computational burden that will increase.

Evaluation methods are not restricted to those measures described above. Another aspect to be applicable for evaluation is statics of the part at the nominal contact state (Mason, 1985). On the assumption that the force and moment acting on the center of gravity of the part during the assembly operation, we can analyze simple statics to get the range of force and moment that will not tumble but slide the part. If there exist large area of this zone, we evaluate the part with high assemblability.

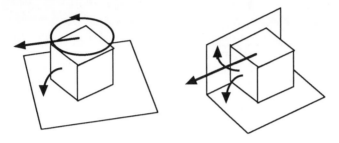

Figure 9 Possible motion of the part with a pair of faces in contact (left) and two pairs of faces in contact (right).

4 EVALUATION BASED ON SIMULATED MOTION OF ASSEMBLY ROBOT

4.1 Basic concepts

Ease of assembly is affected by the tools that assemble the parts. It also differs for the different procedures. We propose to estimate assemblability by evaluating the procedures performed by the assembly robot that assembles the part (Hiraoka, 1993). If the procedure can be easily performed by the robot, we consider the part having high assemblability. The idea is that we can evaluate the assemblability of a product by evaluating the performance of the robot assembling it.

We simulate the motion of the robot for the evaluation. Simple manipulator with 3 degrees of freedom is used. Currently the end-effector and its grasping motion are not considered.

4.2 Simulation of motion of assembly robot

We consider the motion of the robot is given by production engineers or is generated by motion planning system. As we do not have such facilities, we generate simple robot motions as follows.

First we generate the motion of parts that has two portions; one is the motion where the part is mated with other part and the other is the motion where the part is transferred by the robot to other places. We derive mating motion from the assembly procedure mentioned above. For transfer motion we connect the part's feeder and the end of mating motion by a straight line segment. Another point would be added in the middle if it is necessary to avoid some collisions. We give the motion constant velocity along the path for ease of comparison.

Next we generate motion of the robot from the acquired motion of the part. Based on kinematic model and shape model of robot, we convert motion of the part into motion of the robot by using inverse kinematics. Motion is altered if collision is detected. Figure 10 shows a display of the simulation.

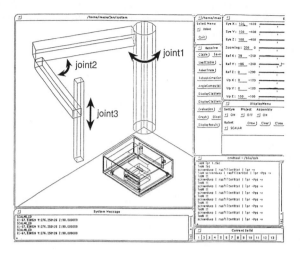

Figure 10 Display of simulated robot motion

4.3 Evaluation of motion as a measure of assemblability

As measures of assemblability we evaluate motion of the robot. If the motion is easy to perform by the robot, assemblability of the part should be high. We use the following measures of performance for evaluation.

· Duration of the procedure: The measure is important as the time duration necessary for the assembly procedure directly affects productivity. In our simulation much difference does not arise for the robot as constant velocity is applied for the motion of parts.

· Joint angular motion: Total change of joint angles and maximum joint angular velocity are used as measures to perform the assembly procedure.

· Operability: Operability (Kotosaka, 1991) is defined as derivative of the joint motion necessary to realize the particular motion of the end-point. Compared to the manipulatability that shows the kinematic performance of robots (Yoshikawa, 1984), operability evaluates kinematic performance of a particular motion for the robot. For the specified motion of end effector r and Jacobian matrix J, we can calculate

$$Oj = \frac{1}{m}\left(\dot{\mathbf{r}}_j^T \left(\mathbf{JJ}^T\right)^{-1} \dot{\mathbf{r}}_j\right) \tag{1}$$

where m is the number of joints. This represents the difficulty of motion of the robot when the end-effector has the motion of r_j in direction j.

4.4 An example and its results

As an example we use two types of framework for small computers. Both are shown in figure 11, one with hard disk drive that is removable sideways and one with hard disk drive removable from above. Simulation was made with the robot shown in figure 10 and the measures described above are compared for these two types. Results are shown in tables 1-4. In tables 1, 2 and 3 results for the motions to mate a part and for the motions to transfer a part are seperately shown.

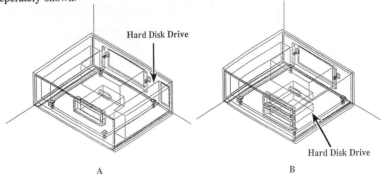

A   B

Figure 11 Two types of framework of small computer used for example.

Table 1 Duration of assembly procedure (simulation steps)

	mating	*transfer*
A	40	200
B	46	196

Table 2 Total changes of joint angles

	joint1		*joint2*		*joint3*	
	mating	*transfer*	*mating*	*transfer*	*mating*	*transfer*
A	0.00	1.52	0.00	1.48	40.00	24.98
B	0.01	1.26	0.24	1.16	0.00	38.93

Table 3 Maximum joint angular velocity (per simulation step)

	joint1		*joint2*		*joint3*	
	mating	*transfer*	*mating*	*transfer*	*mating*	*transfer*
A	0.00	0.02	0.00	0.04	2.00	1.24
B	0.00	0.02	0.01	0.03	0.00	5.00

Table 4 Operability for disassembling operation of hard disk drive

	start	*end*
A	1.00E+00	1.00E+00
B	3.40E-05	2.30E-05

Major differences in the motion are seen in vertical joint 3. For operability shown in table 4 the start point and the end point of disassembling the hard disk drive are compared. The vertical motion that is necessary for mating the drive into the product A can be performed only by vertical joint 3. This makes worst evaluation from kinematic viewpoint. To be fair, we think we should evaluate the motion from other viewpoint such as compliance.

5 CONCLUSIONS

In this paper we propose a new framework for evaluating assemblability of products using simulation of assembly procedures. As the method is based on rational and objective basis, it will be robust for different or changing situations. Two approaches are described. One is based on the evaluation of simulated mating motion of parts. Analysis on the contact configuration space is applied. The other is taking the tools and environments for assembly into consideration. We evaluate the assemblability by way of the performance analysis of simulated motion of the assembly robot.

Further research is necessary to integrate these measures into a single value assemblability. It is definitely necessary to incorporate the expertise and experience of production engineers. More accurate simulation including various aspects of assembly operations and investigations to relate the measures based on simulations with actual productivity are required.

ACKNOWLEDGMENTS

Authors will thank Mr. Kihara and Mr. Mitsuhashi for their basic research work. This research is partially supported by the research program conducted in RIKEN (The Institute of Physical and Chemical Research) that is a part of Cross-over Research Program for Nuclear Base Technology promoted by Science and Technology Agency, Japan.

REFERENCES

Boothroid, G., Dewhurst, P. (1983) Design for Assembly, University of Massachusetts.

Hiraoka, H., Kihara, M., Mitsuhashi, M. (1993) Generation of Indices for Assemblability Evaluation Using Simulation of Robotic Assembly Motion, Proceedings of Design Symposium, 127-133 (in Japanese).

Kotosaka, S., Matsumoto, A., Takata, S., Asama, H., Hiraoka, H. (1991) Computer Aided Design System for Maintenance Robots - 2nd Report: Synthetic Evaluation of Kinematics Considering the Direction of Operations, Proceedings of Spring Conference of Japan Society for Precision Engineers, 119-120 (in Japanese).

Lozano-Perez, T., (1986) Motion Planning for Simple Robot Manipulators, Robotics Research, The Third International Symposium (ed. Faugeras, O.D., Giralt, G.), MIT Press, 133-140.

Mason, M.T. (1985) Mechanics of Pushing, Robotics Research, The Second International Symposium (ed. Hanafusa, H., Inoue, H.), MIT Press, 421-428.

Yoshikawa, T. (1984) Analysis and Control of Robot Manipulators with Redundancy, Robotics Research, The First Symposium (ed. Brady, M., Paul, R.), MIT Press, 735-747.

BIOGRAPHY

Hiroyuki Hiraoka
1978 Graduated from Dept. of Mechanical Engineering, Tokyo University
1983 Received Doctor of Engineering from Tokyo University
1983 Research Associate, Dept. of Precision Mechanical Engineering, Tokyo University
1989 Associate Professor, Dept. of Precision Mechanics, Chuo University

Yoshiaki Takahashi
1993 Graduated from Dept. of Precision Mechanics, Chuo University
1994 Entered Precision Engineering Course of Graduate School, Chuo University

Masayuki Ito
1993 Graduated from Dept. of Precision Mechanics, Chuo University
1994 Entered Precision Engineering Course of Graduate School, Chuo University

50

Dynamic Behavior of Objects in Modelling Manufacturing Processes

José Barata; L.M. Camarinha-Matos
Universidade Nova de Lisboa - Dep. de Engenharia Electrotécnica
Quinta da Torre, 2825 - Monte Caparica, Portugal
Tel +351-1-3500224 / 2953213   Fax +351-1-2957786
E-mail: {jab,cam}@fct.unl.pt

Abstract

Various modelling perspectives for manufacturing systems, both from the structural and dynamic behaviour points of view, are discussed. The utilisation of object-oriented and frame-based paradigms in this modelling context is discussed as well as the connection of models to the real device controllers. The synthesis of control programs from a Petri net model is also presented in this general modelling framework. Finally the concept of object migration is introduced as an approach to deal, in a flexible way, with moving objects in manufacturing systems.

Keywords

Modelling, Object Behaviour, Flexible Manufacturing Systems, Object Persistency, Views.

1. INTRODUCTION

Object Oriented and Frame based (OO&F) techniques have been intensively used in modelling manufacturing systems and processes. OO&F provides a structured modelling approach, allowing for multiple levels of abstraction, a convenient approach for complex systems modelling. However an extended view of objects is necessary in order to capture the dynamic behaviour of such systems.

Particularly in the area of shop floor control, OO&F can be used to model the various manufacturing agents -- robots, NC machines, transportation systems or even continuous processes equipment. An adequate combination of reactive programming and client-server architecture allows for an effective link between the OO model and the local controllers, therefore capturing the dynamic behaviour of the modelled devices and providing a kind of dynamic persistence. Classical real-time aspects, like asynchronous events / interrupts, device drivers, etc., may be adequately modelled/abstracted using such extended OO&F approach.

In this paper various modelling situations are described and experimental results discussed.

2. MODELLING ASPECTS

The Object Oriented paradigm or its "mate" Frame-based/Reactive Programming represent a convenient tool to model the inherent complexity of a manufacturing system. This complexity comes from the amount of relationships among components in association with the diversity of components. On the other hand, the topic of modelling is a pre-requisite for systems integration in CIM. Speaking specifically about system controllers there is a need for a model that supports the interaction between high and low level controllers, and, at the same time, supports the configuration of new systems.

The model should emphasise the relationships among the various components in the cell and hide the specificity's of the hardware. This last item can be easily achieved with the Object Oriented/Frame based paradigm using methods or demons. Methods associated to the component can hide the underlying hardware infrastructure. Another important aspect is the "relation" concept which can provide a flexible way to describe inter-components relationships. This concept has different semantic meanings in the Object and Frame paradigms, which makes modelling a little bit different in these two frameworks.

Before going further in the modelling discussion, it is important to say a few words about which paradigm should be used to model - Frames or Object Oriented. It isn't an easy choice because both have advantages and disadvantages. The fact that a Frame deals with objects as prototypes, allowing dynamic change of the structure of those objects is a good point, especially during the research phase. But this could be also a disadvantage, especially for software engineering production. For instance, a programming environment that doesn't provide strong type checking can be very error prone and lead to software difficult to maintain. On the contrary, these could be the virtues of the Object Oriented paradigm. On the other side, OO systems are quite limited in terms of definition of new relations with customised inheritance mechanisms.

2.1 Structural aspects

The various examples to be discussed in order to introduce the main modelling concepts will use a cell as the basic modelling unit. A cell is a composite entity that is capable of making some transformation, movement or storage related to some product or part. In structural terms, each cell (C) has components to support the input of parts (I), an agent to perform the transforming actions (A) and components to support the output of products/processed parts (O). Therefore, a cell is the tuple: C = (I, A, O).

Parts input and output and the agent will be supported by manufacturing components. Some components can only support one function but there are others components which can support more than one function. Components adapt themselves to the roles they can perform. Some components are more adaptable than others. For instance, the Conveyor is very flexible because it can perform an input, output or agent role, but a CNC machine only can play an agent role.

Components	Input	Output	Agent
Vibrator Feeders	√		
Buffers	√	√	
Indexed Table	√	√	
Gravity Feeder	√		
Conveyor	√	√	√
Robot			√
CNC machine			√
AGV			√
Positioning device	√	√	

Table 1 Components and their possible roles

The generic cell concept can be specialised by activity. There can be cells specialised in assembly, painting, welding, storage, machining, transportation, etc.. A shop floor is just a set of specialised cells.

Metaknowledge should be associated with each specialised cell to represent the specificities of its application domain. For each domain the specific cell has the same structure as the generalised Cell concept (Input Agent, Processing Agent, Output Agent) but the domain and carnality of the implementing components is different in each specialisation. For example, in a Painting or Welding Cell, a vibrator feeder is not a valid Input item, but this component is valid in an Assembly Cell. The Metaknowledge seems to be a very important element at the configuration phase, assuring the validity of cells.

```
FRAME   CELL
name:
base_coordination_system:
processable_products:
input_parts:
connected from:
processor:
connected to:
```

```
FRAME    ROBOT_COMPONENT
is-a: manufacturing_component
Base_coordinate_system:
Controlled by:
Applications: assembly, gluing, ..
DOF: 6
Working_area:
Load:
Repeatability:
Current_position:
Cost:
Cycle_Time:
Next_maintenance:
N_working_hours:
Weight:
Max_speed_by_axes:
```

```
FRAME   ASSEMBLY-CELL
is-a: CELL
val-inp-ag: vibratory_feeder,
            buffer,
            gravitic_feeder,
            Index_Table, agv,
            conveyor
val-out-ag: conveyor, agv, buffer,
            index_table
val-proc-ag: robot
```

Figure 1 Example concepts of cell, assembly cell and component.

At this stage it is convenient to clarify the concepts of <u>agent</u>, input and output, and their relation with the <u>components</u>/manufacturing resources.

Components are entities which participates in the productive process with a specific function and can be controlled by a computational entity. Components models are context independent description of its static and dynamic characteristics. A robot component model, for instance, includes all the characteristics which completely characterise its structural and dynamic aspects.

A robot **agent** (Figure 2) is a model of a robot and associated resources, like tools or auxiliary sensors, when inserted in a particular context. A robot can play different **roles** in different **contexts**. The (expected) behaviour of a robot in an Assembly context is different from its behaviour in a spot welding context.

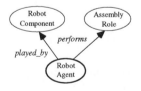

Figure 2 Structure of a robot agent.

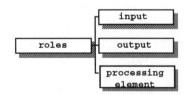

Figure 3 Role taxonomy: main level.

On the other hand, when a robot is performing a given role, it may resort to auxiliary resources, like tools, sensors, buffers, etc., that extend the robot functionality in order to fulfil the functionality required by this role. A robot agent is, therefore, a model of the robot when playing a particular role and extended by selected attributes inherited from the auxiliary resources.

The entity that effectively participates as an assembly robot, for instance, is one which has those characteristics from the robot component model to perform the assembly role.

The agent entity ASSEMBLY_ROBOT is a structure which is supported by two relations: *performs* and *played_by*. The relation *performs* assures the inheritance of the role characteristics to the structure while *played_by* assures the inheritance of those agent relevant aspects, from the component. *Main_attributes* and *component_attrib* are attributes to be used by *played_by* and *performs* relations.

```
FRAME   ASSEMBLY_ROBOT
is-a: agent
performs: ASSEMBLY_ROLE                FRAME   AG_ROBOT_ASSEMBLY_ROLE
played by: ROBOT_COMPONENT             is-a: role
main_attributes:force_sensor,          tools_domain:(grippers,
               current_tool,                          screwdriver)
               available_tools,        aux_res_domain: (buffers)
               available_resources     force sensor:
component_attrib:Base_coord_system,    current tool:
Controlled_by,                         available_tools: gr1, gr2, sd2
Working_area, load,                    aux_resource: buf1, buf2
Current_position                       assembly_device: fixture1
```

Figure 4 Example concept of an agent

The slots, *tools_domain* and *aux_res_domain* represent domain-knowledge that is important during configuration time. The slot *current_tool* is a relation that associates the main player of this role (robot component) to a particular tool. It could be defined as:

new_relation(current_tool, transitive, inclusion(tool_operations, attached_to)

where *attached_to* is the inverse relation. By the "inclusion" restriction, only tool_operations will be inherited by the ag_robot_assembly_role.

Assembly_device is an attribute describing where assembly operations are really done. *Fixture1* is an instance of a component specialized in holding parts.

```
RELATION  PERFORMS                     RELATION  PLAYED_BY
type: intransitive                     is-a: relation
inherit_slot: main_attributes          type: intransitive
inverse_relation: performed_by         inherit_slot: component_attributes
                                       inverse_relation: play
```

Figure 5 Definition of relations *performs* and *played_by*.

A cell is made of entities that are playing different roles.

This modular approach to cell representation facilitates the creation of complex systems by simple "concatenation" of cells. A particular manufacturing unit is made of several subsystems (Transportation Cells, Painting Cell, Assembly Cell, ...). A manufacturing unit could be modelled by a SYSTEM entity, which has access to all characteristics and functionality of all subsystems involved in the Unit.

The way applications see the unit varies with their needs. An application concerned with maintenance activities has different needs from SYSTEM than an application concerned with supervision activities. These differences could be easily supported using the view concept. Using this concept an application only sees the relevant information for its activity.

This is a very convenient concept because it supports information structuring and consistency. Thinking of a robot, an attribute that accumulates its number of working hours is important for a maintenance application, but it could be irrelevant for a direct control application.

We can even think that applications in the same activity area, i.e., accessing the same view, could have different requirements. In this case the access is determined not only by the role of the client but also by its status. Applications may have different status to access a view entity, having conditioned access determined by their status.

These concepts are not easily implemented with current OO&F technologies. The views implementation is different whether it is implemented by frames or by objects. The UNL Robotics group developed some implementations using frames that support this concept. The frame implementations are based on the inheritance and relation mechanisms.

Object Oriented languages should be extended in order to include a create_view constructor that could be related with another new construct VIEW. Using an example in EIFFEL the result could be:

```
CLASSE  robot                          ...
interface                              END configuration
maintenance VIEW                       END robot
 total_hours()                         ...
 hardhome()                            r1,maint_view,op_view: robot;
 ...                                   ...
END maitenance
operative VIEW                         r1 := create(robot);
 hardhome()                            maint_view :=
 move()                                    create_view(robot,
 ...                                                   maintenance);
END operative                          oper_view :=
configuration VIEW                         create_view(robot,
 load()                                                operative);
 max_speed()
```

2.2 Dynamic aspects

Dynamic aspects are related to the components internal state changes. The dynamism presented by components is achieved through controller actions. Every component with dynamics must have a controller associated with it. The main discussion is not centered in the aspects related with the physical components changes, i.e., it isn't important to know what are the inertial conditions associated, for instance, with a robot movement, but it is important to discuss the functional behavihour of the physical component being modelled, i.e., it is important to know what actions should be done in order to move the robot in the most flexible way. The way the model reflects component physical changes and the way physical component reflects model changes is the most important point when discussing dynamic aspects.

Dynamic aspects can also be discussed with two different views: (1) considering the components as isolated entities or (2) considering complex structures, like cells, made of components. In the first view the key point is how components are actuated, without any concerns about their interrelationships. In the second view, aspects related with synchronisation are the most important ones (it will be analysed in the PETRI NETs chapter). In this point the concern is with the first view.

Every component model with behaviour should have a controller model. This model should be like an image of the real controller. Using a frame oriented paradigm, the controllers functionalities could be defined by methods. In this way most of the controller's model is a list of methods, a method for each functionality.

Figure 6 Controlled_by and Controls relations.

A component is related to its controller by a *controlled_by* relation while the controller relates to its component by a **controls** relation.

```
RELATION  CONTROLLED_BY
is-a: relation
type: intransitive
inherits: inclusion (move_wc,
                     move_jc,
                     hardhome,
                     acceleration,
                     speed)
inverse_relation: controls
inverse_relation: controls
```

```
FRAME   ROBOT_CTRL_COMPONENT
is-a: controller
move_wc: method move_wc_fn(x,y,z,q)
move_jc: method
             move_jc_fn(m1,m2,m3,m4)
hardhome: method hardhome_fn
acceleration:demon if write accel_dem
speed: demon if write speed_dem
input: byte demon if needed input_dem
output:byte demon if write output_dem
```

Figure 7 Model of a component and its controller.

One of the most important points in this discussion is the way a controller model is connected to the physical controller. This connection is sometimes not easy because it involves the cooperation of two different computational worlds: the computational world where the model runs and the real controller. To make things even more difficult, sometimes, real controllers have closed architectures. From our experience a lot of effort has usually to be put in trying to open real controllers architecture, and implies the production of an interpreter that runs on it. This interpreter accepts commands from an image that runs in the other world.

The methods of a Controller model implement the actions that are needed to send the right commands to the real controller. The real controller image should be developed using a client-server approach. In this way, implementation methods can ask this server to perform the needed actions. These methods hide the underlying hardware structure from the application, i.e., any application using a robot component doesn't need to know anything about the real robot controller and its image or server. The applications only know what functionalities are provided by the robot component model. This approach could be very suitable to integrate existing controllers, making the integration of legacy systems an easier task.

3. DYNAMIC PERSISTENCE OF OBJECTS

In a manufacturing environment, many information sources -- sensors, state variables of local controllers , etc. -- have their own "life", independent of the computer that is running the general controller model, because they have local processing power. This may lead to the concept of dynamic persistence, that will be introduced and exemplified in modelling manufacturing systems.

Object Persistence is the property of extending the life of an object beyond the running session of the application software that created or changed it. This characteristic is important for applications that may interact with long lifetime objects.

The traditional way of dealing with Object Persistence is storing the objects in secondary memory. In some approaches, classical Database Management Technology has been integrated with OOP languages in order to manage the flow from main to secondary memory and vice versa.

The concept of Dynamic Persistence of Objects is not very different from normal persistence. The basic difference comes from the way persistence is supported: by the local memory of devices' controllers. The use of reactive programming (demons) and methods to "link" the object model to the real cell controllers allows for a permanent update of the dynamic object's model. In this way, a special kind of persistence is achieved - **dynamic persistence**. It is dynamic because the object model reflects, at every time, the status of the physical object. The persistence is assured by the "memory" present in the device controller. There is a tight connection between the object "living" in main memory and the physical controller. We can say that the object virtualizes de physical controller. The physical entity description (object) is connected to the physical entity via demons associated to object attributes. These demons establish a communication link to the physical entity controller.

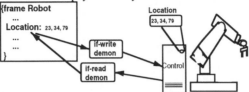

Figure 8 Use of reactive programming to support dynamic persistence

In our work an assembly cell composed by a SONY Scara robot, a fixture, a force sensor, an automatic tool exchanger and two gravity feeders was integrated in a UNIX environment using this approach. However, due to SONY robot's closed architecture a big amount of effort was needed in order to allow this integration. It implied "breaking" the protocol used to down and upload programs, and also the development of an interpreter that runs on top of SONY's own controller. To make error recovery possible (guarded moves) it was also necessary to "break" the teach pendant protocol and to replace it by a PC controller. A controller was also included to drive the force/torque sensor.To connect these external controllers into the UNIX environment it was necessary to develop their "images" on the UNIX side. Each of these "images" are accessible to applications according to a client-server basis.

The same methodology is being used to integrate the several agents belonging to a complex manufacturing cell installed in our facilities. Due to cell's complexity this work is being done by phases. A BOSCH scara robot, a conveyor belt system and an automatic warehouse have been already integrated using this approach.

4. PETRI NETS

Petri Nets are important tools to model the structure and behaviour of controllers and application programs. Complex system dynamics can be described and analysed in a structured way (mathematical methods). There are several types of Petri Nets, which can be used to model distinct types of systems, but Predicate Transition Nets (PTN) seem to be very suitable to model logic controllers. A PTN has predicates associated with transitions which only fire when all input places has marks and the predicate returns true.

The benefits of using a high level modelling tool, like a PTN, shouldn't end in their descriptive characteristics but there should be a direct connection between the description and the real controllers. This means that the model could drive directly its associated physical controller. To assure this connection two different approaches could be used: (1) using PTN to directly program the physical controller, which implies a support by its manufacturer or (2) using a PTN translator which converts PTN to the own language of the physical controller.

The first approach is unrealistic at current stage because the concept of PTN is not well disseminated among controllers' manufacturers, which makes the second approach a better one.

The PTN description should be compiled in order to generate a program which interact with the real system in the way described by the PTN.

The application program or High Level Control Program (HLCP) is generated from a PTN which describes components behaviour and interactions. The HLCP interacts with the execution infrastructure, mainly with the components models. These models are described using an Object/Frame paradigm. As mentioned before, the components behaviour is implemented by methods, which interact with the component's physical controller through a server which supports an image of the physical controller functionalities.

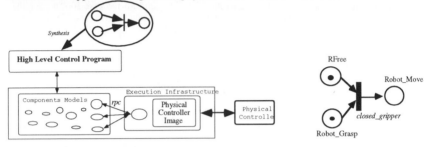

Figure 9 - Interactions between HLCP and Execution Infrastructure.   **Figure 10** A Petri Net example.

The components model associated to a HLCP include only those models relevant for the program that is running in the HLCP. Saying it by other words, only those components which appear in the PTN will be included in the components model associated with HLCP. It should be noted that a Component Model can interact with more than one server, depending on the number of needed controllers to control the components being used. For instance there can be different servers to control de robot, the gripper, gravity feeders, etc.

In order to generate the HLCP from a PTN some considerations should be made about PTN. Components actions can only be done in places with marks. Predicates associated to transitions specify conditions. In order to match places to components actions, places' names include components' name and action name separated by an underscore (for instance, the name for the place that describes the grasp action of a robot is named **robot_grasp**).

The HLCP program generated is a simulator of the PTN being modelled. Different PTNs may have the same kind of simulator, differing only in which order transitions will be fired and which actions will be done.

Program generation was developed using Prolog, and the generated program is also described in Prolog with a frame extension developed in our group- Golog. The generated program can be seen below.

The first section of the generated program is concerned with place definition; every place is defined by an object/frame whose main attribute is the slot mark to store the place's mark value. During this phase the program that contains the components model is consulted.

```
:- consult('models.pl').
:- new_frame(places), new_slot(places, mark).
:- new_frame('RFree'), new_slot('RFree', isa, places), new_slot('RFree', mark, 1).
:- new_frame(robot_grasp),new_slot(robot_grasp,isa,places), new_slot(robot_grasp,mark,1).
:- new_frame(robot_move),new_slot(robot_move,isa,places), new_slot(robot_move,mark,0).
```

After this, transitions are defined. Each transition is defined by checking its enabling condition. When this occurs, input places are updated and transition fires with output place updating and the corresponding method activation: *call_method(robot, move, [true])*. This method's code will send a message to the server which will react by sending the command "move" to the robot.

```
t1 :- get_value('RFree', mark, X0), X0 > 0,
     get_value(robot_grasp, mark, X1), X1 > 0,
     NVal0 is X0 - 1, new_value('RFree', mark, NVal0),
     NVal1 is X1 - 1, new_value(robot_grasp, mark, NVal1),
     call_method(robot, move, [true]),
     get_value(robot_move, mark, VOal0), NVOal0 is VOal0 + 1,
     new_value(robot_move, mark, NVOal0).
```

The main program consists of a forever cycle that continuously apply the existing transition names and randomly choose one which will be checked for its enabling condition.

```
rep_run([]).
rep_run(List) :- length(List, Tam), Pos is ip(rand(Tam)), position(Pos, List, Tr),
             remove(Tr, List, RList), call(Tr, Success), !, fail == Success, rep_run(RList).
run :- repeat, rep_run([t1]).
```

This generated program run with a similar behaviour as the PTN shown in figure 10.

5. OBJECT MIGRATION

In a FMS/FAS System several distinct physical "worlds" may be considered. Assembly cells, transportation systems and automatic warehouses are examples of existing "worlds" in a shop floor environment that have their own controllers (i.e. distinct computational worlds !). These "worlds" are strongly interconnected requiring information exchange, which could be achieved by sharing a centralised repository, by messages or by moving data among "worlds".

Viewing the associated computational "world" as a set of objects which model the physical entities participating in the process, the concept of **object migration** becomes relevant. A computational "world" can include objects which belong intrinsically to that "world". For instance, the object robot belongs intrinsically to the assembly cell "world", but the object pallet doesn't. A pallet migrates between worlds. This object doesn't belong to any specific "world" and can "enter" different "worlds" at different time slots. Taking into account the need of these objects in a FMS/FAS system it would be necessary to develop an infrastructure to support object migration.

Figure 11 SLS and MDT.

```
FRAME  MDT_OBJECT
is-a: migration_object
pallet_id:
kind_of_pallet:( cnc_raw_material,
                    cnc_finished,
                 assembly_raw_material,
                    assembly_finished )
materials_list:
path:
stop_places:
```

Figure 12 Object stored on pallet's memory.

As in our pilot manufacturing system, the BOSCH pallets include an attached memory device - MDT, that can be read/written by various other devices - SLS, located in special places of the system, we have a "physical" support to this object migration (figure 11, 12).

Regarding again the object pallet, it could be seen that while it moves through the "worlds", it can be modified. The modification in the object's structure reflects the changing conditions in the physical "world". At each time slot the object's state reflects the operations done by physical entities, over the moving physical entity, represented by these migration objects.

The attribute *materials_list* includes identifications of those objects carried on the pallet. Pallet's path movement within the cell is controlled by the attribute *path*, which includes a list of conveyor names. The attribute *stop_places* tells the system where the pallet should be stopped.

When a pallet passes in front of an SLS, it reads the contents of the MDT structure in order to determine which action to be done. SLSs are controlled by a PC server which is directly connected to the PLC that controls cell's conveyors. Depending on the MDT's memory contents, the SLS sends commands to the PLC. For instance, if there is a stopper nearby the SLS and the *stop_places* attribute has the name of this SLS, the server commands the PLC to stop the pallet. This is a highly dynamic system, either at the spatial or at the internal structure levels. However, these aspects of migrating objects are still on a developing phase in our system, needing a deeper evaluation.

6. CONCLUSIONS

In this paper we discussed various aspects of modelling manufacturing systems, both from the structural and the dynamic perspectives, resorting to the object/frame based paradigms.

In particular, our experimental results have shown that objects' dynamic behaviour (by means of reactive programming and methods) combined with a client-server approach, provide an effective way to link models with local controllers of manufacturing devices. Therefore this can be a suitable approach for migrating from legacy systems to more integrated high level control systems. The generation (synthesis) of application control programs - directly linked to the above methods from a Petri Net description was also discussed. Current work is addressing the aspects of object migration as a flexible way to "deal" with moving objects in a manufacturing environment.

Acknowledgements
This work has been funded in part by the European Community (Esprit project B-Learn and FlexSys) and JNICT (projects SARPIC and CIM-CASE). We also thank Mr. João Carlos Silva and the students Eduardo Bras, Sandra Gadanho, Luis Fernandes and Nuno Chagas for their contribution to the experimental setup.

REFERENCES

Barata, J.; Camarinha-Matos, L.M.; Rojas Chavarria, J.F. (1994) — Modelling, Dynamic Persistence and Active Images for Manufacturing Processes, *Studies on Informatics and Control* , vol. 3, nº 2-3.

Camarinha-Matos, L.M. (1989) — *Sistema de programação e controle de estações robóticas - Uma arquitectura baseada em conhecimento*, PhD Thesis, Universidade Nova de Lisboa, 12 Jun. 1989.

Camarinha-Matos, L.M.; Negreto, U.; Meijer, G.R.; Moura-Pires, J.; Rabelo, R. (1989) — Information Integration for Assembly Cell Programming and Monitoring in CIM, *21st ISATA*, Wiesbaden.

Camarinha-Matos, L.M.; Osório, A.L. (1990) — Monitoring and Error Recovery in Assembly Tasks, *23ª ISATA*, Viena.

Camarinha-Matos, L.M.; Pinheiro-Pita, H.J. (1990) — Interactive Planning of Motion and Assembly Operations, *Proc. IEEE Int. Workshop on Intelligent Motion Control* , Instambul, Turkey.

Camarinha-Matos, L.M., L. Seabra Lopes, J. Barata (1994) — Execution Monitoring in Assembly with Learning Capabilities, *Proc. of the 1994 IEEE Int'l Conf. on Robotics and Automation*, San Diego.

PART TWELVE

Techniques and Methods for Computer Application

51

O/S.M: An object-oriented approach for software requirements/analysis phase

Louis LASOUDRIS
Institut National des Télécommunications
9, rue charles Fourier - 91011 EVRY - FRANCE
Tel: +33 (1) 60 76 47 44   Fax: +33 (1) 60.76 44 93
E-mail: lasoudri@galaxie.int-evry.fr
Stephen LIAO
City Polytechnic of Hong-Kong
83 Tat Chee Avenue Kowloon HONG KONG
Tel 852 788 7552   Fax 852 78886
E-mail: stephen@hp9000.is.cphk.hk

Abstract

This paper presents Object/System.Modelling (O/S.M) an object-oriented method. Its foundations are based on Object-oriented technologies and on systemic sciences. As others methods, O/S.M has a lot of models for information, behavioural and functional structure, and suggests a guideline for analysis and requirements phases. The use of O/S.M leads analysts to capture requirements and to validate their results.

Keywords

Object oriented methods, system science, modelling technics, information system, object-oriented database.

1   INTRODUCTION

Object/System Modeling (O/S.M: [Liao, 1993], [Lasoudris, 1993]) is an object-oriented method for software requirements/analysis. This method based on the system theory [Von Bertalanffy, 1948] and software object-oriented approaches [Rumbaugh, 1991], [Schlaer 91],[Jacobson, 1990], [Coad&Yourdon90], [Desfray, 1991] uses an object-oriented modeling technics and a guideline for requirements/analysis phase using system theory [Simon, 1962]. The model considers the system to study as an integration of several entities perceptible in real world with their own finalities. The interaction between these entities allows them to collaborate for realizing the system finalities.

The guideline combines the ascending approach (Increasing the level of complexity) and the descending approach (Refinement as in classical methods) [Ross, 1977] permitting respectively first the complexification of the system using conjunction and secondly the simplification using disjunction. Nevertheless, O/S.M gives us a mean to verify the two approaches in a set of coherent and integrated schematas (Information, dynamic, behavioural, functional) which are closed with information model, states model and data flows diagrams as they are defined in O.M.T [Rumbaugh, 1991].

Main software methods use the entity-relationship metamodel [Chen, 1976] for expressing the data semantics, and the Data Flows Diagrams [Gane, 1979] for expressing respectively the data and process (also flow controls) on these datas. The E-R metamodel and D.F.D metamodel will be often respectively implemented by the relational model and by procedural languages at design and implementation levels. These methods, for example SADT [Ross, 1977], have been greatly succeeding since 1980s. Nevertheless today, face to the complexity of the applications and the evolution of the technologies, it is necessary to take into account new software analysis and design methods [Booch, 1991].

System science [Von Bertallanfy, 1948] is a thought method which leads to solve complex problems. This science proves its efficacity in several domains as ecology, medecine, urbanisation, factories organisation,... Since Descartes [Descartes, 1642], we use a refinement thinking mecanism to solve complex problem We have to divide a complex problem to solve it. In software life cycle, requirements/analysis corresponds to the problem space [Booch, 1991]. Most software problems studied are complex.

The bases of Systemic lies on one main idea. All things in real world are complex to understand. So it is not sufficient to divide a problem to understand and to solve it. On the contrary we have to complexify it to understand this problem.   System Science brings some radically different ideas. So we have to think about many problems in terms of complexity (Le Moigne, 1990). Software problems are very complex. Then it seems us that it is possible to bring closer both systemic approach and refinement approach with the goal to increase efficiency in software requirements/analysis tasks.

2   STATE OF ART

The object-oriented approach seems capable to bring some solutions face to the complexity for the applications and the evolution of the technologies.

The first tendency is towards the improvement of the modeling power of the existing models. For example, an E-R model extension ERC+ [Spaccapietra, 1989]. Extensions to E-R model allow us to model with new concepts like generalization/specialization hierarchy and aggregation hierarchy. Nevertheless, encapsulation concept is not proposed.

The second tendency is to propose the object-oriented models for the data bases designs. For example O.R.M. [Pernici, 1990]. Powerful concepts are used like encapsulation which lead us to model static and behaviour structure object. The concept of "inheritance", allowing shared data and methods between class, is one of the more important elements in these models.

Others models like state diagrams increase our thinking about software problem space. For example, state diagrams realized object by object increase our thinking. We are more precise and it is an easier way than the classical way of thought.

An other way to increase thought is to include the environment of the system, to indicate why such event, such message has arrived, and to study the retroaction from the environment to the data base system. Nowadays O.O.S.E [Jacobson, 1990] increases on users cases. With an efficiency manner, Jacobson integrates users needs in requirements/analysis process.

3   O/S.M: OBJECT/SYSTEM MODELLING

Object/System Modelling is an object-oriented method which help analysts in software requirements/analysis tasks. O/S.M proposes a guideline to map the analysis results to object-oriented database O2 [Bancilhon]. O/S.M. based on the system theory and on the object-oriented paradigm, increases the semantic model power and the thought process power. O/S.M model improves the user interface study and the user restitution of the results. Based on these two paradigms O/S.M ensures an homogeneity in the creative thought for understanding and designing a new software analysis.

3.1 O/S.M model.

O/S.M. considers that a real system is composed by a set of entities. An entity takes part in the activities of the system either as an activity executor or being useful for the activity. It activity takes part in the realization of one or several finalities of the system. The evolution of an entity is represented in a space, time, form reference system. An entity exists through the time and through the space. It travels through the time and the space with modifying its form. An entity is associated to a family which determines its structure and behaviour and also its finalities.

Family definition determines constraints on properties, on methods and also on events which could occur in the entity life.

An other originality of O/S.M is an enlarged encapsulation concept. In fact, for O/S.M an entity of a family encapsulates his own entirely history (set of events which occur in the life of the entity). O/S.M defines also constraints on events arrival. A language and a tool built with Graphtalk [Parralax, 1990] is useful to describe constraints, sequence of events. The internal part of entity defined by Family structure and valued for each entity is its characteristic.

Entity has an external part named context. All entities which interfere with a particular entity build up the context of this particular entity. The interference between entities may be static or dynamic.

Static links are classical links like genralization/specialisation, aggregation/part-of, association. Dynamic links are based on shared events, generated events, or characteristic events.

O/S.M increases study of collaboration between entities using the concept of "process of collaboration", which looks like schemas in [Desfray, 1991] or subject in [Coad, 1990], or modules in [Rumbaugh,1991]. This concept allows an analyst to use an ascending approach closely with System science paradigm. After a local study of family object, O/S.M leads to a global analysis by complexification, regrouping family objects in a "process of collaboration" families. Process collaboration is a good mean to look for families in O/S.M process.

O/S.M. classifies the families into three types: the family of agents, the family of passive objects and the family of time (clock, calendar,..).

An agent is an intelligent entity which is able to take decisions. It can initiate at least one "process of collaboration" for reaching the finalities of the system.

A passive object is not an intelligent entity. It undergoes the transformation in the reference system (space, time, form) and is used by the process of the system to study. A passive object can not make any decision, but it propagates the character or context modifications toward the other entities which possess a dynamic link with it.

Entity of time family takes part in the entirely process of collaboration. Its context is composed by all the entities of the real system.

Internal functioning of an entity

A family defines a behaviour for its entities. A function is triggered by event which is produces by the entity itself or by the entity context. So, one or several functions are executed. the result of processing function are one or several events which impact the object itself (we name this event: characteristic event) or its context (by the means of shared events or generated events).

Global functioning of system

When an event has been noted on an entity of a family, it can bring:
• the generation of a other event on one or several other entities.
• the note of the same event on one or several other entities.

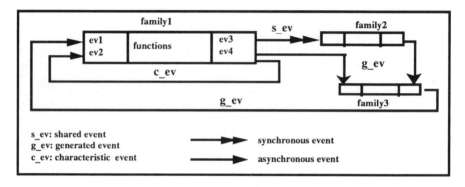

Figure 1 Global and internal functionning of Object/System Modelling.

According to the system theory, a family is a system including a set of entities. It encapsulates a system, a family is itself an entity having properties, functions, events and a context (static links, dynamic links, process of collaboration). This entity can be dynamically bound with other entities through shared or generated events.

According to the system theory, actor is a system This concept permits to take into account organizational aspect of the software system. The actors are the entities of a families of agents. A family of actors is characterized by a name, the finalities, the functions, the events and the knowledge. So we could modelize organization aspects then increasing understanding of software environment or system environment.

Family Models

The first model corresponds to **static structure of family** (Figure 2). For each family we choose to represent Family linked with its roles and attributes. Links between concept are typed.A circle represents a Component Family. A square represents a Family Role. > is the monovaluation symbol while >> is multivaluation symbol. Dotted line represents optional link while full line represents Mandatory link. The couple (x,y) represents for x = F Fixed value, x= L for Free value. For y= P permanent value and y = T Temporary value.

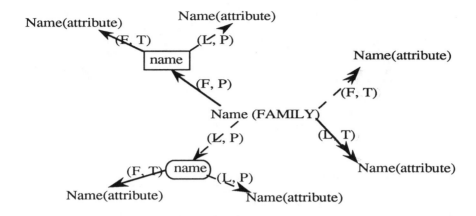

Figure 2 According with the above text representation of static structure of a Family.

Dynamic aspects are represented by state diagram. O/S.M emphasises the concept of role of entity. State entity and role change are represented with the same model.

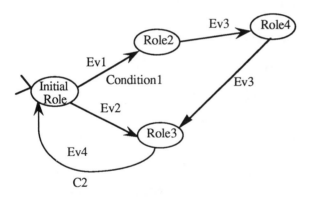

Figure 3 Links representation between roles. We could notice that states of entities are representing with the same model.

Interaction between Family

Static interaction model represents families of the system and links between these families. Monovalued situation link is represented by a full line and arrow, double arrow if the link is multivalued, dotted line if the link is an existence link. The couple (x,y) represnts for x = T Total link x = P partial link y = T Temporary P permanent then we obtain:

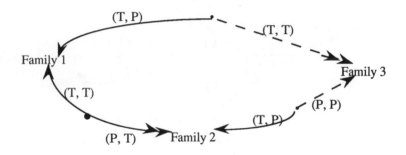

Figure 4: Links Representation between Families.

Dynamic links between entities.
This model presents for a process of collaboration the events sequence.

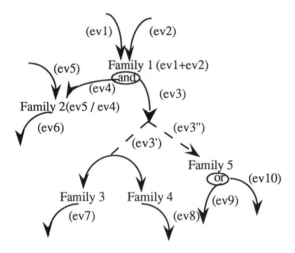

Figure 5: Dynamic links between Families.

3.2 The Requirements/analysis modelling guideline

O/S.M. proposes a modeling guideline permitting analyst to understand and to model
a software system. O/S.M proposes a set of advices to lead requirements and
analysis phasis. It proposes interviews and validation of interviews using the
models. O/S.M uses the concept of perspective to classify the entities, which
permits us to get more homogeneous subset. Face to the life cycle of software,
O/S.M. brings a help for the comprehension and the representation of the existing
and the future software. The guideline comprises four modelling stages, the
descriptive stage, the analysis stage and validating stage and mapping stage. Note
that the process is a non linear process. We have to repeat each step along the
analysis and then we have to build up an application progessively.

The descriptive stage
•Determination of the existing system.

Using process of collaboration, the analyst is able to delimit the space problem. It is when context event associated with process of collaboration has no impact on others non described entities.

•Emergence of the families corresponding to collaboration process.

In an analysis phase it is possible to start with descending decomposition when we meet large system or when there are a lot of families.

•Determination of the fundamental families.

In this step we look for families by grouping entities with similarities in information, behavioural structure.

•Opened prototypes of the fundamental families.

The descriptive stage ends with a gross description of future families. Each families is associated with a prototype which describes static and dynamic structure. At the end of this stage, each entities of real world is instance of one family. Each family describes the attributes and the functions of its entities. The events which could occur on entities of this family are described with their triggers and predicates.

The analysis stage
•Determination of new families by disjunction.

In this step we refine fundamental families using subset theory. So it is a good way to make emergence of new families which are more precise and more homogeneous.

•Determination of new families by conjunction.

It is also possible to obtain new families using fundamental families collected in the first stage. Using conjunction principle we group any fundamental families which particpate in a process of collaboration. In fact we could use this conjunction in complemntarity with disjunction mechanism.

•Construction of the model schemes.

In this step we draw schemes named family information model, behavioural model based on state diagram which apply on entity life and also on role played by the entity. This is very useful to study mechanism of evolution of instanciation link between family and entity.

•Restricted prototypes of the final families.

In this step, we refine constraints on attribute. We specify functions with more details.

We complete also model by additionning some details.

The validating stage
•Application of the validating rules.

O/S.M suggests some rules to be applied on model, or on coherence between models. For example function encapsulated in a family is associated with states of entity of this family. In each state, we have to trigger a function and this function must be declared in final families.

The mapping stage
•Rules to map O/S.M to O2 database.

O2 is an object-oriented database system [Bancilhon,1990]. O/S.M suggests some rules to map attributes and functions of each family in D.D.L and methods described in CO2 language.

4   CONCLUSION

In this paper we present an object-oriented method for analysis software. we use models and guideline suggests by O/S.M in bank application. Using this methods leads us to modelize some applications. nevertheless we have to demonstrate that it is able to be used in others contexts like industrial context. We define also a CASE for this method using Graphtalk metatool.

5   REFERENCES

Bancilhon (1990) The O2 book. Industrial ministry 1990.
Booch (1991) "Object-Oriented Design with application"Cummings Publishing Cy
Bertalanffy (1948) System general theory.
Chen (1976) "The entity-Relationship model. toward a unified View of data"
ACM Transaction Database system Vol 1, N°1, Mars 1976
Coad(1991) "Object-Oriented Analysis" Prentice-Hall International1991
Descartes(1642) Discours de la méthode
Desfray(1991)Le modèle classe-relation (In French) application C++Masson 1991
Gane (1979) Structured Analysis and Structured Design Gane/Sarson Prentice Hall
Jacobson(1990) Object Oriented Software Engineering Addison Wesley 1991
Lasoudris(1993) Object/System Modelling L.Lasoudris and S.Liao I.E.E.E international conference Le Touquet France octobre 93.
Liao(1993)Object/system Modelling Thesis july 93 University of Aix/Marseille 3 France Pr J.L.Le Moigne
Le Moigne (1990) "La modélisation des systèmes complexes" DUNOD 1990 (In French)
Parralax (1990) Graphtalk a CASE tool for metamodelization.
Pernici (1990) "Requirements specifications for object-oriented systems" INFORSID 90 Nouvelles perspectives des systèmes d'information EYROLLES
Rumbaugh (1991) Object modeling technicsPrentice Hall
Ross(1977) "Structures Analysis for requirements definition" IEEE Trans. Software Eng. Vol SE-3, No1
Schlaer(1991) Object-Oriented systems analysis modeling the world with Data. Yourdon Press
Sibertin-Blanc(1991) "Cooperative objects for the conceptual modelling of organizational information systems"IFIP TC8/WG8.1 working conference on the objected-oriented approach in information systems Quebec Canada .
Simon(1962) The architecture of complexity Philosophy of sciences
Spaccapietra (1989)"About entities, complex objects and object-oriented data models"Proceedings of the working conference on information system concepts Namur, Belgium

5   BIOGRAPHY

Louis LASOUDRIS is an information system teacher in Telecommunications National Institute (I.N.T), engineering school of France Telecoms in Evry France. His main interest is Object oriented methodology and more precisely Modelling technics. His research leads him to suggest a new object-oriented method O/S.M. Now, he works on a CASE tool using Case Based Reasoning technologies to allow analysts reusing models.
Stephen LIAO was a student during three years in I.N.T. He worked on a thesis on object oriented method with Pr J.L.Le Moigne. He creates with Louis Lasoudris O/S.M. Now he is a teacher in polytechnic school of Hong-Kong, where he teachs Object-oriented database, language and methods.

52

Expanding Rapid Prototyping Activity: Use It to Develop Real Time System

Jianxin Yan
Dept. of Computer Sci. & Eng. ,
Beijing University of Aeronautics and Astronautics
P.O.Box 1-91#, No.37 XueYuan Road , Peking 100083, China, Tel:086-01-2017251-7605

Abstract

Rapid prototyping is a widely used technique for software development. People usually think it cannot be used in the area of real time system, especially embedded system. In this paper, the author argues that we can effectively apply rapid prototyping technique to real time system, and then we can benefit most in requirement analysis and refinement. The author gives the **'simulation prototype'** concept, which provides a good approach to apply rapid prototype to industrial projects such as real time applications. Furthermore, the author points out that, from the viewpoint of Cybernetics, the essential thought of rapid prototyping is feedback and adjusting, i.e. step refining, which, in fact, has been implicitly or explicitly applied in practical developing activities, though, the developers do so consciously or unconsciously.

Keywords

Prototype, rapid prototyping, real time system, simulation prototype

1 INTRODUCTION

Rapid Prototyping is an effective technique for software development. It has been widely used in various software developing projects, unfortunatetly, except for real time system, especially for real time embedded application. Is there such a big 'gap' between real time system and rapid prototyping method?

The author thinks the answer is NO, this paper is going to discuss the problem.

2 RAPID PROTOTYPING AND ITS ROLE

In computer circle, the developers' experience tell us that one of most important tasks in a system developing process is to get the full requirements of the system correctly from the users. But facts are less smooth than that people image. During requirement definition, many problems often arise. The users describe the system requirements using their specialised knowledge, but the developers must then translate the users' scripts into detailed technical requirements written in computer-specalised terms. In general, the two kinds of engineers(i.e. users and developers) have different technical backgrounds and different technical language. Therefore , it is nature that a communication obstacle lies between the developers and the users of the same system. People has already realised that the earlier an error is made in a project the more catastrophic the effects of that error . Especially when users unconsciously omit the implicit assumptions about the system, the problem which it hence produces may become more serious.

Unfortunately, the tranditional software development methodology based on life cycle model(waterfall model), which provides many requirement analysis and specification techniques though, cannot provide a perfect solution to the above problems. People gradually understand that software development , particularly during its early stages , should be regarded as a learning process in which both the developers and the end-users should participate actively . To be efficient , it requires close cooperation between developers and users. In addition, it will be successful only when it is based on an actual 'working system', which, though, is possibly not mature. Hence, people put forward the technique called as Rapid Prototying(Budde R. et al. 1984, Yeh R.T. et al. 1989). As a better solution, rapid prototyping relies on an idea borrowed from other engineering desciplines, which stresses to produce a cheap and simplified prototype version of the final system rapidly and early during project development. This prototype becomes a learning facility to be used by both the users and the developers and provides essential feedback during the system specification definition. It serves the purposes of experiments and evaluations to guide further development.

Acting as a direct, visible model, prototype's explicit modelling of a system helps both the developers and the users further their understanding of problems, and clarify their misunderstanding about the system. Thus, the users may point out the improper designs quickly, and the developers may also change to eliminate unnecessary , even very costly design requirements.

Prototype is usually constructed at the early stage of the system design, therefore it is not proper to put too many resources, including labor forces and devices, into it. Rapid prototyping emphasizes that the developers should provide a model(prototype) quickly and economically. Usaully, prototype is to be used for a limited period , so quality factors, such as efficiency, maintainability, and full error handling, are little cared. What is important about prototyping is the process but not the prototype itself. Therefore, the major part of the efforts should go into the critical evaluation of the prototype rather than its design.

Used in this way, prototype is often a useful tool for exploring alternative designs and evaluating the appropriateness or feasibility of a new idea. Of course, prototyping and conventional methods are complementary rather than alternative approaches to system development.

3 WHY HAS RAPID PROTOTYPING BEEN REGARDED TO NOT SUITABLE FOR REAL TIME SYSTEM ?

Most engineers support the viewpoint that in areas of embedded software and real time control system, the life cycle model , NOT the rapid prototying, is the most rational approach. They reject rapid prototyping in these areas. Why? The reasons is that they limit their thoughts to that 'prototype' must be actual WORKing model,thus, prototype of real time system, especially embedded system, is too costly and/or impractical to construct. Therefore, in published books and papers about software engineering I have ever read , the authors stated that the aims of rapid prototyping stand out clearly against its application in those backgrounds.

In fact , the major difference between real time system and the other computer applications is time constraint in the former, in which the limitation to reponse time usually is quite strict, failing to this requirement will lead to partial or full failure of the system(O'Reilly C.A. 1985) . Indeed , there is no other essential difference between the two kinds of applications. Except the reason listed above, I cannot find their any other strong reason for rapid prototyping not applicable to real time system at all.

4 SIMULATION PROTOTYPE: APPLYING RAPID PROTOTYPE TO REAL TIME SYSTEM

During the development of systems such as real time applications, it actually exists many difficulties to construct a real working prototype system. Or , even if the prototype was difficultly but successfully created, the benifit produced by it cannot compensate the cost yet (that is to say, 'gain cannot offset loss'). But it does not mean that rapid prototyping cannot be used in real time area at all. In our practices of real time system development, we often become exausted and even ...annoyed by the users' frequent changing their requirements. Though we try our best, tranditional life cycle developing methodology does little help to us. So, we begin to explore and exploit simulation prototyping technique, whose main purpose is still to make user participation easier and then to refine users' requirements , so that we can get specification definition as perfectly as possible.

Based on the prototyping idea, though, simulation prototype is not an actual working prototype version of the final system. Instead, it is a simulation model which reflects the behaviors and properties of that system. Most commonly, prototype does not necessarily reflect all features of the final system. It's enough to describe those key points of the system. In real time system and embedded system, human computer interactions are less important than they in interactive information system. Here, of most important is the system's interaction with its external enviroment, instead. Simulation prototype mainly illustrates the interaction and relationship between the system and its external enviroment, and the logic connections among internal components of the system. In addition, It should show the proper operation sequence of the system, and can help people find where the potential dangers lie. Instead, the technical implementations of some components in the model may be omitted. In simulation prototype, what we notice is the logic and behavior relations among the system, various system components and the external enviroment, NOT their detailed design. So, some system components can be looked as 'black box' with input/output. In general, simulation prototype provides a practicable interactive model for both the developers and the users, which leads to a bettter understanding of the system objectives and operations. Obviously, it will result in the production of specification meeting the need of the users more accurately .

It should be noticed that communication between system model and the users in an easily-understood way is the key point to successful simulation prototyping. According to its implementing approach, simulation prototype can be devided into two categories: 1) prototype on paper, which may be dozens of drawing figures with describing texts; and the better approach 2)amination prototype, which uses computer animated picture to illustrate the key properities of the system. Because 'prototype on paper' lacks dynamic features and continuity, its effect is not as good as that of animation prototype. In practice, we prefer to animation prototype, which can be implemented with simulation software package such as GPSS, SIMSCRIPT II.5(Xiong Guangleng et al. 1991) , even with conventional C or C++ language.

Empirical efforts tell us that when using simulation prototype, in order to get the proposed benefit from it , we must guarantee:

1) Model must be constructed quickly;

2) Both the developers and the users can easily interact with the model, especially the users should need only minimal knowledge of computer techniques to understand it;

3) Model can be easily changed to match the frequently-changing knowledge about the   system.

Furthermore, if we use object oriented analysis and design method , which can increase the flexibility of the model construction and make problem decomposition easier(Booch G. 1986), to construct simulation prototype, our benefit will be still greater.

Our practical facts prove that simulation prototype is a most effective methdology for problem solving when the real world prototyping experiments are too costly and/or impractical to perform.

5 CONCLUSION

This paper has expanded the effectual applicable scope of rapid prototyping technique. The author has argued that rapid prototyping can be used in areas such as real time system and people will benefit most in requirement analysis and refinement from it. Furthermore, the author gives the 'simulation prototype' concept, which provides a good approach to apply rapid prototyping technique to industrial projects whose working prototypes are too costly and/or impractical to be constructed.

In fact, from the viewpoint of Cybernetics(Wiener N. 1961), the essential thought of rapid prototyping is feedback and adjusting, i.e. step refining, which, in fact, has been implicitly or explicitly applied in practical developing activities, though,the developers do so consciously or unconsciously. Of course, that we consciously apply the technique will bring more benifits. Our practices prove that the explicit application of simulation prototype has solved many problems that the life circle model never solved, and simulation prototyping is an effective technique in many industrial computer applications.

6   REFERENCES

Booch, G.(1986) Object-oriented development. IEEE Trans. on Soft. Eng. , Vol.12, No.2 .

Budde, R. et al (1984) Approach to prototyping. Springer Verlang.

O'Reilly C.A.(1985) Fast is not real-time: designing effective real-time artificial intelligence system, in *Application of AI* (ed. Gilnore J. F.), SPIE, Washington.

Wiener, N. (1961) CYBERNETICS Or Control and Communication in the Animal and the Machine(second edition). MIT Press and John Wiley & Sons, Inc.

Xiong Guangleng et al (1991) System simulation of continuous system and discrete event. Tshinghua University Press.

Yeh, R.T., et al (1989) Rapid prototying in software development. IEEE Computer, Vol.22, No.5

BIOGRAPHY

Jianxin Yan *is a second year postgraduate in Beijing Univ. of Aero. & Astro.(BUAA), majoring in computer science & engineering. He received his B.S. degree in Computer Science from BUAA in 1993. He is persuing his M.S. degree. His interests include software engineering, data base, information system, software security. In these fields, he has published over 10 papers. Working for software design and development for years, he has accumulated a fair experience. Now, he is a special invited technical contibutor of* **IDG PCWORLD China** *, the biggest computer magzine in China. It is the height of his ambition to further his studies under eminent teacher's direction.*

53

System integration - an emerging industry

Shi-Ping Hsu
TRW Systems Integration Group
1800 Glenn Curtiss St., Carson, California 90746 U.S.A.
(310) 764-7058 (tel), (310) 764-7059 (fax), shiping@acme.trw.com

Abstract

The trend of "open systems" allows users to build large computer applications quickly, which stimulates the growth of system integration as an industry. However, building large systems requires formal system engineering disciplines. This paper describes the basic system engineering approaches that TRW uses in developing large complex system integration projects.

Keywords

System integration, system engineering, system development cost, speciality engineering

1 SYSTEM INTEGRATION IS A RAPID GROWING INDUSTRY

The computer industry of today has transformed into an "open" industry - layered, standard-based, open systems. The old, proprietary systems are quickly disappearing. What is available now is a wide selection of hardware and software building blocks. By properly fitting these building blocks together, users can create innovative applications quicker and cheaper than ever before. System integration is quickly becoming one of the fastest growing branches in the information market, as shown in Figure 1.

(billion U.S.$)

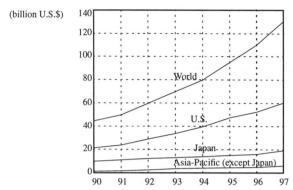

Figure 1. Worldwide system integration market. (Dataquest Japan, February, 1994)

2 SYSTEM ENGINEERING IS THE KEY TO SUCCESS

System engineering is like conducting an orchestra - creating wonder (or disaster) by integrating pieces together (Langley, 1994). A well engineered system is easy to use and simple to understand. A poorly designed system is hard to understand and difficult to change. The following quotes described the characteristics of system engineering:

- System engineering is a branch of engineering that "concentrates on the design and application of the whole as distinct from the parts... looking at a problem in its entirety, taking into account all the facets and all the variables and relating the social to the technical aspects." - Simon Ramo, TRW founder, 1973

- "It is not the single program but the architecture that matters - the way in which every program interacts with every other program in the system; the manner in which it integrates with the system and its environment." (Chorafas, 1989)

- "Although the hardware of computers is still very important in providing new options, the key structure that is being built is generally software." (Rechtin, 1991)

System engineers are like the architects of buildings. They have to understand the bolts and nuts as well as the overall surroundings. They have to look at the project from multiple perspectives - the users, the developers, the managers and the testers. They have to have varied experiences in hardware, software, analysis, standards and user engineering. And above all, they have to be effective communicators.

3 FRONT-END SYSTEM ENGINEERING IS CRITICAL

Successful system engineering usually converts a complex project into a system that is easy to use and simple to understand. Well engineered systems are usually easy to maintain, easy to insert new technology, and flexible in adding new functions. To achieve these goals, system engineering is particularly important in the beginning.

A typical project can be managed in 5 phases as shown in Figure 2. It starts with decomposition and definition (top-down). It ends with integration and verification (bottom-up). The success of definition and decomposition determines the success of integration and verification.

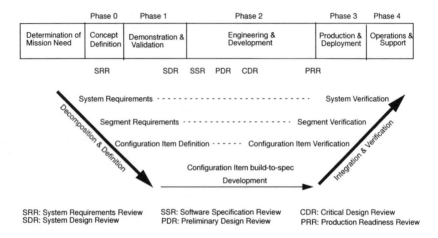

Figure 2. A typical system acquisition life cycle.

Since every project is a subsystem of a larger project in the society, the first step is to define a simple interaction with users and consistent interfaces with the environment. A project can not start until the concept is clearly thought through. Following the concept development is system segmentation. Decomposing a large, complex problem into small, simple segments is a critical step in system engineering. Interfaces among the segments should be designed with minimum complexity. A system's success often can be determined by the segment interfaces. Once the system is segmented, the concept of each segment needs to be translated into precise requirements. A requirement is not a desire, guideline, or objective. It must be testable. Once the requirements are completed, the foundation of the project is set. The rest is development, integration and verification.

We emphasize the front-end system engineering because costs increase dramatically the later problems are found. Figure 3 shows the relative costs of fixing a problem in a project life cycle. The numbers are based on TRW's internal experience which turns out to be quite similar to the prediction by Boehm (Boehm, 1981). Doing it right the first time is the ultimate bargain.

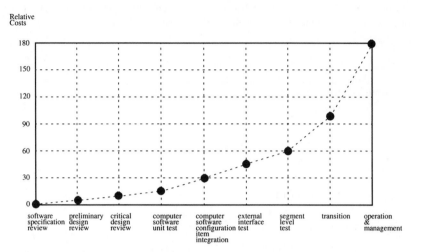

Relative Costs

Figure 3. Costs increase dramatically the later problems are found.
(Based on TRW project experiences)

Experience shows, however, an average of 30% of system development cost is rework or waste. Table 1 shows a typical cost breakdown of a project.

TABLE 1. TRW experience shows that 30% of system development cost is rework. (Table does not include commercial-off-the-shelf hardware/software cost).

Task	Percentage Cost (%)	Rework Cost (%)
Administrative and Support (25%)		
Infrastructure	11	
Human resources	2	
Technology development	3	
Procurement	2	
Marketing & sales	5	
Inbound logistics	1	
Outbound logistics	1	
System Development (35%)		
System engineering	10	2
System integration and test	14	9
Management	7	
Quality assurance, configuration management	4	
Software Development (40%)		
Software requirement	5	1
Preliminary design	5	2
Detailed design	10	4
Code & unit test	10	5
Software integration and test	10	7
Total	100	30

4 TRENDS IN SYSTEM ENGINEERING

More and more system engineering tools are becoming available. These tools help a system engineer in requirements, configuration management, modelling, CASE, and business process reengineering.

In terms of system architecture, open system, hardware independent, and client-server architecture are the major trends. Internet, created under the open system mentality, is probably the most vivid demonstration of the power of an open system. From anywhere in the world, using any hardware platform, running any operating system, a user can access a vast amount of on-line information in multimedia presentation. Today, 20 million users in 75 countries, connected by 48,000 networks, are already accessing information on 7,000 World Wide Web servers. This could not have happened without the open system concept.

5 TECHNOLOGIES FOR SYSTEM INTEGRATION

There are some basic technologies for system integration: computers, local and wide area networks, network management, database management, client/server architecture, mass storage, multimedia, user interface, fault tolerance, and object oriented technologies. While all these technologies are moving fast, the fastest changing technology in 1995 is perhaps ATM networking. Today ATM represents 1% of the market. By 1996, it is expected to grab 50% of the networking market.

In addition to the core technologies mentioned above, a system engineer needs to be equipped with some speciality engineering. These include human factor engineering (Norman, 1988), user engineering, security engineering, reliability/maintainability/availability engineering, life cycle cost analysis, logistics engineering, and performance engineering. The application of speciality expertise can play a key role in the overall engineering decisions.

6 SYSTEM ENGINEERING MANAGEMENT

The ultimate success of an engineering project lies in engineering management (DSMC, 1990). The best management structure may not guarantee results. But a bad one will certainly guarantee non results. A lot friction, frustration, and inefficiency may be created. Work Breakdown Structure (WBS) is a common technique to provide contractual document, financial management and engineering management.

7 SYSTEM ENGINEERING EXAMPLES

Apollo Lunar Landing is an example of good system engineering. This was a 7-year, 7,000 person $2 billion effort resulted in man's first footstep on the moon in 1969. The project management team believed that systems engineering is the design of the design - a reversal of the natural impulse to "run off and do things right away". Every problem that could conceivably occur during the life cycle was considered at the beginning of design. Also, designing the system test was a challenge by itself because the vehicle was to land in a low gravity vacuum. All life cycle phases were considered. The lunar module was sent crashing back to the moon's surface as part of a seismological experiment.

TRW is involved in a few large scale system engineering and system integration projects, including tax modernization, air traffic control, nuclear waste management for the U.S., and fingerprint identification system for England. For each of these projects, TRW uses the system engineering disciplines outlined here to provide a systematic approach to the solution.

8 REFERENCES

(Boehm, 1981) Barry W. Boehm, *Software Engineering Economics,* Prentice-Hall, 1981
(Chorafas, 1989) Chorafas, *Systems Architecture and Systems Design*, McGraw-Hill, 1989
(DSMC, 1990) Defense Systems Management College, *System Engineering Management Guide*, January, 1990
(Langley, 1994) R. J. Langley, *COTS Integration Issues, Risks, and Approaches*, TRW Systems Integration Group Technology Review, Winter, 1994
(Norman, 1988) Norman, *The Psychology of Everyday Things*, Basic Books, 1988
(Rechtin, 1991) Rechtin, *Systems Architecting*, Prentice-Hall, 1991

9 BIOGRAPHY

Shi-Ping Hsu is a Technical Fellow at the TRW Systems Integration Group. His research interest is in database and networking. He developed a high speed pattern matching engine, the TRW Fast Data Finder, for text search, document dissemination and biological applications. Currently, he is developing an ATM testbed for video applications. Shi-Ping Hsu received his M.S. and Ph.D. in Electrical Engineering from California Institute of Technology.

54

An Object-Oriented HCI platform scheme for shop floor

Ping Du, Nianyu Wu, Xueqing Sun
Department of Automation, Tsinghua University
The State CIMS Engineering and Research Center,
Beijing, 100084, P.R. China
Tel. (86-1) 2564179, FAX. (86-1) 2568184

Abstract

The purpose of this paper is try to search an interface scheme that suits for manufacturing environment. First, some problems concerning about how to use Object-Oriented and Direct Manipulation technology in User Interface Management System (UIMS) are discussed, then present a kind of user interface platform scheme for shop floor. Under this scheme user can easily develop his own interface and do some other things such as application system development, data base maintains, network management. Several technical features of this scheme are mainly introduced: application support, visual construct method, interface description method, interface element library, message analysis interface.

Keywords

user interface, UIMS, Object-Oriented, Direct Manipulation

1 INTRODUCTION

The global market competition and the development of new technology have brought challenge to manufacturing engineering throughout the world. Now automatic manufacturing engineering is a very vital field. Among all the new philosophies came out recently both CIM and Concurrent Engineering (CE) pay much attention on Human factor. The Human-Computer Interface (HCI) as a bridge between Human and System has its special position. The feature of HCI has become a very important target to measure the industrial control system, while the more complex the modern controlled object is, the more the requirements to the

Human-Computer Interface the systems ask. Therefore, How to develop a HCI which meets the needs for industrial environment has become a significant task for computer control system designers.

The environment of industrial control is different from the office environment. The various equipment and devices, the variety of the control structure, the distinct operating system and special requirements to interface styles ··· all of these factors cause trouble to design a general and flexible HCI. Therefore, it is some difficulties to directly utilizing the results of the research in UIMS (User Interface Management System) recently. It means that user interface separating from its application software thoroughly is not easy. Authors just based on the analysis of the UIMS feature, present a system configuration interface to Direct Manipulation (Visual Construct Environment) and try to search an interface that suits for manufacturing system in shop floor. Finally, A structure of Object-Oriented HCI platform scheme for the real time is introduced.

2   A SOLUTION FOR HCI SCHEME

In manufacturing environment, different types of application systems have different scheme, their functions and user interface modules are distinct from each other as well. For example, there are various devices from many vendors in shop floor. They should link with computer by their proprietary communication protocol. Each manufacturing system have its own number of equipment and layout and its flow chart. Obviously a proprietary user interface can't satisfy requirements for flexibility and expandability in shop floor. To solve the problem, a HCI platform used for a Real-Time system is presented in this paper.

Two main problems should be solve to implement HCI platform.

●The description and specification of user interface.

●The expandability of application systems and presentation elements of user interface.

As already mentioned, system is different from each other in automatic manufacturing system, the flexibility of the HCI is very important. Users hope that they have the ability to adapt their HCI to the manufacture equipment organization.

One of the approach to improve the flexibility of HCI is UIMS, its main idea is to separate the HCI design from the application design. UIMS offers some method to specify, describe, build and maintain the interface. Thus the user can use UIMS to build what interface he wants.

But current UIMS has two deficiency: one is the description and specification of the user interface is so abstract that the common user can hardly build the HCI themselves.

The other is that it is very difficult to extend a UIMS. The organization of the UIMS is too complex and inter-depended to extend even only a part of UIMS. In the point of UIMS, the function of user interface is limited by the application system. If a new user interface is developed by cutting out of old one, the developer may modify the application design nothing or only a little. But if the developer wants to add some new interactive styles, he will have to modify the application system and to describe new presentation elements of user interface as well.

what we should do is to improve the current UIMS, in other words, to combine the idea of UIMS with some new techniques as follows.

2.1 Using Direct Manipulation interface feature to solve the description problem

By using Direct Manipulation interface the user has the illusion of directly acting upon the objects of interest without intermediary of the system. Many common actions are performed by pointing at or dragging objects on the screen. In addition, by using menu and/or dialogue box. Textual command languages of other obviously syntactic mechanisms are never used (Gerhard, 1987). Advantage of this style is very obvious, because it reflects designer's natural cognition process and cognitive structure to the most extent, and it further shorten syntax distance between user and computer.

In order to realize direct control interface some questions must be discussed. The first, syntax should be minimized. The second, how to realize the flexibility in the presentation component. Finally, how to perform rapid incremental and reversible operations whose impact on the object of interest should be immediately visible.

The most difficult problem is how to hurdle syntax obstacle between user and objects of interest. When user communication with interface, he must utilize some syntax to express his idea. But traditional interface syntax is formed by that which depend on the whole interface, so user must communicate with the whole system, this way is too complex. Only if each object had its own syntax could user interact with it directly. This means the syntax should be minimized.

Although the Direct Manipulation interface has the advantage of easy understanding and operation for novices, it can't meet the needs for constructing complex system and extend user interface. In order to overcome the imperfection, one approach to the disadvantage is that Direct Manipulation and traditional conversation metaphor combine to create a Visual Construct Environment (VCE) for user. In this environment, when direct control is difficult to implement, the dialogue box is used. It means that more interface presentation elements should be prepared. For example, we provide many nonstandard windows and dialogue boxes which make interface easy to be construct as well as some standard windows and dialogue boxes which make interface easy to be extend. In addition, user is allowed to define details of interface description by directly modifying interface description file in VCE, so that he can give full play to his creativity.

2.2 Using Object-Oriented technology to solve the problem about extension

Extension is concerned with two aspects: It is relevant to application system as well as relevant to constructing new style of presentation component.

It is important to provide a programming interface as a bridge between user interface and application system. We called the bridge after the Message Analysis Interface (MAI) for supporting extension of application system.

Considering Object-Oriented technology describes interface presentation component. We regard every screen presentation component as a object, and then establish a standard Object Library. User can easily form new objects with inheritance and polymorphism.

3　USER INTERFACE DESIGN RULES

3.1　Using simplified event model and message queue to achieve real time human computer communication

The description of user interface has several　methods: based on menus , based on state transition network, based on event handler and so on (Gaines, 1988). Event handler model is good for automatic manufacturing systems. Its some features is peculiar to the model, such as: suitable for module design, able to implement application feedback, appropriate for description of parallel control. That is the reason why the model is adopted in this paper.

Automatic manufacturing system interfaces need to manage messages which come from :

●User, such as input of keyboard and move of mouse.

●Application, such as warn message.

●Inter of user interface, such as window management message.

The designer classified message according to above sources. Whenever a message comes, it will be put into a unified message queue. Depending on its arrival sequences, a presenting event can be generated. Then the parallel control of all the messages is accomplished by the event handler.

3.2　Establishing a parallel interface between user interface software and application program

There are three kinds of control pattern of user interface: inter control, outer control and mixed control.　Inter control refers to that the application calls for corresponding　user interface procedure to motivate communication between them. Outer control refers to that it uses corresponding procedure of application to complete the interaction when user interface receive user's control instruction. Mixed control is the combination of these two.

In manufacturing system, generation of message is parallel, user interface and application both have opportunity to be the initiative one in interaction. So mixed pattern is the most convenient way.　Details about how to connect user interface with application by mailboxes is discussed at the part of the Message Analysis Interface.

3.3 Designing system with Object-Oriented technique

Object-Oriented language is very similar to natural language. Data and its behavior is encapsulated together to construct object class. Using objects, the description of user interface becomes very simple and clear. On the other side, the Object-Oriented system architecture is highly flexible and it enhances the reusability of many building blocks (John, 1986).

A user communicates with the system through a world represented on the display screen, which is composed of active objects. Each screen object has its visual representation, a state that defines its appearance on the screen and its relations to other screen objects, and behaves according to its functional role.

The advantages of Object-Oriented are considered as follows (Scott, 1987):
● The creation of subclasses of existing classes allows the designer to create new objects that differ from existing objects in some desired aspects but that inherit almost all of the functionality of their ancestors.
● Designer can use pre-defined components very flexibly, even if most of the behavior of a super class is undesired, he can still use it by overwriting all but the useful properties.
● Extensions can be made on different levels of the hierarchy.
● It has higher programming efficiency compared with traditional module method.

4 THE OBJECT-ORIENTED USER INTERFACE SCHEME

The structure of the system scheme is shown as Figure 1. It includes five main parts.

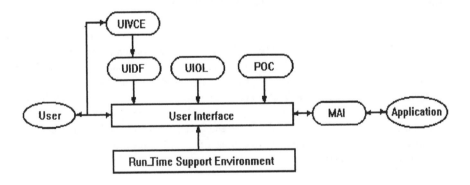

Figure 1 the Object-Oriented HCI scheme.

4.1 Product-Oriented Core (POC)

Although the separating of user interface from application system will bring many benefits, user interface should not be separated thoroughly as it is a outer presentation of application and an element which application uses to interact with human being. In manufacturing system, data transmission and data processing are the basic functions. To achieve these functions application should provide the supports on communication management, machining operation management, data management, etc. In addition, the interface with CAD/CAM and MIS should also be considered. All of these supports we called POC.

Some of the functions are closely related to user interface, such as machining operation control and data management. Here machining operation management refers to operations preparing, NC program checking, tool checking, pallet checking and fixture management. For example, job list starting, emergency stopping, real time supervising, etc. Data management refers to NC program management, job list management and tool data management.

4.2 User Interface Visual Construct Environment (UIVCE)

In Visual Construct Environment, user can construct his own user interface by defining each basic class in interactive task such as window, menu, dialogue or button, etc. Then he also defines real running mechanism, it concerns about recalling action motivated by every input messages. Finally the definition of motivating action, recalling action and data flow will be turned into the description of corresponding operation instance.

Under the visual construct environment, there are some abilities of detecting errors, and can assuring validity of real operation actions through limiting the behavior of objects.

4.3 User Interface Description Method (UIDM)

When constituting interface under UIVCE user should define interface style, class and recalling action (Gang, 1994). All of these will be placed in interface description file automatically. The file is a particular file form, whose language is like the fourth generation language. Therefore, the description is very simple. Besides this file, user needs two files as well, the system configuration file is built by super configuration construct procedure and the resource file is the extended .RC file in Microsoft Windows. Finally, interface platform synthesizes all of the files and builds a project file list. So long as user enters Borland C++ for Windows integrated development environment according to the project file, then compile and link, the executable file user needs is generated.

4.4 User Interface Object Library (UIOL)

In Object-Oriented point of view we regard the user interface as being made up of many presentation elements, such as main window, sub windows, menu, dialogue boxes and icons, etc. We use Object to describe these elements and collects all of the objects in a UIOL. Users can easily pick up any of the objects from the UIOL to build their own interfaces. Moreover, they can also enlarge the library by adding new object into it. The flexibility and expandability of the UIOL fit the situation of current manufacturing systems----it can adapt to the various condition of the manufacture factories.

Objects in the UIOL are divided into two kinds: windows and dialogue boxes, their contents are as follows:

Window: main, supervise, job management,   NC management,   tool management,
             fixture management, file management, list management, graphics management,
             standard management.

Dialogue: standard edit, standard select, job list edit, NC list, tool list, job list, etc.

when using UIVCE , user can select the corresponding icon to activate the description procedure by double clicking on it.

4.5 Message Analysis Interface (MAI)

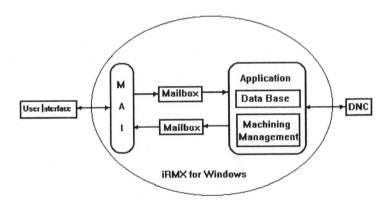

Figure 2 the communication structure of user interface.

The MAI is a bi-directional information channel which links foreground system with background system. The information communication chart is showed in Figure 2. Here foreground means user interface and background means Real-Time machining management system. The interface platform is developed under the following environment: The real time operating system---iRMX for Windows^{TM} is for background and Microsoft Windows 3.x^{TM} as well as DOS is for foreground. By defining some machining messages we extended the message mechanism of Microsoft Windows. The MAI performs the Real-Time communication task by using DOS RTE. The control mechanism of MAI uses parallel control. To transmit message and information by iRMX real time mailboxes makes the application and interface separated physically, but not logically. The MAI process all interactive information between user and application. All these information are classified and code. This makes interface distinguish it's corresponding information, and interpret the code, then send it to main window.

The advantages of using mailboxes are as follows. The common mode of using mailboxes to transmit information is that the communication function module waits for the messages from mailbox to be motivated. Thus the application function module can be executed simultaneously. Moreover message maintaining in mailbox is independent. The communication system consists of several mailboxes, each mailbox process its own message. Therefore, it is very easy to add, delete some messages, which brings more flexibility and more convenience for extension.

5   CONCLUSION

We presented a user interface platform scheme for automated manufacturing system in this paper. This scheme improved current UIMS with Object-Oriented technology and Direct Manipulation feature. According to the scheme we have constructed a template user interface which is reusable and extendible. The result of this work is very significant to realize user interface standardization and quickly development in manufacturing system. The main window of this user interface is as Figure 3. At present, there are still some imperfection in using Direct Manipulation feature and connecting user interface with application system. Our progress in these areas will be reported in future papers.

6   REFERENCES

Gerhard, F. (1987) An Object-Oriented Construction and ToolKit for Human-Computer Communication . Computer Graphics, **21**, 105-9.

Scott, E. H. (1987) UIMS Support for Direct Manipulation Interfaces. Computer Graphics, **21**, 120-4.

Gang, Wu, Jinxiang Dong and Jianxin Ge. (1994) A Model for Automatically Constructing User Interface. Computer Aided Design and Graphics, No 6.

John, L. S, William, D. H and Tereasa, W. B. (1986) An Object-Oriented Interface Management System. ACM SIGGRAPH'86, **20**, No. 4.

Gaines, B.R. (1988) A Conceptual Framework for Person-Computer Interaction in Complex System. IEEE TRANS. on system, man and cybernetics, **18**, No. 4.

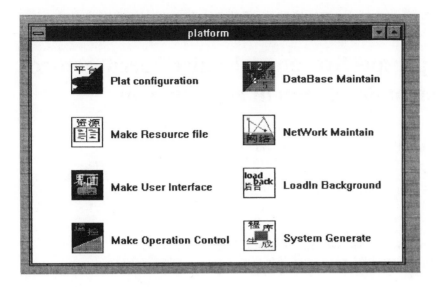

Figure 3 the main window of HCI platform.

7 BIOGRAPHY

Nianyu Wu is a professor in the Department of Automation, Tsinghua University. He was responsible for the establishing the laboratory for hierarchical control system in the state CIMS-ERC. He leads a research group for the Manufacturing Execution System in shop floor. He has designed and implemented more than ten projects of Digital (computer) Control System. He graduated from Department of Electrical Engineering at Tsinghua University, Beijing, 1959.

Ping Du received bachelor's degree of engineering from Tsinghua University in 1993. Now she is a graduate student in the Department of Automation at the same university. Her research interests include: user interface development, computer graphics, UIMS and Computer Integrated Manufacturing.

Xueqing Sun is a graduate student in the Department of Automation at Tsinghua University, where she does her research in the CIMS Engineering Research Center. She received her BE from the same university in 1992. Her research is in the area of Computer Integrated Manufacturing (CIMS), Manufacturing Real-Time Control, Design for Manufacturing.

55

A perspective on a Eunice-based control system development environment

Y. Kawata, A. Yabu, M. Maekawa, T. Kawase, and H. Kozuka
Graduate School of Information Systems, University of Electro-Communications
1-5-1 Chofugaoka, Chofu-shi, Tokyo, JAPAN 182   Ph: +81-424-85-5477   FAX: +81-424-85-5471   E-mail: {yasuro, aki, maekawa, kawasedo, kozuka}@maekawa.is.uec.ac.jp

H. Kobayashi
NTT Data Communications Systems Corporation
Toyosu Center Building, 3-3-3 Toyosu, Koto-ku, Tokyo, JAPAN 135
Ph: +81-3-5546-8963   FAX: +81-3-5546-8990   E-mail: hisa@dc.open.rd.nttdata.jp

Abstract

The long-term objective of our Eunice Project is to identify and develop an integrated environment for Computer-Aided Control System Development, or CACSD. The most salient feature of the project is that all the development lifecycle, from specification to implementation, is to be supported by a single language, Eunice.

This paper discusses the features required of Eunice: early executability by abstract descriptions, refinement mechanism, mixture of discrete models and continuous models, combination of simulation and execution, prespecified adaptive components, support for development aspects that are not directly related to control, extension mechanism for multi-paradigm support, etc.

This paper also states the current status of the project.

Keywords

CAE, CACSD, control system development, executable specification language, simulation, refinement, real-time, Eunice

1   INTRODUCTION

Our research team has been actively pursuing the Eunice Project. This is a research program whose long-term objective is to identify and develop a solid, integrated generic environment for Computer-Aided Control System Development, or CACSD. This paper presents an overview of our approach towards this goal, addressing the technical issues involved. It summarizes the results we have obtained so far. Finally it compares our approach with relevant works.

2 THE EUNICE LANGUAGE AND ITS ENVIRONMENT

2.1 Full-lifecycle-support language

The most salient feature of the Eunice Project is the environment is to be built around Eunice, one single formalism or language. Eunice 'all-in-one' documents supports the full development lifecycle, i.e., from specification to implementation, and also other nontechnical activities involved in development, such as cost estimation. By so doing, in principle it can avoid the following shortcomings inherent in traditional Waterfall Development Lifecycle (Royce, 1970), a still prevalent lifecycle model of information system development in general:

- Formalisms for different phases are different and two formalisms of adjacent phases are often not completely compatible with each other. Some portion of the information obtained in a phase may not be carried on to the next phase properly.
- When an error in a phase is found at a later phase, oftentimes it is corrected only at the later phase and documents/codes of the earlier phase are not updated accordingly. In the end, nothing but source codes correctly reflects the current status of the application. This is more or less a human factor.
- In spite of the popular belief that specifications can be completely separated from implementation, Swartout and Balzer (1982) clearly showed that specification of a system and its implementation cannot be strictly separated, and inevitably they are intertwined.

2.2 Procedure of control system development with Eunice

Eunice-based development is expected to proceed as depicted in Figure 1.

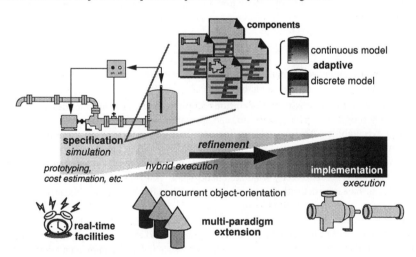

Figure 1 Envisaged development procedure with Eunice

The development begins with acquisition and specification of requirements. For writing specifications, Eunice provides a library of prespecified components and the developer can use them 'as-is' or make necessary modifications by inheritance mechanism in the framework of the standard object-orientation.

These Eunice software components are manipulated, edited, and assembled on a user-friendly multiple window system. For a simple system, when assembled these components automatically work in concert without any adjustments or modifications.

The environment of a control system is the system minus its controller. The controller interacts with, and tries to control it. For a specification of a control system to be complete, its environment also has to be specified. Some prespecified components model objects that belong to environments of control systems: physical devices or appratuses such as motors, tanks, pipes, etc., and such objects as 'room' for a heating system.

The components that model physical devices contain the data of the physical properties of the devices they represent, e.g., nominal radii for pipes. When these components are connected, not only logical component-to-component consistency is validated but also physical component-to-component consistency is validated, e.g., whether nominal radii of connecting pipe components are the same or not.

Components also carry data that are not relevant to control, such as prices for various conditions (bulk purchase, academic discount, etc.), or iconic representation on screens.

Specifications written in Eunice are executable even when they are abstract and not detailed. By executing them, the developer sees whether he has specified what he wants to. By having the customer see the execution, optionally via a visualization tool with pseudo-animation facility, he can make sure whether he understands what the customer wants.

The developer gradually refine and incrementally detail the specifications, and make them closer to implementation, till finally all the control logic has been described, all the physical devices are installed, and the implementation is complete. At any time of this gradual transition, he can execute them.

2.3   Features of Eunice

To attain the aforementioned goal and development procedure, Eunice has the following features, each of which can be a research topic by itself:

Early executability by abstract descriptions Eunice allows more abstract descriptions than ordinary programming languages do, and incomplete descriptions as well. This leads to early executability. Missing information may be supplemented at run time by interaction with users.

This feature lets developers see what they have written represents what they have in mind and also understand correctly what the clients want. If misunderstandings between the two parties surface later in the development, it cost much to fix. Eunice considerably reduces this risk.

As you will see shortly, abstract, specification-level descriptions in Eunice are utilized till later phases, early executability in Eunice is equivalent to capability of rapid prototyping.

Refinement Automatic transformation from a specification to its implementation is indeed ideal (Zave, 1982), but it cannot be done without human assistance; thus, Eunice provides more practical manual refinement mechanism.

Eunice's refinement is not *refinement by replacement*, but instead, *refinement by addition*. In other words, in Eunice, the *pre*-refinement descriptions continue to constrain the *post*-refinement descriptions. The latter can be tested against what the former dictates. After ample testing, it is also possible to switch the semantics of a particular refinement to refinement by replacement, for better performance.

Real-time facilities Real-timeness is usually the major concern of control systems. Eunice provides real-time processing facilities such as deadline specifications on block execution, message sending, and message waiting, and exception handling including timeout handling.

Mixture of discrete models and continuous models The behavior of a controller can usually be described in a discrete model. To model objects that belong to the environment, however, discrete models are not always suitable and continuous models have to be supported. It also applies to analog devices in the controller, if any.

Since some objects need discrete models and others need continuous models, for the whole system the mixture of discrete models and continuous models is supported.

Combination of simulation and execution System descriptions in Eunice goes gradually from the stage of purely specification to the stage of purely implementation. Between these two extreme stages, the descriptions are partly specification and partly implementation, with a varying ratio. We call execution of the system at an intermediate stage *hybrid execution*. Since execution of specification can be seen as simulation, hybrid execution is a combination of simulation and execution.

The difficulty of hybrid execution is how to synchronize simulation part and execution part. The simulation part progresses according to the virtual time while the execution part progresses according to the real time. If done on a single processor, the simulation part, especially when complicated evaluation procedures of continuous variables are involved, will degrade the performance of the execution part and thus inevitably dilute the validity of the obtained results.

Component-based construction with prespecified components Object-orientation, fundamental paradigm adopted in Eunice, is reputed to be effective in class- or object-based reuse. The 'vanilla' object-orientation, however, is not sufficient for component-based construction. For example, name of the messages a server provides have to be known to clients *beforehand* to make them work in cooperation. Eunice provides channel interface between components to enhance component-wise independence.

Adaptive components To facilitate component-based construction, components have to be *adaptive*. It adapts to the context in which it is used. Take a Eunice component that models a tank for example. It might be used to 'pour ten liter of water in' (discrete) or 'pour water in at the rate of one liter of per second for ten seconds' (continuous). Prespecified components such as the one for this tank cannot know beforehand how it will be used in conjunction with other components. Eunice provides framework with which such adaptation is possible.

Support for development aspects not directly related to control Accurate modeling of controllers and their environments is the crucial part of control system development, but there are other factors involved in it. Physical constraints (part of them may be implicitly incorporated into models), cost issues (not likely to be incorporated into models at all), etc., also have to be given consideration from the specification phase. It is well probable that developers weigh their choices in control methods and devices against costs they incur (Swartout and Balzer, 1982).

Since all descriptions are in Eunice, in combination with other Eunice features, developers can try connection of components with other components, overall behavior of the system being constructed, cost estimation, etc., on a user-friendly multiple window system. Thus, Eunice, along with its environment, supports all aspects of decision-making in an integrated manner.

Manufacturers of (physical) components for control systems supply Eunice (software) components because the Eunice components replace conventional catalogues printed on paper. Because of this, Eunice components are circulated at a cheap price, or maybe even distributed for free.

When this perspective is realized, both control system developers and component manufacturers can benefit. Developers can save much of their work in specification validation, design, prototyping, and implementation by utilizing supplied Eunice components. Component manufacturers can convince developers of advantages of their products in an objective manner—Developers can easily see choice of a certain component will improve the overall efficiency of control up to 20 %, or reduce the total cost by 5 %, etc., by executing the corresponding Eunice component with the rest of the system.

Extension mechanism for multi-paradigm support Control systems have vast varieties from 'large' ones in size such as chemical/pharmaceutical plants to 'small' ones such as home appliances. Each minor field has its own paradigm fitted for it, and also long-lived conventions, traditions, etc., specific to that minor field. It is not realistic to try to provide a single paradigm that is suitable for all the minor fields. Instead Eunice provides a fundamental paradigm, namely, concurrent object-orientation, and lets extension of minor-field-specific paradigm.

2.4   Supporting environment for Eunice

The supporting environment for Eunice has the following features:

- It runs under a multiple window system. Information is presented visually, by icons, graphs, etc., in an intuitively easy-to-understand manner. Various kinds of manipulation can be done in a point-and-click mouse interface.
- To facilitate quick and easy searching and browsing of library components, components written in Eunice are stored in a database, not just as files.

3   CURRENT STATUS

Our Eunice project is still in its infancy. As of this writing, Kawata, Kobayashi, Yabu, Onogawa, Kawasaki and Maekawa (1994) discusses gradual, incremental development in Eunice and its supporting visual environment. Kawata, Onogawa, Yabu, Maekawa, Kawasaki and Kobayashi (1994) discusses component-based development, incorporation of physical properties into components, interactive validation of component-to-component connection, and its visual manipulation environment. Kawata, Yabu, Kobayashi, Onogawa, Kawase, Kozuka and Maekawa (1994) discusses adaptive software components that automatically select their behavior from continuous descriptions and discrete ones according to the context in which they are used.

4   RELATED WORK

We have not yet seen any equivalent or similar approaches to ours as a whole. Regarding our individual ideas, however, some parallels can be found.

Open architecture of integrated CACSD environment is discussed in (Andersson, Mattsson and Nilsson, 1992), (Barker, Chen, Grant, Jobling and Towensend, 1993), etc.. They, however, are too concerned with the traditional concept of control system development, i.e., 'control first, other things later', and other aspects of development as pointed out in this paper are not considered. For example, (Barker et al., 1993) bases its discussion on the ECMA/NIST's reference model of integrated CASE environment (so-called 'toaster model') (NIST ISEE Working Group and the EMCA TC33 Task Group on the Reference Model, 1991), and simply replaces its data services with 'modeling services'. Data services in ECMA/NIST model handles *metadata*, i.e., processes involved in development, but the replacing modeling services do not support this.

In the field of simulation, some general-purpose simulation languages support combination of continuous models and discrete models (e.g., SYSMOD   (Smart and Baker, 1984), and COSMOS   (Kettenis, 1992)), and also those dedicated for chemical engineering (e.g., (Barton and Pantelides, 1994) and (Zentner, Elkamel, Pekny and Reklaitis, 1994)). Their one and only goal is simulation itself and the models that have been built for simulation cannot directly be used for later development processes.

In the field of Computer-Aided Design, or CAD, STEP/EXPRESS   (Schenck and Wilson, 1994) (ISO TC184/SC4/WG PMAG, 1992) (ISO TC184/SC4/WG5, 1991), is now an ISO draft of product specification, and thus physical properties can well be described in this framework; however, because of its goal, the EXPRESS language lacks ability to express dynamic, behavioral nature of products. FDL   (Imamura, 1992) supports interactive component assembly on computers, guided by constraints specified for each component. Some works in so-called intelligent CAD incorporate description of dynamic behavior into traditional CAD (e.g., (Nayak, Joskowicz and Addanki, 1992)).

Our hybrid execution is in line with 'hybrid simulation' in the works of Arano *et. al.* (Arano, Chang, Mongkolwat, Liu and Shu, 1993; Arano, Chang, Aono and Fujisaki, 1993). Our approach and theirs differ in that they treat specification and implementation two completely different phases, and use different formalisms for each.

5   CONCLUSIONS

This paper has stated the objective of the Eunice Project, which is to identify and develop an integrated environment for Computer-Aided Control System Development, or CACSD. The most salient feature of the project was that all the development lifecycle, from specification to implementation, is to be supported by a single language, Eunice.

This paper has discussed the features required of Eunice: early executability by abstract descriptions, refinement mechanism, mixture of discrete models and continuous models, combination of simulation and execution, prespecified adaptive components, support for development aspects that are not directly related to control, extension mechanism for multi-paradigm support, etc.

This paper has presented the current status of the project and compared our approaches with others' works.

REFERENCES

Andersson, M., Mattsson, S. E. and Nilsson, B. (1992), On the architecture of CACE environments, in H. A. Baker, ed., 'Computer Aided Design in Control Systems', IFAC Workshop Series, International Federation of Automatic Control, Pergamon Press, pp. 41–46. Selected Papers from the IFAC Symposium, Swansea, UK, 15–17 July 1991.

Arano, T., Chang, C. K., Aono, H. and Fujisaki, T. (1993), A new simulation technique in prototyping development, in 'Proceedings of Summer Computer Simulation Conference (SCSC) '93', pp. 990–995.

Arano, T., Chang, C. K., Mongkolwat, P., Liu, Y. and Shu, X. (1993), An object-oriented prototyping approach to system development, in 'Proceedings of the Seventeenth Annual International Computer Software & Applications Conference (COMPSAC 93)', IEEE Computer Society Press, pp. 56–62. November 1-5, 1993, Phoenix, Arizona, USA.

Barker, H. A., Chen, M., Grant, P. W., Jobling, C. P. and Towensend, P. (1993), 'Open architecture for computer-aided control engineering', IEEE Control Systems 13(2), 17–27. Special Issue on Computer-Aided Control Systems Design.

Barton, P. I. and Pantelides, C. C. (1994), 'The modelling of combined discrete/continuous processes', AIChE (American Institute of Chemical Engineers) Journal 40(6), 966–979.

Imamura, S. (1992), FDL: A constraint based object oriented language for functional design, in G. J. Olling and F. Kimura, eds, 'Human Aspects in Computer Integrated Manufacturing', IFIP Transactions B: Applications in Technology, North-Holland, pp. 227–236. Proceedings of the IFIP TC5/WG 5.3 Eight International PROLAMAT Conference, Man in CIM, Tokyo, Japan, 24–26 June 1992.

ISO TC184/SC4/WG PMAG (1992), STEP part 1: Overview and fundamental principles, ISO CD 10303-1, International Standard Organization. Owner: Howard Mason/John Rumble.

ISO TC184/SC4/WG5 (1991), EXPRESS language reference manual, Document N. 5, International Standard Organization. Release Draft. ISO 10303-11. Owner: Philip Spiby.

Kawata, Y., Kobayashi, H., Yabu, A., Onogawa, K., Kawasaki, A. and Maekawa, M. (1994), Eunice/ITRON: A control system development environment for ITRON machines, in 'Proceedings of the 11th TRON Project International Symposium', IEEE Society Press, pp. 91–105. December 7-10, 1994, Tokyo, Japan.

Kawata, Y., Onogawa, K., Yabu, A., Maekawa, M., Kawasaki, A. and Kobayashi, H. (1994), 'EVE: A graphical specification environment', Submitted to the Transactions of the Institute of Electronics, Information and Communication Engineers. In Japanese.

Kawata, Y., Yabu, A., Kobayashi, H., Onogawa, K., Kawase, T., Kozuka, H. and Maekawa, M. (1994), Eunice adaptive components: Modeling external objects in control systems for better construction and validation of specifications, in C. Mingins and B. Meyer, eds, 'Technology of Object-Oriented Languages and Systems TOOLS 15', Prentice-Hall, pp. 45–55. Proceedings of the fifteenth International Conference TOOLS Pacific 94, Melbourne, Australia, November 28–December 1, 1994.

Kettenis, D. L. (1992), 'COSMOS: A simulation language for continuous, discrete and combined models', Simulation 58(1), 32–41.

Nayak, P. P., Joskowicz, L. and Addanki, S. (1992), Context-dependent behaviors: A preliminary report, *in* D. C. Brown, M. B. Waldron and H. Yoshikawa, eds, 'Intelligent Computer Aided Design', IFIP Transactions B: Applications in Technology, North-Holland, pp. 237–250.

NIST ISEE Working Group and the EMCA TC33 Task Group on the Reference Model (1991), Reference model for frameworks of software engineering environments, NIST Special Publication 500-201, NIST (National Institute of Standards and Technology, United States of Commerce). Technical Report ECMA TR/55, 2nd Edition.

Royce, W. W. (1970), Managing the development of large software systems: Concepts and techniques, *in* 'Proceedings of the IEEE WESCON', IEEE Press, pp. 1–9. Los Angeles, CA, USA, August 25–28, 1970. Reprinted in 'Proceedings of the Ninth International Conference on Software Engineering', pp. 328–338, IEEE Press, 1987. Monterey, CA, USA, March 30–April 2, 1987.

Schenck, D. A. and Wilson, P. R. (1994), *Information Modeling: the EXPRESS Way*, Oxford University Press.

Smart, P. J. and Baker, N. J. C. (1984), SYSMOD—an environment for modular simulation, *in* 'Proceedings of Summer Computer Simulation Conference', North-Holland, pp. 72–82.

Swartout, W. and Balzer, R. (1982), 'On the inevitable intertwining of specification and implementation', *Communications of the ACM* **25**(7), 438–440.

Zave, P. (1982), 'An operational approach to requirements specification for embedded systems', *IEEE Transactions on Software Engineering* **SE-8**(3), 250–269.

Zentner, M. G., Elkamel, A., Pekny, J. F. and Reklaitis, G. V. (1994), 'A language for describing process scheduling problems', *Computers and Chemical Engineering*. To appear.

BIOGRAPHY

Yasuro Kawata received the M.S. degree in computer science from the University of Tokyo and is currently a research associate at the Graduate School of Information Systems, University of Electro-Communications. His research interests include specification languages and methods and environments for computer-aided control system development.

Akifumi Yabu is currently a graduate student at the Graduate School of Information Systems, University of Electro-Communications. His research interests include control system specification, reusable adaptive components, and continuous/discrete combined system descriptions.

Mamoru Maekawa received the Ph.D. degree in computer science from the University of Minnesota. He is currently a professor and the head of the Department of Information Systems Design, Graduate School of Information Systems, University of Electro-Communications. His research interests include software engineering, distributed systems and multimedia.

Hisahiro Kobayashi received the M.S. degree in computer science from the University of Tokyo and is currently employed at NTT Data Communications Systems Corporation. His research interests include specification and programming of control systems, and architecture for distributed online transaction systems.

Both **Tomohiro Kawase** and **Hitoshi Kozuka** are currently undergraduate students at the University of Electro-Communications.

56

Towards More Flexible Parametric Design Support by Introducing a Transformable Object Oriented Language

S. Imamura
Mechanical Engineering Laboratory
1-2 Namiki, Tsukuba-shi 305, JAPAN
Tel: 81-298-58-7107, Fax: 81-298-58-7091
email: imamura@mel.go.jp

Abstract

FDL (Formal Engineering Design Language) has been developed to provide an interactive support environment for basic design. It is an object oriented language that uses constraints for its expressions and information exchanges among objects. The relational data retrieval functionality, which is connected with constraints, is provided. FDL could be a reasonable parametric design support tool for routine design, however, it was not enough to suppoprt more innovative design as its object oriented data structure was too rigid to follow a dynamically changing design context. To overcome this problem, we introduced several meta operations for dynamic data model modification, 'was_a' and 'was' relations for management of modified objects and configuration operators for automatic reconfiguration of design models.

Keywords

parametric design, configuration, constraint based object oriented language, meta operation, structure merging, was_a relation

1 INTRODUCTION

We have been developing FDL (Formal Engineering Design Language), a design language to provide an interactive support environment for basic design[1]. It is an object oriented language that uses constraints extensively for its algorithm representations and information exchanges among objects. The relational data retrieval functionality, which is connected with constraints, is provided. FDL could be a reasonable parametric design support tool for routine design, however, it was not enough to support more innovative design as its object oriented data structure was too rigid to follow a dynamically changing design context. To overcome this problem, we introduced several meta operations for dynamic data model modification, 'was_a' and 'was' relations for management of modified objects and configuration operators for automatic reconfiguration of design models. The features of FDL, the dynamic object modification functionality and configuration operators for re configuring design models will be introduced in the following sections.

2 FEATURES OF FDL

(1) Constraint based object oriented language

FDL is an object oriented language that uses constraints instead of procedural algorithms as its methods. Constraint propagation is used as information exchange among objects. Objects of FDL may express design sub-models, design assembly models, data sheet of standard parts, etc. FDL has two types of slots: part slots to store objects that represent design models and parameter slots to store scalar data types of integer, real, number and string.

Figure 1 shows the FDL description format. Definition of a class object begins by a key word 'class' and ends by 'end'. The data of objects are defined after key words of 'part', 'parameter' and 'database'. Constraints among parameter slots are defined after 'where' and methods are defined after 'method'. Local predicates are defined after a key word 'local' and they are commonly used by methods and constraints. FDL has a multiple inheritance functionality. An object inherits every property except local methods from its super objects specified after a key word 'super'. Instance objects are created from class objects by sending a

```
class <class name> has
[super    <super name> {,<super name>} ;]
[part     <part definition> {,<part definition>} ;]
[parameter <parameter definition> {, <parameter definition>};]
[database  <item definition> {, <item definition>}; [<retrieval formula>;]]
[where
  <slot definition> == <slot definition>;{<slot definition> == <slot definition>;}
  {<constraint formulas>;}]
[method    {:<horn clause>;}]
[local                 {<horn clause>;}]
end.
 <part definition>::= <part name> := #<part class name>
<parameter definition>::= <parameter name> <type> [<parameter constraint>]
<item definition>::= <item name> <type> <unit>
<type>::= real I integer I numberI string
```

> Parts parenthesized by [ ] can be described 0 or 1 time.
> Parts parenthesized by { } can be described 0 or more than 0 times.

Figure 1 FDL description format

```
class shaft has                        class main_shaft has
part   metrial:= #material;            super shaft;
parameter                              parameter
diameter integer [unit(mm), >0],       right_dia integer [unit('mm'), >0],
int_dia integer [unit(mm), >0],        left_dia integer [unit('mm'), >0];
length integer [unit(mm), >0],         where
bearing_dist [unit(mm), >0],           left_dia == diameter;
max_rot integer [unit(rpm), >0],       right_dia = diameter+5;
max_trq real [unit('Kg.m'), .0];       % a tortion equation
where                                  2.334^7*max_trq/ @metrial!sharing_mod+
diameter > int_dia + 5;                @int_dia**4 =X,
end.                                   (X**0.25/5.0 + 1)*5 = ^diameter;
                                       end.
```

Figure 2 FDL Description of shaft and main_shaft Design Models.

message :create(#Class,Instance). Design models are expressed by instance objects.

Figure 2 is an example of FDL program expressing 'shaft' and 'main_shaft' models. The object 'main_shaft' inherits all the slots and constraints described in the object 'shaft'.

(2) Constraints

FDL uses three type constraints, i.e., single attribute constraints, identity constraints and multiple attribute constraints. The single attribute constraints are for specifying data type, existential range, units and default values of parameters. It is written at parameter slots that stay after 'parameter' operator in FDL programs as shown in Figure 2. Other parameter constraints are written after 'where' operator. The identity constraints are denoted by '==' operators and used to express identical relations between two objects in either parameter slots or part slots. The pair of slots applied identical constraints are forced to share same objects. The multiple attribute constraints are used to represent design knowledge, physical lows, standard data, design specifications etc. They are expressed by equations, predicates and inequalities and constraint propagation method[2] is used to solve them. The multiple attribute constraints that exist in different objects are connected by identity constrains and become a global constraint network extending among objects.

(3) Relational database

FDL provides two ways of relational data retrieval functionality. One is using relational formula fully connected with other multiple constraints and the retrieval result is returned to the multiple constraint network. The other is using relational data described as relations between relational attributes and model parameters. Automatic or interactive data retrievals are available through a database window.

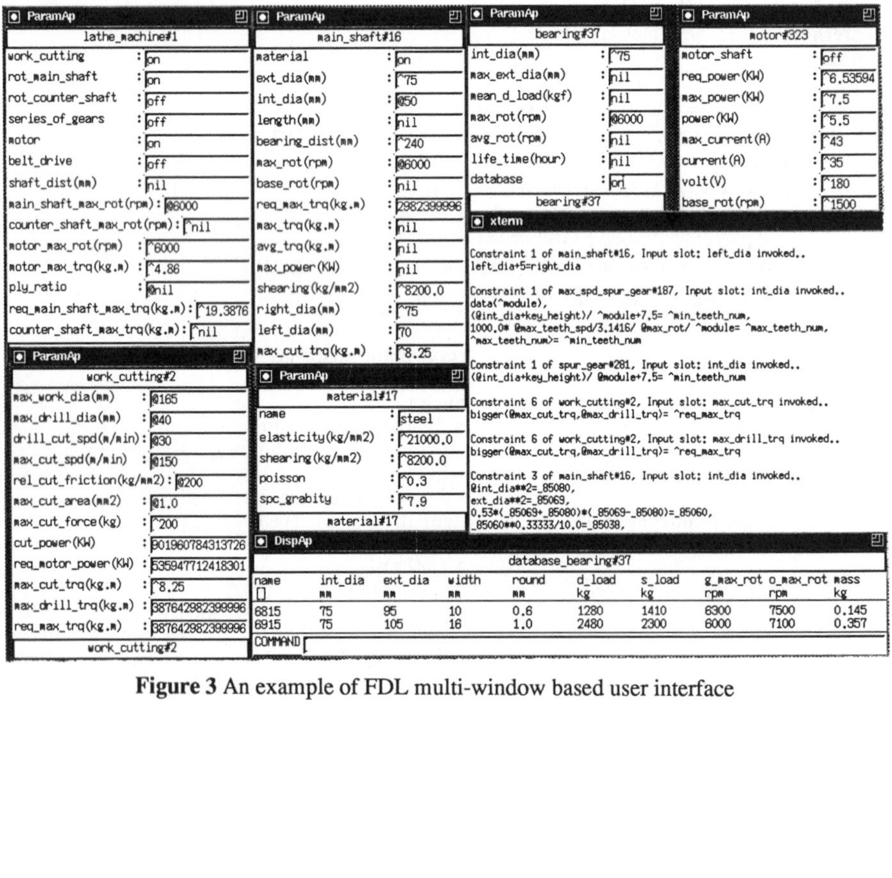

Figure 3 An example of FDL multi-window based user interface

(4) Multi-window based user interface

FDL provides mutli-window based interface. Each window corresponds to a design model and displays part and parameter slots that reflect the data structure of its model. Part slots are displayed as 'on,' 'off' choice items with their slot names, and their initial values are 'off'. If the 'on' is selected, an object bound to its slot is activated and creates its own window. Parameter slots are displayed as variable items with their slot names and units. Their initial values are 'nil'. If a user inputs a value, it activates it's related constraints and results are displayed at parameter slots of the windows.

Figure 3 shows an example of the FDL user interface. Windows representing each component are titled 'ParamAp's and a window for a relational database of a bearing model is titled 'DispAp'. Invoked Constraints are displayed on a xterm window.

3. TRANSFORMABLE OBJECTS AND META OPERATIONS

The object oriented languages generally don't allow structural modification of objects at run time stage, as it violates 'is_a' relations between objects. However the dynamic object modification is often necessary to adapt to unforeseen situations that happen at run time stage of developed programs. Design also often meets unforeseen situations as it doesn't necessarily follow a predefined design context. To allow dynamic modification of objects, we need to introduce meta operations that allow modification of fundamental mechanism of an object oriented languages and management mechanism of modified objects. In this section, we introduce meta operations to assemble or modify design models and 'was_a' and 'was' relations to manage modified objects.

3.1 Structure merging and structure swapping

We introduced a meta operation called *structure merging* for assembling design models. This meta operation creates a new instance by merging two instances, where the set of slots and constraints of the created instance is the union set of those of the two instances. When an instance C is created by a structure merging between an instance A and an instance B, we denote it as:

$$C = A \dotplus B \qquad (1)$$

where the set of slots and the constraints of the instance C are defined by

$$C{\bullet}parameter\_slot = A{\bullet}parameter\_slot \cup B{\bullet}parameter\_slot \qquad (2)$$
$$C{\bullet}part\_slot = A{\bullet}part\_slot \cup B{\bullet}part\_slot \qquad (3)$$
$$C{\bullet}constraint = A{\bullet}constraint \cup B{\bullet}constraint \qquad (4)$$

where $X{\bullet}Y$ denotes the Y set of the instance X. The owner set (set of owners of an object, indicated by inverse relation of 'has_a') of the instance C is defined by

$$owners(C) = owners(A) \cup owners(B) \qquad (5)$$

where owners(X) denotes the owner set of the instance X. The contents of C!X and the owner set of C!X are defined by,

$$C!X = A!X + B!X \qquad (6)$$
$$owners(C!X) = owners(A!X) \cup owners(B!X) - B \qquad (7)$$

where X!Y denotes an object bound to a slot Y of an object X. The formula (6) indicates that the structure merging is a recursive operation, i.e., any substructure of the instance C is created by the structure merging between the corresponding substructures of B and C. If the substructure is a part object, the procedure of the structure merging is described by the formula (6) which is equal to formulas of

$$(C!X) \cdot parameter\_slot = (A!X) \cdot parameter\_slot \cup (B!X) \cdot parameter\_slot \quad (8)$$
$$(C!X) \cdot part\_slot = (A!X) \cdot part\_slot \cup (B!X) \cdot part\_slot \quad (9)$$
$$(C!X) \cdot constraint = (A!X) \cdot constraint \cup (B!X) \cdot constraint \quad (10)$$

The substructure could be a parameter object as well. The structure merging between two parameter objects is allowed only when they are of a same type and their single attribute constraints can be merged; in other words, if the union of single attribute constraints of the two parameters allows solutions. For example, the single attribute constraints of [>2] and [>5] can be merged with a result of [>5], but [>2] and [<1] cannot be merged. The value determined by the parameter merging is given as

$$X + Y = X \quad (11)$$

if the value of X is not nil, otherwise

$$X + Y = Y \quad (12)$$

The structure merging is very useful for assembling parametric design models. Figure 4 shows an example of structure merging used to assemble an additional spur gear pair to a two step spur gear drive. In this case, two sets of corresponding shafts merges and results in a three step spur gear drive. The structure merging is a very simple operation for users: It automatically executes all the necessary and complicated connections of constraints between models. This kind of model structure modification is often required in the design process.

We also introduced another meta operation called structure swapping for exchanging components, such as exchanging types of bearing, motor, v-belt, etc. When an instance C is determined by a structure swapping from A to B, we denote it as:

$$C = A <- B \quad (13)$$

where the structure of the instance C is same as that of the B. If a same part or parameter slot X exists in both A and B, the content of the slot is determined as:

$$C!X = B!X \quad (14)$$
$$owners(C!X) = owners(A!X) \cup owners(B!X) \quad (15)$$

If the X is a parameter slot, the value of the C!X is exchanged to that of the A!X.

Figure 5 shows how the structure merging and the structure swapping work between two objects. In the figure, one hatched object #1!B are merged or swapped by the other hatched object #2!A. By these operations, the object #1 and the object #2 are assembled in different ways.

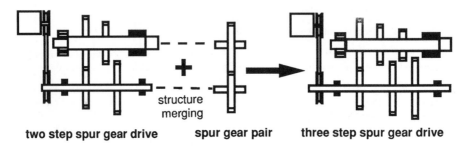

two step spur gear drive   **spur gear pair**   **three step spur gear drive**

Figure 4 Structure merging used for assembling models

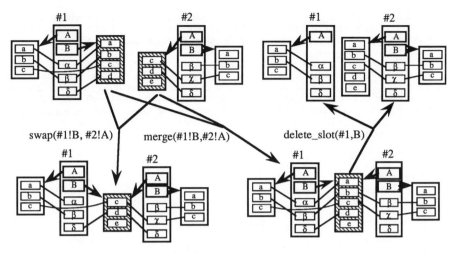

Figure 5 Structure merging and structure swapping

3.2 Other meta operations

There are other meta operations along with the structure merging and swapping for modifying the structure of objects, as shown in Figure 6. The operation (3) in Figure 6 is used for creating a floating class, which will be introduced in the next section, from an ordinary class object. The operation (4) is to transform a floating class to an ordinary class object. The operation (5) is to create a part slot at the 'Location', where an instance created from the class 'PartClass' is stored. (6) is to create a parameter slot at the 'Location', where a 'Type' instance with 'SingleAttriuteConstraint' is stored. The operations (7) and (8) are for deleting part slots and parameter slots. The operations (9) and (10) are for adding and deleting multiple attribute constraints, whereas (11) and(12) are for adding and deleting local predicates.

(1) merge(Obj1,Obj2)   (2) swap(Obj1,Ojb2)
(3) create_floating_class(Obj,FloatingClass)
(4) create_settled_class(FloatingClass,Class)
(5) add_part(Obj,SlotName,#PartClass,Location)
(6) add_partameter(Obj,SlotName,Type,SingleSttributeConstraints,Location)
(7) delete_part(Obj, SlotName)   (8) delete_parameter(Obj,SlotName)
(9) add_constraint(Obj,Constraint)   (10) delete_constraint(Obj,Constraint)
(11) add_local(Obj,Predicate)   (12) delete_local(Obj,Predicate)

Figure 6 Meta operations

3.3 'was_a' and 'was' relations for Managing Modified Objects

We introduced 'was_a' and 'was' relations to enable instance and class objects to be modified without violating 'is_a' relations.

A 'was_a' relation is introduced for managing a modified instance object. Figure 7 (a) shows how the 'was_a' relation works when an instance is modified at run time. We assume that there are a class A and an instance x with an 'is_a' relation between them. In this case, if we add a slot d to the instance x by a meta operation, the 'is_a' relation will be violated. Therefore we delete the 'is_a' relation and introduce a 'was_a' relation with a history of meta operations applied to the instance x. The modified instance does not have its own class, therefore we call it a floating instance.

We can create a class of the floating instance by referring the history written at the 'was_a' relation. By a meta operator (create_floating_class(x,B)), the class B can be created from the floating instance x, as shown in Figure 7 (b). As the class B does not have its super class any more, it is called a floating class and has a 'was' relation with the class A, with a meta operation history attached to it. The instance x is automatically transformed into a conventional type instance with an 'is_a' relation to the floating class B. The concept of the floating class was introduced to enable a class object to be modified without influencing its sub-class objects.

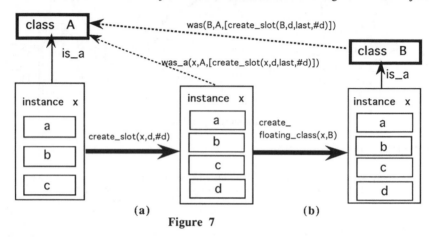

Figure 7

The floating class can be created from either the conventional classes or the floating instances. The floating classes as well as the floating instances inherits some information from its original super class through 'was' or 'was_a' link. This inheritance is different from the ordinary inheritance, as it is controlled by the history written at the link. For example, if a was_a(X,A,[delete_constraint(X,C1)]) link exits between the class A and an floating instance X, the X can still refer all the constraint except the constraint C1 from the class A.

A meta operation to transform the floating class into a conventional class is also provided, as introduce in Figure 6 (4).

4. CONFIGURATION OPERATOR

We can now assemble design models by the structure merging operations etc., however it is still a complicated procedure since we need to apply several structure merging operations and other meta operations to specific parts. To further simplify the assembly procedure, we introduced a configuration operator.

The configuration operator consists of a 'precondition' part and an 'operation' part. The 'precondition' defines the criteria to invoke the configuration operator, whereas the 'operation' defines actual actions of the configuration operator.

Figure 8 shows an example of structural modification of design models by configuration operators: A, B, C and D. Figure 9 is an example description of the configuration operator B, where '|' denotes a delimiter between the precondition part and the operation part and '!' denotes a part of relation. By the configuration operators, the design sub-models assemble by themselves and become an assembly design model that is able to satisfy design requirements.

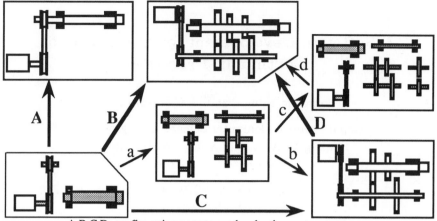

A,B,C,D: configuration operators a,b,c,d: other meta operators

Figure 8 Configuration operators

Obj!motor = Motor, Obj!rotational_main_shaft!main_shaft= Main_shaft,
(Motor!max_rot) / (Motor!base_rot) =A, ((Main_shaft!max_trq)*(Main_shaft!max_rot))/
((Motor!max_trq)*(Motor!max_rot))= B, A =< B, B=< A*A |
create_slot(Obj,gear_drive,last,#gear_train),
create_slot(Obj!gear_drive,second_gear_pair,last,#gear_pair),
merge(Obj!belt_drive!follow_shaft,Obj!rotational_counter_shaft!couter_shaft),
merge(Obj!rotational_counter_shaft!counter_shaft,Obj!gear_drive!follow_shaft),
merge(Obj!rotational_main_shaft!main_shaft,Obj!gear_drive!follow_shaft),
merge(Obj!gear_drive!drive_shaft,Obj!gear_pair!drive_shaft),
merge(Obj!gear_drive!follow_shaft,Obj!gear_pair!follow_shaft);

Figure 9 Configuration operator B

5. CONCLUSION

We have introduced the meta operations of structure merging, configuration and else for model modifications. We also introduced 'was_a' and 'was' relations to manage modified objects. By introducing these new concepts, FDL becomes a more flexible design language that is able to support both parametric design and configuration. However we need more design experiments to improve the mechanism of the meta operations and confirm their effectuality towards more flexible parametric design support.

6. REFERENCE

1) S. Imamura, FDL: A Constraint based Object Oriented Language for Functional Design, Proceedings of the IFIP/WG5.3 Eight International PROLAMAT Conference, Man in CIM, pp.227-236 (1992)
2) G.J. Sussman and G.L. Steel Jr.: CONSTRAINTS - A Language for Expressing Almost-Hierarchical Descriptions, Artificial Intelligence, Vol.14 (1980)

57

A Software Development for Simulation of Continuous Processes

Kyung-Ha Min   and   Chan-Mo Park

Department of Computer Science & Engineering
Pohang University of Science & Technology
Pohang, 790-784 Korea

Tel : +82-562-279-2251
Fax : +82-562-279-2299
e-mail : parkcm@vision.postech.ac.kr

Abstract

This paper presents a user-friendly software which simulates continuous processes which are modeled by use of various differential equations. The simulation programs written in a language developed at the POSTECH are converted into C programs before they are complied and executed. this software is tested by carrying out several simulation runs whose results are already known.

Keywords

simulation, continuous process, differential equation, simulation language

1   Introduction

Most of the continuous processes arising in engineering and manufacturing can be modeled by use of differential equations and continuous system simulation techniques are widely used for analysis and optimization of such processes. With the advancement in computer technology both in hardware and software, electronic prototyping and simulations have also been employed before actual construction for a process is made which enables us to save money and time [1].

In this paper we present a software system developed for simulation of processes which can be modeled by use of differential equations. This software can simulate systems modeled by simultaneous linear and non-linear ordinary differential equations and a partial differential equation. We present a new algorithm to convert a model expressed in differential equations to C program and additionally equip this software with a user-friendliness so that the user can express his model easily. User interface are developed to provide Program Builder and Graphical Display of the simulation results. Using Program Builder, a

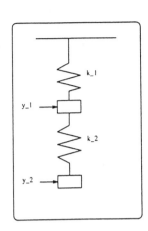

```
BEGIN
VARIABLE
        INDEP : t
        DEP : y_1, y_2
CONSTANT
        k_1 : 3.0
        k_2 : 2.0
RANGE
        R_BEGIN : 0
        R_END : 10
        R_STEP : 0.001
INITIALS
        y_1_INIT_0 : 1.0
        y_1_INIT_1 : 0.0
        y_2_INIT_0 : 2.0
        y_2_INIT_1 : 0.0
MODEL
        "y_1" + (k_1 + k_2)*y_1 - k_2*y_2 = 0"
        "y_2" + k_2*y_2 - k_2*y_1 = 0"
SUBMODEL
END
```

Figure 1: An Example of Simulation Program for a Two-Spring System

user who doesn't know the grammar of the simulation language can make a simulation program for this software. And the graphical display of the simulation results will make visualization very easy[2].

2   Design of Simulation Software

First, we have developed a simulation language with which a user can write a simulation program very easily by representing his/her model in terms of differential equations and initial conditions [2]. This program is then converted into an equivalent C program by a preprocessor of the software. Other necessary routines are also generated automatically by the preprocessor before they are compiled and executed. The simulation results are stored in a file numerically but they can be displayed graphically for easy visualization by a user.

2.1   Definition of Simulation Language

A continuous process is normally modeled by use of differential equations. Therefore, the simulation language should have features to define independent and dependent variables, simulation ranges as well as differential equations representing relationships among constants and variables [3, 4]. Figure 1 shows an example of simulation program written by use of the language developed for a two-spring system.

2.2   Characteristics of the Software Developed

The software developed in this study has the following characteristics.

- The software can be used for simulations of processes which can be modeled by simultaneous linear and non-linear ordinary differential equations which do not contain any singularities. Also, it can handle certain systems which are modeled in terms of limited partial differential equations.

- A simulation program written by a user in accordance with the syntax defined is converted into a C program before it is compiled and executed to make a simulation run.

- The Hamming predictor-corrector method is used to solve an ordinary differential equation [5, 6]. A new algorithm is developed in this study to solve simultaneous ordinary differential equations.

- For an easy input of a model GUI features and interactive editor are provided.

2.3   Organization of the Software System

The software system is consisted of simulation part and user interface part.

2.3.1   Simulation Part

The simulation part is consisted of following modules.

- Preprocessor

  - Translator : This module converts the differential equations specified in the model into an equivalent C program module. Also, other necessary information for simulation is extracted from the model and converted into a C program module.

  - Generator : A complete C program is generated by applying algorithms to solve the differential equations.

- Executor (Simulator) : The C programs generated by the preprocessor is compiled and executed to make a simulation run.

2.3.2   User Interface Part

This part offers following features to make a user input of a model easy.

- Input of differential equations : A GUI scheme is used.

- Input of other information : An interactive editor is implemented so that a user can input all necessary information without failure.

Figure 2 shows the information flow between the modules.

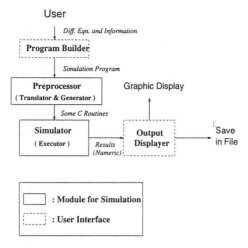

Figure 2: Information Flow between Modules

3   Implementation of the Simulation Software

3.1   Implementation of the Translator

The translator converts differential equations of input model to C modules in order to execute simulation. Types of the differential equations handled by the Translator are single or simultaneous ordinary differential equations and certain types of partial differential equations.

3.1.1   Single Ordinary Differential Equations

A single ODE handled by the Translator is an nth order ODE which is composed of one independent variable and one dependent variable. The preprocessing carried out to convert this ODE to C modules is as follows.

- Substitute a new variable for each order of the differential terms.

- Generate one 1st order ODE by converting given implicit form equation to an explicit form of the highest differentiated variable, and representing with the new variables.

- Generate $n - 1$ new 1st order simultaneous ODE's by using the relations of each substituted variables.

As an example, the followings show the process that converts an nth order ODE to n 1st order simultaneous ODE's.

1. General form of an nth order ODE

$$A_0 y^{(n)} + \cdots + A_n y + A_{n+1} = 0$$

2. Substitution of variables and the relations between new variables

$$y_i = y^{(i)}$$

$$y_i \, (\equiv y^{(i)}) = y'_{i-1}$$

3. For $y^{(n)}$, make an explicit form of the ODE.

$$y^{(n)} = -1 * (A_1 y^{(n-1)} + \cdots + A_{n+1})/A_0$$

4. New n simultaneous ODE's are...

$$
\begin{aligned}
y'_{i-1} &= y_i, \quad \text{for } 1 \le i \le n-1 \\
y'_{(n-1)} &= -1 * (A_1 y_{(n-1)} + \cdots + A_{n+1})/A_0
\end{aligned}
$$

So, the nth order ODE is converted to n 1st order simultaneous ODE's. The process to generate C modules is as follows.

- Convert each ODE to a corresponding C function.

- Convert variables in each ODE as the parameters which are passed to C functions.

- Call converted functions when the value of each ODE is required to know during the execution in main ( ).

3.1.2   Simultaneous Ordinary Differential Equations

The process of changing simultaneous ordinary differential equations into C language is as follows.

1. Differential equations input are analyzed to extract information on variables, coefficients and order of differentiations.

2. The highest differentiated term for each variable is identified and the equation containing the highest differentiated term is marked.

3. As the case of single ODE's, generate new simultaneous ODE's by substituting a new variable for each differentiated term.

4. Instead of converting each equations to its explicit form, the simultaneous equations given are transformed into a linear system of $AX = B$ where A is an $m \times m$ square matrix and X and B are column vectors with m elements, where m is the number of ODE's.

(a) An element $a_{i,j}$ of the matrix A means the coefficient of the highest differentiated term of variable y_j in ith differential equation.

(b) An element x_j of the matrix X means the highest differentiated term (HDT) of variable y_j.

(c) An element b_i of the matrix B means all terms except the highest differentiated term in ith differential equation.

So, the value of new variable corresponding to the highest differentiated term is obtained by solving $AX = B$ and the values of other variables are obtained by solving the simultaneous ordinary differential equations.

As an example, the followings show the process that converts an n simultaneous ODE's to equivalent 1st order simultaneous ODE's.

1. General form of simultaneous ODE's whose variables are differentiated at most 2 times

$$f_i (\ddot{x}_1, \dot{x}_1, x_1, \cdots \ddot{x}_n, \dot{x}_n, x_n, t) = 0, \text{ for } 1 \le i \le n ,$$

Actually, $f_i (\ddot{x}_1, \dot{x}_1, x_1, \cdots \ddot{x}_n, \dot{x}_n, x_n, t)$ is given in a form of

$$\sum_{j=1}^{n}(c_{ij}^2 \ddot{x}_j + c_{ij}^1 \dot{x}_j + c_{ij}^0 x_j) + c_{0j}t = 0$$

2. Substitution of variables and the relations between the variables

$$
\begin{aligned}
x_i &= x_i^0 \\
x_i^1 \ (\equiv \dot{x}_i) &= (x_i^0)' \\
x_i^2 \ (\equiv \ddot{x}_i) &= (x_i^1)', \text{ for } 1 \le i \le n
\end{aligned}
$$

3. Generate new 1st order simultaneous ODE's

$$
\begin{aligned}
(x_i^0)' &= x_i^1 \\
(x_i^1)' &= x_i^2, \quad \text{for } 1 \le i \le n
\end{aligned}
$$

4. Generate Dynamic Inverse Matrix to determine the values of HDT

$$
\begin{pmatrix}
c_{11}^2 & c_{12}^2 & \cdots & c_{1n}^2 \\
c_{21}^2 & c_{22}^2 & \cdots & c_{2n}^2 \\
\vdots & \vdots & \vdots & \vdots \\
c_{n1}^2 & c_{n2}^2 & \cdots & c_{nn}^2
\end{pmatrix}
\begin{pmatrix}
x_1^2 \\
x_2^2 \\
\vdots \\
x_n^2
\end{pmatrix}
= -
\begin{pmatrix}
c_{11}^1 x_1^1 + c_{11}^0 x_1^0 + \cdots + c_{1n}^1 x_n^1 + c_{1n}^0 x_n^0 + c_{10}t \\
c_{21}^1 x_1^1 + c_{21}^0 x_1^0 + \cdots + c_{2n}^1 x_n^1 + c_{2n}^0 x_n^0 + c_{20}t \\
\vdots \\
c_{n1}^1 x_1^1 + c_{n1}^0 x_1^0 + \cdots + c_{nn}^1 x_n^1 + c_{nn}^0 x_n^0 + c_{n0}t
\end{pmatrix}
$$

The process which implements this algorithm to C code is as follows.

1. Apply the methods of single ODE to the variable which are not the highest differentiated term.

2. Apply the current values of each variable of each equations in $AX = B$ and compute the inverse matrix of matrix X in order to find the value of the highest differentiated term.

3. We use Gauss-Jordan method to solve the inverse matrix problem in this software [7].

3.1.3   Partial Differential Equations

We assume the 2nd order PDE as the target of this software. The PDE's are divided into Parabolic type, Elliptic type and Hyperbolic type according to their forms. The process which is required to convert a model expressed by a PDE to C module is as follows.

1. Convert each variable to a finite difference form by applying finite difference method.

2. Convert the given PDE to a finite   difference equation by applying the result of previous step.

3. Correct the difference equation according to the given initial condition and boundary condition.

We execute the simulation of a model expressed by a PDE by solving the converted difference equation.

3.2   Implementation of the Generator

The generator generates C codes automatically to solve the differential equations. The Hamming method is used to solve the differential equations. The form of the differential equations needs not to be considered, because the Translator already translated the differential equations into appropriate C modules. The differentiated order of each variable affects the number of newly generated variables and the repetition times of the loop. These can be easily implemented while the C program is generated.

3.3   Implementation of the Executor

In the Generator, the main program needed in simulation is generated, and in the Translator, the subprograms which are used in the main program are generated. So, the Executor compiles C programs which are generated by the Translator and Generator and carries out simulation run by executing the compiled programs. A user may modify the C program before the simulation is executed if desired. The result of the simulation is stored in a file in the numeric forms.

3.4   Implementation of the User Interface

A graphic user interface is implemented to make the system user-friendly. After a user inputs the differential equations representing his model the user interface system requests other necessary information to the user so that a simulation program is completed without leaving out required information.

4   Examples of Simulation Runs

Two simulation examples are given here. The first is 2-dimensional heat conduction on a thin plate which is modeled by a partial differential equation, $u_t = c^2(u_{xx} + u_{yy})$. The results of simulation is shown in Figure 3. The other example is changes in carbon amount in an LD converter. The differential equation representing the model is $\frac{dC}{dt} = \frac{-1201}{1600} \cdot \frac{RK_o \cdot \rho_{HM}}{W_{HM}} \cdot \frac{P_T}{P_{CO}+2 \cdot P_{CO_2}} \times [O - \frac{P_{CO}}{f_e \cdot C \cdot f_o \cdot K_{CO} \cdot G^2}]$. The result shown in Figure 4 is comparable to a result given in reference [8].

The value of 2D Heat Transfer

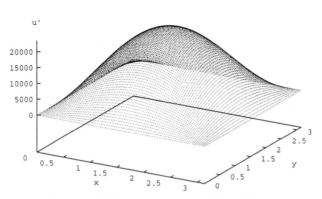

Figure 3: 2-D heat conduct on a thin plate

5   Conclusion

We have developed a software system that simulates continuous processes which can be modeled by use of differential equations. User-friendliness in building a simulation program and graphical visualization of the results are emphasized in our system. The user interface of this software enables users to generate simulation programs without knowing the grammar of the simulation language. Moreover, as is the case of 2-pass simulation method, this software presents more flexibility to the users, because the user can modify simulation programs which are translated to   high level language programs. This system is tested by carrying out several simulation runs whose results are already known. Further study is needed to extend the system to handle a variety of partial differential equations. The treatment of more general objects using this software will lead this study as one of the base technologies of computer animation and virtual reality.

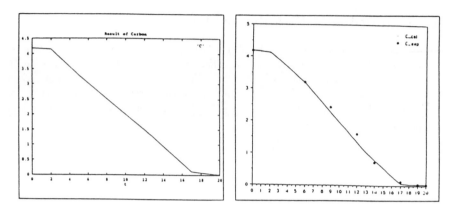

(a) This study   (b) Results in reference [8]

Figure 4: Simulation of Carbon Changes in an LD Converter

References

[1] Aburdene, M. F., " Computer Simulation of Dynamic Systems ", Wm, C, Brown Pub., 1988.

[2] Park, C. M., Yim, C. G. and Kim, M. S., " A Study on Development of User-friendly Continuous System Simulation Language and Application Software ", Proc. of 2nd Beijing International Conference on System Simulation and Scientific Computing, Vol. II, pp. 420-424, Beijing, 1992.

[3] Augustin, Donald C., et al., " The SCi Continuous System Simulation Language ", Simulation , 9, 1967.

[4] Cellier, Francois E., " Continuous System Modeling ", Springer-Verlag, 1990.

[5] Conte, S. D. and de Boor, C., " Elementary Numerical Analysis ", McGraw Hill, 1980.

[6] Park, JaeNyun, " Numerical Analysis ", ChungIkSa, 1986.

[7] Press, William H. et al, " Numerical Recipes in C ", Cambridge University Press, 1988.

[8] Yoon, S.Y., " Dynamic Simulation of LD Converter Processes ", M.S. Thesis, POSTECH, 1991.

Biography

Kyung-Ha Min received his B.S. degree from Korea Advanced Institute of Science & Technology and M.S. degree from Pohang University of Science & Technology in Computer Science, respectively. He is working toward his Ph.D. degree at the POSTECH.

Chan-Mo Park is a Professor of Computer Science at Pohang University of Science & Technology. He received his B.S. degree from Seoul National University and M.S. and Ph.D. degrees from University of Maryland at College Park, Maryland, U.S.A. His experience includes Professorship at the University of Maryland and The Catholic University of America. He was the Dean of Graduate School for Information Technology at the POSTECH.

58
Design Verifications
using
Knowledge Representation Language KOSA

Masamitsu Mochizuki† Isao Nagasawa†† Masanobu Umeda†† Tatsuji Higuchi†††
Takashi Ojima†

Graduate School of Computer Science and System Engineering (Doctoral's programs) †
Kyushu Institute of Technology

Faculty of Computer Science and System Engineering ††
Kyushu Institute of Technology

Olympus Optical Co.Ltd. Camera Development Department †††

Kawazu 680-4 Iizuka-city Fukuoka 820, Japan,
Tel:+81-948-29-7790, Fax:+81-948-29-7760, Email:mochizuk@cad.mse.kyutech.ac.jp

Abstract
We focus on design verifications which can be interpreted as the relationships between inferior and superior characteristics in a model of design object. We are researching on a knowledge representation language and its programming techniques for tolerance analysis which is needed in design verifications. In this paper, the language is applied to some design verification. Design verification systems can be constructed using knowledge representation and programming techniques in the language. A designer who does not have knowledge of computers can easily put their knowledge into this system. A prototype system has been developed and applied to real world problems. The experiments show that this language has enough describability, maintainability and extensibility. We hope that our approach will have a positive impact on design verification systems.

Keywords
CAD, design verification, knowledge representation language

1 INTRODUCTION

The design of products like a camera has become complicated because of variety of markets needs and advancements of qualities. For these reasons, it is important to improve both productivity and quality of product design. We also feel that it important to establish of education systems for junior designers. Generally, design qualities are guaranteed by trial manufactures in these businesses areas. Designers can have a chance to learn new technologies used in the production of new products through the trial manufactures. This helps in the reduction of design mistakes. However, mistakes in design and manufacturing processes still often occur though designers have experiences of the same mistakes in past developments. This is because design technologies are not managed systematically. Mistakes in the design process lead to a wasteful use of resources. Consequently they are a serious problem in periods, cost, and productivity of product development. It is, therefore, better if mistakes in the design of products can found before commencement of production.

Design verification is very important to prevent design mistakes in all stages from conceptual to detail design. The design verification requires heuristic knowledge of skillful designers. However, this knowledge is not generally well-structured, and as a result it is quite difficult to transfer their know-how to junior designers. Therefore, the framework to manage knowledge about design verifications and apply it to design is required to solve the problems.

We developed a language called KOSA and its experimental processing system, which is a framework to represent knowledge about tolerance analysis needed in design verifications. The KOSA is based on object-oriented paradigm and constraint reduction technique like ADL[1, 2]. Then, we also developed programming styles and techniques for representing knowledge about tolerance analysis in KOSA[3].

This paper describes the knowledge representation and programming techniques for applying KOSA to designs verifications. In section 2, we show the overview of KOSA and its fundamental concept. In section 3, we describe how to represent the assembly structure for modeling design objects. In section 4, Fundamental concept of design verifications are described, and we show that KOSA is not only used in tolerance analysis but also in the other design verification problems or design problem solving. In section 5, we describes several programming for design verification.

2   OVERVIEW OF KOSA

An overview of KOSA is introduced using a simple example.

2.1   Object-Oriented Knowledge Representation In Kosa

Knowledge about design verification is represented using object-oriented paradigm in KOSA.

Class
A class represents a object of parts or assemblies with the same characteristics.

```
defclass: gear,                    defclass: gear_pair,
   super: object,                     super: object,
   attributes:                        attributes: rpm_in:real := unknown,
      teeth:int := unknown,                       rpm_out:real := unknown,
      module:real := unknown,
         ⋮                                  ⋮

                                      sub_parts: driver: gear,
                                                 follower: gear,
                                      constraints:
                                         gear_ratio(driver!teeth,
                                                    follower!teeth,
                                                    !rpm_in, !rpm_out),

                                            ⋮
```

(a) A class gear   (b) A class gear pair

Figure 1: The class a gear and a gear pair

Figure 1 is an example of a class description. The class gear describes a part of gear wheel, and the class gear_pair describes a part of assembly. The class gear is a subclass of the class part, and the class gear_pair is a subclass of the class assembly. A

Class has attributes which characterize it with a optional default value or expression in "attributes:" description. The class gear_pair has two gears in "sub_parts:". The gear_ratio constraint in "constraints:" declares relationships of input and output revolutions of a gear wheel. The symbol "!" in the arguments of constraint gear_ratio is reference to attributes of a class itself and/or a sub-class.

Method

A class may have methods, as similar to those found in the conventional object-oriented languages. The followings are methods of the class gear_pair. These methods calculate the distance between centers of two gear wheels (center distance).

```
defmethod(gear, diameter(D)) :-
    D is !teeth * !module.
defmethod(gear_pair, shaft_distance(L)) :-
    send(driver, diameter(D1)),
    send(follower, diameter(D2)),
    L is (D1 + D2) / 2.
```

The send statement calls a method defined in a class. The send statement has the following syntax.

send(<Instance-Name>,<Selector>(<Argument> ⋯ <Argument>))

<Instance-Name> is an instance to which a message is sent. <Selector> is a name of the method.

Instance

An instance generated from a class is an individual part or an assembly. The following statement generates an instance of a class.

make_instance(Ins, $Class$,

attributes($a_1(v_1), \ldots, a_i(v_i), \ldots, a_n(v_n)$),
sub_parts($p_1(i_1), \ldots, p_j(i_j), \ldots, p_m(i_m)$))) The argument

Ins is a parameter, $Class$ is the class name of the generated instance. $a_i(v_i)$ $(i = 1, ..., n)$ means that the initial value of attribute a_i is v_i. $p_j(i_j)$ $(j = 1, ..., m)$ specifies the a part named p_j of the instance Ins is the instance i_j that was generated previously. This predicate assigns a unique name to Ins as an instance of class $Class$.

Following statement deletes an instance named Ins.

del_instance(Ins)

Some of the methods for manipulating instances are shown in Table 1.

Table 1: Examples of methods for manipulating instances

Method	Function
add_attribute(Self,<Attribute>)	add attribute to instance
del_attribute(Self,<Attribute>)	delete attribute in instance
add_sub_part(Self,<Part>,<Class>)	add part to instance
del_sub_part(Self,<Part>,<Class>)	delete part in instance
add_constraint(Self,<Constraint>)	add constraint to instance
del_constraint(Self,<Constraint>)	delete constraint in instance
get_constraints(Self,<Set of Constraint>)	get a set of constraint
get(Self,<Attribute>,<Parameter>)	get attribute value
put(Self,<Attribute>,<Value>)	put attribute value to instance

2.2 Constraint For Representation The Relationship of Attributes

A program is a set of reduction rules having the following form:

$H :- G_1, ..., G_n \mid B_1, ..., B_m.$

H is called a head of rule. $G_1, ..., G_n$ are called guards of rule. $B_1, ..., B_m$ are called bodies of rule. They are represented using atomic well-formed formula. When $G_1, ..., G_n$ is satisfied, constraint H can be reduced to a set of constraints $B_1, ..., B_m$.

For example, a rule of the linear dimension is following:

```
transfer(C1, C2, {Dxt, Dyt}) :-
    true |
    gen(Dxt, Dx),
    gen(Dyt, Dy),
    trans(C1, C2, {Dx, Dy}).
```

In this rule, argument C1 and C2 are coordinates that represent location and orientation. In this body of rule, **gen** is a generator that generates minimum and maximum dimension. **trans** transfers a coordinate C1 to C2.

3 REPRESENTATION OF ASSEMBLY STRUCTURE

This section describes the representation of assembly structure. Representation of assembly structure is very important in representing a model of design objects.

A part has elements which interact each other and realize a specific function , such as a shaft and a hole[4][5]. These elements are called functional elements. The assembly structure is specifyed to be composed of functional elements. Table 2 shows an example of functional elements.

Table 2: Example of functional elements and attributes

Functional element	attributes
point	center
shaft	center, diameter
hole	center, diameter
square_hole	center, diameter, length, width
plane	center

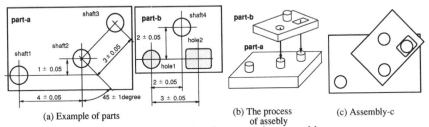

(a) Example of parts   (b) The process of assebly   (c) Assembly-c

Figure 2: Example of parts and a assembly

A part is composed of functional elements. An assembly is composed of parts or inferior assemblies. The assembly object is the general term for the assembly , the part and the functional element . The assembly object has attributes that represent its characteristics. In a part, location and orientation of functional elements are represented by linear and angular dimension. In a assembly, location and orientation of part is represented by the relationship of functional elements.

Figure 2 shows a simple example for representation of an assembly structure.

Assembly-c (see Figure 2(c)) consists of part-a and part-b. The part-a has three shafts (`shaft1, shaft2, shaft3`), the part-b has one shaft (`shaft4`), one hole (`hole1`) and one square-hole (`hole2`). The location and orientation of these shafts or holes are declared by linear and angular dimension (see Figure 2(a)). Let assembly-c be assembled as Figure 2(b) by insert `shaft2` and `shaft3` of part-a into `hole1` and `hole2` of part-b respectively.

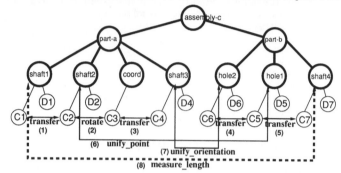

Figure 3: An assembly object representing the assembly

Figure 3 shows the assembly structure of assembly-c. The relationships of attributes of each part is declared by linear or angular dimension rule. Relationship of functional elements are declared by assembly rule like fitting rules or angular fitting rules. In Figure 3, bold circles indicate the assembly object. Bold lines show relationship of part and whole. Lines with arrows indicate constraints of linear dimension rule or assembly rule. Linear dimension (`transfer`) or angular dimension (`rotate`) constraints are declared between some attributes of functional elements to represent a relationship of parts location and orientation. Assembly constraints are declared between some attributes of functional elements to represent a relationship of parts in assembly. Constraints from (1) to (5) are dimension constraints. (6) and (7) are assembly constraints. (8)is measure_ length point. Table3 shows an example of dimension constraints and assembly constraints.

Figure 4 show class hierarchy of the assembly object.

Table 3: Dimension constraints and assembly constraints

A kind of constraint	Name	purpose
Dimension	transfer	linear dimension
Dimension	rotate	angular dimension
Assembly	unify_point	unification of location
Assembly	unify_orientation	unification of orientation

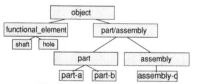

Figure 4: Class hierarchy

4   CHARACTERISTICS OF DESIGN VERIFICATION

Designers define details of implementation structure to satisfy the required specifications as design process goes forward. In that processes, a model specific to individual design activity, such as invention of new ideas, is used. This model is called the model of design object in this paper. It could be, therefore, assumed that a designer verifies his design plans using the model of design object.

The model of design object has two kinds of characteristic values. One is the inferior characteristic represented by attributes such as dimensions, tolerances and geometrical tolerances of parts. Another is the superior characteristic derived from inferior characteristics.

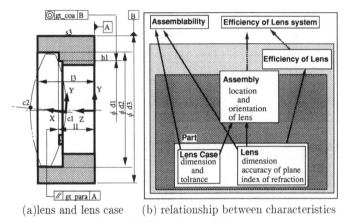

(a)lens and lens case   (b) relationship between characteristics

Figure 5: Lens and its efficiency

Figure 5 (a) is an example of an assembly of a camera lens and a lens case . Figure 5 (b) illustrates the relationship between characteristics that affect the efficiencies of the lens system. Dimensions, tolerances and geometrical tolerances of the parts are inferior characteristics. The locations and orientations of the lens and the fitting plane of the lens are superior characteristics, which are derived from the inferior characteristics. For example, three points are used to fit the lens to the fitting plane of the lens case in this design. The orientation of the plane defined by these three points is variable. The variations of the orientation of the plane are controlled by geometrical tolerances of parallelism of the points. The variations of the location of the points are also controlled by geometrical tolerances of coaxiality of the center of circle defined by the points. Finally, the location and the orientation of the lens affects the characteristics of the zoom mechanism of the lens system, and so the characteristics of the zoom mechanism affect the efficiency of the lens system. In the same way, assemblability of the lens system is derived from inferior characteristics such as dimensions and tolerances of a lens and a lens case.

There are also the same kind of design verification problems that can be interpreted as the relationships between inferior characteristics and superior characteristics. For example, a distribution of dimensions that is a superior characteristic is derived from tolerances of dimensions of parts by tolerance analysis. From these observations, design verification can be seen as a task to verify if superior characteristics derived from inferior characteristics satisfy the specification. Therefore, the language for tolerance analysis is applicable to the representation of this kind of knowledge.

5   PROGRAMMING FOR DESIGN VERIFICATION

In this section, programming techniques for design verification in KOSA are described using several examples. Knowledge base for design verification is constructed by designer using the programming techniques.

5.1   Verify the state of assembly

The following example verifies the state of an assembly. The class in Figure 6 describes an assembly part that consists of a hole and a shaft inserted into the hole.

```
defclass: asse,                                          ··· (1)
   super: assembly,                                      ··· (2)
   attributes: hmin: real := h!d + h!dmin,               ··· (3)
               hmax: real := h!d + h!dmax,               ··· (4)
               smin: real := s!d + s!dmin,               ··· (5)
               smax: real := s!d + s!dmax,               ··· (6)
               drmin: real := (!hmin - !smax)/2.0,       ··· (7)
               drmax: real := (!hmax - !smin)/2.0,       ··· (8)
               state: atom :=                            ··· (9)
               case(!hmax =< !smin -> 'insert',          ··· (10)
                    !smax  < !hmin -> 'can not insert',   ··· (11)
                    true           -> 'other' ),         ··· (12)
   sub_parts: s: shaft,                                  ··· (13)
              h: hole,                                   ··· (14)
   constraints: unify_location(shaft1!c0,hole1!c0),      ··· (15)
                unify_orientation(shaft1!c0,hole1!c0).   ··· (16)
```

Figure 6: A assembly class of shaft and hole

The class `asse` is a subclass of the class `assembly` (see lines (1) and (2)). The lines from (3) to (12) are declarations of attributes of the class. They are a minimum diameter of the hole, a maximum diameter of the hole, a minimum diameter of the shaft, a maximum diameter of the shaft, a minimum clearance, a maximum clearance and verification rule of fitting the shaft and the hole. The `case` is a function to choose a value conditionally. The lines (13) and (14) declare that the class has the shaft and the hole as its parts. The lines (15) and (16) declare constraints over the attributes, which are described declaratively. The constraint `unify_location` and `unify_orientation` restrain location and orientation of the shaft and the hole.

The diameter of the hole have to be adequate to insert the shaft into it. There are three possibilities of fitting condition. In one case, the shaft can be insert into the hole with enough clearance. The second, the shaft can't be inserted into the hole with no clearance. The other is a mixture of these two states.

5.2   Design Verification With Tolerance Analysis

This subsection describes a design verification with tolerance analysis. The following example is the gear design which is an element of mechanical designs. A spur gear has attributes: the number of teeth, module, cutter pressure angle, outer circle diameter, base circle diameter, pitch circle diameter, and addendum modification coefficient. Fig 7 is a description of class `gear_pair` with design verification rule. The attribute a(see line (4)) is a distance between centers of spur gears, the attribute `ab` (see line (5)) is a working pressure angle, and the attribute `sn` (see line (8)) is a backlash. The attributes `chk_module` (see line (12)) and `chk_an_deg` (see line (14)) is assigned to the result of verifying if module and cutter pressure angle of both spur gear are equal to each other. The `transfer` (see line(22)) declares a constraint about a dimension between the origin coordinate c0 of the gear pair and the origin coordinate c0 of the spur gear spur1. The value of the attribute a is derived as a result of tolerance analysis of a dimension between two origin coordinates c0 of the instances spur1 and spur2 using the function measure_length.

```
defclass: gear_pair,                                                ... (1)
    super: pair,                                                    ... (2)
    attributes:                                                     ... (3)
        a: dimension := measure_length(self,spur1!c0, spur2!c0),    ... (4)
        ab: int := acos(( spur1!bcd + spur2!bcd)/2/ !a),            ... (5)
        ans1: real := involute(!ab),                                ... (6)
        ans2: real := involute(spur1!an),                           ... (7)
        sn: real := spur1!m * ( spur1!zn + spur2!zn ) *             ... (8)
                    cos(spur1!an)*(!ans1 - !ans2 - 2 *              ... (9)
                    tan(spur1!an) * (spur1!psx + spur2!psx) /       ... (10)
                    (spur1!zn + spur2!zn)),                         ... (11)
        chk_module: atom  := case(spur1!m = spur2!m - > 'OK',       ... (12)
                        true - > 'unequal modules'),                ... (13)
        chk_an_deg: atom := case(spur1!an_deg = spur2!an_deg - > 'OK',  ... (14)
                    true - > 'unequal cutter pressure angles'),     ... (15)
        dim1: dimension := {10.0, 10.0, 0.0},                       ... (16)
            ⋮                                                       ... (17)
    sub_parts:                                                      ... (18)
        c0: coordinate,                                             ... (19)
        spur1: gear,                                                ... (20)
        spur2: gear,                                                ... (21)
    constraints:                                                    ... (22)
        transfer(!c0, spur1!c0, !dim1),
            ⋮
```

Figure 7: The class gear_pair

The following example shows how the backlash which is a superior characteristic of the attribute a is obtained from these class descriptions.

```
make_instance(I1, spur, attributes: m := 1.0,
                                     z := 12,
                                     dk := {15.0,-0.05,0.05}),
make_instance(I2, spur, attributes: m := 1.0,
                                     z := 41,
                                     dk := {43.0,-0.05,0.05}),
make_instance(I3, spur_pair, sub_parts: spur1 := I1,
                                        spur2 := I2),
Sn := I3!sn.
```

The function make_instance makes an instance of a class. I1 and I2 are variables that hold instances of spur gears. Module, the number of teeth, and diameter are 1, 12, and 15±0.05 in the former instance I1, respectively. They are also 1, 41, and 43±0.05 in the latter instance I2, respectively. The variable I3 holds an instance of a gear pair that consists of two spur gears specified by I1 and I2. The backlash of I3 can be obtained by evaluating the function get_value with arguments I3 and the attribute sn.

5.3   Estimation of Assemblability

Figure 8 illustrates a release button unit of a camera. Fig 8(a) shows an assembly drawing. The part-j is an external surface of a camera. The shaft5 is a release button. If dimensions and tolerances which is inferior characteristics in the assembly have an inappropriate value, the appearance which is a superior characteristic will be spoiled. For example, the central location of `hole3` is not equal to that of `shat5`(Figure 8(c). The release button must stick out of `hole3` to realize the function. Let us assume that the origin coordinate is at the center of `shaft5`, and another coordinate is at the center of `hole3`. Then the problem is measurement of the distance between two coordinates.

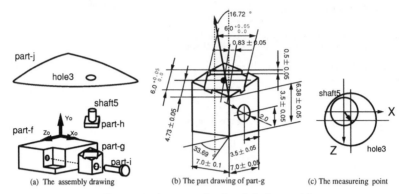

| (a) The assembly drawing | (b) The part drawing of part-g | (c) The measureing point |

Figure 8: An example of real design problem

The result of measurement using worst case analysis is shown in Table 4. The result shows that `hole3` and `shaft5` are fitting without interference, and `shaft5` stick 0.7 ± 0.35 out of `hole3`.

Table 4: Coordinate obtained from the experiment

	X	Y	Z
nominal	0.000000	-0.692726	0.000000
max	0.280278	-0.349070	0.288147
min	-0.280278	-1.036382	-0.288147

6   SYSTEM OVERVIEW

The overview of the system is shown in Figure 9. The system has knowledge-base, data-base and tools such as editors or browsers ,inference engine and user interface.

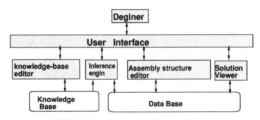

Figure 9: The system overview

The knowledge-base stores tolerance analysis rules and variational modules which are edited using the knowledge-base editor. The data-base stored inference results and assembly structures which are edited using the assembly structure editor. Inference results in data-base can be displayed by the solution browsers. The inference engine evaluates rules in the knowledge-base. The graphical user interface is capable of a three-point perspective drawing because it is necessary for designer to keep continuation of current design work.

7 CONCLUSION

Firstly, we described a framework for the design verification. Most of design verification problems are presumed as evaluation of superior characteristics which is derived from inferior characteristics. Secondly, program techniques for design verification in KOSA were shown using several examples, such as verification of the state of assembly, verification with tolerance analysis and solution of verification of assemblability. Thirdly, We have shown that KOSA for tolerance analysis can be also applied to the design verification problems or design problem solving. The experiments show that knowledge about design verifications can be described and used in practical design fields. Fourthly, the knowledge described in KOSA may be an executable specification or a documentation that it could be useful in the education of junior designers by explaining past design results to them.

Finally, we feel that we need to use the prototype system to analyze more practical designs.

References

[1] Nagasawa, I., Furukara, Y. and Aramaki, S.: ADL: A Designer's Language Based on Logic Programming, *Trans.IPS.Japan*, Vol. 25, No. 4, pp. 606–613 (1984).

[2] Nagasawa, I. and Furukawa, Y.: A Machine Design Calculation Support System, Using the Method of Constraints Reduction, *Trans.IPS.Japan*, Vol. 27, No. 1, pp. 112–120 (1986).

[3] Mochizuki, M., Nagasawa, I., Umeda, M., Higuchi, T. and Ojima, T.: A knowledge Representation Language for Tolerance Analyses and its Programming Techniques, *Trans.IPS.Japan*, Vol. 35, No. 9, pp. 1922–1935 (1994).

[4] Ito, M. and Kono, M.: CONMOTO A Machine Part Description System based on Designers' Mental Processes, *Proc. of IFIP W.G. 5. 2. Working Conference* (1985).

[5] Tomiyama, T. and Yoshikawa, H.: Theory of Design Model, *Journal of The Japan Society For Precision Engineering*, Vol. 49, No. 4, pp. 441–446 (1982).

PART THIRTEEN

Concurrent Engineering

59

Concurrent Engineering with CAM/CAT Systems to Reduce the Production Preparation Lead Time of Personal Computer PCB assembly

Satoru Hashiba

Production Systems Development Laboratory,

NEC Corporation

484, Tsukagoshi 3-chome, Saiwai-ku, Kawasaki, 210 JAPAN

e-mail: hashiba@psdl.tmg.nec.co.jp

Ichiro Kato

Production Systems Development Laboratory

NEC Corporation

Kiyo'o Onodera, Taihei Takeshita, Mitsuo Koga

Production Engineering Department, Personal Computers Division

NEC Corporation

Abstract

The personal computer division of NEC Corporation adopted the distributed development and manufacturing strategy to enable very-short time development and simultaneous multi-plant manufacturing start-up. In order to follow this strategy, CAM/CAT systems to connect design and manufacturing have been developed. In new personal computer product development, these systems contributed significantly to shorten the production preparation lead time of metal mask fabrication, NC data generation, and test fixture board preparation through the concurrent running of these processes. These systems were well adapted to frequent revision of PCBs and helped to maintain the delivery schedule. This report introduces these systems as a successful example of concurrent engineering.

Keywords

CAM, CAT, CAD/CAM linkage, concurrent engineering

1. Introduction

The recent situation in the personal computer business, in which there is a shortening product life cycle, compels manufacturers to reduce product development time, in order to sustain a competitive edge in the market and provide products that satisfy the customers' requirements. We concluded that the challenge of shortening the production preparation lead time had to be solved   by realizing concurrent engineering, which enables parallel processing of all the necessary production preparation jobs. Interfacing the engineering sections and manufacturing sites electronically was also considered to be important in reducing the lead time.

2. Analysis of the previous process

A flow chart of the previous printed circuit board assembly and testing process in our factory is shown in Figure 1.

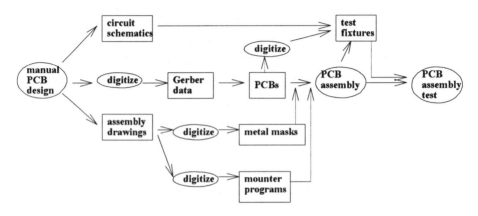

Figure 1. Flow chart of previous PCB manufacturing

(1) PCB design

Printed circuit board (PCB) drawings were drawn by PCB designers and the drawings were again digitized by the PCB designers to provide Gerber data for the PCB manufacturers.

(2) Assembly / Test preparation

Metal masks are important materials used in the cream solder pasting process in SMT mounting. Assembly drawings were digitized to produce metal mask data to obtain the aperture positions for the metal masks. The NC programs for automated SMT mounters were also generated by digitizing the assembly drawings. The question of test fixtures used in the electronic testing process was more serious. The real assembled packages were needed to make the fixtures, and this seriously delayed the start of testing.

3. System concept

To solve the above problems, new computer aided concurrent engineering systems were planned to be built in 1992. The first main strategy in building these systems is to provide measures that enable an early start of production preparation and to make all of the major preparation processes parallel. For example, PCB fabrication, metal mask arrangement, NC program generation, and test fixture preparation should all be done in parallel. It is obvious that this strategy will contribute to shortening the lead time in production preparation.

The second main strategy is to construct a new route of digital data flow connecting the design sections and the manufacturing sites. The personal computer division of our company has distributed design offices and manufacturing plants scattered all over Japan. This distributed arrangement is aimed at responding quickly to sales fluctuations. Therefore, construction of a data way between design offices and manufacturing plants was an urgent need.

To realize these strategies, the three schemes shown below were devised:
(i) definition of standard interface data which connects all the possible CAD systems and the CAM / CAT systems
(ii) circulation of the standard interface data and other library data through *NEC Internet*
(iii) development of the CAM / CAT systems which directly utilize the digital data

4. Construction of the data flow routes

As described in section 3, the key issue was construction of CAD/CAM data flow routes that meet to the real requirements of the production system of our personal computer business.

Our total data base / data communication system layout is shown Figure 2. The main issue to be considered in designing this system was the separation of personal computer division common data and site proper data.

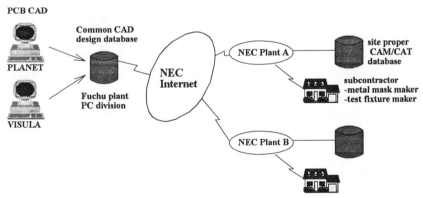

Figure 2. Total Data Base / Data Communication System Layout

Design data is used in common among production plants because the same product may often be manufactured in several plants to follow the market demand. Thus, it is natural that the design data should be concentrated in a common database.

Design data are issued from several CAD systems and they have different data formats. To absorb this difference, a common data format for the CAD/CAM interface was devised and design data originating from various CAD systems are transformed one time into this common format and stored in the database. The format is common for the three CAM/CAT system mentioned below as well as for the CAD systems. Of course the data base is connected to every plant through *NEC Internet* with a sufficient transmission rate, and each plant can directly access the data base. This method allows the plants to extract the desired interface data whenever they want.

On the other hand, site proper manufacturing condition data, e.g., metal mask aperture shapes or individual assembly machine specifications, should naturally be located at each site. Each plant has its own CAM/CAT data base which is maintained separately, reflecting the site's own manufacturing condition changes. Of course, mutual data diversion between two plants is not forbidden if some data can be utilized in another plant.

5. Developed systems

Among the three schemes mentioned in section 3, the most important one is CAM / CAT system development, because these systems generate the detailed output used in the real manufacturing stages, and thus turn the concurrent engineering concept into reality.

These three CAM / CAT systems are:

(1) CAMEL (Computer Aided MEtaL mask design support system):

The main function of this system is to generate Gerber data for the metal masks. The system combines position data from CAD and aperture library data, and generates aperture data for the whole board. The aperture library data should be appropriate for each special situation at the each site, e.g., the direction of cream solder spreading and the type of the solder, and the library is designed to absorb variations in these conditions.

Metal mask manufacturers accept the Gerber data, which can directly be utilized to produce metal masks without any additional processing. The metal mask data is delivered through personal computer communication. This contributes the short data supply lead time.

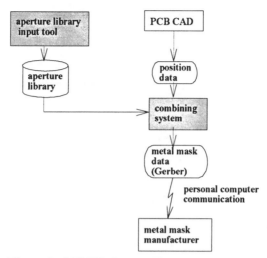

Figure 3. CAMEL System Flow

(2)**CAMOT** (Computer Aided system for surface MOunT assembly):

This system has automatic process planning capability for PCB assembly and NC data generation for mounting / inserting machines. The system flow is shown in Figure 4. The system reads position data from CAD, and generates the fundamental information for PCB assembly, e.g., process sequences, standard times, and part assignment to each process by automatic process planning. This information can be used to prepare the NC data for the automatic machines, the instructions for the operators, and the standard time information used in the production planning.

CAMOT has a sophisticated configuration as a process planning system. CAMOT has three major libraries: a part library, production line library and machine library. These libraries represent the production line model of each plant, and enable the CAMOT system in a sense to *simulate* the production process of the lines. This simulation function allows the generation of accurate line balancing, part assignment to the assembly machines, and the placement sequence on a machine, and the generated NC data has high manufacturing performance through the use of this simulation capability.

CAMOT also has an optimizing function, and the assembly sequence of the parts on a machine can be optimized to minimize the assembly time. This greatly reduces the tact time of the machine assembly, and has a particularly significant effect on mass production.

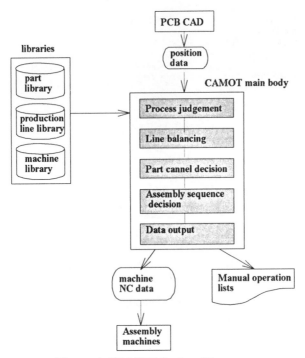

Figure 4. CAMOT System Flow

(3) CATBIRD (Computer Aided Testing by Board fIxtuRe Design):

This system, whose system flow is shown in Figure 5, is a computer aided electronic test support system. This system is actually embedded into the PCB CAD system, and CATBIRD operates as a command level function of the PCB CAD system. This system has two main functions. One is *prior test pad appending*. This means that before the layout design of the PCB, test pads expected to be required are added as dummy parts of the PCB assembly, and the PCB designer arranges them at the appropriate positions on the PCB so that the electronic testing can be done easily.

In a high density PCB, it is difficult to directly probe the part pads, but addition of unnecessary test pads seriously damages the PCB routability. This system adds only the minimum necessary test pads so that the PCB routability is kept high.

The other function is *electric test fixture board design support*. This system generates a priority of test easiness of the pads on the same net, and the designer interactively selects the pads to be touched by the test probe, taking account of this priority. It is also required that in some areas where the fixture board may collide with parts already placed, the fixture board should be hollowed out to allow the parts to escape. The part height and size are put into the testing library, and the system notifies the designer of places where collisions may occur. This function helps the fixture board designers to decide where the test probes should be arranged and where the fixture board should be hollowed out.

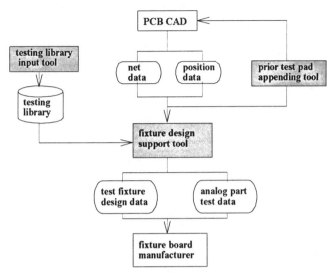

Figure 5. CATBIRD System Flow

5. Results

After the introduction of these three CAM / CAT systems, the new job flow for PCB manufacturing is significantly different. The resulting flow chart is shown in Figure 6.

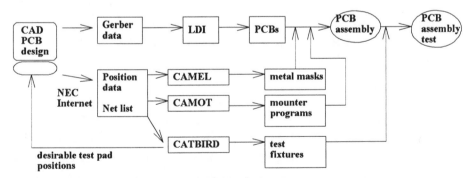

Figure 6. New Flow for PCB Manufacturing

As shown in the figure, all the preparation jobs can be performed in parallel. Every new CAM / CAT system works without paper drawings, real PCBs, or assembled cards. This parallel processing is achieved because each CAM / CAT system can work independently once the standard CAD interface data is released. These CAM / CAT systems also make it possible for the PCB card manufacturing itself to be done independently, because real PCBs are no longer necessary for any other preparation process. For instance, a test fixture board can be made without a real PCB or PCB assembly, and the fixture board and the PCB assembly are finished at almost the same time. The fixture board and the PCB assembly fit exactly without any correction, even though it is the very first time they have been joined. This surprised the fixture board engineer and he got a realistic feeling of concurrent engineering. The lead time effect of the CAM/CAT systems is summarized in Figure 7.

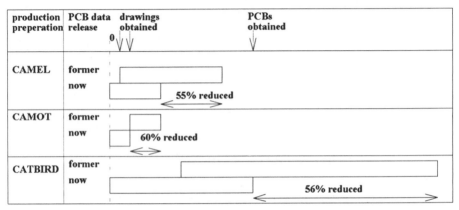

Figure 7. Lead Time Reduction Results

Besides the CAM / CAT systems, PCB manufacturing itself made great progress. PCB manufacturers equipped themselves with Laser direct image exposure machines (LDIs). An LDI can read Gerber data, which is directly output from CAD and can draw a PCB photo master immediately without a film. This made the PCB manufacturing lead time remarkably short.

6. Conclusion

These three CAM / CAT systems have been adopted for NEC's new series of personal computers, and they contribute meaningfully to short time production preparation and simultaneous setup at all the manufacturing sites in the personal computer division, guaranteeing delivery of shipments.

7. Acknowledgments

This remarkable job was accomplished not only by the five authors of this paper but by many people in charge of the development and manufacturing of personal computers in our company. Especially, we wish to show our gratitude to the engineers of each plant, who made great efforts to introduce the CAM / CAT systems into practical use.

60

Global concurrent engineering approach for production systems

D.BREUIL, M ALDANONDO

Ecole Nationale d'Ingénieurs de Tarbes - Laboratoire Génie de Production
Avenue d'Azereix - BP 1629. 65016 Tarbes FRANCE
Tel : (33) 62 44 27 16 - Fax : (33) 62 44 27 08

Abstract

Basic studies are the first phase of an engineering project. They aim to determine solutions in order to improve performances of production units, to evaluate their feasibility, their profitability, and to establish the planning of their installation.

They concern the three domains of the production systems, the product design, the process definition and the management. Such studies may be decomposed in six functional steps, dealing with the components of these domains

This paper focuses on the physical organisation design (of production and flows activities) and the management design (decision system), and their relations.

The goal of this communication is to present an approach for Global Concurrent Engineering which improve the co-ordination between those two domains. The main points of the proposed approach deal mainly with :

- the links and influences between decision system, physical organisation and resources design (what, when, how they must be taken in account during the whole design process)
- the identification of constraints for these design steps, nature of which may come from technical aspects, implementation environment, or financial objectives
- the evaluation of the consequences of design alternatives

1- INTRODUCTION

Production systems engineering aims to design or renew all the components which participate to the production. These components include mainly:
   - the physical elements, ie all the resources like equipments, workers,..required for manufacturing, maintaining, preparing, distributing..
   - the management which can be split in decision and information systems with their own organisation, mechanisms and resources
   - the interfaces between the production system and its environment which can be internal or external to the company

Such projects are normally decomposed in several phases, from the very first global studies to the launching of exploitation This decomposition is made according to a hierarchical approach which allows at each level, to evaluate the feasibility and the profitability of the proposed solutions. Criteria for this decomposition may be different from one project to another; they may consider the production system according to different points of view, either the company domains or departments (production, maintenanc, information,...) or the functions to be modernised (civil engineering, automatisms, assembly, logistics,..) or else the manufacturing organisation (workshop 1, workshop 2, warehouse,...).

Basic Studies are the first phase of an engineering project. Their aim is to determine global solutions in order to improve perfomances of production units, to evaluate their feasibility, their profitability, and to establish the planning of their installation.
During these studies, designers must choose between various alternatives upon the different components of the production system . However for many different reasons (like time and costs allocated to these studies, process definition,...), their decisions are based on global views, even though they must get clear enough precision on the future in order to find a realistic cost balance.

Many research works propose to help developers during design phases. Quite often they are oriented towards detailed studies activities and they address only few aspects of them. However, some of them may be used at global level as we will see later..
Two generic ways have been the subjects of several works which are not global studies beginning oriented, but their principles or concepts may be adapted for this purpose :
re-engineering and concurrent engineering.
a) re-engineering concerns the fundamental transformation of the processes inside the organisation or the structure of a company, from the top management to the bottom; it is aimed to prepare the organisation to permanent evolution.
Several methods have been proposed in that domain which are oriented towards the management of the company changes from human resources or company culture point of view supported by the different possibilities offered by information technology.

However, re-engineering approaches do not take into account the production (or manufacturing) processes.

b) The definition for Concurrent Engineering stated by US Institute for Defence Analysis (report R-338, 1988) considers Concurrent Engineering as "a systematic approach to the integrated, concurrent design of products and their related processes including manufacturing and support. This approach is intended to cause the developers, from the outset, to consider all elements of the product life-cycle from conception through disposal, including quality, cost, schedule and user requirements".

The parallelisation of the design processes (proposed in this definition) is quite necessary for the management of Basic Studies; however, approaches presented in the majority of research works are too much detailed for their global level.

2- BASIC STUDIES

Basic Studies concern all the domains of the production systems, the product design, the process organisation, the management and the information systems. Such studies may be decomposed in functional steps, dealing with the components of these domains. During such studies very strong interactions occur between the steps.

Results coming out of Basic Studies include the global description of the solution which concern mainly:

- the type and characteristics of all the resources that will be used (from buildings to containers for products transportation inside workshops),
- the different flow operating rules,
- the layout of resources and various charts adapted to different points of view (ex civil engineering, circulation of people, or energies,...)
- the structure of decision system,
- the organisation and resources allocated to information system

The following simplified diagram (figure 1) shows the main relations between the different steps of Basic Studies.

As this diagram shows, there are loops between the steps, especially at the interfaces between the domains. If design process is considered as a sequence, each step may lead to modify the results of one or more previous ones because, from its point of view, some impossibility to meet required performances occurred.

This may come for instance from technical difficulties in equipments design which implies to modify the production activities or from the optimisation of management which requires to re-organise workshops.

More over, design at each step must take into account constraints coming from the general environment (market trends, expected output flows,...), from the company strategies on its own domain and from forecasted exploitation of the future production system.

The main steps are concerned with :

a) The design of products concerns the general definition of the products and the associated production activities.

b) The design of the physical organisation concerns :
  - the definition and the dimensioning of equipments as well as their layout
  - the definition of all the facilities for the support of production activities, from the buildings to the energies,
  - the flow regulation rules adapted to the required performances for the production systems
  - the required resources for all the internal logistics

c) The design of decision and management system deals with the organisation of :
  - the definition and organisation of decision functions and decision centres in each function,
  - the management organisation of human resources and the definition of required competencies.

d) The design of information system considers the organisation and the resources which may be implemented to support all the other components of the production system. This starts with the definition of data organisation and processing and leads to the definition of required types of equipments, packages, networks,...

e) The last functional step concerns the planning of the implementation of the proposed solutions and the estimation of the costs and profitability of the total project.

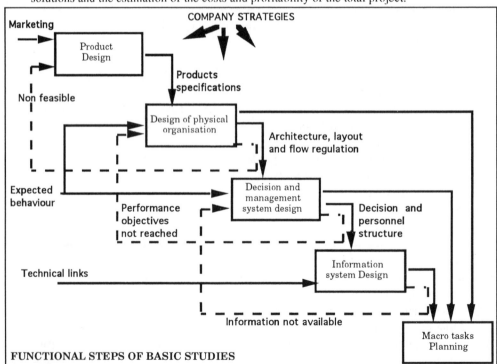

Figure 1

For all those steps, different competencies among the designers are involved in order to handle efficiently all the point of views. One way to bring together all correct informations consist in grouping designers in teams to obtain proper co-operation level; but meetings cannot stand full time; so the need of concurrency in design activities implies the distribution of coherent and right informations for all partners in a project. Since influences from one point of view may be of interest for all others and reciprocally, the network of informations exchange is quite large. Further more, data used at this design level are global and aggregated data, representing several aspects of the future production system.

3- DESIGN RELATIONS BETWEEN PHYSICAL AND MANAGEMENT ORGANISATION

3.1- organisation of physical system
The design process may be decomposed in 6 main activities (figure 2):
- the first one deals with the flow analysis; this is a critical point since these flows are not always clearly identified at the Basic Study level for different reasons which are related to the difficulty to get representative flow (and products) data inside the company and the fuzziness of company objectives regarding the expected or forecasted product flows out of the company.

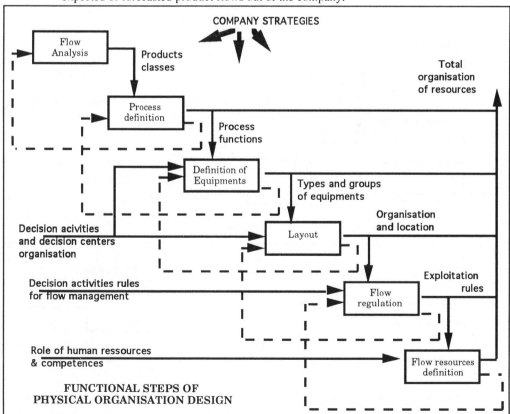

Figure 2

- then the process definition will point out the functions required for manufacturing the products
- the definition of equipment will lead to the definition of the nature and global performances of equipments that will be required to perform the production activities.
- layout activity aims to organise and structure workshops and inside them major groups of equipments according to the different possible choices offered by the previous activity
- flow regulation activity is concerned with the architecture of the flow management
- flow resources definition activity is in charge of the determination of the resources associated to the product flow and storage.

3.2 Management organisation

These activities aim to determine the decision structure according to the required performances for the production system and the organisation of human resources associated to this structure.

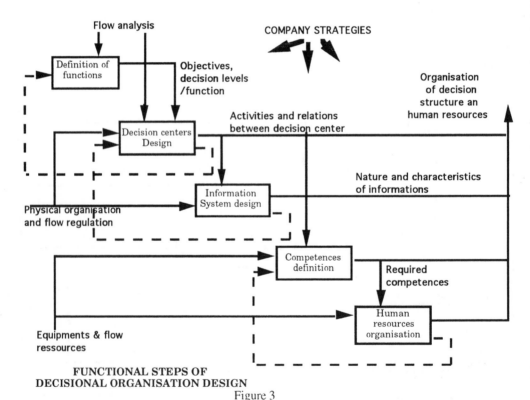

**FUNCTIONAL STEPS OF
DECISIONAL ORGANISATION DESIGN**

Figure 3

The decision structure covers all management functions of the considered production system (that may include sometimes logistics functions, suppliers relations,..) from the top level of production management to the decision taken in real time inside workshops. Five main activities may be identified at the level of Basic Studies:

- the definition of management functions and of the global architecture of decision centres
- the definition of the activities inside each decision centres and the informations they need
- the design of the architecture of the information system
- the definition of the required competencies for managing each decision centres; this is particularly important when the improvements of decision structure introduce new concepts, new decision rules or new relations between decision makers.
- the organisation of the human resources for identifying groups of personnel, relations between groups and the structure of this organisation.

3.3 Relations between the design processes

On these diagrams, the main links between these processes are shown and they pointed out the strong relations that exist between those organisations.

In fact, quite often, the stating of a physical organisation determines flow regulation decisions; but such a choice depends on management principles of the flows (decision aspect) which depend themselves on the nature of the process which have been designed and such reasoning could go on again,...

If design processes are conducted separately, there will be of course many chances to spend a long time for adapting each one at the end when designers come together.

On the other hand, if the different steps are simply integrated, sequenced just one after the other or if designers met at some times, there will be also a lot of time lost in discussions or in re design

4- STRUCTURE OF THE PROPOSED APPROACH

The optimisation of the objectives of the designed production system depends mainly on the coherence of all the different flows which will exist in this system and above all the physical and decisional ones.

So, in order to minimise the backward loops we mentioned above or the loss of time, conflicts, etc,... it is necessary to :

- design simultaneously those flows
- to co-ordinate the other design functions from the design decisions taken upon the flows

Then the functional steps will be linked as it is shown in the figure 4. In this diagram, the activities dedicated to the specific design of process, equipments or management remain almost the same as previously.

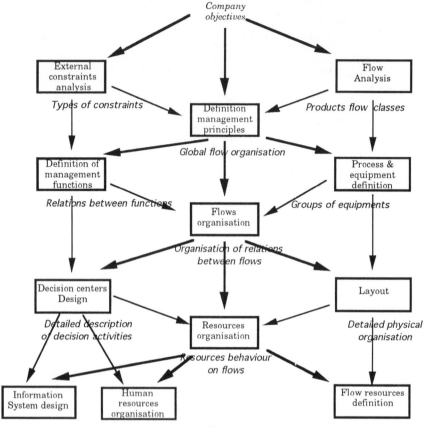

Figure 4

However, co-ordination design activities are the skeleton of all the design processes. The main co-ordination activities are:

- **definition of flows management principles**: this activity aims to define and adapt the general management principles to each class of product flows, to determine the types of management rules (technical or production management aspects) which will be applied to the products and decision flows according to the performance objectives and responses to external constraints. This will allow us to determine on one hand the necessary management functions and on the other hand the organisation of process functions.

- **organisation of flows**: this activity is concerned with the design of the architecture of flows. Once management functions and groups of equipment have been defined, it is necessary to examine their internal and external cross links and especially the interfacing points between the different flows. This will lead to the determination of the co-ordination between the different functions regarding the flows and to verify the coherence of the rules applied to these flows.

- **resource organization**: this last important co-ordination activity first defines the relations between the resources associated with the different flows and their nature according to the performances they must reach. The objective is to establish the activity frame of these resources inside the production system and their possible influences on the different flows.

5- CONCLUSIONS

This approach for Basic Studies may be extended to the whole phases of an engineering project. Then we get a structured hierarchical approach over these phases; the criteria used for the structuration of the hierarchy is the degree of detail of the data used during these phases.

This methodology has been used for several companies (11 full approach, 4 partially, some more are under changes) in different market domains (food industry, wood industry, mechanical or chemical). The only common point between them was the size, no none was greater than 600 people and was challenged on costs and response time to customers. The small size simplified the relations between the global co-ordination process and others.

The results were the creation of small independent entities (20 to 50 people) or production units (sometimes subsidiaries) in which people were quite involved in the success of their activities. Most of them have their own independent management structure which includes commercial or purchasing functions even though they are co-ordinated by company management. In all cases the weight and complexity of centralised management was avoided (sometimes with difficulty) in order to let as large as possible autonomy to the units.

In these companies, objectives towards environment were reached (response time, costs, production volume increase...) and up to now they are still alive while some of their challengers disappeared in the mean time.

REFERENCES

AHMED (1994), M. S. AHMED, J. A. G. KNIGHT, H. S. ABDALLA, Global Concurrent Engineering, Intelligent Manufacturing Systems, IMS'94 (IFAC) Vienne, 1994.
ARVINDE (1994), B. ARVKNDH, S. A. IRANI, Cell formation: the need for an integrated solution of the subproblems. International Journal of Production Research, V 32, No 7, 1994.
BREUIL (1984), D. BREUIL, Outils de conception et ed decision dans les organisations ee gestion de production, These d'etat, Bordeaux, 1984.
BREUIL (1984), D. BREUIL, Schemas Directeur: episodes 1 et 2 Logistique et Management, V 1, No 1, 1993, V 2, No 1, 1994.
BURBIDGE (1989), J. L. BURBIDGE, Production Flow Analysis for Planning Group Technology, Oxford University Press, Oxford, 1989.

BURBIDGE (1992), J.L. BURBIDGE, Changes to group technology: process organisation is obsolete, International Journal of Production Research, V 30, n° 5, 1992

BURBIDGE (1993), J.L. BURBIDGE, Group technology, where do we go from here?, Advances in Production Management Systems (APMS 93), Athens, Greece, September 1993

DAS (1993), S. K. DAS, A facility layout method for flexible manufacturing systems, International Journal of Production Research, V 31, n° 2, 1993

ESPRIT 7752 (1994), Architecture for Global Concurrent Engineering System, Deliverable 3.2

FERRAND (1994), D. FERRAND, G. PAQUET, Apprentissage organisationnel et Re-engineering, Logistique et management, Vol 1, n°2, 1994

GALLOIS (1989), P-M. GALLOIS, Typologie des entreprises industrielles, Revue française de Gestion Industrielle, V 8,n°1, 1989

GUHA (1993) S. GUHA, W. J. KETTINGER, J.T. TENG, Business Process Re-engineering: Building a comprehensive Methodology, Information Systems management, 1993

HALL (1993), G.HALL, J. ROSENTHAL, J. WADE, How to make re-engineering really work, Harvard Business Review, 71,6, 1993

M. HAMMER (1990), Re-engineering work: Don't automate, Obliterate, Harvard Business Review, 68,4 1990

M. HAMMER, J.CHAMPY (1993), Le Reengineering, DUNOD, 1993

Mac LEAN, 1993, Computer-Aided Manufacturing System Engineering, Advances in Production Management Systems (APMS 93), Athens, Greece, September 1993

BIOGRAPHY

Dominique BREUIL (age 44) is Associated Professor at the Ecole nationale d'Ingénieurs de Tarbes since october 1993. Previously he worked as manager of Organisation Department in several engineering companies where he was in charge of the industrial organisation of the production systems designed by engineers of these companies.

Micehl ALDANONDO (age 34) is Maitre de Conférences at the Ecole nationale d'Ingénieurs de Tarbes

61

Methods and Tools for Technological Databases to Support Concurrent Engineering

Dr.-Ing. A. Nestler, Dr.-Ing.Ch. Schöne
Institute for Production Engineering, Dresden University of Technology
Mommsenstr. 13
01062 Dresden, Germany, telefon: (0351) 463 7088 , fax:(0351) 463 7159

Abstract

Modern production concepts requires new approaches. Better than before must succeeded the integration man in production process. Therfore fexible goals for the problems with alternativ possibilities and strategies for specific solutions are important. Because of more decentralized decisions local experts needs more global informations about the processes. Technical information systems are in this connection most important. Technological data in particular represent a special bottleneck, in this case. That means, we have to find out creative approaches especially for this data type. There is demonstrated to solve this problem by means of tools and methods for a technological database.

Keywords

Concurrent engineering, technological database, optimization model, neural network, cutting values

1 INTRODUCTION

Present trends of development influence the production engineering and the often existing conventional methods are insufficient. The flexible marked demands, the increasing complexity of products, processes and structures are the cause of this development. This leads to concepts of development, planning and production with decentralization, team work and parallel sequences of activities and tasks. Such changes make high demands on information systems. Especially the intelligent usage of resources are in great demand. Knowledge and experience have to be made available for a wide range of experts out of different working disciplines in an efficient manner. Mostly ensured and updated information to be taken from various business-oriented- and business-independent data stocks; standardized data access; rapid information processing and easy-of-use data supply. Thereby it is possible to cope with changing requirements on market immediately. Additionally, essential suppositions for cooperative problem solving can be gained in the stages of product design and product manufacturing.

2   STATE-OF-THE-ART AND PRESENT DEMANDS

To tackle with these problems, there is no doubt about the recent market - and research and develpment dominance of *tool management- and resource management systems*. Those systems accepted to be efficient means of rationalization are planning- or process-oriented. From their content, tool data management activities are linked with efforts *to make available cutting values* supported by firm databases summarizing best cutting parameters. Trends towards *CAD/NC based on manufacturing features* are underlined by data linkage combining geometric and technology via product models.

All of these development subjects are strictly oriented to technological data. Technological data are available *concurrently*, on the one hand. On the other hand, those data are *mostly oriented to a predefined user system* which is the only one determining data structure, - content and application alternatives for this range of users. Even the very sensitive technological data type are more complicated concerning integration. Tools and methods of present technological databases results from functionalities of relational data base management systems.

Especially the technological data area based on firm-dependent and standardized ordinal systems for instance number systems, thesauri, classification methods and so on. In addition to data bases also used other conventional and knowledge based methods in practical operation for data storage and -supply: macro techniques (higher progamming languages C++, FORTRAN, EXAPT, ...), table techniques, decision tables (decision rules, IF-THEN-conections, ...). All systems needs data exchange with strange systems. Standardized data interfaces for export/import and programmable interfaces for the adaption are important evaluation criterions.

The following demands for information processing results from present working methods. The general characteristics (Kimura 1992, Krause 1993) are applicable for technological data area:

- Informations are incomplete, uncertain and heterogeneous
- Flexible solutions fields (enviroments, constraints) desirable, allow scope for variants
- Distributed problem solving and simultaneously processes data keeping.

In consideration of creative aspects tools and methods for operativ and decentral user support makes available for different user demands in different business departments.

3   CONTENT AND INTEGRATING FUNCTION OF TECHNOLOGICAL DATABASE

Technological data type is an essential part of computer aided manufacturing. It is very difficult to make available technological data. To fulfill system integration requirements, there is a special need for novel systems especially considering this data type.

A joint project carried out by the Dresden University of Technology and the Research Association for Programming Languages Aachen*) which title is *Technological database* was first of all, directed to find out appropriate tools for production planning and NC manufacturing. As to be demonstrated following, the gained results - methods and data support - are able to contribute entire system integration.

*)The paper presents results gained by project 92D promoted by the Working Group of Industrial Research Associations (AIF). The project's title is "Development of a production planning - and NC manufacturing database considering the needs of small- and medium-sized enterprises in the East German Federal States".

Using this technolological data, design to manufacturing is influenced in the planning stage, CAP processors are supplied for production planning and Requirements for production equipment and -facilities in manufacturing can be derived, too. To cope with different user demands out of the production planning departments, we can apply flexible data structures and standardized methods of access.

In this case, the understanding of a *technological database* (production planning- and manufacturing database) is focussed on a software tool for storage, handling and management of data, necessary or generated by production planning and NC manufacturing. Additionally, selected algorithms and methods to process, link, calculate or visualize the managed data are a part of this database, too.

Similar software systems put the main emphasis on management functions, for instance - resources management. However, the target of the technological database consists in supplying the necessary planning data *near to the user*. To carry out CAP- and CAM functions even by advanced programming systems, we have to make available data on manufacturing process. Relevant data ranges to be solved for technological data are structuring, collection, processing and supply. Information out of technological data are covering subjects as in tab 1.

Machine tools, Attachments	Clamping devices, Fixtures, Chucks	Tools, Tool holder, Measuring instrumets, Testing instruments	Workpiece materials, Cutting materials, Cutting values	Manufacturing features, Manufacturing operations, Cycles

Tab. 1: Technological data areas

If those data are input into a user system, for instance an *NC programming system*, we can set the user free of everyday activities. Simultaneously, the user is supported in essential cases of manufacturing decision making, for instance: selection of tools and clamping devices, determination of cutting values, subdivision of cuts and manufacturing strategies, simulation in connection with collision avoidance check, and so on. To reduce lead times in production planning and NC programming is one of the targets. Simultaneously, security and manipulation of planning results should be very flexible.

These data are an essential foundation stone *not only for production planning and NC-programming*. If there are made available these data out of different business departments, it is possible to update resources and generate production plans concurrent with manufacturing sequences planning. Following this strategy, *early* planning stages can integrate up-to-date information out of *late* planning steps. As described, a technological database is an essential part of business' manufacturing data management carried out by distributed information systems (fig. 1).

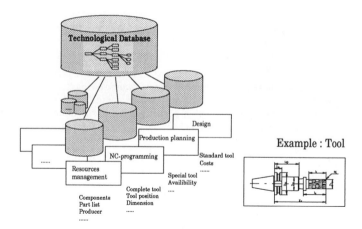

Fig. 1: Technological data in different business departments

4   DESCRIPTION OF SELECTED TECHNOLOGICAL DATABASE MODULES

The technological database is to be shaped as an autonomous software tool in order to be as independent as possible of any kind of user systems. The software tool is focussed on the EXA-TDO (*TechnologyDataOrganization*) system (N.N., 1993) used for technological data management, as an example (fig.2). Delivering necessary interfaces there are available functions as data exchange with import and export, easy-of-use communication and integrated access.

The user is enabled to carry out data records and - updates in a centralized mode, where as data can be used in a decentralized one. The technological data can be transferred to various already existing business software environments for instance databases or file management systems. In addition to pure data handling functions, a technological database has to solve another sophisticated task directed to differentiated data supply and data evaluation at various using levels - design, production planning, NC programming, resources management, and so on.

The necessary data volumes are to be supplied depending on the use of available functionalities of a technological data management system or a user system as, for instance, an NC programming system. To cope with *complex correlations among data*, it is useful to enable allocation of technological database areas. Linkage via *identifyers* is one method to be applied. Basing on a machine tool, there can be linked accompanying resource components. Additionally, there can be predetermined cutting values coping with the foreseen material which can be allocated to manufacturing features. Updated manufacturing data can be concurrently used by the NC programming staff as well as resource management - and design departments during process planning. Already recorded data can be supplied for business process chains; those data can be added, if necessary.

A new technological database quality can be gained if there can be considered *various data sources* by means of different data processing methods alternatively - for instance *tables, mathematical models, knowledge based methods* etc. Alternative data recording and data supplying functions oriented to an NC programming system should be demonstrated for the example of the *technological data area c u t t i n g   v a l u e s.*

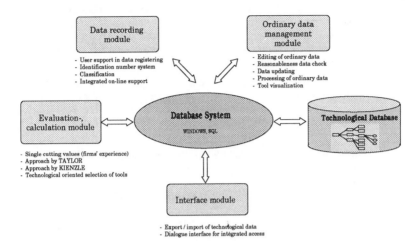

Fig. 2: Structure of a technological database (example, represented for EXA-TDO)

Alternative data sources can be shaped by *discrete cutting values* tested under practice demands and cutting parameters derived from empiric series of tests (wear- and force measurement) to be made available for user via production technological data baseas. Firm-independent recommended values have to be adapted according business-oriented to machine tools and materials in the majority of cases. Data collections can be added by *mathematical (calculation) models* based on primary data. This is another possibility to determine starting values for the manufacturing process. Being part of a technological database, these models are available for an extended group of users - in comparison with an internal module realization integrated into a manufacturing processor of a special NC system. As an essential supposition, workpiece- and cutting material data as well as machine tool- and manufacturing element data must be available for all alternatives determined before.

In the following the trend from complex, rigid and exactly models towards a clearly, flexible system with new characteristic features should be demonstrated. Rough technological models with a high ability to adaption is an integrated part of the developed database.

5   INTEGRATION OF MATHEMATICAL MODELS

A module on complex cutting values' determination shaped for any milling tasks (Kochan, 1992) has been applied for manufacturing features. This module SAFRAE is an integrated part of a technological database. Within the *technological module for milling*, determination of cutting values is based on technical constraints. First of all, these are force-determined constraints acting for maximum feed, as, for instance, shank load, cutting edge strength, moment load, tool dislocation a.o. technical constraints.

From the user's point of view, to apply the module for determination of cutting values can be advantageous in two cases: In the first case cutting values predefined for a special manufacturing subject the loads of technical constraints (tool dislocation, roughness,...) are calculated. As a result, we are enabled to evaluate the recommended values' validity resp. feasibility. Furthermore, there can be found out critical constraints in the case of selected manufacturing features (grooves, pockets,...). In the second case of a manufacturing subject there are calculated cutting values (maximum values) based on technical constraints. As a result, these values can be used as initial data for manufacturing.

Load of technical constraints is visualized by numerical and graphic means. Percentage loads of discrete technical constraints give some information on their influences' priority. Help-texts are integrated to support the results' interpretation. Each technical constraint is represented by its essential influencing parameters. Additionally, the user is proposed alternative activities to handle the influence parameters (fig. 3).

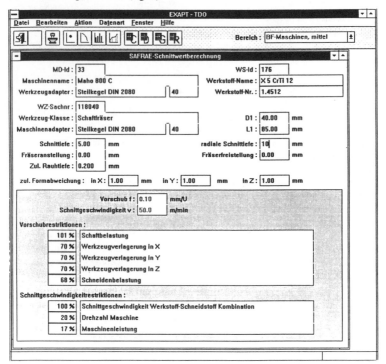

Fig. 3: Usage of EXA-TDO with access to the intergrated optimization modul for milling

Before taking over one feasible variant (one cutting value, for instance) into the application (the NC program,...), the user is enabled to improve the result by considering and comparing with other variants (other tools, cutting materials and so on). Evaluating this methodology for the example above mentioned we see, that the user is enabled to call in question the given variants; he can find out *creative results* by means of system's support.

6   ADDITION OF KNOWLEDGE BASED METHODS

The use of experience knowledge within a technological database is a rather necessary alternative for the determination of cutting values. Applications of mathematical models for technological data determination give useful start parameters for the using in different technological software systems. Nevertheless it is necessaray to specify the cutting values after practical tests, for instance by running in a new NC-program on the machine tool. Also, for technological problems, which are too difficult to describe by mathematical formular, it is better to use suitable knowledge based methods or use them in addition.

To acquire knowledge on technical and manufacturing correlations, the triple assignment among object-feature-feature value is the foundation. First and most difficult part of knowledge processing process is characterized by knowledge acquisition methods.

With the knowledge akquisition the complexity of mathematical models are better available. The technological influence parameters can prepared for declaration components as one possibility for an *intelligent help*. Especially for planning variants and the application of different equipments the knowledge of the relationsship between different parameters is indispensable (Kochan, 1992).

Another way is to reduce the technological models for specific form features to get a 'rough model' applications with specific restrictions. At the same time it is interesting to add specific knowledge to the technological rough models. *Neural networks* are applied for knowledge acquisition and - processing in rising extent (Barschdorff, 1991). The computerized model 'neural network' is directed to automated knowledge processing, being similar to the function of human brain. At present, neural networks are especially used to solve sophisticated problems; influence of various input parameters on the system can not be found out at all resp. incompletely, on the one hand. On the other hand, this influence is too sophisticated to be modeled. Unknown correlations among input- and output parameters can be analyzed and assigned to by means of neural networks.

Sequence of operations necessary for practical use of neural networks can be subdivided in the steps:

- Supply of learning database for the neural network

- Determination of network structure and learning parameters

- Train a neural network

- Application of the neural network.

In the following, two manufacturing problems will be demonstrated. The primary use is the determination of cutting values. The second use is the decision support for specific working conditions in manufacturing section. It is necessary to get useful parameters to describe a technological situation and to make a evaluation. Now ist is possible to chosen out the necessary parametrs for a learning database. This data represents the praxis relevant database. In this case, it must be guaranteed, that there are no data lacks in records of learning database. The Development is based on the state of the art of the technological database, their cutting values and describing attributes.

With the technological database the neural network will be traind on base of known *individual cutting values*. Using network for new manufacturing operations cutting values are not known yet for, describing attributes are defined. In this case, a new manufacturing operation may be to manufacture a workpiece material characterized by special mechanical properties or to use a tool being of general new cutting edge geometry or consistency. That means, input data for neural network are not part of the learning database (fig.4).

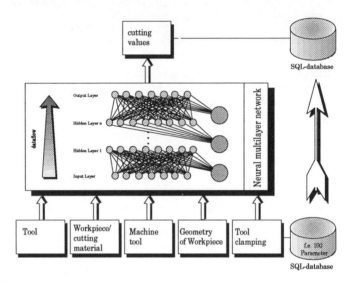

Fig.4: Use of the neural network for cutting value determination

Applicability of cutting values determined via neural network are to be checked according to their reasonableness by an expert. If these values have been evaluated to be appropriate, corresponding practice investigations on a test are carried out to verify theoretical values feasibility. The results of practice work are useful for a rather training and qualification of neural network. So we have a flexibility part of a database by using actual technological data from monitoring.

Another example is the *determination of cutting edge mismatch* in 5-axes milling. Tool change is necessary before finishing, partially for tools characterized by large operating times. This behaviour becomes a special problem, if the tool is to be changed within a tool path. In this case, it is to be guaranteed, that both tools - one to be taken off and the other to be set in - cause no free cutting scores at the workpiece surface.

At the beginning, significant features are evaluated out of the set of possible parameters influencing cutting edge mismatch in 5-axes milling. Tests have been carried out for these significant parameters. In addition to tool parameters there are varied milling mode (climb - or cut-up milling), leading angle and cutting values. Test specimens have been evaluated and classified acoording to resulting axial- and radial cutting edge mismatch. Input- and output values for the neural network are summarized in fig.5.

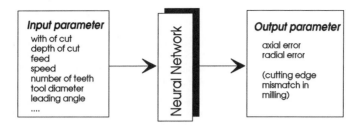

Fig.5 : In- and Outputparameter of neural network for determination of cutting edge mismatch in 5-axes milling

This application shows, how neural network can used for decision support before manufacturing. So it is possible to plan the practical operation of tool, the selection of cutting cinematics for special working conditions more effektive.

7   EFFECTS AND APPLICABLE FIELDS

The presentation illustrates the realized access to efficient flexible technological modules. To use this result in combination with a technological database, there can be gained effects which are variously contributing to the target of Concurrent Engineering.

The chosen example of the technological data area in connection with the complex problem cutting values illustrates approaches to be used for team-based methodologies. User's creative decision making is improved caused by simultaneous availability of discrete cutting values as firm-dependent best values, on the one hand , and cutting values which have been determined by computations, optimization and knowledge processing on the other hand. Computation modules and neural networks are enabled to cope with specific problems to be solved meeting the demands of user groups:

- Using technological information in design
- Precalculation
- Computing criterical loads
- Using knowledge about manufacturing conditions
- Reducing tests
- Flexible resource allocation
- Case studies.

That means, within appropriate information systems there we can derive approaches on generally applicable fields for decision making in design, resources' application and - supply as well as quality assurance a.o. business departments only by supplying information originally destined for NC programming.

8   SUMMARY

Further tasks ranging in alternative technological data supply out of various sources have to be solved by simultaneous development of system development. Further methods and tools - genetic algorithm for optimization and simulation, multimedia in connection with natural linguistic access - have an importend influence on future demands.

In addition to continuous development of new technological database areas, existing areas can be made more qualified by means of advanced methods and extended manufacturing data contents. To cope with maximum functional reliability, process security and a better utilization of Know-how much better, advices on application, machine tool influencing parameters and manufacturing situations have to be made more transparent.

An essential supposition to coincide conventional data (resource data, computation models,...) and additional alternative knowledge sources (experience,...), on the one hand, and to cope with the demands for supplying data on new manufacturing techniques (High-Speed Cutting, Solid Freeform Manufacturing,...) and resources (cermets, ceramics,...), on the other hand, is represented by an integrated application of conventional CAx-solutions and knowledge processing.

Concurrent Engineering in the field of technological data is based on novel approaches including *alternative methods for various using levels*. A technological database represents an essential link with other data management systems of technical production planning.

9   REFERENCES

Kimura, F. u.a. (1992): Interim Report. The first int.CIRP Workshop on Concurrent
        Engineering for Product Realization. 27. u. 28.6.1992, Tokio. In: Anals of the CIRP41
        H.2,S.743-746
Krause, F.-L. (1993): Produktentwicklung mit Simultaneous Engineering. FACTS, Mai 1993.
        Springer Verlag
        September 12-16
N.N.(1993): Einführung in die EXAPT-Technologiedatenorganisation EXA-TDO. EXAPT NC
        Systemtechnik GmbH, Aachen
Kochan, D., Nestler, A., Schöne, Ch. (1992): Intelligent software support for cutting process.
        in Olling, G.J.; Kimura, F.: Human Aspects in Computer Integrated Manufacturing.
        Proceedings of the IFIP TC5/WG 5.3, Eight International PROLAMAT Conference,
        Man in CIM, Tokyo, Japan, 24-26 June 1992, Amsderdam: Elsevier, 1992
Barschdorff, D., Monostori,L. (1991): Neuronal networks- Their applications and perspectives
        in intelligent machining, Computers in Industry 17 (1991) S.101-119, Amsderdam:
        Elsevier Science Publishers

62

CIM-FACE:
A Federated Architecture
for Concurrent Engineering

L.M. Camarinha-Matos; A. L. Osório
Universidade Nova de Lisboa
Quinta da Torre - 2825 Monte Caparica - Portugal
Tel +351-1-2953213   Fax +351-1-2957786   E-mail cam@fct.unl.pt

Abstract
The requirements for a platform for Manufacturing Systems integration and Concurrent Engineering are discussed in parallel with an analysis of the evolution of manufacturing and organizational paradigms in industrial companies. A prototype system combining both information integration and cooperation support functionalities is presented and particular emphasis is put on the execution supervision of the enterprise's plan of business processes. Finally open questions and directions for further research are summarized.

Keywords
Systems Integration, Concurrent Engineering, CIM, Federated Architecture, Modeling.

1. INTRODUCTION

A realistic approach to design an architecture to support Concurrent Engineering has to take into account results and tendencies emerging from various research sub fields of the advanced manufacturing area.

Information based Integration. Information integration has been recognized, since long ago, as a basic requirement for the implantation of advanced manufacturing systems and for concurrent engineering in particular. This has been an important research topic in many international projects. The ISO STEP group (ISO STEP 1991) is a particularly notorious example in this area.

Integrated Manufacturing Systems are complex systems that can be analyzed from various perspectives, not limited to the technologic aspects -- in their many facets -- but including also organizational and social views. From a software engineering perspective, a CIM system can be seen as an integrated federation of multiple heterogeneous software modules (Camarinha 1993) (Osório 1994). This heterogeneity comes from a diversity of reasons, such as the use of different development and data management technologies, heterogeneous computational platforms and operating systems, and even from being based on different underlying "cultures" in terms of the target users. These software modules typically run in a distributed computational infrastructure and show a considerable degree of autonomy, either because:

i) they were developed as stand alone or loosely integrated components, specially in the case of legacy systems; and

ii) the decision making process is, to a large extent, based on the humans that use such tools or on the characteristics / behavior of the machines being controlled, in spite of the increasing level of intelligence of computer aided tools.

A common Information System (CIM IS), including shared concepts / data models, provides, therefore, a basic kernel to support the integration of heterogeneous and distributed functional modules in an engineering and manufacturing environment.

Concurrent Engineering. The increasing globalization of the economy and openness of markets is imposing tough challenges to manufacturing companies, leading to the concept of lean / agile manufacturing. One of its manifestations is the recognition of the product, and thus product data, in its entire life cycle, as the main "focus of attention" in CIM IS.

Product Data Management may be considered an essential set of tools for tracking products from conception / design to retirement / recycling. The concept of Concurrent Engineering (CE) has become more and more popular in recent years as a result of the recognition of the need to integrate diversified expertise and to improve the flow of information among all "areas" involved in the product life cycle. Team work based on concurrent or simultaneous activities, potentially leads to a substantial reduction in the design-production cycle time, if compared to the traditional sequential "throw it over the wall" approach.

Evolving from earlier attempts, represented by the paradigms of "Design for Assembly / Design for Manufacturing", Concurrent Engineering is a consequence of the recognition that a product must be the result of many factors, including:

-Marketing and sales factors
-Design factors
-Production factors
-Usage factors (intended functionalities / requirements)
-Destruction / recycling factors.

For all these areas there are hundreds of computer-aided tools (CAxx) on the market that help the human experts in their tasks. At a particular enterprise level various of these tools are normally available, together with some proprietary software developments. A platform that supports the integration of such tools (information and knowledge sharing) as well as the interaction among their users (team work) is a computational requirement for CE.

New organizational structures. On the other side, observing companies' evolution in terms of organization, a strong paradigm shift towards team-based structures is becoming evident. Team work, as a practical approach to integrate contributions from different experts, by opposition to more traditional hierarchical / rigid departmental structures, is being extended to all activities and not only to the engineering areas.

Complementarily there is a tendency to establish partnership links between companies, namely between big companies and networks of components' suppliers. Similar agreements are being established between companies and universities. Such network structures may be seen as extended or virtual enterprises.

This tendency creates new scenarios and technologic challenges, specially to Small and Medium Enterprises (SMEs). Under classical scenarios, these SMEs would have big difficulties -- due to their limited human and material resources -- to have access or use state of the art technology. Such partnerships facilitate the access to new technologies and new work methodologies but, at the same time, impose the use of standards and new quality requirements.

In terms of the IS, this new situation requires the definition of common models (sometimes the use of common tools). Standards, like STEP, are expected to play an important role in such inter-enterprises cooperation.

The efforts being put on the implantation of high speed networks (digital high ways), supporting multimedia information, open new opportunities for team work in multi-enterprise / multi-site networks. But this new scenario also brings new requirements in terms of control: access rights to the information, scheduling of access, control of interactions, etc. (distributed information logistics infrastructure).

Integration perspectives. Taking into account the scenario described above, the definition of a platform for Concurrent Engineering involves, in our opinion, three related sub-problems:

i) Definition of common models. This is a basic requirement in order to enable communications between members of the engineering team. The adoption of common modeling formalisms is a first requirement. Formalisms like IDEF0, NIAM, Express/ Express-G, Petri nets are being widely used. The consolidation of STEP may help in terms of product modeling, but many other aspects not covered by STEP have to be considered, like process and manufacturing resources modeling. MANDATE seems still far from offering usable results. Business Processes modeling, as proposed by CIMOSA (Esprit 1989), is also contributing to facilitate dialogue.

ii) Engineering Information Management. Definition of integrating infrastructures and information management systems able to cope with the distributed and heterogeneous nature of CIM, has been the subject of many research projects from which various approaches and prototypes have been proposed in last years. Management of versions, a difficult problem in engineering data management, is even more complex when different versions may be produced / explored in parallel / concurrent way. Various centralized and decentralized solutions have been experimented, the concept of federated architectures developed and the issue of interoperability between different data management technologies and standards has been pursued. The need for a more mature technology for Engineering Information Management, combining features from Object Oriented and Knowledge Based Systems, Concurrent / multi-agent systems, is becoming evident.

iii) Process supervision. To build a platform that supports concurrent engineering it is not enough to guarantee that the various computer-aided tools used by a team are able to communicate and share information. In other words, it is not enough to provide an integrating infrastructure and to normalize information models. Even though these aspects are essential, there is also the problem of coordination. It is necessary to establish a supervision architecture that controls or moderates the way and time schedule under which computer-aided tools (team members) access the infrastructure and modify shared information. In other words, it is necessary to model the engineering processes and to implement a process interpreter or supervisor.

The platform for integration and concurrent engineering -- CIM-FACE: Federated Architecture for Concurrent Engineering -- being developed at New the University of Lisbon addresses these three issues.

2. THE CIM-FACE ARCHITECTURE

Figure 1 illustrates the main blocks of the CIM-FACE architecture. Application (computer-aided) tools are integrated via the integrating infrastructure, which provides access to the common IS.

One important part of the CIM-FACE prototype is the EIMS (Camarinha 1993) (Osório 1994) (Figure 2) subsystem which provides basic information management functionalities as well as an integrating infrastructure to support the connection of a federation of heterogeneous software tools. The implemented EIMS prototype, inspired on the developments of STEP / Express (Schenck 1994), is based on a hybrid and distributed programming environment supporting the connection of tools implemented in UNIX and MSDOS environments. The integrating infrastructure supports two connection modes: tight and loose connection. For tight connection a library of Information Access Methods (IAM) is linked to each tool, thus hiding the communication details (RPCs, messages format, etc.). As this part of the system was developed in an earlier stage, before the availability of the STEP SDAI specification, the IAM methods don't follow the standard, although they are quite similar, as inspired by EXPRESS concepts.

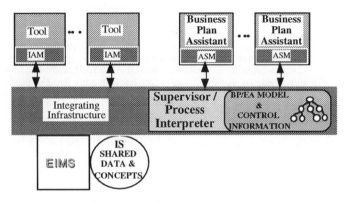

Figure 1 CIM-FACE Architecture.

For loose connection mode, a STEP-port based on neutral files interchange is included. Reactive programming was used as an effective mechanism to implement interoperability between different data management technologies. The "visible" part of EIMS is implemented on a KBS (Knowledge Craft), but links to a RDBMS or CAD DB were established to offer object's persistency. We think the interoperability mechanisms will play a role in the migration of legacy systems to more advanced Engineering Information Management Systems.

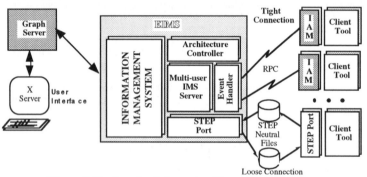

Figure 2 Engineering Information Management System.

Another major component of CIM-FACE is the Supervisor or Process Interpreter. The objective is to provide a framework that allows for a set of autonomous agents -- computer aided tools and engineers -- to cooperate in solving complex tasks.

Special tools - Business Plan Assistants - provide a front end to the human experts allowing them to take part in the execution of the business processes.

Each agent may have its own data models (partial views of the world) and its decision making capabilities. For instance, the time window within which a CAxx tool is active, interacting with the federation, and the kind of interaction is decided by the couple "CAxx tool - human user". A global supervision system -- "federal government" -- can impose some rules regarding the interaction, like refusing it if some pre-conditions are not satisfied, but it cannot consider the agents as obedient "slaves". On the other side, as agents are supposed to cooperate, they are not completely independent from each other. For instance, the actuation of a CAPP agent depends on the existence of a product model generated by a CAD agent.

3. THE ENTERPRISE FUNCTIONAL VIEW

Business processes. In order to implement a control strategy it is necessary to be able to model the dynamic behavior of the system being controlled. In our current experiment, the dynamic behavior of the enterprise is modeled by hierarchies of Business Processes and Enterprise Activities, according to CIM-OSA. In each level of the hierarchy, a Procedural Rule Set (PRS) defines precedence constraints between BP or EA of that level as well as their starting (firing) conditions. This hierarchical structure are called business plan.

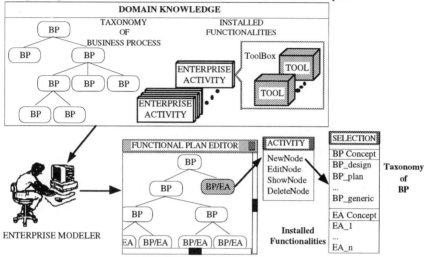

Figure 3 Edition of the enterprise's business plan.

To support the edition of business process plans in the operational phase, we consider some domain knowledge, including a taxonomy of classes of Business Processes and a Catalog of installed functionalities (Enterprise Activities, i.e., functional model of a particular enterprise).

```
{{ PlanNode:
  Identifier :
  Name :                          {{ BusinessProcess:
  InputConcepts :                   is-a : PlanNode
  OutputConcepts :                  SubObjectives :
  PrevActivities :                    . . .
  NextActivities :                }}
  UserRules :
  StartActivity :      ; methods
  EndActivity :
  ShowStatus :                    {{ Enterprise-Activity:
  ShowSubPlan :                     is-a PlanNode :
  SendMsg :                         ToolBox :
  ReadMsg :)                          . . .
  ... }}                          }}
```

Figure 4 Basic concepts in a business plan.

An EA model includes the list of supporting tools, i.e. Implemented Functional Operations in the CIM-OSA terminology. In current stage, the business plan is edited by a human, resorting to the domain knowledge previously defined for the particular enterprise.

```
{{BP012:
   Type : BP
   Instance : BP_design
   Name : 'Design P1'
   Input :
   Output : Product
   PrevActivities :
   NextActivities :
   UserRules : [all_concluded(Self),
            version(Self, less_than(5)), ...]
   SubObjectives : [EA123, EA124, EA125]
   Initializer : [create_instance(Product,
            Name, NewVersion)]
}}
```

```
{{EA123:
   Type : EA
   Instance : Product_Rough_Design
   Name : 'RoughDesign_P1'
   Input :
   Output : RoughDesign
   PrevActivities : []
   NextActivities :  [EA124]
   UserRules : [attribute_value(RoughDesign,
                  material, metal), ...]
   ToolBox: [CAD1]
   Initializer : [create_instance(RoughDesign,
                  Name, NewVersion)]
}}
```

```
{{EA124:
   Type : EA
   Instance : Product_Design_Analysis
   Name : 'DesignAnalysis_P1'
   Input : RoughDesign,
   Output : DesignConstraints
   PrevActivities : [EA123]
   NextActivities : [EA125]
   UserRules : [attribute_value(RoughDesign,
                  material, metal), ...]
   ToolBox: [PDA1]
   Initializer : [create_instance(DesignConstraints,
                  Name, NewVersion)]
}}
```

```
{{EA125:
   Type : EA
   Instance : Product_Design
   Name : 'Design_P1'
   Input : [RoughDesign,
            DesignConstraints]
   Output : Product
   PrevActivities :  [EA124]
   NextActivities : []
   UserRules : [attribute_value(Product,
                  weight, less_than(23)), ...]
   ToolBox: [CAD1, FEA1]
   Initializer : []
}}
```

Figure 5 Examples of BP and EA concepts.

The system also has a model of each tool; describing its inputs, outputs, functionalities, etc.

Petri nets. Another important formalism to model system's dynamic behavior, being extensively used in manufacturing systems is Petri nets (in its various derivations) (Zhon 1993). In order to take advantage of tools and methods available for Petri nets, namely for qualitative and quantitative analysis, attempts to derive PN models from BP/EA/PRS are justifiable. In a first experiment, we started from a hierarchy of BP/EA, as a "global plan" (external view) that describes the intended behavior of the agents with respect to the federation, and automatically derived a PN representation from it (Camarinha 1994) (Osório 1994). The control structure, i.e. the process interpreter was then synthesized from this PN. Our system implements a kind of *loose control* that acts more in terms of preventing/enabling situations, but the actual events (completion of actions) are decided externally. The user is assumed to "inform" the control system, via the special tools "business plan assistants", about the conclusion of activities, as it will be explained in next chapter. Nevertheless, some control rules can then be used to check whether "typical" consequences of that activity have been achieved or not.

However, this experiment showed that event condition Petri nets cannot easily represent the full semantic of PRS. A more appropriate tool could be Predicate Transition Nets, but this was not tested in CIM-FACE.

4. MODEL INTERPRETER

The model interpreter is responsible for the "execution" of the various enterprise's BP/EA plans. It is responsible for:
    -keeping track of the execution status of each plan (and each node in the plan)
    -supporting consistency maintenance (by verifying pre- and post-conditions for each BP/EA)
    -helping or directing tool selection for each EA
    -providing a common access to the enterprise information models
    -providing a platform for progressive improvement of the control structure (addition of new rules).

An important part of the execution environment is the set of CA-tools that - in cooperation with the various human experts - actually implement the EAs. A "protocol" is necessary to specify the interactions between each couple tool-user and the integrated federation. The "performer" of this protocol can be a layer separated from the applicative part of the tool. For legacy systems it is quite hard or nearly impossible to modify their control architectures. For new tools this "protocol performer" can be seen as a common script (library) that can be linked to the tool.

As a first attempt, a tandem structure (Figure 7) was implemented, separating the tool itself from the protocol performer, here called *tool assistant.*. From the implementation point of view, this tool assistant can be a module linked to the tool or even a parallel (detached) process. The second alternative is more suited to legacy systems.

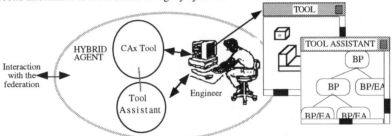

Figure 7 Tool assistant.

With basis on this initial approach, but taking into account that:
    -the "tasks" to be realized in a given business plan are hierarchically decomposed
    -"real" application tools appear only associated to the leaves of this hierarchy (EAs)
    -the human experts may interact with (inspect, start, finish, etc.) activities at various levels
    of the hierarchy
it seems more reasonable to have not tool assistants but ***business plan assistants***.

When a user "joins" the federation (logs in), a business plan assistant is launched. Through a graphical interface (Figure 8), the user has access to a specific business plan (or part of it, according to his access rights). For each node in the plan he can inspect its status, start or stop the BP/EA. When starting an EA, the associated application tools are launched.

Therefore, the human expert plays an important role in the evolution of the business plan execution.

As the control events are decided externally, by the human, how can the process interpreter be sure the decision was appropriate? For instance, lets suppose that current user of tool T_i informed the system that he has finished a generation of a process plan for a given production task. Should the control system simply accept such information as a fact or should it be cautious and try to investigate the accuracy of the information? Therefore different "kinds" of control systems can be defined, ranging from a totally confident system to a cautious one.

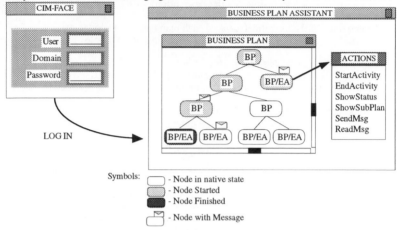

Figure 8 Business plan assistant interface.

In some cases, and for some application domains, it will be possible to define a set of verification rules to test the validity / accuracy of each access protocol action issued by business plan assistants. In other cases that might be difficult. Therefore, our proposal is to have an architecture that can start from a level of total confidence and progress towards a more cautious system once verification rules are added to its control knowledge base.

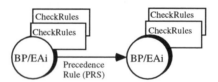

Figure 9 Check rules associated to BP/EA.

An example for a minimal checking level is a rule that implements a CIM-OSA PRS rule:

```
IF <all previous activities are finished> and <Self was requested to start>
THEN   <enable execution of Self>
```

A more specific example can be associated to the BP_Design_P1 of type BP_design:

```
IF <MarketEvaluation_p1 of type MarketEvaluation was started> and
     <BP_Design_P1 was requested to start>
THEN <enable execution of BP_Design_P1>
```

For instance the following rule could be attached to the EA124 (Figure 5):

```
IF <RoughDesign.verified of EA123 Output is true> and
     <EA123 is inactive> and <Self was requested to start>
THEN <enable its execution>
```

It shall be noted, however, that rules associated to the nodes (BP/EA) of a business plan are related to the coordination / interaction level of the plan execution. Other kind of rule, related to information consistency, are more appropriately defined in association to the concepts present in the IS. See, for instance, the following example using Express notation. This class of rules are supposed to be checked by the Information Management System and not by the Process Interpreter.

```
ENTITY manufacturing_step;
   identification : IDENTIFIER;
   manpower_time : INTEGER;
   machine_time : INTEGER;
   cycle_time : INTEGER;
   . . .
WHERE
   rule_1: cycle_time >= machine_time;
   rule_2: cycle_time >= manpower_time;
END_ENTITY;
```

Messages. As various BP/EA nodes may be active (in parallel), "operated" by different humans, possibly in different geographical locations or a change in one may affect the progress of others, it is important to provide a message exchange mechanism. This can be seen as a particular e-mail facility associated to the plan nodes. From one BP/EA node, the human agent can send a message to another BP/EA (or to a group). See (Figure 8). At current stage, these messages are intended for communication among the human experts, but we may even think of messages to be processes by high level (autonomous) process execution. In this way, if an agent modifies the model of a part, he can notify the agent performing the process plan, which may need to revise his plan.

Contexts and versions. As mentioned before, the various agents involved in the "execution" of a business plan have access to a common IS. As there might be various business plans evolving concurrently, it is convenient to organize the population of object instances in the IS by groups associated to the respective business plan. The notion of context is implemented to support this structure. A context is a "subspace" containing all object instances created / manipulated by a business plan (Figure 10).

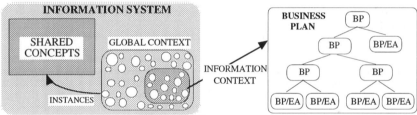

Figure 10 Context concept.

A more difficult topic is the representation of versions. If one agent (human) decides to re-start a BP and create a new version (of the product model, for instance), which information should be "copied" into the new context? How to identify the potential inconsistencies in the results generated by the following activities, and that are based on the previous version?

This is still a topic being developed in CIM-FACE.

5. CONCLUSIONS

CIM-FACE represents a prototype federated architecture to support systems integration and concurrent engineering activities.

The implemented parts include:

i) an infrastructure for information integration and management that was successfully evaluated in the context of an European Esprit project.

ii) A process interpreter that provides basic functionalities for concurrent execution and coordination of a set of business processes / enterprise activities, driven by a team of human experts. The proposed architecture supports the definition of multiple checking levels, allowing progressive degrees of robustness.

An intensive evaluation of this control strategy in a real application has still to be done. CIM-FACE is not a finished system, but an ongoing research. For instance, the integration of different versions in this architecture is a topic of current research.

Acknowledgments

The work here described received partial support from the European Community -- the Esprit CIM-PLATO, ECLA CIMIS.net and FlexSys projects -- and from the Portuguese Agency for Scientific and Technologic Research (JNICT) -- the CIM-CASE project.

6. REFERENCES

Barata, J. A.; Camarinha-Matos, L.M.; Rojas, F. -Dynamic persistence and active images for manufacturing process, Journal Studies on Informatics and Control , vol.3, n. 2-3.

Camarinha-Matos, L.; Sastron , F. - Information Integration for CIM planning tools, CAPE'91 - 4th IFIP Conference on Computer Applications in Production and Engineering, Bordeaux, 10-12 Sep, 1991.

Camarinha-Matos, L.M.; Osório, A. L. (1993) CIM Information Management System: An Express-based integration platform, IFAC Workshop on CIM in Processes and Manufacturing Industries, Espoo, Finland - published by Pergamon Press.

Camarinha-Matos, L.M.; Pinheiro-Pita,H.J. (1993) Interactive planning in CIM-CASE, *Proceedings of IEEE Int. Conf. on Robotics and Automation*, Atlanta, USA, 2-7 May93.

Camarinha-Matos, L.M.; Pita, H.; Osório, L. (1993) Hybrid Programming Paradigms in CIM-CASE, *Proceedings of the IFIP/IFAC Worksing Conference on Knowledge Based Hybrid Systems in Engineering and Manufacturing* [IFIP Transactions B-11, North-Holland], Budapest, Hungary, 20-22 Apr 93.

Camarinha-Matos, L.M.; Pita, H.; Rabelo, R.; Barata, J. (1995) Towards a taxonomy of CIM Engineering Activities, Int. Journal of Computer Integrated Manufacturing [to appear in 95].

Camarinha-Matos, L.M.; Osorio, A.L. (1994) An integrated Platform for Concurrent Engineering, Proc.s 3rd CIMIS.net Workshop on Distributed Information Systems for CIM, Florianopolis, Brazil. To appear in the Journal of the Brazilian Society of Mechanical Science.

Camarinha-Matos, L.M.; Afsarmanesh, H. (1994) Federated Information Systems in Manufacturing, Proceedings of EURISCON'94, Malaga, Spain.

Esprit Consortium AMICE (1989) Open System Architecture for CIM, Springer-Verlag.

Osório A. Luis; Camarinha-Matos, LM. (1994). Information based control architecture, Proceedings of the IFIP Intern. Conference Towards World Class Manufacturing, Phoenix, USA, Sep 93, edited by Elsevier - North Holland.

Schenk, D.A.; Wilson, P.R. - Information modelling: The EXPRESS way, Oxford University Press, 1994.

Steiger-Garção A.; Camarinha-Matos, L.M (1992) Design of a knowledge-based information system, in Integration of Robots into CIM [Ed. Bernhardt, Dillman, Hörmann, Tierney], Chapman & Hall, Cap. 21 & 22, 1992 .

Schenck, Douglas; Wilson, Peter (1994), Information Modeling the EXPRESS Way, Oxforf University Press, New York.

STEP, ISO (1991). Reference Manual, ISO/TC 184 /SC4.

Welz, B. G. et al. (1993). A toolbox of integrated planning tools - a case study, IFIP Workshop on Interfaces in Industrial Systems for Production and Engineering, Darmstadt, Germany, 15-17 Mar 1993.

Zhon, Mengchu; DiCesare, Frank (1993) Petri net Synthesis for Discrete Event Control of Manufacturing Systems" - Kluwer Academic Publishers.

63

Graphic environment for virtual concurrent engineering

Fei Zheng   Shanghui Ye   Mei Chen
Department of Electromechanical Engineering
Xidian University, Xi'an, P.R. China

Abstract

A product description model for CE is proposed from the point of view of life cycle of product in this paper. A graphic environment is set up based on the product description model. A method to combine shell model and voxel model through sweeping transformation is achieved in technical realization, so as to supply strong and convenient graphic supports for the software environment.

Keywords

Concurrent engineering, graphic environment, product development

1 INTRODUCTION

Concurrent engineering (CE) is a system method for integratively and concurrently designing product and various processes corresponding to it, including manufacturing processes and supporting processes. Such a method demands that all factors through the life cycle of product, from formation of concept to treatment of scrapped product, including quality, cost, schedule and requiments of customers, are considered by product developers at the begining of design (Karwowski,1990). Since its rising from 1980s, CE has been paying great attention to both military circles and industrial circles. CE is a complex system engineering, it is more complicated, more active and more uncertain than CIMS (Kusiak,1990). Therefore there are no concrete theories and ripe modes in studies of CE at present. CEs put into effects by some enterprises are still local and limitary. Further and deeper studies in CE are expected for Carring out.

Scientific visualization and virtual reality are two new thchnologies developing rapidly at present. In today's science and technology, computer experimenting method is going to be the third scientific field which is parallel to the traditional theory method and experimenting method (Greenwood, 1992). So, scientific visualization and virtual reality, which are the two direct computer experimenting thchnologies, should be effectively used in even wide and concrete fields.

It should be necessary and effective to study such a complex system as CE by using computer experimenting method. But such studies are not much. An attempt is taken in this paper.

2 PRODUCT DESCRIPTION MODEL

To study CE effectively, some concepts must be understood clearly, such as space–time, stage, level, virtual and practical. CE cannot change actual space positions of product for too much, but can fairly short times for useless transmissions in certain degrees. There are many stages in CE that cannot be changed in sequence. Many levels exists which cannot be described clearly and easily in simple and short manners, either using top–down method or bottom–up method. CE can only avold many mistakes and repeatations by simulating in virtual stages, so as to guarantee the success of practical manufacturing in only once.

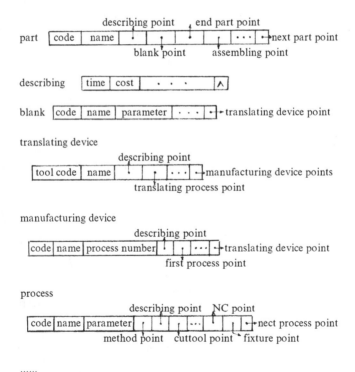

Figure 1   A product description model.

From the point of view of mathematics, CE is a multi—object, multi—value and multi—level dynamic nonline system. It would take a long time to study such a complex system. Based on the above concepts, described from processes, which would be fundamental and effective by our consideration, a product description is proposed as figure 1 (Zheng,1994).

The establishment of such a product model is corresponding to the processes of practical product development, which is taken from general to specific, from obscure to clear. It is expended dynamically according to the practical development processes of product. Sequent design stages and concurrent design stages are naturally divided. Level data structures are used in sequent design stages and network data structures are used in concurrent design stages. Dynamic chain—table forms are used in parallel levels, multi—point forms are used in different levels. Therefore it is corresponding to processes of practical refining and information expending.

Such a product model can be expended dynamically by various specific developers according to the practical processes of product development. It is the inner description form of the graphic environment to be discussed below. It effectively supports the outer description form of the graphic environment.

3 GRAPHIC SOFTWARE ENVIRONMENT

Though computer graphics have being widely used in CAD, these applications are limited in shape designing of parts with determined schemes, or in designing purely from the point of view of shapes (Requicha,1992). This is the fundamental reason that there exist gaps among CAD, CAPP and CAM.We here attempted a graphic environment from the point of view of life cycle of product, so as for developers of product to be able to cooperate at a unitary environment with unitary informations.Such a graphic environment takes the general graphic layout of the corresponding enterprise for its main menu.It uses icons to indicate relative items, each item represents a pop—up window. Such a window describes detail contents of the item. It may also a general graphic layout of a department, its items are also responsable for other pop—up windows, and so on.

The graphic environment supports decomposing processes of function and structure in product developing. For example: regarding a product as a general function block. After it has been decomposed to sub—function blocks as power block, transmission block, fixture block, cycle block and so on. Each sub—function block can also be decomposed to next lower level function blocks. At the same time all the corresponding developers can take part in the detail designs, and so on. Decomposition can be carried out till to designs of specific parts. Multi—level computers connected with network could be used to support gradual decomposing of tasks of the product development. Then personals as many as possible could commonly participate in the product's design, evaluation and checking.

The graphic environment also supports designs of parts with manufacturing simulating. It uses objective graphic informations to contact relationships of arrangements and managements of materials, equipments, personals, man—hours and so on, to share informations of process operations. Then developers of CAD, CAPP and CAM can be connected naturally to cooperate. Therefore, various problems, which would probably arised in the practical processes of developing and manufacturing, could appeared as early as possible, and could be solved just in time. Thus concrete and effect utilization of CE could be given security. To fit for various conditions in engineering and to give full plays to developers' creatives, design with free–form sculpture and reverse operations are also provided.

Multi–level assembling operations, which are reverse to decompositions of function and structure, are also supplied in such a graphic environment. Hence decompositions of function and structure can be checked and revised, relative informations for arranging assemblies and managements are supplied. The following figures are realized by such a graphic environment using solid model. Figure 2 (a) corresponds to the assembled state of a belt–whell structure, Figure 2 (b) corresponds to its disassembled state.

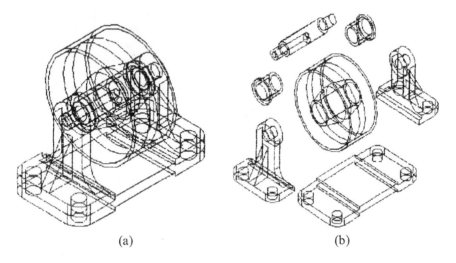

(a)   (b)

Figure 2   A belt–whell structure.

Such a graphic environment would supply objective and convenient operations for virtual CE.

4   TECHNICAL REALIZATION

To practically build up a prototype system of the graphic environment for virtual CE proposed above, from the point of view of technical realizations, two problems must be solved ahead: firstly to set up a suitable general model of product description from the point of view of life cycle of product development, such a problem is stated above; secondly to deal with the corresponding relationship between operating of graphics and dynamic expending of such a model. Here analysis of graphic treating method of such corresponding relationships is given emphasis to.

Traditional primitive mixing modeling method has no direct corresponding relationship to the description method of practical design and manufacturing of product. There is also no direct corresponding relationship between processes of primitive mixing modeling and processes of manufacturing of products. That is to say solid model is not enough.

We analysised various manufacturing methodes such as maching, milling, grinding, drilling, cutting, shaping, boring and so on. They can be classified to two methods as reciprocation and rotation. They can be corresponded to sweeping and swinging bodies from the point of view of solid modeling. Moreover, if 3D bodies are described by voxel, then processes of manufacturing of product can also be correponded to modeling processes of product. Objectively speaking, voxel is the unitary atom to form 3D bodies, it is generally described as unitary cube. Concrete contents of volume graphics kerneled with voxel can be obtained in reference (Kaufman,1993).

We call traditional geometric model which is based on shell description as shell model, so as to be distinguished with voxel model. Both of them exist advantages and disadvanges. They all are necessary in practical realization of the graphic environment for virtusl CE. We are developing a method of combining shell model and voxel model through sweeping transforming, so that both of their advantages can be given full play to. Then effective graphic support and convenient interface operation can be supplied for the practical realization of such a software environment for virtual CE.

We have made deep researches in sweeping transforming methods. Various effective and convenient sweeping transforming methods have been realized. Many simple or complex 3D objects can be formed quickly and simply through only one sweeping (Zheng,1992).

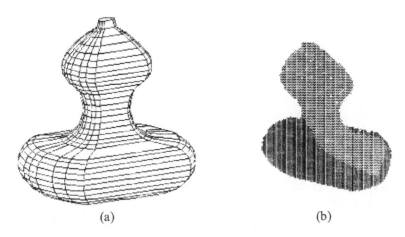

(a)   (b)

Figure 3   Shell model and voxel model.

We realize corresponding relationships between shell graphics and voxel graphics by changing sweeping formed shell primitive to voxel primitive: simple or complex shell primitive can be formed conveniently and quickly through sweeping transforming, shell primitive can be easily transformaed to voxel primitive. Shell primitive can be used to realize various functions of traditional geometric modeling, and can be used to precisely describe artifacts; voxel primitive can be processed operations such as cutting, pasting, translating, rotating, zooming and so on, with the aim of its corresponding shell primitive, and to describe inner parts or to simulate processes. Then strong and convenient graphic environment can be obtained.

Figure 3(a) is a shell model formed by just one sweeping through tracing transforming method. Figure 3(b) is the corresponding voxel model of Figure 3(a), it is displayed with simple lighting mode. All of them are realized in personal computer environment.

5   CONCLUSIONS

CE concerns with various factors in all the life cycle of product. The product description model set up above is only a basic and primary model considered from the point of view of product design. The graphic environment based on it is also only a re—study experimenting environment for CE. There are still no concrete theory and ripe mode in studies of CE. So it would be significant to set up such a product description model and to process computer experimenting with corresponding graphic environment. Moreover, a method to combine shell model and voxel model through sweeping transforming are proposed from the point of view of technical realization, so as to supply strong and convenient graphic supports for its practical realization. Further researches are still under developing.

6   ACKNOWLEDGEMENT

This work has been supported by Fundation of Xidian University.

7   REFERENCES

Greenwood,A. et al. (1992) Science   at the frontier,   National Academy Press, Wassington, DC USA.
Karwowski,W. and Rahimi,M. (1990) Ergonomics of hybrid automated system II, Elsevier Science Publishers B.V.
Kaufman,A. Cohen,D. and Yagel,R. (1993) Volume graphics. Computer,July.
Kusiak,A. (1990) Intelligent manufacturing system, Prentice Hall.
Requicha,A.A. (1992) Solid modeling and beyond. IEEE computer graphics and applications, vol.12, no.9.
Zheng,F. (1992) Study of sweep representation modeling in computer 3D graphics, Xidian University, MS thesis (Chinese).
Zheng,F. Ye,S.H. and Chen,M. (1994) Computer aided graphic environment in product development. The third international conference in system integration. Brazil.

8 BIOGRAPHY

Fei Zheng is a lecturer at Xidian University. His research interests include computer graphics, computer aided design, scientific visualization and modern industrial design. **Shanghui Ye** is a professor at Xidian University. His research interests include structural optimization, geometrical modeling and computer aided mechanical design. He is a fellow of Chinese Institute of Electronics. **Mei Chen** is an engineer of Research Institute of Space Radio Technology at Xi'an. Her research interests include computer aided design, digit image processing and scientific visualization.

PART FOURTEEN

Neural Network Applications

64

Constructing Efficient Features for Back Propagation in Visual Inspections

Y.H. Chen
Department of Mechanical Engineering
The University of Hong Kong
7-29, Haking Wong Bld, Pokfulam Road
Hong Kong
Tel: (852) 859-7910 Fax: (852) 858-5415

Abstract

Back propagation(BP) is one of the most widely used artificial neural networks(ANNs) because of its powerful problem solving capability and simplicity. However, one of the inherent drawbacks of the BP algorithm is its slow convergence speed. When applied to visual inspections in machine vision, the large amount of the image data may make the convergence of the BP algorithm unpredictable.

This paper presents an application of the BP algorithm for automatic IC mark inspection in the semiconductor assembly environment. Before applying the BP algorithm, feature construction starts by preprocessing the IC mark image through run-length coding, or the quadtree subdivision methods. By rearranging the constructed features to a form appropriate to the input of BP algorithm, the convergence speed can be increased significantly. This in turn makes the automatic in-line IC mark inspection possible.

Keywords

Computer vision, neural network, visual inspection

1 INTRODUCTION

Machine vision systems have been used in a wide range of visual inspection applications (Gonzalez,1986). Recent applications employ advanced techniques such as Artificial Intelligence (Perner,1994) and Fuzzy Logic (Chen,1994). This paper discusses an application in the semiconductor industry where ICs (Integrated Circuit) need to be marked after the moulding process. A typical mark as shown in Figure 1 normally includes company logo, company name and the IC model number. The usual mark defects are broken characters, contaminated mark, smeared marks, upside-down, mis-aligned mark, etc. In-line visual inspection is practised in marking process because defective marks can be easily rectified before being cured in the next process. To automate the in-line mark inspection, a machine

Figure 1 A typical IC with marks.

vision system has been used. Back propagation is implemented for the mark inspection.

Because of the slow convergence speed of the BP algorithm, it is not practical to use the pixel data as direct input to the BP. Researchers have approached this problem through disparate means ranging from modifications to the BP algorithm (Jacobs,1988) to hardware implementation of the algorithm in VLSI component (Kung,1990). Another method of increasing the convergence speed of the BP algorithm is through feature construction (Hecht-Nielsen,1989). Feature construction is often used to preprocess the data before being used by BP. Feature construction is a methodology by which newer features can be constructed from the original set of features, resulting in a less complex representation. The number of features that are required for learning concepts is effectively reduced, while also reducing the number of peaks that are present in the feature space. The reduction in the complexity of the feature space is very beneficial for learning algorithms such as BP. As the number of peaks are reduced in the feature space, the number of hyper-planes that are required to separate examples belonging to different classes are drastically reduced, thus improving the learning process.

2 FEATURE CONSTRUCTIONS

In automatic visual inspection, a part image is usually taken under a constrained condition. For inspection, one normal approach is by matching a part image to a template image pixel by pixel (Gonzalez,1986). This method requires the process of a large amount of data which means a long lead time. In order to solve this problem, data compression techniques are used in the image preprocessing stage to construct an input pattern for a neural network.

In the proposed application, inspection systems are required to be fully automatic and in-line with marking machines. A flowchart of the inspection process is shown in Figure 2. IC mark image is captured immediately after the marking process. The system set-up of the inspection system is similar to conventional visual inspection systems (Hedengren,1989). The principle contribution of this paper is the study of image feature construction through two potential image compression techniques namely run-length coding and quadtree method. Further more, the constructed image features are used as the input for a BP network which is trained to give inspection results. Since run-length coding and quadtree methods are applicable to all image data compression applications, the proposed method has the potential to be used as a basis for implementing an automatic visual inspection system which can be used for all 2D visual inspection purposes.

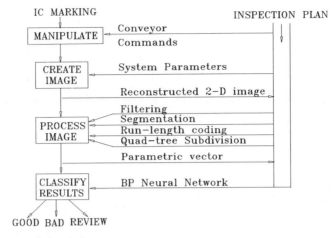

Figure 2 The automatic inspection process.

Feature Construction through Run-Length Coding

Run-length coding is normally used to compress a binary image data. This method exploits the fact that along any particular scan line there will usually be long runs of zeros or ones. Instead of transmitting the individual bits, we record numbers indicating the length of such runs. The run-length code for the image line

0	1	1	0	1	1	1	0

is stored as a linked list {1,2,1,3,1}. In the inspection system presented in this paper, each scan line starts with a zero. For an image under inspection, there may be **n** scan lines which are represented as linked list L_1, L_2,...,L_n. The linked lists with maximum length is found by $N_{max}=\max\{\text{len}(L_1), \text{len}(L_2),...,\text{len}(L_n)\}$. During network teaching, N_{max} is stored as the number of BP input features. For linked lists whose lengths are less than N_{max}, zeros are inserted at the head of lists to ensure that all lists have a uniform length. This addition does not affect the performance of the network. It is easily seen from the feed forward summation

$$S_j = \sum_i a_i W_{ji}$$

where a_i=activation level of unit i, and W_{ji}=weight from unit i to unit j.

The run-length code is a complete representation of an image. Let r_{ik} is the kth run of the ith line. Suppose further that there are m_i runs on the ith line. The area of the image is just the sum of the run lengths corresponding to ones in the image:

$$A = \sum_{i=1}^{n} \sum_{k=1}^{m_i} Y_{i,2k}$$

The center of image can be found through horizontal and vertical projections. Suppose h_i is

the ones in a scan line, then

$$h_i = \sum_{k=1}^{m_i} Y_{i,2k}$$

From this we can easily compute the vertical position y of the center of area using

$$A.x = \sum_{i=1}^{n} i\, h_i$$

It is difficult to compute the vertical projection directly from the run lengths. Instead consider the first difference of vertical projection.

$$\overline{v_j} = v_j - v_{j-1} \text{ with } \overline{v_1} = v_1$$

From this first difference we can compute the vertical projection itself using the simple summation

$$V_j = \sum_{l=1}^{j} \overline{v_l}$$

Given the vertical projection, we can easily compute the horizontal position x of the center of area using

$$A.x = \sum_{j=1}^{m} j\, v_j$$

200, 200, 200, ... 35 , 11 , 108, 11 , 35 ... 29 , 19 , 104, 19 , 29 ... 11 , 41 , 96 , 41 , 11 ... 11 , 178, 11 ... 11 , 41 , 148 ...	200, 0, 0, 0, 0 200, 0, 0, 0, 0 200, 0, 0, 0, 0 ... 35 , 11 , 108, 11 , 35 ... 29 , 19 , 104, 19 , 29 ... 11 , 41 , 96 , 41 , 11 ... 11 , 178, 11 , 0 , 0 ... 11 , 41 , 148, 0 , 0 ...	

(a) Original image   (b) Run-length code   (c) Rearranged code
Figure 3 Feature construction by run-length coding

Figure 3(a) shows a partial image of a PCB routing. The image is preprocessed to enhance the quality. Run-length coding starts to build a series of line codes as shown in Figure 3(b).

However, the lines of code do not have uniform length. Therefore, the run-length code need to be rearranged. In the proposed implementation, the code is nomalized by adding zeros in the front of all scan lines where the code length is less than the maximum code length as shown in Figure 3(c). The normalized code is then used as input features for a BP neural network.

(a) Image subdivision by quadtree method

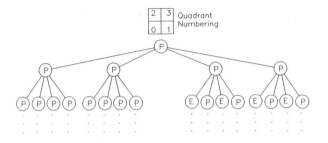

(b) A partial quadtree data structure for the mark in (a)
(E=Empty, P=Partially ful)

Figure 4 Feature construction through quadtree method

Feature Construction through Quadtree Subdivision
The fundamental idea of quadtree is the divide-and-conquer power of binary subdivision. A quadtree is derived by successively subdividing a 2D image in both directions to form quadrants. Each quadrant may be full, partially full, or empty, depending on how much of the quadrant intersects the image area. A partially full quadrant is recursively subdivided into subquadrants. Subdivision continues until all quadrants are homogeneous, or until a predetermined cutoff depth is reached. Figure 4(a) shows a mark image under 2^6 subdivisions. The internal data structure of a quadtree is shown in Figure 4(b). For visual inspection problems, image compression must be achieved without any lost of details. In order to meet this requirement, the recursive subdivision process of the Quadtree method can only be terminated at pixel level. The construction of such a Quadtree is time consuming. Further more, the Quadtree structure is not the desired format for neural network input. Therefore, Quadtree method is not chosen for this application.

3 IMPLEMENTATION BASED ON RUN-LENGTH CODING

For any visual inspection, reliable image acquisition is critical. In the mark inspection example, a window enclosing a mark image must be defined. This window is also called the Area of Interest(AOI) which is shown in Figure 5(a) as dashed line. All further image processing will only be performed within the AOI.

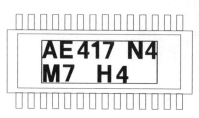

Max Column=25
Row=80

11,8,16,24,107,9,12,9,28
10,10,15,24,23,7,17,5,8,21,26,10,11,9,18,7,3
......
5,7,5,7,11,24,14,5,6,7,16,6,19,6,31,6,3,7,3,6,8,5,6,7,3
......
224
224
......
224
3,11,11,7,22,46,6,13,6,88
......
158,7,59

(a) The sample image   (b) Partial run-length code
Figure 5   A sample applicaton

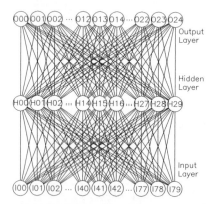

Figure 6 Network structure for the sample problem

In order to locate the mark image, a bounding rectangle box is defined as shown in Figure 5(a) as a solid rectangle within the AOI. This box is drawn based on a major axis (the major axis of an equivalent ellipse that has the same second moments of area as the mark image) and a minor axis(the minor axis of an equivalent ellipse that has the same second moments of area as the mark image). Run-length code is acquired by scan from the upper left corner of the box to the lower right corner of the box. Figure 5(b) shows part of the run-length code for the mark image in Figure 5(a). The run-length code is normalized by first adding zeros to the end of some rows so that all rows have equal length. Second, all the code is divided by the maximum code. After these two operations, a three layer neural network is constructed as shown in Figure 6. The number of neurons in the input layer is equal to the number of rows of the run-length code. In the output layer, the number of neurons is equal to the number of columns of the run-length code. The number of neurons in the hidden layer is chosen arbitrarily. In this example, 30 neurons are used for the hidden layer.

In the output layer, the target outputs are associated to the columns of the run-length code. When the first column is presented, the target output of the first neuron in the output layer is 1 and other neurons in the same layer must be 0. The second column of the run-length code requires the target output of 1 from the second neuron of the output layer and 0s from all other neurons. Similarly, the network responds with a 1 in the position of the column being presented to the network. All other values in the output layer should be 0.

When training the network, both learning rate and momentum are used. They are set to 0.55 and 0.45 respectively. After 3085 epochs, the sum-squared error of the network is reduced to 0.001 which is the preset value. After training, the network can easily identify inverted marks. In order to correctly identify all other defects, a proper threshold value of the output neurons must be defined.

4 CONCLUSIONS

This paper has presented a novel development of an automatic visual inspection system. Feature construction of an image is achieved through run-length coding. Quadtree subdivision is also discussed. The constructed features have a significant reduction of data as compared with the original image data. For inspection, a three layer BP neural network is constructed based on the feature pattern. After training, the neural network is capable of identifying major mark defects such as inverted mark and broken characters etc. The definition of threshold values of the output neurons is critical to achieve consistent inspection results.

5 REFERENCES

Chen, Y.H(1994), Computer vision for industrial inspection through linguistic fuzzy variable input. *20th International conference on industrial electronics, control, and instrumentation*, Bologna, Italy, **Vol. 2**, pp 1349-1353

Piramuthu, S.(1990) Feature construction for back-propagation. *Proc. internat. workshop on parallel problem solving from nature*, Springer-Verlag, Heidelberg, pp 264-268

Gonzalez, R.C. and Safabakhsh, R.(1986) Computer vision techniques for industrial inspection: a tutorial overview. *IEEE Computer*, pp 400-420, May, 1986

Hecht-Nielsen, R.(1989) Theory of the back-propagation neural network. *Proc. internat. joint conference on neural networks*, **Vol.1**, pp 593-606, June, 1989

Hedengren, K.(1989) Methodology for automatic image-based inspection of industrial objects. *Advances in machine vision*(J.L.C. Sanz Ed.), Springer-Verlag, pp 160-190, 1989

Jacobs, R.A.(1988) Increased rates of convergence through learning rate adaptation. *Neural Networks*, **Vol.1**, pp 295-307

Kung, S.Y., Vlontos, J. and Hwang, J.N.(1990) VLSI array processors for neural network simulation. *Journal of neural network computing*, pp5-20, **Vol. 2**, 1990

Perner, P.(1994) A Knowledge-based image-inspection system for automatic defect recognition, classification, and process diagnosis. *Machine vision and applications* **Vol. 7, No. 3,** pp 135-147,

Piramuthu, S.(1990) Feature construction for back-propagation. *Proc. internat. workshop on parallel problem solving from nature*, Springer-Verlag, Heidelberg, pp264-268

6 BIOGRAPHY

Yong-Hua Chen received the B.Eng degree in Mechanical Engineering from Southwest Jiao-Tong University, P.R.China in 1985. In 1991 he received Ph.D degree in Mechanical Engineering from the University of Liverpool, U.K. Since then Dr. Chen has worked in Motorola Electronics Pte (Singapore), Asia Matsushita (Singapore) and Swire Technologies Ltd (Hong Kong) as Sr. Automation Engineer, Sr. Development Engineer and Automation Manager respectively. Since late 1993, Dr. Chen has been a lecturer in the Department of Mechanical Engineering, The University of Hong Kong. His research interests include computer aided design, 3-D model-based vision and visual inspections.

65

A Neural Network Algorithm for Solving the Job-Shop Scheduling Problem with Priority

Jianfei Chen and Shaowei Xia
Dept. of Automation, Tsinghua University, Beijing 100084,China
National Laboratory of Pattern Recognition, Beijing 100080,China

Abstract

Hopfield-type neural networks have shown part success in solving hard optimization problem. As problem becomes more and more complex, it is increasingly difficult to find a suitable encoding of constraints into an energy function, which will effectively guide the network to a desired solution. This paper describes how the job-shop scheduling problem with priority is encoded into a modified Hopfield neural network, and solved. The neural network algorithm is easy to be virtually realised on digital computers. At the end an example is given.

Keywords

NP-hard, neural network, job-shop scheduling problem, priority

1 INTRODUCTION

The job-shop scheduling problem(JSSP) is a resources to tasks allocation problem. The resources are usually called machines, the tasks jobs. Each job may consist of several subjobs (refered to as operations) which subject to some precedence restrictions. The job-shop scheduling is a classical operations research problem with numerous applications but very few practical solution approaches. Due to the large number of constraints the problem is known to be NP-hard one, so that even a good (which may be not optimal) feasible solution is acceptable in most applications. The scheduling problem with priority is one important class among the scheduling problems, and this paper especially introduces the neural network algorithm for solving it.

2   FORMAL DESCRIPTION OF GENERAL JSSP

Let S_{ik} denote the starting time and t_{ik} the processing time for operation k of job i. The cost function will be the sum of the starting time of each job's last operation: $\sum_{i=1}^{n} S_{ik_i}$, where n is the number of jobs, k_i is the last operation of job i. The total waiting and processing time for all jobs equals $\sum_{i=1}^{n} S_{ik_i} + \sum_{i=1}^{n} t_{ik_i}$ where $\sum_{i=1}^{n} t_{ik_i}$ is a constant.

The precedence constraints are given by the following inequality:

$$S_{ik} - S_{i,k+1} + t_{ik} \leqslant 0, \qquad i=1, \cdots, n; \ k=1, \cdots, k_i - 1$$

The starting time should be positive, so

$$S_{i1} \geqslant 0, \qquad i=1, \cdots, n.$$

For any two operations (i, k) and (j, p) assigned to the same machine (where the first index denotes the job number and the second index the operation number), the following constraints need to be satisfied in order to avoid the overlap in time between these operations: $S_{ik} - S_{jp} + t_{ik} \leqslant 0$, if operation (i, k) is performed first; or $S_{jp} - S_{ik} + t_{jp} \leqslant 0$, if operation (j, p) is performed first.

In conclusion the mathematical formulation for the job-shop scheduling is described as follows (Zhou and etc. , 1990):

$$\min \sum_{i=1}^{n} S_{ik_i}$$

s. t.
$$S_{ik} - S_{i,k+1} + t_{ik} \leqslant 0, \qquad i=1, \cdots, n; \ k=1, \cdots, k_i - 1 \tag{1}$$
$$S_{i1} \geqslant 0, \qquad i=1, \cdots, n \tag{2}$$
$$S_{ik} - S_{jp} + t_{ik} \leqslant 0 \quad \text{or} \quad S_{jp} - S_{ik} + t_{jp} \leqslant 0 \tag{3}$$

In addition, there is another constraint,

$$S_{ik_i} \leqslant \text{expect}_i, \qquad i=1, \cdots, n \tag{4}$$

If job i has due date expect_i equals the due date, otherwise expect_i may take an enough large number.

3 PROPOSED NEURAL NETWORK FOR GENERAL JSSP

3.1 Energy function

The energy function of the neural network is chosen as follows:

$$E = \sum_i S_{ik_i} + \frac{1}{2}\sum_i\sum_k H_1 F(S_{ik} - S_{i,k+1} + t_{ik}) + \frac{1}{2}\sum_i H_1 F(-S_{i1}) + \frac{1}{2}\sum_i H_1 F(S_{ik_i} - expect_i)$$
$$+ \frac{1}{2}\sum_{ik}\sum_{jp} H_2 F(\min(S_{ik} - S_{jp} + t_{ik}, S_{jp} - S_{ik} + t_{jp})) \tag{5}$$

where H_1, H_2 are all big positive parameters, $H_1 \gg H_2$. $F(x) = \begin{cases} x^2, & x > 0 \\ 0, & x \leqslant 0 \end{cases}$

3.2 Motion equations

The motion equations of the neural network are stated as follows:

$$\frac{dS_{ik}}{dt} = -\frac{\partial E}{\partial S_{ik}} \tag{6}$$

$$\triangle S_{ik}(t) = \begin{cases} \min(2^q, LS_{ik}(t) - S_{ik}(t)), & \text{if } \frac{dS_{ik}}{dt} \geqslant 2^{q-1} w_{ik,ik} \\ \max(-2^q, -S_{ik}(t)), & \text{else if } \frac{dS_{ik,ik}}{dt} < -2^{q-1} w_{ik,ik} \\ 0, & \text{else} \end{cases} \tag{7}$$

$$S_{ik}(t+1) = S_{ik}(t) + \triangle S_{ik}(t) \tag{8}$$
$$i = 1, \cdots, n; \ k = 1, \cdots, k_i; \ q = 0, \cdots, b_m - 1$$

The processing times are all integer numbers; even not, they may be converted into integer numbers. S_{ik} can also be thought of as integer numbers, and b_m is the maximum bits of binary digit needed to represent S_{ik}. $LS_{ik_i} = expect_i - t_{ik_i}$, $LS_{ik} = LS_{i,k+1} - t_{i,k+1}$, $i = 1, \cdots, n; \ k = k_i - 1, \cdots, 1$.

The initial values of S_{ik} are selected as follows:

$$S_{i1} = 0, \ S_{ik} = S_{i,k-1} + t_{i,k-1}, i = 1, \cdots, n; \ k = 2, \cdots, k_i$$

Determination of $w_{ik,ik}$: Assuming the coefficient matrix of (1) is A, the coefficient matrix of (3) A_1,

$$W = H_1 A^T A + \frac{1}{2} H_2 A_1^T A_1$$

and $w_{ik,ik}$ is the ikth diagonal element (corresponding to S_{ik}) of W.

It may be proved that the function E in (5) will monotonously decrease if S_{ik} change according to the motion equations (6), (7) and (8).

3.3 'Hill-climbing' term

Rather than use a stochastic technique to enable the neural network to escape from local energy minima (which may correspond to infeasible solutions), a simple but effective 'hill-climbing' term was added to the motion equations, that is,

$$\frac{dS_{ik}}{dt} = -\frac{\partial E}{\partial S_{ik}} - \sum_{jp} H_3 h(\min(S_{ik}-S_{jp}+t_{ik}, S_{jp}-S_{ik}+t_{jp}))$$
$$\times select(S_{ik}-S_{jp}+t_{ik}, S_{jp}-S_{ik}+t_{jp})$$

where H_3 is a positive parameter, $H_3 > H_1$.

$$h(x) = \begin{cases} 1, & x > 0 \\ 0, & x \leqslant 0 \end{cases}, \qquad select(x,y) = \begin{cases} 1, & x \leqslant y \\ -1, & x > y \end{cases}$$

3.4 Implemented method

The value of H_3 is selected by trial and error, and is hard to get a suitable amount. If H_3 is too small the 'hill-climbing' term can not make effectiveness, if H_3 is too big the neural network will oscillate. Therefore in the implemented method we avoid the selection of the value of H_3, taking adjustment instead of 'hill-climbing' term. The rules for adjustment are:

if $S_{jp} < S_{ik} < S_{jp} + t_{jp}$
   if $LS_{ik} \geqslant S_{jp} + t_{jp}$ then $S_{ik} = S_{jp} + t_{jp}$
   else if $LS_{jp} \geqslant S_{ik} + t_{ik}$ then $S_{jp} = S_{ik} + t_{ik}$
if $S_{ik} < S_{jp} < S_{ik} + t_{ik}$
   if $LS_{jp} \geqslant S_{ik} + t_{ik}$ then $S_{jp} = S_{ik} + t_{ik}$
   else if $LS_{ik} \geqslant S_{jp} + t_{jp}$ then $S_{ik} = S_{jp} + t_{jp}$
if $S_{ik} = S_{jp}$ and $t_{ik} \leqslant t_{jp}$
   if $LS_{jp} \geqslant S_{ik} + t_{ik}$ then $S_{jp} = S_{ik} + t_{ik}$
   else if $LS_{ik} \geqslant S_{jp} + t_{jp}$ then $S_{ik} = S_{jp} + t_{jp}$

if $S_{ik} = S_{jp}$ and $t_{jp} < t_{ik}$

if $LS_{ik} \geqslant S_{jp} + t_{jp}$ then $S_{ik} = S_{jp} + t_{jp}$

else if $LS_{jp} \geqslant S_{ik} + t_{ik}$ then $S_{jp} = S_{ik} + t_{ik}$

In conclusion $S_{ik} (i = 1, \cdots, n; \ k = 1, \cdots, k_i)$ evolve in light of the motion e-quations $(6), (7)$ and (8), and are sequently adjusted by the above-mentioned rules. The evolution and adjustment are repeated until S_{ik} freeze not only in evolution but also in adjustment.

It may be proved that the implemented method will certainly converge and the solution obtained by it must be feasible.

4 ALGORITHM FOR THE JSSP WITH PRIORITY

In the JSSP with priority, the jobs have different priorities. The jobs may be scheduled in order according to their priorities (from the highest to the lowest). Moreover, when scheduling the jobs with low priority, the time table of the jobs with higher priorities is not admitted to be delayed. For the consideration, the algorithm for the JSSP with priority may consists of the following steps:

1. Assumpting LS_{ik} is the latest start time of operation (i, k), let $LS_{ik} = \infty$.
2. Schedule the job(s) i with the highest priority, and then let $LS_{ik} = S_{ik}$, $k = 1, \cdots, k_i$.
3. Acording to priority, schedule the job(s) i with lower priority, and then let $LS_{ik} = S_{ik}$, $k = 1, \cdots, k_i$.
4. If there is(are) job(s) unscheduled, return to step 3.
5. End.

5 NEUMERIAL RESULTS

A program to simulate the neural network algorithm was writen in C language on VAX 785. We tested many problems, and found that the neural network algorithm had been quite successful in finding valid schedules, and the solutions had been superior to those found by heuristic algorithms.

Table 1 gives a JSSP with priority. Figure 1, 2, 3 are the results found by the neural network algorithm, and the results have been known to be very good through comparision.

Table 1   A JSSP with priority

Jobs	Machine Process time			Priority
1	$\dfrac{M_1}{5}$ $\dfrac{M_2}{8}$ $\dfrac{M_3}{2}$			3
2	$\dfrac{M_3}{7}$ $\dfrac{M_1}{3}$ $\dfrac{M_2}{9}$			3
3	$\dfrac{M_1}{6}$ $\dfrac{M_2}{3}$ $\dfrac{M_3}{2}$			2
4	$\dfrac{M_2}{4}$ $\dfrac{M_3}{3}$ $\dfrac{M_1}{4}$			2
5	$\dfrac{M_3}{6}$ $\dfrac{M_2}{4}$			2
6	$\dfrac{M_1}{2}$ $\dfrac{M_2}{3}$			1
7	$\dfrac{M_3}{5}$ $\dfrac{M_1}{6}$ $\dfrac{M_3}{2}$			1

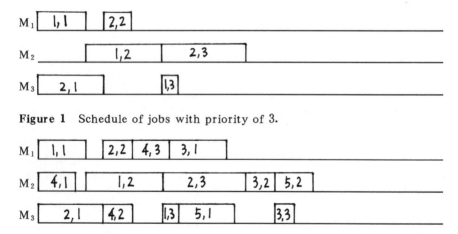

Figure 1   Schedule of jobs with priority of 3.

Figure 2   Schedule of jobs with priorities of 3 and 2.

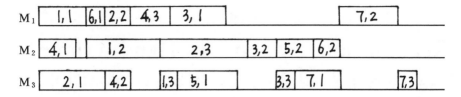

Figure 3 Schedule of all jobs.

6 CONCLUSION

The job-shop scheduling with priority is a kind of NP-hard problem. Although it has many application situations, there is few efficient method to solve it. After many tests we think that the neural network algorithm is a good candidate and merits further studies.

7 REFERENCE

Zhou, D. N. , Cherkassky, V. , Baldwin, T. R. and Hong, D. W. , (1990) " Scaling Neural Network for Job-Shop Scheduling", Proc. IJCNN' 90, Vol. 3, 889-92, Washington, D. C.

8 BIOGRAPHY

Jianfei Chen is a Ph. D. student in system engineering, department of automation, Tsinghua University. He has published several papers on discrete optimization, scheduling, multi-objective decision system, object-oriented analysis, and etc.. He is also temporarily a research fellow of National Laboratory of Pattern Recognition, Chinese Academy of Sciences.
Shaowei Xia is a professor in the department of automation, Tsinghua University. She graduated from the department of electrical engineering, Tsinghua University in 1953. Since then, she has worked in this university in the area of automation and system engineering. In 1980-1981, she visited the Nomura Institute in Japan as a visiting scholar. Her current research interest includes dynamic input-output models and regional linkage models, economy-energy-environmental system analysis; neural networkks and its applications. Shaowei Xia is a council member of system engineering federation of Chian, and a vice chairman of the social-economic system engineering society of China.

66

Recognition System of Hand-written Figures by Using Neural Networks and Genetic Algorithm

Hidehiko YAMAMOTO
Dept. of Information Science
Wakayama University
930, Sakaedani, Wakayama, 640
Japan
Tel: +81-734-54-0361
Fax: +81-734-54-0386
Email: tiger@wusun.center.
wakayama-u.ac.jp

Abstract

This paper describes the development of the computer system to recognize hand-written ten figures from 0 to 9 by learning synaptic weights of neural networks including genetic algorithm. The genetic algorithm process is (1) To generate initial population, (2) To calculate fitness for each individual based on the recognition results of hand-written teaching data, (3) Genetic operations such as crossover and mutation, (4) To select high fitness individuals and (5) To judge conditions. After the recognition simulations, it is ascertained that the developed system has the highest recognition rate.

Keywords

Hand-written figures, Neural networks, Genetic algorithm.

1. INTRODUCTION

Some researches for hand-written recognition systems by neural networks have been carried out. This paper describes the computer system to recognize hand-written ten figures from 0 to 9 by using genetic algorithm (Goldberg, 1989) and neural networks learning (D.E.Rumelhart, 1986).

There are some researches for neural networks learning by genetic algorithm. These researches are applied to comparatively simple problems such as XOR problem (Harp, 1989, Whitley, 1989, Kitano, 1990). My research is aimed at the development of the image processing system practically used in production systems. In such practical image processing systems, complicated neural networks construction has to be used.

In this paper, the decisions of the complicated neural networks and the synaptic weight values are made by using the genetic algorithm including learning. The developed hand-written recognition system is applied to 180 pattern figures and the good recognition rates are acquired.

2. OUTLINE OF FIGURE RECOGNITION SYSTEM

This chapter describes the outline of the developed hand-written figure recognition system. The input information of the system is the figures which are written, from 0 to 9. The ways to input the figures are capable of inputting figures directly written in with a mouse and of inputting figures

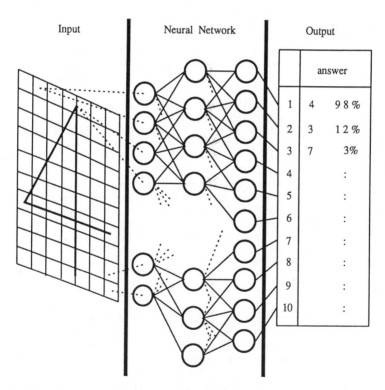

Figure 1 Outline of Figures recognition system

written on a white board with a felt-tip pen. The latter way
needs a CCD camera and the image processing such as
binalization. Finally, which figure was input is recognized
as the output results of the neural networks shown in Figure
1.

The output 10 neurons of Figure 1 networks construction
correspond to the 10 figures, 0, 1, $\cdots$, 9 in turn. As the
recognition result, the figure corresponding to the neuron
which has the maximum output value is shown.

The construction of the networks has three layers and each
neuron number of input layer, hidden layer and output layer
is 71, 70 and 10.

3. CONSTRUCTION DECISION BY GENETIC ALGORITHM

3.1. Basic Operation of Genetic Algorithm

The decision of the three layered neural networks construction described in Chapter 2 is made by the following
algorithm. The construction decision I am discussing here
means to decide the existence of the combination between each
neuron and to decide the synaptic weight values if the
combination is existent. The algorithm to decide the networks
construction is based on genetic algorithm shown below.
Figure 2 shows the algorithm.
(1) To generate 90 individuals of initial population which
    consist of randomly coded genes.
(2) To calculate fitness for each individual based on the
    recognition results of hand-written teaching data.
(3) To carry out genetic operations such as crossover and
    mutation.
(4) To repeat step(2) and to select individuals that have
    high fitness values in turn. The number of the selected
    individuals is equal to that of the initial population.
(5) To judge the system finish condition. If one of the
    condition is satisfied, the algorithm comes to an end. If
    any conditions are not satisfied, the algorithm keeps to
    carry out from step(3).

3.2. Coding

The contents to memory neural networks construction as gene
phenotype are synaptic weight values. The information of the
synaptic weight values is expressed with binary number with
two bites. Because of the expression method, the values from
0 to 65536 are alloted to the binary notation from -3.0 to
3.0. The sequence of 0 and 1 is considered as a single
individual of genotype. Figure 3 shows the coding example

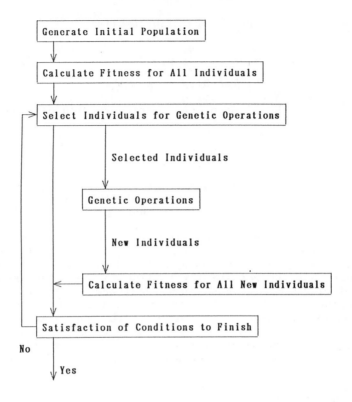

Figure 2 Network construction algorithm

used the genotype. If the synaptic value is less than 0, it is considered that the combination between the neurons is not existent.

3.3. Fitness

After the recognition of hand-written figures, the fitness is

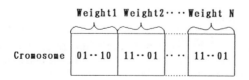

Figure 3 Cromosome model

calculated with the recognition results.

$$F = N_T \times N_{OUT} - \sum_{n=1}^{N_T} E(n) \quad \cdots\cdots\cdots\cdots (1)$$

N_T : Teaching data number
N_{out}: Output neuron number
$E(n)$: Mean square error for teaching data

$$E(n) = \frac{1}{2} \times \sum_{c=1}^{N_{OUT}} (y_C - Y_C)^2 \quad \cdots\cdots (2)$$

y_C: Recognition result
Y_C : Output value of teaching data
c : Output layer neuron

In back-propagation learning, the network that has low mean square error is considered as a good network (Yamamoto, 1993). In genetic algorithm, the individual that has high fitness has the possibility to live in the future generation. In order to have a good individual whose mean square is low possessed a high fitness, the mean square error is converted by using equation (1). Because of the equation (1) and (2), the fitness covers real numbers from 0 to (Teaching data number) × (Output neuron number). As 51 teaching data are used in the example, real numbers from 0 to 510 are capable in use.

3.4. Crossover

As a crossover method, the basic one-point crossover is used. The method is to select two individuals randomly from among the population and to exchange a part of chromosome to each other. As a result, new two individuals are generated.

3.5. Mutation

Mutation is to select one individual randomly from among the population and to forcely exchange the gene at the randomly selected locus.

3.6. Finish Conditions

When the system meets the following situation, the algorithm shown in Chapter 3.1. comes to an end. The situation is when the generation number becomes 2000 or the fitness become over 509.

4. SIMULATION

4.1. Neural Networks Learning

As the hand-written recognition simulations, two kinds of simulation conditions are used and the learning for the neural networks construction by using 51 teaching data is carried out. The conditions are (1) to carry out ten times back-propagation learning just before calculating fitness and (2) to carry out just genetic algorithm (No back-propagation learning). The learning results show in Figure 4. In the Figure, best fitness curves and average fitness curves for each condition are shown. As can be seen in Figure 4, the best fitness curve converge of the condition (2) that is no back-propagation learning is worse than that of the condition (1).

4.2. Recognition Results

The hand-written recognition process is to draw 180 kinds of figures on a white board and to recognize them in the following three condition.

<Condition 1> To carry out ten times back-propagation
            learning just before calculating fitness
<Condition 2> To carry out just genetic algorithm (No back-
            propagation learning)
<Condition 3> To carry out just back-propagation learning
            (No genetic algorithm)

  The recognition results for each condition show in Table 1, 2 and 3. Average recognition rates indicate 0.905 in Table 1, 0.437 in Table 2 and 0.794 in Table 3. Judging from the results, the construction decision just by genetic algorithm indicates a low recognition rate. The recognition rate of Condition 1, the method by genetic algorithm including back-propagation learning is better than that of Condition 3, the method by just using back-propagation learning. The construction decision method by Condition 1 abbreviated about 50 % combination number between neurons.

5. CONCLUSIONS

This paper reported the genetic algorithm method including back-propagation learning to recognize hand-written figures. By simulating hand-written figures recognitions in the three conditions, genetic algorithm including back-propagation learning, just genetic algorithm and just back-propagation learning, the recognition differences among the three

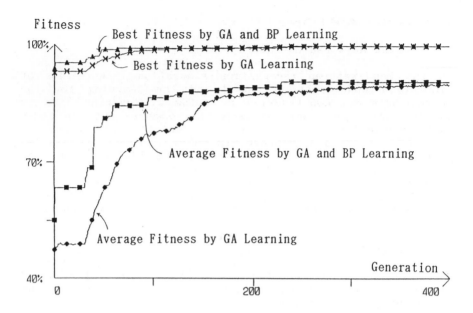

Figure 4   Simulation results

Table 1   Result of recognition rate (GA & NN)

Number	0	1	2	3	4	5	6	7	8	9
Recognition Rate (%)	95.3	82.0	74.7	98.1	93.2	95.1	94.7	99.3	78.7	74.7
									Average	90.3

Table 2   Result of recognition rate (GA)

Number	0	1	2	3	4	5	6	7	8	9
Recognition Rate (%)	52.9	40.2	38.2	48.1	50.6	39.3	29.1	48.6	39.5	51.2
									Average	43.7

Table 3 Result of recognition rate (NN)

Number	0	1	2	3	4	5	6	7	8	9
Recognition Rate (%)	94.6	88.7	60.7	66.9	95.5	73.3	92.8	72.0	69.1	79.5

	Average	79.4

conditions have been acquired. As a result, the recognition rate of the condition using genetic algorithm including back-propagation learning had the highest value.

6. REFERENCES

D.E.Rumelhart,J et al., Parallel Distributed Processing (MIT Press, 1986).

Goldberg,D., Genetic Algorithm in Search, Optimization, and Machine Learning (Addison-Wesley, 1989).

Harp,S. et al., Towards the Genetic Synthesis of Neural Networks, Proceedings of ICGA-89, (1989), pp.360-369.

Kitano,H., Designing Neural Networks Using Genetic Algorithm with Graph Generation Systems, Complex Systems, Vol.4, (1990).

Whitley,D. and Hanson, T., Optimizing Neural Networks Using Faster, More Accurate Genetic Search, Proceedings of 3rd ICGA, (1989), pp.391-396.

Yamamoto,H. and Fujimoto,H., Multiobjective Evaluation Expert System with Neural Networks Assisting Production System Design, Journal of Decision Systems (Editions HERMES, 1993), Vol.2, pp.213-224.

7. BIOGRAPHY

Dr. Hidehiko Yamamoto joined Wakayama University in 1992 after a ten year career in automotive industry, Toyota. He developed many automatic production systems using robots, NC machine tools and JIT system and also developed the expert system assisting for production line design. He has been an associate professor of Department of Information Science at Wakayama University. He is currently involved in a broad range of research for an intelligent factory automation such as a knowledge learning system of production system design and an intelligent production simulator for flexible transfer production systems.

67

Automatic knowledge acquisition for adjustment of radio frequency amplifier

Yasuhiro Hirayama, Akio Ukita
Production Engineering Development Laboratory, NEC Corporation
484, Tsukagoshi 3-Chome, Saiwai-ku, Kawasaki, 210, Japan
telephone +81-44-548-8823
fax +81-44-548-8817
email lkt60738@pcvan.nec.co.jp

Abstract

Radio frequency amplifiers used in microwave telecommunications amplify modulated microwave signals for feeding to the antenna. Since the elements that make up this amplifier have different electrical characteristics, the amplifiers need to be adjusted after being assembled to meet specifications. Generally, the adjustment process is performed by highly skilled workers. However, since the parameters to be adjusted are inter-dependent, the adjustment procedure becomes all the more complex. when new products or functions are introduced, even an experienced worker will require a fairly long time to learn the adjustment process. Thus, an adjustment support system may prove helpful in improving the adjustment efficiency, especially when knowledge acquisition is performed in the designing stage.

This paper proposes the use of an adjustment support system utilizing a microwave linear simulator and an artificial neural network. After circuit design is completed, the circuit output pattern is simulated by microwave linear simulator. During simulation, the values of the circuit elements are changed at random within prescribed limits. This paper permits the neural network to learn by collecting pairs of circuit output patterns and sets of circuit elements.

Keywords

Microwave telecommunication, field effect transistor, microwave integrated circuit, microstrip line, stub, impedance matching, adjustment process, microwave linear simulator, neural network

1 INTRODUCTION

Radio frequency amplifiers used in microwave telecommunication equipment within the GHz band amplify modulated microwave signals for antenna transmission. Since these amplifiers amplify signals within the microwave band, sensitive elements of the amplifier affect the amplifier's characteristics. Thus, the amplifiers need adjustment after being assembled to meet prescribed specifications.

Radio amplifiers have several adjustment specifications. Since these specifications interfere with each other, the adjustment process is fairly complex. Moreover, since the adjustment process must be performed 100% manually, it costs higher than other manufacturing processes. In the case of new products, since obviously there are no experts with whom to consult with, several problems regarding productivity arise.

Thus, production personnel hope that a system for supporting adjustment work, including tools that indicate the need for adjustment or automatic adjustment devices, be developed for new products. In the development of this system, an adjustment software must be created since knowledge for the adjustment of the radio frequency amplifiers must be processed on a computer.

An expert system is a possible solution. However, an expert system entails several problems, including development costs, need for human experts, etc. An effective system may be created based on design data obtained after circuit design. One of the most effective ways of acquiring adjustment knowledge is the use of a microwave linear simulator. Thus we are proposing an adjustment system using a microwave linear simulator and a neural network. An empirical study was performed in actual application to prove the effectivity of this system.

2 OUTLINE OF RADIO FREQUENCY AMPLIFIER

This document studies radio frequency amplifiers used in microwave telecommunication equipment. These amplifiers consist of microwave elements, such as microwave integrated circuits (MIC), field effect transistors (FET) and microstrip lines. Generally, MICs and FETs are connected series.

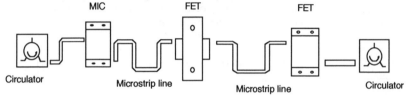

Figure 1 Radio frequency amplifier

MICs amplify the modulated microwave signals at low noise, low-distortion level. The FETs amplify the modulated signal into high power output signals for feeding to antenna. Microstrip lines connect the MICs to the FETs. The circulator is a circuit that sends the input signal from the terminals to designated terminals only in the direction specified by the arrows.

One of the characteristics of microwave signals is that they are reflected if the output and input impedances are not equal. If the signals are reflected, they are returned to the signal generator. Therefore, power cannot be transported efficiently. Consequently, microwave signals cannot be sufficiently amplified if the impedance between elements in the amplifiers is mismatched (Yoshida, 1985). Radio frequency amplifiers must therefore be adjusted by applying the principle of stub for impedance matching (Ogawa, 1992).

3 DESIGNING AND ADJUSTING RADIO FREQUENCY AMPLIFIERS

3.1 Designing

A radio frequency circuit simulator is used in designing radio frequency amplifiers. This circuit simulator simulates the actual operation of circuit based on its high frequency characteristics (White, 1985). For instance, it is possible to express the characteristics of the elements using scattering parameters, parameters widely used in analyzing radio frequency circuits, and use the parameter for simulating radio frequency circuits. For the microstrip line, the width and length of the strip can be entered from the radio frequency circuit to permit the analysis of the characteristics of the elements. Using this kind of radio frequency circuit simulator, the characteristics of a circuit can be studied.

However, at present, information provided by designers to the manufacturing group is limited to the adjustment specifications which contain the items for the adjustment. Moreover, data and information obtained from the radio frequency circuit simulator are not efficiency used. If only information obtained at the design stage can be efficiently used, ordinary operators, not only the skilled ones, can easily perform adjustment work contributing to reduced adjustment time.

3.2   Present adjustment method

Figure 2 shows the present method used in adjusting radio frequency amplifiers. The adjustment process is performed by operators. The sweep oscillator generates microwave signals within the frequency band required for the adjustment process. The scaler network analyzer measures the gain of the signal passing through the radio frequency amplifier. Using the signal generated by the sweep oscillator and the waveform obtained by the scaler network analyzer, the operator performs adjustment of the radio frequency amplifier.

To measure the observed waveform, the operator touches the microstrip line with a slim bar to which a small copper plate is attached. The operator looks for the characteristics that produce the optimum waveform. After finding the areas to be adjusted, the operator solders a small copper plate to those areas. In this process, stubs are attached to parts that are assumed to improve the characteristics. Thus, the kind of adjustment process is more or less a trial-and-error process. Through this process, the operator gradually learns by himself.

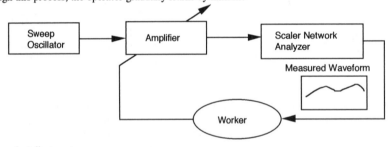

Figure 2 Adjustment

4 AUTOMATIC ACQUISITION OF ADJUSTMENT KNOWLEDGE

4.1   Proposed method

As explained in chapter 3, a system that can acquire knowledge even without skilled operators must be built. In this case, it is assumed that the design stage has been completed. The proposed system makes use of design data (which was not efficiently used as described in 3.1), inputs it to the circuit simulator, and obtains the knowledge necessary to perform the adjustment process.

First, design values of the circuit elements are entered in the circuit simulator. Some of the elements, though physically identical, may show variations in their electrical characteristics. Thus, these variations in electrical characteristics are set within a limited range. The outputs of the circuit are then calculated by the circuit simulator and then stored. Next, the length of the stub, which is used for adjustment, is varied so as to produce circuit outputs that meet the adjustment specifications. Once the circuit outputs meet the adjustment specifications, the stub length and the circuit outputs before adjustment are paired to form one "learning" data. In the same manner, the values of the circuit elements are varied within the prescribed range to compute learning data with the circuit simulator. These learning data are used to train the neural network about the relation between the circuit's outputs and stub's length.

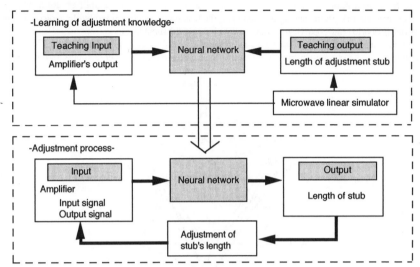

Figure 3 Adjustment using neural network

As the neural network learns, the amount of adjustment (which is the stub's length) required for a circuit to produce a prescribed output can be obtained. In the actual adjustment process, the output of the circuit to be adjusted can be fed to the trained neural network to obtain the amount of adjustment required (See Figure 3).

In the case of radio frequency amplifiers, there are several parts to be adjusted. In this case, a neural network can be built for each adjustment item and subsequently trained.

This proposed system enables the generation of computer-analyzable software at the completion of the design stage, without assistance from skilled operators. Acquisition of adjustment knowledge using this system has the following advantages:
· Eliminates the need of skilled workers
· Permits the automatic acquisition of adjustment knowledge
· Permits computer analysis

However, the implementability of this proposed system lies on the possibility of training the neural network on the relation between the circuit outputs and the amount of adjustment required for the circuit elements. In principle, the neural network can be trained if the circuit output patterns and the circuit element to be adjusted behave within the scope of a definable relationship (one-to-one, two-to-one, etc.). The selection of data from the circuit outputs to be entered into the neural network and the architecture of the neural network are important factors in acquiring adjustment knowledge. These two factors can be determined by using a circuit simulator in place of an actual circuit in the adjustment process shown in Figure 5.

This is because unless the circuit model can be adjusted on the simulator, it also follows that the actual circuit cannot be adjusted in the actual adjustment process.

4.2   Architecture of the neural network

The neural network proposed in this document has a feed forward architecture with a three-layered perceptron, as shown in Figure 4, and learns through a method called error back propagation learning (Matsumoto, 1992) and (Uesaka, 1993).

Scattering parameters, that are measured from a radio frequency amplifier, are used as the input of this network. In the actual input process, the frequency band required by the radio frequency amplifier is divided into p parts. p values are then fed into the input layer.

The number of output layers is equal to the number of stubs used for adjustment, The length of a stub will be an output neuron. The hidden layer depends on the number of inputs and outputs.

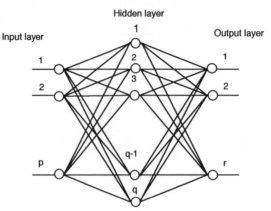

Figure 4   Architecture of the neural network

4.3   Learning data

Learning data is produced by varying all variable parameters and pairing circuit output pattern with corresponding stub lengths. First, the value of parameter is kept within $\pm a\%$. The parameter is then varied form its design value within $\pm a\%$ and the corresponding outputs are analyzed with the circuit simulator. The output data are then stored. Next, analysis is performed with the parameter while a stub is attached to the microstrip line to be adjusted. Various conditions are analyzed until the output meets the specifications. The length of the stub which produced the desired output is then obtained. The circuit output prior to adjustment and the adjustment stub are then paired from a learning data. The process described above is performed for all variable parameters to obtain various learning data.

Assuming that the number of variable parameters is k and that error range of the parameters is divided into n equal parts to produce $n+1$ values within the error range, the number of possible combinations will be $(n+1)^k$. As the number of parameters and divisions increase, the learning data also increase. Thus, the neural network becomes more complex making it more difficult to implement.

If the variation range of a parameter is small, the amounts by which the adjustment and output are varied can be converted linearly to train the network on the resulting change in output corresponding to the amount by which the adjustment is changed(see Figure 5). Using this concept, the k described earlier will equal the number of adjustment stubs thereby resulting in reduced learning data.

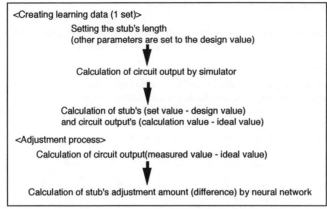

Figure 5 Training the neural network on the effects of the stub length

5 SIMULATION

5.1 Simulation model

To test the effectivity of the proposed method, we simulated the FET radio frequency amplifier shown in Figure 6. Impedance matching open stubs were connected to the input and output sections of the amplifier.

After measuring the characteristics of the module in the adjustment process, the length of the adjustment stub is adjusted based on the results of measurement to produce a constant gain level within the prescribed frequency range.

The neural network described in chapter 4 is then built for this circuit model. Among the scattering parameters, the S21 parameter (forward transfer coefficient) is used. This parameter express the characteristics of FET amplifiers. The frequency band (11 to 14 GHz) used by the FET amplifier is divided into 6 equal parts to produce seven values. The outputs will consists of two neurons. These two neurons output the length of stub 1 and 2. The number of hidden layers were determined depending on learning progress.

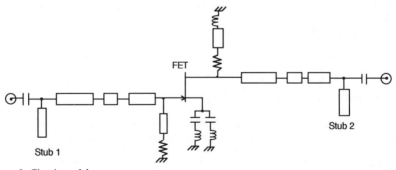

Figure 6 Circuit model

5.2 Results of simulation

After training the neural network using the learning data generated with the steps shown in Figure 5, we varied the parameters of the FET amplifier from the design values on the simulator and performed adjustment for the model shown in Figure 6.

Figure 7 shows example of measured values before and after adjustment. The objective is to produce a gain of +10 dB over the entire frequency range. After adjustment, the gain at each frequency approached the target value indicating that the adjustment process succeeded.

Ideally, it is desirable to meet the specifications by adjusting the stubs once only. Learning data cannot be created for all cases. Thus, the learning process of the neural network cannot be made perfect and adjustment must therefore be performed a couple of times.

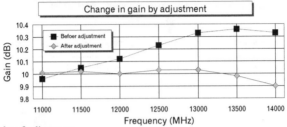

Figure 7 Results of adjustment

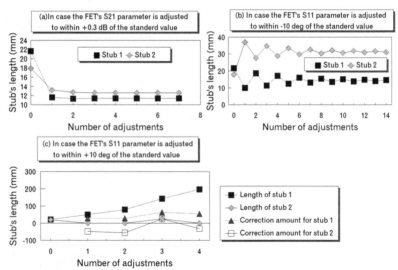

Figure 8 Number of adjustments and stub length

Figure 8 shows the number of times adjustment is performed and the stub length. If the FET amplifier's variation pattern is (a), the adjustments converge. If the variation pattern is (b), the adjustments converge but somewhat show some fluctuations. For pattern (c), adjustment does not succeed and the stub length diverges resulting in negative values. We were therefore able to find out through simulation that there are some patterns like pattern (c), that cannot be adjusted. In pattern (c), the correction value is indicated by the neural network. The length of stub 2 before the first adjustment was 19.7mm. However, after adjustment, the correction value indicated -48.0mm. Figure 9 shows the result of changing the target adjustment value for pattern (c). The target adjustment value was changed to +9.6dB. In this case, the adjustment process succeeded.

The cause of this problem was not the poor performance of the neural network but the physical impossibility to adjust the stub. However, the number of adjustments may increase. Thus the neural network need to be improved.

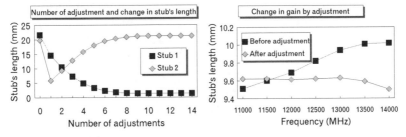

Figure 9 Result of adjustment after changing the target value

6 CONCLUSION

As proposed above, we are proposing a neural network to reduce adjustment related problems and make the adjustment process more efficient without adding too much work load. The simulations and evaluation of the effectivity we conducted prove that adjustment knowledge can be acquired using neural networks and circuit simulations.

However, an actual radio frequency amplifier has several parts and parameters to be adjusted and that each of these parts and parameters are interdependent with each other.

Thus, it is still uncertain if adjustment knowledge can be effectively acquired for every adjustment part and parameter. To solve this problem and lead to a conclusion taking into consideration these adjustment parts and parameters, combination of adjustment knowledge and cooperative deduction must be performed. Moreover, an actual radio frequency amplifier consists of several stages. Thus, simulation is more complicated than the simulation explained in this document. Further study is therefore necessary.

7 REFERENCES

White, J.F. and Kounosu, T. (1985) Microwave semiconductor application. CQ-Publication, Tokyo
Yoshida, T. (1985) Know-how of microwave circuit design. CQ-Publication, Tokyo
Ogawa, T. (1992) Basic knowledge for microwave circuit design. Transistor technology. **29-8**, 270-9
Matsumoto, G. and Ohtsu, N. (1992) Neuro-computing. Baifukan, Tokyo
Uesaka, Y. (1993) Mathematical basis of neuro-computing. Modern science, Tokyo

8 BIOGRAPHY

Yasuhiro Hirayama was born in 1966. He received his B.E. and M.E. degrees in electrical sciences from Kyushu University in 1988 and 1990, respectively. he joined NEC Corporation in 1990, and has been engaged in the development of some test and tuning equipment.

adjustment was 19.7mm. However, after adjustment, the correction value indicated -48.0mm. Figure 9 shows the result of changing the target adjustment value for pattern (c). The target adjustment value was changed to +9.6dB. In this case, the adjustment process succeeded.

   The cause of this problem was not the poor performance of the neural network but the physical impossibility to adjust the stub. However, the number of adjustments may increase. Thus the neural network need to be improved.

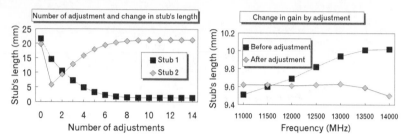

Figure 9 Result of adjustment after changing the target value

6 CONCLUSION

As proposed above, we are proposing a neural network to reduce adjustment related problems and make the adjustment process more efficient without adding too much work load. The simulations and evaluation of the effectivity we conducted prove that adjustment knowledge can be acquired using neural networks and circuit simulations.

 However, an actual radio frequency amplifier has several parts and parameters to be adjusted and that each of these parts and parameters are interdependent with each other.

   Thus, it is still uncertain if adjustment knowledge can be effectively acquired for every adjustment part and parameter. To solve this problem and lead to a conclusion taking into consideration these adjustment parts and parameters, combination of adjustment knowledge and cooperative deduction must be performed. Moreover, an actual radio frequency amplifier consists of several stages. Thus, simulation is more complicated than the simulation explained in this document. Further study is therefore necessary.

7 REFERENCES

White, J.F. and Kounosu, T. (1985) Microwave semiconductor application. CQ-Publication, Tokyo
Yoshida, T. (1985) Know-how of microwave circuit design. CQ-Publication, Tokyo
Ogawa, T. (1992) Basic knowledge for microwave circuit design. Transistor technology. **29-8**,
      270-9
Matsumoto, G. and Ohtsu, N. (1992) Neuro-computing. Baifukan, Tokyo
Uesaka, Y. (1993) Mathematical basis of neuro-computing. Modern science, Tokyo

8 BIOGRAPHY

Yasuhiro Hirayama was born in 1966. He received his B.E. and M.E. degrees in electrical sciences from Kyushu University in 1988 and 1990, respectively. he joined NEC Corporation in 1990, and has been engaged in the development of some test and tuning equipment.

68

Neural networks in new product development

J. Bode[*], *S. Ren*[*], *S. Luo*[*], *Z. Shi*[†], *Z. Zhou*[†], *H. Hu*[†], *T. Jiang*[†], *B. Liu*[†]
[*]*Dept. of Automation, Tsinghua University*
Beijing 100084, China, Tel:+86-1-2564179 Fax:+86-1-2568184
[†]*Institute of Computing Technology, Academia Sinica*
Beijing 100080, China

Abstract

Presents the results of the application of back propagation three layer perceptrons to cost estimation problems in design. Typical requirements of neural networks to be used in new product development are derived. Network construction has to consider the small number of training sets as a major characteristic of design problems.

Keywords

Design, concurrent engineering, product development, cost estimation, neural network

1 INTRODUCTION

New product development is obtaining increasing attention after it has been realized that it plays a long underestimated role in modern industries. Examples show that product design, although only constituting 5% of total product cost, can determine 70% of production cost (Grady et al., 1991). Moreover, large efforts are made to reduce the time between conceptualization of a new product and its final commercialization (time-to-market) because it is expected to realize higher first mover market shares and margins (Perry, 1990).

Most current computer systems to support design (like CAD, CAE, CADAM) focus on the well structured late phases of development. However, many important decisions are made in the early phase of conceptual design. There, computer support is poor, the major reason being the uncertainty of knowledge, difficulties in formalizing design procedures, and lacking information about the impacts of conceptual design decisions on downstream activities like manufacturing, marketing, maintenance, or disposal and recycling.

The PENDES project of the national Chinese CIMS-ERC (Computer Integrated Manufacturing Systems Engineering Research Center) investigates the possibilities to support decisions in the early phases of new product development (PENDES is the reverse acronym of Support of early phases of the development of new products).

In this paper we present results of our research on applying neural networks to cost estimation problems in conceptual design.

2 OPPORTUNITIES OF NEURAL NETWORK APPLICATIONS IN NEW PRODUCT DEVELOPMENT

Among the properties of neural networks their ability to generalize functional relationships among example data is of utmost importance for design. This feature is valuable wherever these relationships are assumed, but not known. This is the case, for example, for some dependencies between design decisions (e.g. to detect incompatible solutions), or some impact of design decisions on downstream activities (e.g. to identify manufacturing or maintenance problems of a given design). Moreover, neural networks are adaptable. This gives rise to the hope that changes in the connection between design-relevant data (e.g. the reduction of development effort as experience with a technology grows) are recognized without explicitly reprogramming the system. For design applications of neural networks and discussions of machine learning approaches see (Reich et al., 1993), (Ivezic/Garrett, 1994), (Bratko, 1993, p.162), (Ehrlenspiel/Schaal, 1992, p.409f.), (Bahrami/Dagli, 1992), (Becker/Prischmann, 1993).

This paper reports on the first results of our research. Section 3 describes experimentation with a three layer perceptron, section 4 generalizes the outcomes, derives requirements for neural network applications in the field of new product development and suggests promising solution methods to tackle these requirements.

3 EXPERIMENTS

3.1 Layout of experiments

General approach

The course of our research project is split into two phases:

1. **Feasibility study and conceptual design:** In this first phase we experiment with different network architectures and test selected approaches to learning. The goal is to become acquainted with the behavior of neural networks in typical design situations.

   It certainly has an appeal to study the behavior of neural networks with data extracted form real world situations. The fact that the relationship between input and output values of real data is not known seems to be one of the main reasons to apply machine learning principles. However, this same fact is a major drawback for performance measurement and control as the standards to be reached by the neural networks are unknown.

   Generally, the performance of a neural network depends on two factors: topology, node characteristics, and learning algorithm on the one hand (**network properties**), and the set of training data values and their interrelationships on the other hand (**external factors**). In this first phase we concentrate on the response of selected neural networks on controlled changes of the external factors. This will improve our knowledge on the suitability of networks with different properties given a certain design situation and a set of training data.

   We therefore do not want to be restricted to real world data and confront neural networks with **artificially generated data**. (Becker/Prischmann, 1993) and (Ivezic/Garrett, 1994, pp.154ff.) follow a similar approach.

2. **Pilot application:** It is in this second phase that we experiment with **real world data** from a pilot application. As the system's behavior is well understood research can concentrate on

calibrating the network parameters, accommodating external factors (e.g. selection of training sets) and investigating the suitability of neural networks to practical design problems.

The first phase comprises five steps: generation of training data, construction of neural network, training, generation of test data, and testing. The second phase follows a similar structure except that training and test data is not generated but elicited from a pilot application situation.

For the first phase we have selected a cost estimation problem from computer design. A pilot application for the second phase is being developed with a manufacturer of molds for injection-molding of plastic parts, again for cost estimation. This paper mainly reports on the first phase.

Application example outline

Cost estimation - the problem selected as application example - is helpful in the design of new products to make sure that target cost are met and competitive prices can be realized (Bode et al., 1995). As conventional costing methods generally require quite detailed information about the product in question it is only in a late phase of the development process that the available knowledge suffices to base cost-related decisions on it. However, the designers' influence on cost decreases over time because decisions with substantial cost impact have been made in the early phases. In case the estimated cost do not satisfy the requirements earlier commitments must be revised which results in delays and creates new unplanned cost.

It is therefore desirable to estimate cost already in early phases of development. However, design decisions have been made only for a small amount of important conceptual product variables then, and their relationship to cost often is not clear. Neural networks can help to detect these functional relationships.

We selected a cost estimation problem from computer design. The new product under development is a portable 'palmtop' computer. It will be considerably smaller and lighter than a notebook computer, to a large extent independent from external power sources, and as easy to carry as a folder for paper files.

Generation of training and test data

In an early phase of development only few attribute values about the product in question are available. They fall into three classes: **component-related data** describe modules and parts which are readily available and need no or only little further development; **development-related data** specify attributes of the design process of those components which need further development; **production-related data** focus on cost drivers during manufacturing.

Table 1 shows the attributes for the above mentioned cost estimation task. In the application example we suppose that the screen needs further development in order to reduce its power consumption.

In the selected (artificial) case the attribute values are connected through the following cost function:

Table 1 Cost estimation problem: input and output data

	Component-related input data		Development-related input data			Production related input data	Output data
Attributes	*Memory [MB]*	*CPU*	*Size*	*Improve-ment target*	*Technical difficulty*	*Production volume*	*Cost*
Values (examples)	0.64; 4; 16	386; 586	486;	palmtop; notebook; desktop	0%; 50%; 100%	0 (very simple) to 10 (very difficult)	5000; 10,000; 15,000	very low; low; medium; high; very high; unfeasible

$$C = [\sqrt{mem} + m(CPU)] \cdot 250 + \underbrace{\frac{\overbrace{[(2^{2 \cdot (\frac{improvement}{100})^2} \cdot 2^{2 \cdot (\frac{technicaldifficulty}{10})^2}) - 1] \cdot 5 \cdot 10^6}^{\text{development cost}} + \overbrace{20 \cdot 10^6}^{\text{fixed cost}}}{productionvolume}} \quad (1)$$

$$\underbrace{\phantom{C = [\sqrt{mem} + m(CPU)] \cdot 250}}_{\text{variable production cost}}$$

with function *m* mapping a CPU to a real number between 1 and 9. The three additive components represent the variable production cost, development cost, and fixed production cost, respectively. The result of this equation is a real numbered cost figure which ranges between 1,783 and 22,250 cost units if input data ranges are as shown in Table 1.

An analysis of the equation as well as a numerical simulation proved the function to be sufficiently complex, i.e. to be nonlinear over the domain of discourse, and not to be linearized without oversimplification.

The cost range is partitioned on the real line into 5 approximately equally sized classes labeled as shown in the cost column of Table 1. A sixth class represents all combinations of input data which are technically unfeasible. In the selected setting we suppose that a CPU type '386' is not feasible in combination with a memory of 16 MB, regardless of the real circumstances for the reasons of experimentation named above.

Note further that the attribute *size* has no impact on cost. This is for testing if the neural network is able to detect that there are input data without any relationship to the output. Network construction in the real world would start with assuming all factors with impact on cost, those factors being provided by an expert. However, as relationships are uncertain in the real world the expert might name irrelevant factors. In such a case the neural network must be able to isolate these and feedback this discovery to the user.

3.2 Construction of neural network

We use a three layer perceptron applying the back propagation algorithm (for an introduction see (Lippmann, 1987)). Six nodes in the first layer and six nodes in the third layer represent input attributes and output cost classes, respectively. Nine nodes in the hidden layer are applied because first tests showed comparatively good performance. However, more research needs to be done on the investigation of the network topology.

Descent step and momentum are dynamically adapted according to the shape of and the position on the hypersurface. As output function an ordinary sigmoid function is used for each node in hidden and output layer. A compress factor applies for cases where the network is trapped in a local minimum. In deep local minima a jump algorithm helps to prevent oscillations.

A winner-takes-it-all approach regards the output node with highest activation value as representing the output class assigned to the input data.

3.3 Results

Training with 150 randomly generated sets produced an ambiguous result. The neural network clearly identified the irrelevance of the attribute *size* to cost by assigning very low weights to the outgoing arcs of the respective input node. However, classification of data into cost ranges was not satisfying.

We stopped experimenting with this kind of setting as it is not realistic in the domain of product development. Firstly, such large amounts of training data usually are not available (see below). Secondly, past case examples and the experience of designers in the real world are not randomly distributed throughout the input space.

In a second experiment we therefore partitioned the input space. The attribute values of one partition remained constant for all training and test sets (memory, CPU, and size). The attribute values of the other partition (improvement target, technical difficulty, and production volume) varied among the data sets. Furthermore, we reduced the number of training sets to more realistic levels.

The results are shown in Table 2 and Figure 1. Three different neural networks with identical topologies but differing weight matrices were generated by training them with 27, 18, and 12 training sets, respectively. Reiteration of training data (sweeps) was stopped when the network's error fell below a threshold, or when the error values started to oscillate and no other measures could prevent oscillation.

As performance attributes average square errors and correct classifications (i.e. the number of sets that were assigned to the correct cost class by the neural network) were evaluated. Naturally, the systems perform better when assessed with training data itself. Moreover, the small number of training sets makes average values and percentages obsolete. This probably accounts for the reduced performance at greater training set sizes which is counterintuitive. Thus, performance on test data is more meaningful.

Correct classifications of 62% to 69% during testing appear low. However, it should be

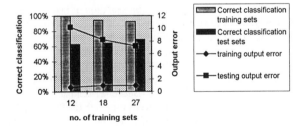

Figure 1 Results of neural network training and testing

Table 2 Results of neural network training and testing

No. of training sets	No. of iterations (sweeps)	Training error (10^2)	Correct classification of training sets	No. of test sets	Testing error (10^2)	Correct classification of test sets
27	3064	0.98	25 (92.6%)	64	7.25	44 (68.7%)
18	1398	0.99	17 (94.4%)	64	8.27	41 (64.1%)
12	398	0.71	12 (100.0%)	64	10.22	40 (62.5%)

considered that common applications of neural networks work with many hundreds, thousands, sometimes far more than ten thousand training sets in order to achieve satisfying results (Ivezic/Garrett, 1994), (Pal/Mitra, 1992). From this point of view the performance is better than we expected. Nonetheless, the main focus of further research should be directed to measures that help to improve performance despite the small number of training sets as will be discussed below.

4 REQUIREMENTS OF NEURAL NETWORK APPLICATIONS IN NEW PRODUCT DEVELOPMENT

From our experience with the above described experiments we derive some requirements of neural network applications in new product development. Future research should concentrate on these topics.

4.1 Ability to deal with small number of training data

We regard the small number of training data as the major point requiring attention. Several possibilities can be assessed to improve network performance.

1. Based on design cases from the past experts are asked to alter case data systematically and estimate the outcome. Consequently, training data can be multiplied. Problems with the reliability of the simulated data and the fact that simulated data is likely to be situated close to the base cases in the input space must be considered.

2. Often experts have substantial, yet approximate, background knowledge about the relationships between certain data. If this information is inserted into the neural network before training (prewiring; e.g. by removing connections between nodes if a relationship can be excluded) the learning process could be accelerated. However, common multi layer perceptrons usually do not allow **explicit** representation of knowledge which is essential for the input of background knowledge. Exceptions can be found especially in the field of fuzzy neural networks where topologies have been suggested that represent the rule structure of a fuzzy controller (Horikawa et al., 1992), (Lin/Lee, 1991).

3. Neural networks have problems learning relationships between very unevenly distributed data. It has been reported that transformation of data to an even distribution prior to learning can improve performance (Becker/Prischmann, 1993).

4. Certain input data from a training set can be left out on a random basis to use this incomplete sample as a new set. This would multiply the number of training sets without much effort. (Ivezic/Garrett, 1994, p.153) report of encouraging results. This approach is

only possible with suitable network topologies. Common back propagation multi layer perceptrons would interpret missing data as zero values which is not desirable.

5. Depending on the application neural networks might be segmented, each segment being trained separately with the same training sets. Segmentation results in a smaller number of input and/or output nodes and might speed up learning. This approach necessitates independence between the segments of input or output data, i.e. a segment of input nodes determines the activation of a segment of output nodes regardless of the values of input attributes from other segments.

6. The input attributes are concentrated, or clustered, into transformed attribute classes (e.g. by functional analysis) before constructing the neural network, or classification networks (e.g. Kohonen feature maps) cluster the input data automatically in real time. This method might be able to reduce the number of input dimensions and therefore requires less training.

4.2 Further requirements

Neural networks should be able to handle **fuzzy data** as much knowledge cannot be provided in a precise manner in early development phases. A lot of research has been initiated to construct fuzzy systems (e.g. adaptive fuzzy controllers) by using connective methods (Kosko, 1992), (Romaniuk/Hall, 1992), (Lin/Lee, 1991), (Horikawa et al., 1992).

It should be noted that fuzzy decision problems in design mostly have a structure different from fuzzy controllers. Fuzzy controllers usually obtain crisp inputs form their environment (e.g. the sensors on a moving vehicle). The input is then fuzzified, processed using fuzzy rules, and the fuzzy rule output finally needs defuzzification as most technical systems need crisp control data. The rationale to use fuzzy systems is the simple structure of fuzzy rules compared to complex analytical equations.

Problem solving in design, unlike common fuzzy control, has to deal with both crisp and fuzzy input data (e.g. it might be uncertain which type of component should be applied in the product under development). The output (e.g. cost of the new product) should not be defuzzified because fuzzy inputs can never lead to crisp outputs. A fuzzy output reflects the fuzziness of the information available at present. A fuzzy neural network to support new product development should take this into consideration.

In addition to that, data describing past cases mostly is crisp which means that the neural network is trained with crisp data. After training, however, it should be able to be applied using fuzzy values.

Moreover, **incomplete data** should lead to meaningful results as not all parameters of a design problem are known from the outset. This demand requires suitable topologies as common multi layer perceptrons would interpret missing data as zero values. In contrast, missing data means that **all** values of an attribute are possible.

Finally, the designer should not be forced to a rigid **separation of input and output data** as required by common neural networks. All data play an equal role, and it is their interplay that is of interest. Thus, one point of view might regard cost as a function of the product attributes, while another sees certain product attributes as an output of a preset cost decision.

5 REFERENCES

Bahrami, Ali / Dagli, Cihan H. (1992) Design retrieval by fuzzy neurocomputing. *Journal of Engineering Design*, **3**, 4, pp.339–356.

Becker, Jörg / Prischmann, M. (1993) Supporting the design process with neural networks - a complex application of cooperating neural networks and its implementation. *Journal of Information Science and Technology*, **3**, 1, pp.79–95.

Bode, J. / Hu, Y. / Liu, P. / Shu, B. (1995) Cost estimation in new product development: Knowledge pluralism, knowledge granularity, and agent based representation, in *Computer Applications in Production and Engineering*, Proceedings of CAPE '95 (ed. Q.N. Sun), Chapman & Hall, London.

Bratko, Ivan (1993) Machine learning in artificial intelligence. *Artificial Intelligence in Engineering*, **8**, pp.159–164.

Ehrlenspiel, K. / Schaal, S. (1992) In CAD integrierte Kostenkalkulation (Cost estimation integrated into CAD). *Konstruktion*, **44**, pp.407–414 (in German).

Horikawa, Shiu-ichi et al. (1992) On fuzzy modeling using fuzzy neural networks with the back-propagation algorithm. *IEEE Transactions on Neural Networks*, **3**, 5, pp.801–806.

Ivezic, Nenad / Garrett, James H., Jr. (1994) A neural network-based machine learning approach for supporting synthesis. *Artificial Intelligence for Engineering Design, Analysis and Manufacturing*, **8**, pp.143–161.

Kosko, B. (1992) Neural networks and fuzzy systems - a dynamical systems approach to machine intelligence. Prentice Hall, Englewood Cliffs.

Lin, Chin-Teng / Lee, C.S. George (1991) Neural-network-based fuzzy logic control and decision system. *IEEE Transactions on Computers*, **40**, 12, pp.1320-1336.

Lippmann, Richard P. (1987) An introduction to computing with neural nets. *IEEE ASSP Magazine*, April 1987, pp.4–22.

O'Grady, Peter / Young, Robert E. / Greef, Arthur / Smith, Larry (1991) An advice system for concurrent engineering. *Int. J. Computer Integrated Manufacturing*, **4**, 2, pp.63–70.

Pal, Sankar K. / Mitra, Sushmita (1992) Multilayer perceptron, fuzzy sets, and classification. *IEEE Transactions on Neural Networks*, **3**, 5, pp.683–697.

Perry, Tekla S. (1990) Teamwork plus technology cuts development time. *IEEE Spectrum*, October 1990, pp.61–67.

Reich, Y. / Konda, S. / Levy, S.N. / Monarch, I.A. / Subrahmanian, E. (1993) New roles for machine learning in design. *Artificial Intelligence in Engineering*, **8**, pp.165–181.

Romaniuk, Steve G. / Hall, Lawrence O. (1992) Decision making on creditworthiness, using a fuzzy connectionist model. *Fuzzy Sets and Systems*, **48**, pp.15–22.

6 BIOGRAPHIES

Jürgen Bode (associate professor, DAAD), Shouju Ren (professor) and Shaowu Luo (Ph.D. student) are with the Computer Integrated Manufacturing Systems Engineering Research Center (CIMS-ERC) at the Department of Automation of Tsinghua University, Beijing. Zhongzhi Shi (professor), Zhonglin Zhou (assistant professor), Hong Hu (assistant professor), Tao Jiang (student) and Bianlan Liu (graduate student) work at the Institute of Computing Technology at Academia Sinica (Chinese Academy of Sciences), Beijing.

PART FIFTEEN

AI Applications

69

Cost estimation in new product development: Knowledge pluralism, knowledge granularity, and agent based representation

J. Bode, Y. Hu, P. Liu, B. Shu
Dept. of Automation, Tsinghua University
Beijing 100084, China, Tel:+86-1-2564179 Fax:+86-1-2568184

Abstract

Presents a distributed model based on the contract net approach to support cost estimation during all phases of product design. Several methods are available for cost estimation (knowledge pluralism), each one represented by one agent. Agents compete and cooperate to solve costing problems. Unlike common multi agent architectures agent construction follows data accuracy and data availability (knowledge granularity) of the method to be applied.

Keywords

Multi agent architecture, contract net, design, product development, cost estimation, concurrent engineering

1 INTRODUCTION

Cost estimation is helpful in the design of new products to make sure that target cost are met and competitive prices can be realized. As conventional costing methods generally require quite detailed information about the product in question it is only in a late phase of the development process that the available knowledge suffices to base cost-related decisions on it. However, the designers' influence on cost decreases over time because decisions with substantial cost impact have been made in the early phases (Figure 1). Thus, conventional cost estimation is merely stating facts fixed in the past; it does not serve as a tool for an effective controlling of the development project (Becker/Prischmann, 1993). In case the estimated cost do not satisfy the requirements earlier commitments must be revised which results in delays and creates new unplanned cost.

In order to reduce the need for repeated iteration cycles between design and cost estimation support is required to provide cost information throughout all phases of the development process. This should be carried out in the following steps:

- decompose the product into components,
- for each component select among the different methods of cost estimation,

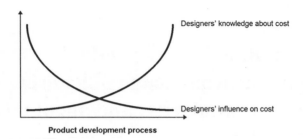

Figure 1 Knowledge about, and influence on cost during product development
(Becker/Prischmann, 1993).

- execute the estimation process, and
- combine the partial solutions.

The PENDES project investigates the possibilities to support decisions in the early phases of new product development (PENDES is the reverse acronym of Support of early phases of the development of new products). We suggest in this paper a novel architecture to realize a computer-based cost estimation of new products along all phases of the development process. Using the market paradigm a module, or entity, representing a cost estimation method can be regarded as a relatively autonomous "organization" offering a service on the cost estimation "market". Different "organizations" might cooperate if they have complementary capabilities, or compete if they offer the same type of service. This point of view allows to solve the cost estimation problem in a distributed and parallel manner.

2 KNOWLEDGE PLURALISM AND KNOWLEDGE GRANULARITY

Several cost estimation methods have been developed and are in use at present (for an overview see (Stewart/Wyskida, 1987)). They differ widely in their properties (see Figure 2):

- the **accuracy** of a method depends on the selection of cost drivers, the precision of the chosen mapping between cost drivers and cost, and the fuzziness of the input data;
- in general, accurate methods require more data and more precise data than simple rules-of-thumb (**knowledge granularity**);
- sophisticated statistical fitting methods (e.g. data analysis methods, neural networks) often require extensive computation (**computing complexity**);
- the representation of cost estimation knowledge might be more (e.g. rules-of-thumb) or less (e.g. neural networks) **transparent**, allow **explanation** of results (e.g. expert systems), and be **available for further analysis** (e.g. sensitivity analysis).

Although several methods are available for the task of cost estimation (**knowledge pluralism**) they cannot be classified into "good" or "bad" ones. Instead, each method is more or less suitable for a given design situation, depending on the fit between the methods' properties and the goals of the estimation problem.

Among all properties of a method knowledge granularity is of foremost interest. It is mostly the data available at the point of time of cost estimation which determines the selection of an

Source of input data
Default values
Databases
Case search
Simulation
Machine learning

Methods	Properties						Phase of development process
	Accuracy	knowledge granularity	computing complexity	transpa-rency	explanation ability	analysability	
Parametric	low	low	low	med	low	high	early
Activity-based	high	high	high	high	high	high	late
Neural networks	med	low - med	very high	low	low	low	early
Value Engineering	...	...	...	...	...	...	...
Similarity factors	...	...	...	...	...	...	...
Life cycle costing	...	...	...	...	...	...	...
Discounted cash flow analysis	...	...	...	...	...	...	...
Learning curve	...	...	...	...	...	...	...
Sequential block	...	...	...	...	...	...	...
...	...	...	...	...	...	...	...

Figure 2 Examples of cost estimation methods, their properties, the suitable development phases to be supported, and sources of input data for cost estimation.

appropriate method. Especially at early phases of product development the small amount of knowledge available makes the use of detailed methods obsolete.

Furthermore, for a product under development there is not just one best estimation method. Each fairly complex product can be subdivided into components, with each component at a different level of development maturity. Thus, one component might be readily available beforehand, or be subject to only minor changes, whilst another one is in the phase of conceptual design. Therefore, different cost estimation methods might apply to different components in parallel.

Another selection problem is the **source of input data** (see Figure 2). For detailed estimation methods a large amount of data is used, mostly provided by a corporate database, as described in (Fischer et al., 1993; Becker/Prischmann, 1993). Becker/Prischmann suggest that missing data be replaced by default values. Another procedure in case of missing information is the use of data from similar cases of the past (case search). Sometimes input data required by cost estimation methods must be generated by simulation. Machine learning and data analysis methods can provide the functions which relate the cost drivers (i.e. the independent variables) with cost (for the application of neural networks and data analysis on cost estimation see (Bode et al., 1995), (Becker/Prischmann, 1993), and (Mileham et al., 1993)). The accuracy of input data apparently has an impact on the precision of the cost estimation method.

3 TASK SHARING IN AN AGENT-BASED SYSTEM

3.1 Introduction

The above described knowledge pluralism shows similarities with a **market for consulting services**. In our model a set of consulters specializing on cost estimation (cost estimators) offers its services on a marketplace where they meet on the demand for solutions on cost

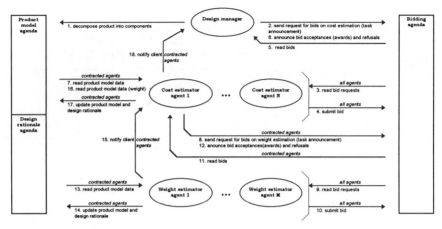

Figure 3 Agent communication for cost estimation (example).

estimation problems. All cost estimators are able to predict the cost of a given product under development. However, as each consulter uses a different cost estimation method their results differ, not only with regard to the calculated cost figure, but with regard to the quality attributes of the results (i.e. properties like accuracy, data availability etc.) which are closely connected with the properties of the estimation methods as depicted in Figure 2.

When a cost estimation problem appears the cost estimators compete, and a manager has to select the appropriate one. He will do so according to the required quality attributes of the expected result. As these quality attributes depend on the design stage (e.g. a late design stage generally allows for higher accuracies) they are not fixed over time but need to be evaluated at real time.

As cost estimators might not have sufficient information to perform their task they have to cooperate with other consulters specializing on providing the missing data (product attribute estimators).

We have modeled this problem solving approach using distributed, partially autonomous, cooperative agents. The model will be described below in more detail. For other approaches on distributed problem solving in design see (Park et al., 1994), (Sriram/Logcher/Fukuda, 1991), (Mani, 1994).

3.2 Agent communication

We define an agent as a partially autonomous unit of information processing which is able to perform a certain task within a knowledge domain (similarly (Park et al., 1994); for a discussion of different concepts of agents see (Zelewski, 1993)). Figure 3 depicts an example of the message-passing process between agents during cost estimation. Data is passed via two kinds of media: a pair of agents can communicate directly by exchanging messages (e.g. arc 18 in Figure 3), or information is made available to more than one agent by storing it in a blackboard-like agenda (for details on the blackboard architecture see (Nii, 1989)).

Two kinds of agendas can be distinguished. The product model and design rationale agendas store input data, intermediate and final results of the agents' problem solving process.

The bidding agenda serves as the communication platform to prepare and execute the process of selecting the appropriate agents for problem solving.

The process of task allocation follows the **contract net** approach (Davis/Smith, 1983), (Parunak, 1988) which is based on three principal roles of agents: the **managers** which identify tasks and assign them to other agents for execution, the **bidders** which offer to perform a task, and the **contractors** whose bids have been accepted by the managers.

The communication process in Figure 3 is to be explained in more detail. The design manager decomposes the product under development into its components and stores this information in the product model agenda which contains design-relevant data in a standardized form (e.g. using STEP/EXPRESS; see (STEP, 1992)). Thereafter, the need for cost estimation is identified and bids from cost estimator agents solicited (arcs 1, 2). Each cost estimator agent is a specialist for one method of cost estimation. All cost estimator agents evaluate the bid request (i.e. check if, and under which conditions, it is possible to perform the task) and submit a bid in case they are able to contribute to the problem solution (3, 4). Each bid must contain as major quality attributes the estimated accuracy of the result and the time required to perform the task. Based on this information the design manager evaluates the bids (i.e. compares the submitted quality attributes with its own goals) and announces awards for the successful bidders and refusals to the others (5, 6). An award constitutes a contract to perform the task.

As products are made up by several components, and different cost estimation methods might be suitable for each component, several agents can be contracted to perform the task on several components concurrently.

It should be noted that in early phases of design the product model is uncertain in two ways: it is incomplete because not every single feature of the product in question is known yet, and it is fuzzy because the value of certain product attributes might not be known precisely.

We now suppose that one cost estimator (agent 1 in Figure 3) has been contracted by the design manager to perform the task. This agent is able to perform the parametric cost estimation method (Stewart/Wyskida, 1987) using a given simple cost function $C = f_c(product\ weight)$ as applied for instance in the estimation of cost of injection-molded components (Mileham et al., 1993). The agent checks the product model for the required input data, product weight (7). This variable is still missing at this point of time. The cost estimator, trying to obtain the data, now himself solicits for bids by whoever might be able to provide a weight estimation (8). This is where other than cost estimator agents come into play.

The bidding and contracting process of the weight estimator agents is similar to the one previously described (9–12). Here, the cost estimator agent, formerly a bidder, now takes over the role of the manager for the weight estimation task. In Figure 3 weight estimator agent 1 is awarded the contract. It proceeds by reading data stored in the product model, say, the volume and the material type of the product (13). Then the agent's problem solving procedure is applied, in the example the function *product weight = product volume * material density*. The product model is updated (14) with the resulting weight and the estimation rationale as part of the design rationale (described below). After being notified that the task has been completed successfully (15) the cost estimator abandons its role as manager and performs its task. Reading the product data (weight), updating product model and design rationale, and notifying the design manager is executed similar to the weight estimation procedure (16–18).

The appeal of this contract net approach is its similarity to ordinary markets. Each agent can be illustrated as an organizational entity (e.g. a person, a company) in its role as (potential)

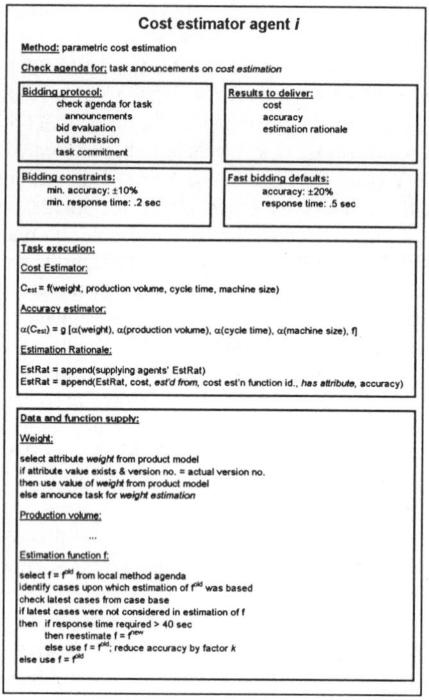

Figure 4 Basic structure of selected cost estimator agent.

supplier or buyer in a market. This specific point of view makes multi agent systems of this kind relatively easy to design and understand. Agents contain all knowledge to execute bidding and perform their task, and thus are (partly) autonomous. Like in ordinary markets they compete if they are able to deliver similar services. The succeeding bidder is selected according to the goals of the buyer, here the manager agent. Furthermore, agents can cooperate if their services are of complementary nature (here the cost estimator and the weight estimators). Note that cooperation might also occur among cost estimators in case they use complementary approaches (e.g. life cycle costing uses and extends traditional costing methods which requires both methods to be applied in combination).

Finally, it should be mentioned that it is not difficult to add or remove agents from the system as knowledge in general is represented locally and therefore easy to encapsulate.

3.3 Agent structure

Figure 4 and Figure 5 show the major parts of the internal structure of a cost estimator agent and several weight estimator agents respectively. For the ease of reading the figures only show those agents' components which are necessary for the understanding of the multi agent model. The cost estimator agent represents knowledge about the application of the parametric cost estimation method (in contrast to the above example extended to four independent variables).

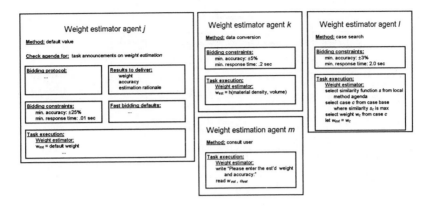

Figure 5 Basic structure of selected weight estimator agents.

Other agents on this level might be specialists in detailed cost estimation, life cycle costing, neural networks, or other costing methods (see the examples in Figure 2).

An important step in the bidding phase is **bid evaluation**. This requires the estimation of the quality attributes accuracy and response time which have to be submitted along with the bid as they constitute the decision variables for the design manager to award the appropriate agents. The ordinary procedure is to gather information about the quality attributes of the input data and the estimation function because they determine the accuracy of the result. This can involve soliciting judgments from agents with knowledge on the estimation of this data (not represented in Figure 3). However, if the required response time is limited preset bidding constraints and fast bidding defaults allow for quick bidding.

The **estimation rationale** links the result of the estimation process with its prerequisites. As a single attribute of the product model (say, cost) might have different values (depending on the cost estimation method and the input data used) it is important to be able to identify the history of an estimated value. One might call this "versioning" of attribute values. Even if, in the example of Figure 4, a value is assigned to the attribute *weight* in the product model, the cost estimator must check if this value is still up to date by comparing the information in the rationale with the information currently available. The estimation rationale is part of the design rationale which captures the history of the whole development process (i.e. the underlying intent and logical support of all design-related decisions, (Klein, 1993)).

Competition is illustrated with the weight estimator agents displayed in Figure 5. Four methods to generate a weight value are available, each method represented by one agent. The most simple one (agent *j*) just provides a default value which is a rather common method in case of missing attribute values (Becker/Prischmann, 1993), (Park et al., 1994, p.53). On the other hand, case search methods (agent *l*) can provide more valuable and accurate information. It might even be necessary to solicit outside information by consulting the user (agent *m*).

After bidding, the cost estimator in his role as (temporary) manager awards the contract to the agent whose quality attributes accuracy and response time appear most suitable to his goals.

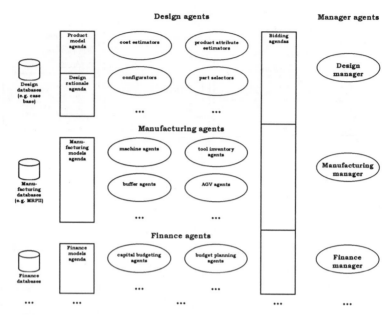

Figure 6 Cooperating agents beyond the design function.

The cooperation between two levels of specialist agents described here (cost and weight estimators) can be extended to any number of levels. For instance, weight estimator agent k in Figure 5 might require information about the value for the attribute *volume* and will announce this task to volume estimator agents. Thus, the **depth** of cooperation varies according to the problem to be solved (**vertical cooperation**).

Another form of cooperation extends beyond the scope of design specialists (Figure 6). For example, during cost estimation it can be necessary to check if sufficient manufacturing capacity for the product under development will be available. This falls into the competence of manufacturing agents, and the cost estimator will solicit their bids to obtain the needed information. Another important address for cost estimators are finance specialists: if, for example, new equipment has to be installed for the manufacturing of the new product cost figures need to be calculated by using capital budgeting methods (e.g. net present value method). Evidently, the **breadth** of cooperation is another attribute of the multi agent problem solving process (**horizontal cooperation**).

Both forms of cooperation are particularly attractive if collaborating agents originally were designed for other purposes, e.g. to support manufacturing resources planning (MRPII) and control (Lin, 1992), (Zelewski/Bode, 1993). This shows that the multi agent approach can contribute to the issue of **software reuse**. Further note that it is not by accident that economic theory also employs the terms vertical and horizontal cooperation as contract net systems are based on the market paradigm.

4 DISCUSSION

Several aspects of the described application can be compared with conventional multi agent models.

- Unlike agents in, say, manufacturing systems an agent does not always represent knowledge associated with a real (i.e. material) object like a machine, an inventory, a part. Instead, an **ideal object** (here: a method) constitutes an agent. This does not violate design rules for multi agent systems. Ideas compete, and ideas cooperate, facts which make a multi agent approach particularly useful. Similar applications are reported from the domain of financial services where different investment opportunities compete for an investor's money (Buhl/König/Will, 1993).
- An important criteria for agent construction is the level of **knowledge granularity** that is required to perform the task. Different agents, yet capable of solving similar problems, are able to deal with different knowledge granularities. The multiplicity of knowledge concerning a single type of problem like cost estimation (**knowledge pluralism**) is the motivation to represent this knowledge in a multi agent structure.
- The distribution of roles is dynamic, as it is in ordinary markets. A specialist agent (e.g. a cost estimator) can adopt the role of a manager when it requires information to which it has no direct access. Thus, a single problem to be solved by the system might result in a veritable **bid cascade** from one specialist to the other.
- The multi agent approach to cost estimation is **transparent** and **easy to understand** because of its similarity to organizational structures outside the computer world. It allows **incremental system development** as agents can be added, modified, or removed easily. Several components of the product under development can be dealt with in **parallel** because of the partial autonomy of the agents. Agents originally designed for other than design tasks can provide their knowledge for cost estimation thus realizing benefits from **software reuse** and **concurrent engineering**.
- Some multi agent systems exhibit a sophisticated **goal structure**. Agents can attempt to maximize capacity usage, minimize cost, bargain for prices, etc. (Lin/Solberg, 1992). Here, cost estimators have no specific goals besides participating in bidding wherever possible. Management agents, however, have requirements concerning the accuracy of a result, and, to a lesser extent, the response time.

5 REFERENCES

Becker, Jörg / Prischmann, M. (1993) Supporting the design process with neural networks - a complex application of cooperating neural networks and its implementation. *Journal of Information Science and Technology*, **3**, 1, pp.79–95.

Bode, J. / Ren, S. / Luo, S. / Shi, Z. / Zhou, Z. / Hu, H. / Jiang, T. / Liu, B. (1995) Neural Networks in new product development, in *Computer Applications in Production and Engineering*, Proceedings of CAPE '95 (ed. Q.N. Sun), Chapman & Hall, London.

Buhl, H.U. / König, H.-J. / Will, A. (1993) ALLFIWIB: Distributed knowledge based systems for customer support in financial services, in *Proceedings 17th Symposium on Operations Research* (ed. A. Karmann et al.), Springer, Heidelberg, pp.537–540.

Davis, R. / Smith, R.G. (1983) Negotiation as a metaphor for distributed problem solving. *Artificial Intelligence*, **20**, pp.63–109.

Fischer, Joachim / Koch, Rainer / Schmidt-Faber, Bastian / Szu, Kou-I (1993) Konstruktionssynchrone Kostenprognose als CIM-Komponente (Design-synchronous cost forecast as component of CIM systems), in *Wirtschaftsinformatik '93* (ed. Karl Kurbel), Physica, Heidelberg, pp.378–390 (in German).

Klein, Mark (1993) Capturing design rationale in concurrent engineering teams. *Computer*, **26**, 1, pp.39–47.

Lin, Grace Yuh-Jiun / Solberg, James J. (1992) Integrated shop floor control using autonomous agents. *IIE Transactions*, **24**, 3, pp.57–71.

Mani, N. (1994) Towards modelling a multi-agent knowledge system in engineering design, in *1994 IEEE International Conference on Systems, Man, and Cybernetics* (Conference proceedings), Vol. 1, pp.160–164.

Mileham, A.R. / Currie, G.C. / Miles, A.W. / Bradford, D.T. (1993) A parametric approach to cost estimating at the conceptual stage of design. *Journal of Engineering Design*, **4**, 2, pp.117–125.

Nii, H. Penny (1989) Blackboard systems, in *The Handbook of Artificial Intelligence* (eds. A. Barr / P.R. Cohen / E.A. Feigenbaum), Vol. IV, pp.1–82.

Park, Hisup / Cutkosky, Mark R. / Conru, Andrew B. / Lee, Soo-Hong (1994) An agent-based approach to concurrent cable harness design. *Artificial Intelligence for Engineering Design, Analysis, and Manufacturing*, 8, pp.45–61.

Parunak, H.V.D. (1988) Distributed artificial intelligence, in *Artificial Intelligence - Implications for CIM* (ed. Andrew Kusiak), IFS/Springer, Kempston/Berlin, pp.225–251.

Sriram, D. / Logcher, R. / Fukuda, S. (eds.) (1991) Computer-Aided Cooperative Product Development, Springer, Berlin, pp.253–333.

STEP Part I: Overview and fundamental principles (1992). ISO TC 184/SC 4 Committee Draft.

Stewart, Rodney D. / Wyskida, Richard M. (eds.) (1987) Cost estimators reference manual. Wiley, New York.

Zelewski, Stephan (1993) Agenten. *IM Information Management* (München), **8**, 2, pp.79–81 (in German).

Zelewski, Stephan / Bode, Jürgen (1993) Koordination von Produktionsprozessen - Ein Ansatz auf Basis von Multi-Agenten-Systemen (Production process control - an approach based on multi agent systems). *IM Information Management* (München), **8**, 2, pp.14–24 (in German).

6 BIOGRAPHIES

Jürgen Bode is associate professor (DAAD) with the Computer Integrated Manufacturing Systems Engineering Research Center (CIMS-ERC) at the Dept. of Automation of Tsinghua University. He earned his doctoral degree in management science at Universität zu Köln, Germany. Yongtong Hu, Ping Liu and Bo Shu are students of automation engineering at Tsinghua University.

70

The Determination of AGV's Traffic Control Model by ID3 through an Implicit Knowledge Learning

Dan JIN
610, FMS
Mechanical Engineering Department
Shanghai Jiaotong University
No. 1954, Huashan Road, Shanghai, China

Wenwei YU, Yukinori KAKAZU
Department of Precision Engineering
Hokkaido University
West-8, North-13, Sapporo, Hokkaido, Japan
Tel: +81-11-706-6443
Fax: +81-11-700-5070
Email: kakazu@hupe.hokudai.ac.jp

Abstract

Traffic control problem of AGV becomes more and more important in flexible manufacturing systems when mechanical and automated production plants are much more expanded. In this paper, we provide an information based evaluation method—Induction of Decision Tree (ID3) for learning the implicit experience like human expert to determine the traffic control model. We evaluate the method on several artificially generated data sets. The significance and advantage of ID3 can be seen through the results.

Keywords

Decision making, Induction tree, ID3, Machine learning, AGV, FMS, Traffic control problem

1. INTRODUCTION

As mechanized and automated production plants are expanded, the problem of transport, handling and storage become increasingly important, because of which AGV is used in a considerable scope. As the number of AGV is increasing, the problem of conflict and congestion in the intersections in a working area becomes larger[2]. It is useful to consider this kind of problem as urban traffic problem in intersections.

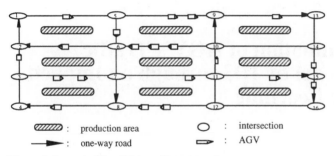

⬛▨▨▨ :　production area　　　　○ :　intersection

➤ :　one-way road　　　　⇨ :　AGV

Figure 1 Layout of the 4×4 working network.

　　　Let us consider the traffic problem of twenty AGVs in the 4×4 network working area in a factory. Figure 1 shows one instance of the distribution of AGVs.

　　　In the 4×4 network, there are 16 intersections. AGVs walk along the one-way roads, go through the intersections and arrive to the production area to finish their works respectively. As Figure 1 shows, for each intersection, the number of AGVs which will go through this intersection are different. In Figure 1, for the 6th intersection, 4 AGVs will go through, which seems to cause the conflict and congestion problem. But for the 3rd, 12nd, 14th intersection respectively, no AGV will go through. From this, we can see that in the certain working area, for the different intersection, the volume of the traffic-flow of AGVs has big difference, which leads to the necessary by using different traffic control model to avoid the problem of conflict and congestion.

　　　In order to select a suitable traffic control model, except the different intersection, we should consider many other elements simultaneously, which is very important. But it is also very difficult because between these elements, there are no explicit knowledge having been known. The expert systems which use if-then rules is not suitable for a complex task because it will lead to the bottleneck problem. On the other hand, the expert systems are powered by knowledge that is represented explicitly rather than being implicit in algorithms. So we would be interested in studying the learning behavior of human expert (which is called machine learning). When a human expert determines to use which kind of traffic control model, he uses his implicit knowledge in his brain. Therefore, in order to make computer learn the implicit knowledge like human expert and not suffer from the bottleneck problem, we provide the approach called induction of decision tree (ID3) to solve the traffic control model problem of AGV which can also been reckoned as classification problem.

　　　In this paper, we discuss the decision problem for two kind of control models[3]:

(1) delay model—make the AGVs run as fast as possible;

(2) volume model—make traffic-flow's volume of AGVs through intersection maximum.

2. ID3

2.1 Structure

ID3[1] is a Learning From Examples (LFE) method that constructs decision trees to represent concepts. Given a training set of examples with their features and classes, it produces a discrimination tree whose nodes indicate useful feature tests and whose leaves assign classes to examples.

Table 1 Table of Examples

Examples	Features				Classes
	f_1	f_2	...	f_m	
e_1	v_{11}	v_{12}	...	v_{1m}	c_{k_1}
e_2	v_{21}	v_{22}	...	v_{2m}	c_{k_2}
e_3	v_{31}	v_{32}	...	v_{3m}	c_{k_3}
$\vdots$	$\vdots$	$\vdots$	$\ddots$	$\vdots$	$\vdots$
e_n	v_{n1}	v_{n2}	...	v_{nm}	c_{k_n}

Consider a set of examples $E=\{ e_1, e_2, \cdots\cdots, e_n \}$, the features $F=\{ f_1, f_2, \cdots\cdots, f_m \}$ and the classes $C =\{ c_1, c_2, \cdots\cdots, c_p \}$. Here, we assume that the examples are exhaustive, and that the domain for each feature f_j ($j= 1, 2, \cdots\cdots, m$) is a finite set $Dom(f_j)$, the domain for each class c_k is also a finite set $Dom(c_k)$. For each example e_i, a value v_{ij} ($\in Dom(f_j)$) for each feature f_j and the class of each example c_k ($k = 1, 2, \cdots\cdots, p$) is given (see Table 1).

One example decision tree that correctly classifies each example in the example set is like the structure shown in Figure 2.

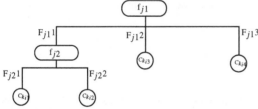

Figure 2 Decision Tree for the example set in Table 1.

As shown in Figure 2, a decision tree is a top-down structure, and consists of several branching nodes and leaves. A feature f_j corresponds to each branching node, and subset F_{jk} of $Dom(f_j)$ corresponds to each branch from the branching node. Here, each F_{jk} must satisfy the following conditions:

$F_{jk} \subset Dom(f_j)$;   (subset)

$F_{jk} \cap F_{jl} = \phi$; $(k \neq l)$;   (disjoint)

$\cup_k F_{jk} = Dom(f_j)$;   (exhaustive)

Each leaf in a decision tree can be regarded as a classification rule. The path from the root to the leaf determines the rule's conditions, and the class at the leaf represents the rule's action (namely, the result of the classification). For the decision tree shown in Figure 2, there are 4 rules corresponding to each leaf l ($l = 1, 2, 3, 4$), which are shown as follows:

$(v_{ij_1} \in F_{j1}1) \wedge (v_{ij_2} \in F_{j2}1) \rightarrow C_{k_{i1}}$;

$(v_{ij_1} \in F_{j1}1) \wedge (v_{ij_2} \in F_{j2}2) \rightarrow C_{k_{i2}}$;

$v_{ij_1} \in F_{j1}2 \rightarrow C_{k_{i3}}$;

$v_{ij_1} \in F_{j1}3 \rightarrow C_{k_{i4}}$.

Decision tree can be rewritten into rules, but it is different with the if-then rule system. It is much more simpler than the if-then rule system, and very easy to be understood. Also from decision tree, we can know the importance of each feature for the classification task, which will be discussed in the learning algorithm of ID3.

2.2 Learning algorithm

The learning algorithm of ID3 uses an information-theoretic measure to determine which feature to branch on at each node, and then makes recursive calls to build subtrees for each created branch.

Imagine selecting one example at random from a set S of training examples, and announcing that it belongs to some class C_i. So the average amount of information (also known as the entropy of the training set S) needed to identify the class of a case in S is:

$$I(S) = - \sum_{i \in \{C\}} p(C_i) \times log_2[p(C_i)] \qquad \cdots\cdots\cdots\cdots\cdots \quad (1)$$

where, $p(C_i)$ stands for the probability of examples in S that belongs to class C_i.

After S has been partitioned in accordance with the n outcomes of a test F (namely feature), the expected information requirement can be found as the weighted sum over the subsets, as:

$$I_F(S) = \sum_{i=1}^{n} p(F=V_i) \sum_{j \in \{C\}} -p(C_j|F=V_i) \times log_2[p(C_j|F=V_i)] \quad \cdots\cdots\cdots\cdots \quad (2)$$

Then the information that is gained by partitioning S in accordance with the test F can be measured as:

$$gain(F) = I(S) - I_F(S) \qquad \cdots\cdots\cdots\cdots\cdots \quad (3)$$

ID3 then selects the feature that maximizes the equation 3, and the algorithm is applied recursively to each child node until the training examples at that node are either of one class or have the same class values for all features. A summary of the ID3's learning algorithm is presented as follows.

From the learning algorithm of ID3, we can see that the simplicity in knowledge formalism of ID3 makes the learning methodologies considerably less complex than semantic networks and avoids the bottleneck problem in the if-then knowledge-based export system. Furthermore, using ID3 the implicit knowledge can be learned. Namely, the information provided by a feature for the classification purpose can be achieved and the importance of each feature in classifying also can be understood from the decision tree, although the example objects from which the decision tree is constructed are only known through their values of a set of features.

LEARNING ALGORITHM OF ID3

Let E be a set of classified training examples.
Let F be a set of features for describing examples.
Let $GAIN(F_i)$ be the function shown in equation 3 , where $F_i \in F$.

INDUCETREE (E)
IF all training examples at node are one class, or have the same class values for all features
      - return a leaf node for TREE, labelled with the most common class of examples in E.
ELSE
    - let $F_{best} \in F$ be the feature with the largest value of the gain $GAIN(F_{best}, E)$.
        For each value V_j of feature F_{best},
          Generate subtrees using *INDUCETREE (E_j)*,
            where E_j are these examples in E with value V_j for feature F_{best}.
          Return a node labelled as a test on feature F_{best} with these subtrees attached.

---

3. USING ID3 TO DETERMINE TRAFFIC CONTROL MODEL OF AGV

3.1 Advantages

The characteristics of decision tree and the ID3's learning algorithm lead to several advantages of using ID3 to determine the AGV's traffic control models.

    In the past, the knowledge-based expert systems were usually used to select the traffic control model of AGV, which required hundreds or even thousands of such rules. But when using decision tree to represent the acquired knowledge, the knowledge formalism is relatively simple. Furthermore, the learning methodology is considerably less complex than those employed in expert systems. So the bottleneck problem can be solved.

    On the other hand, using ID3, the implicit knowledge for traffic control model determination can be learned. The information provided by a feature for determination task can be known. That is although there are many elements affecting the determination of traffic control model simultaneously, there must be some one which is the most important feature for the determination, and should be checked at first during the traffic control model selection. Similarly, there should be some feature which is the second important , and will be examined at second during selection, and so on. This is just like the human expert's behavior in determining the control model. Using ID3 , this kind of behavior can be learned, but it can not be learned by the other methods, such as Artificial Neural Network (ANN).

    From these, it can be said that using ID3 in the determination of traffic control model of AGV seems very effective.

3.2 Problem of traffic control model

In this paper, we try to investigate the effectiveness of determining two kinds of control models (delay model and volume model) for each intersection to avoid the conflict and congestion problem of AGVs in a small (4 ×4) network working area in a flexible manufacturing system as shown in Figure 1.

    For applying ID3 in the determination of control model, selecting the features for the problem is necessary and also very important. Here, we considered four elements which affect the determination. Table 2 shows the four features, and their meaning and values.

Table 2 Features used in ID3

No.	Feature	Value of feature		Meaning of value
F_1	season	0	$f11$	busy
		1	$f12$	normal
		2	$f13$	leisure
F_2	production quota	0	$f21$	large
		1	$f22$	medium
		2	$f23$	small
F_3	working time range in one day	0	$f31$	day time (8:00-17:00)
		1	$f32$	night (17:00-22:00)
		2	$f33$	midnight (22:00-8:00)
F_4	number of AGVs entering the intersection	0	$f41$	number<=1
		1	$f42$	1<number<=3
		2	$f43$	number>=4

Here, we assume that when the production quota is "large" and "medium", in 4×4 network area, 20 AGVs are necessary, while only 10 AGVs are needed when the production quota is "small".

4. EXPERIMENTAL RESULTS AND DISCUSSIONS

To investigate the effectiveness of determining the control model by ID3, some simulation experiments were carried out, from which 66 experimental data groups were achieved. Each data group has four values corresponding to four features respectively, and also has one model value which was the experimental result that is the suitable control model which should be used in the situation determined by the values of features. So let D be the data group, for the 66 data groups, there is:

$D = \{ D_1, D_2, D_3, \cdots\cdots, D_{66} \}$.

Let F be the set of features. Because four features have been used, so:

$F = \{ F_1, F_2, F_3, F_4 \}$.

We assume the value of a feature $F_j \in F$ by $Dom(F_j)$. According to the values for each feature shown in Table 3, that:

$Dom (F_1) = \{ f_{11}, f_{12}, f_{13} \}$; $Dom (F_2) = \{ f_{21}, f_{22}, f_{23} \}$;
$Dom (F_3) = \{ f_{31}, f_{32}, f_{33} \}$; $Dom (F_4) = \{ f_{41}, f_{42}, f_{43} \}$.

where, the meaning of f_{ij} are showed in Table 3.

Let C be the set of traffic control models to be selected. In this research, there are two model—volume model and delay model—should be determined. So, the class set for this research is:

$C = \{ C_1, C_2 \}$.

where, C_1— delay model;

C_2— volume model.

Therefore, the data group set D is the combination of the feature set and the class set:

$D = F \cup C = \{ F_1, F_2, F_3, F_4; C_1 \text{ or } C_2 \}$.

For inducing the decision tree, we selected 36 data groups randomly from the gotten 66 data groups as the training data set (training data set 1). Based on this training set, by applying the information entropy measure, a decision tree (DT1) for determining the AGV's traffic control model was induced as shown in Figure 3. From here, we can determine which traffic control model should be used for each intersection.

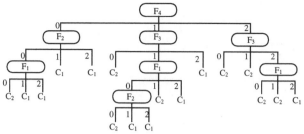

Figure 3 Decision tree 1 (DT1) based on training data set 1 for determination.

Figure 4 shows one distribution figure (distribution 1) of AGVs, when the busy level of production is normal ($F_1 = 1$), the production quota is medium ($F_2 = 1$) and its time is at 10:00 in the morning ($F_3 = 0$). There are 20 AGVs working simultaneously in this case. The number of AGVs to go through each intersection is different, for which different traffic control model should be used. We use the decision tree shown in Figure 3 to select the traffic control model.

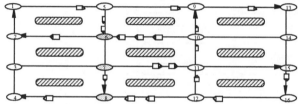

Figure 4 Distribution 1 of AGVs under ($F_1 = 1$, $F_2 = 1$, $F_3 = 0$).

For indicating how to use decision tree to determine the control model, we choose the 6th, 7th and 8th intersections as examples. As shown in Figure 5, "_____" line is the determining path for the 6th intersection, "..........." line is for the 7th intersection, and "—.—.—" line is for 8th intersection. For the 6th intersection, when the test data $t = \{ 1, 1, 0, 2 \}$ is inputted into the decision tree, the feature at the root is tested at first. Since the root of the decision tree is the feature "F_4", the feature "F_4" of the test data is examined firstly. That the value of F_4 for the 6th intersection equals "2" leads the test to go to the subtree on the right side. Then test this subtree's root feature which is feature F_3. Since the F_3's value of the test data is "0", the path goes to the left side which is a leaf marked with the kind of class "C_2". So the determination's result is "C_2" which means to use the volume model. This result is as the same with the result of the simulation experiments.

Using the decision tree shown in Figure 3, we can determine the control model for each intersection in the working area shown in Figure 4. The results are shown in Table 3.

Table 3 The control models for all intersections in distribution 1 selected by ID3

No.	1	2	3	4	5	6	7	8	9	10	11	12	13	14	15	16
no. of AGVs	0	1	0	1	1	5	0	3	2	1	3	1	1	0	0	1
F_4	0	0	0	0	0	2	0	1	1	0	1	0	0	0	0	0
model	C_1	C_1	C_1	C_1	C_1	C_2	C_1	C_2	C_2	C_1	C_2	C_1	C_1	C_1	C_1	C_1

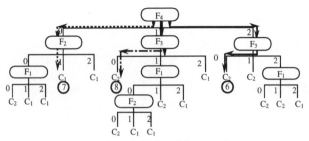

Figure 5 Decision paths in DT1 for all intersections.

Figure 6 shows another distribution (distribution 2) of AGVs, when the features $F_1 = 2$, $F_2 = 2$, $F_3 = 2$. In this circumstance, according to our assumption, 10 AGVs are enough for the working task. The result of the determination of control model is shown in Table 4.

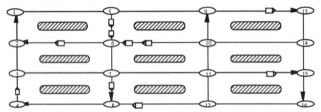

Figure 6 Distribution 2 of AGVs under $(F_1 = 2, F_2 = 2, F_3 = 2)$.

Table 4 The control models for all intersections in distribution 2 selected by ID3

No.	1	2	3	4	5	6	7	8	9	10	11	12	13	14	15	16
no. of AGVs	0	1	1	0	0	4	0	2	0	0	0	0	1	0	1	0
F_4	0	0	0	0	0	2	0	1	0	0	0	0	0	0	0	0
model	C_1	C_1	C_1	C_1	C_1	C_1	C_1	C_1	C_1	C_1	C_1	C_1	C_1	C_1	C_1	C_1

From these results, we can see that the learned knowledge are all included in the decision tree which is an extremely simple knowledge formalism. When we will determine to use which traffic control model, we can get the determine rule (the path) easily and quickly based on the known feature values. Since the way of determination using ID3 is completely different with the expert system, the bottleneck problem of the if-then rule based expert system can be solved.

And it also can be seen that when using ID3 to select the traffic control model, we test the feature in turn according to the value of information supplied by each feature. This is like the expert's behavior that is he first test the most important element for the determination of the traffic control model, and then test the others in turn based on his experience which is implicit. Therefore, we can say that using ID3, the importance of each feature in determination of traffic control model can be received, and the implicit knowledgement can be learned.

In order to examine the effectiveness of determining the control model of AGV using ID3, we used the decision tree 1 (DT1) shown in Figure 3 to detect the test data groups which are the remained 30 data groups (except the 36 training data groups used for inducing the decision tree) in the total 66 data groups received from the simulation experimental results. The correct rate of determination for this test data is 80.00% which can be considered high. The determination errors occurred in 6 test data groups which are shown in Table 5.

Table 5 Test data groups misdetermined by DT1 in test data groups 1

No.	F_1	F_2	F_3	F_4	results in simulation experiment	model selected by ID3
1	0	2	2	2	C_1	C_2
2	1	0	1	1	C_1	C_2
3	1	2	1	1	C_1	C_2
4	1	2	2	2	C_1	C_2
5	2	2	0	1	C_1	C_2
6	2	2	1	2	C_1	C_2

We put these 6 test data groups into the training data set 1 and got the training data set 2 which has 42 training data groups. Based on the training data set 2, the decision tree 2 (DT2) was induced shown in Figure 7.

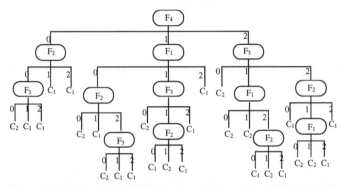

Figure 7 Decision tree 2 (DT2) based on training data set 2 for determination.

Using the DT2 to determine the traffic control model for each intersection of the distributions shown in Figure 4 and Figure 6, we got the same results as shown in Table 4 and Table 5. The reason of this is that the new decision tree (DT2) learned all knowledge included in the decision tree 1 (DT1). So if using DT1 can correctly determine the control model for each intersection under a distribution, then using DT2 can also get the same results.

Using the DT2 to examine the same test data groups as used in DT1, we got the correct rate of determination as 93.33%, which has raised. Table 6 shows the data groups in which error happened.

Table 6 Test data groups misdetermined by DT2 in test data groups

No.	F_1	F_2	F_3	F_4	results in simulation experiment	model selected by ID3
1	0	1	0	1	C_2	C_1
2	2	1	0	1	C_2	C_1

As the same, we again put these two test data groups into the training data set 2, and got the training data set 3 in which there are 44 training data groups. The decision tree (DT3) in accordance with the training data set 3 is constructed as shown in Figure 8. The correct rate of determination using DT3 for the same test data groups is 100.00%.

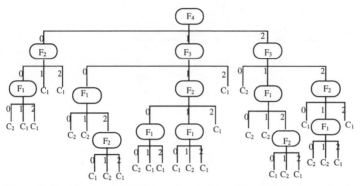

Figure 8 Decision tree 3 (DT3) based on training data set 3 for determination.

From these results we can understand that the constructure of decision tree and the correct rate of determination using decision tree is different in accordance with the different training data set. The more the correct information contained in the training data set and the more the number of the training data is, the more precise the decision tree is, which can more correctly determine the control model.

5. CONCLUSIONS

In this paper, we presented a example-based learning method — ID3 for AGV's traffic control model determination. The preliminary results presented in this paper suggest that the ID3 is useful for the decision of traffic control model. Even though the various sorts of elements influencing the determination of AGV's traffic control model will change at anytime and will be different with the training data set, using the decision tree constructed from the training data set can also select a suitable control model correctly.

At the same time, using ID3, the implicit knowledgement can be learned, and the bottleneck problem can be solved, which can make the machine to determine using which traffic control model for AGV very quickly like the human expert's behavior.

On the other hand, the results also indicate that the constructure and the correct rate of decision tree are sensitive to the number and the value of the training data sets. In order to get a perfect decision tree, the adequate and correct training data set should be used.

It can also be seen in our results that when the correct rate arise, the complexity of decision tree became bigger. In fact, the simple decision tree is expected under the same correct rate. So, it is our future work to get the simpler decision tree.

6. REFERENCES

[1] J.R.Quinlan: "Induction of Decision Tree", Machine Learning, 1,1,1986;
[2] A.Raouf & S.I.Ahmad: "Flexible Manufacturing", Amsterdam,Oxford,New York, 1985;
[3] William.R.McShame & Roger P.Roess: "Traffic Engineering", Prentice-Hall, Englewood Cliffs, New Jersey, 1990.

71

A tool of automatic generation of fault trees for complex systems

Zhiming Jian, Dongcheng Hu, Shibai Tong
Dept. of Automation, Tsinghua University, Beijing, PRC 100084

Abstract

AGFT(Automatic Generation of Fault Trees) is a problem with which system reliability analysts are concerned very much. This paper introduces a development environment of AGFT expert systems--TEE. As a tool for developing AGFT expert systems, Expert systems developed by TEE can automatically construct fault trees not only for ordinary complex systems, but also for those including control loops. The paper discusses TEE's structure, the knowledge base, the knowledge representation, and the inference algorithm.

Keywords

Fault tree analysis, automatic generation of fault trees, expert system development environment.

1 Introduction

In recent decades, fault tree analysis has been extensively applied in many fields such as aviation, space flight, nuclear energy and chemical industry to analyze system reliability and determine causes of unsafe or hazardous events. However, at any cases, the construction of fault trees is a work requiring the bulk of time unless the analyzed system is very simple. On the other hand, if two analysts analyze the same system or one analyst analyzes a system at different time, the results usually aren't the same. So, in order to reduce cost and standardize results, to develop an AGFT method is a very attractive work. Since 1973, many algorithms for AGFT have been presented, and corresponding programs have been coded. Because of some reasons(Poucet, 1990; Xie, 1992), these algorithms and programs haven't been extensively applied as their earlier proponents predicted. In recent years, some people presented utilizing knowledge-based method to carry out AGFT, and developed several expert systems, e.g., A. Poucet did in 1990, G. Xie did in 1992. However, difficulties still exist, especially when their software tools are applied in control systems. This is mainly caused by the tangled feedforward or feedback control loops in these systems.

In order to solve this problem,   we design and implement an expert system development environment--TEE. TEE is a powerful tool not only for field expert to build expert systems, but also for analysts to analyze complex practical engineering systems.

TEE is a powerful expert system development environment for field experts. Field experts can use TEE to build AGFT expert systems of generating fault trees for their field systems. The only work of field experts needs to do is to describe his related field knowledge into TEE's knowledge base by means of TEE's friendly interfaces. TEE's inference engine is specially   designed and solidified, in the other words, it is not necessary to design and construct inference engines when field experts build future expert systems. When we designed TEE's inference engine, we took in some research results of fault tree synthesis. This makes expert systems built by TEE automatically generate fault trees not only for normal complex systems but also for those containing feedback or feedforward loops.

After field experts' works(build expert systems using TEE) finish, TEE is a powerful expert system for system analysts. System reliability analysts can use TEE to construct fault trees for systems need to be analyzed.

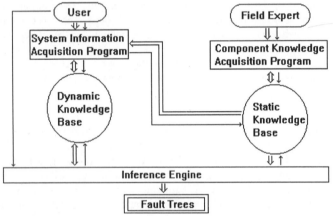

Figure 1 TEE's structure.

2 Overview of TEE

TEE's structure is shown on figure 1.

TEE takes the knowledge base of a future expert system into two parts: the dynamic knowledge base, the static knowledge base. There are frames of the two knowledge bases in TEE. The static knowledge base stores knowledge of components consisting of a kind of systems. For each component, there are three classes of knowledge: the component name, the component icon, the component I/O function and failure mode descriptions. In the dynamic knowledge base, the stored is CID(the Component Interconnection Diagram), which describes component interconnection relationship of analyzed systems. The component knowledge acquisition program mainly consists of an icon editor and a knowledge description program. The icon editor is a powerful bitmap establishment and modification program, which is used to set up icons of components. The knowledge description program provides an interactive environment, which makes field experts conveniently establish, modify the I/O function and failure mode descriptions of components and glance over them. The system information acquisition program is a diagram editor that can be used to establish, modify CID of analyzed systems, and define top events.

TEE provides two interfaces: the field expert interface and the user interface. The field experts describe their field knowledge(related to generation of fault trees) in the static knowledge base by the former, and build AGFT expert systems for automatically generating fault trees. After an expert system has been built by field experts, it can be provided to users(analysts). Users can conveniently construct fault trees for systems need to be analyzed by means of the user interface.

Considering it is possible that a user is a field expert, the two interfaces can be substituted each other easily.

3 The knowledge base and the knowledge representation
3.1 The static knowledge base and the static knowledge representation

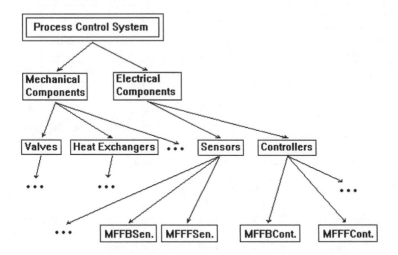

Figure 2 The classification structure of the static knowledge base.

The static knowledge base contains characters of all components(in function and failure cases) consisting of a kind of systems. In the static knowledge representation, semantic network, frame and rule knowledge representation models are applied. Semantic network represents the layer and classification structure of the static knowledge base; The frame is used to represent the attributes of each component; Rules describe the I/O function and failure mode of the attributes.

Generally, semantic network is applied in representing the knowledge of some related objects and concepts. In TEE, components constituting systems are organized by classifying components. Semantic network is suited to represent this classification and inheritance relationship. Taking process control system as an example, figure 2 shows the classification structure of the static knowledge base(It is similar to Poucet's component KB1.). The classification structure is interactively constructed by field experts using TEE's expert interface. It is different according to different kinds of systems, and for the same kind of systems, it is possible that the classification structures constructed by different experts are different. Obviously, the more reasonable the structure organization is, the faster the searching for components in the course of inference is. So, to some extent, the reasonability reflects the level of an expert.

NAME	ICON	RULES	
		IF	THEN
process_control_system. electrical_component. controller. MMFBcontroller	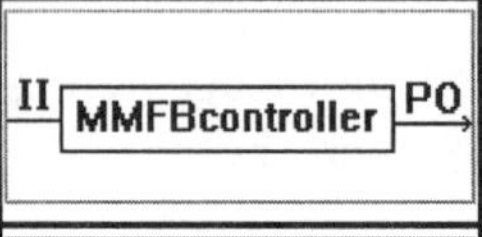 II $\rightarrow$ MMFBcontroller $\rightarrow$ PO	II	PO(+1)
		MMB Cont.RV **AND** II	PO(-1)
		MMB Cont. BR **AND** II	PO(0)
process_control_system. electrical_component. sensor. MSsensor	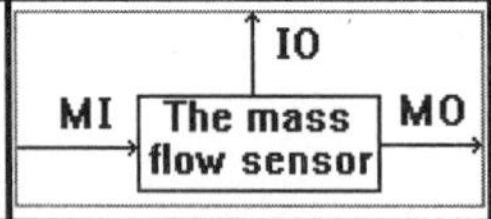 IO / MI $\rightarrow$ The mass flow sensor $\rightarrow$ MO	MI	MO(+1)
		MI	IO(+1)
		MSen. BR **AND** MI	IO(0)
prosess_control_system. mechanical.component. valve. Pcontrol_valve	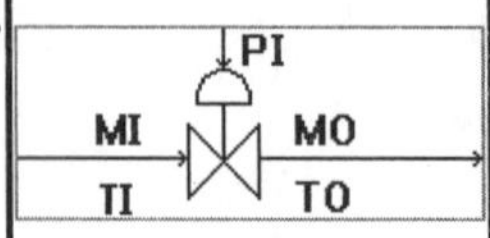 PI MI $\rightarrow$ MO II TO	MI	MO(+1)
		TI	TO(+1)
		PI	MO(-1)
		PVa. RV **AND** PI	MO(+1)

Figure 3 The structure of the component knowledge frame.

As mentioned before, the attributes of a component include three parts: name, icon and rules. In TEE, the three parts are organized by a frame, that is to say, a frame with three slots is defined to describe each component. Figure 3 shows the structure of the component knowledge frame. In the component name slot, in order to keep the completeness and uniqueness and to make it convenient to search, components use their classification names. For instance, the classification name of a mass flow feedforward controller is:

process_control_system.electrical_component.controller.MFFFcontroller

The icon is also an important slot in the component knowledge frame, because the icon is a main tool for identifying each component in CID, and the inference engine also acquires system constitution and component interconnection information by icons from CID. Icons are described and stored by bitmaps.

The most important slot in the component knowledge frame is the rule slot, it includes all rules describing I/O functions and failure modes of a component. Generally, a logic system composed of numbers -10, -1, 0, +1, and +10 is used to describe the I/O functions and failure modes(Lapp&Powers, 1977, 1979; Andrew, 1990). '+' and '-' denote directions of the I/O gains. '0', '1' and '10' denote the magnitude. The inputs and outputs of a component are defined by some system variables: 'M' means mass flow; 'T' means temperature; 'P' means pressure; 'I' means electric current; and etc. 'I', 'O'(as the second letter) and numbers '1', '2', '3', ... are used to separate different input ports and output ports, e.g., "MI2" expresses the mass flow of the second input port.

Figure 3, taking several components that ever appear in process control systems as examples, shows the structure of the component knowledge frame.

3.2 The dynamic knowledge base and the dynamic knowledge representation

What stored in the dynamic knowledge base is CID of analyzed systems and top events assigned by analysts. The reason utilizing CID to represent the analyzed system is that, CID of the analyzed system includes the system information needed by constructing fault trees of systems:

    (a) Components consisting of the analyzed system;

    (b) Interconnection relationship of the components.

In order to represent and save this information, TEE defines several general data structures respectively for lines, texts, components of CID. For example, the general data structure of components is defined as follow:

```
struct component {
    int x, y;
    char *classname;
};
```

(x, y) is the coordinate of a component icon in CID. It indicates the position of a component in CID. *classname* indicates the type of the component. CID is interactively drawn on screen by users using the diagram editor. By means of searching the static

knowledge base, the diagram editor can provide a menu of component icons for users. When a component is selected, the component's icon is popped out and shown on screen. Users can flexibly move the icon   on screen. So, the data above are generated automatically.

When CID of a system has been edited, the analysis to the system may begin. The first step of the analysis is to assign a top event by means of the top event description program. Its course is: At first, the top event description program traces   CID of the analyzed system, defines the I/O lines of the system and the connection lines among components as nodes. Then, a top event can be described by the system variable disturbance on a node. Figure 4 shows the CID of a mass flow control system(For a process control system, CID is its P&ID--Pipe&Instrument Diagram. In figure 4, nodes have been defined.). One of its top events--"a moderate increase of the mass flow on node 1" can be described as "M1(+1)".

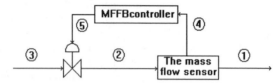

Figure 4 A mass flow control system(CID).

4 TEE's inference engine

As mentioned previously, TEE's inference engine is designed specially and fixed. This is the main difference between TEE and general expert system development tools. Because of this, TEE is only applied in developing AGFT expert systems. In order to make expert systems built by TEE can generate fault trees not only for ordinary complex systems but also for those with tanglesome control loops--complex control systems, the mechanism that identifies loops(negative feedback loops, negative feedforward loops) and carries out special operators when these loops are encountered is introduced into when designed TEE's inference engine.

CID of the analyzed system cannot be taken as the basis of inferring to generate fault trees, because CID does not include the I/O functions and failure modes of components consisting of the analyzed system. So, before inferring to generate fault trees, the inference engine must get a system representation which includes the related knowledge of the system such as the top variable(the system variable describing the top event), the system constitution, the I/O functions and failure modes of components consisting of the system. In TEE, the representation is a table named the relation table.

Figure 5 shows the inference algorithm of generating fault trees. STACK_F is a stack used to store the developed mid-events, it abides by the FILO law. Sometimes, a system variable exists in a NFFL and exists in a NFBL at the same time, so, after having performed the NFFL, NFBL operators, it is possible that some fault tree branches are developed repeatedly. Therefore, in order to avoid the repetition, to inspect the developed fault trees and to cancel the duplicated fault tree branches are necessary.

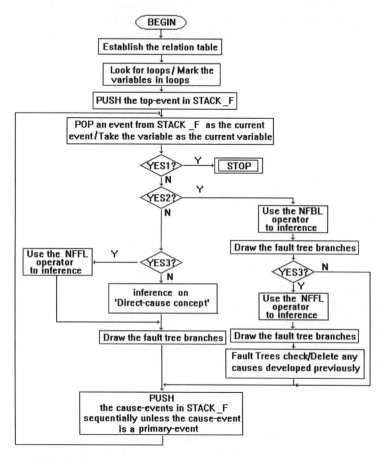

YES1:STACK_F empty;
YES2:The current variable is the first variable of a NFBL;
YES3:The current variable is the terminating variable of a NFFL;

Figure 5 The inference algorithm of generating fault trees.

5 Conclusion

TEE is a tool developed in MS-Windows environment. With friendly interfaces, TEE can be applied conveniently in building AGFT expert systems for AGFT of electrical systems, process systems, process control systems and other similar systems. After TEE having been finished primarily, we developed an experimental expert system "FTCES" generating fault trees for process control systems. The results are satisfactory.

At present, TEE still is a prototype system. There are some perfective works need to be done.

Acknowledgment

This work is being supported by the Chinese Natural Science Foundation .

References

Poucet, A.(1990), STARS: Knowledge based tools for safety and reliability analysis, *Reliability Eng. &Sys. Safety,* 30, 379-397.

Xie,G.(1992), The study on the expert system for fault tree construction(in Chinese), Doctoral dissertation, Institute of Nuclear Energy Technology, Tsinghua University, Beijing.

Lapp, S. A.&Powers, G. J.(1977), Computer-aided synthesis of fault trees, *IEEE Trans. Reliability,* R-26, 2-13.

Lapp, S. A.&Powers, G. J.(1979) , Update of Lapp-Powers fault tree synthesis algorithm. *IEEE Trans. Reliability,* R-28, 12-14.

Andrew, J.(1990) , Application of the digraph method of fault tree construction to a complex control configuration, *Reliability Eng. & Sys. Safety,* 28, 357-384.

Shi, Zhongzhi(1987), Knowledge engineering, Tsinghua University Press.

Biography

Mr. Zhiming Jian: Earned Bachelor of Engineering from the Huazhong University of Science and Technology(Wuhan, P. R. China) in July 1989. Earned Master of Engineering from Wuhan Institute of Technology(Wuhan, P. R. China) in Sept. 1991. At present, is a Ph.D. Candidate(Dept. Automation, Tsinghua University, Beijing, P.R.China). Current research interests include: Reliability Engineering, Human Factors, Expert System, and Object-oriented Method.

72

A Knowledge-Based Advisor for Determination of Limits and Fits

C K Mok and S Y Wong
Department of manufacturing Engineering
City University of Hong Kong
Tat Chee Avenue
HONG KONG
Fax : (852) 788 8423

Abstract

The determination of limits and fits exists in nearly all engineering design works which involve mating parts. Their selection is a crucial process in achieving functional requirements and optimum cost of manufacture. This paper describes the development work of a prototype knowledge-based system (KBALF) capable of determining limits and fits. When relevant information for a given application is inputted, the system can help engineering designers through the whole process of determination of fits, ranging from recommendations of fits for assemblies, calculation of limits of sizes, to production of fit specification documents. By incorporating database of standard nominal sizes, the system also facilitates the process of standardization which in turn can reduce the cost of manufacture.

Keywords

Knowledge base system, limits and fits, CAD, standardization.

1. INTRODUCTION

The need for limits and fits for engineering parts was brought about mainly by inherent inaccuracy of manufacturing methods, coupled with the fact that `exactness' of size was found to be unnecessary for most workpieces. In order that functions should be satisfied, it was found sufficient to manufacture a given part so that its size lay within two permissible limits, i.e. a tolerance, this being the variations in size acceptable in manufacture. Similarly, when a specific fit condition is required between mating parts, it is necessary to ascribe an allowance, with positive or negative, to the basic size to achieve the required clearance or interference.

Almost all engineering design tasks involving assembly of parts necessitate the determination of limits and fits. The decision is an important process in achieving functional requirements and optimum cost of production. Thus the incorporation of an knowledge-based system in this aspect in a CAD system is deemed necessary.

2. FUNDAMENTALS OF LIMITS AND FITS

The fits between cylindrical parts are designated by custom as ` holes ' and `shafts ', in which case the term ` size ' refers to the diameter of the mating parts; whereas for non-cylindrical parts, it refers to a length width or other dimension.

The disposition of the tolerance zone for a HOLE relative to the basic size is denoted by a capital letter and the magnitude, or grade, of the tolerance by a suffix number, e.g., H7; whereas for a SHAFT , by a small letter and the magnitude, or grade, of the tolerance by a suffix number, e.g., p6. A fit is describe by a combination of symbols : H7-p6 or H7/p6.

3. SYSTEM DESIGN

3.1 Modular Approach

The present system is designed with an objective to provide the following functions related to limits and fits determination :-

(I)   recommendations and explanations for fits for mating parts;
(ii) calculation of limits of sizes;
(iii) production of fit specification documents; and
(iv) auto-draughting routine to specify fits in tabular form as well as on assembly drawings.

In order to have a more flexible structure, make easier for modifications and facilitate for future expansions to cover other areas related to selection of limits and fits, such as in-house standardization, process capability, process selection, etc., all the aforesaid functions are designed under a modular approach. Fig. 1 shows the block diagram of KBALF.

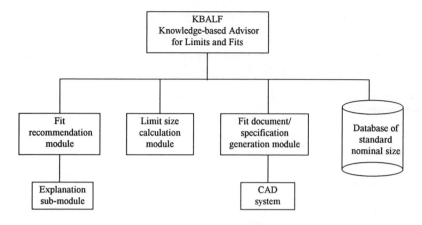

Fig. 1 Block diagram of KBALF

3.2 International Standard Used

The present must be designed on basis of an international standard as most companies adheres to some international standards.

The ISO Limits and Fits System was chosen as the basis for developing the system in view of the facts that it is the most widely used international standard fits can be selected with regard to part size. The ISO system provides a comprehensive ranges of limits and fits for engineering purposes. It is based on a series of tolerances graded to suit all classes of work from the finest to the coarsest. These tolerances are intended for most general applications and should be used whenever a graded series of tolerances is needed; and they are not restricted solely to diameters, although only cylindrical parts are referred to explicitly in the standard. They may also be applied to the width of a slot, the thickness of a key, etc., and to lengths, heights and depths. The range of fits provided by the standard is very comprehensive, but a guite small selection will satisfy most normal requirements.

3.3 Knowledge Domain

With a view to design a generalized system, various sources of knowledge have been taken into account for developing the knowledge domain of the knowledge-based system. The sources considered are as follows:-

(i)   Publications
      a. International standards

b. Books on mechanical engineering design and engineering tolerances

c. Journal papers

(ii)   Both authors have more than ten years experience in engineering design works.Their extensive experience and knowledge in this aspect are captured in the present system.

(iii)   Other relevant knowledge acquired through discussions with designers of some major local companies.

4. SYSTEM DEVELOPMENT

4.1 Fit Recommendation Module

The four components of fit specifications are considered in the present system. Rules are then applied to suggest hole type, hole tolerance grade, shaft type and shaft tolerance grade individually. For example, hole type `H' is always used for the hole basis system. Shaft types are chosen according to the functional requirements of a given application. Tolerance grades are selected based on the precision level required.

BS4500 provides 28 types of standard Holes and Shafts and a series of tolerances grades, so as to cater for a wide range of applications. To use all 28 types of standard holes and shafts results in 784 (28 x 28) combinations of fits. Nevertheless, the fit conditions required for the majority of engineering products can be met by a quite limited selections of fit combinations. Thus it would cause serious problems in design, planning, purchasing, production, repair and maintenance ...etc resulting high cost on the products; if one uses all the fit combinations. In view of this the BS4500A recommends some selected Holes and Shafts as indicated below for general engineering applications :-

Selected Hole: H7, H8, H9, H11

Selected Shaft: c11, d10, e9, f7, g6, h6, k6, n6, p6 s6.

Standard fits are achieved by combination of the above selected holes and shafts. All the data of those holes and shafts are stored in the module in form of lists. Incorporated within this module, there is an explanation sub-module which can provide explanations for the reasoning behind a recommended fit and tolerance specification. This sub-module is important for training inexperienced designers, while experienced designers can skip it.

4.2 CAD Interface Module

In most cases, it is necessary to specify a number of fits of a given product or assembly on an assembly drawing as well as in a tabular form. This module was set up to connect the fit knowledge-based system with a CAD system to realize the aforesaid purposes. This module can automatically generate the fit specification drawings.

The present system is implemented on an IBM personal computer which is installed with the AutoCAD system. An interface program is needed to convert the results of the knowledge-based system to the CAD software for generating the fit specification drawings on location of the assembly drawing as specified by the designer. The AutoLISP which is specially provided for use with the AutoCAD system is adopted for writing the interface program. It is a modified version of the common A.I. language `LISP`.

4.3 Limit Size Calculation Module

The limit size calculation module is used for calculating the limits of size of the mating parts for a fit such as maximum and minimum limits of size of the male part, maximum and minimum limits of size of the female part, and maximum and minimum clearance or interference of the fit based on data of standard tolerances and limit deviations included in ISO 286-1 & 2 : 1988.

5. SYSTEM OPERATIONS

5.1 Main Menu

The menu system is developed through a system of pull-down menus. The designer can pull down the menu from the tie bar and select the appropriate command from the pull-down menu. The layout of the menu system is shown in Fig. 2. Details of these menus are described in the subsequent paragraphs.

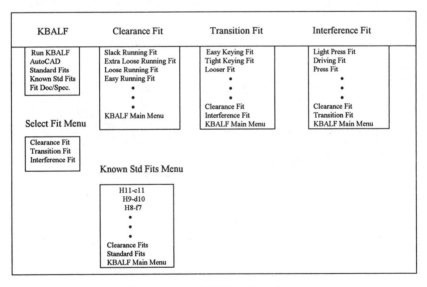

Fig. 2 Layout of KBALF Menu System

From left to right, the first pull down menu consists of two menus, KBALF, the main menu and Select Fit, its sub-menu. KBALF menu is the main menu to activate KBALF, flick back to AutoCAD menu and selections of fits menus. Select Fit menu enables the designer to select a standard fit among clearance fits , transition fits and interference fits.

The second pull down menu consists of Clearance Fit menu and Known Std Fits menu. Clearance Fit menu consists of a series of standard clearance fits such as slack running fit, extra loose running fit,..., etc. The last item KBALF Main Menu of this menu enables the designer to go back to KBALF main menu for other choices if he cannot find any suitable fit in this menu. The sub-menu 'Known Std Fits menu' provides a series of known standard fits in BS4500A system, such as H11-c11, H9-d10, H9-e9, ..., etc. for the designer to choose standard fits with which he is familiar. The last three items let the designer page between other menus.

The third pull down menu is Transition Fits which is for the selection of transition fits. The fourth full down menu is Interference Fits which is used for the design of interference fits. Through the menus system a designer can run the KBALF software and select appropriate fits according to BS4500A hole basis system.

5.2 Example

An example of the application of KBALF for a hydraulic cylinder is presented in this section. For this assembly the designer can determine the five fits with the help of KBALF. After the determination of fits is completed, a fit specification drawing can be produced on the assembly drawing as shown in Fig. 4. At the same time a fit document list can also be generated as shown in Fig. 5.

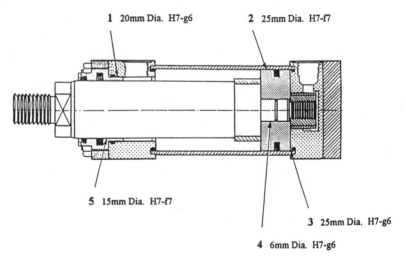

1 20mm Dia. H7-g6   **2** 25mm Dia. H7-f7

5 15mm Dia. H7-f7

3 25mm Dia. H7-g6

4 6mm Dia. H7-g6

Fig. 4 A fit specification drawing for a hydraulic cylinder

5	Rod	Bush	H7-f7
4	Rod	Piston	H7-g6
3	Cap	Tube	H7-g6
2	Piston	Tube	H7-f7
1	Bush	Head	H7-g6
Item No.	**Male/Shaft Part**	**Female/Hole Part**	**Fit Specification**

Fig. 5. A fit document list for a hydraulic cylinder

6. CONCLUSIONS

A CAD-integrated 'knowledge-based system' capable of determining limits and fits has been developed using LISP in the AutoCAD environment. The present system can help engineering designers through the whole process of determination of fits, ranging from recommendation of fits for assemblies, calculation of limits of size, to generation of fit specification documents. This advisory system can also facilitate the process of standardization which in turn can reduce the cost of manufacture.

The modular approach provides a more flexible structure which can allow for easier modifications and knowledge enhancement. It also facilitates future expansions to cover other areas related to limits and fits, such as in-house standardization, process capability analysis, process selection, etc.

REFERENCES

[1] Autodesk, Inc., 'AutoCAD Release 12 Programmer's Reference Manual'.

[2] Autodesk, Inc., 'AutoLISP Release 12 Programmer's Reference Manual'.

[3] British Standards Institution, 'BS 4500 : ISO limits and fits', 1990.

[4] Oyvind, B., 'Computer-aided Tolerancing', ASME Press, NY, 1989.

[5] Greenwood, D.C., 'Product Engineering Design Manual', McGRAW-HILL.

AUTHORS' BIOGRAPHY

Mr Chiu-Kam MOK is Lecturer in the Department of Manufacturing Engineering.

Mr Seung-Yin WONG is former research graduate in the Department of Manufacturing Engineering.

73

Integrating manufacturing database and equipment

Z.K. Lu**, E.H.M. Cheung* & K.B. Chuah*
*Department of Manufacturing Engineering, City University of Hong Kong,
HONG KONG   Tel:852-7888420; Fax:852-7887660
**Department of Computer Science, Nanjing University of Science &
Technology, CHINA   Tel:86-25-432333;   Fax:86-25-431622

Abstract

One of the important features of a manufacturing database is that it must be able to communicate effectively with the different elements at different levels of the manufacturing system. For instance, we need to transfer schedule instructions to certain equipment or machines to control the sequence of their operation; download NC programs to appropriate machine tools; capture real-time production data like the status of certain equipment or work-in-progress etc. Therefore, in manufacturing database design we need to consider its method of communication with the equipment or other system elements and the related data structure in the database physical model. This paper presents the method of communication between an Rdb/VMS manufacturing database and the equipment or other elements in a CIM system via BASEstar and discusses how the data related to the equipment and elements are analysed. It also describes the mechanism for ensuring data integrity in the communication between the database system and equipment or elements of CIM system.

Keywords

Manufacturing database, communication, CIM system, BASEstar

1   INTRODUCTION

CIM is essentially about effective communication between the manufacturing database and the different elements or equipment of the manufacturing system it serves. The operation of a manufacturing system requires for example, the issuing of production schedules to shopfloor equipment or machines to control the sequence of their operation; the downloading of NC part programs to appropriate machine tools for the manufacturing of certain parts; the capturing of useful production data such as equipment/machine status or work-in-progress etc. We must organize and manage the data in the manufacturing database in such a way that meets these communication requirements (Chryssolouris, 1992). Thus, early in the manufacturing database design stage, we need to consider and decide with care its method of communication with the equipment or other system elements and the related data structure in the database physical model.

VAX/VMS provides the BASEstar software integration platform which can help to bridge the communication gap between equipment and the various application systems of which the manufacturing database system is one. In short, the BASEstar software is designed to facilitate the integration of manufacturing applications and equipment, accelerate the development of applications and device connections, and provide an architecture for the consistent development of manufacturing systems. It also provides the callable services modelled after the VMS system services which provide the building blocks of application and device integration. The manufacturing database can make use of these callable services to define and manipulate data in the BASEstar system to communicate with the various system elements and equipment. The manufacturing database we are talking about here is a relational database system based on the Rdb/VMS. To access the data in this database system, we must use the SQL which provides the programming environment of Rdb. We also need to link this database system with BASEstar via SQL/Services. The SQL/Services then provides application access to Rdb/VMS database from remote computers running a wide variety of operating systems following the client/server model.

2   COMMUNICATION OF MANUFACTURING DATABASE AND EQUIPMENT

The key BASEstar component used to integrate manufacturing applications and equipment is its application programming interface (API). The API is a shareable image containing entry points known as callable services. The BASEstar callable services offer developers the ability to:

- Share data among applications
- Control and monitor device operations as well as collect data from plant devices
- Communicate between programs using event-driven messaging
- Control application processing
- Configure some elements of the manufacturing environment

The BASEstar data management callable services provide developers with a common mechanism for defining, organizing, and accessing data in an integrated manufacturing environment. It includes the ability to collect, manage, and distribute current value plant data, automatically notify applications of critical changes in plant information, and export data to an Rdb/VMS database for permanent storage.

Based on a CIM environment of VAX/VMS system and application softwares, the proposed manufacturing database is to have a structure as shown Figure 1. There are three levels of data in this database. It is designed to serve the CIM system currently being developed (Cheung et al, 1994 $_{a\&b}$) at the City University of Hong Kong. The first level is the main database which supports all the manufacturing functions of this CIM system. The second level is a host of equipment controller bases which are distributed on the different services to control the respective equipment. It is the bridge between the main database and equipment. It serves to accept instructions from the main database, send them to the appropriate equipment, record the status and other information coming from equipment and maintain or update the related data in the main database. The third level provides the equipment interface based on the BASEstar software. The data at this level are used to describe the current status of the manufacturing system; control and monitor the equipment; and control other application processing.

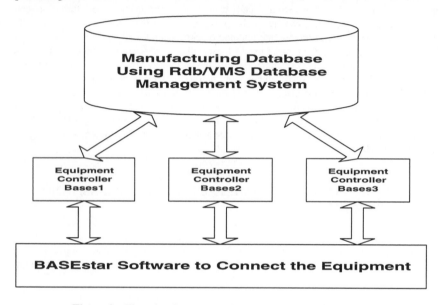

Figure 1   Three-level structure of proposed manufacturing database

3   DATA ANALYSIS RELATED TO THE EQUIPMENT

The information required by the management from the manufacturing database would normally be the manufacturing system's main performance indicators like: (1) the status of current jobs; (2) the status of equipment; (3) the setup and processing operation times etc. The computation of these indicators should take into account the requirements or specifications of schedule, quality levels, equipment maintenance, safety, personnel training, and the effects of ongoing efforts such as setup time and operating-time reductions. A distribution of these indicator data among various equipment is to be provided in order to enhance performance. These indicator data are maintained decentrally and the overall data management is performed by a system-wide, global database system.

In the manufacturing database, the data related to the equipment in the shopfloor are such items as: PartL/ULStation, ToolL/ULStation, AGV, Robot, MeasuringEquipment, Operation, MeasuringRecord, AGVSchedule, Equipment, EquipmentStatusReport. In Figure 2, when an instruction in Operation is translated to the equipment, it would effect the relations:

Equipment, Workpiece, Pallet and FixtureElement. When the status of an equipment is be recorded in the database, it would also effect the relations: Equipment, Workpiece, Pallet, FixtureElement and EquipmentStatusReport. For the same reason, the status of the robot would effect the relations: Robot, Workpiece, Pallet, Fixtureelement, Tool_Using and ToolMagazine. When the AGVSchedule is to be translated to the AGV, it would effect the relations: Workpiece, Warehouse, Buffer, Tool_Using and CentralToolBase. The status of AGV would also effect the Workpiece, Warehouse, Buffer, Tool_Using, CentralToolBase and AGV relations. The result of Measuring Equipment would effect the relations: Workpiece, Tool_using. The status of part load/unload (L/UL) station would effect the Workpiece and Warehouse relations, and the status of tool load/unload (L/UL) station would effect the Tool_Using, ToolMagazine and CentralToolBase relations.

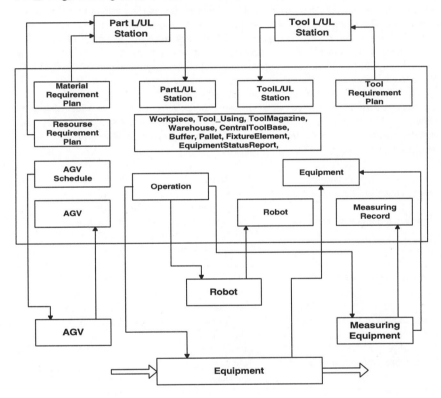

Figure 2 Relationship of data in database and equipment on shopfloor.

Data are defined and referenced in a BASEstar system using logical points. A logical point represents a data item derived from the manufacturing environment. It is stored for immediate retrieval in the VAX memory. The value of a logical point can be derived from any of the following sources:

- User input
- Application program
- Mathematical expression
- Plant device
- Point mapping

Users can create logical point definitions and set their value using the command line interface and the menu interface. The BASEstar's data management callable services perform these functions from application programs. In addition to deriving the value of a point from a user or application input, a point value can be the result of a mathematical calculation called expression. Logical points can also be equated to values stored in the memory of a plant device. Device data defined as physical point is collected upon request. Logical points can be connected to physical points to integrate data from plant equipment into the BASEstar application environment.

Once the data is defined as logical and physical points, a BASEstar software can automatically distribute data values when changes occur; equate the value of a logical point to another logical point through point mapping; and export data to other systems such as other databases, applications or computer systems. Examples of these logical points in the manufacturing system are:

• Data related to Operation:	instruction	text(20);
• Data related to Equipment:	status	text(1)
	instruction_no	text(6)
• Data related to AGVSchedule:	time	number(6)
	holding_thing	text(8)
	from_buffer_no	text(4)
	to_buffer_no	text(4)
• Data related to AGV:	position_z	number(4)
	direction	3*number(3)
	location	2*number(5)
• Data related to Robot:	position	3*number(4)
	direction	3*number(3)
	location	2*number(5)
• Data related to MeasuringRecord:	choices	text(6)
	precision	text(6)
• Data related to MaterialRequirementPlan:	date	date
	time	number(6)
	material_no	text(8)
	plan_qty	number(4)
	input_buffer_no	text(4)
• Data related to ResourceRequirementPlan:	date	date
	time	number(6)

	resource_no	text(6)
	plan_qty	number(4)
	input_buffer_no	text(4)
• Data related to	date	date
ToolRequirementPlan:	time	number(6)
	tool_no	text(6)
	equipment_no	text(4)
	plan_qty	number(4)
	input_buffer_no	text(4)
• Data related to	status	text(1)
PartL/ULStation:	send_instruction	text(6)
	finished_instruction	text(6)
• Data related to	status	text(1)
ToolL/ULStation:	send_instruction	text(6)
	finished_instruction	text(6)

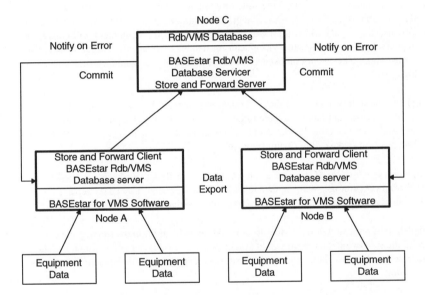

Figure 3 Method of exporting data from equipment.

As mentioned earlier, we can create logical point. We can also connect the logical point to a shopfloor device through BASEstar Device Connect Management (DCM) or to another device connection system. In this case, we must convert the ASCII logical point name into internal binary format which can then be used in any BASEstar Kernel callable services to reduce processing overhead. We can define a logical point as a data structure. This will allow efficient processing of complex information by way of adding a data structure definition and then defining the fields within the structure.

Production or shopfloor data collected by BASEstar software can be sent to the manufacturing database. To export data, an external map is created to determine a format for the data. The external map matches the format of BASEstar data that is necessary in the database system. BASEstar data is exported to database using a database server that services as a link between BASEstar software system and the database system as shown in Figure 3. An external map is created so that data elements in a set of logical points can be mapped into a corresponding set of fields in a database table. Each element in an external map can then be added to an external map by the BASEstar callable service.

4   DATABASE INTEGRITY CONSTRAINTS

The manufacturing database system described above allows for different users or programs to access the data. The information collected from equipment or other system elements in the manufacturing system can also change the state or content of the database. Therefore, we need to provide a mechanism to maintain the integrity constraints in the manufacturing database system. Because the events which are invoked by the state of the equipment will change the state of manufacturing database, the manufacturing database will automatically maintain the integrity constraints. The events invoked by the equipment in the manufacturing system are listed below:

- Completion of an instruction by robot and other equipment- This event would effect such data as: instruction_no in the data table Equipment; instruction_no, equipment_no, position, status in data table Workpiece; location, position, rotation in data table Pallet; and location, position in data table FixtureElement.

- Completion of a task by AGV- This event would effect such data as: buffer_no in data table Workpiece; storage_status in data table Warehouse; buffer_no in data table Tool_Using; and storage_status in data table CentralToolBase.

- Completion of an instruction by Measuring Equipment- In addition to effecting the data listed in the first event, this event would also effect such data as: weight, volume in data table Workpiece; and volume, tipradcorr, lworkmax, diamaximum in data table Tool_Using.

- Completion of a task by Part L/UL Station- This event would effect such data as: buffer_no in data table Workpiece; and storage_status in data table Warehouse.

- Completion of a task by Tool L/UL Station- This event would effect such data as: buffer_no in data table Tool_Using; and storage_status in data table CentralToolBase.

These events are invoked by different equipment or other system elements which may be located physically at different sites. We must process these events in proper time to maintain the states of database and ensure the integrity in the manufacturing database which is concurrently being used by several users or application softwares. To do this, we need to introduce the distributed transaction concept which uses the two-phase commit protocol to ensure the consistency of the database system. This is further elaborated elsewhere (Cheung et al, 1994 $_{a\&c}$).

5   CONCLUSION

This paper reports the result of the final phase of a CIM system database project carried out at the Department of Manufacturing Engineering, City University of Hong Kong. Earlier efforts on the database's conceptual design, logical and physical modelling etc. were presented elsewhere. In this paper, a method of communication between an Rdb/VMS manufacturing database and the equipment or other elements in a CIM system via the BASEstar platform is described. The paper also discusses the data integrity constraints adopted in this database system.

6   ACKNOWLEDGEMENT

The authors would like to thank the Croucher Foundation for providing a research grant for this project. They would also like to acknowledge the support provided by the Department of Manufacturing Engineering of City University (formerly City Polytechnic) of Hong Kong and Nanjing University of Science & Technology.

7   REFERENCES

Cheung, E.H.M., Chuah, K.B. and Lu, Z.K. (1994$_a$) Logical Design of A manufacturing Database in CIM Environment, Croucher Foundation Internal Report, Manufacturing Engineering Department, City Polytechnic of Hong Kong, March, 1994

Cheung, E.H.M., Chuah, K.B. and Lu, Z.K. (1994$_b$) The Conceptual Design of a Database System in an Educational CIM Environment, Proc. International Conference on Data and Knowledge Systems for Manufacturing and Engineering, Hong Kong, May 2-4, 1994, 443-448

Cheung, E.H.M., Chuah, K.B. and Lu, Z.K. (1994$_c$) Physical Design of A Manufacturing Database in CIM Environment, Croucher Foundation Internal Report, Manufacturing Engineering Department, City Polytechnic of Hong Kong, June, 1994

Chryssolouris, G. (1992) Manufacturing Systems: Theory and Practice, Springer-Verlag, New York

Steudel, H.J. (1992) Manufacturing in the Nineties, Van Nortrand Reinhold, New York

VAX/VMS Workstation series: BASEstar Kernel User's Guide; BASEstar Callable Service Reference; Distributed Transaction in the RDB/VMS.

8   BIOGRAPHY

- *Z.K. Lu*

Mr Lu graduated with BSc degree in Science in 1982 and later obtained his MSc degree in Computer Science in 1991 from Nanjing University of Science and Technology. He has joined the Department of Computer Science in Nanjing University of Science and Technology since 1982. He worked on a CIM development project funded by the Croucher Foundation in City University of Hong Kong between 1993 and 1994. His current research interests are CIM system analysis, design & development and Database system design.

- *E.H.M. Cheung*

Dr Cheung obtained his MSc in mechanical engineering from University of Manchester Institute of Science and Technology and PhD from University of Manchester, U.K. His PhD research was in design of textile machinery. He is a member of the Institution of Mechanical Engineers, a Fellow of the Institution of Electrical Engineers, a Chartered Engineer and a Fellow of the Hong Kong Institution of Engineers. He joined the Department of Manufacturing Engineering, City University (formerly City Polytechnic) of Hong Kong in 1987 when the Department was established and is now the Associate Head of Department. His current interest is CIM system development and implementation, flexible automation and laser material processing.

- *K.B. Chuah*

Dr Chuah graduated with a mechanical engineering degree from Loughborough University, U.K. His PhD research was in Shiphull Roughness and Hydrodynamic Drag, a project funded by the US Navy. Since then, he has worked in several U.K. universities as research fellow and lecturer before joining the City University (formerly City Polytechnic) of Hong Kong in 1990. His main research areas are surface metrology, CIM system development & implementation and project management.

74

Intelligent evaluation in FMS simulation

E.H.M. Cheung & K.B. Chuah
Department of Manufacturing Engineering,
City University of Hong Kong, Hong Kong.
Fax: 852-788-8423   Phone: 852-788-8407 & 852-788-8437
E-mail: meedche@cityU.edu.hk & mebchuah@cityU.edu.hk

X.N. Li
Department of Manufacturing Engineering,
Nanjing University of Science & Technology, Nanjing, P.R. China.
Fax: 025-443-1622   Phone: 025-443-1512 ext 2615

Abstract

Traditional evaluation process of FMS simulation mainly depends on users reading and analyzing the statistical data lists or status data files. This is time-consuming and rather inefficient. Moreover, different interpretations on a FMS performance and problems such as where is the real bottleneck of a FMS may be deduced by different users of a FMS simulator who have different levels of understanding of FMS and simulation theory and practice. This paper introduces an artificial intelligent (AI) method to aid the evaluation process. The application of AI technique for the evaluation shows that efficiency can actually be increased in the evaluation process and more accurate conclusions of the analysis can be obtained.

Keywords

Artificial intelligent, FMS simulator, intelligent evaluator, relationship tree.

1   INTRODUCTION

FMS simulator can be a very powerful and effective tool to aid the design and analysis of a FMS. Performance of a FMS can be evaluated by analyzing the statistical data recorded from dynamic simulation running of the simulator. This evaluation process after the simulation is very important and useful for finding potential problems, and for modifying the design of a FMS, or for scheduling strategy to achieve ultimate economical benefits. However, in normal practical simulation exercises, it is usually found that the evaluation process of FMS simulation is rather inefficient. This is because the immediate outputs of a simulation exercise are often presented in the form of statistical data lists or status data file, rather than clear and crisp information. To draw accurate and correct evaluation conclusions from the statistical data, such as whether the efficiency of a FMS is high enough, which equipment is the bottleneck, what are the real causes to the bottleneck, and whether the scheduling strategy or rules in FMS operation are efficient, users often have to sieve through and analyze many data lists manually. This is very time-consuming and results in low efficiency in FMS simulation.

Another problem is that accurate and correct evaluation requires the users to have sufficient theoretical and practical knowledge about FMS concept and simulation technique. In other words, because of the difference in ability of simulator users to conduct evaluation, different results may be deduced from the same sets of data obtained from a simulation exercise. For example, if simulation statistical data show that a machining centre has low utilization and its queuing length is too long, some inexperienced users of the simulator may deduce that this machining centre is a bottleneck. Sometimes this may not be really the case. For experienced users, they will probably further analyze other data lists such as the data concerning AGV and tool-changing robot. This is because the transportation speed, capacity and scheduling strategy of an AGV and tool-changing robot or other devices will directly or indirectly affect the working efficiency, thus resulting in decreasing utilization of the machining centre and increasing its queuing length.

The shortcomings mentioned above not only reduces the efficiency, reliability and validity of FMS simulation exercises, but also impedes the wide and practical applications of simulation in FMS design and analysis. This paper deals with the investigation of an *Intelligent Evaluator* which is embodied in a FMS simulator to aid the evaluation of FMS simulation. The *Intelligent Evaluator* has been implemented in an effective and practical FMS simulator called SAIL1SIM, which is being developed by the authors at City University of Hong Kong [1] [3] [4].

2   KEY TECHNIQUES FOR THE INTELLIGENT EVALUATOR

Some key techniques adopted for the *Intelligent Evaluator* are:
- knowledge acquisition from FMS domain and simulation evaluation;
- knowledge base design.

2.1   Relationship tree and FMS domain knowledge acquisition

The necessary FMS domain knowledge can be directly acquired from FMS and simulation domain experts. To do this, it is necessary to first analyze the dynamic logical relationship between devices and equipment and their evaluation criteria. For example, the utilization of a machining centre is concerned with loading capacity, tool-changing robot loading capacity, buffer capacity, last or next operating machine (e.g., another machining centre or washing machine) utilization, quantity of spare tools etc. In this process, a new method called *Relationship Tree* method is employed. Through analyzing and identifying the dependency relationships between a specified device and other devices in a FMS physical environment and dynamic operation path, some *Relationship Trees* can be drawn. These tree-like diagrams are very useful and effective for acquiring the required and accurate domain knowledge: facts and rules. Figure 1 shows an example of the *Relationship Tree.*

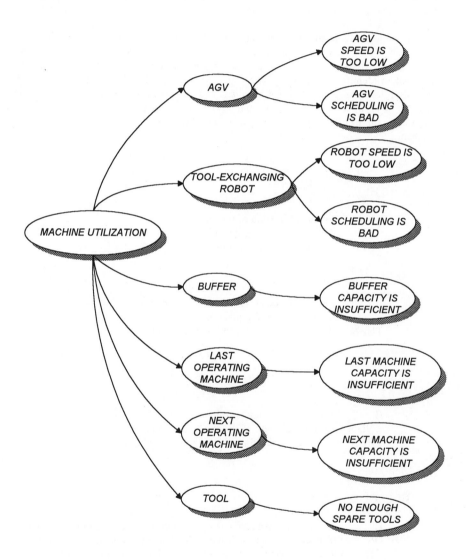

Figure 1 An example of the *relationship tree* for rule acquisition.

2.2   Knowledge base

In the knowledge base, formal logic predicates are used to represent facts such as the physical data of a FMS and operational statistical data of a FMS simulation. For example, **utilization(Device_name,X)** represents the **utilization** of the **Device_name** (a machining centre or other equipment) is **X**, **max_queue(Device_name,Y)** represents the **maximum queue length** of the **Device_name** is **Y**.   Here the **Device_name** can be an operating machining centre, a ROBOT or an AGV.

In addition to formal logic predicates, rules are used to represent the dependency relationships among device and equipment in the FMS, such as conditions, conclusions and actions. These diagnosis rules and the information contained in the knowledge base are used to identify bottlenecks. The diagnosis rules can be divided into three classes: analysis rules that guide a preliminary analysis of simulation results, local rules that identify local operational problems and global rules that identify FMS bottlenecks. Following are examples of analysis rules, local rules and global rules (expressed in English):

(1)   Analysis rule:

ARULE1

IF	average utilization of the machine<
	desired minimum utilization
THEN	the machine is under-utilized

ARULE2

IF	average queue length of the machine>
	desired maximum queue length
THEN	the queue length of the machine is too long

(2)   Local rule:

LRULE1

IF	the machine is under utilized
AND	the queue length of the machine is too long
THEN	the under-utilized machine operates with possible loading problem

(3)   Global rule:

GRULE1

IF	the under-utilized machine operates with possible loading problem
OR	the machine operates with possible loading problem
OR	the over-utilized machine operates with possible loading problem
AND	the utilization of AGV loading for the machine>
	desired maximum utilization

THEN   the machine is diagnosed to be impaired by a low speed or
              over-utilized or badly scheduled AGV
DO   report that the AGV is the possible bottleneck

GRULE2

IF   the under-utilized machine operates with possible loading problem
OR   the machine operates with possible loading problem
OR   the over-utilized machine operates with possible loading problem
AND   the utilization of ROBOT for tool-loading for the machine>
              desired maximum utilization
THEN   The machine is diagnosed to be impaired by a low speed or
              over-utilized or badly scheduled ROBOT
DO   report that the ROBOT is the possible bottleneck

2.3 Inference Strategy

The *Intelligent Evaluator* is implemented and programmed with Turbo Prolog and other
modules of the SAIL1SIM simulator are programmed in Turbo C. In Turbo Prolog, the
inference process can be described as follows: given a goal, Prolog searches the
knowledge base(database) for a fact that matches the goal; when Prolog finds a match and
substantiates the appropriate variables, it leaves a pointer where the match has occurred;
when a goal matches the head of a rule rather than a fact, the atoms within the body of the
rule are treated as subgoals that must all be satisfied in order to prove that the head is
satisfied.

3   STRUCTURE OF THE INTELLIGENT EVALUATOR

Figure 2 shows the structure of the *Intelligent Evaluator*. The part of modelling & data
inputting shown in Figure 2 is a very user-friendly interactive graphical modelling
environment in the SAIL1SIM simulator that has been described in literature[2]. After
modelling for an FMS and inputting the needed processing plans including types and
quantity of parts, the processing routes, operating time etc., the simulator can begin its
simulation running. In the course of the simulation running, the simulator records all the
needed data concerned with time and status into the data base. When simulation is ended,
the simulator reads and calculates the statistical data. Then the *Intelligent Evaluator*
starts the evaluation process by means of getting statistical data (currently the utilization,
maximum queue length and operating, part-loading, tool-loading and tool-unloading
time). It uses the data as the facts to match heads of *Analysis Rules* in the knowledge
base. If a match happens, the *Intelligent Evaluation* will get a preliminary diagnosis
and proceed with intention to find out a local problem by testing the *Local Rules*. If a
local problem is found, the *Intelligent Evaluator* will go to the final diagnosis by testing
the global rules and then output an evaluating conclusion.

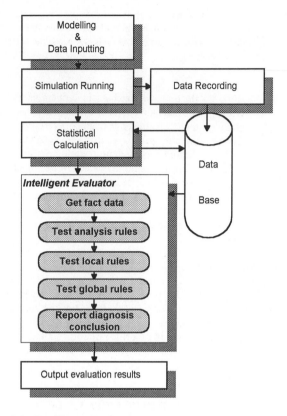

Figure 2   The structure of the *Intelligent Evaluator*

For example, if there is a fact that the average utilization of MC2 (identification of Machining Centre No.2 in the simulator) is 0.52 and it is less than 0.80 which is the desired average utilization of machines, the testing of *Analysis Rules* will get a diagnosis of "MC2 is an under-utilized machine". Next step is to test the *Local Rules* and it is found that the maximum queue length of MC2 is 8 which is bigger than the desired queue length of 4. The diagnosis of "the under-utilized machine MC2 operates with possible loading problem" is then obtained. Finally, if there is another fact that the utilization of ROBOT is 0.94 which is bigger than 0.90, the desired utilization of a ROBOT, then the testing of *Global Rule* will get a diagnosis of "the under-utilized machine MC2 is impaired by a low speed or over-utilized or badly scheduled ROBOT".

Following the procedures described above, this *Intelligent Evaluator* has been applied to aid the evaluation of a FMS which consists of two machining centres, a washing machine, twelve buffers, a part-loading/unloading station and a robot for tool-changing. Diagnosis of the evaluation after the FMS simulation has revealed that the robot is a possible bottleneck. Further analysis of the model parameters and scheduling process has also revealed that there are some problems with the scheduling process or with rules of the robot for tool-changing. Improvement has since been made for the scheduling of the robot and the next simulation and evaluation has shown that this bottleneck has been removed.

4 CONCLUSION

Traditional evaluation of FMS simulation results mainly depends on reading and analyzing many data lists manually by users. This not only reduces the efficiency, reliability and validity of FMS simulation exercises, but also impedes the wide and practical applications of simulation in FMS design and analysis. The *Intelligent Evaluator* introduced in this paper has been proved successfully as an effective and practical tool to aid the FMS simulation evaluation. This will improve the efficiency and accuracy of the evaluation process. Furthermore the conclusions drawn from the simulation exercise will be more consistent and less dependent on the users' ability to analyze large quantity of data.

5 ACKNOWLEDGEMENT

The authors would like to thank the Croucher Foundation for providing a research grant for this project. They would also like to acknowledge the support by the Department of Manufacturing Engineering of City University of Hong Kong and of Nanjing University of Science and Technology.

6 REFERENCES

[1] E.H.M. Cheung et al, "Developing SAIL.1 within a CIM environment", Croucher Foundation Project Reports, Department of Manufacturing Engineering, City University of Hong Kong, June 1993.

[2] E.H.M. Cheung, K.B. Chuah and X.N. Li, "A new architecture for FMS simulators", Proceedings of the International Conference on Data and Knowledge Systems for Manufacturing and Engineering (DKSME'94), Chinese University of Hong Kong, 2-4 May 1994, pp 485-490.

[3] K.B. Chuah, E.H.M. Cheung and X.N. Li, "A user-oriented interactive FMS simulation modelling environment", Proceedings of the International Conference on Advanced Manufacturing Technology (ICAMT'94), Universiti Teknologi Malaysia, Johor Bahru, Malaysia, 29-30 August 1994, pp 369-375.

[4] K.B. Chuah, E.H.M. Cheung and X.N. Li, "A study of fast modelling techniques for FMS simulation", Proceedings of the 1994 Pacific Conference on Manufacturing, The Institution of Engineers, Jakarta, Indonesia, 19-22 December 1994.

7   BIOGRAPHY

*   *E.H.M. Cheung*

Dr. Cheung obtained his MSc in Mechanical Engineering from University of Manchester Institute of Science and Technology and PhD from University of Manchester, U.K. His PhD research was in design of textile machinery. He is a member of The Institution of Mechanical Engineers, a Fellow of The Institution of Electrical Engineers, a Chartered Engineer and a Fellow of The Hong Kong Institution of Engineers. He joined the Department of Manufacturing Engineering, City University (formerly City Polytechnic) of Hong Kong in 1987 when the Department was established and is now the Associate Head of the Department. His current research interest is CIM system development and implementation, flexible automation and laser materials processing.

*   *K.B. Chuah*

Dr. Chuah graduated with a mechanical engineering degree from Loughborough University, UK. His PhD research was in Shiphull Roughness and Hydrodynamic Drag, a project funded by the US Navy. Since then, he has worked in several U.K. universities as research fellow and lecturer before joining the City University (formerly City Polytechnic) of Hong Kong in 1990. His main research areas are surface metrology, CIM system development and implementation and project management.

*   *X.N. Li*

Dr. Li obtained his MSc and PhD in Mechanical Engineering from Harbin Institute of Technology, P.R. China. His PhD research is "Computer Aided Design on a Hydraulic Pump". Then he worked on the research of simulation and control of FMS. He joined the Department of Manufacturing Engineering of Nanjing University of Science and Technology in 1989, and is now the Assistant Director of the Department and the Head of the Electromechanical Control Research Section of the Department. He worked in City University of Hong Kong from 1993 to 1994 for the Croucher Foundation funded research project on FMS simulation. His main research areas are Fluid Power Transmission and Control, CAD, FMS control and simulation.

PART SIXTEEN

Computer Application Techniques

75

Mobile Visualization:
Challenges and Solution Concepts

J.L. Encarnação, M. Frühauf, T. Kirste
Computer Graphics Center (ZGDV)
Wilhelminenstr. 7, D-64283 Darmstadt, Germany
Email: {jle|fruehauf|kirste}@igd.fhg.de

Abstract

Relatively cheap and widely available wireless data communication services are available to the mass-market. Within the coverage of the cellular network used, the vision of information access for 'everyone, anytime, anywhere' has become reality. A *Global Information Visualization* environment can be imagined where the vision of 'All information at your fingertip' is reality. In this environment, every information, data, or tool available on any computer connected to the global network is directly accessible to the user.

However because of the low bandwidth of wireless networks and the limited resources of mobile hardware, the handling of highly interactive distributed multimedia applications faces severe problems.

The objective of 'Mobile Visualization' is to provide effective solutions to these problems. This paper discusses the relevant aspects of systems and data models for Mobile Visualization and tries to provide an integrative view on the field based on the vision of Global Information Systems. In addition, we describe some of the research projects in this field from our institute, the Computer Graphics Center (Darmstadt/Germany).

Keywords

Mobile Computing, Global Information Systems, Distributed Systems, Visualization.

1 INTRODUCTION: WIRELESS COMMUNICATION AND GLOBAL INFORMATION VISUALIZATION

Long-range wireless data communication is available and effectively usable today. It is a matter of a few minutes to connect a portable PC to a cellular phone with data transmission capability and log on to any mailbox using standard communication software. No specific knowledge is required for this operation.

Relatively cheap and widely available wireless data communication services are therefore available to the mass-market. Within the coverage of the cellular network used, the vision of information access for 'everyone, anytime, anywhere' has become reality – at least as far as the basic technology is concerned.

Using stationary sites granting access to the Internet, global connectivity is available to the experienced user. Likewise, data transfer between mobile systems is quite simple (and cheaper than between mobile and stationary systems, looking at the tariff structure of some network providers).

So it appears as if a *Global Information Visualization* environment could be constructed in which the vision of 'All information at your fingertip' is realized. In this environment, every information, data, or tool available on any computer connected to the global network is directly accessible to the user.

However, the mobile world is far from perfect once more complex applications and services are considered. File transfer and mail reading are essentially non-interactive applications, for which a transfer rate of about 9.6 Kbps (Kilobit per second) poses no real problem (cf. Table 1 for bandwidth data of typical cellular networks) – after all, serial communication with character terminals provides usually the same bandwidth. But today's stationary display and interaction systems differ substantially from the early 'glass teletypes'. High-resolution displays and highly interactive applications based, *e.g.*, on the paradigm of direct manipulation or even immersive interfaces using virtual reality, require massive bandwidth between application and display.

Network	Switching technology	Bandwidth	Availability
GSM	connection	2.4, 4.8, 9.6 Kbps	exhaustively 1995
	packet	9.6 Kbps	exhaustively 1996
Modacom	packet	9.6 Kbps	since 1993
UMTS	packet	$n \times 64$ Kbps – 2 Mbps	$\sim$ 1998

Kbps = Kilobit per second
Mbps = Megabit per second
GSM   = Global System for Mobile Communication
UMTS = Universal Mobile Telecommunication System

Table 1: Communication bandwidth of mobile communication networks

Today, users have become accustomed to comfortable, easy-to-use and interactive and systems based on the concepts of direct manipulation. A step backwards in interactivity will cause a serious acceptance problem for such largely non-interactive applications. So in order to build 'everyone, anytime, anywhere' information applications that will be *used*, the problem of how to make these services interactive across a slow data link has to be solved.

A directly related problem is the communication of time-dependent data (sound, video). Users of mobile systems will expect to deliver access to all of the multimedia data making up modern information applications – at least within the limits defined by the mobile system's input and output capabilities.

Finally, the problems of scalability have to be tackled when considering that thousands of users may simultaneously access the same data or service from all over the world. Suitable concepts for caching, replication, and migration of data and services have to be identified which guarantee the response times required for interactive applications as well as the transfer rates required for the rendering of time-dependent data.

The objective of *Mobile Visualization* is to provide effective solutions to these problems. This paper discusses the relevant aspects of systems and data models for Mobile Visualization and tries to provide an integrative view on the field based on the vision of Global Information Systems. In addition, we describe some of the research projects of our institute in this field.

The remainder of this paper is organized as follows:

Section 2 gives an overview over the fundamental problems of mobile visualization within the integrative context of a global information infrastructure.

In Section 3, we discuss aspects of possible solution strategies which are based on the concepts of object migration and fragmentation in heterogeneous environments.

Section 4 describes a number of experimental systems and concepts being under development at the Computer Graphics Center (ZGDV), which rely on the concepts discussed in the previous section.

Finally, a discussion of the ideas presented in this paper is given in Section 5.

2   CHALLENGES FOR MOBILE VISUALIZATION AND INTERACTION

2.1   Global information management systems

When T. Nelson coined the term 'docuverse', he envisioned a global hypertext-like structure, containing and interlinking the entire human knowledge (cf. [16]). Global information management systems based on this vision – such as the World-Wide Web (WWW) [1] – show that on-line access to vast amounts of distributed information is possible not only for the expert, but also for the end user. The ultimate goal of these activities is the construction of a unified, globally accessible repository for arbitrary information and services[1] which is available to everyone. (The term 'docuverse' will be used to denote this repository).

Based on this scenario, the following assumptions hold for the docuverse:

1. The docuverse contains not only multimedia documents, but also services which may actively communicate with the user or other services and might even *change* the docuverse. (*E.g.*, a subscription service adding a user to an object representing the list of subscribers.)
2. The set of object and service types present in the docuverse is constantly changing and expanding.

[1] Such as the WWW Pizza ordering service available for Santa Cruz residents (URL: http://www.pizzahut.com).

Taking Pt. 1 to its extreme, the docuverse appears as a set of uniquely identifiable *active* objects which refer to each other through mechanisms such as messaging and static associations ('hyperlinks'). In conjunction with Pt. 2, the notion of the docuverse thus implicitly contains a dynamically extensible class system which allows the definition of an object's message response behavior.

This concept informally describes the notion of a docuverse as it is understood in the course of this paper. It is this dynamic set of heterogeneous, communicating active objects, which ultimately will be accessed by mobile and stationary data terminals.

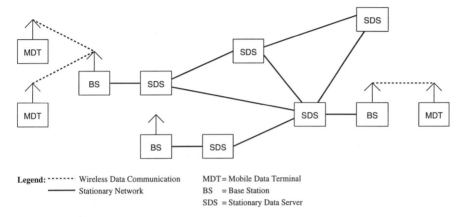

Figure 1: Typical communication infrastructure for mobile data access

It is now an interesting challenge to *combine* the concept of a 'docuverse' with wireless data communication into a system model granting any user ubiquitous, mobile access to the global information repository. The idea is to use portable computing devices as mobile 'windows' into the docuverse. These *mobile data terminals* (MDT) use wireless data communication to access *stationary data servers* (SDS) which implement the docuverse (cf. Figure 1).

A simple straightforward realization of this *Global Information Visualization* scenario is inhibited by the following contradictory properties of the different architectural components:

1. Limitations of the communication services such as low bandwidth and temporary disconnection.
2. Limited local resources (in terms of storage capacity, computing power, data entry and data output capabilities).
3. Complexity of the docuverse (in terms of the available data formats and the display/interaction requirements of the individual data objects and services).

Although many more problem areas for mobile information systems can be identified besides Pts. 1 and 2 (cf. *e.g.* [5, 7, 19, 23]), the issues identified above are the the most obvious and pressing ones when trying to build a Global Information Visualization System (GIVS). An in-depth discussion of these aspects is given in Section 3.

2.2 Managing content data

Owing to the openness of the docuverse where any information provider may offer his own specific data and services, a large – virtually unbounded – number of different data types and formats have to be managed by a user's MDT. Typical standardized formats for graphical and multimedia data are listed in Table 2.

In addition, there are a lot of more proprietary formats used for transfer and storage of graphical data. Especially in networked information systems, this fact raises many problems during transfer and presentation of the data. Therefore, a widely agreed understanding on standards to be used for data exchange is desirable.

Data type	Format
Raster image	JPEG, TIFF, IIF and many others
Audio	AIFF
Video, Motion picture	motion JPEG, MPEG, MPEG-2, MPEG-4, H.261
Text, Documents	RTF, Postscript, PDF
Document structure	HTML, SGML, HyTime
Presentation and Visualization	QuickTime, MHEG

Table 2: Data formats for multimedia data

The main challenge in visualization of information on a mobile computers, however, is the transfer and presentation of non-textual information. Table 3 lists the data rates to be transferred and processed for visualization of one screen (480 x 640 pixels) of information using different types of data.

Data	Format	Size/Data rate	quality
Text	ASCII	~ 5 KByte	n/a
Graphics	2D-/3D-Graphics	10 KByte – 100 KByte	n/a
Image	B/W (1Bit/Pixel)	38 Kbyte	n/a
	Color (8Bit/Pixel)	300 KByte	n/a
	True color (24Bit/Pixel)	900 KByte	n/a
	JPEG True color	60 KByte	15:1 compression 'virtually original'
Video	MPEG-1	1.5 Mbps	$\frac{1}{4}$ VCR quality
	MPEG-2	4–10 Mbps	VCR quality
	MPEG-3	5–20 Mbps	HDTV quality
	MPEG-4	9–40 Kbps	interactive multimedia, video telephony
	H.261	$p \times 64$ Kbps (64 Kbps–2.048 Mbps)	video telephony
Audio	8-Bit μlaw	64 Kbps	speech monaural
	2 channel, 16 Bit PCM at 44.1 kHz	~ 1.4 Mbps	CD stereo

Table 3: Data rates for multimedia data

Considering the limits of mobile computing devices such as processor speed, short-term and long-term storage, and network bandwidth, it becomes obvious that a highly sophisticated software architecture for networked mobile information systems is required.

2.3   Practical usability issues

Besides the more fundamental problems discussed in the previous subsections, the 'docuverse' concept introduces numerous issues concerning an effective usability of the global information repository. Some of them are [4]:

- The docuverse is a large information space containing complex information structures. Helping the user to orient himself is a primary challenge for such scenarios. Tools are required which support the visualization of these structures even on small, low-resolution displays.
- Support for navigation through huge information spaces has to be provided even if the size and structure of the information space change dynamically – *e.g.*, because of changing network connections. Orientation and navigation support tools must be able to dynamically adapt to changing environments.
- Access to external data sources has to be provided. Tools for data format conversion and compression and decompression of data have to be accessible on the network.
- Mobile computing devices have to be able to down-load tools for visualization of specific data formats from the network, even if the user is not aware of the physical location of the appropriate tool.
- Adaptable and configurable user interface have to be provided. The adaptation to the needs of specific users should be accomplished by the evaluation of a specified user profile.

- New input devices such as pen or touch screen have to be supported by providing new interaction techniques.
- Hand-writing recognition is absolutely essential for hand-held systems without any keyboard.
- Speech recognition is an alternative to hand-writing recognition as an input mechanism for hand-held systems without any keyboard.
- Real-world metaphors or physical-world metaphors make the design of access mechanisms for large electronic information spaces, e.g. electronic libraries etc., much easier.

In the next section, we will outline a data model and a system architecture which is able to provide a solution to these problems.

3 SOLUTION STRATEGIES

3.1 A simple model for interaction with multimedia objects

The 'docuverse' consists of *information objects*, abstract active entities receiving requests and responding with replies. Individual requests and replies are elements of suitable alphabets. As far as this paper is concerned, refined concepts such as classes and inheritance are not relevant.

Communication between user and an information object is a cyclic four-step process: (1) The object sends a value to the user's display and the value is rendered appropriately. (2) The user may then interact with the display's input devices which eventually (3) results in an event being sent to the original object. This event may cause (4) state changes, which again (1) cause reply values to be transmitted to the display for rendering.

When looking at this processing model, one can easily discover that there are several levels involved, at which data processing occurs (cf. Figure 2). Consider, *e.g.*, an object representing some graphics drawing. Rendering is ultimately by drawing pixels, interaction events are button press/release notifications and coordinate change notifications. This level next to the user's sensory and affectory system may be called 'physical' or 'iconic' level.

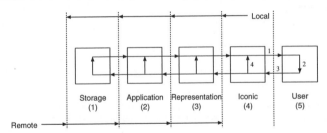

Figure 2: Simple layer model for graphics applications

On the next lower level, graphics primitives such as circles, lines etc. may be used; interaction events may be 'picking' and 'dragging' of these primitives. This may be termed the 'representation' level. It contains an intermediate representation of the data which abstracts from both the concrete rendering device and the application semantics. Finally, the 'application' level captures the application semantics, the concepts and functions it uses for realizing its behavior. In the context of a CAD system, these might be 'nuts' and 'bolts' resp. 'adding an axle' etc. Usually one has also to provide a 'storage' layer which provides long-term storage for object data (using another representation).

Clearly, this simple model – and the ideas behind it – is quite similar to the layered models of communication, such as the OSI-Model. Although one could easily identify additional layers (cf. *e.g.* [6] for the image case), this is matter of detail discussion. What we want to assert here is that such layers clearly *exist* – and that the existence of this layer can be exploited for optimizing *mobile* visualization. This issue will be elaborated in the following sections.

3.2   Locating the interface

Since each pair of layers constitutes an interface, the interesting question is: *which* interface coincides with the separation between mobile system (MDT) and data source (SDS)?

Besides the simple model, where all layers execute on the mobile system and no communication occurs at all (*i.e.*, a local application with local storage), there are two central models currently used for distributed graphics applications:

- Remote application with local display server (Interface between layers 4 and 3).

    One alternative possible for high-bandwidth networks is to define a front-end server (a la X [20]), which provides the necessary primitives for realizing an object's interface behavior. The behavior itself is realized by the object at an SDS site, where it can exploit all resources of a stationary computer. Only primitive requests are sent to the MDT. Within the expressive power of the front-end server, any possible interface behavior could thus be realized by a MDT, without exceeding the available resources.

    However, since only primitive requests or event notifications are transmitted between MDT and SDS, a substantial amount of bandwidth is required in order to reach acceptable response times for interactive applications. Anyone who has tried to run the X-protocol over a slow serial line has experienced this (cf. *e.g.* [8]). Furthermore, in the case of temporary disconnection (= very low bandwidth), interactivity is not given at all.

    Therefore, this approach is generally not feasible considering the properties of wireless data communication. The MDT must be able to realize object behavior with more autonomy from the SDS[2].

- Local application with remote storage (Interface between layers 1 and 2).

    Using this approach, an object's complete interface behavior is migrated to the MDT as program executable on the MDT's operating system. (For today's global information systems, this is typically done manually via ftp.) This strategy clearly avoids the disadvantage of the high communication bandwidth required for the dumb server scenario and guarantees very high autonomy. But here too, severe disadvantages prohibit this approach:

    - If heterogeneous MDTs are used (which is to be expected), a separate executable version of the interface behavior for every possible MDT needs to be available.
    - Because MDT storage capacity is limited, the number of executables to be down-loaded at the same time is limited (note that even in the age of shared libraries, executable programs are quite substantial in size). Assuming a large number of different object types (*i.e.*, different object behaviors) in the docuverse, this requires a caching scheme with preemption. This, however, causes problems in the case of a cache miss: because of the large size of the executables, transfer time will be at least several minutes, making information access unpleasant at best. Exploratory information access – which may cause a large number of executables to be loaded – will be avoided by the user.

Clearly, for a mobile window to the docuverse a concept is required which provides a kind of down-loading of interface behavior, while at the same time avoiding the problems of heterogeneity and size.

3.3   Fundamental solution concepts

Based on the discussion in the previous section, two fundamental solution concepts can be identified:

Object migration: The first idea is to allow for an on-demand migration of object behavior. Because heterogeneous environments have to be considered, a well-defined *representation* for denoting object behavior is required. In order to render the docuverse open to any possible information object, this notation has to fulfill certain completeness properties. One aspect is definitely *computational completeness*, which allows any algorithm to be part of an object's behavior. Another aspect is *resource completeness*, which gives an object control over any input or output device (this is the tricky part).

Object fragmentation: As a rule of thumb, allocating a layer to the MDT results in additional storage and processing requirements. Likewise, moving a layer from the MDT to the SDS results in higher bandwidth requirements. Therefore, an optimal choice, considering the interaction with a large amount of different objects, would be to use the separation between levels 3 and 2.

---

[2]Because sending data from MDT to SDS via wireless is a power consuming job, the limited battery capacity of mobile systems is another reason for requiring more autonomy of the MDT.

Because different layers of the same object are executing on different machines, one can speak of *object fragmentation* in this case. Clearly, once the concept of well-defined object representations has been introduced, it may be left for a dynamic, run-time decision, *how many* of the different layers of an object are migrated to the user's site. This supports dynamic load balancing considering the various costs for computation and communication with respect to the currently available computing machinery and network technology: If the user's 'MDT' is a powerful stationary workstation connected via FDDI, the SDS will migrate the complete object. Accessed via a small mobile device using slow wireless communication, the SDS may decide to migrate only layers 1 an 2.

Next, we will give a short discussion of the fundamental architectural components required for *implementing* migration and fragmentation of active objects.

3.4 An architecture for a distributed realization of object behavior

A basic architecture supporting object migration in heterogeneous environments is quite simple to identify: It consists of an abstract machine (AM) which provides computation, storage, input, and output services for the realization of object behavior. This AM is implemented on the different hosts, on MDTs as well as on SDSs. Object behavior is then defined in terms of the 'programs' this machine is able to execute and the 'process states' it maintains for programs currently executed.

An object or parts thereof may then be represented by 'closures' which consist of a program and a process state. Object migration and replication is realized by moving or copying closures from one AM to another.

The idea of 'function shipping' or 'remote programming', which is basically equivalent to the architecture outlined here, exists for quite a while now. A prime example is the well-known 'Postscript' language which is used to remotely program printing devices. More recent examples are 'Telescript' [24] or 'Safe-TCL' [2] (used for enabled mail applications).

Remote programming enables the dynamic extension of remote functionality through data communications, without requiring human intervention. This contrasts with the usual RPC (remote procedure call) based approach. Using RPCs, a only fixed set of remote functions is available. Extending this functionality requires changing the source code of both client and server.

However, the basic idea of remote programming requires a substantial amount of detailing in order to be of any use to the problems of mobile visualization. At least three problem areas exist:

General distributed process management: Imagine the replication of one object at several MDTs. In order for the object to present a consistent state to every user, there must be some kind of coordination between the individual replicas. However, which mechanism used (*e.g.*, primary copy or 'gossip' architectures; pessimistic, optimistic, or time-stamp based concurrency control; . . .), depends on the function performed by the object. Therefore, the system has to provide a comprehensive set of tools for managing and coordinating distributed processes.

Layer identification: Consider a layered graphics object o, whose behavior is described by a program p written in any typical programming language, be it declarative, procedural functional or object-oriented. From an analysis of p's syntactic structure alone, it is usually not clear which of o's layers are described by which of p's parts. This, however, hinders an effective automatic partial migration of o: In order to determine which parts of p constitute a layer of o requires a thorough semantic analysis of p!

So, in order to allow for practical automatic object fragmentation, languages based on *extended object models* (such as described in [12, 18]) are required which allow for a syntax-based layer identification. Such object models add components like 'interaction behavior' etc. to the basic slot/method model.

Input/output functionality: In Section 3.3, we called for *resource completeness* of the object representation. By this we mean that the representation allows to describe any possible input/output operation on the available devices. Because future devices may provide arbitrary enhancements to existing functionality, the only definite limits to useful input and output operations are the capabilities of the user's sensory and affectory system. Although eventually the goal, this value domain clearly is somewhat removed from a satisfying solution in the near future.

As intermediary solution, the fundamental concept of a 4D (x, y, z, t) space – as proposed by the HyTime standard [17] – is a more viable approach. Output is defined in terms of the content of sub-volumes of this space; basic input is captured by the selection of points or point-sets in this space. Dynamic reactions to input are reflected by the introduction of multiple spaces connected by paths to be traversed upon certain events.

The really interesting question is, how to define a concrete language for defining such 'interactive spaces'. Some of the issues involved here are:

- What is the underlying language concept (*e.g.*, declarative a la HyTime or imperative like MHEG)?
- What are the 'within-volume' primitives (*e.g.*, lines, circles, cubes, spheres, textured spline surfaces)?
- How to provide for an efficient 'fast path' implementation for typical interactive spaces (*e.g.*, windowed $2\frac{1}{2}$D spaces)?
- How to cope with a wide range of display capabilities (*e.g.*, provide capability information or use 'normalized' descriptions).

3.5   Coping with multimedia content data

While the previous section dealt with the abstract description of an object's overall input/output behavior, the management of concrete multimedia content data to be presented within individual (z, y, z, t)-sub-volumes is the most obvious problem in low-bandwidth networks with resource-poor display terminals. Underlying solution strategies fall into the following categories:

- Reducing the size of data itself (Compression).
- Getting the most interesting part first (Within-content-anchoring).
- Allowing for the transfer of detail data later on (Detail on Demand; Progressive Refinement).
- Fetching data to be accessed in future while network is idle (Prefetch).
- Keeping frequently used data online (Caching).

Ideally, the data exchange architecture should support all of these features within a unified model giving user and implementor of information services a wide range of tradeoff alternatives. Furthermore, most of the individual categories influence each other, giving rise to numerous optimization possibilities.

For example, low-resolution displays are usually available on mobile hardware, thus coinciding with low-bandwidth networks. A general use of progressive refinement technique for information transfer thus automatically addresses both problems of display heterogeneity and low bandwidth within the same approach (see Table 4).

Refinement		Display-Res.	Bandw.		Terminal	Network
low	$\xrightarrow{\text{requires}}$	low	low	$\xleftarrow{\text{provides}}$	Mobile	Wireless
medium		medium	medium		Stationary PC	Ethernet
high		high	high		Graphics Wks.	FDDI

Table 4: Progressive Refinement requirements vs. terminal capabilities

Furthermore, the computational complexity of progressive refinement algorithm should match with the computing power available. Low refinement stages should be computationally simple as to allow execution on comparatively slow mobile hardware. Higher refinements can be more expensive. For raster images, simple algorithms such as approaches based on Binary Condensed Quad-trees (BCQ, [3]) can be used for the first refinement levels. Furthermore, the MDT's programmability can be employed to support *arbitrary* algorithmic image descriptions (*e.g.*, fractal or state-machine based), because the rendering algorithm can be transferred with the image itself.

For single linear multimedia contents – such as sound, video, text – prefetching is straightforward (the only question being the scheduling of simultaneous prefetch operations when accessing more than one such content at the same time). However, the static associations between individual docuverse objects ('hyperlinks') hold the potential of extending the prefetch across multiple objects, maybe exploiting usage information and user preferences for predicting objects to be accessed next. One very simple example is the concept of 'guided tours' found with hypertext models, where the user has the choice between several predefined paths through the same set of documents. Clearly, once a path has been selected it is an ideal 'oracle' for predicting the user access behavior.

Another hypermedia feature useful for optimizing content access are within-content anchors, Anchors which do not refer to the complete destination object but only to a part thereof. This allows for a quick access of a hyperlink destination by transferring only the *context* of the anchor, a content region surrounding the anchor point (resp. the anchor region, if region anchors are supported[3]). The remaining parts of the document content may then be treated by prefetch or detail-on-demand mechanisms.

Finally, the simultaneous transmission of multiple content data must also be considered. The communication between consumer and producer of content data (*e.g.*, between display volume and network) must be handled by parallel processes. Several of these processes which render data arriving at the communication port directly into the respective display volumes may execute in parallel, so that negotiation between them, the underlying transport system, and the server is necessary in order to observe the current focus-of-interest of the user. How these processes may be described in the object's behavior definition is another issue.

Likewise, one has to consider the subjectively optimal scheduling of multiple simultaneous refinements – *e.g.*, by giving priority to the volume with the current input focus.

4 EXPERIMENTS AND PROJECTS

In this section, we describe some of the of research projects at our institute which are concerned with the construction of Global Information Visualization Systems (GIVS).

4.1 The HYPERFUNK system

HYPERFUNK [10] is an experimental prototype of a GIVS, based on the concepts discussed in Section 3. In addition to creating a platform allowing an evaluation of the viability of these concepts, the following objectives have been pursued by the HYPERFUNK implementation:

- Enable the demonstration of the potential of ubiquitous mobile information access to expert and non-expert users (information consumers as well as information providers).
- Demonstrate the 'docuverse' concept of a unified global information and service management and the idea of a mobile windows to the docuverse.

For this purpose, an example 'micro' docuverse (DV/0) has been created for HYPERFUNK, demonstrating the following facilities:

- Access of private data.
- Access to public data (a simple electronic newspaper).
- Transparent migration of specialized object behavior.

Architecturally, HYPERFUNK consists of multiple MDTs (Notebooks running MS-Windows) and a single SDS (a stationary UNIX-Server). The Notebooks are equipped with GSM (= Global System for Mobile communications) mobile phones for wireless data communication; they use the 9.6 Kbps data transmission service provided by the GSM standard. The SDS is equipped with a standard analog PSTN (Public Switched Telephone Network) modem. The transition from GSM to PSTN is the responsibility of the GSM service provider. (cf. Figure 3.)

High-level communication between SDS and MDT is based on stream sockets relying on SLIP (Serial Line Internet Protocol) and TCP. This reduces available net bandwidth, but substantially simplifies the implementation of data communication – *e.g.* through providing multiple logical channels across one physical channel, access within the complete Internet, and error recovery.

Within HYPERFUNK, the language HCL is used as language for representing object behavior. HCL[4] [9] is a LISP-like dialect with small-footprint interpreters available for UNIX and Windows. It contains a run-time extensible object system (HCLOS), socket-based communication mechanisms, and functions for creating graphical user-interfaces (using X/Motif resp. MS-Windows). So, a kind of resource completeness as well as computational completeness can be granted for object representations.

Both SDS and MDT employ an HCL-Interpreter to realize object behavior. With respect to behavior definitions, the HCL running on the SDS operates as master, containing all known object behaviors (*i.e.*, classes and methods). If the slave HCL on the MDT encounters objects with unknown behavior, it requests the necessary classes and methods from the master and augments its local object system.

---

[3] For a generalized mechanism supporting server-based anchor region computation, see [13].

[4] HCL is an acronym for 'HyperPicture Command Language'; it has initially been developed for the description of active objects in the computational hypermedia system 'HyperPicture' [15].

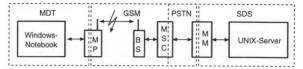

Legend: MP = GSM Mobile Phone with data transmission capability.
            BS   = GSM Base Station.
            MSC = GSM Mobile Switching Center with signal conversion to analog
                    PSTN modem transmission.
            MM = Stationary analog PSTN modem.

Figure 3: Communication architecture of HYPERFUNK

The user interface to DV/0 is created by the interface behavior of an object in the docuverse itself, the 'entry-point'. It is highly interactive and metaphor-based (therefore intuitive, aiming at non-expert users), demonstrating the wide range of interface behavior available for objects in the docuverse. (For a general discussion of metaphor-based user-interface construction see *e.g.* [21].)

The entry-point object presents itself as a two-story 'shopping & service center' (the entrance hall of this center is shown in Figure 4). The ground floor contains public services (*e.g.*, the newsstand providing access to newspapers – cf. Figure 5), the first floor gives access to private services (*e.g.*, the user's office with a cabinet containing folders and documents, Figure 6) Based on this metaphor, a user may navigate through DV/0 by 'walking' around the building, 'knocking' at doors, 'entering' rooms etc. Besides this primary navigation mechanism, a selection of classical navigation support tools as can be found in hypertext systems is available: Various 'information booths' provide orientation and descriptive access through overview maps and keyword search; a history-list and user-configurable hot-lists allow for backtracking and shortcuts. (See [22] for a complete discussion of the entry-point.)

Finally, 'Persons' can be charged by the user to perform more complex tasks. One example is the vendor at the newsstand, who will perform the necessary modifications of the metaphor when subscribing to a newspaper.

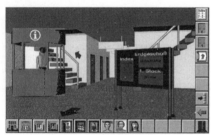

Figure 4: Entrance hall   Figure 5: Newsstand

4.2   Active Mail

Besides the communication of object representations between two active AMs (abstract machines), one can also look at the idea of instantiating the destination AM upon the user's *access* of the transferred object layers. This fits with the idea of electronic mail (Email): Active objects are a specific kind of Email, the mail reading tool (= AM) is started once the user tries to read a mail containing an object representation. Thus, active mail is another incarnation of the object migration concept introduced for mobile visualization.

An active mail can for instance build up a form on the receivers site, consisting of text fields (for editing purposes), multiple choice elements etc. It may accept and evaluate the answers of the receiver (i.e. check the consistency) and send back automatically a reply to the originator of the mail.

| Figure 6: User office | Figure 7: Active mail tool |

Of course, besides reading Email one should also be able to *write* Email. Therefore, an easy-to-use tool for creating the behavior definitions which describe an active object is required. At the ZGDV, a prototype of such an *authoring tool* for active mails has been developed. The central feature of the system is a user-friendly mail handler, which allows one to compose active messages interactively supported by a graphical feedback. The active messages appear as a form consisting of editable text fields, multiple choice elements, inline pictures and anchor buttons etc. Anchor buttons are the source for linking multimedia elements to the active messages. Figure 7 shows the composing tool and a sample active mail.

Active mails are – just like active objects in HYPERFUNK– represented using the LISP-dialect HCL. Furthermore, the composing tool itself is realized with HCL and can therefore be transmitted as its own active mail.

The format for the interchange of mails is based on the Multipurpose Internet Mail Extension (MIME). The reading of delivered active mails is supported by a standard mail reading tool supporting MIME; *e.g.*, elm 2.4.

Abstracting from the concrete application of creating active mails, the composing tool may be regarded as a first step towards an authoring tool for the interactive creation of fully programmable active objects. This enables information providers in the docuverse to create information objects without conventional programming.

4.3 The MoVI project

In the previous sections, we have indicated numerous open issues regarding the functionality required on a mobile system for the global visualization of information. These topics are investigated in-depth within the MoVI project [14, 11]. MoVI (Mobile Visualization) is a basic research project funded for two years by the German Science Foundation (DFG). Besides the ZGDV, the Interactive Graphics Systems Group of the Darmstadt Technical University, the Computer Science Dept. of the Rostock University, the Fraunhofer Institute for Computer Graphics Rostock, and the Computer Graphics Center Rostock participate in this project.

The primary research aims of the MoVI project are:

- the access,
- the visualization and
- the interaction

of multimedia information on distributed mobile computer systems. A main task will be to refine and detail the solution concepts outlined in this paper. Some aspects of refinement are:

- On the architectural level, a distributed *information agent* will be responsible for performing object location, replication and migration. *Presentation servers* are responsible for implementing the semantics of an object's behavior description on the respective display. Finally, *user agents* mediate between the docuverse as such and the view preferences of the individual users.

- On the level of the object model, *contexts* are proposed as a central mechanism for filtering the vast amount of information available in the docuverse to sizes manageable by the user. Possible contexts are the semantics of the information, aims and requirements of different users as well as location, time and technical environment. Queries for information objects are then evaluated against the current user context specification.

5   CONCLUSION

In this paper, we have introduced the main task of Mobile Visualization as providing mobile windows to a globally unified information repository, the 'docuverse', to everyone.

We have observed that the docuverse is a complex domain, containing and undetermined set of data formats and information services. From this observation, we have concluded that a system concept supporting the *dynamic* configuration of a mobile data terminal is a fundamental requirement for creating a mobile information system based on the docuverse vision. As areas of primary importance, we have the investigated the aspects of:

- Object migration and fragmentation based on a well-defined external representation for active objects.
- Extended object models as fundament for an effective automatic object fragmentation.
- Resource completeness of the representation with respect to a suitable value domain (*e.g.*, the 4D (x, y, z, t) space).
- Techniques for an efficient management multimedia content data in resource-poor environments.

As illustration for these concepts, some of the current research & development projects at our institute have been described briefly. From the experiences with our prototype systems, the following conclusions can be drawn so far:

- The concept of object migration and object fragmentation is very powerful once supported in heterogeneous environments. Furthermore, a naive mechanism supporting object migration can easily be implemented on interpreted languages providing external representations for all denotable values, such as LISP.
- In order to construct a system with general applicability, the fundamental solution concepts of migration and fragmentation require a substantial amount of detailing. The various areas of interest have been discussed throughout this paper.

As final remark, we want to emphasize that mobile visualization is a world of *compromise*. Due to the substantial limitations of the various system components, every non-trivial application will suffer in one way or the other from these limitations. Therefore, it is essential for a Global Information Visualization System using mobile data terminals to provide the user with a *subjectively* optimal compromise. On the other hand, every user and application has its own rating for the different aspects of the 'Quality-of-Service'. No single model will be able to capture all tradeoff possibilities *unless* it is essentially a dynamically programmable system. Hence, a concept supporting the migration of active objects will eventually be inevitable for mobile visualization applications.

ACKNOWLEDGMENTS

The authors would like to thank Kaisa Väänänen for designing and implementing the user interface metaphor of HYPERFUNK, Jian Zhang for doing a great job in building the low-level communication software of this system, and Jürgen Schirmer for building the prototype composing tool of our active mail system.

The work on HYPERFUNK has been supported by DeTeMobil.

REFERENCES

[1] Berners-Lee, T., Cailliau, R., Groff, J., Pollerman, B. WorldWideWeb: The Information Universe. *Electronic Networking: Research, Applications and Policy*, 1(2):52–58, Spring 1992.

[2] Borenstein, N.S. Email With A Mind of Its Own: The Safe-Tcl Language for Enabled Mail. Internet Draft, Proc. ULPAA'94, 1994.

[3] Dürst, M.J. Progressive Image Transmission for Multimedia Applications. In Magnenat Thalman, N., Thalman, D., editors, *Virtual Words and Multimedia*. John Wiley and Sons, 1993.

[4] Encarnação, J.L., Frühauf, M. Visualization and Interaction: The Enabling Technologies for Applications and Services of Mobile Computing Systems. In *TENCON'94*, 1994. (Keynote Address).

[5] Forman, G.H., Zahorjan, J. The Challenges of Mobile Computing. *IEEE Computer*, 27(4), April 1993.

[6] Hildebrand, A., Magalhães, L.P., De Martino, J.M., Seibert, F., Strack, R., Tozzi, C., Wu, S.T. Towards a Visual Computing Reference Model. *Computers and Graphics*, 19(1), 1995.

[7] Imielinski, T., Badrinath, B.R. Mobile Wireless Computing: Challenges in Data Management. *Communications of the ACM*, 37(10):18–28, October 1994.

[8] Kantarjiev, C.K., Demers, A., Frederick, R., Krivacic, R.T., Weiser, M. Experiences with X in a Wireless Environment. In *Proc. USENIX Symposium on Mobile & Location-Independent Computing (August 2–3 1993, Cambridge, MA)*, pages 117–128. USENIX Association, 1993.

[9] Kirste, T. HCL Language Reference Manual, Version 1.0. ZGDV-Report 68/93, Computer Graphics Center, 1993.

[10] Kirste, T. An infrastructure for Mobile Information Systems based on a Fragemented Object Model. Submitted to Distributed Systems Engineering Journal Special Issue on Mobile Computing, January 1995.

[11] Kirste, T. Mobile Visualization – Project Home Page. Available via WWW (URL: `http://www.igd.fhg.de/www/zgdv-mmivs/~movi/`), January 1995.

[12] Kirste, T. Some issues of defining a user interface with general purpose hypermedia toolkits. In Schuler, W., Hannemann, J., editors, *Proc. Workshop on Methodological Issues on the Design of Hypertext-based User Interfaces (July 3–4 1993, Darmstadt, Germany)*. Springer, 1995.

[13] Kirste, T., Frühauf, M. On the Use of Extents in Distributed Multimedia Computing Environments. In Herzner, W., Kappe, F., editors, *Multimedia/Hypermedia in Open Distributed Environments (Proc. Eurographics Symposium, June 6–9 1994, Graz, Austria)*, pages 247–263. Springer, 1994.

[14] Kirste, T., Heuer, A., Kehrer, B., Schumann, H., Urban, B. Concepts for Mobile Information Visualization – The MoVi-Project. Submitted to Eurographics Workshop on Scientific Visualization 1995, January 1995.

[15] Kirste, T., Hübner, W. An Open Hypermedia System for Multimedia Applications. In Kjelldahl, L., editor, *Multimedia: systems, interaction and applications (Proc. 1st Eurographics Workshop on Multimedia, April 18–19 1991, Stockholm, Sweden)*, pages 225–243. Springer, 1992.

[16] Nelson, T.H. All for One and One for All. In *Proc. Hypertext '87 (November 13–15 1987, Chappel Hill, North Carolina)*, pages v–vii. The Association for Computing Machinery, 1987.

[17] Newcomb, S.R., Kipp, N.A, Newcomb, V.T. The "HyTime" Hypermedia/Time-based Document Structuring Language. *Communications of the ACM*, 34(11):67–83, November 1991.

[18] Rettig, M., Simons, G., Thomson, J. Extended Objects. *Communications of the ACM*, 36(8):19–24, August 1993.

[19] Satyanarayanan, M. Mobile Computing. *IEEE Computer*, 26(9):81–82, September 1993.

[20] Scheifler, R.W., Gettys, J. *The X Window System*. Digital Press, 1992.

[21] Väänänen, K. Interfaces to Hypermedia: Communicating the Structure and Interaction Possibilities to the Users. *Computers & Graphics*, 17(3):219–228, 1993.

[22] Väänänen, K. Hypermedia im Mobilfunk: Spezifikation der Anwendersicht mit der Metapher "Dienstleistungszentrum". Project Deliverable R.1, Computer Graphics Center, July 1994.

[23] Weiser, M. Some computer science issues in ubiquitous computing. *Communications of the ACM*, 36(7):75–85, July 1993.

[24] White, J.E. Mobile agents make a network an open platform for third-party developers. *IEEE Computer*, pages 89–90, 1994. November.

76

Experience and potential of AutoCAD based piping applications for the Shipbuilding Industry

Njål Vikanes, Naval Architect , MRIH
Vikanes Consulting
6065 Ulsteinvik , Norway , Tel. +47 70 01 25 45

Abstract

This presentation emphasis the potential for PC-based piping applications in the shipbuilding industry. With the various working pattern for each yard and contract, flexible CAD tools with low learning curve is essential for a successful concurrent engineering. With the recent developments these systems represent the most cost effective solution available. This paper will go throw the different stages from conceptual design to final fabrication drawings.

Keywords

Shipbuilding , CAD engineering ,Pipe design, 3D pipe models, Pipe arrangement plans, ISO drawings, Pipe stress, Fluid flow analysis, AutoCAD®, SUNRISE SYSTEMS Ltd., COADE®, Rebis™ .

1 INTRODUCTION

As a Naval Architect, I have followed the introduction of Computer Aided Design in the shipbuilding industry. From my first years of experience with all manual drafting, throw the period of UNIX based proprietary systems, to the introduction of DOS based PC-systems and the present powerful 32-bits WINDOWS systems.

The potential for designers and builders with these new flexible tools are enormous even if you compare with what were available only two years ago, but the implementation of these systems has been slow for the traditional industries such as shipbuilding.

With AutoCAD® being the world most used CAD software with over a million users and Rebis™, as the marked leader for pipe design applications I believe the potential is strong for the global industry of ship designers, builders and ship operators. With its drawing format being the industry standard it gives a unique flexibility to utilize the work from conceptual design all the way throw the different stages , also during the operational life of the ship.

2   DESIGN AND ENGINEERING

2.1 Conceptual design

System drawings.
For a shipdesign the pipe systems plays an important role as for the cargo-, utility-
and ship systems. In the ship conceptual design phase the pipe schematics for pipe
system process is developed with PRO-FLOW. Based on owner/trade demands and
classifications rules, the pipe systems optimization can be tailored with a Fluid Flow
analysis program as PIPENET . With such a tool, it's more flexible to decide on
pump capacities, pipe sizes etc. By running Microsoft windows version of these
programs they can easily be dynamically liked by using the industry standard **O**bject
Language **E**xtension.

Arrangement pipe drawings.
For ship project where pipe systems plays an major role in the ships functions a 3D
pipe model study in critical regions like cargo area, pump and engine rooms will be
a functional way to determine the space needed for the various systems. By using
PRO-PIPE it is convenient to route the different pipes on to the ship model. The
material take off from PRO-FLOW and PRO-PIPE is essential information for cost
estimates and purchasing. Values for weight and center and gravity is also possible
to get from PRO-PIPE.

Figure 1 Information flow during conceptual design phase.

2.2 Detail Engineering

For the detailed engineering the usage of the following modules is described:
- PRO-FLOW   for Pipe flow and instrument diagrams .
- PRO-PIPE   for Pipe 2D drawings and 3D models.
- PRO-ISO   for Pipe pre-fabrication drawings.
- PRO-PLANT/STEEL   for Structural Steel 2D drawings and 3D models.
- SA II link   for to-ways link between ISO drawings and the pipe
                        stress program CAESAR II.

System drawings.
As a part of the classifications drawings, the pipe system drawings describing the
different systems as: cargo systems, fuel and lube oil systems, cooling systems, air
systems, loading systems, utility systems etc. To effectively design these type of
drawings and to utilize their information I've used PRO-FLOW .

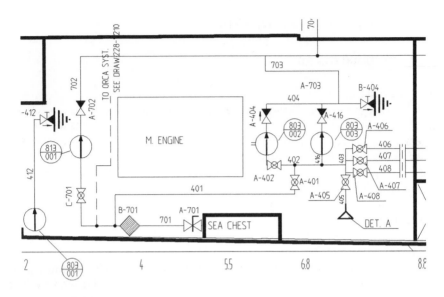

Figure 2   Example of a schematic drawing.

This program is a specification driven drawing program with automatic routines to draw valves and equipment , pipeline etc. There are also routines to set up relations to group parts of a systems into vendors or gather al components belonging to a pipeline number. There is a two-way dynamic link that makes it possible to do global editing in a central material take off base. This can the be updated in the respective drawings.

The program also have full blown Material Manager that can produce any type of report and material take off from the drawings..

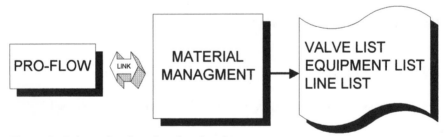

Figure 5   Information flow for pipe drawings.

Pipe drawings in 2D and 3D

The pipe layout and arrangements plans is the most labor intensive in the pipe design process. With the program PRO-PIPE it is possible to make a pipe arrangement and a pipe 3D model at the same time. Actually your pipe plan is your model. You design in a "double line mode" with three coordinates. So you actually builds a model with each component only drawn as it would have been drawn manually seen from above. Each component have a intelligent center line containing all information related to the component. in order to have a complete model you have to model al pipe in the vertical directions, but then you have a complete model to generate sections and ISO drawings from. You can also convert the model to 3D wireframe representation. When working with a pipe arrangement it is possible to use information from the schematic drawings trough a link.

The program is specification driven that means that you have to select the components from a pipe specification

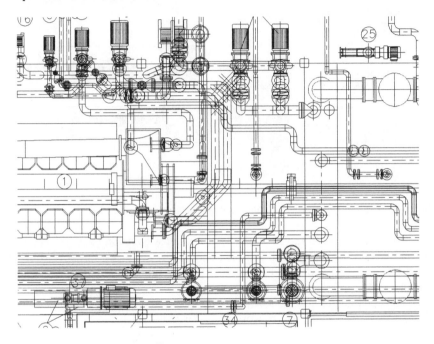

Figure 4   Example of a "raw" 2D & 3D pipe arrangement drawing.

With such a drawing as shown abovek, you can make transverse and longitudinal sections. You can also make plan sections alt different elevation.

Each component contain information as size, line no., material, component description, measure standard, cost and weight. there are routines for automatic routing. You just rout the centerline and the use automatic routines to put on pipe, elbows and other components.

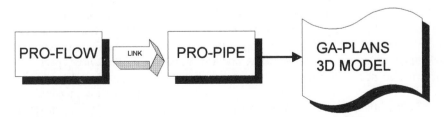

Figure 3   Pipe drawing production .

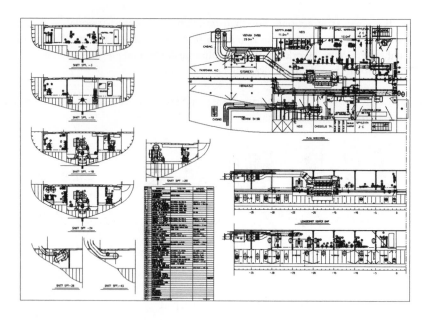

Figure 6   Typical "flat" steel arrangement with equipment placed.

For the interference with the structural steel it is most common to us the flat steel plan and sections as background and then only model the pipes, however with the recent power increase of the PC-systems its now practical to include the steel structure in 3D.

It is possible to link the steel structure to the pipe drawing , and in that way it is possible for somebody else to work parallel with the steel.

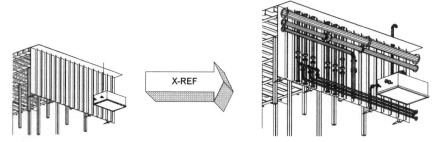

Figure 7   Linking steel structure to a pipe model.

The program have routine to run systematic clash check pipe to pipe and pipe to steel. The program is specification driven. This means that you select a pipe spec and size and the program takes the information as measurements, descriptions etc. from this defined spec. When you modify the spec during the project you can go back and run spec checker on the earlier drawings to update to the revised specs.

There is also powerful filter function witch makes it possible to do selections on sizes, line numbers , layer and specs. For example if you have several pipelines on top of each other, it is possible to isolate one by line number. Also for material takeoff does the filter give the designers good flexibility.

The conversion to 3D for pipe presentation is also a powerful function. Regardless if which drawing mode a pipe component is inserted in it can be converted into several different modes as double line, single line, wire frame, solids and clean outer contours.

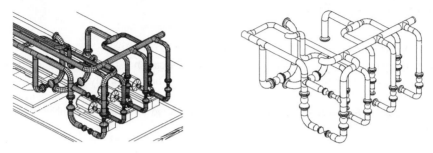

Figure 8   Example of a wire frame model and clean outer contours drawing.

The clean outer contours is the best presentations for plans and sections giving a clean picture with lots of space for measurements and annotations. The wire frame models is ideal for plotting a model drawing and rendering . With rendering it is possible to apply different materials on the model. In that way it is possible to easily present clear models with any pipeline color schemes. With the AutoVisjon program one can set up walking paths for a walk-throw presentation.

2D sections
With the shape of a ship hull the sections puts constrains on a pipe systems. If you only have the steel structure in 2D it is possible to generate 2D sections from the pipe model and place it on the steel sections.

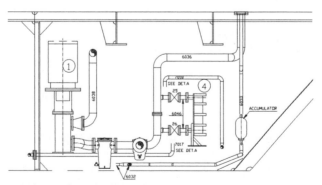

Figure 9   A steel and pipe section.

Equipment.
The equipment such as pumps and heat exchangers is arranged in separate drawings and then linked to the pipe drawings in the same way as the structural steel. The equipment can be a 2D or a 3D presentation with pipe nozzles attached. The pipe nozzles contains the pipe sizes, line numbers and pipe specifications so it is easy to start to drawing pipes as the program reads this settings from the nozzles.

Steel structure

Working in an AutoCAD environment the hull shape can be imported from any fairing program supporting DXF or IGES formats into AutoSurf. Steel plate parts can be defined with the parametric based AutoCAD Designer and steel profiles with the PRO-PLANT/STEEL .

Automatic generation of prefabrication ISO drawings

From the pipe model its possible to generate prefabrication ISO-drawings. First one have to set up the preferences for the ISO-drawing. There are many options on how you want the final layout. Sheet size, measurements, annotation, component graphics etc. Also weather you like to run ISO's based on pipe line number or any other criteria.

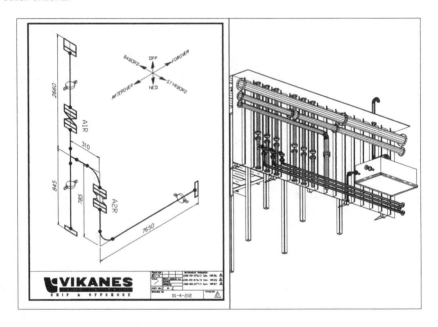

Figure 10   Automatic generation of ISO drawings is a powerful feature.

By generating the ISO-drawings from the model were eliminates the risk of human errors by alternatively redrawing of each pipeline. In general we calculates that the generated ISO-drawing is 80% completed . To complete the ISO-drawings we use the PRO-ISO program. This is a stand alone non-scale ISO program that use the same pipe specifications as used for the arrangement plans.

Pipe stress analysis

If required the ISO-drawing can be imported to a pipe stress program such as CAESARII for either static or dynamic stress analysis. Modifications to the pipeline can be done in the stress program and this can also be imported back to the ISO-drawing.

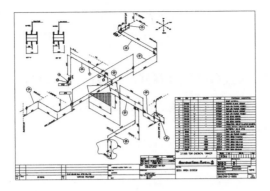

Figure 11   A example of a system ISO-drawing.

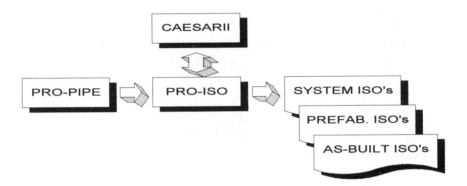

Figure 12   A schematic layout of the ISO-drawing production.

1   TOTAL CONCEPT

As a total concept for pipe design the document and drawing manager like the Autodesk WorkCenter program can keep track of all drawing revisions and related documents as well as electronic distributions of drawings .

Vikanes, Njål.(1994) Experience and potential of AutoCAD based piping
applications for the Shipbuilding Industry.
Born 1957 , Norwegian,
1975-78 Naval Architecture (MRIH) ,
1979-81 Business Administration   (NDH)
1981-82 Global Fishery Services Ltd.,
1983   Kvaerne Rosenberg AS
1984-85 Fishing Vessel Unit ,FAO, Rome , Italy.
1986-90 Bård Vikanes, Naval Architecture office,
1991- Vikanes Consulting (CAD implementation) .

77

Computer simulation for stress analysis in a discontinuous composite

H. G. Kim
Department of Mechanical Engineering, Jeonju University
Jeonju, 560-759, Korea, Tel: 0652-220-2613

H. G. Park
Department of Mathematics, Jeonju University
Jeonju, 560-759, Korea, Tel: 0652-220-2521

S. H. Chang
Department of Industrial Engineering
Kumoh National University of Technology
Kumi, 730-701, Korea, Tel: 0546-467-4319

Abstract

A computer simulation has been performed for the application to the stress analysis in a discontinuous composite solid. To obtain the internal field quantities of composite, the micromechanics analysis and finite element analysis (FEA) were implemented. As the procedure, the reasonably optimized FE mesh generations, the appropriate imposition of boundary condition, and the relevant postprocessing such as elastoplastic thermomechanical analysis were taken into account. For the numerical illustration, an aligned axisymmetric single fiber model has been employed to assess field quantities. It was found that the proposed simulation methodology for stress analysis is applicable to the complicated inhomogeneous solid for the investigation of micromechanical behavior.

Keywords

Stress analysis, composite, fiber, matrix, micromechanics, constraint, plasticity, deformation, MMC, RVE, FEA

1 INTRODUCTION

Composite is one of the strongest candidates as a structural material for many aerospace and other applications (Divecha et al., 1981; Nair et al., 1985). Among composites, metal matrix composite (MMC) has been under development for more than 20 years. However, the initial emphasis was on continuous filament MMCs. They were first developed for applications in aerospace followed by applications in other industries (Gibson, 1994). The expansion into non-aerospace and non-military fields came about slowly as the price of MMC was coming down. This is due mainly to the development of new low-cost fibers (Nair et al., 1985). In recent years, short fiber reinforced metal matrix composites (SFMMCs) have been extensively investigated because it is more economical to produce the production of SiC fibers (whiskers), which has also led to the use of platelet or particulate SiC in MMCs. One of the advantages of discontinuous composite is that they can be shaped by standard metallurgical processes such as forging, rolling, extrusion, and so forth (Taya and Arsenault, 1989).

In these MMCs, where the matrix and reinforcements are well bonded, thermally induced significant residual stresses can arise due to Coefficient of Thermal Expansion (CTE) mismatch between two constituents (Taya and Mori, 1987; Derby and Walker, 1988). In fact, residual stresses are the system of stresses which can exist in a body when it is free from external forces. They are sometimes referred to as "internal stresses" or "locked-in stresses". Therefore, it can be mentioned that accurate prediction of the magnitude and distribution of residual stress is crucial to the design and analysis of MMCs. In recent numerical studies (Levy and Papazian, 1991; Povirk et al., 1991), it was shown that the magnitude of thermal residual stress is significant, adequate to result in substantial plastic yielding around fibers after cooling from the processing temperature though the age hardening effect was neglected.

In this paper, the overall procedure to investigate micromechanical deformation behaviors considering temperature dependent material properties as well as precipitation hardening effect was studied using micromechanics approach. An axisymmetric FEA based on incremental plasticity theory using *von* Mises yield criterion and Plandtl-Reuss equations was implemented to evaluate properties of the representative volume element (RVE) with constraint condition. Some results of numerically simulated thermomechanical behaviors were demonstrated using elastoplastic analysis.

2 MODELS AND PROCEDURES

Figure 1 shows an overall procedure for composite analysis. From the given information, such as geometric and material properties, a designer can choose the analysis types and RVE depending on geometric and material characteristics. For an appropriate RVE, the boundary conditions should be considered rigorously. In the FEA preprocessing, it is quite efficient to set an coarse mesh in order to find out the correctness of overall results.

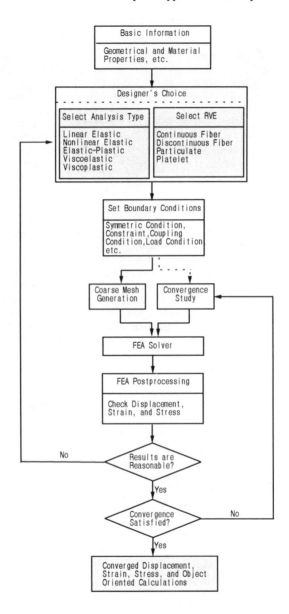

Figure 1 Overall computational Procedure for composite design.

The micromechanical model to describe a short fiber reinforced composite is a single fiber model as shown in Figure 2. In this model, a uniform fiber distribution with an end gap value equal to transverse spacing between fibers was selected. The fibers were assumed as uniaxially aligned with no fiber/matrix debonding allowed for, in keeping with the actual situation in many MMCs (Arsenault and Pande, 1984). Figure 3 describes the relevant meshes.

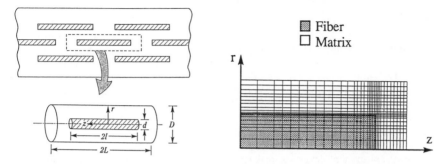

Figure 2 RVE of discontinuous composites.   **Figure 3** FE mesh for RVE.

3 ANALYSIS

3.1 Thermomechanical Formulation

The FEA formulations in this work are centered on the thermo-elasto-plastic analysis with small strain plasticity theory (Cook et al., 1989) using an axisymmetric single fiber model. The model is based on incremental plasticity theory using *von* Mises yield criterion, Plandtl-Reuss equations and isotropic hardening rule. The strains here are assumed to develop instantaneously. To solve nonlinearity, Newton–Raphson method has been implemented. Based on the thermomechanical theory (Cook et al., 1989),

$$\{d\varepsilon^{el}\} = \{d\varepsilon\} - \{d\varepsilon^{pl}\} - \{d\varepsilon^{th}\} \qquad (1)$$

where $\{d\varepsilon\}$, $\{d\varepsilon^{el}\}$ and $\{d\varepsilon^{pl}\}$, $\{d\varepsilon^{th}\}$ are the changes in total, elastic, plastic and thermal strain vectors, respectively. The thermal strain vector $\{d\varepsilon^{th}\}$ is

$$\{d\varepsilon^{th}\} = \{CTE\}\{\Delta T\} \qquad (2)$$

According to von Mises theory, yielding begins under any states of stress when the effective stress σ_e exceeds a certain limit, where

$$\sigma_e = \left[\frac{1}{2}\{(\sigma_x-\sigma_y)^2+(\sigma_y-\sigma_z)^2+(\sigma_x-\sigma_z)^2\} + 3\left(\tau_{xy}^2+\tau_{yz}^2+\tau_{xz}^2\right)\right]^{1/2} \qquad (3)$$

The stress increment can be computed via the elastic stress–strain relations :

$$\{ d\sigma \} = [D] \{ d\varepsilon^{el} \} = [D] (\{ d\varepsilon \} - \{ d\varepsilon^{pl} \} - \{ d\varepsilon^{th} \}) = [D_{ep}] \{ d\varepsilon \} \tag{4}$$

where $[D]$ is the elastic stress–strain matrix and $[D_{ep}]$ is the elastoplastic stress–strain matrix which is given by

$$[D_{ep}] = [D] \left(1 - \left\{ \frac{\partial Q}{\partial \sigma} \right\} \{C_\lambda\}^T \right) \tag{5}$$

where Q is the plastic potential and $\{C_\lambda\}$ is the factor influencing to the plastic multiplier. Elastoplastic stress–strain matrix can be solved iteratively, in which the elastic strain vector is updated at each iteration and the element tangent matrix is also updated. The preset criterion for convergence, ie,. plasticity ratio was used as 1% at all integration points in the model. Detailed procedures to solve material plasticity is described in the reference (Cook et al., 1989).

3.2 Stress analysis for a composite

In FEA, component stresses are calculated for each element at its integration points (or Gauss points). The stress values are then extrapolated to the nearest node using element shape functions, resulting in a nodal component stress for that node due to that element. At a node shared by two elements, therefore, we have two nodal stress values, one from each element. In general, the nodal stresses in the entire model are averaged by the stress contributions from all elements shared by a particular node, as shown in Figure 4.

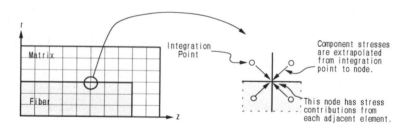

Figure 4 General process of the interfacial stress calculation in FEA.

This averaging scheme is acceptable in most cases, but there are some instances where the scheme becomes quite inappropriate, as **discontinuities** in element stiffness. In this case, stress averaging does not make sense at nodes shared by elements with different material properties or different geometric properties. In such case, the calculation processed by elements of the same material or geometric property individually can be a good choice. Therefore, geometric or material mismatch at the interface can be evaluated. This scheme is especially available for stress analysis of inhomogeneous materials, such as fiber or whisker reinforced composites.

On the other hand, we can implement the concept of volume average method for regional stress analysis (Kim and Choe, 1994). The overall stress in a domain can be calculated through a simple averaging scheme given by the following equation :

$$\langle\ \sigma_{ij}\ \rangle_{\Omega}\ =\ \frac{\int_{\Omega}(\sigma_{ij})_k\ V_k\ d\Omega}{\int_{\Omega}\ V_k\ d\Omega} \tag{6}$$

where $(\sigma_{ij})_k$ is the stress in element k and V_k is the volume of that element. Hence, equation (6) is used to group each domain stress. Hence, the average stress–strain response can be obtained in each domain, which represents regional RVE stresses. By employing this stress grouping approach, a representative domain stress–strain curve can be delineated. In a composite, for instance, the composite domain Ω_c can be decomposed into the whisker region Ω_w and the matrix region Ω_m, and in the same fashion, the field quantity in the matrix region Ω_m can also be decomposed to surrounding matrix region Ω_{m1} and matrix region between whisker ends Ω_{m2}.

4 NUMERICAL ILLUSTRATION

The composite and unreinforced Al 2124 was processed in identical fashion, namely, by a powder metallurgy (PM) process involving hot processing above the solidus followed by hot extrusion. The SiC whiskers were 0.5-1.0μm in diameter with an average aspect ratio of 4 and tended to be aligned in the extrusion direction which corresponds to the longitudinal axis of the tensile samples. After machining, the samples were heat treated for the T–6 condition. From the matrix test data, a bilinear representation of the matrix stress–strain curve was obtained for FEA simulation. Thus, stress–strain characteristics of the matrix were defined by the elastic modulus, yield stress and work hardening rate (tangent modulus). These characteristics were measured at room temperature on the PM 2124 Al alloy and were found to be E_m=70GPa, σ_{my}=336MPa and E_T=1.04GPa, respectively.

The material properties of high temperature behavior were implemented by the documented data (Phillips et al, 1978; Frost et al, 1982). Other material properties chosen were ν_m=0.33 and α_m=2.36x10^{-5}/K for the matrix and E_f=480GPa, ν_f=0.17 and α_f=4.3x10^{-6}/K for the the reinforcement (Kim et al., 1994; Kim and Choe, 1994). Here, E is Young's modulus, E_T is tangent modulus, σ_{my} is matrix yield stress, ν is Poison's ratio and α is CTE.

FE computations were performed using four noded isoparametric elements. The schematic of mechanical (tensile) loading case is shown in Figure 5(a) and (b), and the thermomechanical (cooling) deformation behavior is shown in Figure 5(c) and (d), respectively. Hence, the constraint boundary condition enforces elastic and plastic constraint by requiring that the radial and axial boundary of RVE is maintained in the straight manner during deformation (Nair and Kim, 1991).

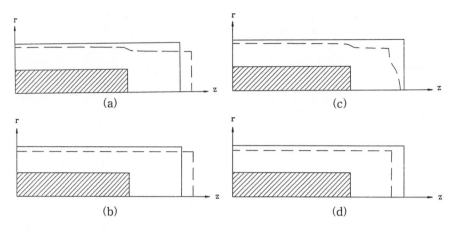

Figure 5   The schematics deformation shape (a) without and (b) with constraint condition for mechanical loading, and (c) without and (b) with constraint condition for thermal loading. (Hatched regions indicate the fiber and dashed lines indicate the distorted shape).

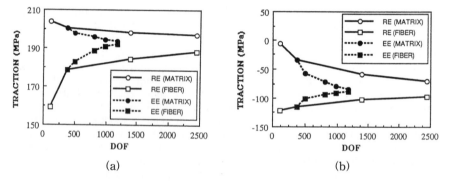

Figure 6   Results of local convergence computed by traction differential approach as a function of degree of freedom (DOF) for (a) the mechanical loading (b) the thermal loading.

Mesh convergence for composite stress analysis can well be obtained using traction differential approach as reported in the previous work (Kim et al., 1994). Figures 6(a) and (b) show the results of convergence study indicating the efficiency of the mapped mesh generation.

Results of simulation to predict the accurate thermal stresses induced by CTE difference are shown in Figures 7 to 9. Figure 7 shows a typical behavior of thermal deformation for cooling down ($\Delta T = -200K$). The simulation results in the displaced vector delineating thermomechanical behavior. In Figure 7, note that the displaced magnitude is actually exaggerated to understand the behavior easily. To understand the microscopic behavior, the magnitude and distribution of thermal stress is of interest to investigate the internal stresses. For instance, the axial component of thermal matrix stresses for $\Delta T=-200K$ are shown in Figure 8. For axial thermal stress contour, compressive stresses are found in the region between fiber ends whereas tensile stresses are found as expected by axial constraint effects. Likewise, the region between fiber ends shows an extensive deformation because of the combined effect of tensile and compressive constraint conditions.

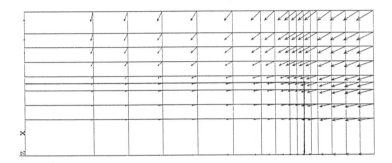

Figure 7 The displacement vector for thermal loading of cooling down ($\Delta T=-200K$).

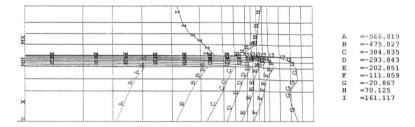

A	=-566.819
B	=-475.827
C	=-384.835
D	=-293.843
E	=-202.851
F	=-111.859
G	=-20.867
H	=70.125
I	=161.117

Figure 8 The typical axial thermal stress contour in the matrix (Unit:MPa). ($\Delta T=-200K$).

Figure 9 shows the the vector plot of principal stresses for the thermal loading case. The direction and magnitude of principal stresses are important in the standpoint of yielding prediction and plasticity evolution. The principal components in the fiber region show all compressive and likewise those in the matrix region show all tensile. Thus far, yielding would start in the surrounding matrix region near the fiber center.

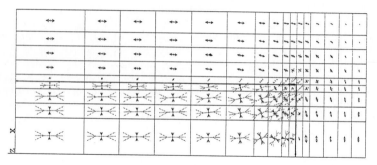

Figure 9   The principal stress contour for the thermal stress case.

5   CONCLUDING REMARKS

A simulation methodology for thermomechanical stress analysis was investigated to predict the internal quantities. It was found that the local deformation behavior in SFMMCs can be simulated depending on the designer's choice. The numerical examples show that the thermal stresses are generated substantially and they are sufficient to provide a plasticity around the stiff reinforcement.

6   REFERENCES

Divecha, A.P., Fishman, S.G. and Karmarkar, S.D. (1981) Silicon carbide reinforced aluminum – A formable composite, *Journal of Metals*, 12-17.

Nair, S.V., Tien, J.K. and Bates, R.C. (1985), SiC reinforced aluminum metal matrix composites, *International Metals Review*, **30**, 275-290.

Gibson, R.A. (1994) McGraw-Hill Inc. Principles of composite material mechanics, New York.

Taya, M. and Arsenault, R.J. (1989) Pergamon Press, Metal matrix composites, Thermomechanical behavior, New York.

Taya, M. and Mori, T. (1987) Dislocations punched-out around a short fiber metal matrix composite subjected to uniform temperature change, *Acta Metallurgica*, **35**, 155-162.

Derby, B. and Walker, J.R. (1988) The role of enhanced dislocation density in strengthening metal matrix composites, *Scripta Metallurgica*, **22**, 529-532.

Levy, A and Papazian, J.M. (1991) Elastoplastic finite element analysis of short fiber reinforced SiC/Al composites: Effects of thermal treatments, *Acta Metallurgica*, **39**, 2255-2266.

Povirk, G.L., Needleman, A. and Nutt, S.R. (1991) An analysis of the effect of residual stresses on deformation and damage mechanisms in Al-SiC composites, *Materials Science and Engineering*, **A132**, 31-38.

Arsenault, R.J. and Pande, C.S. (1984) Interfaces in metal matrix composites, *Scripta Metallurgica*, **18**, 1131-1134.

Cook, R.D., Malkua, D.S. and Plesha, M.E. (1989) John Wiley & Sons. Concepts and applications of finite element analysis, New York.

Kim, H.G. and Choe, G.H. (1994) Role of whisker Stresses in the deformation and fracture of whisker reinforced metal matrix composites, *Proceedings of Korean Society of Composite Materials*, 25-30.

Phillips, W.L. (1978) Elevated temperature properties of SiC whisker reinforced aluminum, *Proceedings of ICCM/2*, Edt. by Norton, B., 567-576.

Frost, H.J. and Ashby, M.F. (1982) Pergamon Press. Deformation mechanism maps, New York.

Kim, H.G., Chang, S.H., Chang D.S. and Chung, S.K. (1994) A numerical study using micromechanics model for metal matrix composites, *Proceedings of Korean Society of Composite Materials*, 25-30.

Kim, H.G., Grosse, I.R. and Nair, S.V. (1994) Finite element mesh refinement for discontinuous fiber reinforced composites, ASME *Journal of Engineering Materials and Technology*, 116, 524-532.

Nair, S.V. and Kim, H.G. (1991) Thermal residual stress effects on constitutive response of a short fiber or whisker reinforced metal matrix composite, *Scripta Metallurgica*, **25**, 2359-2364.

7 BIOGRAPHY

Hong Gun Kim received MS degree from Hanyang University, Korea in 1984 and Ph.D degree from University of Massachusetts at Amherst, USA in 1992. He further studied on the computational solid mechanics as postdoctoral researcher in the Pennsylvania State University, USA during 1992-1993. From, 1994, he has been working as full time lecturer in the department of mechanical engineering in Jeonju University, Korea. His interesting research areas are numerical analysis, solid mechanics, composite materials and CAD/CAM/CAE.

Hong Goo Park received MS degree from Wayne State University, USA in 1984 and Ph.D degree from University of North Texas, USA in 1989. He studied on the isomorphism problems of combinatorial objects such as graphs, digraphs, and designs in the combinatorial theory as well as the generalization of the Bays-Lambossy theorem to arbitrary Cayley objects. From 1990, he has been working as assistant professor in the department of mathematics in Jeonju University, Korea. Currently, he is interested in an open problem related to the above topics through the use of numerical analysis and theory of permutation polynomials as a new approach.

Sung Ho Chang received MS and Ph.D degrees from University of Michigan, USA. He further studied on the factory automation and numerical control as postdoctoral researcher in the Seoul National University, Korea. From 1991, he has been working as assistant professor in the department of industrial engineering in Kumoh National University of Technology, Korea. His interesting research areas are numerical analysis, factory automation and precision measurement.

78

A PLC-based pseudo-servo hydraulic actuator system

S. T. Chen.   B. Z. Chen.   J. J. Tsai.
Department of Industrial Education
National Chang-Hua University of Education
Chang-Hua, TAIWAN. Tel. 886-4-7232107 Ext. 7215,
Fax. 886-4-7211097. Email:Twnet@Chnut028.edu.tw.

Abstract

In this paper, a new linear pseudo-servo hydraulic control system is introduced. In the hydraulic control system, we use conventional solenoid valves, hydraulic cylinders and a Programmable-Logic-Controller (PLC) to construct a translational actuators. With the PLC controlling instantaneous signal change, i.e., on and off in solenoid valves, we can simulate a pseudo servo motion for the hydraulic cylinders. Although, the performance of the actuator system is not quite ideal compared to the general hydraulic servo valves and servo-actuator system, the price of the actuator system is much lower and the installation of this control system is much easier. The control system is discussed in this paper and the control characteristics of the actuator system are presented fromexperimentalresults.

Keywords

Hydraulic Control System, Programmable-Logic-Controller

1 INTRODUCTION

Usually, there are two servo control systems used in automation (D'azzo and Houpis,1986), one is the conventional servo-hydraulic(pneumatic) control system, the other is the electrical servo-motor control system . The former is suitable for the applications where higher speed and larger actuation forces are needed. The later is appropriate for higher precision applications. However, both of the controller systems are expensive and complicate to implement. Therefore, in this paper, we propose a low-cost PLC-based hydraulic control system to simulate a pseudo servo translational motion. Several experiments are conducted to test the characteristics of the control system. Although the precision of our system can not compete with the two servo-control systems, our control system is easy to implement and is suitable for various industrial applications where precision factor is not major concern.

In this paper, the design of the PLC-based control system is described in Section 2. Section 3 presents the experimental setup for testing the actuator characteristics. The experimental results are collected and discussed in Section 4. A brief conclusion and recommendation are given in Section 5.

2 THE DESIGN OF THE PLC-BASED CONTROLLER

In this section, the design of the PLC-based controller is discussed in details. We use conventional solenoid valves, hydraulic cylinders and a Programmable-Logic-Controller (PLC) to construct a translational actuators as shown in Figure 1.

The system consists of

(1). A Programmable-Logic-Controller (PLC) made by OMRON company, model No. C-20H-C6DR-DE (0mron,1991). The controller programs can be edited and saved in EPROM.

(2). A set of hydraulic units consists of a hydraulic cylinder, where the specification is ϕ 20x300 mm, and two single-ended-two -positioned solenoid valves, where the models are SV08-20 and V08-24.

(3). A Linear potentiometer, where the maximum cylinder displacement is 300 mm.

(4). An A/D, D/A converter.

(5). A personal computer.

The input and output of the hydraulic fluid use the same outlet of the hydraulic cylinder as shown in Figure 2. Therefore, we can control the two solenoid valves to adjust the amount of the hydraulic fluid to or from the hydraulic cylinder. With the PLC controlling instantaneous signal change, i.e., on and off in solenoid valves, we can simulate a pseudo servo motion for the hydraulic cylinders. The actual displacement is sensed by the linear potentiometer . The signal can then be fed into the personal computer through an A/D converter. The result is compared with a reference input to form a closed-loop system. Although the performance of the actuator system is not quite ideal compared to the general hydraulic servo valves and servo-actuator system, the price of the actuator system is much lower. In addition, the installation of this control system is much easier.

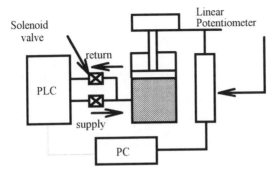

Figure 1 The Design of the PLC-based controller.

Since we would like to test the characteristics of the controller and actuator, different conditions, such as the open and close time of the solenoid valve, has to be adjusted by the PLC. An example program for the PLC is shown in Figure 2 as a ladder diagram. If we want to change the open or close time for the solenoid valve, we can just adjust the value shown in TIM000 and TIM001. If we want to change process continuation time, we can adjust the value shown in CNT003. For example, In the diagram, the supply solenoid valve is closed for 0.5 second after activated for 0.2 second, and this process is continued for five time. Therefore , the symbols #0002 in TIM 000, #0005 in TIM 001, and #0005 in CNT 003, represent 0.2 second, 0.5 second, and five times, respectively.

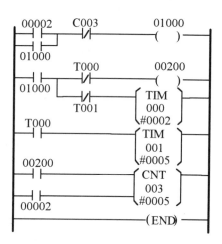

Figure 2 The Program of the PLC in the experiment.

3 THE EXPERIMENT OF THE TRANSLATIONAL ACTUATOR

The experiment setup is shown in Figure 3. We use the design ideas discussed in previous section. On the top of the hydraulic cylinder, different loading is placed to simulate various loading conditions. For the solenoid valve V_2, we design another set of motor and pump to retrieve the hydraulic fluid out of the cylinder to force the descending of the piston. The displacement of the piston can be measured through the linear potentiometer and an A/D converter. In order to test the characteristics of the translational actuator, different controller conditions and loading conditions are given in the experiment.

The experiments can be classified into three groups:

(1). For different loading and hydraulic pressure, the displacement   and speed change of the piston are tested.

(2). For different open and close time of the solenoid valves, the displacement and speed change of the piston are tested.

(3). The repeatability is tested through a defined control condition.

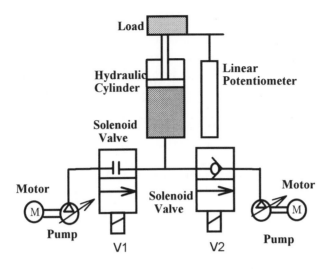

Figure 3 The Design of the experimental setup.

The solenoid valve are on and off for 5 times in all the experiments. The displacement are measured through the linear potentiometer. The parameters used in the experimental results are defined as,

P_n : the pressure of the hydraulic power defined as n kg/cm^2.

F_m ; loading units equal to m, where 3 kg per unit.

T_{r-s}: the on-off time interval of the solenoid valve shown in Figure 4, where value r represents the time when the solenoid valves on and value s represents the time when the solenoid is off.

Figure 4 The On-off time interval of the solenoid valve.

4 THE EXPERIMENTAL RESULTS

For different loading units and hydraulic pressure, we test the displacement and speed of the piston. The results are as follows,

4.1 Various pressure and loading conditions vs. displacement

The results are shown in Figure 5 -(a) to (d). The time interval for the solenoid valve is $T_{0.2-0.5}$, Y-axis represents the displacement of the piston, and X-axis represents the sequence of 5 on-off operation of the solenoid valve.

Displacement(mm)

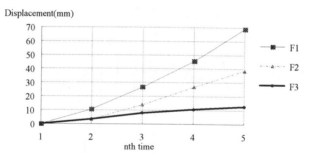

Figure 5-(a) The pressure is 7 kg/cm².

Displacement(mm)

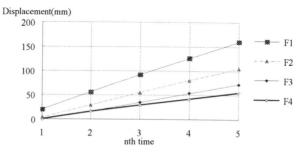

Figure 5-(b) The pressure is 8 kg/cm².

Displacement(mm)

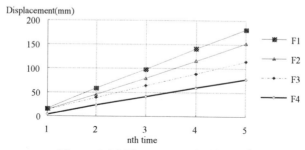

Figure 5-(c) The pressure is 9 kg/cm².

Displacement(mm)

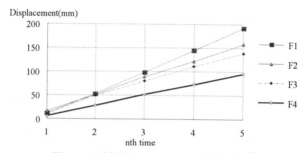

Figure 5-(d) The pressure is 10 kg/cm².
Figure 5 The Various loading and pressure conditions
Vs. the cylinder displacement.

From Figure 5-(a), it is noticed that the pushing force is not enough for the cylinder when the pressure is 7 kg/cm^2 and for the first on-off operation in the solenoid valve. The situation does not exist when the pressure is larger than 8 kg/cm^2. The resolution for different pressure and loading conditions can be calculated by dividing the final displacement by 5. In addition, From Figure 5, it is observed that the first displacement is instable when the starting point is at the origin because of the static friction. Therefore, the design of the controller should avoid the origin.

4.2 The characteristics of the descending hydraulic cylinder

Because the static friction is so large that the hydraulic cylinder will descend only when the loading is greater than F3 or F4. The displacement characteristics curves are shown in Figure 6. The starting point is at the maximum displacement of 300mm. X-axis represents the sequence of 5 on-off operation of the solenoid valve. In Figure 6, X-axis 2 stands for the first on-off operation of the solenoid valve. The descending speed is fast since the pulling force contains self weight and the force inducing by retrieving the hydraulic fluid from the cylinder. The displacement is 171.37mm when the loading is F4 and 5 on-off operations in the solenoid valve. The position control for descending is more difficult than for ascending.

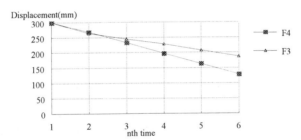

Figure 6 The displacement of the descending hydraulic cylinder.

4.3 Different operation time of the solenoid valve vs. the displacement and speed change of the piston.

(a) T$_{r\text{-}s}$ v.s. speed

The purpose of this part of experiment is to teat the relationship between the on-off time and the ascending speed of the hydraulic cylinder. The pressure is set at P11 and the loading is set at F1. The results are shown in Table 1. The final displacement is measured as X value.

Table 1 T$_{r\text{-}s}$ vs. displacement and speed of the cylinder

	Displacement (mm)	Velocity (mm/s)
T$_{0.1\text{-}0.5}$	51.37	102.74
T$_{0.2\text{-}0.5}$	150.37	150.37
T$_{0.3\text{-}0.5}$	270	180

where Velocity = x/5r. The speed increase as the time for opening the solenoid valve increase.

(b) Tr-s, S value v.s. the displacement

The pressure is set at P11 and the loading is set at F1. In order to avoid the origin influence, mentioned in previous experiment, the starting point is chosen at 10mm from the origin. The displacement is shown in Figure 7. It is noticed that the displacement for $T_{0.1-0.1}$ is twice different from the other displacement. Although the motion is stable, the motion displacement is too large and is difficult to control accurately. Hence, we conclude the closing time for the solenoid valve do influence the control stability and displacement.

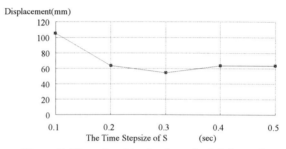

Figure 7 The solenoid valve is on for 0.1 Second.

A defined control condition v.s. the repeatability. The pressure is set at P11 and the loading is F2. The solenoid valve on-off time is set as $T_{0.2-0.5}$. This process is repeated 5 times. The final values is measured through the AD/DA converter. The results are shown in Figure 9. The maximum and the minimum deviation value compared with the average value is 12.75mm and 2.25mm, respectively. The repeatability for this controller need further improvement.

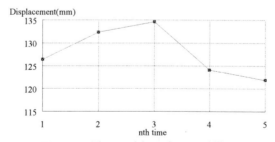

Figure 8 The precision of repeatability.

From the experimental results, we know that the pressure in the cylinder, loading condition, the on-off time of the solenoid valve, and the static friction are the influence factors for the control characteristics. The pressure should not too small and the S value of T_{r-s} condition should avoid 0.1 second.

5 CONCLUSION

In this paper, a new linear pseudo-servo hydraulic control translational actuator is introduced. We use conventional solenoid valves, hydraulic cylinders and a Programmable-Logic-Controller (PLC) to construct the translational actuator. With the PLC controlling instantaneous signal change, i.e., on and off in solenoid valves, we do simulate a pseudo servo motion for the hydraulic cylinders. From the experimental results, when the loading is below 12 kg and the pressure is above 8 kg/cm^2, we can obtain a linear displacement results. Therefore, the minimum pressure in the cylinder should not less than 8 kg/cm^2. In addition, the nonlinear effects such as the static friction between the piston and the wall of the hydraulic cylinder do influence the performance of the actuators. Although the performance of the actuator system is not quite ideal compared to the general hydraulic servo valves and servo-actuator system, the price of the actuator system is much lower and the installation is much easier. Hence, the control system remains valuable and can be applied to those industrial applications where the accuracy is not major concern. Further research should be concentrated on improving the repeatability and overcoming the influence of the static friction.

ACKNOWLEDGMENTS

The author wishes to acknowledge the support of the National Science Council in Taiwan, grant number NSC83-0422-E-018-008.

REFERENCE

D'Azzo, J. J. and Houpis, C. H., (1986) *Linear Control System Analysis and Design,* McGraw Hill, New York.

Omron,(1991), SYSMAC Mini H-type PCs: C20H, C28H, C40H ,Installation Guide,Operation Manual, Vol.1,Japan.

Omron,(1991), SYSMAC C-Series PCs Ladder Support Software, Operation Manual,Vol.1,Japan.

BIOGRAPHY

Su-tai Chen is an associate professor in the Department of Industrial Education, National Chang-hua University of Education, Taiwan. He received the B.S. degree in Mechanical Engineering from National Taiwan University in 1984, and the M.S. and Ph.D. degrees in Mechanical Engineering from the University of Iowa, U. S. A., in 1988 and 1991, respectively. His current research interests include kinematics and control of Robot, industrial controller and actuator, computer modeling and simulation, automation and educational software development, and multimedia application in industrial education.

Bau-zo Chen and **Jia-jun Tsai** are research assistants and students in the Department of Industrial Education, National Chang-hua University of Education, Taiwan.

79

Computer–Controlled Temperature System of Polymerization Reaction Autoclave

Authors: Li Shou–cheng, Guo Cai–xia, Zhang Ning
Wang Mei and Liang Jun
Affiliation: Professor, Lecturer, Associate Professor
Engineer and Engineer
Address: Departement of electrical engineering
Northern Jiaotong University
Beijing, China
Telephone: 3240346
Postal number: 100044

Abstract

This paper presents the making of the polymerization reaction autoclave temperature control system based on computer, giving its hardware structure drawing, software flow chart graph, experimental graphics and evaluation. It has been proved that the artificial intelligent control is a efficient way to chemical production process.

Keywords

Computer controlled temperature system, polymerization reaction autoclave, pearl producing process, Quick–Basic language, data acguisition, D–A converter

1 INTRODUCTION

The intermittent polymerization reaction autoclave is an equipment that produces anion–cation–exchange resin pearl. The crux of the pearl producing process is the cooling control mode in which thermal energy is shot up in homothermal stage, as ab, cd, ef stage is shown in Figure 1.

The polymerization reaction autoclave—the object controlled has characteristics of nonlinearity, time dependent, random variation, etc. The designed targets cannot often be attained by adopting classical controlling theory and method to set up the control system.

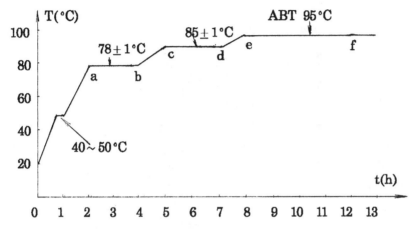

Figure 1 Process curve.

2 NEW CONTROLLING AND REGULATING SYSTEM

2.1 System diagram of controlling and regulating

The acquistition and control of temperature signals are achieved by means of data acquisition and control boards, PC –1216C and PC–1612. Temperature is detected by copper thermal resistance. The executive device consists of D–A converter, V–I converter, air compression system, electro – pneumatics transducer and pneumatic diaphragm regulating valve. The system diagram as shown in Figure 2. Stored into the computer as recorded data are the technological curves made in producing anion – cation – exchange resin pearls. Manual experiences can be summarized into four sections of 'Generation Model Regulations'. The temperature in the autoclave is regulated according to the temperature detected in the autoclave and by means of controlling the steam valve and the water valve.

The structure sketch polymerization reation autoclave is shown in Figure 3.

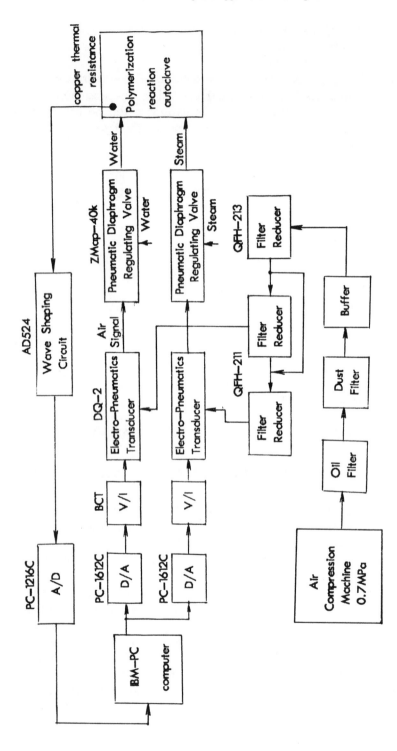

Figure 2 System diagram of controlling and regulating.

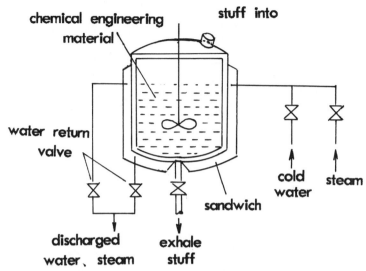

Figure 3 Structure sketch of polymerization reation autoclave.

2.2 Structure of system software

The new control system is developed through summing up the polymerization reaction long term manual control experiences and applying the 'Generation Model Regulations' (i.e 'IF⋯ ⋯ THEN⋯ ⋯ ELSE' form statement), used in artificial intelligence and micro –computer and the techniques of data acquisition, etc. The purpose of the system is to meet the production requirements conveniently.

Software is programmed in Quick–Basic language.

Whole software consists of main program and five subroutines. The structure of main program is shown in Figure 4.

2.3 System has other features

The system realizes the self–acting and real time marking and recording of the data of production. The computer can check and monitor the system. After operation starts, first , check all the I/O interfaces.

If any problems arise, the system will give an alarm and prompt and stop working for check and repair. When the system is in operation, the computer practises real time monitoring. Once the system is found out of control, it will give an alarm and prompt and change to hand operation.

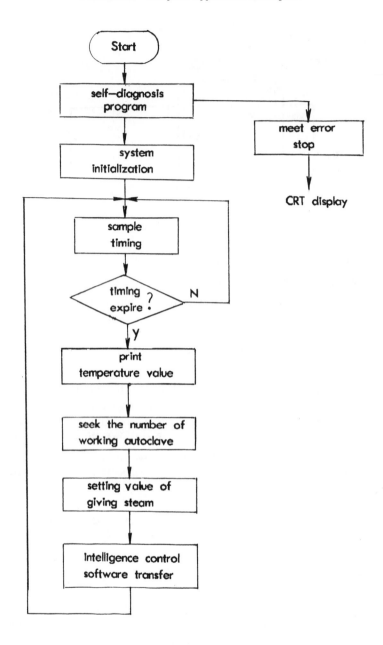

Figure 4 Structure of main program.

3 USER ESTIMATION

The system is simple and various control can be carried out easily. The beneficial result of production is 10% higher.

4 REFERENCES

蒋慰孙 俞金寿，(1988)，过程控制工程，烃加工出版社，北京。

沈平，(1987)，时间滞后调节系统，化学工业出版社，北京。

郭锁凤，(1987)，计算机控制系统，航空工业出版社，北京。

王学慧 田成方，(1987)，微机模糊控制理论及其应用，电子工业出版社，北京。

周明德，(1990)，微型计算机硬件软件及其应用，清华大学出版社，北京。

李守成，(1992)，电子技术（电子学Ⅱ），高等教育出版社，北京。

天津自动化仪表四厂，(1990)，常用调节阀的选择，气动薄膜调节阀说明书，天津。

上海自动化仪表厂，(1990)，QFH系列空气过滤减压器，上海。

北京祥云数据设备公司，(1990)，PC–1612C16通道12位D/A板用户手册；PC–1216C12位16通道AD/DA版用户手册，北京。

J.M.Smith, (1959), No.2, A controller to Overcome Dead-time, ISAJL.

R.F.Giles, (1977), Vol.16, No.1, Gain-adaptive Deadtime Compensation, ISA Trans.

5 CHIEF COMMENTATOR BIOGRAPHY

Li shou cheng was born in 1934 in Tianjin,China. I am working in the department of electrical engineering, Nothern Jiaotong University, Beiing, China. I am a Professor and have compiled 5 books as ' Electronics (Electrolechnics Ⅱ) ' 、 ' Digial Electronics' , and have published 20 articles.

I am the permanent members of a council of board of directors of Electrotechnics of Universitys, China, Beijing and China railway.

Prof. Li shou cheng
  Department of Electrical Engineering
  Northern Jiaotong University
  Beijing 100044
  China
Tel: 2256622(Rail 40500) — 2440(H)
  3240346(Rail 40346)   (0)

PART SEVENTEEN

Robotics

80

Rapid prototyping of large complex shapes with a fully automatic robot system

J.S.M. Vergeest, J.J. Broek and J.W.H. Tangelder
Delft University of Technology, Faculty of Industrial Design Engineering
Jaffalaan 9, NL 2628 BX Delft, Netherlands
Telephone +31 15 783765, Fax +31 15 787316
Email j.s.m.vergeest@io.tudelft.nl

Abstract

Rapid prototyping has become an increasingly prominent technique for the product development process. In several situations CNC milling is among the favourable prototyping methods. We present a system based on a 7 degrees-of-freedom mechanism including software to mill a prototype in a robust and fully automatic way given a preliminary CAD model. Practical results of this system are shown. Further, we describe an approach to extend the domain of manufacturable shapes based on an analysis of the 7-dimensional configuration space of the mechanism.

Keywords

Computer-aided design, rapid prototyping, robot path planning, voxel representation, configuration space

1 INTRODUCTION

In industry product prototypes have always been essential at critical points of the design process, as a basis on which design decisions can be made. Both virtual prototypes and physical models are widely used for very different purposes from mechanical testing to assessing consumer acceptability of the designed product. Presently, as companies get pressed to accelerate the product development process and to enhance their design practices into a concurrent engineering approach, the need for methods to produce prototypes very rapidly from preliminary CAD models has grown (Echave and Grau 1993). This has stimulated the development of new prototyping methods and it has also encouraged the research for improving existing principal techniques.

This paper describes the development and application of an automatic robot milling facility that produces physical replica of CAD-defined objects. The advantages of this rapid prototyping process compared to e.g. stereolithography, are significant in situations that the prototypes are large ($>\approx 300$ mm) and at the same time must be inexpensive. However, robotic prototyping can be economic only if path planning proceeds fully automatically. This holds for several other types of robot applications as well, and it is the time consuming task of manual or semi-automatic path planning that causes some of the robot potentials to be unexploited in industry, in particular for small-batch applications.

In Section 2 we briefly list the pros and cons of robot (CNC) milling as a means to obtain product prototypes and we summarize the specific technical requirements. The hardware, algorithms and software of the current system are presented in Section 3. In Section 4 we propose a configuration space method to solve some particular problems of robot motion in the milling task. Conclusions about the method are drawn in section 5.

2   RAPID SHAPE PROTOTYPING USING A MILLING ROBOT

There are many different ways of rapid prototyping. Sometimes a graphical presentation of a CAD model (either on a workstation or in virtual reality) is sufficient to make preliminary design evaluations (Smets 1993). Physical prototypes can be made either using stereolithografic techniques and other incremental methods, or with conventional CNC milling techniques which is often supported in a CAD/CAM environment (Barkan 1993). The best type of prototyping depends, among other factors, on the intended purpose; visual inspection, mechanical testing, ergonomic studies, functional testing etc. In a comparison of four prototyping technologies (Wall 1992) it was concluded that in many situations CNC machining is the best choice.

The benefits of CNC are:
1. Especially for large objects it is relatively fast
2. A wide range of materials can be applied
3. A high accuracy can be obtained
4. The costs are relatively low
5. Manufacturing experience can be gained from the prototyping process
6. The necessary hardware is often already available at the company.

There are also drawbacks:
1. The shape domain is restricted due to limited accessibility by the machining tools
2. Most current systems are far from automatic as they are just intended for final CAM
3. Computation and the physical process both are still too slow.

Our research aims at the development of CNC-based rapid prototyping with emphasis on objects with free-form shapes. Based on experience (both from education and from industrial collaboration projects) we obtained a number of requirements:
1. The system must be safe
2. Any object must be correctly processed, as far as it is reachable
3. Inaccurate and, to some extent, even incomplete geometric input must be handled

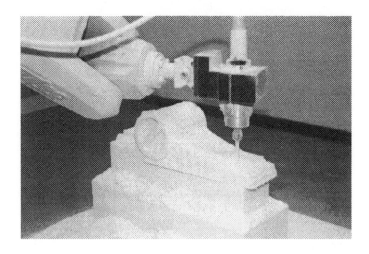

Figure 1: The Sculpturing Robot System performs rapid prototyping of CAD-defined objects.

4. The process must be fully automatic
5. Prototypes with dimensions up to 1m³ must be manufacturable without refixturing
6. The process should not last longer than 4 to 48 hours, depending on the product size
7. The user must be able to make a trade-off between speed and accuracy of the prototyping process
8. Each milling process must be performed in such a way that a subsequent, more accurate machining process can be made. It is acceptable that excess material remains but no material may ever be removed from the nominal part.

3 THE SCULPTURING ROBOT SYSTEM

Taking account of the above listed 8 requirements we have developed the

Sculpturing Robot (SR) system. We based this development on more than 10 years experience at the Faculty of Industrial Design Engineering in CAD-defined prototype production with CNC milling (Lennings 1992). The SR system accepts considerable inaccuracies in the geometry from the CAD model. Mostly the geometry is defined by a (possibly large) number of Nurbs surfaces that either exactly, or within some tolerance enclose one or multiple volumes. The system also accepts contour data, e.g. from medical scan devices. All the CAD user needs to do is specifying which model should be fabricated and selecting the required spatial accuracy.

The hardware consists of a 6 degrees-of-freedom industrial robot. All 6 joints are of the revolute type. The stock of material is placed on a horizontal turn table, rotatable around its vertical axis. A milling device consisting of a toolholder and a milling tool

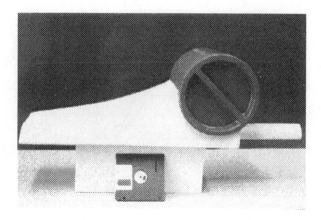

Figure 2: Typical prototype manufactured by the SR system before finishing.

is mounted on the end-effector of the robot, see Figure 1. Both the robot and the turn table are computer-controlled via a precomputed path file. This file essentially contains move-to instructions in terms of position and orientation of the toolholder relative to the robot base, and as 7th parameter the orientation of the turn table.

The system is currently running the Srplan1 software (Tangelder 1994). This path planner is based on a strategy where only 5 different tool orientations are allowed relative to the stock, one vertical and 4 horizontal directions, mutually perpendicular. With this strategy it was feasible to represent both the target CAD model and the stock-in-progress, which encloses the CAD model, by 6-grid voxel data structures. These data structure capture the process of material removal due to intersection of the tool with material, and also support efficient collision avoidance (Tangelder 1994). Figure 2 shows one of the prototypes obtained with the SR system. The length of this object is approximately 40cm and it took about two-and-a-half hours to machine it out of foam at a spatial accuracy of 1mm. The precomputed path file consisted of approximately 31 000 robot/table movement instructions.

The development of the path planning software has been supported by a new graphical simulation system Srsim3 which visualizes the mechanism's motions but also the material removal process (Walstra 1994). In a related research project this simulation will be extended through virtual reality to give the CAD user direct and intuitive control over the rapid prototyping process.

4   PATH FINDING IN CONFIGURATION SPACE

It is quite obvious that a milling strategy with only 5 different access directions, will in general not meet requirement 2 (see section 2). Certain portions of the model may be inaccessible by the tool in all 5 directions, whereas they would be accessible when the

Figure 3: Milling slots of this type cannot be done with a small number of tool access orientations

tool has other orientations. One approach to solve this is to search for those alternative orientations or, more generally, to determine for a given target CAD model a minimal number of orientations such that the model can be completely milled. For each of these orientations a Z-buffer data structure can be built suited for the Srplan1 algorithms (Tangelder 1995).

For some situations the number of access directions may get too high (see Figure 3) and a different path find method is needed. The example geometry shown in Figure 3 suggests that a continuous milling path is possible if the tool orientation varies continuously too along the path. For this type of situations direct search in configuration space of the mechanism may be the most effective strategy.

For the SR system the configuration space C^7 is the set of all possible 7-tuples $(q_1 \ldots q_7)$, where q_i is the angle that parameterizes joint i. It can be easily demonstrated that building up a representation of C^7 and performing a direct search for collision-free milling paths in this 7-dimensional space is computationally very expensive.

We formulate the problem of path planning by using a general formalism for the definition of all objects (mechanism, auxiliary devices, obstacles) including the stock-in-progress S, which is an object that changes during the process; material is removed from the volume S due to a penetrating milling tool T. This latter effect of volume removal is extremely important; if a point in Eucledian space changes from being "inside material" into a point in empty space, a subspace in configuration space will be created that extends the total set of allowed configurations. The dimension of this subspace is related to the redundancy of the task. Hence the existence of a collision-free trajectory at a given moment during the process depends heavily on the trajectory realized up to then.

This dynamic aspect is important for the research into new trajectory planning methods. Whereas most theories start from a problem definition in terms of start situation and goal situation, a material removal process allows a multitude of start/goal situations

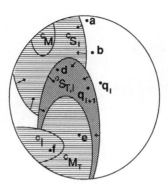

Figure 4: Schematic overview of the configuration space C^n at the beginning of subtask i. The subregions are described in the text. The arrows indicate shrinkage of a subset.

to be considered simultaneously. This provides for a larger and richer space in which solutions are available.

We introduce a number of subsets in the configuration space C^n of the mechanism in consideration, where n is the number of degrees of freedom of the mechanism. Each n-dimensional point $\mathbf{q}$ is located in zero or more of these subsets. The three most important subsets are:

1. $^C I$ is the subset of C^n containing all configurations $\mathbf{q}$ that would cause intersection (collision) of the mechanism (including T) either with any of the fixed obstacles in the work cell or with itself. Another forbidden set is $^C M_T$, the configurations for which T would damage the target CAD model M.

2. $^C J_i = {}^C I \cup {}^C M_T \cup {}^C S_i$, where $^C S_i$ are the configurations for which the mechanism (not T) intersects S_i, the stock-in-progress at the time the milling process has come to point i. Due to the material removal process we have $^C S_{i+1} \subseteq {}^C S_i$.

3. $^C S_{T,i}$ is the set of configurations representing actual material removal, i.e. for which T intersects S_i. For this set it holds that $^C S_{T,i+1} \subset {}^C S_{T,i}$.

When the milling process has proceeded till step i with $\mathbf{q} = \mathbf{q}_i$, then the next point $\mathbf{q}_{i+1}$ in C^n can be reached by a path through $^C Q_i$, where

$$^C Q_i = (C^n - {}^C J_i) \cap {}^C S_{T,i}.$$

If the path $\mathbf{q}_i \rightarrow \mathbf{q}_{i+1}$ does not intersect $^C S_{T,i}$ (but remains inside $C^n - {}^C J_i$) then no material gets removed during this step, i.e. $^C Q_{i+1} = {}^C Q_i$. $^C Q_i$ is the overlap of an increasing set $(C^n - {}^C J_i)$ and a decreasing set $^C S_{T,i}$. Therefore a general statement cannot be made on the change of the size of $^C Q_i$ (see Figure 4). However, for a particular mechanism and a

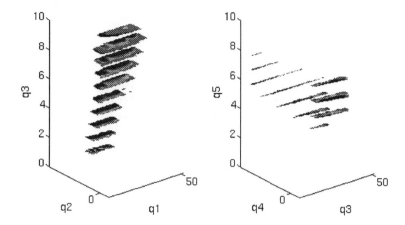

Figure 5: Projections of C^7 onto the space spanned by q_1, q_2, q_3 (left) and onto the space spanned by q_3, q_4, q_5. The 3D volumes are visualized as parallel slices. The colors (gray values in this picture) indicate which regions of C^7 contain relatively many configurations that are allowed and that contribute to the execution of the task. All units are arbitrary.

given target model M, it is possible to analyse the behaviour of ^CQ_i at any i.

The Srplan1 algorithm (outlined before) is based on maximal depth machining with 5 different tool orientations. The resulting path in C^7 therefore is, for the most part, located either on the hypersurface defining the boundary of ^CS_i or ^CM_T. As stated before, in a typical milling process a path planning strategy directly based on geometry in Eucledian space will most of the time work well. However, if these strategies fail to remove parts of ^CQ then a search for permitted corridors through ^CQ and $C^7 - ^CJ$ could be effective. For $n = 7$ this is feasible only when a small fraction of C^7 is actually considered, e.g. only the part for which T is near the difficult spots on the boundary of M. Even with this restriction the search space may still be too large. Then a statistical analysis will help in finding permitted corridors. An example of such a statistical analysis is shown in figure 5. The analysis produces projections of C^7 on 3D spaces spanned by triplets of coordinates. To every point in this 3D space is assigned a scalar (or color) representing the probability that $\mathbf{q}$ is located in $^CI, ^CM_T, ^CS, C^7 - ^CJ$ or in ^CQ after integration over the remaining 4 coordinates. This allows navigation in multiple 3D spaces rather than in a single 7D space. For a simple case (M being a cube enclosed by a larger cube S) we generated a number of 3D distributions (two of which are shown in figure 5). It turned out that for all coordinate triplets plotted so far, regions with $\mathbf{q} \in^C Q$ (after integration) showed up. These regions are valid starting points for milling paths. We refer to a similar approach in (Faverjon 1989).

5 CONCLUSIONS

We have described the hardware, software and results of the Sculpturing Robot system. Using the current path planning strategy with only 5 predefined tool access directions, shape prototypes can be rapidly made for a significant shape domain. This domain is being extended by allowing more tool orientations. For certain types of geometries it will be necessary to perform search directly in (a limited part of) configuration space. We have described an approach to achieve this and we demonstrated the feasibility of this approach by an initial analysis. In this analysis, the entire robot workspace was considered. The next step will be to take a complex object (e.g. as in Figure 3) and to restrict the analysis to the region which could not be handled by the heuristic strategies.

6 ACKNOWLEDGEMENTS

We very much thank Bram de Smit and Adrie Kooijman for their contributions to this work.

7 REFERENCES

Barkan, P. and Iansiti, M. (1993) Prototyping: a tool for rapid learning in product development. *Concurrent Engineering: Research and Applications*, **1**, 125-134.

Echave, I. and Grau, S. (1993) Stereolithography in small and medium enterprises, in *Proc. of the third FAIM Conference* (Eds. M.M. Ahmad and W.G. Sullivan), CRC Press, Boca Raton, 200-209.

Faverjon, B. and Tournassoud, P. (1989) Motion planning for manipulators in complex environments, in *Geometry and Robotics, Lecture Notes in Computer Science*, (Eds. J.-D. Boissonnat and J.-P. Laumond, Springer-Verlag, Berlin, 87-115.

Lennings, L. (1992) CAD/CAM integrations in practice: Two cases of computer aided toolmaking. *Computers in Industry*, **18**, 127-134.

Smets, G.F.J. and de Ruwe, P. (1993) Rapid prototyping using virtual reality and robot milling. Project description, Delft University of technology (in Dutch).

Tangelder, J.W.H. and Vergeest, J.S.M. (1994) Robust NC path generation for rapid shape prototyping. To appear in *Journal of Design and Manufacturing*.

Tangelder, J.W.H. (1995) Tool access orientations for milling 3D shapes based on a spatial 3D visibility data structure. To be published.

Wall, M.B. Ulrich, K.T. and Flowers, W.C. (1992) Evaluating Prototyping Technologies for Product Design. *Research in Engineering Design*, **3** 163-177.

Walstra, W.H., Bronsvoort, W.F. and Vergeest, J.S.M. (1994) Interactive simulation of robot milling for rapid shape prototyping. To appear in *Computers & Graphics*.

81

New developments in exchange of robot controller models

T. Sørensen, E. Trostmann, and U. I. Kroszynski
Control Engineering Institute,
Technical University of Denmark.
Building 424, DK-2800 Lyngby, Denmark
Tel. +45 45 934 419 Fax +45 42 884 024

Abstract

With the increasing capabilities of advanced computer based systems for design and simulation of robotics systems, the complexity of the data formats increases, which are used to store the product design results. Standardisation of the product description interface between these design systems, and robotic simulation systems, is a pre-requisite for the implementation of CIME systems in order to exchange such comprehensive design results freely between different program package. Standardisation will allow to replace software components and production machines by more effective ones as they appear in the market, leading to the application of customised CIME systems based on multiple vendors.

Much work has been done under ISO-STEP (ISO10303-STandard for the Exchange of Product model data) to define standardised resource- and application models for shape/geometry, and for kinematics of mechanisms, but the dynamic description of the robot arm as well as the servo-loops, and the robot controller has not been addressed. New STEP-proposals for standardised information models that include these missing data are under development in ESPRIT project 6457 InterRob 'Interoperability of Standards for Robotics in CIME'.

This paper highlights some issues in the InterRob development work of a new information model for robot controllers. The information model will be used to specify a proposal for a standardised product description interface for the exchange of robot controller models between advanced design systems and robot off-line programming and simulation systems.

Key Words

CAD/CAM, CACSD, CIME, STEP, Products Specification, Information Modelling, Robotic Controller, Robot Off-line Programming, Simulation.

1. INTRODUCTION

Computer Integrated Manufacturing and Engineering (CIME) need intelligent communication of product definition data as a basis. Information about the products must flow back and forth between a large number of senders and receivers beyond the boundaries of a single enterprise in an intelligent, reliable, complete, and unambiguous fashion. Any bilateral agreement about this topic, even national standards will fail to meet the needs of industry (Schlechtendahl, 1989).

The evolving international product data standard STEP has demonstrated the feasibility of neutral product data representation. It provides a methodology for product identification and specification for product modelling throughout the product life cycle of any type of mechanical product. The identification of product related concepts (or reference models) is used to establish a definition of what information constitutes a complete computer-interpretable definition of a product. The specification phases in STEP include a specification of the functionality of the formal product data model, a specification of the protocol that describes the context in which the product model is used, and a specification of how the information can be accessed.

Product data technology currently still relies on existing national and de facto standards within specialised areas. Such initiatives like SET, VDAFS, IGES, and EDIF are widely used, but in the years to come a migration towards STEP is expected to take place and thus facilitating full integration based on the open systems concept. An early presentation to the open systems concept was given in (Kroszynski et al. 1989).

In order to achieve a more universal application of the STEP methodology for neutral product modelling and data representation, and for exchange of product data between different systems, STEP has to be extended so more complete product models can be represented by including data for geometry, kinematics, dynamics, control, and technology; so far no other international project activities than InterRob, are known to address this challenge.

The remaining part of this paper highlights some issues in the InterRob development work of a new information model for robot controllers (Sørensen, 1994). The information model will be used to specify a proposal for a standardised product description interface for the exchange of robot controller models between advanced design systems and robot off-line programming and simulation systems.

2. THE ROBOT CONTROLLER INFORMATION MODEL

2.1. Motivation

The success of a robotic design and simulation systems is highly correlated with the degree to which the systems are capable to model and predict the behaviour of physical robotic systems by simulation of real executed tasks.

A realistic simulation of the robot tasks calls for a very detailed model of the robotics system in question, and the free exchange of such detailed models demands a well developed product description interface, as shown in Figure 1, that is general enough to carry the data and functionality of the models used in the simulation.

The product information model is used as a reference model for the development of this product description interface, and one of the main properties of the controller information model is therefore that it has the capacity to define a STEP based robotics interface general enough to carry the data and functions reflecting a large diversity of robot control systems. The control systems addressed should not 'only' reflect the present situation of the real industrial robotics world, but a robotics interface must be identified and specified such that main issues from current and future scientific developments in the area can be mapped onto the proposed schema.

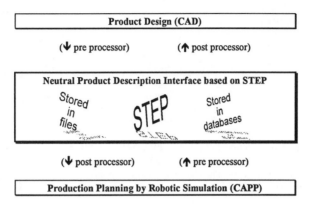

Figure 1 The STEP based product description interface between CAD and CAPP. The pre- and post-processors are programs that communicate design and production planning data and functions via the interface.

The same kind of control-scheme is implemented in several different ways from manufacturer to manufacturer, and the implementation details are most often concealed as unreachable micro codes stored in microchips situated at different places in the controller's integrated circuit. The robotics control interface must therefore also be designed so it is open for whatever controller, hidden or not, that is going to be transferred via the interface.

The total sum of properties (functionality and data) of the future robotics systems will increase relative to the properties of the currently existing robotics systems. Traditional robot simulation systems only cover a minor subset of the properties of the existing robotics simulation systems. The properties that need to be covered by realistic robot simulation systems of tomorrow will therefore also increase. Figure 2 shows a qualitative picture of this situation.

The 'traditional' robotic simulation systems are typically based on a shape model and a kinematic model of the robot manipulator, as well as some very simple general interpolation algorithms for moving the robot arm. The real robot, however, is a dynamic entity governed by a complex controller adapted to the specific robot.

The application of 'traditional' robotic simulation systems in the automotive industry have shown a considerable mismatch in timing between simulation and real-time execution of the off-line programmed robot task. Furthermore, the simulated path differs from the path traversed by the real robot. A more realistic real-time simulation is needed to achieve a more reliable and useful prediction of robot task to be executed.

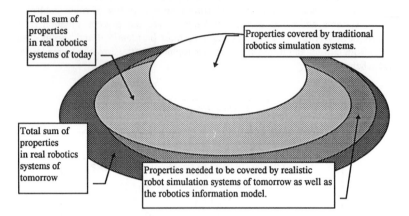

Figure 2 A qualitative, schematic overview of the relation between the properties of current and future robotics simulation systems compared with the properties of the real robotics systems of today and tomorrow. The robotics information model must be able to carry all the properties (functionality and data) needed by a realistic simulation system of tomorrow.

A much more realistic simulation can be achieved by extending the capability of the robotic simulation systems to cope with robot dynamics, and more realistic robotics motion planning and control.

Robotic simulation systems, that are able to cope with this extended description, are still in their infancy, and very few in number, one example is a 'Robot Programming and SIMulation system' called ROPSIM (Trostmann, 1992) which is a prototype system under continuing development at the Control Engineering Institute, Technical University of Denmark.

The 'old' robotics models that 'only' include shape/kinematic information can be modelled in a traditional CAD system. The requirements for the extension of the robotics model call for including the modelling of manipulator dynamic and of robot controllers that are not found in such CAD systems; these facilities have to be modelled in dedicated systems for Computer Aided Control Systems Design (CACSD).

Introducing CACSD systems as an addition to traditional CAD systems <u>does not eliminate</u> the need for a neutral, standardised interface between design and robotic simulation, on the contrarily, it stresses the need for an extended interface that includes dynamics and control, as shown in Figure 3.

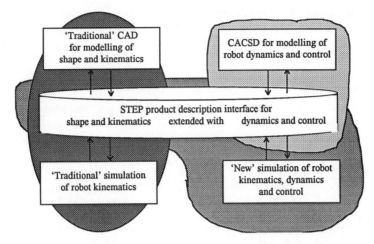

Figure 3 The extended STEP interface provides a neutral interface for the transfer of dynamic and control data and functions between CACSD systems and advanced simulation systems, as well as 'traditional' shape and kinematics CAD data between CAD and simulation.

2.2. The approach used in InterRob

The information model of robot controllers addressed here has been developed utilising a so-called 'model based' approach. In this approach a controller model is developed by model synthesis based on a detailed analysis of all components of the real robot system with respect to their data and functionality. The analysis leads to a *model identification*, followed by a *model specification* to be used in the synthesis of a model that includes all data and functionality that are relevant for the performance of the real robot system.

Model identification
Figure 4 shows a generic model of the motion controller applied in the InterRob project. The general model is used as a reference model for the development of the STEP based information model.

The model in figure 4 has been developed by a detailed analysis of common features of different robot controllers.

The two major tasks of a robot controller are, roughly speaking, to transform a robot program from program statements into robot movements and to control these movements in a desired manner dictated by the program.

Robot program statements can generally be divided into execution commands and movement commands.

Execution statements are used to control the execution flow of the program, these statements typically comprehend the well-known loop structures such as while-do, do-until, and for-do, as well as conditional statements controlled by Boolean expressions based on sensor signals. These sensor signals control the signal-flow of communication between the robot and its peripherals (part manipulators, gantry cranes, belt conveyors, etc.).

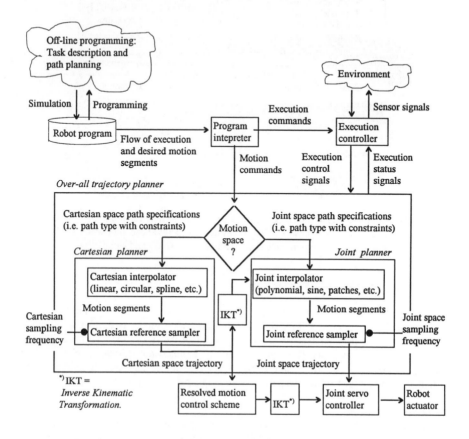

Figure 4 The generic motion controller used in InterRob.

Motion commands are typically move ptp, move linear, move circular, move uncoordinated, etc. These motion commands are read from the source program by a program interpreter and transformed into so-called motion trajectory segments by low-level motion planning (or trajectory planning). The motion segments are converted finally converted into reference signals for the motion controller by the robot controller.

Model specification
All the model entities that have been carefully identified need a detailed specification of their data and functionality. Nearly all definitions used for the controller specification are new to STEP in the sense that they are not defined in any STEP resource so far (December 1994).

The STEP methodology utilises the so-called EXPRESS information modelling language (EXPRESS, 1992) and the EXPRESS-G diagram techniques to do this data modelling and specification. Figure 5 shows a simplified Application Reference Diagram (ARM) in EXPRESS-G for a vital part of robot controller.

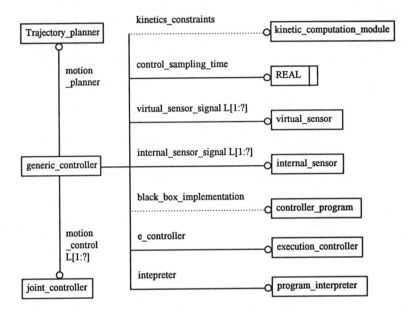

Figure 5 Simplified ARM diagram in EXPRESS-G for the robot controller.

The corresponding specification written in the EXPRESS language is shown in figure 6.

```
EXPRESS specification:

*)
ENTITY generic_controller;
  e_controller                : execution_controller;
  interpreter                 : program_interpreter;
  motion_planner              : trajectory_planner;
  motion_control              : LIST [1:?] OF joint_controller;
  control_sampling_time       : REAL;
  virtual_sensor_signal       : LIST [0:?] OF virtual_sensor;
  internal_sensor_signal      : LIST [1:?] OF internal_sensor;
  kinetic_constraints         : OPTIONAL kinetic_computation_module;
  black_box_implementation    : OPTIONAL controller_program;
END_ENTITY;
(*
```

Figure 6 EXPRESS specification for the simplified the robot controller in figure 5.

The data structure of product models is defined by the entities and their attributes (i.e. member data), written in EXPRESS, but a description of the product's functionality is not included directly in the EXPRESS language. A *service* entity to define so-called member functions has therefore been defined in the InterRob specification to extend the classic EXPRESS definitions to also include a functional description to the entities of the robot controller.

This way, the entity definition becomes very close to the definition used for objects in object-oriented languages. Therefore, this approach not only adds the description of the product's functionality into the specification, but it also provides the benefit of a smoother integration between the specification and schema-driven processors based on object oriented computer programming language such as C++, Eiffel, and SmallTalk.

Each of the general member entities of the generic_controller shown in figure 5 and 6 consists of several member entities and sub-entities used to form the detailed product information model of the robot controller. It is beyond the scope of this presentation to go into a detailed discussion of the total information model. Instead, a brief description of some of the vital entities is given in the next section.

2.3. Short description of the generic controller

Controller_program

The generic motion controller, shown in Figure 4-6, is defined either by a controller program, or by several different entities.

The controller_program contains all the functionality of the controller, it is stored in a separate file outside the STEP file. The separate file consists of object code (black box), or source code, and the neutral STEP file will contain a pointer to this program file (i.e. the file name and path). The description of the procedural interface for the controller_program has been done in the RRS project (Bernhardt, 1994) and this work can be included in the STEP interface defined here by the controller_program entity.

Generic_controller

The generic_controller is open for sensor integration via the virtual_sensor entity ((Boysen, 1989) and (Clausen, 1993)), and, of course, internal_sensors such as position sensors based on encoders, velocity sensors based on tachometers, and force_sensors (Clausen, 1993). Also the control_sampling_time is specified within the generic_controller.

The generic_controller works on the basis of a program_source that consists of a robot program with execution commands (e.g. program flow commands and motion commands) and motion data. The execution commands are read into a so-called execution_controller, that controls the data-flow from program command to real-time execution of the program in the generic_controller. The motion data are read into the trajectory_planner also controlled by the execution_controller.

The program_interpreter converts the statements listed in the program_source to internal program commands understandable for the execution_controller.

The general motion related data-flow in the generic_controller is as follows:

Input to the generic_controller are path specifications in Cartesian space or joint space as well as input variables (in joint or Cartesian space) which indicate the constraints of the path that are read into the trajectory_planner. Path constraints are kinematic/geometric limits in the workspace of the manipulator. Also dynamic constraints such as force or torque limits of the actuator, inertia, and friction in the manipulator, etc. may be considered here. These path constraints can be written explicitly in the program_source, or they can be computed by the kinetic_computation_module.

The trajectory planner

The trajectory_planner consists of a joint_space_trajectory_planner and a cartesian_space_trajectory_planner, as well as an optional kinetic_computation_module as shown in Figure 7.

```
EXPRESS specification:

*)
ENTITY trajectory_planner;
  joint_space        : joint_space_trajectory_planner;
  cartesian_space    : cartesian_space_trajectory_planner;
  kinetic_constraints : OPTIONAL kinetic_computation_module;
END_ENTITY;
(*
```

Figure 7 EXPRESS specification of the trajectory planner.

The program_source containing Cartesian path specifications with constraints given in Cartesian space is read into the cartesian_space_trajectory_planner and joint path specifications with constraints given in joint space is read into the joint_space_trajectory_planner. The cartesian_interpolator interpolates the desired Cartesian curve (line, circle, spline, or whatever) in accordance to the given path constraints. These interpolated Cartesian motion segments are sampled to a series of discrete time_based_cartesian_space_vectors describing the Cartesian trajectory (i.e. positions, velocities, and accelerations of the TCF along the path). The sampling is done by the cartesian_reference_sampler at a given cartesian_sampling_frequency.

The discrete time_based_cartesian_space_vectors (expressed by Cartesian space knot points) are converted to corresponding time_based_joint_space_vectors by the kinetic_computation_module.

The joint_interpolator interpolates the desired joint curve (4-1-4'order polynomial, cubic polynomial, 5'th order polynomial, spline, or whatever) between two subsequent 'start' and 'end' time_based_joint_space_vectors. The interpolated joint motion segments are sampled to a series of discrete time_based_joint_space_vectors describing the joint trajectory (i.e. positions, velocities, and accelerations of the joints along segment) between the 'start' and 'end' time_based_joint_space_vectors. The sampling is done by the joint_reference_sampler at a given sampling frequency.

The discrete time_based_joint_space_vectors are used as set points for the joint_controller utilising a joint-based control scheme.

A motion controller with a resolved-motion control scheme will use the time_based_cartesian_space_vectors instead of the time_based_joint_space_vectors. In each case, output from the generic controller are control signals to the actuator drives (i.e. the servo motor drives) of the manipulator (Refer to Figure 4).

The joint controller

A joint_controller is a topological entity that contains the model description of the joint controller used in the generic_controller.
The joint_controller is defined by a list of control_elements.

```
EXPRESS specification:

*)
ENTITY joint_controller;
   implementation : LIST [1:?] OF control_element;
END_ENTITY;
(*
```

Figure 8 EXPRESS specification of the joint controller.

A control_element is a topological entity that defines model elements of the joint controller. One control_element could define a proportional controller, other elements could be used to define classical PI, PID, and PD controllers, other more advanced controllers such as adaptive controllers could also be defined by control_elements. The control_element reflects the control_system_type (linear_system, or non_linear_system) and the control_domain (digital, or analogue), the joint_controller can consist of several different control_elements with different control_system_types and control_domains. That way, a digital (linear or non-linear) system can be coupled with an analogue (linear or non-linear) system.

Furthermore, the list of control_elements can model the joint_controller as a Single-Input, Single Output (SISO) system, Multi-Input, or Multi-Output (MIMO) system.

It is implicitly assumed in InterRob that the joint_controller can be described exclusively by control_elements that are time-invariant and deterministic with concentrated parameters.

The inter-connection of control_elements that model the joint_controller, is done by the control_line entity, each control_element has one input and one output control_line attached to it.

The definition of the joint controller in STEP, as proposed and presented here, has been developed by the use of modelling technique similar to the techniques suggested by (Schmid, 1993) for defining CACSD data in STEP, and this proposal is being studied in further details in the near future with reference to further integration of the data formats.

3. CONCLUDING REMARKS

Each time data is transferred from one system to another it must be translated. This is an inefficient and costly process. It is ironic to note that CIME technology that is expected to reduce costs actually increases the costs because of incompatibility between the various subsystems. The use of neutral interfaces reduces the actually burden of translating between different systems. The robotics area has been presented as an example in this paper. However, the neutral file concept is not restricted to this particular field.

The use of standard interfaces allows to establish CIME implementations based upon multiple vendor software systems and production equipment. This results in a more rational and transparent use of design, manufacturing, and production resources, and in the end, in more reliable and better designed and produced products. The user can choose the CIME elements that fit him best from a vendor base, since neutral interfaces make possible interconnection.

The InterRob project has contributed to STEP in the area of kinematics (Input/105, 1994), and by the end of 1995 the project is supposed to come up with two new draft proposals for STEP information models in the area of robot arm dynamics (multi-body dynamics), and robotics control as discussed here.

4. REFERENCES

Bernhardt, R. et al. (1994). The Realistic Robot Simulation Interface Specification. Version 1.0. January 19, 1994.(Ed. Dr. R. Bernhardt), IPK-Berlin, Pascalstrasse 8-9, 109 587, Berlin Germany.

Boysen, N.P. (1989). A Robot Controller Concept for Sensory Controlled Motion. Ph.D. Dissertation. June 1989 Control Engineering Institute, Technical University of Denmark, Rap. no. S89.52. 1989.

Clausen, T.G. (1993). A Generic Sensor Interface for Motion and Force Control in Robotics. Ph.D. Dissertation. June 1994 Control Engineering Institute, Technical University of Denmark, Rap. no. S93.53. 1993. ISBN 87-89961-04-08.

EXPRESS (1992). ISO DIS 10303-11; EXPRESS Language Reference Manual. ISO TC184/SC4/WG5

Input/105 (1994). Input to the ISO 10303 Part 105: Integrated Application Resource: Kinematics. Note of April 5, 1994. Rapport no. InterRob.WP1.DTU.005.94.

Kroszynski, U.I., Palstrøm, B., Trostmann, E., and Schlechtendahl, E.G. (1989). Geometric Data Transfer between CAD Systems: Solid Models. Computer Graphics and Applications Journal, IEEE, September, 1989, pp 57-71.

Schlechtendahl, E.G. (1989). Intelligent communication of product definition data. 11th World Computer Congress, IFIP '89, San Francisco. Pre-print pp. 1-5.

Schmid, Chr., Schumann, R. (1993). Standardised data exchange between CACSD systems. International Federation of Automatic Control. 12th world congress. Sydney, Australia, 18-23 July 1993. Pre-print of Papers .Vol. 6 pp. 121-126.

Sørensen, T. (1994). Pilot specification of a STEP based reference model for exchange of robotic models, Chapter 8 K4: Robotics Control. InterRob Deliverable no D1.1.2 June 94.

Trostmann, S. and Nielsen, L. F. (1992). Model Driven Simulation of Robot Systems Ph.D. Dissertation. December 1992. Control Engineering Institute Technical University of Denmark ISBN 87-89961-00-5

5. ACKNOWLEDGEMENT

The InterRob project is partly financed by the Community of the European Commission (CEC) under the European Strategic Program for Research in Information Technology (ESPRIT). The authors wish to thank the CEC and the Technical University of Denmark for their support in the project. We are also most grateful to our colleagues in the project for showing their enthusiasm, motivation, and spirit of co-operation in the daily project work.

6. BIOGRAPHY

M.Sc.E.E. Torben Sørensen

received his M.Sc. degree in electrical engineering from the Technical University of Denmark (DTU) in 1986. He then joined the Control Engineering Institute of DTU where he was appointed as a research fellow in the CAD and Robotics areas. Now he is appointed as a Senior Engineer in the CAD and Robotics areas. He has published several papers in CIME.

Mr. Sørensen has been co-supervisor and external examiner for several M.Sc. students in the CAD and Robotics areas, and has also participated in industrial consulting activities.

In ESPRIT project 2614/5109 (NIRO) he was responsible as Work Package leader (Working Group A). In ESPRIT project 6457 (InterRob) he is currently leader in Workpackage 1 'The STEP Interface for Robotics Applications'.

Torben Sørensen plans to finish his Ph.D.-dissertation in the CAD and Robotics areas in spring 1996.

Professor, Ph.D. Erik Trostmann

received his M.Sc. degree in mechanical engineering from the Technical University of Denmark in 1954 and his Ph.D. in control theory and engineering at Case Western Reserve University in Cleveland in 1963.

Since 1970 he has been a full professor in Control Engineering at the Technical University of Denmark. He has participated in many design and development projects in the areas of automatic machines and plants such as CNC machine tools, foundry machines food production machinery.

He is currently engaged in research within CIME, CAD, CAM, FMS, CNC and Robotics. He has been a project member of the ESPRIT CAD*I project, no. 322, and NIRO project no. 2614/5109. At present he is engaged in the ESPRIT InterRob project 6457. He has published many papers in the above fields of interest.

Dr. Sc. Uri Israel Kroszynski

is since, 1985, an Associate Professor at the Control Engineering Institute of the Technical University of Denmark, where he specialises in CAD integration in CIME environments.

Originally, a ground water hydrologist his interest shifted first to coastal hydrodynamics and computational hydraulics and then to computational geometry.

Dr. Kroszynski started his Engineering education at the Universidad de la República in Uruguay, and continued at Technion in Israel, where he graduated, receiving his Bachelor, Master, and Doctor of Sciences Degrees in Civil Engineering in 1969, 1971, and 1974 respectively.

He has participated as Senior Researcher in large national and international projects in both Israel and Denmark, and published many papers in the above fields of interest.

82

Robot measurement system in Car—Body Assembly

Min Li Zheng Zhou jing Rua Motor Leaf Spring Co, Ltd
Zheng Zhou City Henan province
(450001)China
Tel:(0371)6255751   Fax:86. 371. 6221590

Abstract

The procedure of car—body assembly incorporates fitting—up and welding of varied parts and units. In view of the fact that it is difficult to measure and adjust the seams between parts and body skeleton and embedment, three—dimentions system is required in the procedure of assembly. The introduction of Robot 3—D vision system in car—body assembly makes it possible to position the parts in their accurate place, and measure the distance between parts and units. Therefore high quality of assembly and the flexibility of the car—body will measure up our expectations, and cost of design and assembly will be reduced, design procedure simplified and the to—be—supplied parts can be standarized.

keywords

Robot, 3—D vision system

1. Background of the Research Project

Car—body assembly incorporates fitting—up and welding of car doors, wings, windows and other parts. The surface feature of these parts is varied with the colour. So the robot must have the capability of image processing. In case that the distance between a certain part and the car skelton is given, a robot of 2—D vision system is good enough for the purpose,while the position of a part in three dimentions is to be discribed,a 3—D vision system is needed.

2. 3—D vision system

Sense of vision is one of the most important organic functions of human being and a great part of our in—formation comes from the sense of vision. The key technology of designing a brain—robot is how to make the robot have the sense of vision,with which the robot,like human,can receive most of its information three dimentions by (or mainly by)the binocular parallax. A high—intelligent robot should have a 3—D vision system which is now an important research in the image processing and artificial intelligence fields.

The information that human's eyes process is mainly the brightness,colour and distance of objects within the visibility. Human's vision system is composed of the following sections:photoelectric transformation(acted on part of the retina),optical system,eyeball motion system(horizontal,vertical and circular ,motion),information processing system(the nervous system connecting retina and cerebrum).

Fig. 1 shws the hardware system of a computer which can perform the above functions.

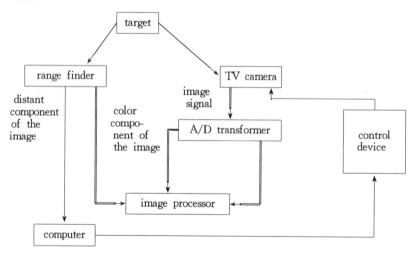

Fig. 1. The hardware system of a robot's vision function

The imput image is changed into digital data through the A/D transformer so that a digitized image is displayed. Usually an image can be displayed with 512x512 or 256x256 scanning lines, and the brightness of the 256 scanning spot can be shown with eight binery digits.

Once the image is input, a number of processes are conducted to distinguish and comprehend the image, the information of distance found through a range finder, the position and orientation of the object defined through a computer, the information of colour received through a colour analyzing filter, etc. Again, the above information are treated in an image processor and the vesult is put into the robot which will operate in the light of the information.

In order to get the information of distance, the following techniques can be applied: optical triangulation, stereoscopic vision, focusing and ultrasonic wave.

If the technique of optical triangulation is applied. the choice of appropriate light is a problem. Due to the variety of the performance, it is hard to control the intensity of illumination on the objects. The desirable plan is to choose different illuminants according to the different tasks, such as spot light source, parallel light, laser beams, movable spot light or flat light, etc. The correct choice of light mode can simplify the problem, and even play a key role in the performance. In operation, the angle beween light source and camera must be accurately fixed. Optical triangulation technique is the preference because this technique can be applied to measure the distance to a spot, to a line and to several spots in a network.

It is rather difficult to control simultaneously the performance of the robot if a universal computer is incorporated to process the information.

It takes time to process information, especially the information of fringes. In order to solve the problem, an image processor is introduced, which is a special purpose processor capable of obtaining information of tringes and part of the object. This special image processor is also a flexible one which is adaptable to customer's programs. The feature parameters from the processor are analysed in the computer and the feature and the relative position of the fitting parts can be defined. so an image processor involved in the hardware system is really a time—saving device. As the surface feature of the metal parts is variable, the data of contrast and brightness from the image processor in the software is useful and practical. Images from up to four cameras are composed so that the accurate position of an object can be calculated. The calculation involves the data from motion of the six—freedom—direction arm. This system has the ability to distinguish and divide the outlines of objects, which makes it possible to locate the lines drawn on the car—body or the protruding lines.

The vision system of car—body robot is a 3—D grey scale system, a universal system applied in a wide field. Adopting the technique of optical triangulation, the system is integrated with a robot, having the capability of man — to — machine communication.

3. Principle and its Application

The problem of fitting—up falls into two categories In the first category, the opening of the fitting socket is big enough for the plug—in unit. The only thing to do is to adjust the relative positions of the fittings and the sockets. In this case the friges of the opening are exposed to the camera. The width of the seams is measured and adjusted to a proper value. The measure is relative.

In the second category, the first thing to be done is to position the opening, then plug in the fittings.

In the later case the measure should be perfect, and in order to meet the high precision, it is necessaty to use multi—angle camera to orientate the socket. In doing so the distance between the camema and three or four spots on the fringe of the opening should be measured so that the plain of the opening can be found. After the coordinates of the three or four spots are measured, the robot's movement to fit the parts can be calculated.

with the aid of the measurement error of the three spots and the seams, the motion vector of each spot can be calculated. And the horizontal movement and the rotation of the robot's arm is on the hasis of these vectors so that the fittings can be plugged in the openings with standard error.

The procedure is composed of bearing calibration, measurement specification and robot procedure. Each time before the robot assembles a part, a calibration is needed. Such calibration mainly refer to the position adjustment of the camera, finding the matrix and the proportional coefficient between camera's coordinate system and robot's coordinate system.

The proportional coefficient can be calculated in the image processing so long as a drawing of three circles with even space in between is put in the scene and the data of distance is put in by artificial system.

The robot has an arm with three grabs which are positioned artificially. The pins are parallel to the normal line of the scene's plane. Then a faint circle is marked at the spot to which the pin is aiming so that the orientation of the robot is stored. Next the robot moves to a position from which the camera can see the objects. The robot with a camera moves parallelly along the objects while image processing is conducted. Based on the data obtained the position of the objects and the matrix of the robot's coordinate can be calculated.

The range finder is calibrated by the manufacturer in view of the desired measuring range which is put in the system artificially. The proportional coefficient of distance measurement is calculated by the computer. when the crabs of the robot are in the position ready for measurement, the car body should be in the best position and the reference line for measuring distance can be found through image processing. The calibration can be conducted on the calibration trestle.

The range finder can supply intensive information directly so that the calculation of the data from the vision system is simplified, because the 3—D information can be obtained directly upon the 3—D sense and separation and feature obtaining can be carried out without a complicated 2—D image processing.

The method of definition includes model definition and measuring procedure definition. The term's of definition are shown as in the following table;

Table 1.   Terms of Definition

Method of Definition	Terms of Definition
Mode Definition	Distinguish the measur— ing spots on the Object
	Position the measuring spots on the object
Measuring Definition	The code of the camera
	The code of model machine
	The real value of measurement
	Measuring difference
	The statistic date of measurement

In order to lower the undue presicion demand and make the assembly oporation flexible, a vision system with a camera fixed on the machanical arm is the way out. In the operation of assembling the engine cowling , it is necessary to measure the seams width and the relative distance between partition board and car body(that is, the degree of leveling). In the above operation a 3—D vision system should be introduced.

In the process of windowshield assembly, on sensing units, which conduct the measurement of four points along units, together with a projective line crossing the middle of the scene emitted by an infrared laser, make it possible to measure the distance triangularly. As the windowshield is sucked up by the suction cups, the robot is just out of the sight of the windowshield and the vision system. Then the four spots along the frings of the window openings are located and measured. The result of the measurement is the coordinates of the spots in the 3—D space.

If the accurate position of the windowshield in the clamping apparatus is known, it is possible to make out the motions of the robot to assemble the windowshield

Besides the assembly of windowshield, the robot we have designed can be used to assemble wheels engine covers, car doors and wings.

4. summery

The adoption of robot 3—D vision system in car body assembly has a number of advantages, such as simplification of the designing procedure, reduction of the designing cost and assembling cost, and standarization of the supply system. Moreover, this system is easy to operate and programmable. It also has the function of automatically measuring and adjusting its parts and units. In a word, such a system is truly beneficial and practical.

PART EIGHTEEN

Manufacturing System

83

A joint perspective for the implementation of small flexible manufacturing system

Hongyi Sun & Jan Frick

Department of Business Administration, Stavanger College
P.O.Box 2557 Ullandhaug, 4004 Stavanger, Norway
Telephone +47 51831518, Fax: +47 51581550
e-mail: h-sun@hauk.hsr.no

Abstract

In this paper, a joint perspective for the implementation of FMS will be discussed. The main theme of the joint perspective is that design and implementation of FMS is not only a technical issue, and that both technical and organisational issues should be considered when adopting FMS. An FMS case from Denmark will be used to illustrate the joint idea. The intention of this paper is to raise the issue to technical engineers and technical managers who design and implement the FMS.

Keywords

Flexible Manufacturing System, implementation, organisational issues

1 INTRODUCTION

Flexible Manufacturing System/Cell (FMS/FMC) is a key member of computer application in production. In the past ten years, FMS application had experienced rapid increase. An FMS database (Tchijov 1990) revealed that there were at least 1000 FMS in the world in 1988. The number must be much more than that now. A recent survey of 600 manufacturing companies from 20 countries indicates that 18% of the companies have adopted FMS. For many advanced industrialised countries every manufacturing company will have an FMS on average.

Although its application increased, FMS did not perform as good as expected. A UK survey (Voss, 1988) found that most (86%) of the companies with FMS achieved technical success in terms of increase in productivity, a little over half (57%) realised other benefits such as reducing throughout times, quality improvement, increasing flexibility, reducing inventory, while only 14% managed to improve their competitive position in terms of increase market shares, sales and profits. In summary, these studies revealed that although most FMS systems are technical success, i.e. operating on the shop floor without breaking down, the business performance in terms of sales and profits were not improved much.

Many researchers (Voss 1988, Bessant 1990 etc.) have found that one of the main reasons of the business failure of FMS is that the organisational and managerial issues were considered for FMS systems. In this paper, the author will try to propose a joint perspective. The theme of the joint perspective is: technical and organisational issues related to FMS should jointly be considered. In the rest of this paper, the idea will be elaborated and examples will be discussed to provide details.

2 TRADITIONAL VERSUS JOINT DESIGN AND IMPLEMENTATION

The traditional perspective of FMS design and implementation is partial (i.e., only consider technical issues) or sequential (i.e., considering mainly technical issues in design stages and organisational issues later in implementation stage). The practice represents a technology-centred perspective. Joint design and implementation proposes a another perspective, i.e., joint, and a parallel approach.

Theoretically, the joint perspective is supported by the Sociotechnical theory (Kidd *et al* 1990). The sociotechnical theory deals with integrating technical and human considerations in the development and implementation of systems, either manufacturing or others. The assumption of sociotechnical theory is that any manufacturing system is a combination of a technical sub-system (i.e., physical layout and equipment) and a social sub-system (relationships and organisation of people who perform the work). The technical and social sub-system are in mutual interaction with each other. The two sub-system co-exist, depends upon each other, and mutually influences each other. When one of them changed, advanced for example, the other will also change in some way. The relationship is what as the relationship between husband and wife, who should mutually love, fit and help each other. This means, very naturally, that the analysis, design and implementation of the manufacturing system must seek to a combination of the technical and social sub-systems. If one side of the two were not considered in advanced, the two may not fit each other in the implementation. The intention is to produce two sub-system that fit each other, and operate in harmony. The main differences of the two perspectives are summarised in table 1.

Table 1 The differences of traditional and joint design and implementation of FMS

	Traditional	Joint
Theoretical Perspective	Technology-centred	Joint technology and organisation
Practical Approach	Sequential: technology then organisation	Parallel: technology and organisation

Practically, however, there are still many questions regarding the joint approach. We will focus on those organisational/managerial questions since we come from this field. These questions are as follows:

- What organisational issues should be considered in the joint approach?
- When (at which stages) should these issues be considered?
- Who should consider these the organisational issues?

In the next section, a real case company implementing FMS will be described. Then we will try answer the above questions with the evidences from the case study as well as references from other research.

3. A CASE FMS

Although there are disagrees on the definition of FMS and FMS, FMS can be used as a general term of flexible manufacturing technologies including FMS and FMC as long as the technological configuration is mentioned clearly. The research reported in this paper was based on previous research from literature and one of my case studies. Both sources are the main activities of my Ph.D. studies in the past three years. The case company had been studied through interviews and observations. The total time used for the empirical study are more than two full days. The persons interviewed include production technology managers and foreman in charge of the FMS.

The case company is a manufacturer of engines for ships. It sells their products world-wide and is pretty competitive. In 1986, the milling shop started to use a small FMS. It has two milling machining centres, each of which has a tool magazine with about 60 tools. There is a tool warehouse behind the system which has 200 tools. A robot was used to exchange the tools both between the magazines and the tool warehouse. There are two loading/unloading stations. The transportation system is rail and pallets. The pallets are sent into the machining centres by another robot. A central computer controls all the operations.

The implementation started in 1986 and lasted about 5.5 years. Suppose that the whole implementation process is divided into four steps: initialisation and justification, planning and design, installation and training, and routinisation and learning. It took 3.5 years to go through the first three steps and to get the equipment installed in the shop. It took another 2 years before the system really worked satisfactorily. The last two years was a typical example of the paradigm shift and transition.

At the beginning the system did not work as expected, the utilisation rate was lower than 70%, breakdowns was too often and the productivity was not high. It looked as if the problem was technical and much efforts were devoted to repair and maintain the physical equipment and computer system. Few changes were introduced in the organisational aspect. The performance did not improve. Both the top management and the employees were not satisfied with its operation. Other companies also knew its failure.

Meanwhile, a professor visited the company as a Danish representative of the FAST programme. He found that the way this FMS was implemented and operated was not correct. The employees were kept out of the implementation team and were asked only to load and unload the part. The professor managed to give a seminar to all the staffs involved in the team, except for the technology manager who was the champion and project manager of this system.

Table 2 The organisational issues along the implementation process of FMS

Four stages	Organizational issues
Initiation & justification	• Champion
	• Company goal
	• Conditions
	• Labor Union
	• Management support
	• Communication
Preparation & design	• Goal specification
	• Work organization
	• Project team
	• Cooperation with vendor
Acquisition & installation	• Training
	• Staff selection
	• Participation
Routinization & learning	• Learning
	• Uncertainty
	• Auditing
	• Organization changes

4.1 The Initiation & Justification Stage

In SME (Small and Medium Sized) companies, it is champion (or initiator) who starts and manages the projects of FMS implementation, normally production manager. He should be a multi-skilled expert, dealing with issues regarding both the organisational and technological issues related to the implementation. Many companies are lack of such a champion (Beatty 1990).

The champion should make a proposal (a technical report) to the top management so that official support form top management will be achieved. The report should elaborate the company goals and potential contribution of the recommended FMS. Additionally, the organisational conditions of the company should also be considered at this stage. Conditions refers to staffs and experts who implement and operate the FMS. The labor force consequences have also to be covered and clearly stated. One of the critical issues related to labor force is the deployment of the replaced employees. This issue is especially critical in countries where labor force is strong. The champion should consider and give answers to the following questions. Will the current human resources support or fit the FMS? If not, how to cope with it? Will the labor union cooperate with the suggestion? Is there any resistance from labor forces and other parties?

Communication is the only approach to disseminate FMS information to all parties. The decision to adopt the FMS will be made based on the opinions of the top managers, the middle managers from other functions and the representatives of the labour union. If there is resistance from them, it will be hard for the decision to be made. Poor communication and lack of the necessary information are the main reasons for resistance. If the operators understand well about the FMS, they will be in favour of FMS.

4.2 The Preparation and Design Stage

The design of the FMS (systems) in SME will be mainly conducted by the vendor. AT this stage, the main technical tasks of the champion is to make preparation and provide enough information for the design.

First, the company's goal will be broken down to operational variables such as product, volumes and the variants of parts. Champions should consider the following issues. What products is the company producing now and in the future? What are the company's manufacturing missions? What are the production tasks under these missions? What is the technology strategy in the near future?

Additionally, the champion should also consider the following organization questions. What is the work organization for the FMS? What is the qualification of the staffs needed in the FMS? How is the division of labor? What are the responsibility and authority of the operators? How could the people replaced by the FMS be rearranged? How could the FMS be integrated with the existing systems and other functions? It is widely believed that FMS needs a new kind of work organization, which is characterized as autonomous working group, i.e., self-managing group.

A project team should be organized at the beginning of this stage. The members should be selected by the champion and officially assigned by the top management. In our cases, there are three to four technical persons in the team. They do have quite frequent communication and connections with people in the marketing, design and other functions regarding integration issues. But it seems not necessary to have a full-time member from other areas in the group.

Co-operation with the vendor covers all the stages, but mainly the design one. The experiences of champions in our cases can be summarised as follows. (1) Cooperate sincerely with the vendor as long as possible, and use every chance to learn from the vendor. (2) Don't believe all of their promises. Everything has to be made sure. (3) And keep your technical requirements strictly, for vendors may try to make do with similar ready-developed technologies.

4.3 The Installation and Training Stage

Installation is the very traditional and basic activity of adopting new technologies. Intuitively, equipment will be installed at this stage. However, it goes beyond that. The following issues should be considered and implemented, too.

The selection of operators is also an important issue. From our investigation, not everybody would like to work in FMS. The personality should be considered. The general rule is that the people who would like to take more responsibility will be proper for FMS, while people who would like to focus on narrow job should be considered to be transferred to other areas where less responsibilities are required. Usually young people would like to take the challenges of more responsibilities and participation. The training of the operators is carried usually in training center of the vendor and then on the shop of the user.

The participation of operators starts since the installation stage. Before the stage starts, it should be decided who are going to work in the FMS so that they can participate as early as possible. Participation is crucial to the success of FMS implementation, because it is the operators, rather than the champion or the designer, who will operate the FMS afterwards. It is important for them to express their opinion about the layout, the organization and so on.

4.4 The Learning and Routinization Stage

Uncertainty occurs both in technology and organisation. It seems as if the uncertainties are unpredictable and inevitable. This proved the necessity of a learning and routinization stage. The suggestion is that don't be scared and get disappointed when uncertainties (or problems) occur, just try to solve them. The unexpected problems may not be recognised as misfortune and disappointments.

Audit of objectives of the FMS or the uncertain problems occurred in technology and organisation will help the company to improve the performance of the FMS. It must be mentioned that both technical performance in terms of shop floor productivity and quality and business performances in terms of sales and profit should all be audited.

Organizational changes are implemented at the final stage after the equipment is on the shop floor, then the work organization can be tested. Organization design starts since the preparation and design stage, however, organizational issues could not be prepared completely. Many organizational dimensions have to be tested and modified at the learning stage. All of the three cases have experienced organizational changes to certain extents. The main dimensions of the organizational changes we observed include (1) the job enrichment, (2) the new work organization, (3) the direct integration with other department, and (4) the new culture.

These organizational changes will not happen automatically or spontaneously. The case illustrates the importance of the new culture and new attitude of the managers. If the new manager did not have a new thinking regarding the work organisation and the new attitude towards workers, it might be impossible to implement the new work organisation in this company.

5 CONCLUSIONS

This paper has proposed and discussed a joint perspective for the implementation of FMS. The main theme is that organisational issues should be considered in parallel of technical issues. The detailed dimensions of the organisational and managerial issues have been

During the seminar, he introduced the idea of the new work organisation, and hoped that the workers would actively participate in managing and operating the system.

When the project manager heard about the seminar, he got very angry with the professor. He said that this was not the way he was going to manage this system. He asked the professor to leave the factory. The project manager himself left the company and moved to another place in 1989. The management found a new technology manager which was also in charge of the FMS system. Since the new manager took over the management of the FMS, series of organisational changes were implemented, without further investments in technology.

The new work organisation around this system can be characterised as an autonomous working group. In this group, there are seven persons, including an implicit foreman. The other six are all skilled workers. Although they have joined training courses given by the German vendor. The company has decided that every other year, they will be sent to the vendor for a new course.

Normally, it needs three people to operate the system. The workers are very autonomous. They receive the week plans every Monday from the planning department. And they themselves make the daily schedules together. If they like, they can work in the weekends and holidays. Everybody in the system has a key to the shop. That is something special for workers in FMS. If a guy is ill, he can just make a call to his colleagues. And they will find a substitute. Their salary system is also different from other workers. They get monthly salary and the bonus which depends upon the group's performance such as the utilisation rate. Other workers are paid by the piece. If there is no order, other workers get no payment, but the workers in FMS still get their salaries.

Another feature is their interaction with maintenance department. If some serious problems occur and the workers on job cannot solve them, they can directly call the maintenance department, rather than report to the foreman. The foreman is mainly responsible for the economic evaluation and improvement of the system's performance. The foreman said:" I did not like it at the beginning, I felt I lost something. After a consultant took me to visit companies in Sweden, I believed it functions. Now it is OK for me. But not everybody can work like this."

And as a result of the above organisational changes, the system has turned out to be very efficient and successful. At the time this interview was conducted in July 1991, the utilisation rate of the FMS is more than 90%. The system can produce more than 60 different milling parts.

4. DISCUSSIONS

In this section, the organisational and managerial issues that should be consider for the implementation of FMS will be discussed based on the evidences of this case and other research. The issues are highlighted in table 2 and be discussed briefly below. For details of these issues and the implementation stages, please see other publications (Sun 1993, Sun & Riis 1994).

discussed along the implementation process from initiation to rountinisation. It was concluded that the champion is the key figure integrating technical and organisational issues. This implies that companies needs experts who know both technology and organisation. Additionally, it also suggests that technical people should cooperate with managers. Finally, the engineering and management departments should cooperate in both research of FMS implementation and education of technical staff with managerial knowledge.

REFERENCES

Beatty, Carol A., 1990, Implementing Advanced Manufacturing Technology, *Business Quarterly*, Autumn, 46-50.

Bessant, John, 1990, Organization adaptation and manufacturing technology, *Proceedings of the final IIASA conference on CIM: Technologies, Organizations, and People in Transition* by Haywood (eds.), Luxembourg, Austria, 351-360.

Kidd, P. T., B. Hamachrt, G. Lane, H. Bolk, E. Havn, T. Klevers and L. Klein, 1991, Joint technical and organizational design of CIM system for SME's in *Proceedings of The 7th ESPRIT CIM-Europe*, May, Tulin, Italy, 149-161.

Sun, Hongyi and J O Riis (1994) "Organisational, Technological, Strategic, and Managerial Issues along the Implementation Process of Advanced Manufacturing Technology: A General Framework of implementation guide", *International Journal of Human Factors in Manufacturing*, Vol. 4 (1), pp.23-36.

Sun, Hongyi, (1994) "Patterns of Organisational Changes and Technological Innovations", *The International Journal of Technology Management*, Vol 9, No.2. pp.213-226.

Tchijov, Iouri (1990). The diffusion of Flexible manufacturing systems (FMS), in *the Proceedings of the final IIASA conference on CIM: technologies, Organisations, and people in transition,* ed. by Haywood, Austria

Voss, C., 1988, Implementation: A Key issue in manufacturing technology: the need for a field of study, *Research Policy*, 17, 55-63.

84

A computational method for system status change of manufacturing systems

Deng, Z.
Narvik Institute of Technology
Teknologiveien 10, Narvik, Norway • tel: 47 76 92 21 81
fax: 47 76 94 48 66 • e-mail: Ziqiong.Deng@sin.no

Bjørke, Ø.
Norwegian Institute of Technology, Trondheim University
Trondheim, Norway
tel: 47 73 59 37 82 • fax: 47 73 59 71 17

Abstract

This paper describes a computational method for status calculations for manufacturing systems at discrete time moments. Conventionally, simulation methods are used to analyse discrete event system based on a diagram model. This paper reveals a computational method instead of simulation methods. It is considered that this new method gains more efficient and effective than the conventional method.

Keywords

Computer integrated manufacturing, system analysis, system modelling, discrete event system, manufacturing system theory

1 INTRODUCTION

The status changes of manufacturing system happen in a series of discrete time moments. Therefore the analysis features of it are different from that of continuous system. Conventional simulation methods are used to analyse discrete event system based on a diagram model. This paper reveals a computational method instead of simulation methods but also based on a diagram model. It is considered that this new method gains more efficient and effective than the conventional method.

2 DIAGRAM MODEL

In conventional method, a diagram model can be built for example as an Activity Cycle Diagram(ACD) (Deng 1988) as shown in Figure 1 where a manufacturing system consists of two activity cycles with one for the **machine tokens flow** called machine activity cycle starting, say, from a queue MIDLE(machine in idle status), to an activity SETUP(machine being set up), to a queue MREADY(machine ready for machining), to an activity MACHINING(machine doing machining) and approaching back to the queue MIDLE. The other cycle for **worker tokens flow** called worker activity cycle starting, say, from a queue WIDLE(worker in idle status), to an activity SETUP(worker doing the work of setting up) and approaching back to the queue WIDLE. This model expresses the behaviour of a manufacturing system which consists of a group of workers taking care of a group of semi-automated machine tools.

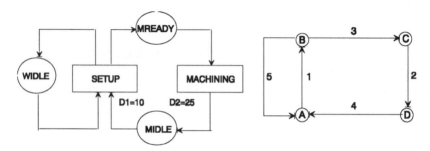

Figure 1 ACD diagram model   Figure 2 MST diagram

In our case, for the convenience of computational analysis, we redraw the diagram model as shown in Figure 2 where branches 1, 2, 3, 4, and 5 are linked by nodes A, B, C and D referred to Manufacturing System Theory(MST) (Bjørke 1994). Branch 1 is equivalent to activity SETUP in Figure 1, branch 2 to activity MACHINING, branch 3 to queue MREADY, branch 4 to queue MIDLE and branch 5 to queue WIDLE.

3 SYSTEM STATUS AND VARIABLES

System status can be expressed via system status variables $E_{(b,t)}$ as follows:

$$
E_{(b,t)} = \begin{array}{c}
 \\
b_1 \\
\vdots \\
b_i \\
\vdots \\
b_m
\end{array}
\begin{array}{c}
t_1 \quad \cdots \quad t_j \quad \cdots \quad t_n \\
\left[
\begin{array}{ccccc}
e_{11} & \cdots & \cdots & \cdots & e_{1n} \\
\vdots & \vdots & \vdots & \vdots & \vdots \\
\vdots & \vdots & e_{ij} & \vdots & \vdots \\
\vdots & \vdots & \vdots & \vdots & \vdots \\
e_{m1} & \cdots & \cdots & \cdots & e_{mn}
\end{array}
\right]
\end{array}
$$

where rows $b_1 \ldots b_m$ of the matrix stand for branches, columns $t_1 \ldots t_n$ stand for individual tokens.

This is an 1/0 matrix. If $e_{ij} = 1$, it means there exists a token t_j which is now resided in branch b_i, otherwise t_j is not resided there. However, if we only consider the token amount distributed in individual branch of the system other than consider tokens one by one, then the system status can be expressed briefly as:

$$E_{(b)} = \begin{matrix} b_1 \\ \vdots \\ b_i \\ \vdots \\ b_m \end{matrix} \begin{bmatrix} e_1 \\ \vdots \\ e_i \\ \vdots \\ e_m \end{bmatrix}$$

where the value of e_i is the token amount in branch b_i.

From the discussion above, we know that a system status means a dedicated distribution of all tokens in individual branches of the system at a certain time moment. With the time progressing to another discrete time moment, the distribution may be changed, i.e. the system status may be changed to a new status.

4   COMPUTATIONAL METHOD FOR SYSTEM STATUS CHANGE

The status change of manufacturing system happens at following discrete time moment:

- **activity starting(firing)** by due tokens, i.e. if there exist due tokens in previous queue(s) of an activity (or activities), then this activity (or activities) can start;
- **activity ending(firing)** by due tokens, i.e. if there exist tokens flowing through the activity(or activities) which posses different ending time moments, then system status change happens at earliest time moment by earliest arrival tokens.

The computational method for system status change can be as follows.

4.1   Calculating firing nodes, $F_{(n)fr}$ and active tokens, $E_{(b,t)act}$

It is found that the branch-head-node incidence matrix $AN_{(b,n)}$ (Bjørke 1994) can help to find the firing nodes. Therefore, based on the existing system status, the firing node matrix can be found by the equation of

$$F_{(n)fr} = AN^T{}_{(b,n)} \quad \langle \wedge . m \wedge / \leq \rangle \quad E_{(b)}$$

where in the operator $\langle \wedge . m \wedge / \leq \rangle$, $m \wedge$ means firstly masking the '0' elements in $AN^T{}_{(b,n)}$, then logically ANDing the nonmasked '1' elements in $AN^T{}_{(b,n)}$ with the corresponding elements in $E_{(b)}$. If all nonmasked elements of former matrix is equal to or less than($\leq$)

corresponding elements of latter matrix (i.e. this is an '$\wedge$' operation), then the solution is '1'(it means a firing happens at this node), otherwise '0'.

After a firing happens, the system status is changed. After we find the firing nodes $F_{(n)fr}$, we can further calculate the new status of the system.

We know that not all of tokens in branches will always join the firing, some of them may be still stayed idly in the branches. Therefore we first have to calculate active tokens as follows:

$$E_{(b,t)act} = AN_{(b,n)ti} \quad \langle whole\,/\,column{:}\vee.\wedge{:}i=1,t\rangle \quad E_{(n,t)act}$$

where the operator $\langle whole\,/\,column{:}\vee.\wedge{:}i=1,t\rangle$ means using the whole former matrix corresponding to t_i, i=1,t, one by one to do operations of Boolean inner product with the corresponding column of the latter matrix (Deng 1994).

The $E_{(n,t)act}$ is the matrix of individual active tokens in relation with firing nodes which can be derived from $F_{(n)fr}$ (Deng 1994).

The $AN_{(b,n)ti}$, i=1,n, are the set of branch-head-node incidence matrices for every token (Bjørke 1994).

4.2 Calculating idle tokens against the firing nodes, $E_{(b,t)idle}$

Due to some of tokens may not join the firing, it is necessary to calculate idle tokens as

$$E_{(b,t)idle} = E_{(b,t)} - E_{(b,t)act}$$

where $E_{(b,t)}$ is the old status of the system.

4.3 Calculating firing-node arrival tokens, $E_{(b,t)act..arr}$

We have also to distinguish another situation in where although tokens are active to flow to firing nodes, but the firing nodes arrival times of them may be different between them. In this case, we need to take the firing of earliest arrival batch of tokens as the due firing. Therefore, we can calculate the firing nodes arrival tokens as

$$E_{(b,t)act.\,arr} = \Delta T \quad \langle \Delta T\,/\,elmt{:} = /1\,;\neq /0\rangle \quad D_{(b,t)act.\,to.\,nd}$$

where $D_{(b,t)act.\,to.\,nd}$ is the matrix in where elements contain duration value of flowing to firing nodes for each active token (Deng 1994), and $\Delta T = \min_{ij}\{\ D_{(b,t)act.\,to.\,nd}\ \}$, and the operator $\langle \Delta T\,/\,elmt{:} = /1\,;\neq /0\rangle$ means that ΔT value compares with each element in the latter matrix, if it is equal each other, then set the resulted element in the resulted matrix to be '1', otherwise '0'.

4.4 Calculating firing-node nonarrival tokens, $E_{(b,t)act.non.arr}$

It can be calculated straightforward as

$$E_{(b,t)act.\,non.\,arr} = E_{(b,t)act} - E_{(b,t)act.\,arr}$$

4.5 Calculating the tokens of crossing over firing-node, $E_{(b,t)act.arr.over}$

Note that even we have calculated the $E_{(b,t)act.\,arr}$, but it reflects still the active arrival tokens in relation with the ***original*** branch. For the purpose of calculating new status of the system, we need to know which branches these tokens will enter after crossing over the firing nodes. Therefore

$$E_{(b,t)act.\,arr.\,over} = AP_{(b,n)ti} \quad \langle whole\,/\,column: \vee .\wedge : i = 1, t \rangle \quad AN^T_{(b,n)} \ E_{(b,t)act.\,arr}$$

where $AP_{(b,n)ti}$ are the set of branch-tail-node incidence matrices for every token (Bjørke 1994), and the operator $\langle whole\,/\,column: \vee .\wedge : i = 1, t \rangle$ means using the whole former matrix corresponding to t_i, i=1,t, one by one to do operations of Boolean inner product with the corresponding column of the latter matrix (Deng 1994).

4.6 Calculating new system status, $E_{(b,t)}$

It can be calculated as follows:

$$E_{(b,t)} = E_{(b,t)idle} + E_{(b,t)act.\,non.\,arr} + E_{(b,t)act.\,arr.\,over}$$

5 CONCLUSION

In comparison with conventional simulation methods, this computational method revealed above gains advantages of freeing from tedious simulation programming and reprogramming from time to time whenever a new simulation model is created, and freeing from time-consuming searching during the simulation program running.

Further works will be done by authors in the direction of finding more efficient mathematical method in future.

6 REFERENCES

Bjørke, Ø. (1994) Manufacturing system theory - a geometric approach to connection
Deng, Z. (1988) A study of modelling part and tool flow for FMS, proceedings of PROLAMAT '88, IFIP Working Group 5.3, Dresden
Deng, Z. (1994) An approach for analysing discrete event systems based on manufacturing system theory, scientific report on working in NTH, NTNF visiting senior scientist program, Norway

85

Re-STRUCTURING
THE MANUFACTURING PROCESS

G. Halevi
Technion -- Israel Institute of Technology
Dobnov St. 20a Tel-Aviv 64369   ISRAEL
Fax. + 972 3 696 2833
Tel. + 972 3 696 2833

ABSTRACT

The manufacturing process is flexible by nature. However, the structure of the manufacturing employed today makes it rigid. Today competitive environment is compelling companies to be the best at the work they does. These competitive fields have brought up the need for re-evaluation of manufacturing methods and restoring its embedded flexibility.

In this paper we propose a new manufacturing structure, one that will respond to all today's manufacturing objectives, and have computer programs to implement and support it.

The basic idea of the proposed system is to concentrate on the objectives of manufacturing and not on how to improve today's manufacturing structures. The method to achieve it is to evaluate the pro and cons of the existing structure and find methods to overcome the problematic areas, which are:
1. Decisions are being made too early stage
2. Decisions are being made by managers that are not always experts in the decision matter
3. Data transfer between stages is of decisions and does not reflect the decision maker
    intentions, alternatives, ideas,   etc.
The paper will describe in details an operational system that overcomes these obstacles.

Keywords

Manufacturing systems, product specification, product design, process planning, production management

1 INTRODUCTION

The manufacturing process is flexible by nature. However, the structure of the manufacturing is a result of evolution over many years. Evolution that utilized, the available resources of that time, to the most. Therefore the manufacturing process is rigid and based on human efforts and skill. Therefore, it became rigid.

The introduction of computers in industry had an effect in two different aspects:

1. Manufacturing hardware -- most of today's machines are computerized, efficient, accurate and powerful. With this respect the computerized revolution is a big success.
2. Manufacturing software -- Most of today computerized production management programs are based on the manual concepts, i.e., chain of activities, and are not very satisfactory.

Today's competitive environment is compelling companies to be the best at the work they do. The competition on conventional basis has lost its spurs as most companies are using similar machinery, similar quality control, similar inventory control, etc., Therefore the competitive manufacturing objectives has changed to: Customer service and satisfaction, quick response to market demands, new product sophistication and options. These competitive fields have brought up the need for re-evaluation of manufacturing methods. Many new methods such as: Concurrent Engineering, TQM, BPM, WCM, VE, QFD, Groupware, Mission statements, etc., is proposed today. Most of the proposed methods emphasize the need for each stage and discipline of the manufacturing process to consider the objectives and problems of the others. They differ in how to implement these objectives. The common suggested method is based on team-work, committees and group discussions. Very few methods offer, beside ideas, a computerized method of implementation.

In this paper we propose a manufacturing structure that will respond to all today's manufacturing objectives, and have computer programs to implement and support it.

The basic idea of the proposed system is to concentrate on the objectives of manufacturing and not on how to improve today's manufacturing structures and keeping it unaltered. The method to achieve it is to evaluate the pro and cons of the existing structure and find methods to overcome the problematic areas i.e., solution by elimination. It calls for understanding the meaning and causes of each phenomenon and then to evaluate it, and decide further actions.

Very good reasons led to the present day manufacturing structure. However, the situation, working tools, conditions and objectives has since changed, and there is no reason to keep the old faithful solutions, to problems that do not exist any more. New problem needs new solutions, employing the best available tools and procedures.

As an example one may consider the basic production management procedure of: MPS - Master Production Schedule; MRP & MRPII Material Requirement (Resource) Planning; and CPS - Capacity Planning System. Each one of these stages has a unique specific defines objective. Basically they all are working with the same logic and technique. As it was very time-consuming to perform an accurate computation of loads and on hand inventory by manual means, it was a very wise decision to separate them into three unique stages, each stage with its own objective. It was a compromise between accuracy of data and elapses time to compute it. Therefore:

MPS -- ignore available inventory, work in process, and compute work load, with a rough attempt to balance it. As MPS has to supply a long term data, these liberties are tolerable, and would not significantly distort the overall decisions.

MRP -- consider available inventory, and purchased order, but ignores available finite capacity. As MRP plans for the intermediate periods, working with infinite capacity may be excused. In cases that it does not provide accurate enough data rough cut capacity planning is adding in order to simulate finite capacity.

CPS -- consider inventory, purchasing, work in process, and finite capacity. As CPS is a short term period planning, it must consider all parameters in order to plan a practical production program.

Today, with computers, that have the speed, and may handle large data bases, there is no actual need to sacrifice planning accuracy to computing speed and tediousness. One accurate program may serve the purpose of all three stages, and they may be combined to one stage in the manufacturing structure. There are many more cases like this in the present day manufacturing structure and technique.

In our studies we arrived at the conclusion that the most critical drawbacks of present day manufacturing philosophy are the methods that:

1. Decisions are being made too early.
   Therefore they must ignore market fluctuations and demands and shop floor on-line
   situation. Thereby a fictitious constraints, such as: limited product options, bottlenecks,
   machine overload and underload, meeting due dates etc. occurs
2. Decisions are being made by managers that are not always experts in the decision matter.
   They do not always know the implication of their decision on the overall plant strategy.
3. Data transfer between stages is of decisions and does not reflect the decision maker
   intentions, alternatives, ideas, etc.,

In order to have good performance, the manufacturing process must consider points of view of many disciplines, as each discipline considers the problem on hand from a different angle. Given a manufacturing problem each experts will consider a different problem:-

- The product designer will evaluate methods of achieving product functions.
- Marketing will evaluate its seductiveness to the customers
- Finance will evaluate the required investment
- Manpower will consider the workforce demands
- The manufacturing engineer will consider floor space and material handling
- Purchasing and shipping will consider how to store the product
- and so on.

A compromise of demands of all disciplines will result with a good manufacturing decisions. This desired situation is commonplace currently. The question is how to arrive at such a decision? The commonplace method is by creating team works, committees. Very few propose a computerized system to handle this task.

A team-work and a committee must have a chairman. As each discipline has its own criteria of optimization, no doubt that discussions and arguments will take place. It would be a blessing if a decision will be reached unanimously. It is more probably the chairman, on the basis of the discussions, will have to dictate his decision. Personal favoritism, gift of speech in many cases might bias the decision.

These many disciplines may be divided into three groups according to their impact and involvement on the manufacturing process.

Group A - Includes the disciplines that make the decisions.
Group B -- include all disciplines that are affected by the decision taken.
Group C-Includes the technical disciplines whose decisions are based upon mathematical computations.
Accordingly, the propose organization is composed of 4 stages that are indicated in group A. The disciplines in each group are marked in the table 1.

Research has shown the following observations:--
- More than 60% of the product committed cost is established at the product definition stage.
- The total committed cost after the design stage is 60 to 80%.
- The total committed cost after the process planning stage is 90 to 95%
- 70% of the features in a design are "fillers" i.e. they are not essential for the product performance
- There are many ways to produce a part and product.
Therefore our structuring proposal considers and assigns the appropriate weight to these observations. The method is by the following concepts:

Table 1 Arrangement of disciplines in groups

GROUP III	GROUP II	GROUP I
BILL OF MATERIAL	MARKETING SALES	PRODUCT
ORDERS	COSTING ECONOMICS	SPECIFICATION

MATERIAL
  REQUIREMENT
  PLANNING
	SHAPE
	COLOR
	STORAGE
		PRODUCT
		DESIGN

	COLOR	PROCESS
PRODUCTION SCHEDULING	MATERIAL HANDLING PACKING	PLANNING
	SHIPPING PURCHASING	PRODUCTION MANAGEMENT
	BOOKKEEPING CUSTOMER RELATION	

1. Product definition will be supported by a special computer program, that guides the user, and advice him on any problematic feature, and consult, by computer databases, with other disciplines (instead of discussion group).
   A computer trace program will keep designer reasoning.

2. Each stage will make decisions **ONLY** in the area of its expertise, and to product functionality.
3. Data transfer between stages may be incomplete. A note stating   "I DO NOT CARE" or "IT SHOULD BE WITHIN LIMITS X,Y,Z" should be allowed. Each stage will supplement data according to its expertise field.

4. Data transfer between stages will be by matrix of all possible solutions and not concrete decisions. The matrix will include the penalty value of each selected decision path.

5. The final operation decision (one path of the matrix) will be made at the instant (on-line) that it should be performed.

6. Decisions on manufacturing methods made at the planning stages will not oblige the production stages. The production stage must, however, deliver all the scheduled orders.

A description of the proposed system will be given in the following sections. The description will concentrate only on the four stages of group I as shown in table 1 .

2   PRODUCT DEFINITION

Any manufacturing activity start with a product, with no product there is nothing to manufacture. The activities in the product design field might be of three types:-

1. The company manufacture to order. In such cases the product is defined by the customer. However, it will be a good idea to consult with the manufacturer the design in order to reduce manufacturing costs.

2. The company manufactures a line of existing products. In such cases it is a good idea to keep track on design changes that might reduce manufacturing costs, increase product appeal to customers, and introduce options and new models of the same product.

3. The company would like to enter a new field of activities.

It is a natural tendency for the one who specifies product characteristics to aim for the best, and rightfully so. However, he is not always aware of the costs and manufacturing implications of the specifications. In many cases, reducing the specified values by as little as 5% may result in cost reduction of more than 60%. We assume, that the product specifier may change the specifications if he will be aware on this effect. Lets take for example a bathroom scale. The accuracy has to be specified. It is nice to have a scale with an accuracy of

+/- 1 gram. It can be done. However, it calls for costly accurate sensors and supports circuit. If the accuracy will be specified as +/- 100 grams, a simple sensor and low cost product will result.

Another extreme example might be when a electronic product specifies a monitor of $20^{1}/4$". It can be done. However, it is a special monitor and calls for developing (not research). It will be costly and take several month to develop. However, if the product specifier will be aware of it and will change the size to 20", than an off the self monitor can be used at a very low cost.

Any product comes to fill a certain needs. To perform its objectives, certain functions must be available, and have to be specified accordingly. However, there are many features that do not contribute to the main objectives of the product, but come to serve as supporters. Following the previous example, the monitor is essential for the product performance, but the method by which it is held is irrelevant to the product performance. We suggest to call such features "fillers." Research has indicated, that in many cases, the product specifier does not pay much attention when specifying fillers. Although, the fillers cost is not negligible.

The objective of the product definition module is to assist the product specifier in defining a product, that will meet all product objectives, at a minimum cost and lead time. The philosophy behind this proposal is that product definition is an innovation process, and as such must leave freedom and judgment to the individual that perform this task. The role of the computer will be to draw the attention of the user to the meaning of his decisions, and in some case to propose alternatives. However, the final decision is left to the user.

It has been realized, that the computer system that assists in defining new product, may be used to follow up existing product, and suggests options and alternatives in cases of market demand and technology changes.

The basic concept behind this system is that each manufacturing company is in a specific line of products or business. A line of products usually has many common features. Instead of designing each product and its features from screech, it is advisable to find the commonalty of the features, and attempt to design a feature that will serve many products. Moreover, when new technology or material appears in the market, to evaluate its use on the feature basis, and if economic to apply it to all relevant products.

This concept employed and tested In one electronic company. It proved that savings of $11 million could have been saved in a project of $15 Million, just by not inventing the wheel over and over again. Further more, a study of their line of products was carried out, and a computer program prepared to assist and guide the product specifier. The computer program, constructed as a dialog system, includes: check lists, databases of suppliers off the self products, inventory, dead stocks, tools, jigs and fixtures, test equipment, market research, competitive products, group B personal notes, etc.

It proved that within a few days an optimum product that gets the approval of all disciplines was be specified. The difference between this approach and the Value

Engineering, or the TQM is that it handles the initial design, and not the post-mortem improvements.

3 PRODUCT DESIGN

The objective of the product designer is to meet the product functions, as specified. It is an innovation process, but it will be very helpful to employ techniques and designs that have proved themselves in the past. Therefore, a CAD system that can retrieve designs according to product functions, by any attribute and key will be used. Auxiliary files referring to customer satisfaction and service, costs, complexity, etc.,. and any other information that might assist the product designer in making a sound effective decision will be part of the CAD system. These auxiliary files, will be continuously updated by the personal performing topics of group B, by meeting decisions, by publications in technical literature, seminars and conferences. Thus CAD will really be a Computer Aided Design.

By these means a series of alternative designs will be generated, and a sound decision will be made in selecting the best alternative.The selected alternatives is a concept design decision, and based upon it a detail design will commence.

It is advisable to use as many standard components in order to reduce product cost and design lead time (Outsourcing). For these reason standard components file will be part of the CAD system.

Any product or a component has a specific task or a function that it should perform. Some of such tasks are dictated by the product objectives. They will be refereed as primary functions. However in the design there are features that their function is to support the primary functions. As an example, a slide projector objective is to display a transparent slide in a clear, focus manner on a screen. To accomplish this task it is important to have an appropriate light source, lens of appropriate size and shape and a mean of adjusting the distance between the slide and the lens. These are the primary features. However, the method by which to adjust the distance does not contribute directly to the functionality of the slide projector. The designer have the freedom, based on criteria of optimization, to select one of many design options. Such as: gear and rod, worm gears, bolt and nut, friction rod, slots. The shape of the holder and its size have no bearing on the slide projector functions. Such parts are auxiliary features, or "fillers". Any chosen design might meet the design objective. Another example, electromagnets have to supply a required force. The force is a function of the magnetic flux, i.e., depends on ampereturns of the exciting coil and on the permeability of the core. Any combination of ampereturns can supply the required force. The designers have freedom in making this decision without effecting product performance. Similarly with hydraulic or pneumatic design, the force is a function of selected pressure and cylinder diameter. Any combination will do. If there are any design constraints, it will restrict the freedom, but not eliminate it.

Therefore the design features can be divided into two categories:
A. Functional features. They are dictated by the product functionality.

B. Filler features. They are to fill a space, or there are many methods to arrive at the design
objective, or a wide range of dimensions can serve the design function.
As an example: a peep hole can be of any diameter in the range of 20 to 60 mm.
It is important to the functionality to keep two shafts at an X mm apart. It is a type A category.
However, the method and shape of the part that connect the two shafts are meaningless as
long as it keeps them apart as required. Therefore it is a type B category.

A research indicates that about 75% of part dimensions and features are of category B, i.e.,
that the designer specified them not because they contribute to the functionality of the
product, but because the designer is obligated to specify a "complete" design and furnish a
"complete" drawing. He must specify shapes, dimension, tolerances even to those portions of
the part, that he does not care. The drawing is transferred to the next stage of process
planning. The process planners have no indications whether the indications are of category A
or B, and he must meet the drawing demands. Some minor modifications might reduce
manufacturing cost by a significant amount, for example corner radius in a pocket, flatness of
a bottom of a hole. To over come extreme cases, the process planner will call the product
design or a review committee after parts have been rejected to discuss the proposed changes.
From my experience, usually the product planners respond by I doesn't care and approves
such requests.
   In the proposed system, the CAD system (and drafting rules) will allow notes such as "I
don't care", or "any dimension between X and Y is good enough", or "make your own shape
and size". These notes will come to serve as transferring the process planner intentions to the
other disciplines. Additional data will reside in a "note" files of the CAD system. Any one of
the group B experts might add notes, constraints, requests, ideas as to the design and the
liberty and boundaries of allowed changes.
   By this method, the expert in his field may change or supplement the design in order to
reduce its cost without having to call for meeting and committees.

4 PROCESS PLANNING

The task of this phase is to determine what operations, on what machines and tools, in what
sequence, and details are needed in order to transform raw material to the desired shape. The
product designer has knowledge on processes, but he is not and expert in this field. The
process planner is the expert, therefore, let him make the decisions of all those shapes and
dimensions that the product designer doesn't care. By knowing the product planner intentions
the process planner may supplement the drawing to assure DFM (Design For Manufacturing)
without effecting product performance. There are two methods of implementing this idea:
1. A CAPP (Computer Aided Process Planning) programs will reside in the CAD system
     and be automatically called during the design process drawing the attention of the designer
     to any feature or dimension the increase the product cost, and suggest modifications. The
     decision whether to accept or reject the suggestion still lies with the designer.
     Although it is an ideal approach, it is not practical today as CAPP systems of today are not
     advanced enough to play such a role.

2. The product designer will design only those features that are cardinal to the performance of the product, and leave for the process planner the task of designing the "fillers" according to the dictated constraint.

There are many methods to produce a part. The process planner considers many alternatives. He selects the "best" alternative and transfers it as his decision (routing) to the next stages, i.e. production management. Naturally, the process planner will optimize the process, and select the best machines for the job. However, in the shop there are new modern machines and old machines. This situation brings up the problem of overload and underload of machines. Scheduling theories are available to handle with such cases.

Moreover, shop floor control is a dynamic task as: machine fails, tool breaks, employees are missing, orders' changes, parts   reject and rework, power failure, etc. CPS (Capacity Planning System) and scheduling is done in the office, before the jobs are released to the shop floor. CPS task is to balance the load. To accomplish some times alternative machine or operations are defined. It is done in the office at a planning stage and its decisions is transfer shop floor. Situation on shop floor is dynamics, and when on-line information is available, it might be that the alternative process is not needed. Alternative, usually means, less effective process, it will take longer and cost more. If the decision, or the knowledge of the alternatives, will be transferred to shop floor control, the foreman may employ his judgment to decide, based on actual shop floor state what alternative to employ.

The process planner is not a production management expert, however he makes the decisions that affect production control and efficiency. A process planner might go down to shop floor and consult with shop personnel. His decisions serve as   long term decisions. Routings are not being changed for every batch of production. Meetings and committees might be called to decide if to chance routings.

The proposed system philosophy is that each decision should be made by the most qualified person and at the moment of execution. Therefore, process planning is deviled into three stages. The first stage is the Absolute Theoretical Optimum (ATO) process. It means that only technological constraints are being considered. It disregards shop resources and order quantities, therefore the ATO may be considered as transferring process planner ideas.

The second stage builds an operation-resources matrix, where the operations are those that were specified into ATO stage, and the resources are those available at a specific shop. Solving the matrix for the minimum path of all operation (with priority) including penalty , which depends on batch quantity, result with an optimum practical routing to be used for MRP, MPP, and CPS.

The matrix will be transferred to shop floor for scheduling. It represents almost unlimited methods of producing the components, without making a decision. The most qualified person to make such a decision, is the person on shop floor, at the moment of need. The decision on shop floor will be of a single production step, taking into account the immediate shop floor situation, and be repeated whenever needed. It is allowed for the practical production method to deviate from tje planned one. It suppose to be this way. A computer program will assist the foreman in making his decisions.

By regarding the routing as a variable instead of a fix rigid routing. experimental program indicates that productivity can be increased by 30%

5 CONCLUSIONS

The existing manufacturing organization structure were good in the manual era. In the computer era, flexibility can be restore to the manufacturing process.
In this paper we propose a new manufacturing structure, one that will respond to all todays manufacturing objectives, and have computer programs to implement and support it.
The basic principles of the proposed system are:
1. Transfer of data between disciplines will be of thought, ideas, alternatives, objectives and not merely decisions.
2. Decisions will be made by the most qualified personnel.
3. Decisions will be made at the latest moment possible, i.e. at the execution time.
    Methods and tools to implement the proposal is detailed.

6   REFERENCES

- Halevi, G. (1980) The Role of Computers in Manufacturing Process, John-Wiley,NY
- Halevi, G. (1988) All-Embracing Production Control, & Integration of PM into CIM, in IFIP State-of-tee-Art Computer-Aided Production Management (ed. A.Rolstadas)
- Halevi G. (1993) The magic matrix as a smart scheduler, Computers in Industry 21(1993)245-253

7.   BIOGRAPHY

Gideon Halevi received his M. Sc in Mechanical Engineering from the University of Pennsylvania (1959) and his Doctor of Science in Technology from the Technion (1973). He is as adjunct professor at the Technion (1981), teaching mostly at the graduate school on CAD/CAM and Robotics. For more than 30 years he has been active in industry in production development, Combine Technical Operations, heading the corporate computing center and as manager and director of the CAD/CAM Research and Development Center. He has developed the All Embracing Technology, which was published in his book "The Role of Computers in Manufacturing Processes (Wiley 1980). He is an active member of the CIRP, Israeli representative to IFIP TC5, a member of IFIP WG5.3 and WG5.7, past chairman of SME chapter 319 and chairman of IPICS.

86

Product model sharing: A formal approach and applications

J.S.M. Vergeest
Delft University of Technology, Faculty of Industrial Design Engineering
Jaffalaan 9, NL 2628 BX Delft, Netherlands
Telephone +31 15 783765, Fax +31 15 787316
Email j.s.m.vergeest@io.tudelft.nl

Abstract

Product model interchange is an increasingly essential process. Due to differences between the systems involved this process is possible only under severe conditions. It is important to be able to formulate these conditions exactly for each given application, in order to predict or to verify that model transfer or model sharing proceeds correctly. We propose to describe these conditions using the formalism for the finite automata. We have analysed data transfer conditions for two specific design systems and we present the resulting conditions. The formalism seems attractive for analysis of model sharing environments and it is helpful to develop tools that are easier to integrate.

Keywords

Product data interchange, product model sharing, formal description

1   INTRODUCTION

Product data interchange is still facing severe problems, practical as well as theoretical problems. Malfunctioning communication, both within an enterprise and between different commercial parties leads to high direct and indirect costs and form a real threat for the industry. In addition, the need for data communication between different companies is growing due to the increasing specialization which is required for the development and manufacturing of complex products.

At first glance it seems surprising that the data interchange problem is far from being solved, given the excellent digital communication facilities that are available today at low cost. However, as we will see, these facilities help to make the existence of some fundamental issues very clear. Whereas data transfer into the traditional downstream directions (e.g. from product design to process planning, manufacturing or documentation) is well under control, this is not the case for most other types of product data interchange such as communication within the design stage. This latter application however, is of vital importance since more and more effort is put into the early stage of product design, where multiple engineering disciplines meet.

Technically, information transfer is particularly difficult if the metadata is non-persistent or even ill-defined. This is exactly the case during product design; at almost any level of detail, the data structures and the data constraints are dynamic and hardly predictable. It has turned out extremely hard to agree internationally on product information models, even for specific industrial application areas (Owen 1993). And even within this restricted scope the use of STEP (ISO 1994) takes away only a small portion of the problems. As has been pointed out recently, genuine information exchange requires the knowledge and explicit representation of the semantics of the data (Eastman 1994). This is directly related to the type conversion problem; if data is transferred from one system (be it a computer tool or CAD system) to a different system, then a conversion method must be applied to map the data representation of the first system to the representation form of the second. This cannot be accomplished without full knowledge of the internal data structures and the possible data operations of each of the two systems (Vergeest 1994). This issue falls outside the current scope of STEP, but it must be addressed.

Product model sharing is often depicted as a set of tools that have read and write access to one single product model, with a set of rules and procedures to maintain the consistency of the model. To realize this picture three issues that must be addressed include:

1. How to find those rules. Can they be determined in advance or will they originate and change as the product model evolves.

2. Would these rules restrict the usage of the tools compared to the situation that each tool has to deal only with its internal model.

3. Given a set of tools and rules, how must these rules be represented.

In this paper we analyse some of the interaction between points 1 and 2. Some of the rules can be determined in advance. They can be derived from properties of the involved tools. In section 2 we introduce the description method for an individual tool (or system). In section 3 we formulate data sharing using this description method. With a practical example we show in section 4 how some consistency rules can be derived from the descriptions of the involved systems. Here we assume that the semantics of the descriptions is limited to regular sets in 2D and 3D spaces.

2 DESCRIPTION METHOD FOR THE INVOLVED SYSTEMS

In order to define the rules for data interchange between different systems A, B, ...we need to take account for their information structures as well as for the possible state transitions within each system.

The notions of state and state transition are well captured by the finite automaton (Perrin 1990). System A (any computer-based design tool) is represented by the finite automaton $A = (Q^A, q_0^A, \Sigma^A, F^A, \Delta^A)$ where Q^A is the set of all possible states of A and hence represents the design space of system A. If a design space contains continuous parameters (as is e.g. the case for geometric systems) it is not finite. However, we assume that in reality Q^A is based on a discrete representation and hence is finite. The initial state $q_0^A \in Q^A$ corresponds to the empty design model. The alphabet Σ^A is a finite set of symbols representing all atomic commands of the tool. Such a command changes the current design model into the next, possibly the same model. If A is in state q_1^A then upon reading symbol $a \in \Sigma$ a transition takes place to (possibly the same) state q_2^A.

For example, if we wish to describe all 3D and 2D manipulations of a common CAD system these atomic commands can be translation, rotation, scaling, copying etc. The symbol a representing such a command must specify the type of manipulation, parameters of the manipulation (e.g. the translation vector) and identifiers of the entities on which the manipulation is applied. A geometric operation can be applied to the entire model q^A or to one or more of its constituents. On the CAD system these constituents can be selected using an identifier or they can be designated with a cursor. In the latter case a includes the specification of screen coordinates or similar parameters. It is important to notice that the effect of a command a on state q^A can be dependent on such properties as the actual screen location of q^A.

A string $s^A \in \Sigma^{A*}$, where Σ^{A*} is the set of all possible strings made with symbols of Σ^A, represents (a part of) a design session. $F^A \subset Q^A$ are the final states. If, in the above example, $q_2^A \in F^A$ then, on reading symbol a, a non-empty output symbol $y^A \in \Delta^A$ is produced, where Δ^A is the output alphabet of A. We can write $y^A = \lambda^A(q_1^A, a)$, where $\lambda^A : Q^A \times \Sigma^A \rightarrow \Delta^A$ is the output function of A. We impose that non-empty output is produced if and only if the input symbol corresponds to the unique "transmit design model" command on the system. We call this input symbol t^A. Then y^A is an external representation of design model q_1^A. Although t^A causes A to change to a different state we assume that the new state will, from the point of view of the user of system A, be the same as the old state. We therefore make no distinction between the two states in our notation.

The definitions for automaton A apply also to finite automata representing the tools B, C, ...

We briefly mention some aspects:

1. The automaton A must be deterministic; to each give pair (q_1^A, a) belongs exactly one resulting state q_2^A. We therefore can introduce the function $\delta^A : Q^A \times \Sigma^{A*} \rightarrow Q^A$ to denote transitions, e.g. $q_2^A = \delta^A(q_1^A, a)$.

2. A is complete; each state in Q^A can be reached from q_0^A along some path, on reading some $s^A \in \Sigma^{A*}$.

3. y^A cannot represent the design history of q_1^A, i.e. it cannot be derived that s^A was read to obtain $q_1^A = \delta^A(q_0^A, s^A)$. To proof this we note that if y^A would represent the design history, so would each q^A. But then Q^A would not be finite.

3   MODEL TRANSFER AND MODEL SHARING

We first consider the transfer of design models from A to B, where A and B are of exactly the same type. In terms of the finite automata A and B, the transfer can be described starting from the situation that A is in state q^A. This state has been reached from q_0^A by reading $s^A \in \Sigma^{A*}$. Subsequently $t^A \in \Sigma^A$ is accepted by A and $y^A = \lambda^A(q^A, t^A)$ is produced. Now it holds that

$$y^A = s^A \Rightarrow \delta^B(q_0^B, y^A) = q^A$$

i.e. if the output of A is fed into B, then B gets in exactly the same state q^A, and we would call the transfer process successful. However, in general it will be impossible to derive s^A from the output y^A (see point 3 in the previous section). It can be shown that it is only feasible to obtain some string s'^A with the property $q^A = \delta^A(q_0^A, s'^A)$. This string does bring B into the required state q^A, but via a different path. This may lead to problems for design model sharing.

If A and B are different, we have to define very accurately what we mean by a successful transfer process, and particularly what it means if A and B are "in the same state". To do this we use an evaluation function $V : Q^A \cup Q^B \rightarrow S$. A transfer process that translates q^A into $q^B = \delta^B(q_0^B, \lambda^A(q^A, t^A))$ is successful if and only if $V(q^A) = V(q^B)$. Without loss of generality we assume that the input alphabets have been extended such that $\Sigma^A = \Sigma^B \equiv \Sigma$. We also assume that the empty models on A and B are equivalent, i.e. $V(q_0^A) = V(q_0^B)$.

For some applications it may suffice, as a criterion for successful model exchange, that the geometric extent of the model is conserved. The description in the rest of this section is made in terms of these geometrical extents. We define the geometric extent designated by a state q as a subset $Z \subseteq I\!R^3$. For CAD systems it is essential to take account of the 2D graphical presentation of the design, since Σ may include cursor positions as we mentioned earlier. Therefore we extend the definition of q with the so-called screen image $I \subseteq I\!R^2$ of the model obtained by a map $P : I\!R^3 \rightarrow I\!R^2$ which maps Z onto I. P may involve affine 3D transformations such as translation, rotation and scaling, as well as a projection of the transformed object onto a plane. We adopt the definitions in right-handed coordinate systems as in (Fiume 1989). To simplify the discussion we restrict the projections to the orthographic projection that maps points (x, y, z) to (x, y). P can be defined using a 4×4 homogeneous matrix, followed by the projection $(x, y, z) \rightarrow (x, y)$.

It is important to realize the implications of this choice for V. In effect the semantics that CAD systems usually possess is significantly reduced by this V. In particular V is insensitive to the CAD system's internal geometric representation form.

In some respects V induces a stricter condition for equivalence than often applied in practice. Especially the condition that the screen images must be the same, as must be the actual coordinate systems in $I\!R^3$, are rather stringent. This as opposed to applications where two models are even called geometrically equivalent if there exist a homogeneous transformation that can transform the two states into each other (Torres 1994).

We distinguish three type of strings in Σ^* (corresponding to three types of commands on the design system):

1. Affine transformations $\tau : I\!R^3 \rightarrow I\!R^3$ of the form STR, where R and T compose a rigid body transformation in $I\!R^3$ and S is a non-uniform scaling operation.

2. Viewing transformations $P : I\!R^3 \rightarrow R^2$ as described above, which leave Z of the model unchanged, but determines the image I.

Sequential CAD							
Time →	T_1	T_2	T_3	T_4	T_5	T_6	T_7
Input $\in \Sigma^{A*}$		s_1^A	t^A	s_3^A		t^A	
State $\in Q^A$	q_0^A	q_1^A	q_1^A	q_2^A	q_2^A	q_2^A	
Output $\in \Delta^A$			y_1^A			y_2^A	
Model transfer from A to B			$\downarrow$ $\downarrow$			$\downarrow$ $\downarrow$	
Input $\in \Sigma^{B*}$			y_1^A	s_1^B	s_2^B	y_2^A	s_3^B
State $\in Q^B$	q_0^B	q_0^B	q_1^B	q_2^B	q_0^B	q_3^B	q_4^B
Output $\in \Delta^B$							

Simultaneous CAD							
Time →	T_1	T_2	T_3	T_4	T_5	T_6	T_7
Input $\in \Sigma^{A*}$		s_1^A	t^A	s_3^A		t^A	
State $\in Q^A$	q_0^A	q_1^A	q_1^A	q_2^A	q_2^A	q_2^A	
Output $\in \Delta^A$			y_1^A			y_2^A	
Model transfer from A to B			$\downarrow$ $\downarrow$			$\downarrow$ $\downarrow$	
Input $\in \Sigma^{B*}$			y_1^A	s_1^B	s_2^B	$y_2^A s_1^B$	
State $\in Q^B$	q_0^B	q_0^B	q_1^B	q_2^B	q_0^B	$q_4'^B$	
Output $\in \Delta^B$							

Figure 1: Model transfer process without and with support of simultaneous use of design tools. The interpretation of a column in the figure is as follows. T_i represents a point in time or a time interval. At time T_i the system receives the indicated input symbol and changes into the new state written in the column. At the same time an output symbol may be produced. A downward arrow indicates that the output symbol is transferred to another system.

3. Modeling operations m which affect Z (and hence in general I), where for a given m there exists no operation of type τ such that $\delta(q, m) = \delta(q, \tau)$ for all $q \in Q$. Examples of m are creation, deletion, Boolean and other non-linear operations on geometry.

If we denote the space of possible m by $\mathcal{M}^A$ then Q^A can be represented by the set

$$((I\!R^4 \times SO(3)) \times (I\!R^6 \times SO(3)) \times \mathcal{M}^A$$

and Q^B with a similar expression.

Still taking V as a geometric evaluator we can distinguish between model transfer as a unique event and model transfer in a framework of design model sharing. We illustrate this in Figure 1. At time T_3 system A is in state q_1^A which is, upon acceptance of symbol t^A transferred to system B through the external representation y_1^A. This symbol is able to bring system B from its initial state into q_1^B. A necessary condition is that $V(q_1^A) = V(q_1^B)$. Suppose that the purpose of this model transfer is to perform further modeling on B as done at time T_4. However, at the same time T_4 the user on system A decides to change q_1^A into q_2^A In preparation of the next model transfer process, B is returned to its initial state q_0^B at T_5. At T_6 the updated model q_2^A is transferred to B through the symbol y_2^A causing B to be in state q_3^B.

Then, if model sharing is *not* supported the user on B has to redo all its operations (s_3^B at time T_7); the modeling process is sequential and the work s_1^B is lost.

Model sharing does support simultaneous work on A and B with reuse of the work s_1^B as depicted in Figure 1. At T_6 the string $y_2^A s_1^B$ automatically brings B from its initial state to $q_4'^B$. The key condition is that $V(q_4'^B) = V(q_4^B)$.

An objection to this procedure is that s_1^B will not be available at time T_6 since we know that the automaton does not store such information. Therefore, an external system is needed to keep track of design histories.

4   A PRACTICAL EXAMPLE

For two commercial design tools we have analysed the requirements for model sharing, taking into account a number of types of system commands (geometric transformations, surface filleting and trimming, and NURBS degree changing). The two systems have complementary functionality (global versus detailed modeling) and it could therefore be useful to let them simultaneously act on the same model. A picture of the functionality of the two systems is shown in figure 2. In table 1 it is summarized which operations on system A and B cause inconsistencies. The experimental evaluation proceeded as in Figure 1. On system A a model q_1^A was created. q_1^A was transferred to q_1^B and it was verified that $V(q_1^A) = V(q_1^B)$ (despite the different internal representation forms of the two design models). Then two activities took place simultaneously:
1. On system A q_1^A was further designed to q_2^A and
2. On system B q_1^B was further designed to q_2^B.

Then q_2^A was transferred into q_3^B. Finally all design commands under 2) above were re-executed on B, resulting in q_4^B. This re-execution was done using a standard log file procedure available on system B. No specific utilities were used.

q_4^B was compared with the model $q_4'^B$ that would have been obtained if the two design

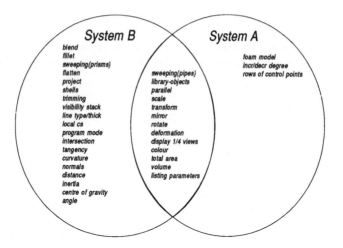

Figure 2: Symbolic picture of the overlapping alphabets Σ^A and Σ^B.

Simultaneous use of systems A and B			
	b_1 fillet/bl.	b_2 holes	b_3 trimming
a_1 changing degree	−	$+^1$	$+^3$
a_2 deformation	+	$+^2$	+
a_3 scaling	+	+	+
a_4 translation	+	$+^2$	$+^4$
a_5 mirroring	+	$+^2$	$+^4$
a_6 rotation	+	$+^2$	$+^4$

Table 1: Operation b_j can be automatically repeated on a model after this model is revised by operation a_i (+) or it cannot be (−).

[1...4] There are circumstances that the automatic repeat fails (Vergeest 1994).

activities were done sequentially. If $V(q_4'^B) = V(q_4^B)$ then the + sign was put in table 2.

As shown in table 2, some operations (corresponding to strings in Σ) can take place simultaneously on A and B for all possible states of A and/or B. More often, the + sign holds only for certain subsets of $Q^A \times \Sigma^*$ and $Q^B \times \Sigma^*$. Details are given in (Vergeest 1994).

For the tests we did not yet take into account the actual screen positions of the projected models. We rather relied on the CAD user; if he or she was able to designate the correct constituent with the cursor, we counted this action as being automatic. This is motivated by the expectation that affine transformations and viewing transformations can be included relatively easily in such a way that deviations in 3D and 2D spaces are automatically compensated.

5 CONCLUSIONS

Table 2 has resulted from some ad hoc tests for each listed type of operation as an approach to determine the $Q \times \Sigma$ subsets empirically. However, in general the sets Q and Σ are strongly structured and it can be expected that much properties of the $Q \times \Sigma$ subsets can be derived from those structures. It is therefore very important that computer tool vendors make documentation about the $Q \times \Sigma$ structures explicitly available; this will help in the development and standardization of data type conversion algorithms. Using the finite automaton description method, it can be precisely described which information can reside in the subsystems and which information must be represented externally in order to manage product model sharing.

6 REFERENCES

Eastman, C. (1994) Out of STEP? *Computer-Aided Design*, **26**, Nr. 5, 338-340.

Owen, J. (1993) STEP: An Introduction. Information Geometers, Winchester.

Perrin, D. (1990) Finite Automata, in *Handbook of theoretical computer science* (ed. J. van Leeuwen), Elsevier Science Publishers, Amsterdam, 3-57.

Torres, J.C. and Clares, B. (1994) A formal approach to the specification of graphic object functions. In *Eurographics '94* (eds. M. Daehlen and L. Kjelldahl), Blackwell Publishers, *Computer Graphics Forum*, **13**, Nr. 3, C-371-380.

Vergeest, J.S.M. (1994) Managing simultaneous design activities using shared models, in *Proceedings 4th Int. FAIM Conference, FAIM'94* (eds. W.G. Sullivan et al.), Begell House, New York, 390-398.

87

A New Modeling and Analysing Approach to Material Flow and Productivity - An Application of Manufacturing Systems Theory (MST) to Production Systems

Kesheng Wang
Professor, Dr. ing., Department of Production and Quality Engineering
NTH, N-7034 Trondheim, Norway
Tel: 47 73 597119 Fax: 47 73 597117 Email: keshengw@protek.unit.no

Abstract

This paper describes an application of Manufacturing Systems Theory (MST) to the modeling and analysis of material flows and productivity of a batch-processing engineering company which produces a large variety of electrical motors. In MST, the various physical and economical systems within manufacturing engineering may be modeled and analyzed by one abstract geometrical system. The transformation from physical variables (concrete) to geometrical variables (abstract) has a dramatic reduction on the number of modelling approaches. The modeling approach is more general, systematic, unified and effective than any other approach.

Keywords

Modeling and analysis, graph theory, systems theory, manufacturing systems, production planning and control

1 Introduction

Manufacturing engineering is one of the most important fields in the world today. The modeling of manufacturing systems helps in analyzing various designing, planning and operating decisions. In recent years, there has been a rapid growth in interest in the basic theoretical and methodological issues in the modeling and analysis of a variety of complex large-scale systems. The modeling approaches for manufacturing systems reported in literature include linear programming, dynamic mathematical programming, markov method, simulation model, network model, bond graph, queuing model, pertrinets, etc. [4,5,6,7,8,12,13] Most of these approaches are domain-dependent, that means none of them can be suitable to model all kinds of problems in manufacturing systems.

With the increasing technological complexity and scale of manufacturing systems which spans from design, manufacture to management, the request to derive solutions of problems more objectively, systematically and logically could never be overemphasized. Theory of Technical system, [9] Physical System Theory, [10] System Science, and Computer aided systems theory [11] are all developed for meeting this request.

The Purpose of developing Manufacturing Systems Theory [1,3,15] is also exactly to meet the request for the modeling and analysis of large-scale and complex manufacturing systems. In Manufacturing Systems Theory, the large amount of physical and economical systems within manufacturing engineering may be modelled and analyzed by one geometrical system. All these physical variables within the systems can be interpreted as a set of geometric variables. This transformation leads to a dramatic reduction on the number of modelling approaches.

In this paper we will present how to use a new modeling and analyzing approach, i.e., by Manufacturing Systems Theory (MST), to model and analyze flows of material, cost of unit weight and productivity in a multi-stage production system. The system is a batch-processing engineering which produces a large variety of electrical motors. The approach of modeling and analyzing is described in the following sections in details.

For the reason of comparisons and simplifications, the case study is exactly the same as the one of reference 12. The production system consists of 12 production stages. There are two production lines; one for manufacturing of rotors and other for stators. These components are finally assembled. A set of standard functions written in APL (A Programming Language) has been used for the automatic computer-aided analysis. A complete program list is proposed for the illustrated example.

2   Manufacturing Systems Theory

The general procedure of modeling and analyzing in MST for any kind of system is briefly described as the following :

(1) Set up primitive system (Y-object);
> In this stage, a whole system will be disconnected into several logically-independent units, which are called elements or primitive systems. The physical variables (dual variables) and govern equations (the relationship of the physical variables) of each element are defined. Then the characteristics of all elements are represented in mathematics by a Y-object.

(2) Set up the connected system (V-object);
> The element can be modeled by a directed line and two boundary nodes, and the connection of all these elements can be modeled by a directed linear graph, which is called a behavior graph of a system. The topology of a connected system can be represented as V-object algebraically.

(3) Determine the given source (I_N-object);
> The influence to a system from environment may be selected as sources.

(4) Solve the problem (e_N-object).

After the Y-, V-, and I_N-object has been established, the solution process is just as simple as to transform the primitive systems into the connected system.

$$e_N = V^t \ Y \ V \ I_N$$

The procedure described above shows that the approach is general, systematic, unified and effective compared to the other approach.

3 Modeling of a transformation process

The different components of a production system under consideration are conceptualized to perform any one, or a combination, of the following basic processes:

(a) Transformation process: the transformation process can be defined as a transformation of resources to achieve a well-defined change in their physical, chemical, technological, biological or functional characteristics.
(b) Transportation process: This is a special type of transformation process in which material is simply moved from one geographic location to another at a cost. This includes collection, translocation and distribution of various kinds of materials.
(c) Storage processes: This is a special type of transformation process in which the input and output are identical in form, and the resources are carried over time.

The MST has been successfully applied to modeling and analysis of the transportation process and storage process.[13] This paper will focus on the material transformation processes taking place within a production system. The free body diagram of the ith transformation process is shown in Fig. 1.

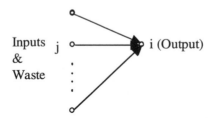

Figure 1. Free body diagram of ith transformation process

The response-stimulus relationships for a transformation process are as follows:

(1) Material flow equation:

$$I_j = Y_j \ I_i$$

(2) Cost equation:

The associated parameters and variables are as follows:

$$E_i = \Sigma Y_j \, E_j + f_i(I_i)$$

> I_j = flow of material input, or waste, of the jth type,
> I_i = material output, or design capacity, of the process,
> E_j = cost per unit material input, or waste, of the jth type,
> E_i = unit cost of output of ith process,
> Y_j = process technological coefficient for input, or waste, of jth type.
> f_i = the cost function, which represents the variation in processing cost with design capacity which accounts for labor, capital and overhead costs by the technological coefficient at various stages.

(3) Sensitivity

The sensitivity S_j for the change in output cost for a corresponding unit change in some factor at any stage can be carried out by partial-differentiating the output cost, E_{out}, with respect to the technological coefficient, Y_j, at various stages, as follows:

$$S_j = \frac{\partial E_{out}}{\partial Y_j}$$

(4) Production index

In the system, productivity P_i is defined as the ratio of the ideal output cost per unit weight to the actual cost per unit weight, i.e.,

$$P_i = \frac{ideal \ cost \ per \ unit \ weight}{actual \ cost \ per \ unit \ weight} \cdot 100$$

Production index P_i will help in identifying potential areas of improvement. It is worth nothing that the productivity index for this system will always be less than 100 except in the rare case when it could be equal to 100. For a fully, or highly, automated industry, a high index number, approaching 100, could be regarded as satisfactory, while in a labour-intensive industry a much lower value could be considered good. The interpretation of these indices will depend on individual managements.

4 A case study

4.1 Description of the production system

The purpose of the application of MST is to analyze the inter-relationships between the various process of the existing system in order to identify areas of productivity improvement. The given production system consists of 12 production stages. There are two production lines; one for the manufacture of rotors, and the other for stators. These components are finally assembled together into a product.

4.2 Primitive systems

According to the concept of MST, a complete and whole system can always be separated into a number of subsystems and a subsystem may be divided further into several elements which are no longer interconnected each other, i.e., are logically independent. The collection of elements are called primitive systems. The given production system can be viewed as a whole system, the various production stages as subsystems and each transformation process as an element. The present production system consists of 7 subsystems (production stages), i.e., transformation processes, are: (1) cast iron foundry, (2) machining, (3) press working, (4) winding, (5) die casting, (6) balancing, and (7) assembly.

4.2.1 Elements in primitive systems

The basic components in the primitive systems are elements. In other words, a primitive system is a physical element which has terminal pairs at which different measurements are made, that is, all measuring procedures (instruments) have, at least conceptually, two terminal points. A measurement therefore always has to be associated with a topological line having two boundary points coinciding with the points in which the instrument is connected. Now let's explain how to use a directed topological line with two boundary nodes to model a transformation process. The first transformation process in the production system, foundry, is considered. In foundry, the flow of material input, Pig iron, is transferred into the flow of material output, raw casting, and the waste or rejects. The foundry process may be viewed as a subsystem, and the subsystem consists of two primitive systems, i.e., two elements. One element is the basic transformation process that pig iron is transformed into raw casting, and another element is one that pig iron is transformed into waste and rejects. Each element is modeled as a directed topological line and nodes represents total flow of material input, output and waste or rejects respectively. The direction of line is based on the requirement of computation. The rule for determining the direction of a arrow is that in the case of computing material flow, the direction of an arrow is against the direction of the transformation process, and in the case of cost calculation, the direction of an arrow is same as the direction of the transformation process. The diagram of the foundry process is shown in figure 2.

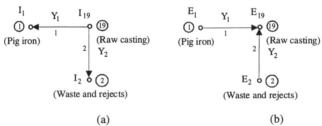

(a)   (b)

Figure 2. The diagram of primitive system of the Foundry.
(a) for the flow of material, (b) for the cost analysis.

4.2.2 Dual variables

In transformation process system, there are two different kinds of physical quantities to characterize the properties of an element : the unit cost of material is defined as a transvariable denoted by E, while the flow of material is defined as a intervariable denoted by I. The technological coefficient representing the input requirement per unit of output is denoted by Y.

4.2.3 Relationship between the dual variables

In foundry process, material flow equation are as follows:

$$I_1 = Y_1 I_{19}$$
$$I_2 = Y_2 I_{19}$$

and cost equation is:

$$E_{19} = Y_1 E_1 + Y_2 E_2 + f_{19}(I_{19})$$

Where

I_1 = total input of pig iron, I_2 = total foundry scrap or waste, I_{19}= total output of raw casting, E_1 = unit cost of pig iron, E_2 = unit cost of disposing foundry scrap, E_{19} = unit cost of raw casting, Y_1 = the pig iron required unit of raw casting, Y_2 = the scrap generated per unit of raw casting.

4.2.4 Y-object

All kinds of physical variables are represented with N-object in MST. The representation of N-object is easy to be carried out in a computer. In the present example, Y-object is a 2-object, i.e., a diagonal matrix, which may be derived from the diagrams of primitive systems shown in Figure 3. And Y-object may be set up as followings:

$$Y = \text{diag } [Y_1, Y_2, \ldots\ldots Y_n]$$

where n = 29 and is the number of the directed lines of the system.

4.3 Interconnected system

The important stage in MST is to set up the connection of the individual primitive systems. The complete interconnected system is modeled by a directed linear graph which characterizes the topological property of the system.

4.3.1 Behavior graph

The behavior graph for computing material flow is found just by joining the diagram of primitive systems together. In a complete behavior graph, all nodes have to be connected to a common reference. But for simplification, we will never really bring the corresponding lines and the reference points into the behavior graph. We have to keep it in mind that an identity matrix, YI, is always added to a transformed YN-object.[2] The simplified behavior graph is shown in Figure 4.

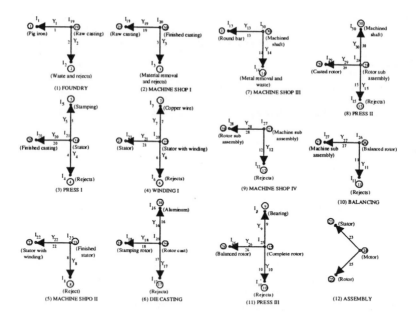

Figure 3. The diagram of primitive systems.

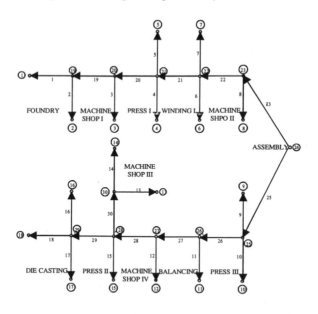

Figure 4. The simplified behavior graph of the production system

The behavior graph of the production system for cost computation is obtained just by turning the direction of the arrows in the behavior graph for flow of material computation. So we will not show that diagram.

4.3.2 Topological structure

The behavior graph above enables us to directly derive the corresponding topological matrices. The total incidence matrix VU, the positive incidence matrix VP, and the negative incidence matrix VN are defined:

$$VU_{ij} = \begin{cases} +1 & \textit{with tail of an arrow} \\ -1 & \textit{with head of an arrow} \\ 0 & \textit{(else)} \end{cases}$$

$$VP_{ij} = \begin{cases} +1 & \textit{with tail of an arrow} \\ 0 & \textit{(else)} \end{cases}$$

$$VN_{ij} = \begin{cases} +1 & \textit{with head of an arrow} \\ 0 & \textit{(else)} \end{cases}$$

where i = the number of lines and j = the number of nodes.

4.4 Given sources

In MST, sources are defined as the influence between a system and environment. We only need to consider node sources for the production system. There are two kinds of node sources :

(a) Node source for material flow: The analysis of material flow is to help a manager to make the bill of material for producing necessary products. The easiest way is to compute the necessary products for one final product, ie, one motor in the present example. The node source can be represented as a 1-object form (vector):

$$\begin{array}{cccccccc} & 1 \ 2 \ 3 & & 22 \ 23 \ 24 \ 25 \ 26 \\ I_N = [0 \ 0 \ 0 & ... & 0 \ 0 \ 1 \ 0 \ 0 \ ...] \end{array}$$

The number of elements in the IN is equal to the number of nodes. One motor is produced in the 24th node (see Fig. 4).

(b) Node sources for the cost of unit weight: The data collection mainly consists of the material, labour, capital and overhead costs incurred as well as the quantity of material waste generated at various stages. The data for material cost per unit weight are derived from available accounting data as given as followings:

$E_1 = 1028,$ $E_2 = -250,$ $E_3 = -150,$ $E_4 = -150,$ $E_5 = 13632,$ $E_6 = -10000,$
$E_7 = 40972,$ $E_8 = -200,$ $E_9 = 113119,$ $E_{10} = -200,$ $E_{11} = -200,$ $E_{12} = -200,$

$E_{13} = 332,$ $E_{14} = 50,$ $E_{15} = -200,$ $E_{16} = 13667,$ $E_{17} = -400$ $E_{15} = 13632$

The minus sign represents that the waste or rejects can be reused.

The average estimated values of the factor f for different stages are given as following:

$f_{19} = 5.32,$ $f_{20} = 4.71,$ $f_{21} = 0.231,$ $f_{22} = 2.81,$ $f_{23} = 0.84,$ $f_{24} = 8.63,$ $f_{25} = 5.46,$
$f_{26} = 0.44,$ $f_{27} = 1.84,$ $f_{28} = 0.46,$ $f_{29} = 0.44,$ $f_{30} = 0.57$

The node source I_{NC} may be set up based on the data which given above:

$$\begin{matrix} 1 & 2 & 3 & 4 & 5 & & 28 & 29 & 30 \end{matrix}$$
$$I_{NC} = [1028 \ -250 \ -150 \ -150 \ 13632 \ ...... \ 0 \ 0 \ 0]$$

The factor f may be represented as :

$$\begin{matrix} 1 & 2 & 3 & 4 & & 27 & 28 & 29 & 30 \end{matrix}$$
$$f = [0 \ 0 \ 0 \ 0 \ ...... \ 1.84 \ 0.44 \ 8.63 \ 5.46]$$

4.5 Formulation and solution

4.5.1 Roth's diagram for ordinal scale of measurement system

The elements in a production system is characterized by ordering elements which has inherent orientation.[1,2]. The corresponding system is called a system of ordinal scale of measurement. The change of reference frame in the case of ordinal scale of measurement is performed by the positive part of V-object, VP, and the negative part, VN. Roth's diagram in Figure 5 shows the transformation process graphically. All the formulation processes are carried out based on the Roth's diagram routinely.

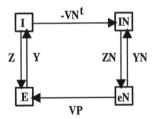

Figure 5. Transformation diagram of the ordinary scale system

4.5.2 Formulation

(a) For the computation of material flow, formulation is as the follows

$$YN = -VN^t \ Y \ VP$$

$$YNN = YN + YI$$
$$ZNB = YNN^{-1}$$
$$eNB = ZN \ INB$$

(b) For the cost calculation, formulation become

$$ZNC = ZNB^{\ell}$$
$$F = eNB \times f$$
$$INNC = INC + F$$
$$eNC = ZNC \ INNC$$

(c) Sensitivity analysis

During the calculation of sensitivity S_j, it is not necessary to derive the symbolic functions and make partial differentiation. The approximate formula is

$$\frac{\partial E_{out}}{\partial Y_j} = \frac{\Delta E_{out}}{\Delta Y_j}$$

The procedure of cost calculation in (b) can be used for sensitivity analysis and the algorithm is:

(1) Give an increment $\Delta Y_j = 0.01$
(2) Substitute $Y_j + \Delta Y_j$ in Y-object
(3) Calculate E_{nout}
(4) $\Delta E_{out} = E_{nout} - E_{out}$
(5) $S_j = \Delta E_{out}/\Delta Y_j$

(d) Productivity analysis

The procedure of cost calculation in (b) can also be used for productivity analysis and the algorithm is:

(1) Set up Y-object using general technological coefficient value
(2) Calculate E_{gout} using Algorithm (b)
(3) Set up Y-object using ideal technological coefficient value
(4) Calculate E_{iout} using Algorithm (b)
(5) $P_i = (E_{iout}/E_{gout}) \times 100$

4.6 Solutions

In order to make the computer-aided computation, APL programming language has been chosen as a programming tool. A set of different functions has been developed for making the programming easy. The program list of analysis of material flow, and unit cost of output material with ideal and general technological value and productivity index are shown in Appendix.

5 Conclusions

This paper has presented a new modeling and analysis approach to material flow and productivity by the use of Manufacturing Systems Theory. The general procedure by which the production system can be modeled and analyzed is: (1) to establish primitive systems; (2) to determine the behavior graph and topological object; (3) to select sources; and (4) to formulate and compute solutions via Roth's diagram. The case study has proved that MST can be fruitfully utilized for the purposes of costing and production planning of multi-stage manufacturing systems and it is useful in the practical implementation of this methodology to the modeling and analysis of manufacturing systems. According to author's knowledge, the modeling approach is more general, systematic, unified and effective than any other approach. We conclude that Manufacturing Systems Theory (MST) may be a most prospective approach to modeling and analysis of manufacturing and engineering systems.

6 References

(1) Bjørke, Ø., Towards a manufacturing systems theory - Applications so far, Proceedings of Conference Manufacturing International '88. Atlanta, Georgia. USA., 1990.

(2) Bjørke, Ø. and Wang, K., Manufacturing systems theory - Examples, Institute report, Department of production engineering, NTH, Trondheim, 1990.

(3) Bjørke, Ø. and Wang, K., Manufacturing systems theory, Lecture notes, Norwegian Institute of Technology, Trondheim, 1993.

(4) Buzacott, J. A., Prediction of efficiency of production systems with internal storage, International Journal of Production Research, Vol. 6, p. 173   , 1968.

(5) Davis, R. P., and Kennedy, Jr. W. J., Markovian modelling of manufacturing systems, International Journal of Production Research, Vol. 25, p. 337, 1987.

(6) Davis, R. P., and Miller, D. M., Analysis of the machine requirements problem through a parametric Markovian model. SME technical paper MS79-69, 1979.

(7) Dobois, D., and Stecke, K., Using petrinets to represent production process. Proceedings of 22nd IEEE conference on decision and control, December, 1983.

(8) Egbelu, P. J., Davis, R. P., Wysk, R. A., and Tanchoco, J. M. A., An economic model for machining of cast parts, Journal of Manufacturing Systems, Vol.1, p. 207, 1982.

(9) Hubka, V. and Eder, W. E., Theory of Technical Systems - A total concept Theory for Engineering Design, Springer-Vorlage, Berlin, 1988.

(10) Keening, H. E., Toad, Y, and Kesavan, H. K., Analysis of Discrete Physical System, McGraw-Hill, New York, 1967.

(11) Pichler, F., and Diaz, R. M., (Eds.), Computer aided systems theory - EUROCAST'93, Spring-Vorlage Berlin Heidelberg, 1994.

(12) Stack, K., and Solberg, J. J., The optionality of unbalancing both workloads and machine group sizes in closed queuing networks of multi-server queues, Operations Research, Vol 33, p. 882, 1985.

(13) Sushil, Singh, N. and Jain, B. K., A physical system theory approach to material flow and productivity analysis, Engineering Costs and Production Economics, Vol. 13, pp.207-215, 1988

(14) Wang, K. and Bjørke, Ø., Manufacturing systems theory in Scandinavia, International symposium on manufacturing science & technology for 21st century, Beijing, 1994.

(15) Wang, K. and Bjørke, Ø., A general, efficient approach to dynamic analysis of interconnected mechanical system using the theory of connections, International journal of system science, Taylor & Francis Ltd., Vol. 25, No. 7, pp. 1157-1178, 1994.

7 Appendix

1. Material flow with general technological coefficient value

YG←TYPE_DIAGONAL
1.176 0.176 1.25 0.25 0.67 0.01 0.34 0.893 0.02 0.127 1.22 0.22 0.705 0.385 0.031 0.005 0.974
0.001 0.053 1.053 0.32 0.01 0.69 1.54 0.54 0.28 0.22 0.94

VU←TYPE_TOPOLOGY
TYPE INCIDENCE PART OF BRANIN TABLE
19 19 20 20 21 21 21 22 22 22 23 23 24 24 25 25 25 26 26 27 27 28 28 28 30 30 2
29
1 2 19 3 20 4 5 21 6 7 22 8 23 25 9 10 26 11 27 12 28 30 15 29 13 14 16 17 18

TYPE TRANSFORMER PART OF BRANIN TABLE

YNG← - (TRA VN) MUL YG MUL VP
YDIG←TYPE_DIAGONAL
1 1 1 1 1 1 1 1 1 1 11 1 1 1 1 1 1 1 1 11 1 1 1 1 1 1 1 1

YNNG←YNG + YDIG
ZNBG←INV YNNG
INB← 0 0 0 0 0 0 0 0 0 0 0 0 0 0 0 0 0 0 0 0 0 0 0 0 1 0 0 0 0 0 0
eNBG←ZNBG MUL INB
eNBG
0.7565 0.1132 0.1287 0.007681 0.2611 0.172 0.1092 0.1551 0.01194 0.001925 0.000375 0.01989
0.1948 0.0683 0.003953 0.07636 0.06 0.2564 0.6433 0.5146 0.7681 0.8601 0.705 1 0.385 0.375
0.3754 0.3953 0.2727 0.1265

(2) Unit cost of output material with general technological coefficient value

ZNCG←TRA ZNBG
f← 0 0 0 0 0 0 0 0 0 0 0 0 0 0 0 0 0 0 0 5.32 4.71 0.231 2.81 0.84 0.57 0.44 6 1.84 0.44
8.63 5.46
FG←eNBG×f
INC←1028 -250 -150 -150 13632 -10000 40972 -200 113119 -200 -200 -200 332 50 -200 13667
-4000 13632 0 0 0 0 0 0 0 0 0 0 0
INNCG←INC + FG
eNCG←ZNCG MUL INNCG
eNCG
1028 -250 -150 -150 1.36E4 -1.0E4 4.097E4 -200 1.131E5 -200 -200 -200 3323 -150 -200
1.367E4 1168 1425 5589 9996 1.215E4 1.485E4 1.632E4 1.315E4 1.314E4 1.249E4 1.576E4 5037

(3) Unit cost of outputmaterial for ideal technological coefficient value

YI←TYPE_DIAGONAL
1.08 0.08 1.12 0.12 0.67 0.01 0.34 0.893 0.003 0.11 1.08 0.02 0.705 0.295 0.031 0.005 0.974
0.001 1.001 0.03 1.03 0.32 0.005 0.685 1.24 0.24 0.21 0.08 0.87

YNI← - (TRA VN) MUL YI MUL VP
YNNI← YNI + YDIG
ZNBI← INV YNNI
eNBI← ZNBI MUL IN3

ZNCI← TRA ZNBI
FI← eNBIxf
INNCI← ZNCI MUL INNCI
eNCI← ZNCI MUL INNCI
eNCI
1028 -250 -150 -150 1.363E4 -1.0E4 4-09E4 -200 1.131E5 -200 -200 -200 3323 -150 -200 1.36E4
-4000 1.36E4 1168 1425 5589 9996 1.215E4 1.485E4 1.63E4 1.315E4 1.314E4 1.249E4 1.576E4
5037

(4) Computation of productivity index PI

□←PROINDX←eNCI ÷ eNCG × 100
100 100 100 100 100 100 100 100 100 100 100 100 100 100 100 100 100 93.55 84.77
97.4 93.43 82.98 77.13 90.26 87.59 87.59 89.52 91.43 81.1

PART NINETEEN

Manufacturing Technology

88

Model of Root-Bead Welding for Off-line Programming and Control

O. Madsen, J. Lauridsen, H. Holm, J. Boelskifte, I. Hafsteinsson.
Department of Production, Aalborg University.
Fibigerstraede 16, DC-9220 Aalborg East, Denmark. Tel: +45 98 15
85 22, Fax: +45 98 15 30 30, email: I9OM@iprod.auc.dk.

Abstract

In this paper, a model is presented which, based on a description of the shape of a welding seam (the gab size and bevel angles), can be used to select variables controlling the welding of a root-bead. The model can be used in an off-line programming system as well as in a geometry-sensor based control system.

The model is a stationary model and it is established on an empirical basis. Tests indicate that the model can be used to select control variables for seams with root gabs between 2 and 5 mm. The allowed deviation from the nominal value of the gab is app. 0.5 mm.

The model has been implemented in an off-line programming system generating robot programs for welding thick walled large diameter nozzles perpendicular onto pipes. Tests with this off-line programming system indicate that if the workpiece geometry is defined sufficient well, a satisfactory weld quality is obtained when executing the control variables computed by the model.

Keywords

Robotics, root-bead welding, off-line programming, process modelling, sensor based control.

1 INTRODUCTION

In many industries, welding fabrication involves many labour intensive manual operations which impose high demands on the skills and concentration of the operator. Because of the significant requirements to the humans involved and the high demands to welding flexibility, automation by robots offers a great potential for many industries to improve welding rates and in particular welding quality.

However, one of the obstacles preventing many industries from using robots to perform welding tasks is the cumbersome work associated with programming the robots. This obstacle is of particular importance when the robots have to be re-programmed often.

One way to solve this problem is to automate the programming task by using a computer aided off-line programming system. In order to perform computer aided off-line programming of robots for welding tasks (and also other tasks), a number of models are needed: models of the robot and its environment, workpiece models, and inverse welding process models. An inverse welding process model is here defined as: a table or a mathematical equation which transforms the workpiece geometry into appropriate control variables, which, when executed, yield a weld seam fulfilling the required weld quality criterion.

For a large number of welding applications, sufficiently accurate models of the robot and its environment can be designed in off-line programming and simulation systems such as ROBCAD and GRASP. Furthermore, geometrical models of the workpieces to be welded can be designed in CAD-systems.

However, sufficiently reliable inverse welding process models are rarely available. Some models have been developed for the root-beads in seams with backing (Andersen et al (1989), Harrits (1990), Galopin (1991) and Lauridsen (1991)). But for welding root-beads without backing, no inverse welding process model is available.

Because of this, and because such a model is needed in a computer aided off-line programming system used at Odense Steel Shipyard Ltd, Denmark, an inverse welding process model for welding root-beads without backing has been developed at Department of Production, Aalborg University, Denmark. The objective of this paper is to present this model.

2   THE WELDING TASK

The inverse welding process model presented in this paper has been developed so that it can be used at Odense Steel Shipyard, Ltd, Denmark, for welding thick walled large diameter nozzles onto pipes (in the following called pipe branches). Pipe branches of various dimensions are being welded at Odense Steel Shipyard, however, the work presented here is focused on pipe branches having the following characteristics (see figure 1):

- The diameter (D) of the main pipe is larger than the nozzle diameter (d) (D/d > 1.5)
- The nozzle centre axis intersects and is perpendicular to the centre axis of the main pipe.
- The diameter of the nozzle is relatively large (d > 200mm).
- The nozzle and the main pipe are made of carbon steel (St. 37).

The weld seam is produced by preparing the pipe and the nozzle in a numerically controlled flame-cutting machine. With a vertical cut, a hole is made in the main pipe, and the nozzle is cut such that the resulting weld groove angle (Ω) is app. 45° along the entire seam. Measurements on pipe branches have showed that the variations of Ω usually are less than +/- 2°. The nozzle is tack welded onto the main pipe. It is attempted to have a constant root-gab, but due to various production tolerances (e.g. ovalities of the pipes), the resulting root-gab (d_{root}) varies between 2.0 mm and 5.0 mm. The resulting weld bead must have adequate fusion and convexity of the back face of the weld seam.

The cell in which the robot based nozzle welding is performed consists of a workpiece positioner having one degree-of-freedom, a welding robot and a welding machine. The basic outline of the cell is shown in figure 1. As it appears of this figure, the workpiece is fixed on the workpiece positioner so that the degree of freedom of the positioner (θ) coincides with the centre axis of the main pipe.

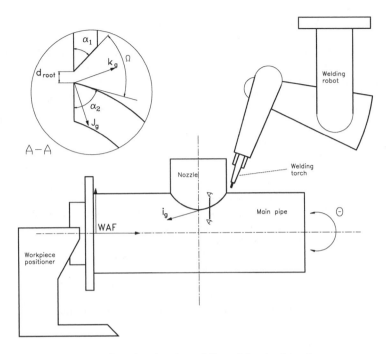

Figure 1 A sketch of the cell performing the welding of the pipe branches

3 MODEL SPECIFICATION.

In this section the input/output to/from the inverse welding process model are defined. However, first it must be stated that the model has been developed for stationary (steady-state) operations. Stationary inverse welding process models can be used to compute trajectories of control variables if the conditions of the weld process is such that it can be assumed that the weld process runs through a number of quasi-stationary states during operation. Experiments indicate that for pipe branches having the characteristics listed in section 2 this is a valid assumption (see Lauridsen et. al. (1995)).

Secondly, it must be stated that the model applies for MAG-welding (gas metal arc welding) only and for the equipment parameters shown in table 1.

Table 1 Equipment parameters used for developing the inverse welding process model

Welding machine:	MIGATRONIC KME400
Wire:	FILARC PZ 6102 (d=1.2mm)
Gas:	$ArCO_2$ 82/18

3.1   Output of the Inverse Welding Process Model.

The output from the inverse welding process model are the values of the variables controlling the weld process in a certain pre-determined point along the welding seam.

There are a large number of variables involved in MAG-welding, and these variables can be defined and selected in various ways depending on the specific welding task. For the welding task presented in this paper, the following variables are used:

U	:	The welding voltage.
W	:	The wire feed speed.
θ	:	The positioner angle (see figure 1).
v_T	:	The travel speed (the speed of the torch relatively to the workpiece).
$^{WAF}T_{tnom}$	:	A transformation matrix specifying the nominal location (position and orientation) of the welding torch relatively to the workpiece attachment frame (WAF). WAF is located on the workpiece attachment plane of the positioner (see figure 1), and the nominal location is defined as the location of the torch if no oscillation is specified.
OSC	:	The oscillation pattern specified by (see figure 2):

	OW :	The oscillation width.
	OV :	The oscillation vector represented relatively to WAF.
	OF :	The oscillation frequency.
	OT :	The hold time in percentage of total oscillation time.

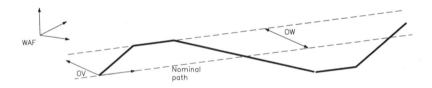

Figure 2 Illustration of the variables specifying the torch oscillation.

How a trajectory of a vector containing the control variables listed above can be transformed into a trajectory of equipment control vectors controlling the equipment shown in figure 1, is described in Madsen&Holm (1994).

3.2   Input to the Inverse Welding Process Model.

The inverse welding process model must be able to compute control variables in all points along the welding seam. For welding of pipe branches, this means that the inverse welding process model must be able to cope with the varying seam profile shapes found when moving along the seam.

As indicated by figure 1, the groove profile can be described by means of the following shape parameters:

α_1 = The bevel angle of the nozzle.

α_2 = The bevel angle of the main pipe.

Ω = The groove angle.

d_{root} = The root gab.

If it is assumed that Ω is constant $45°$, and that: $\alpha_1 + \alpha_2 + \Omega = 180°$, then α_1 is dependent of α_2, and the seam profile can be described by two parameters: α_2 and d_{root}. For the pipe branches in this investigation α_2 and d_{root} belongs to the following intervals: $\alpha_2 \in [50° ; 90°]$ and $d_{root} \in [2 \text{ mm} ; 5 \text{ mm}]$

The position and orientation of the groove is represented by a transformation matrix ($^{WAF}T_{groove}$) which specifies the location of a groove frame relatively to the WAF-frame. In every point along the welding seam, the groove frame is defined as follows (see also figure 1): The origin of the groove frame is located on the edge of the main pipe. The X-axis of the groove frame (represented by the unit vector $\mathbf{i}_g$) is tangent to the spatial curve formed by the groove edge, and the Z-axis (represented by $\mathbf{k}_g$) is parallel with the bisector for the groove. The direction of the Y-axis (represented by $\mathbf{j}_g$) can be found using the right-hand rule.

The orientation of the welding seam relatively to gravity should in principle also be included as input to the inverse welding process model, since gravity has a significant effect on the resulting weld quality. Usually, if the orientation of the welding seam with respect to gravity changes, a change in control variables is needed. However, by using a workpiece positioner as shown in figure 1, it is possible to keep the orientation of the seam with respect to gravity constant along the entire welding seam. How to control the workpiece positioner so that this is obtained is described in Lauridsen et. al. (1994).

4   THE INVERSE WELDING PROCESS MODEL.

In the following, the inverse welding process model will be presented, i.e. it will be described how the transformation shown in figure 3 is carried out.

Figure 3 Definition of the task of the inverse welding process model.

Experiments have shown that some of the control variables need not be varied as a function of model input. These constant control variables are shown below in table 2.

Table 2 Constant control variables.

Welding voltage (U):	15.3 V
Wire feed speed (W):	2.5 m/min
Oscillation frequency (OF):	1.8 Hz
Oscillation holding time (OT):	35 %

The transformation from the WAF-frame to the torch frame ($^{WAF}T_{tnom}$) is computed as:

$$^{WAF}T_{tnom} = {}^{WAF}T_{groove} \cdot {}^{groove}T_{tnom} \tag{1}$$

where $^{WAF}T_{groove}$ is known from the model input and $^{groove}T_{tnom}$ is a function of the desired tip-to-workpiece distance (D_w), the desired travel angle (α_t), and the desired torch work angle (β_t) (see figure 4 for a definition of α_t and β_t)

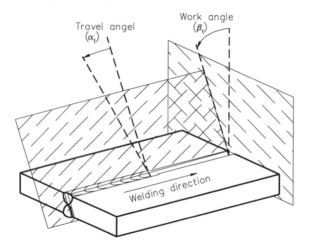

Figure 4 A definition of the angles (α_t and β_t) specifying the torch orientation.

In the model the tip-to-workpiece distance (D_w) is constant = 15 mm, and the travel angle (α_t) is constant = -27°. Experiments have shown that the torch work angle β_t must be varied as a function of α_2 in order to cope with the varying mass distribution along the seam. Experiments have shown that β_t (in degrees) can be computed as:

$$\beta_t = 0.104 \cdot \alpha_2 - 5.351 \tag{2}$$

The oscillation width (OW) is computed as a function of the root-gap:

$$OW = \frac{d_{root}}{2} \tag{3}$$

The oscillation vector (OV) is parallel with the root-gab. This means that OV represented relatively to the groove frame can be computed as:

$$OV = \left(0 \quad -\sin(\alpha_2 + \frac{\Omega}{2}) \quad -\cos(\alpha_2 + \frac{\Omega}{2}) \right)^T \tag{4}$$

By means of $^{WAF}T_{groove}$ this vector can be transformed into a representation relatively to the WAF frame.

Experiments have shown that the travel speed V_t must be varied as a function of both α_2 and d_{root}. In order to establish this relationship experiments were made in test pieces where α_2 were constant and the root gab varied. Welding was performed with constant travel speed, and the gab sizes at which the welding quality changed from being satisfied to being unsatisfied were determined. By performing this experiment for a number of different travel speeds a tolerance box was determined, inside which the travel speed must be selected. A curve was then drawn inside the tolerance box and, finally, a 3rd order polynomial was fitted to this curve. Such experiments were made for three different α_2-values resulting in the following three 3rd order polynomials:

$$\alpha_2 = 90.0^\circ: \quad V_t = 4.828 - 1.550 \cdot d_{root} + 0.160 \cdot d_{root}^2 - 0.002 \cdot d_{root}^3$$

$$\alpha_2 = 71.0^\circ: \quad V_t = 5.397 - 2.155 \cdot d_{root} + 0.344 \cdot d_{root}^2 - 0.019 \cdot d_{root}^3 \tag{5}$$

$$\alpha_2 = 51.5^\circ: \quad V_t = 4.930 - 1.439 \cdot d_{root} + 0.107 \cdot d_{root}^2 - 0.003 \cdot d_{root}^3$$

Given a certain pair of α_2-d_{root}-values, the travel speed is computed as follows: First the root-gab d_{root} is inserted into the three 3rd order polynomials shown above. The result of this are three travel speeds $(V_{t90}, V_{t71}, V_{t51.5})$ corresponding to the root gab and $\alpha_2 = 90^\circ$, 71° and 51.5°. A 2nd order polynomial is then fitted to these three travel speeds and their corresponding α_2-values. Finally, the travel speed is computed by inserting the α_2-value into this 2nd order polynomial.

5 MODEL TESTS.

A number of tests have been made in order to test the model. In these experiments linear test pieces with constant α_2 and d_{root} values were produced. The inverse welding process model was then used to compute control variables. These control variables were then executed in a test cell. Figure 5 summarises the results obtained from these tests.

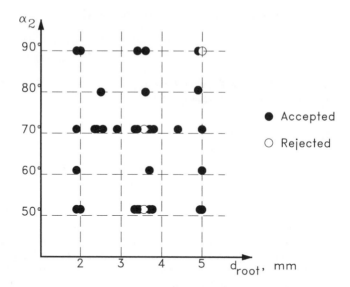

Figure 5 A summary of the experiments made in order to test the inverse welding process model. A total of 37 experiments was carried out.

As it appears from this figure more than 37 test experiments have been performed. Of these 37 experiments only 3 failed to fulfil the quality demands whereas the remaining 34 experiments were accepted. The rejected experiments were repeated, this time resulting in a satisfactory weld quality. Hence, it is evaluated that the rejected experiments failed due to factors outside the scope of the inverse welding process model (e.g. that the thermal distortion of the workpiece was larger than expected).

In order to test the robustness of the model, a number of experiments were made where the inverse welding process model was used to compute control variables for a given model input. These control variables were then executed in seams which had a shape which differed from the shape after which the control variables were computed. The results of these experiments indicated that a satisfactory weld quality can be obtained as long as:

- The misalignment between the two pipe edges is less than ±1 mm.
- The root gap is known with an accuracy of ± 0.5 mm.
- The welding torch is guided in the middle of the weld groove with an accuracy of less than ± 0.3 mm.
- The groove angle (Ω) is 45° ± 5°.

6   CONCLUSIONS.

In this paper an inverse welding process model has been presented which can compute the variables controlling the welding of a root-bead in a seam with a gab. Input to the model is a description of the shape of the welding seam. Experiments indicate that the model is valid as long as the deviations between the modelled (expected) workpiece shape and the shape of the physical workpiece are below certain thresholds.

The inverse welding process model has been implemented in an off-line programming system for welding pipe branches. Here the model input is computed based on a computer representation of the pipe branch. The off-line programming system is described in Lauridsen et. al. (1995).

The main conclusion from experiments with the off-line programming system is that if the workpiece shape is known sufficiently well, an execution of the control variables computed by the inverse welding process model results in a satisfactory weld result.

However, the experiments also indicated that it is difficult and costly to keep the shape deviation on an acceptable low level. Instead, it is proposed to use a geometry-sensor system to measure the actual seam shape. The architecture of such a system is presented in Madsen&Holm (1995).

7   REFERENCES

Andersen, K. et al. (1989) A Novel Approach Towards Relationships between Process variables and Weld geometry. *In Proceedings of ASM International Conference on Trends in Welding Research;* Gatlingburg, USA, 14-18 May, 1989.

Harrits, P, (1991) *Automation of Multiple Pass Welding Using an Industrial Robot.* PhD-Thesis, The Control Engineering Institute, Technical University of Denmark; 1991.

Galopin, M. et. al. (1991) Design of Optimum Adaptation Tables for Robotic Arc welding Using Vision Sensing. *In proceedings of EUROJOIN 1;* Strasbourg; France; 5-7 November 1991.

Lauridsen, J. (1991) *Computer Aided Off-line Robot Programming of Multi-Pass TIG-Welding.* PhD-Thesis, Department of Production, Aalborg University, Denmark, Dec. 1991.

Lauridsen, J, Madsen O, Holm, H, (1995) An Off-line Programming System for Welding the Root Bead in Pipe Branches. Paper summited to the *4th International Conference on Trends in Welding Research*, Gatlingburg, USA, 5-9 June, 1995.

Madsen, O, Holm, H, (1994), A Real-time Motion Control Interface for 3D-Tracking in Welding. *In Proceedings of the 9th IPS-Research Seminar,* Fuglsø, Denmark, 18-20 April.

Madsen, O, Holm, H, (1995), Control System Architecture for Robotic Welding of Tubular Joints. Paper summited to the *4th International Conference on Trends in Welding Research*, Gatlingburg, USA, 5-9 June, 1995.

89

Enhanced Mark Flow Graph to Control Autonomous Guided Vehicles

Santos Filho, D. J.;   Miyagi, P. E.
Escola Politécnica da Universidade de São Paulo
Departamento de Engenharia Mecânica/Mecatrônica
Av. Prof. Mello Moraes, 2231
05508-900 - São Paulo - S.P. - Brazil
tel: +55-11-818-5491
fax: +55-11-813-1886
email: diolinos@cat.cce.usp.br

Abstract

The aim of this work is the evaluation of the modeling power of Enhanced Mark Flow Graph (E-MFG) when approaching manufacturing plants whose transportation systems are based on autonomous guided vehicles (AGV). We present main concepts regarding E-MFG and its application in the description of the control part as well as the operative part of production systems. Through the information associated with the structural components of E-MFG it is possible to represent complex control strategies involving different routes of AGV systems in the production of a mix of parts in a highly flexible environment.

Keywords

Discrete event dynamic system, Petri nets, Mark Flow Graph, manufacturing system, autonomous guided vehicles, autonomous system

1   INTRODUCTION

The difficulty in controlling a transport system composed of several Autonomous Guided Vehicles (AGV) in a flexible manufacturing environment is strongly related to its level of automation. On the other hand, the concept of automatic process control can be generalized to the case in which control object is a dynamic production system driven by events, i.e., a discrete event dynamic system - DEDS (Santos FⁿO, 1991). Therefore, this work approaches the problem of controlling an AGV transport system with high functional flexibility that can be treated as DEDS.

Recently the technological evolution of the equipaments used in automated manufacturing systems has increased substantially the degree of automation of these systems. In this way, increases in the functional flexibility of manufacturing system meet the changes of the market dynamics with high degree of autonomy.

Mark Flow Graph - MFG (Hasegawa, 1984) is a technique derived from Petri nets (Reisig, 1992) that is effective to represent and control the dynamic behavior of DEDS in special, cases where the system must be described and controlled in a distributed way. The major property of the MFG is that it explicitly describes the control strategy of independent, sequential and concurrent events. However, there are some problems related to the modelling of systems with complex dynamic behavior involving non trivial control rules. To represent the control rules in a consistent manner it is necessary to consider several sub-graphs that need to be synchronized. The resultant graph is difficult to be interpreted and analysed. This aspect motivated the proposition of extensions of MFG to increase its modeling power and a methodology to represent Integrated Manufacturing Systems (IMS) by subsystems that operate cooperatively. In this context, the Enhanced Mark Flow Graph - E-MFG (Santos F$^{\underline{o}}$, 1993) is based on individualized marks and additional rules to control the transitions firing.

E-MFG technique intends to combine the advantages of a higher level representation and the properties of condition-events nets (Valette, 1986) to effectively realize the control. By using the individualized marks and specific rules for transition firing it is possible to define more complex control strategies for the combination of pre-determined activities and the representation of appropriate knowledge to assure a greater degree of autonomy and more efficient solution for cases where the system reaches abnormal states .

This work presents the E-MFG technique emphasizing its ability to model and control AGV transport systems with high level of flexibility and autonomy. The technique is applied to a system composed of several AGV that must attend production units with high operational flexibility. E-MFG elements are combined to control the movements of AGV in a consistent manner, supervising the flow of information that assures the desired dynamic behavior of the production system.

2   ENHANCED MARK FLOW GRAPH - E-MFG

The E-MFG is a modeling technique that is capable of working with individualized marks with attributes. An attribute is used to represent and control the changes in the marks' flows (alternative routes) and select the tasks (conditions) associated with boxes. The basic structure of the graph and its main elements is maintained. The concept of marks with attributes is based on the theory of Tagged-MFG (Yoshida, 1988). In this sense, the basic rules related to the attributes are not changed. New rules for transition firing are added based on the theory of ϕ-net (Fuji Electric, 1989).

Next, the basic structural elements of E-MFG and the dynamics of transitions firing and marking updating are described.

2.1   Individualized Marks

In E-MFG the marks have an attribute vector that defines their individuality. These attributes can be associated with information related to the product, process and control. If the attribute is zero, it means that the attribute does not exist. This concept is similar to the idea of label used in Tagged-MFG. The Figure 1 illustrates an example of the structure of an individual mark.

$$\text{Mark} \quad \bullet \quad = \langle a1, a2, a3, a4 \rangle$$

$$\text{where,} \quad \begin{cases} a1 = \text{piece type} \\ a2 = \text{batch} \\ a3 = \text{origin} \\ a4 = \text{destination} \end{cases}$$

Figure 1 Example of the structure of an individual mark.

2.2   Composed Individual Mark

Taking advantage of the modular nature of MFG, macro-elements were introduced to simplify the representation of several common devices in manufacturing systems. The resulting Functional MFG (Miyagi, 1988) is therefore a combination of basic elements of MFG. The functional boxes named capacity, grouping and dispersing handle only indistinct marks, i.e., capacity, grouping or dispersing boxes treat only the amount of items (workpieces, parts, etc.) when we are modelling the storing in buffers, the packing and unpacking tasks in manufacturing systems. By using individualized marks, it is also possible to describe the input and output sequence of items and specify the item contents. When the palletizing (assembling) and depalletizing (disassembling) process is modelled by grouping and dispersing boxes in an E-MFG, the mark that represents the set of items is named *composed individual mark*. This denomination is justified by its structure that stores the information about each item before the grouping or distributes the information for each item after the dispersion process. The stored information is represented by an additional attribute in the form of a control code. Figure 2 illustrates the representation of functional boxes known as grouping, dispersing and capacity and an example of the composed individual mark.

2.3   Handling the Marks Attributes

The problem of handling attributes associated with marks can be interpreted in two forms, i.e., the attributes can be modified by "conditional changes" or "selective filtering".

Conditional changes mean that an attribute state can be modified depending on the current state of the proper attribute, i.e., we have a function that updates the state of the system. This task is executed by another functional box named "controller box". This box performs the function of controlling the attribute states, i.e., this box updates the global state of the system based on the current local state described by each mark. Production rules of "if-then-else" type are applied to verify and update the attributes previously specified. The Figure 3 illustrates an example of controller box changing the attributes of the marks.

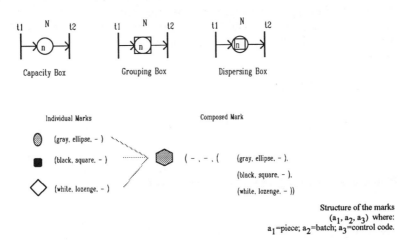

Figure 2 Basic functional boxes and an example of composed individual mark.

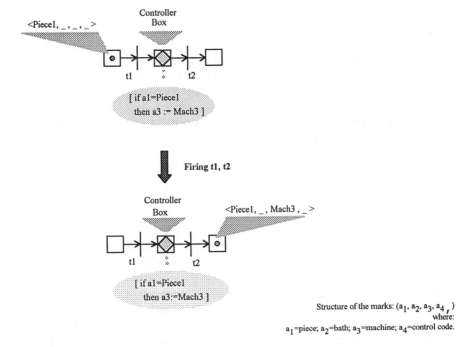

Figure 3 Controller box changing the state of an individual mark.

The *selective filtering* establishes fields corresponding to attributes that must follow in the mark after the firing of transition . This specification is accomplished through inscriptions in the directed arcs. It is important to note that the inscriptions do not represent restrictions on firing of transition connected by the directed arc. The inscriptions limit the attribute fields to be transmitted. The Figure 4 illustrates the selective filtering process. The variable inscription "a1" above the input arc of transition "t1" specify that only the attribute corresponding to the first element of the attribute vector of individual mark must be maintained. Other attributes are not transmitted.

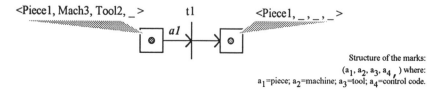

<Piece1, Mach3, Tool2, _ >   t1   <Piece1, _ , _ , _ >

Structure of the marks:
(a_1, a_2, a_3, a_4) where:
a_1=piece; a_2=machine; a_3=tool; a_4=control code.

Figure 4 Representing a selective filtering process.

2.4   Dynamics of Firing Rules

The dynamics of transition firing are established by decision rules according to a specific hierarchy. As the same case of ϕ-nets, we have three levels of hierarchical decision.

The first level corresponds to the additional restriction rules of the type "if-then-else", necessary to represent specific control strategies. The specification of these rules is done by inscriptions in the transitions. If the transition does not have inscription, there are no additional rules to consider in its firing. A transition that satisfies the additional restriction rules is named *"transition in readiness condition"*. The second level corresponds to the enabling rules for transition firing. A transition is called *"enabled transition"* if it is in the readiness condition and also satisfies the following conditions:

• There are no boxes with marks in the output side.
• There are no boxes without marks or with marks with restrictions in the input side.
• There are no internal or external gates in the disabled state.

The third and last level corresponds to the proper *"execution rules"* of firing. These rules are related to the arbitration rules in cases of existing conflicts despite the rules for selective filtering of attributes defined by the inscriptions above the directed arcs. An enabled transition where all the execution rules are satisfied is called *"fireable transition"*. A fireable transition fires immediately, accomplishing the flow of marks above the graph that models the dynamic behavior of the system. The attributes that are not transmitted are assumed to be zero and we have a combination of attribute information in the marks, i.e., the attribute vector in the mark that appear in the output side of the transition is the result of a logic "or" operation of the attributes in the corresponding input side. The Figure 5 illustrates an E-MFG graph where the firing of a transition and the resulting changes in the involved marks occur.

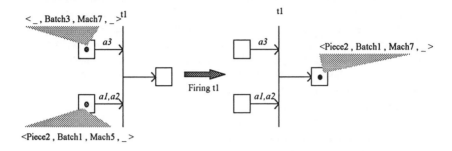

Structure of the marks: (a_1, a_2, a_3, a_4) where:
a_1=piece; a_2=bath; a_3=machine; a_4=control code.

Figure 5 Changes in the marks attributes resulting the transition firing.

3   APPLICATION OF E–MFG TO AGV CONTROL IN IMS

3.1   Object Control and Control System

AGV transport systems can be represented by two fundamentals and interconnected subsystems that operate cooperatively. These subsystems represent the operative part and control part respectively, as illustrated in Figure 6. This approach is aimed at establishing a distinction between the execution sub-system (operative part) and the coordination sub-system (control part). The operative part sends to the control part information above its state and the control part processes this information to send commands to operative part.

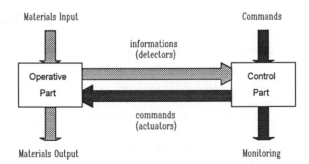

Figure 6 Operative part and control part of production systems.

The control object model that represents the plant of the production system corresponds to the operative part of the IMS with the AGV transport system. Considering the hypothesis that the qualitative control of IMS can be treated as a combination of sequential subsystems, the devices that compose the system can be described as elements with a finite number of discrete states. Therefore, controlling the system means the synchronization of several elements (that represents the physical devices) involved in the accomplishment of a certain set of activities.

The use of MFG to model the control object generates an explicit representation of local states reachable for each device and permits through its structural elements (external gates and output arcs) an effective exchange of signals between the controller and the control object. Here, the control object receives information from the controller by the external gates.

The control part that represents the controller can be modeled by E-MFG. Conditions for evolution of the system considering activities to be realized involve the synchronization and ordering of operations to be executed by several devices that compose the physical system.

This methodology permits the analysis of the model that in fact represents the IMS, i.e., there is not the semantic gap between the model for simulation and the model for control. The Figure 7 illustrates an example of E-MFG model where the information exchange between control part and operative part is explicitly shown.

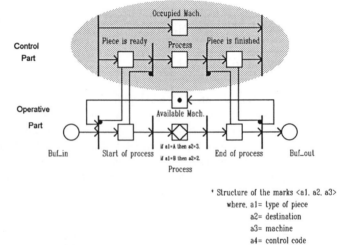

Figure 7 Example of the E-MFG model for the control part and operative part.

3.2   Control of AGV Transport System

To control the traffic of AGV it is necessary to control material flow between production units in parallel with movement of vehicles. The difficulty is in the specification of an activity sequence for the AGV, i.e., to define in advance the sequence of production units that the AGV must follow we face the problem of indeterminism of DEDS and the problem of stochastic nature of transportation systems (Brand, 1988). To control the AGV traffic we consider that the transport is accomplished from the operations sequence control of each item (workpiece) in each production unit. When a production process of a certain item is finished the AGV is requested to transport this item to the next production unit.

Through the attributes associated with individual marks we can represent the source and next production unit. Then, the AGV control systems identify the next production unit of the item in process and specify a vehicle to do the transportation task. If individual marks are not used it is necessary to represent all possible routes for the AGV. Therefore, using E-MFG we can apply the following procedure to develop the control system of IMS with AGV system. First, definition of the E-MFG model of the control object.

• Model of the transport via;
• Model of each production unit (a workstation or a manufacturing cell)
• Model of the items input, output stations and the parking stations of the AGV.

Second, definition of the E-MFG model of the control system to be developed in two stages.

• Stage one. Specification of the operation sequence control related to the production process of each type of product (item).
• Stage two. Establishment of the AGV control to execute the item transport between the production units. The function of control system is to update the destiny of the AGV in agreement with the production step.

Applying this methodology we obtain the model of the controlled system that can be evaluated using an E-MFG simulator for example.

3.3   Application Example

Lets consider a flexible production system composed of four work stations, one input station, one output station and parking station for AGV. This system is supposed to produce a mix of three different products with the following operations sequences:

• Product A: EST.in → EST.1 → EST.2 → EST.3 → EST.4 → EST.out.
• Product B: EST.in → EST.2 →EST.1 →EST.3 →EST.out.
• Product C: EST.in → EST.3 → EST.2 → EST.1 → EST.4 → EST.out.

The attribute vector composition is "a1=order type", "a2=order lot", "a3=route (specify the station to be visited)" and "d1=identification of the AGV".
In the model of control object we specify one transport activity for each section visited by the AGV to connect one station to another. Concerning workstations, we define several bypass points to permit the entrance and exit of the AGV. The bypasses depend on the attributes of marks that specify the route and destiny of the workpieces/AGV.

In this text is not possible to show the several steps of the modeling procedure to construct the E-MFG graph (Santos F$^{\underline{o}}$, 1993). Figure 8 illustrates one part of resulting E-MFG model of the AGV control system. The marks in boxes *B.in*, *B.1*, *B.2*, *B.3* and *B.4* represent the request of transport from the stations *EST.in* , *EST.1*, *EST.2*, *EST.3* and *EST.4* respectively. From boxes *B.x*, the control system verifies if there is any AGV available in the stations. If so, it is moved to the corresponding station. Obviously there are priorities among the stations where the available AGV must be stopped.

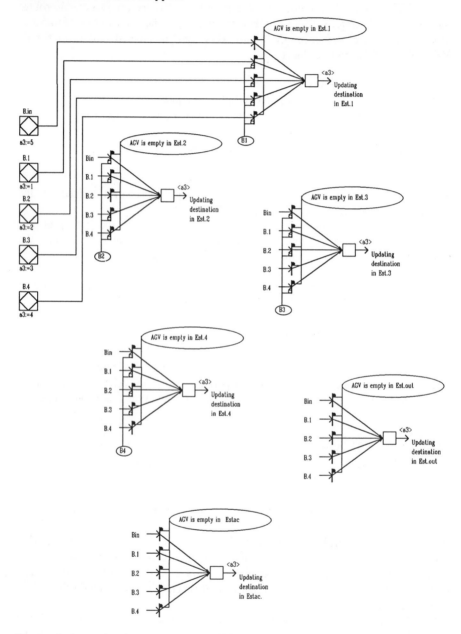

Figure 8 Example of E-MFG of the decision procedure for requests of the system.

4 FINAL COMMENTS

The case of AGV was used to show the potential of this technique for the specification and implementation of additional control rules. The capability of E-MFG to handle individualized marks is very important to describe the complex information flow necessary to coordinate the activities of AGV systems that transport materials in production systems.

The next step is the study of a generalized modeling procedure, development of tools to analyze systems and test of alternative control strategies.

5 REFERENCES

Brand, K. P. and Kopainsky, J. (1988) "Principles and Engineering of Process Control with Petri Nets", IEEE Transactions on Automatic Control, IEEE, Vol.33, n° 2, pp.138-149.

Fuji Electric Ltd. & Fuji FACOM Control Ltd. (1989) "Material Flow System Control (ϕ-net)", Journal of the SICE, v.28, n°9, pp.826-827, Tokyo, Japan.

Hasegawa, K.; et al. (1984) "Proposal of Mark Flow Graph for Discrete System Control", Transactions of the SICE; Vol.20, n° 2, pp.122-129.

Miyagi, P.E. (1988) "Control System Design, Programming and Implementation for Discrete Event Production Systems by Using Mark Flow Graph", Doctor Thesis, Tokyo Institute of Technology, Japan.

Reisig,W.(1992) "A Primer in Petri Net Design", Springer-Verlag,Berlin-Heidelberg

Santos F°, D. J. (1991) "Discrete Event Systems and their Control", Proc. of 1st International Congress of Instrument Society of America, ISA, São Paulo, SP. Brazil, pp.2.1-2.12 9 (in Portuguese).

Santos F°, D.J. (1993) "Proposal of Enhanced Mark Flow Graph for Modeling and Control of Integrated Manufacturing Systems", Master Thesis, University of São Paulo - Escola Politécnica, SP, Brazil (in Portuguese).

Valette, R. (1986) "Nets in Production Systems", Lecture Notes in Computer Science, Springer-Verlag, Berlin, Germany, pp.191-217.

Yoshida, S. (1988) "Tagged MFG Based Control of Materials Flows in Job Shop Production Systems", Master Thesis, Tokyo Inst. Tech., Japan (in Japanese).

SANTOS F°, D.J.S., born in 1961. Graduate in Eletrical Eng. from Escola Politécnica da USP - Brazil in 1988. Received the M.Sc. from the Department of Mechanics/Mechatronics Engineering of Escola Politécnica da USP - Brazil in 1993. Assistant of Department of Mechanics/Mechatronics Engineering of Escola Politécnica da USP since 1993.

MIYAGI, P.E. Ph.D., born in 1959, graduated in Eletrical Eng. from Escola Politécnica da USP - Brazil in 1981. Received the M.Sc. and Ph.D. degrees from Tokyo Institute of Technology - Japan in 1988 respectively. Associate Professor of Department of Mechanics/Mechatronics Engineering of Escola Politécnica da USP since 1993.

90

An Architecture for Robot Off-line Programming of Filling Welds for Multipass GMA-Welding of Pipe Branches

J.K. Lauridsen, O. Madsen, H. Holm
Department of Production, Aalborg University
Fibigerstraede 16, DK 9220 Aalborg East
Denmark
Phone: +45 98 15 42 11 / 2959
Fax   +45 98 15 30 30
Email: i9jkl@iprod.auc.dk

Abstract

This paper presents an architecture of an off-line programming system for generation of robot programs for welding the filling welds in pipe branches. The paper outlines the need of process models and how the structure of such process models must be in order to obtain the required quality to the pipe branch welding. Furthermore it is shown, how the process models are used to generate the control variables, so that the required quality to the pipe branch weldings is obtained.

Keywords

process modelling, off-line programming, robot welding

1.   INTRODUCTION

In a large number of industries the joining of nozzles attached perpendicular on large diameter (see figure 1) pipes involves many labour intensive manual operations which impose high demands to the skills and concentration of the operator. Because of the significant requirements to the humans involved and the high demands to welding flexibility, automation by robots of this type of welding offers a great potential for many industries to improve welding productivity and in particular to improve welding quality.

However, there are only few reports on automatic robot based methods for this kind of welding. The reasons behind this are:

1. The path to be followed by the welding torch forms a complex so called "saddle type" curve.
2. Multi-pass welding is necessary.

3. The welding position and the welding seam profile constantly change along the welding seam.

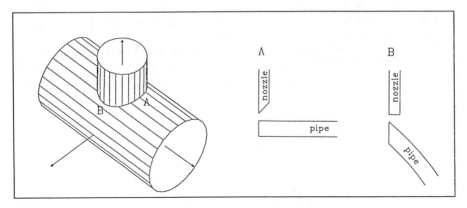

Figure 1.   Sketch of a pipe branch. The diameter of the pipe is approximately 400 mm and the length is approximately 800 mm, and the wall thickness is approximately 8 mm.

Due to these reasons programming of robots with conventional on-line techniques becomes very cumbersome, and makes practical application almost impossible. One way to conquer this problem is to automate the programming task.

Automation of the programming of such a complex welding task implies a sensor based programming system, which must contain the following elements:

- Computer based models of the physical welding equipment.
- Computer based models of the physical welding process
- Computer based models of the physical workpieces to be welded.
- Computer based strategies and algorithms for planning the welding task.
- Computer based models of sensor control.

The computer based models are capable of matching the reel world into a computer world, by which the information needed for off-line generation of the robot program are extracted. The models of the welding equipment (i.e. geometric and kinematic model of the welding robot and workpiece manipulator, geometric model of the welding torch) and the workpiece can relatively easy be established in commercial available software packages.

This is, however, not the case with the computer based models of the physical welding process. These are models, that describe the relations between the control variables (i.e. the torch position, orientation, welding velocity, wire feed speed etc.), the joint geometry, and the desired weld quality.

Furthermore, the computer based strategies and algorithms needed for planning the welding task and off-line generation of the robot program, are not presently available.

Consequently, considerable effort must be placed in the following fields:

1. Establishment of computer based welding process models, that are capable of describing the relations between the control variables, the joint geometry, and the desired weld quality. Since the dynamic structure and the quality requirements to the root bead in the pipe branch weldings differs significantly from the filling welds, it is necessary to operate with two different type of welding process models.
2. Elaboration of computer based strategies and algorithms needed for planning the welding task and off-line generation of the robot program.

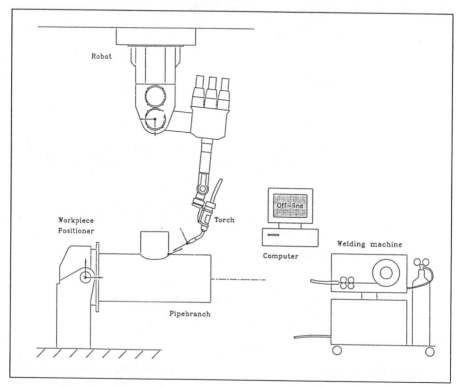

Figure 2.   Outline of welding cell.

In corporation with Odense Steel shipyard, Denmark, the Department of Production, Aalborg University, has started the development of the mentioned computer based programming system (see figure 2). The work is primarily concentrated on the above mentioned fields.

In the last year, considerable results have been obtained in research in the above fields, (see. Lauridsen 1991, 1995, Madsen 1992, 1993, 1994 and 1995). Particular, results have been obtained in the field of root bead welding, so that it now is possible automatically to off-line program the welding of the root bead in a pipe branch.

This paper presents the results obtained so far on the establishment of an off-line programming system for robots for carrying out welding of the filling welds of nozzles attached perpendicular on large diameter pipe.

2.   DESCRIPTION OF THE WELDING TASK

Since the nozzle is attached perpendicular to the main pipe, the intersection curve between the inner radius of the nozzle and the outer radius of the pipe forms a saddle curve. Normally, when a specific welding task is considered, a number of cross sections of the workpiece along the welding axis is constructed. In this case particular case, the cross sections are created along the intersection curve between the nozzle and the pipe, so that the tangent to the intersection curve are normal to the plane, in which the cross sections are constructed.

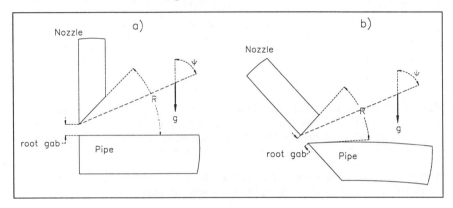

Figure 3.   Cross sections along the intersection. a) is a cross section in position A of figure 1 and b) is a cross section in position B of figure 1 .

By using the equipment setup as shown in figure 2, and in particular the use of a workpiece manipulator one important advantage is obtained. If it is assumed that the centreline of the main pipe is parallel with the axis of rotation of the manipulator, then for any position on the intersection curve, a value of the axis of rotation of the manipulator can be determined, so that the tangent to the intersection curve is perpendicular to gravity. (see Lauridsen, 1994). Furthermore, if the angle R in figure 3 is constant alone the intersection curve (this is at Odense Steel Shipyard obtained in production by using a numerical controlled flame cutting machine), then the angle Ψ between the groove bisector and gravity is constant along the entire intersection curve. This means, that using a workpiece manipulator, the welding orientation along the entire intersection curve is constant.

However, the welding task cannot be considered as constant, because the mass distribution along the intersection curve changes, and changes from 2 to 5 mm in root gab must be expected. Furthermore, from figure 4 it is seen, that the free groove area, in which the filling beads are to be deposited changes along the intersection curve.

3.   THE OFF-LINE PROGRAMMING TASK

Since off-line programming of the root bead welding has been realised, the programming task to be described here concerns the filling welds only. This off-line programming task is to generate a program for robot, manipulator and welding machine, so that the required welding quality of the finish weld is obtained, when the program is executed. The program is a trajectory of position, orientation and velocity of the end-effector for the robot, arc voltage and wire feed speed for welding machine and rotation angle of the workpiece manipulator.

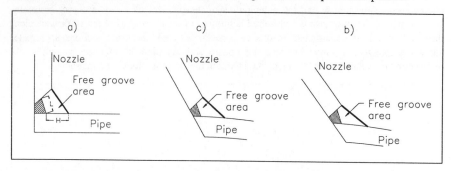

Figure 4. Free groove area along the intersection curve. a) is a cross section in position A of figure 1. b) is a cross section in position B of figure 1 and c) is a cross section somewhere in between. The hatched area indicates the root bead, and the filled line indicates the required quality of the finish weld.

In order to generate the trajectory is it necessary to be able to compute the 4x4 transformation matrices of equation (1), and the corresponding velocity, arc voltage and wire feed speed.

$$^{base,robot}T_{torch}=^{base,robot}T_g \cdot {}^gT_p \cdot {}^PT_{torch} \tag{1}$$

Where $^{base,robot}T_{torch}$ represents the location of the torch relatively to the base of the robot.

$^{base,robot}T_g$ represents the location of the g-frame (see figure 5) relatively to the base of the robot.

^gT_p represents the location of the p-frame relatively to the g-frame.

$^PT_{torch}$ represents the location of the torch relatively to the p-frame.

The first matrix on the right hand side of equation (1) is a function of the workpiece shape and work cell layout. How this is computed is described in Lauridsen (1994).

The second matrix, which describes the location of the p-frame relatively to the g-frame, is a function of the root bead shape only. Consequently, this matrix is a result of the root bead welding, and can be derived here from (Madsen (1994) and Lauridsen (1995)).

The third matrix describes the position and orientation of the torch relatively to the p-frame. In order to compute this matrix, the control variables for welding the filling beads are needed. These control variables are torch position, orientation and velocity (including weaving variables) relatively to the p-frame, wire feed speed, and welding voltage. In figure 5 and figure 6 the definition of a number of the torch position and orientation control variables is shown.

In order to generate these control variables, welding process models of the filling beads are required.

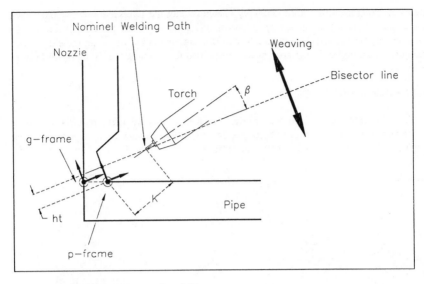

Figure 5. Definition of control variables.

4. THE PROCESS MODELS

As mentioned earlier, a process model describes the relation between the control variables and the joint geometry and the desired weld quality. In a mathematical way, this can be expressed as:

control variables = f(joint geometry,weld quality)

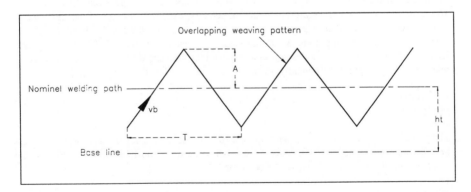

Figure 6. Definition of weaving pattern.

As indicated in figure 4, the joint geometry changes along the intersection curve. However, since the resulting root bead geometry is known, and since the desired quality is known, the free groove area (FGA) and the variables L and H in figure 4 is known along the entire intersection curve. Consequently, the process models needed for planning the filling beads become functions as:

control variables = f(FGA,L,H)

Given the dimensions of the pipe and nozzle as indicated in figure 1, the required quality of the finish weld can only be obtained by depositing two filling beads into the free groove area. (see figure 7). This means, that a process model for bead 2 in figure 7 and a process model for bead 3 is required.

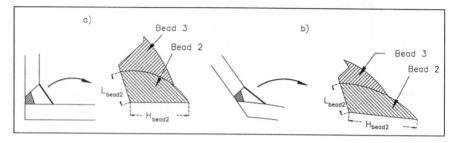

Figure 7.   Free groove area with 2 filling beads.

The requirement to the quality of bead 2 must satisfy equation (2).

$L_{bead2} \leq \varepsilon_1 \cdot L$, where ε_1 is approximately 0.7   (2a)

$H < H_{bead2} < \varepsilon_2 + H$, where ε_2 is 2 mm   (2b)

The requirement given i equation (2a) is determined because the resulting geometry of bead 2 cannot be considered isolated, but must be determined so that the it is possible to weld bead 3. The requirement given i equation (2b) is determined by the required surface of the finish weld.

Experiments show, that it is possible to obtain the required quality for bead 2 only by changing the weaving amplitude A and the value of h_t, as a function of L (see figure 4) and keeping all other control variables constant. h_t describes the distance of the center of weaving relative to the p-frame. Consequently, the process models for bead 2 is on the form

$A = f_1(L,H)$ (3)
$h_t = f_2(L,H)$

The process model for bead 3 is more complicated, because the expected shape of bead 2 must be taken into consideration. This means, that two conditions must be satisfied. Firstly, the cross section area of bead 3 must satisfy equation (4)

$Area_{bead3} = FGA - Area_{bead2}$ (4)

This equation can be satisfied by proper selection of the relation between wire feed speed (W) and travel speed (v). Secondly, the area of bead 3 must be distributed properly in order to reach the required quality of the finish weld. This can be obtained by proper adjustment of A and h_t. Consequently, the process models for bead 3 is on the form:

$$h_t = f_3(L,H)$$

$$A = f_4(L,H) \tag{5}$$

$$W,v = f_5(\text{Area}_{bead3})$$

5.   USE OF MODELS

As indicated in section 3, considerable amount of the information used and generated during the off-line programming of the root bead can be reused when doing off-line programming of the filling beads. In figure 8 is shown, how the information is reused, how off-line programming of the filling beads takes place, and how the process models of the filling beads are used.

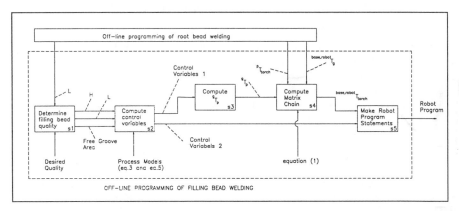

Figure 8   Off-line programming of the filling beads.

The top box of figure 8 illustrates the off-line programming of the root bead, whereas the bottom dashed box indicates off-line programming of the filling bead.

In s1, the free groove area and H is determined based on L (see figure 4), which is derived from the root bead and the desired quality of the filling bead. When the free groove area, H and L are known, the process models in equation (3) and equation (5) are used for generation of the control variables for bead 2 and 3 respectively. This is done in s2. The output of s2 is two sets of control variables. The first set is variables related to position and orientation of the welding torch and are used in s3. The second set is the wire feed speed (W), the travel speed (v), the welding voltage, and the value of the axis of rotation of the manipulator.

In s3 the matrix $^PT_{torch}$ is computed based on the control variables, and in s4 the matrix chain in equation (1) is completed. In s5, $^{base,robot}T_{torch}$ is together with the second set of control variables used for making robot program statements.

Doing the sequence (s1 - s5) for a number of discrete steps along the entire intersection curve, yields the complete trajectory for robot , manipulator and welding machine, and the off-line programming of the filling beads has been completed.

6.   CONCLUSION

In this paper an architecture for off-line robot programming of filling welds for multipass welding of pipe of pipe branches has been suggested. The architecture is characterised by:

- The trajectory of a control vector, that controls the equipment (the workpiece manipulator, the robot and the welding machine) is generated off-line, based on the use of a model of the workpiece, which is going to be welding and on welding process models of the filling beads.
- A workpiece manipulator is used to position the workpiece so that the orientation of the welding seam with respect to gravity during welding is kept as described in a given control criterion.

Experiments performed so far indicates, that the suggested process models can be established, and consequently the suggested architecture forms an important contribution towards the realisation of and automatically programming system for welding pipe branches.

Currently, the architecture of the off-line programming system is being implemented in a prototype system at the Department of Production, Aalborg University, Denmark, and considerable efforts are put into the establishment of the suggested process models.

7.   REFERENCES

Boelskifte, J. Hafsteinsson, I. (1994) Kinematic control for off-line programmed robotic root-pass welding of pipe branches. Department of Production, University of Aalborg, Denmark.

Galopin, M. Dao, T.M. Boillot, J.P. (1991) Design of optimum adaptation tables for robotic arc welding using vision sensing. In Proceedings of EUROJOIN, First European Conference on Joining Technology, Strasbourg, France, 5-7 Nov. 1991.

Holm, H. (1998) Off-line programming of robots for high strength welding of tubular joints. Main report from the ROPS-project, (In Danish).

Koguchi, T. (1990) Development of automatic welding method for large diameter nozzle attached to piping with GMAW. IIW Doc. XII-1162-90. July, 1990.

Lauridsen, J.K, Madsen, O. Holm, H. Hafsteinsson, I. Boelskifte, J. (1994): Model Based Control of a One Degree of Freedom Workpiece Manipulator for Welding of Nozzles. In Proceedings of EURISCON '94, European Robotic and Intelligent Systems Conference, Malaga, Spain, August 22-26, 1994

Lauridsen, J.K. (1991) Computer Aided Off-Line Programming of Multipass TIG-Welding. Ph.D. dissertation, Department of Production Aalborg University, Denmark, December.

Lauridsen, J.K. Madsen, O. Holm, H.(1995) An Off-line Programming System for Welding the Root Bead in Pipe branches. Department of Production, Aalborg University.

Madsen, O. Lauridsen, J.K. Holm, H. Nielson, J. Schluter, F (1993) Model Based Control of a Workpiece Manipulator for Welding of Nozzles. In of Proceedings of: International Conference of Modelling and Control of Joining Processes, Orlando, USA, 8-10 December, 1993.

Madsen, O. (1992) Sensor Based Robotic Multi-Pass Welding. Ph.D. dissertation, Department of Production Aalborg University, Denmark.

Madsen, O. Lauridsen, J.K. Holm, H. Hafsteinsson, I. Boelskifte, J. (1994) Model of Root-Bead Welding for Off-line Programming and Control. Department of Production, Aalborg University

Madsen, O. Lauridsen, J.K. Holm, H, (1995): Control System Architecture for Robotic Welding of Tubular Joints. Department of Production, Aalborg University.

Miyake, N. (1980) Multipass welding of nozzles by an industrial robot. In Proceedings of 10th International Symposium on Industrial Robots, Milan, Italy.

Neerland, H. Aune, D. (1986) Evaluation of systems for robotic multi-pass welding". SINTEF-Report no. STF17 A86085, Norway. (In Norwegian).

8.   BIOGRAPHY

Name: Jan Kirkegaard Lauridsen
                   M.Sc., Mechanical Engineering, Ph.D.,
                   Assistant Professor.
Affiliations: Department of Production
                   University of Aalborg
                   Fibigerstraede 16
                   DK-9220, Aalborg East
                   Denmark.

I am employed as an Assistant Professor at the University of Aalborg, where I teach students in Mechanical engineering, primarily in the science of welding. Furthermore I am researching in control of welding, laying my main effort in the field of off-line programming of robots for carrying out multipass welding. I received my M.Sc.-degree at the University of Aalborg in 1988 in the field of welding High Strength Low Alloyed Steels. In 1992 I received my Ph.D.-degree at University of Aalborg in the field of off-line programming of welding.

91

Integrating Solid Freeform Manufacturing with Relief Creation Software

Chua C.K.
Lecturer, School of Mechanical and Production
Engineering, Nanyang Technological University,
Singapore 2263, 65-7994897 (P), 65-7911859 (F),
MCKCHUA@NTUVAX (EM)

Abstract

The integration of a colour scanner, 3-dimensional CAD/CAM system and Stereolithography Apparatus (SLA) provided a powerful means of building art to part. Artwork such as Chinese characters, human faces or flowers can be input by a scanner, converted into 3-D reliefs within a CAD/CAM system and built using the Stereolithography Apparatus (SLA). In this article, the effectiveness of such an integrated system is illustrated with 3 examples.

Keywords

Art-to-Part, SLA, Scanner, CAD/CAM, Reliefs

1   INTRODUCTION

There are presently numerous commercially-available software for product design for a particular range of industries which include ceramics, glassware, bottle making, both plastic and glass, jewellery, packaging, food processing, for moulded products and products produced from forming rolls and badges, and embossing rollers (Chua, 1993 and Lee, 1992). All of these industries share a common problem: most of their products have elements of complex engraving or low relief on them. Traditionally, such work is carried out by skilled engravers either in-house, or more often by a third-party sub-contractor, working from 2D artwork. This process is costly, open to unwanted misinterpretation of the design by the engraver and most importantly, lengthens the time of the design cycle.

The CAD/CAM revolution has boosted the production and the performance of many industries everywhere for the past 15 years (Lee, 1993). However, its applications in the above industries are still at their infancy stage. Prototyping is still very much a manual process which relies largely on the skills of an experienced craftsman who uses handtools such as a small chisel to carve and shape the model out of a plaster block. Little attention has been focused on the use of quick and accurate rapid-prototyping equipment for building prototypes in this industry.

The use of CAD/CAM and Stereolithography Apparatus (SLA) reduces the time required for design modifications and improvement of prototypes. The steps involved in the art to part process include the following:

- Scanning of artwork
- Generation of surfaces
- Generation of 3D relief
- Wrapping of relief on surfaces
- Converting triangular mesh files to STL file,
- Building of model by the SLA

2   SCANNING OF ARTWORK

The function of scanning software is to automatically or semi-automatically create a 2D image from 2D artwork. It would normally be applied in cases where it would be too complicated and time consuming to model the part from a drawing using existing CAD techniques. Figure 1 shows the 2D artwork of a series of Chinese characters and a roaring dragon.

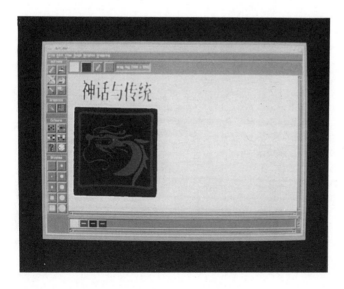

Figure 1 2D artwork.

3   GENERATION OF SURFACES

The shape of a part is generated to the required shape and size in the CAD system for model building. A triangular mesh file is produced automatically from the 3D model. This is used as a base onto which the relief data is wrapped and later combined with the relief model to form the finished part.

4   GENERATION OF 3D RELIEFS

The next stage in creating the 3D relief is to assign to each colour in the image a shape profile. There are various fields which control the shape profile of the selected coloured region, namely, the overall general shape for the region, the curvatures of the profile (convex or concave), the maximum height, base height, angle and scale. There are three possibilities for the overall general shape; a plane shape profile will appear completely flat, whereas a round shape profile will have a rounded cross section and lastly, the square shape profile will have straight angled sides. Figure 2 illustrates the various shapes of the 3D reliefs. For each of these shapes, there is an option to define the profile as either convex or concave. The square and round profiles can be given a maximum height. If the specified shape reaches this height, it will 'plateau' out at this height giving in effect a flat region with rounded or angled corners, depending on whether a round or square shape was selected for the overall profile respectively. Figure 3 illustrates the 3D relief of an artwork.

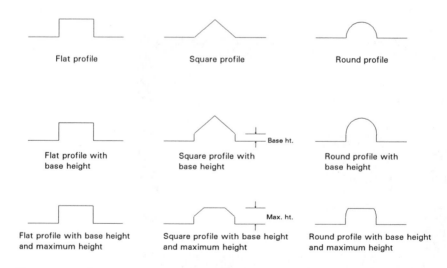

Figure 2 Various shapes of the 3D relief.

Figure 3 3D relief of an artwork.

5   WRAPPING OF RELIEF ON SURFACES

The 3D relief is next wrapped onto the triangular mesh file generated from the object's surfaces. This is a true surface wrap and not a simple projection. The wrapped relief is also converted into triangular mesh files. The triangular mesh files can be used to produce a 3D model suitable for colour shading and machining. The two sets of triangular mesh files, of the relief and the part shape, are automatically combined. The resultant model file can be colour-shaded and used by the SLA to build the prototype.

6   CONVERTING TRIANGULAR MESH FILES TO STL FILE

The STL format is originated by 3D System Inc. as the input format to the SLA, and has since been accepted as the *de facto* standard of input for Layered-manufacturing systems (Fidoora, 1991, Mueller, 1992 and DeAngelis, 1991). Upon conversion to STL, the object's surfaces are triangulated, which means that the STL format essentially consists of a description of inter-joining triangles that enclose the object's volume. The triangular mesh files are also triangulated surfaces, however, of a slightly different format (see figure 4). Therefore, an interface programme written in Turbo-C language was developed for the purpose of conversion. The converted triangular file adheres to the standard STL format as in figure 5. It has the capability of handling triangular files of huge memory size.

DUCT 5.2 TRIANGLE BLOCK P 18 AUG 1993 21.43.28
*

1
@1
1
GREEN
Paint Duct @1

1	0	4	2	0
0.00000	10.00000	0.00000	20.00000
0.00000	0.00000	0.00000	0.00000
0.00000	0.00000	1.00000	1.00000
0.00000	1.00000	0.00000	1.00000

0	0	0	0
0.00000	0.00000	0.00000	0.00000
0.00000	0.00000	0.00000	0.00000
0.00000	0.00000	0.00000	0.00000
0.00000	0.00000	0.00000	0.00000
0.00000	1.00000	0.00000	0.00000
10.00000	0.00000	0.00000	0.00000
0.00000	1.00000	0.00000	0.00000
0.00000	20.00000	0.00000	0.00000
0.00000	1.00000	0.00000	0.00000
10.00000	20.00000	0.00000	0.00000
0.00000	1.00000	0.00000	0.00000

3	1	4
4	1	2

Figure 4 The original triangular file format.

solid print
 facet normal -0.00000e+00 2.00000e+02 -0.00000e+00
   outer loop
     vertex   0.00000e+00 0.00000e+00 2.00000e+01
     vertex   0.00000e+00 0.00000e+00 0.00000e+00
     vertex   1.00000e+01 0.00000e+00 2.00000e+01
   endloop
 endfacet
 facet normal 0.00000e+00 2.00000e+02 0.00000e+00
   outer loop
     vertex   1.00000e+01 0.00000e+00 2.00000e+01
     vertex   0.00000e+00 0.00000e+00 0.00000e+00
     vertex   1.00000e+01 0.00000e+00 0.00000e+00
   endloop
 endfacet

Figure 5 The converted triangular file to follow the STL format.

7   BUILDING OF MODEL BY THE SLA

Californian company 3D System Inc., pioneered the Solid Free Manufacturing (SFM) technologies when they released their commercial SFM system in December 1988 - the SLA-250 model of their StereoLithography Apparatus (SLA) (Fidoora, 1991 and Mueller, 1992). Stereolithography works by using a low-power Helium-Cadmium laser or an Argon laser to scan the surface of a vat of liquid photopolymer which solidifies when struck by a laser beam. The SLA makes use of a variety of photopolymers with different properties suited for different requirements. The properties of the cured photopolymers should allow SLA prototypes to be used for making soft tools like rubber moulds for mass production. Research has shown that feasible rubber moulds can be made from SLA-produced jewellery rings (Lee, 1993). The SLA is capable of a 0.125-mm minimum layer thickness and an accuracy of within 0.5%.

8   CASE STUDIES

Three case studies are selected to illustrate the significant advantages of using the proposed art to part technique over the conventional tools and processes. These case studies are done to cover various types of artwork designs including animals, a human face, flowers as well as Chinese characters (see figure 6), and at the same time to show the feasibility of replacing the current plaster mould prototype with the resin model prototype. Alongside the advantages obtained in adopting the use of the proposed prototyping technique, the case studies also revealed shortcomings which provide scope for future work.

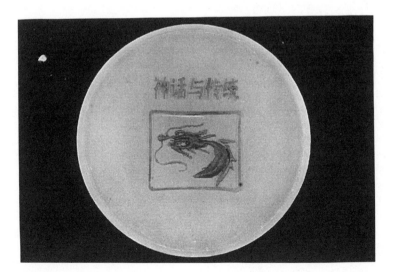

Figure 6 Resin prototype of the Chinese characters and roaring dragon.

9   ADVANTAGES OF THE INTEGRATED PROCESS

The integration of the scanner, the CAD/CAM system and the SLA provides a list of specific advantages to the art to part process as given below:

9.1   It saves time

The existing technique of hand-carving takes about two weeks to complete a plaster mould. However, relief can be created in the CAD/CAM system in two hours' time and the prototype will be ready for examination in the next morning after going through the SLA.   The time to market has become a competitive issue in the need to prototype quickly (Wood, 1990, Brown, 1991 and Poindexter, 1991).

9.2   It is easy to amend

There is often a need to amend the design of the prototype. Serious amendments will result in discarding of the plaster mould and doubling of the time needed to produce a model. The CAD/CAM system allows changes to be done quickly and easily and rebuilding of the model is also a simple task.

9.3   It is easy to master and apply

The whole system is relatively user friendly and the procedures for generating relief are short and simple. There is also a high potential in further extending the application into other industries.

10   CONCLUSION

The CAD/CAM software allows the formation of complicated and time consuming reliefs on models such as jewellery, ceramics tableware, pewter ware, etc. to be semi-automatically or automatically created. The software provides realistic viewing function to see the colour shaded final model and permits amendments to be made easily. Experiments on building models using the SLA have been carried out to study the application of the relief generating software system. Three models, the Chinese Legend and Tradition, the human face and the orchid, were built and examined. It was found that substantial amount of polishing work is needed to improve the surface finish of the resin models. The major advantage of this prototyping technique is the ability to create more prototypes for less time and cost.

11   ACKNOWLEDGEMENT

The author acknowledges the industrial sponsorship of Delcam (Singapore) for this project.

12   REFERENCES

Brown, A. S. (1991), Rapid prototyping. Part without tools, Aerospace America, 29 (8).

Chua C. K., Hoheisel W., Keller G. and Werling E. (1993), Adapting decorative patterns for ceramic tableware, *Computing and Control Engineering Journal*, 4 (5).

DeAngelis, F. E. (1991), Laser-generated 3D prototypes, *Laser in Microelectronic Manufacturing*, San Jose, CA, USA, Sep. 10-11.

Fidoora, J. C. (1991), Stereolithography models - the keystone for limited production models, *49th Annual Technical Conference - ANTEC '91*, May 5-9.

Lee H. B., Ko M. S. H., Gay R. K. L., Leong K. F. and Chua C. K. (1992), Using computer-based tools and technology to improve jewellery design and manufacturing", *International Journal of Computer Applications in Technology*, 5 (1), 72-80.

Lee H. B. (1993), Computer-aided and manufacturing for the jewellery industry, *a thesis for the degree of Master of Engineering in Nanyang Technological University*.

Mueller I. (1992), Stereolithography. A rapid way to prototype die castings, *Die Casting Engineer*, 36 (3).

Poindexter, J. W. (1991), Rapid prototyping in an integrated product development environment, *1991 SAE Aerospace Atlantic*, Dayton, OH, USA, Apr. 22-26.

Wood, L. (1990), Rapid prototyping. Uphill, but not moving, *Manufacturing System*, 8 (12).

13   BIOGRAPHY

Chua Chee Kai is a lecturer with the School of Mechanical and Production Engineering in Nanyang Technological University, Singapore. He is a member of WG5.3 of TC5 of IFIP and is the Singapore Representative for TC5. He is also currently the President of the Institute of Industrial Engineers, Singapore.

92

A Decision Aid System for Manufacturing : FlexQuar

P. Régnier[1], G. Doumeingts[1], M. Bitton[2]
[1]GRAI/LAP, Université Bordeaux 1, 351 Cours de la Libération,
33405 Talence Cédex, FRANCE - Tel : (33) 56 84 65 30,
Fax : (33) 56 84 66 44, Email : doumeingts@lap.u-bordeaux.fr
[2]Clemessy - 18 rue du Thann BP 4499, 68057 Mulhouse, FRANCE

Abstract

This paper is to propose a generic architecture to help decision makers to cope with decision and organisation related changes (manufacturing events, modification of variables, redistribution of roles,...). We show how these problems can be partly solved adopting the FlexQuar approach. The recent results obtained in the FlexQuar ESPRIT III CIME project, sponsored by the European Communities, are described here. This three year project started in September 1991, and has four industrial application sites in order to test and integrate the project results. One demonstrator has been implemented and tested in the end of the first year (Pirelli company) and another one implemented, in a second application site, for validating the second year of the project. This paper focuses on the second demonstrator.
First, we formulate the problems linked to the integration and evolution needs of the Production Manufacturing Systems. Second, the FlexQuar architecture is described in order to define the methodology which has been used for the project (GIM), and the two main components of the FlexQuar architecture i.e. the Information Server (called IS) and the GraphicaL User Environment tool (called GLUE). Third, we present the FlexQuar second year demonstrator, which has been used to validate the proposed architecture in a car components manufacturer application site. Four, we describe the scenario which has been used, in order to give ideas on the practical interest of the FlexQuar architecture.

Keywords

Car components manufacturing, CIM systems modelling & design, GRAI Integrated Methodology, Information Server, Manager Advisor, Scheduling.

1   INTRODUCTION

This paper intends to provide an answer to the basic issue that presently prevents the success and diffusion of CIM in many industrial organisations : the accessibility and the availability of the information for the various decision makers of the Manufacturing System.

2   THE PROBLEMS OF COMMUNICATION IN ORGANISATIONS

2.1   The integration related problems

The majority of organisations adopting CIM technology base their communication on specific information systems, that have been built in the past according to previous and specific application requirements. This approach allows the solution of "local" issues but is unable to provide a global and dynamic view of business requirements. Therefore, the Manufacturing Information Systems implemented in many industries consist of heterogeneous hardware and software platforms, and incompatible data structures. Filling the information gaps between different departments, and between different responsibility levels within the same department, is difficult and expensive, if not impossible. Often, some of the adopted components are obsolete, no more fully supported by vendors and difficult to maintain due to the lack of skills within the organisation and on the market. However, the large investments in these systems, their highly valuable information contents and the reliance of present business operations upon them, make very difficult to justify their replacement. These legacy systems have to be included into any information system evolution, taking care of information system integration.

To face this problem, industries are presently focused on a more human based approach inside their Information Systems. The possibility of retrieving, structuring and processing the needed data in the more appropriate way to solve particular problems, while adopting personal approaches, should be given to each person. Any member of the organisation would then contribute to make the Information System more flexible and performing, by including his own experience and solutions to his needs on specific problems areas. Nevertheless, information visibility should also be restricted, avoiding both information access / modification from unauthorised persons, and accidental damage to information completeness / coherency due to improper operations (FlexQuar, 1994).

2.2   The evolution related problems

The CIM Systems have to continuously evolve in order to face the evolution of the environment and technology. Thus, there is a specific need to design a CIM Information System able to closely follow the evolution of the CIM System.

To answer this problem it is necessary to be able to elaborate the model of the Manufacturing System taking into account all the features including the integration and the dynamic evolution aspects. Such design needs to use a Reference Conceptual Model and a structured approach defined with precise steps. From the model of the Manufacturing System, we will deduce the CIM Information System. GIM (Doumeingts, 1993), the GRAI Integrated Methodology, which is based on the GRAI Model (Doumeingts, 1984), has been elaborated in order to fill these specific requirements. It uses an integrated set of modelling techniques and a structured approach allowing to represent and to design the CIM System in an integrated way and to derive from it, the Conceptual Data Model (CDM) of the CIM Information System. From this Conceptual Data Model, the Implementation Data Model (IDM), which is implemented in the IT System, can be further derived.

This approach brings many benefits and enables the generation of a complete model of the CIM System, giving with precise modelling techniques the various views of the System : the functional, the physical, the decisional and the informational views. The set of models obtained is an excellent documentation on the CIM System. If the CIM System evolves, the modularity of GIM allows to modify only the models concerned by this evolution, while maintaining the links of integration. As the Information System Model is directly linked to the other models, one modification in the functional, physical or decisional models automatically triggers the evolution of the Information System Model. Moreover, a mapping exists allowing to translate easily the evolution of the CIM System into changes in the various models, and more especially in the Information System Model.

An example of evolution in the CIM System could be a change of responsibility of a manager. Here, the CIM Information System must be adapted in order to provide the manager a new Conceptual Data Model adapted to his new responsibilities. For instance, if the manager has a new production unit to control, the Conceptual Data Model will have to be adapted giving information linked to this new area. Due to the high level of integration of the various models of the CIM System, the adaptation of the Conceptual Data Model will be made more easily. Moreover, the mapping rules will allow to get the right Implementation Data Model related to the new Conceptual Data Model.

Due to these advantages, the Implementation Data Model can be coherent with the reality, whatever kind of evolution of the CIM System. This coherence is guaranteed by the fact there exists precise mapping rules describing how to go from the Conceptual Data Model to the Implementation Data Model. This is really a key point for the continuous adaptation requirements of the CIM Information Systems.

The human based approach should be improved by giving the ability to cope with the evolution of the organisation and of course the subsequent redistribution of roles and tasks. The information systems will have to be quickly adapted to new situations, and people will have to be supported in order they can take into account the experience and expertise accumulated by previous persons contributions (FlexQuar, 1994).

2.3   Conclusion

The application of FlexQuar is expected to meet the users needs which motivate this project, adopting an innovative approach based on :
• easy development and integration of CIM architecture with strong emphasis on organisational and human issues,
• easy and fast conversion of CIM systems to the evolving environment of the factories,
• easy and economical integration of the various information sources existing in the factories in coherence with both the organisational needs and the organisation evolution,
• improvement of the quality of the information supplied to users, developing their "sense of ownership" to the CIM applications,
• full testing on relevant industrial sites using FlexQuar tools (Information Server and GLUE) to implement the GIM modelling and design results, producing improvements on the application sites, and receiving extensive feedback from them.

3   THE FLEXQUAR ARCHITECTURE
(FlexQuar, 1994)

From the analysis of decision processes through to systems development, FlexQuar addresses the previously identified industrial needs. The FlexQuar architecture uses non proprietary Client-Server tools supported by low-cost hardware (workstations and PCs), and is based on three main elements described below.

3.1   A CIM systems modelling & design methodology - GIM

GIM (GRAI Integrated Methodology) (Doumeingts, 1993) allows to structure and specify CIM applications taking into account the global need of information exchanges between the different levels of an organisation. This methodology has been elaborated for improving the efficiency of CIM systems modelling and design. It is based on the results of various ESPRIT projects (ESPRIT Project 418 - Open CAM System, ESPRIT Project 2338 - IMPACS) and it has been refined within the current FlexQuar project. GIM is supported by the PROGRAI software tool, which allows to perform easier GIM applications.

3.2   An high performance Information Server Module - IS

It is in charge of collecting, handling and dispatching the information flows characterising the CIM reality. This server will be supplied with configuration facilities, to take into account the legacy information architecture, and to make it adaptable to dispatch consistent information as requested by the organisation context and evolution. The Information Server Module can support several information types such as production process real time data, data stored in data bases, images, drawings, etc. It provides functionality to access, transmit, convert and process data stored in different environments (e.g. relational or hierarchical data bases, expert systems, control systems) and on different hardware platforms (e.g. mainframes, minicomputers, workstations, PCs).

3.3   A GraphicaL User Environment tool - GLUE

This tool supports the requirements for the fast development of the interfaces required to properly present the information needed by "advice giving" applications. GLUE directly allows the end users to develop their own applications (typically Management Advisors), without asking for help from software developers. The GLUE tool is supplied with a high level interface that permit the processing of heterogeneous information made available by the Information Server.

3.4   Conclusion

The aim of this project is the demonstration of FlexQuar architecture and concepts on the practical problems of improving the responsiveness of a production system to the quality and maintenance feedback.

As a result the following modules are being developed, integrated and tested :
• Information Server,
• GLUE,
• a set of Management Advisors for production planning, maintenance and quality,
• a set of generic knowledge bases tools for the solution of maintenance and quality problems.

4   ARCHITECTURE OF THE SECOND FLEXQUAR DEMONSTRATOR
(FlexQuar, 1994) (Humbert, 1994)

All the previous modules will be integrated in three successive demonstrators according to GIM (GRAI Integrated Methodology). Two demonstrators have already been tested, one in the tyre production for validating the first year of FlexQuar project (Pirelli application site), and the second one in a car components manufacturer application site, for validating the second year of the project.

Software prototypes are being tested and integrated in the four FlexQuar application sites. In this paper, we focus on the second demonstrator, i.e. on the car components manufacturer application site, as it relates to the more recent results obtained in the FlexQuar project.

This application site (and the derived demonstrator) shows the ability of the FlexQuar Architecture to support legacy system interfacing and modular adding of Information Server Clients into a full Client/Server schema.

4.1   The scope of the demonstrator

The FlexQuar demonstrator, based on this specific application site, is described in the following figure.

Figure 1 The car components manufacturer Application Site

The <u>Standard User</u> is the set of EISs (Executive Information Systems) implemented to support the everyday operations of the plant. They are intended to provide standard and predefined access to the shared information, and parametric control on planning and scheduling EISs algorithms.
The <u>Advanced User</u> is a GLUE based Manager Adviser allowing a manager to acquire and manipulate the information he is interested in.
The <u>Information Server</u> is the server application allowing Standard and Advanced Users to retrieve and exchange information. Both Standard and Advanced Users modules have internal components which are Information Server Clients.
   In this application site, the Information Server supports the integration with the already in place Information system, mainly through an Oracle DB support, updated from the legacy systems. Within the demonstrator, the Oracle DB is filled with "frozen" information from the legacy systems, but the rest of the module is working exactly as in the real conditions. The demonstrator is limited to one Standard User and/or one Advanced User at the same time (depending on the type hardware configuration used). The real situation allows multiple Standard Users and multiple Advanced Users to be active at the same time.

4.2   The objectives and functionality of the demonstrator

The demonstrator, implemented at the car components manufacturer application site, is typical of the requirements expressed by the users (people in charge of production, logistics, maintenance) in the moulding shop. There are two types of users (the Standard User and the Advanced User), although they correspond to two similar requirements : what has to be done is, using a PC, locally formulate series of questions concerning the information spread out in the various remote systems which exist, bring it back to the local station, process it, and/or edit it on this station (see following figure on next page).

The <u>Standard User</u> wants the system to take into account the most complex aspects of what has to be thought about during his workshop control activities, mainly : (1) according to the references produced, size the batches, (2) optimise the recombinations of the production resources, (3) organise the basic manufacturing sequences, and (4) determine which periods are free for maintenance actions.
For each of these aspects, the user would like, as often as he likes and at any time :
- the necessary information to be automatically fetched from the existing systems,
- these information to be brought back to his local PC,
- the pre-processing to include predefined and set options, but which can be deactivated by the operator dialogue,
- the result to be displayed in the synthetic form which is best adapted to the decision he will have to make.

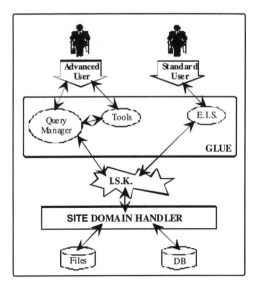

Figure 2 Two types of potential users

The <u>Advanced User</u> wants to glance through and "freely" consult any information likely to help him to know : (1) how production is progressing, (2) the current combination of resources (the machines, the tools, the operators,...), (3) any information able to help him build up a new "Standard User" type functionality.
In order to do so, the user would like :
- the aspects of the information contained in the remote systems to be presented according to the language he commonly uses in his activity, and not computer language,
- the activity-oriented language to formulate requests,
- the resulting information to be automatically fetched from the existing systems,
- this information to be brought back to his local PC,
- this information to be stored in list form in a spreadsheet, in order to be able to reorganise it freely.

The FlexQuar functional modules make possible to build up an application which meets these two types of requirements (Advanced User type requirements and Standard User type requirements). The Information Server (FlexQuar general module) is used to connect the user's local station to the remote systems which have the information to reach. GLUE (FlexQuar general module) is used to create a graphic environment adapted to each kind of user. The domain handler, interfaces each of the information bases involved, and manages the technical or computer side of the data.

4.3 The architecture and configuration of the demonstrator

The various computer modules to be implemented are used to build up various architectures : one or several clients, one or several servers, one or several ISK (Information Server Kernel) regardless of the number of servers. The following figure gives an example of the possible combinations : 2 clients (2 PCs), 2 servers (2 systems), each of them including an Information Server Kernel.

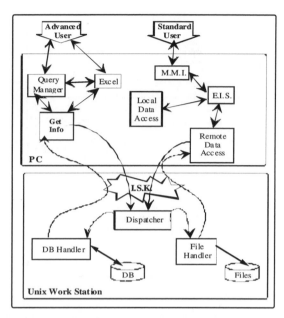

Figure 3 An example of possible combination : compute modules and flows

All these modules have been implemented for the demonstrator in 1 PC, local system capable to support both types of users, integrating the client modules, and 1 UNIX station, materialising the existing systems, containing the information to be reached, and integrating the server type modules.

• *The Standard User type client modules are :*
- MMI : managing the operator dialogues,
- Local Data Access : managing the accesses to the local information, information held by the user himself,
- Remote Data Access : managing the accesses to the remote information, linked to the server modules,
- EIS : "decision making" type pre-processing of the information.

• *The Advanced User type client modules are :*
- Query Manager : managing the MMI aspects, making the user able to build his own CDM (Conceptual Data Model), and then to use it to express his queries based on the obtained model,
- Excel : used to display the results of the queries, making the user able to reshape them, or to insert them in some other local reports,
- Get info : managing the accesses to the remote information, according to the orders submitted by the Query Manager.

• *The Server modules are :*
- ISK : Kernel of the Information Server,
- Dispatcher : receives the queries from the clients, and distributes them to the handler concerned,
- DB Handler : according to the query, fetches the information from the dedicated data base, and returns it to the client,
- File Handler : according to the query, fetches the information from the files, and returns it to the client.

5   THE SCENARIO FOR THE DEMONSTRATOR
(Humbert, 1994)

The global scenario used with the demonstrator is shown in the following figure.

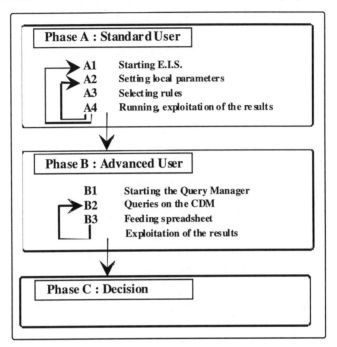

Figure 4 Block diagram scenario for demonstrator

5.1   Phase A - Standard User

Following a change in the production context, e.g. a change in the volume to be produced for two references, the person in charge of production must reorganise the production, and first of all, has to redefine which batch size is best adapted to the new conditions. He shall have to proceed as follows :

A1 - Using his PC, he activates the FlexQuar application, and chooses among the functionality offered by the EIS "help in defining the sizes of lots".

A2 - To redefine the new production conditions. For this, he wants several types of parameters to be taken into account :
• Purely local variables, as they concern his own field of action, and are never used by others : e.g. the re conversion time / production time ratio for the 2 references which have changed. For this, he activates the "variable" function in the menu, selects the variable, and put the new values he wants. The new values are stored locally.

• Conditions under which the information is taken into account, i.e. validate or not validate its pre-processing rules. For example, he deactivates the rule "provide the progress made in production".
The remote information involved in the EIS's logic (e.g. the volume to be produced or the stock status) will be automatically identified, located and collected by the FlexQuar modules, without anymore operations performed by the user.

A3 - To view the new EIS proposals, which will therefore take into account the latest conditions set, the operator initiates the execution, and obtains the complete list of batch sizes for each of the references. The demonstrator's various screens highlight the route taken by the information flow :
        - messages from the client to the dispatcher,
        - messages from the dispatcher to the file handler,
        - search in the files,
        - reply from the file handler to the client,
        - EIS pre-processing,
        - viewing the pre-processed information.

The manager will consider the results obtained and either apply them or not. He can also change the local parameters and the conditions again, and then observe in what way the given values act upon the previous results.
The production manager, or the logistics manager, or even the maintenance manager, could in the same way activate the 3 other EISs if he wants to make sure that all the aspects of the current organisation, relating to local production, are consistent with the new situation.

5.2   Phase B - Advanced User

According to the second phase of the scenario, before implementing the new measures, the production manager wants to know how the current production is progressing. In order to do so, he does not have a set of tools available, which cover all the objects to be taken into consideration, and for all the new situations.

B1 - He first of all wishes to gain access to the Domain Handler which covers the production control information, and view the associated Conceptual Data Model. In order to do that, he initiates the FlexQuar Query Manager application.

B2 - Once he has located the "lot" entity, he asks questions concerning the status of the lots in progress relating to the 2 references considered. To view the new proposals made by the EIS, which will take into account the latest conditions fixed, the operator starts the execution , and obtains the complete list of lot sizes for each of the references. The demonstrator's various screens highlight the route taken by the flow of information.

B3 - He wants the result of the query to be stored directly in a calculation sheet, as for making the final decision the list will have to be reorganised, and some basic calculations carried out beforehand (e.g. sum of the batches in progress, reference by reference, to be multiplied by a production time coefficient).

5.3   Phase C - Decision Making

The production manager then has all the information allowing him to make the right decision, i.e. to decide when the size of the batches will have to be readjusted.

6 CONCLUSION

The FlexQuar project ends on August 1995 and the third and final demonstrator to be provided by the FlexQuar consortium is a portable demonstrator at the end of the project. The major work is actually performed on the testing and integration aspects of the FlexQuar architecture. The final FlexQuar product will be composed of GIM Tools (GRAI Integrated Methodology for CIM modelling & design), Information Server, GLUE and specific Management Advisors. These four main components should be then very useful to cope with the evolution and integration needs of Production Manufacturing Systems.
At the moment, the feasibility of the FlexQuar product has yet been proved. Moreover, a market study has shown the strongly request from the industry for such systems.

7 REFERENCES

Doumeingts, G. (1984) La méthode GRAI : méthode de conception des systèmes de productique, Thèse d'État en Automatique, Université Bordeaux 1.
Doumeingts, G. Régnier, P. Fénié, P. (1993) FlexQuar Consortium ESPRIT III CIME 6408 FlexQuar project - Adaptive System for Flexible, High Quality and Reliable Production, Deliverable M12T11 - GRAI Integrated Methodology. August 1993.
FlexQuar Consortium, (1994) ESPRIT III CIME 6408 FlexQuar project - Adaptive System for Flexible, High Quality and Reliable Production, Edited Periodic Progress Report for second year. September 12, 1994.
Humbert, A. and al. (1994) FlexQuar Consortium ESPRIT III CIME 6408 FlexQuar project - Adaptive System for Flexible, High Quality and Reliable Production, Deliverable M24T52 - Interim Report on Demonstrators. September 16, 1994.

8 BIOGRAPHY

Guy DOUMEINGTS is professor at the Université Bordeaux 1. He is director of the LAP (Université Bordeaux 1) and head of the GRAI (CIM group of the LAP). His research activities are the modelling techniques for manufacturing, CIM and Production management. He is expert for European Commission and French Ministry of Research, co-ordinator of several international projects and chairman of IFIP working group "C.A.P.M.".
Pascal RÉGNIER is PhD Researcher in the GRAI/LAP Laboratory of the Université Bordeaux 1. His thesis is focused on the Production Management Area and he is involved in the FlexQuar project as Technical Responsible for the GRAI.
Moïse BITTON got his PhD (Automatic control, CIM Systems) in 1990 at the Université Bordeaux 1. Since 1985, he has been working in the field of CIM. He is now a CIM consultant at Clemessy. He has been in charge of all European projects in the field of organisation and CIM systems at Clemessy company. His current activity focuses on shop floor monitoring and control systems.

9 ACKNOWLEDGEMENTS

This paper has benefited from the comments of the persons involved in the FlexQuar project : G. Squellati (ARS[3]) , M.Corti (ARS), L. Cagnana (ARS), C. Piddington (BAe[4]), D. Copson (BAe), M. Wing (BAe), J. Mc Mullin (BAe), L. Drummond (BAe), G.Basaglia (PIRELLI[5]), L. Treanor (PIRELLI), M. Guida (PIRELLI), A. Concialini (PIRELLI), A. Odasso (PIRELLI), E. Moglia (PIRELLI), C. Iannella (EFESO[6]), J. Colom (CLEMESSY[7]), A. Humbert (CLEMESSY), C. Lambert (CLEMESSY), P. Fénié (GRAI[8]).
[3]ARS / IT company, [4]British Aerospace / aircraft manufacturer, [5]Pirelli / Tyres and Cable manufacturer, [6]EFESO Consulenza / Consultancy Company, [7]Clemessy / CIM and automation company, [8]GRAI / Groupe de Recherche en Automatisation Intégrée du LAP de l'Université Bordeaux 1.

INDEX OF CONTRIBUTORS

KEYWORD INDEX